International Conference on Cloud Technology and Communication Engineering (CTCE 2017)

Journal of Physics: Conference Series Volume 910

Guilin, China
18 - 20 August 2017

Editors:

Yu-Chen Hu
Wen-Tsai Sung

ISBN: 978-1-5108-5108-5
ISSN: 1742-6588

Printed from e-media with permission by:

Curran Associates, Inc.
57 Morehouse Lane
Red Hook, NY 12571

Printed by Curran Associates, Inc. (2017)

For permission requests, please contact the Institute of Physics at the address below.

Institute of Physics
Dirac House, Temple Back
Bristol BS1 6BE UK

Phone: 44 1 17 929 7481
Fax: 44 1 17 920 0979

techtracking@iop.org

Additional copies of this publication are available from:

Curran Associates, Inc.
57 Morehouse Lane
Red Hook, NY 12571 USA
Phone: 845-758-0400
Fax: 845-758-2633
Email: curran@proceedings.com
Web: www.proceedings.com

Table of contents

Volume 910

The 2017 International Conference on Cloud Technology and Communication Engineering (CTCE2017)

18–20 August 2017, Guilin, China

Accepted papers received: 11 October 2017
Published online: 1 November 2017

Preface

Papers

Preface

2017 International Conference on Cloud Technology and Communication Engineering [CTCE2017] has been held in Guilin, China from August 18-20, 2017. Aiming at providing an excellent international academic forum for all the researchers and practitioners, the conference enjoys a wide spread participation among all over the universities and research institutes. Any paper and topic concentrates on related fields of Cloud Computing, Big data, Signal Processing and Wireless Communication are broadly welcomed. Besides that we also look forward to making the conference perfect circumstances to share resources, exchange opinions and inspire studying.

CTCE2017 proceeding tends to collect the most up-to-date, comprehensive, and worldwide state-of-art knowledge on Education and Development. All the accepted papers have been submitted to strict peer-review by 2-4 expert referees, and selected based on originality, significance and clarity for the purpose of the conference. The conference program is extremely rich, profound and featuring high-impact presentations of selected papers and additional late-breaking contributions. We sincerely hope that the conference would not only show the participants a broad overview of the latest research results on related fields, but also provide them with a significant platform for academic connection and exchange.

The Technical Program Committee members have been working very hard to meet the deadline of review. The final conference program consists of 72 papers divided into 4 sessions. The proceedings would be published in a volume by IOP Conference Series: Journal of Physics: Conference Series.

We would like to express our sincere gratitude to all the TPC members and organizers for their hard work, precious time and endeavor preparing for the conference. Our deepest thanks also go to the volunteers and staffs for their long-hours work and generosity they've given to the conference. The last but not least, we would like to thank each and every of the authors, speakers and participants for their great contributions to the success of CTCE2017

CTCE2017 Organizing Committee

IOP Conf. Series: Journal of Physics: Conf. Series **910** (2017) 011002 doi:10.1088/1742-6596/910/1/011002

Keynote Speaker 1: Prof. Shien-Kuei Liaw, National Taiwan University of Science and Technology, Taiwan

Speech Title: WDM bidirectional Free Space Optical Transmission in 1550 nm Band

Abstract: In this talk, high-speed free space optics communication (FSO) technologies will be reviewed and introduced. Then we will design and demonstrate two proposed FSO schemes. The first scheme is bi-directional short-range free-space optical (FSO) communication with 2x4x10 Gb/s capacity in wavelength division multiplexing (WDM) channels short transmission distance. The single-mode-fiber components are used in the optical terminals for both optical transmitting and receiving functions. The measured power penalties for bi-directional, four-channel WDM FSO communication are less than 0.8 dB and 0.2 dB, compared with the back-to-back link and uni-directional transmission system, respectively. The second scheme is hybrid optical fiber and FSO link in outdoor environments such as cross bridge or inter-building system. A sensor head is used for monitoring the condition of bridge, and in the case of the bridge being damaged the transmission path could be changed from fiber link to FSO link to ensure data link connectivity. In both cases, the single-mode-fiber (SMF) components are used in the optical terminals for both optical transmitting and receiving functions. The influences of environmental factor including window glasses, air turbulence and rainfall will also be addressed. The colorless and colored window glasses introduce losses under various incident angles, but did not induce substantial power penalties. The air turbulence induces extra transmission loss and instability in the received power. Raindrops are the most influential environmental factor. The bit error rate (BER) test shows that raindrops result in a seriously impaired BER to interrupt the transmission instantaneously. After appropriate performance improvement, these proposed transmission structures show potential applications for outdoor transmission under various natural weather conditions.

Biography

Shien-Kuei (Peter) Liaw received double PhD degrees from National Chiao-Tung University in photonics engineering and from National Taiwan University in mechanical engineering, respectively. He joined the Chunghua Telecommunication, Taiwan, in 1993. Since then, he has been working on optical communication and fiber based technologies. Prof. Liaw joined the department of Electronic Engineering, National Taiwan University of Science and Technology (NTUST) in 2000. He has ever been director of the Optoelectronics Research Center and the Technology Transfer Center, NTUST. He was a visiting researcher at Bellcore (now Telcordia), USA for six months in 1996 and a visiting Professor at University of Oxford, UK for three months in 2011.

He owned six U.S. patents, and authored or coauthored for 250 journal articles and international conference presentations. He earned many national honors such as Outstanding Professor of the Chinese Institute of Electrical Engineering in 2015; the winner of The 7th Y. Z. Hsu Scientific Paper Award in 2009; the best project award of National Science and Technology Program for Telecommunication in 2006; the outstanding Youth Award of The Chinese Institute of Electrical Engineering; and the outstanding Youth Academic Award of the Optical Engineering Society of the Republic of China.

He has been actively contributing for numerous conferences as technical program chair, international advisory committee, session chair and invited speaker. He serves as an Associate Editor for Fiber and Integrated Optics. Currently, Dr. Liaw is a senior member of IEEE, OSA and SPIE. Currently, Dr. Liaw is a distinguished Professor of National Taiwan University of Science and Technology (NTUST), Vice President of the Optical Society (OSA) Taiwan Chapter.

Keynote Speaker 2: Prof. Wen-Tsai Sung, National Chin-Yi University of Technology, Taiwan

Speech Title: Innovative IoT System view based on Wireless Sensors Networks technology

Abstract: IoT (Internet of Things) System is a rapidly developing area, a combination of Network, mathematics and computing technology, in order to enhance the complex sensors network and data aggregation. Traditional Wireless Sensors Networks method does not have the ability to process hung amounts sensors signals that is why the Wireless Sensors Networks design often only one cluster or one layer framework. This speech issue brings together some of the optimal fusion of innovative information technology and methods and it provides to the listeners on this issue have further improved System Integration and Applications in Wireless Sensors Networks. This will allow scientists to develop smarter process strategies for multi-sensors signals and data.

Biography

Wen-Tsai Sung is working with the Department of Electrical Engineering, National Chin-Yi University of Technology as a professor and Vice-Dean of Academic Affairs. He received a PhD and MS degree from the Department of Electrical Engineering, National Central University, Taiwan in 2007 and 2000. He has won the 2009 JMBE Best Annual Excellent Paper Award and the dragon thesis award that sponsor is Acer Foundation. His research interests include Wireless Sensors Network, Data Fusion, System Biology, System on Chip, Computer-Aided Design for Learning, Bioinformatics, and Biomedical Engineering. He has published a number of international journal and conferences article related to these areas. Currently, he is the chief of Wireless Sensors Networks Laboratory. At present, he serves as the Editor-in-Chief in three international journals: International Journal of Communications (IJC), Communications in Information Science and Management Engineering (CISME) and Journal of Vibration Analysis, Measurement, and Control (JVAMC), he also serves as the other international journals in Associate-Editor and Guest Editor (IET Systems Biology).

Keynote Speaker 3: Dr. Weidong Huang, University of Tasmania, Australia

Speech Title: Understanding How Network Performance Affects User Experience

Abstract: Much research has been done to support remote collaboration on physical tasks. However, the focus of the research has been mainly on system and interface design and their impact on collaboration. Relatively less attention has been paid to investigating how network performance can affect user experience and task performance. In this talk, I will present a research work on this issue in which participants were asked to work collaboratively in pair using a remote mobile tele-assistance system we developed. In this study, five network scenarios were examined and network performance (QoS) was measured using four metrics including delay, jitter, bandwidth and packet loss. User experience (QoE) was measured using both objective and subjective metrics. The formal included time taken and number of instructions repeated for task performance while the latter included user ratings of quality of audio experience, quality of video experience and overall quality of experience. The results indicated that the packet loss rate in QoS is the biggest contributor to loss in QoE. I will also discuss implications of the research and possible directions of future work.

Biography
Dr. Huang is Director of Collaboration and Visual Analytics Research Lab at the University of Tasmania, Australia. He received his PhD in computer science from University of Sydney. His main research interests lie in human-centred computing in general, and Visual Analytics, Human-Computer Interaction and Visual Perception in particular. He is the author of over 100 publications in these areas including a book entitled "Handbook of human centred visualization". His research is supported by Australian government funding agencies and commercial partners with a total value of over 3 million dollars, and has achieved impact both in academia and in industry. He gives keynote/plenary speeches, invited talks and seminars at conferences and universities worldwide.
Dr. Huang is the Executive Editor of Journal of Visual Languages and Computing (SCI indexed Q1 journal) and a guest editor for several prestigious SCI indexed journals. He is a founding chair of the technical committee on visual analytics and communication for IEEE SMC Society. He has served as a conference chair, a PC chair, or an organization chair for a number of international conferences and workshops.

IOP Conf. Series: Journal of Physics: Conf. Series **910** (2017) 011002 doi:10.1088/1742-6596/910/1/011002

Keynote Speaker 4: Dr. Wen Cheng Lai, National Taiwan University of Science and Technology, Taiwan

Speech Title: IoT Design for Cloud and Big Data Application

Abstract: The Internet of Things (IoT) has accumulated a lot of attentions in past a couple years. IoT, wearable devices, cloud computing, big data, and data mining are current buzzwords in the new wave of technologies. According 5G communication trend, low power, massive connection and low latency can support smart home, smart city, wearable device, wireless surveillance, smart grids, wire-area wireless sensor network, remote diagnostic, autonomous car intelligent transportation and Industrial 4.0 which makes the IoT, Cloud, and Big Data are enabling this trend the next big wave of gold rush. It is believed that wireless communication are essential to the success of enabling real values of IoT-based applications, thus imposing many research challenges on the integration from IoT chip to cloud. This talk primarily examines the problems of current IoT applications and supported cloud trend by big data. Integrated chip design to enable local intelligence and responsive data analysis for many big data applications while maintaining very low energy consumption. With the scale of Internet to connect everything, it's obvious and evidentially true that Big Data and IoT will be closely knit together. Any solution that intended to address IoT will need to cope with the amount of data generated from millions of un-attended devices connected remotely.

Biography
PhD of Electrical Engineering, National Taiwan University of Science and Technology
PhD of Engineering Management, Nueva Ecija University of Science and Technology
Certification in MFP/ PMP/ CFA/ CPA/ DMP
Doctor of Management, University of Peking and University of Maryland University College
Research Interest: Analog, Mixed Signal IC and SOC design technology, RF devices and integrated circuits, MEMS Sensor, Microwave,.....
Working Experience: National Taiwan University of Science and Technology, ASUSTek Computer Inc, Toshiba Corp , Foxconn (Hon-Hai Technology Group), Quanta Computer Inc, Micro Star Inc, Compal Electronics Inc,

Chair

Prof. Shien-Kuei Liaw, National Taiwan University of Science and Technology, Taiwan

Co-chair

Prof. Hanmin Jung, University of Science & Technology, Korea

Editors

Prof. Yu-Chen Hu, Providence University , Taiwan
Prof. Wen-Tsai Sung, National Chin-Yi University of Technology, Taiwan

Technical Program Committee

Prof. Basel A. Mahafzah, The University of Jordan, Jordan
Prof. Mu-Chun Wang, Minghsin University of Science and Technology, Taiwan
Dr. Murat Dener,Gazi University, Turkey
Prof. Cheng-Yuan Tang, Huafan University, Taiwan
Dr. Noran Azizan Cholan, Universitiy Tun Hussein Onn Malaysia, Malaysia
Prof. Wen-Tsai Sung, National Chin-Yi University, Taiwan
Dr. Wei-Liang Wu, University of Taipei, Taiwan
Dr. Hongsheng He, University of Tennessee, China
Dr. Huey-Ming Lee, Chinese Culture University, Taiwan
Dr. Chien-Hung Yeh, Feng Chia University, Taiwan
Dr. Ray-Hwa Wong, Hwa-Hsia University of Technology, Taiwan
Prof. Shian-Chang Huang, National Changhua University of Education, Taiwan
Dr. Asif Iqbal Hajamydeen, Management and Science University, Malaysia
Dr. Ozgur Ozdemir, Selcuk University, Turkey
Prof. Mu-Song Chen, Da-Yeh University, Taiwan
Dr. Muh-Tian Shiue, National Central University, Taiwan
Dr. Guodong Wang, South Dakota School of Mines and Technology, USA
Dr. Mustafa Shakir, COMSATS Institute of Information Technology, Pakistan
Dr. Tayeb Brahimi, Effat University, Saudi Arabia
Dr. Hui Yang, Beijing University of Posts and Telecommunications, China
Dr. Chiang-Chi Liao, National Kaohsiung First University of Science and Technology, Taiwan
Dr. Qingshan Wang, Central South University, China
Prof. Shukui Zhang, Soochow University, China
Dr. Mingxing Luo, Southwest Jiaotong University, China
Dr. Mohammad Maifi Hasan Khan, University of Connecticut, USA
Dr. Guoqing Xiao, Hunan University, China
Dr. Zhenyu Wang, North China Electric Power University, China
Dr. Cheng-Kuan Lin, Soochow University, China
Dr. Jianwei Liu, China University of Petroleum, China
Dr. Yousra Aafer, Purdue University, USA
Dr. Sibel Birtane, Istanbul Arel University , Turkey
Dr. Pankaj Kumar Mishra, CSIR-Central Institute of Mining & Fuel Research, India

IOP Conf. Series: Journal of Physics: Conf. Series **910** (2017) 011003 doi:10.1088/1742-6596/910/1/011003

IOP Conf. Series: Journal of Physics: Conf. Series **910** (2017) 011004 doi:10.1088/1742-6596/910/1/011004

Peer review statement

All papers published in this volume of *Journal of Physics: Conference Series* have been peer reviewed through processes administered by the proceedings Editors. Reviews were conducted by expert referees to the professional and scientific standards expected of a proceedings journal published by IOP Publishing.

IOP Conf. Series: Journal of Physics: Conf. Series **910** (2017) 012001 doi:10.1088/1742-6596/910/1/012001

Research on Network Scanning Strategy Based on Information Granularity

Futong Qin[1], Pengfei Shi[2], Jing Du[3], Ruosi Cheng[4] and Yunyan Zhou[5]

Luoyang Electronic Equipment Test Center of China, Luoyang, China
Email: [1]ftqin@163.com, [2]spf123000177010@163.com, [3]jdstarry@aliyun.com, [4]lincoln_cheng@foxmail.com, [5]54zyy@sina.com

Abstract. As the basic mean to obtain the information of the targets network, network scanning is often used to discover the security risks and vulnerabilities existing on the network. However, with the development of network technology, the scale of network is more and more large, and the network scanning efficiency put forward higher requirements. In this paper, the concept of network scanning information granularity is proposed, and the design method of network scanning strategy based on information granularity is proposed. Based on single information granularity and hybrid information granularity, four network scanning strategies were designed and verified experimentally. Experiments show that the network scanning strategies based on hybrid information granularity can improve the efficiency of network scanning.

1. Introduction

With the continuous development of computer and network technology, the computer network technology is widely used in people's lives and work in all aspects. However, the hidden dangers of the network itself and the flaws of the network communication protocols make the network vulnerable to various types of attacks [1][2]. Moreover, the openness of network architecture, the sharing of resources, the diversity of access methods, the irregularity of terminal distribution and the unpredictability of network boundary make the network technology to provide people with many conveniences but also bring many problems and hidden dangers.

Study the security of computer network, especially how to improve the security of computer network must first know both the enemy and us. Network scanning is an important way to understand the status of computers in network security research. Through network scanning, it can obtain some characteristics of the computer on the network, such as the operating system running on the computer, the vulnerability information in the system and other fingerprint information, which is very valuable for network security researchers. Therefore, network scanning plays an important role in network security research.

Network scanning as the basic means to obtain system and network information, the size of the target scan and the level of information they want to obtain, largely determines the time required to scan the network. If the scanning strategy is configured improperly, it may result in a small amount of information but a long scanning time, greatly reducing the scanning efficiency. Common network scanning strategies generally include random scanning, target list-based scanning [3], routing scanning [4] and partition scanning [5] and so on. Random scanning is randomly select computers on the network to scan. Target list-based scanning first collect a list of potentially vulnerable host addresses before scanning, and then scan the hosts in a specific target list. Routing scanning using BGP route prefix to reduce the scan space and selectively scan the IP address space. Partition scanning makes the target list

IOP Conf. Series: Journal of Physics: Conf. Series **910** (2017) 012001 doi:10.1088/1742-6596/910/1/012001

divided into several sub-lists before scanning, and then conducts multiple scanning tasks according to the sub-list at the same time. All of these scanning strategies are to improve the efficiency of network scanning from the point of reducing the network scanning IP address space, and the influence of information granularity on scanning efficiency is not considered.

The purpose of this paper is to analyze the network scanning principle and define the granularity of network scanning information, so as to discuss the method of designing efficient network scanning strategy, and to provide guidance for computer network scanning. Aiming at the problem of network scanning efficiency, we first propose a network scanning strategy design method based on information granularity. The experimental results show that this method can significantly improve the network scanning efficiency.

2. Network Scanning Technology

Network scanning is a kind of information system security protection means, it uses the method of simulated hacker attack to detect and evaluate the security of the target system. The basic principle of network scanning is through a series of means to simulate the attack behavior of the system and get the attack results. Based on these, analyzing the results so as to understand the security configurations and the application services in the target, analyze the existing security vulnerabilities, assess the network risk level objectively, guide the network administrator corrects the network security vulnerabilities and system errors in the configuration according to the results, make prevention before the launch of hacker attack.

Network scanning technology can be divided into three categories according to a complete scanning process. The first is target survivability scanning technology, that is to determine whether the target exists in the network, and in the on-line state. The second is target information collection technology, including detection of the target host open ports, the target operating system type, open services, etc. The third is vulnerability scanning technology, according to the information collected to determine or further test the system for the existence of security risks.

2.1 Target Survivability Scanning Technology

Target survivability scanning is to discover whether the target host survives, the basic principle is to use packets generated by a variety of network protocol and the inherent nature of the network protocol itself for scanning. A commonly used scanning method is PING scanning.

PING scanning usually adopts methods that send various types of ICMP or TCP, UDP request message, through the results of the message sending to determine whether the target is alive. Several major approaches include ICMP broadcast, ICMP sweep, ICMP non-echo, TCP and UDP sweep. The ICMP broadcast speed is fast but cannot find the Windows system host, ICMP sweep is easy to use, but slower. Compared with the detection of ICMP protocol, TCP sweep is more effective. It uses the famous "three-way handshake" to scan, in the handshake process to send SYN packets to the target, regardless of the target host is to respond to SYN/ACK message or RST message, you can determine the target host survival. It is now the most effective method of target discovery.

2.2 Target information scanning technology

Target information include the information of the target host port open, operating system type and open information service, etc. The main technologies of target information scanning are port scanning technology, operating system identification technology and system service identification technology.

2.2.1 Port scanning technology. Port scanning [6] is mainly divided into full connection, half connection and secret scanning, etc.

Full Connection Scanning: Full connection scanning is the most basic TCP scanning method, also known as the "TCP connect scanning", which principle is to use the connect() system call provided by the operating system to connect to a specific TCP port on the target host. If the target port is listening, connect() can connect successfully, else the connection will be prompted to fail. The advantage is that any user in the system can call the connect() function without any special permissions, and scanning speed. The disadvantage is easy to be detected and filtered by the security software, but also in the target host's log file will produce a series of error message about the service connection is established and immediately disconnect.

a) Half Connection Scanning: Half connection scanning also known as "TCP SYN scanning". TCP SYN scanning sends a SYN packet, if return to the SYN/ACK packet said target port in the listening state, if return to RST packet indicates the port is not open. If a SYN/ACK packet is received, the scanner then sends an RST packet to terminate the connection. This method has the advantage of being covert because it does not need to establish a complete TCP connection, even if the log has been recorded on the scanning, but the record on try to connect is much less than full scanning. The disadvantage is that in most operating systems, the sending host needs to construct the IP packets, this is too much trouble, and constructing SYN packets needs super or authorized users to access the special system call.

b) Secret Scanning: A typical secret scanning technique is the TCP FIN scanning. TCP FIN scanning is subtler than TCP SYN scanning, its packets can through the firewall and packet filtering systems without any hassle. TCP FIN scanning method is based on the principle that the port that is not opened will give RST response to the FIN packet, but the open port will ignore the FIN packet directly, do not answer. This method is associated with the type of the target system, some system reply RST regardless of whether the port is opened or not, at this point, this scanning method can be used to distinguish the target system is a UNIX or Windows.

2.2.2 Operating System Identification Technology. Each operating system, or even each kernel revision, has small differences in the TCP/IP stack, and these differences make the response to the corresponding packet not the same. Some network scanning tools can provide a response list, by comparing with it, the operating system running on the target host can be identified. Operating system identification technology mainly contains active identification and passive identification.

a) Active identification: Active contract, many times test, screening of different information, for example, according to the returned ACK value, some systems will return the sequence number of the acknowledged TCP packet, while others return the sequence number plus 1. Some operating systems use some fixed TCP window or set the DF header of the IP header to improve performance, which is the basis for judging the corresponding operating system. Active identification technology to determine the

b) accuracy of Windows is relatively poor, can only determine the approximate range, it is difficult to determine the exact version, but UNIX, CISCO network equipment more accurate. However, the more hops between the target host and the source host, the worse the accuracy. This is mainly because many of the eigenvalues in the data packet have been modified or blurred in the transmission process.

c) Passive identification: Passively monitoring the network data packets, using the data in the packet header of DF, TOS, window size and the TTL value to determine the operating system. Because this method does not need to actively send detection packets, only need to crawl the network packets, so called passive identification.

2.2.3 System Service Identification Technology. The detection of system service is mainly based on the scanned port to determine, or the use of HTTP response analysis, binary information detection and other means to achieve.

According to the port to identify system services is to directly use the port and the corresponding services relationship, this approach is an earlier way, for a wide range of assessment, it has some value, but its accuracy is low, it often produces injustice in actual use. For example, it thinks that as long as open port 80 is to open the HTTP protocol. However, this is not the case, this is the fundamental flaws of using port scan technology in the service decision.

Relatively speaking, the binary information detection method is relatively accurate, access to the service banner, is a more mature technology, can be used to determine the current operation of the service, the judgments are more accurate. Moreover, it not only can determine the service, but also can determine the specific version of the service, but it is easy to be cheated by banner camouflage methods and produce false positives.

2.3 Vulnerability Scanning Technology

Vulnerability scanning technology is a kind of security technology developed to make system administrators keep abreast of the security vulnerabilities in the system and take corresponding precautions to reduce the security risk of the system. Using vulnerability scanning technology, you can scan the security vulnerabilities of the LAN, Web sites, host operating systems, system services, and firewall systems, etc.

Vulnerability scanning can be mainly divided into passive and active scanning according to the scanning strategy. Passive scanning based on the hosts, the general system architecture adopts C/S mode, through installing agents in the hosts to collect vulnerability information. There are certain limitations in this mode, which are caused by the software compatibility problems. Active scanning is network-based, sending specific packets through the network to detect the existence of security vulnerabilities of the target system, this vulnerability scanning system more widely applicable, the use of such scanning system is more common especially in internet companies.

3. Design Method of Network Scanning Strategy Based on Information Granularity

3.1 The Definition of Information Granularity

3.1.1Definition 1: Information Granularity. Information granularity refers to the extent to which the information obtained by network scanning can clearly reflect the security problem of the target system. The smaller the granularity of information, the more clearly the security problem of the target system.

According to the degree of detail of information obtained by network scanning, we defined four information granularity levels, as shown in figure 1.

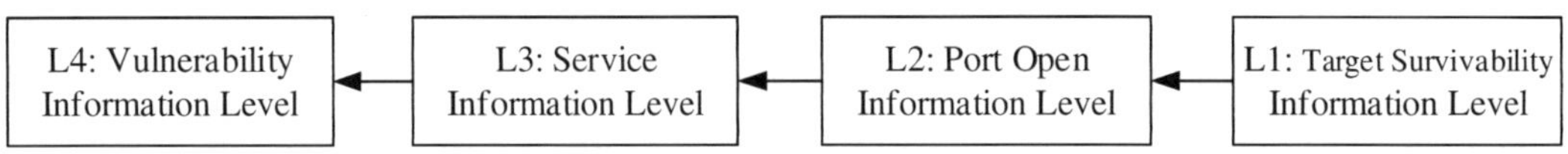

Figure 1. The levels of information granularity of network

3.1.2Definition 2: Target Survivability Information Level. The information obtained by network scanning is the survivability of the hosts in the target network, and the information granularity is the largest, which can be used to statistic the distribution of surviving hosts in the network. Using the same scanning tool, the time spent in this granularity scan is minimal.

3.1.3Definition 3: Port Open Information Level. The information obtained by the network scanning is the open ports of the target hosts. The information granularity is smaller than the target survivability level. It is based on the target survivability detection and further obtains the host open port information.

3.1.4Definition 4: Service Information Level. The information obtained by network scanning is information such as open service, application version and host operating system. The granularity of the information is smaller than the port open level. The port open level scanning can only obtain the status of the port open, while the service information scanning can determine whether the service corresponding to the port is the default service that the port has, and can obtain information such as the application version of the service and the target host operating system.

3.1.5Definition 5: Vulnerability Information Level. The information obtained by the network scanning is the vulnerability information of the target host, and the information granularity is the smallest. Vulnerability information level scanning results can be directly used by the attacker to attack the target host, or as a "security scanning", can directly find the system security option setting error, in order to improve system security.

3.2 Design of Network Scanning Strategy Based on Hybrid Information Granularity

Network attack and defense has always been a game problem. For the attacker, the sooner the problems found in the network, the more able to break through the target system. For the defender, the sooner the problems found in the network, the more timely to prevent possible attacks.

By the second section of this paper on the analysis of network scanning, it is not difficult to find that the larger the scanning network scanning information granularity is, the less data is needed to judge the scanning result. The smaller the network scanning information granularity is, the more data the scanning result needs to be obtained. If the network scanning information granularity is the target surviving level, the scanning host only needs to send the SYN message to the target. The feedback information obtained is that the target host responds to the SYN/ACK or RST message. Regardless of the message information, the time is only the sum of the sending, feedback and normal network transmission times. If the network scan information granularity is the service information level, such as obtaining the system service information, regardless of whether the target host is alive, the scanning host will send detection information to it and wait for the detection target feedback result, and then judge the target system service information according to the feedback result. The time of the whole process takes is the sum of the time to send probe packets, feedback packets, packet analysis and network delay. If the target host survival, the network delay is the normal network delay, if the target host is not alive, the network delay consists of the normal network delay and the timeout period of the host to wait for detection of target feedback. From the above analysis, we can see that the smaller the granularity of network scanning information, the more complex the analysis and judgment of the scanning results, the more time is needed, especially when the target host is not alive.

When designing a network scanning strategy, in the same size of the scanning IP address space conditions, if we first using the coarse-grained scanning to reduce the IP address space, and then using the fine-grained scanning, these will greatly reduce the proportion of network overtime in the whole scanning time-consuming, will certainly be able to improve network scanning efficiency. In view of this, this paper designs the following four network scanning strategies.

C1: For the scan target IP address list A, performing L1 target survivability scanning.

C2: For the scan target IP address list A, performing L2 port open scanning.

C3: For the scan target IP address list A, performing L3 service information scanning.

C4: For the scan target IP address list A, first performing L1 target survivability scanning to obtain the survive hosts' IP addresses, then make these IP addresses into a new list A1, and performing L3 service information scanning on A1.

Among them, scanning strategy C1, C2 and C3 are strategies based on single information granularity, C4 is a strategy based on hybrid information granularity.

4. Experiment Analysis

In this section, we will analyze the four C1~C4 network scanning strategies to verify the effectiveness of the network scanning strategy based on information granularity in improving the scanning efficiency. The target network is built in laboratory environment.

4.1Experimental Setup

The target network consists of four subnets. Each subnet host sets the IP addresses to the address segments as in table 1.

Table 1. The IP addresses corresponding to subnets

Subnet	**IP address segments**
Subnet1	192.168.11.0/24
Subnet2	192.168.12.0/24
Subnet3	192.168.13.0/24
Subnet4	192.168.14.0/24

IOP Conf. Series: Journal of Physics: Conf. Series **910** (2017) 012001 doi:10.1088/1742-6596/910/1/012001

Each subnet has five hosts, and each host is deployed the same FTP and Web services with the ports 20 and 80 are opened. Set up a scanning host connected with each subnet, and deploy the network scanning tool Nmap [8] in the scanning host. The specific experimental network topology as shown in Figure 2. In the experiment, we set up two groups of scanning targets. The group 1 includes two subnets, subnet 1 and subnet 2, the group 2 includes all four subnets in the target network.

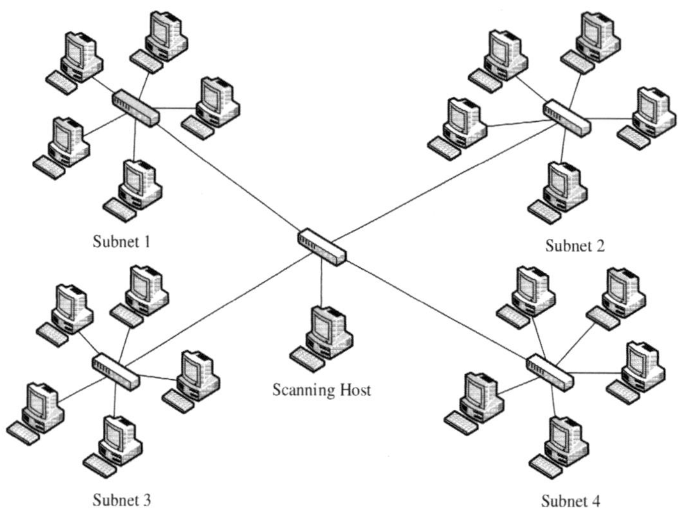

Figure 2. The specific experimental network topology

4.2Experimental results and analysis

Set the scanning host scanning the two groups of targets respectively in accordance with the scanning strategies C1 - C4. For each group of targets, five scans were performed using each scanning strategy. The time spent for each scanning was recorded separately. The average of five scanning times was taken as the experimental result. All the results were shown in table 2.

Table 2. Scanning time results (unit: second)

Scanning Target	C1	C2	C3	C4
group 1	25.886	72.124	137.744	56.306
group 2	112.608	200.778	223.206	129.188

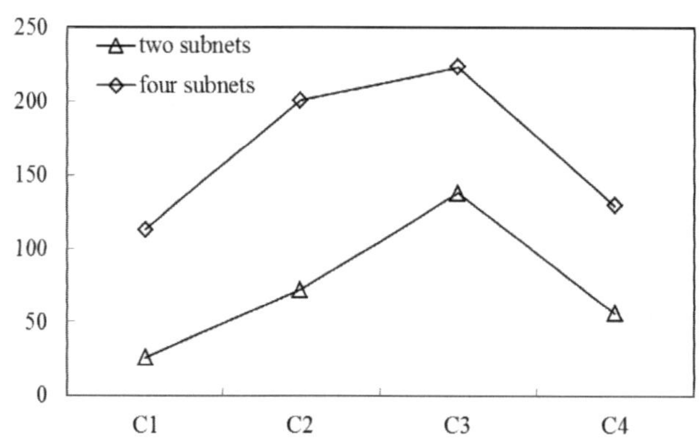

Figure 3. The influence of IP address space on scanning time

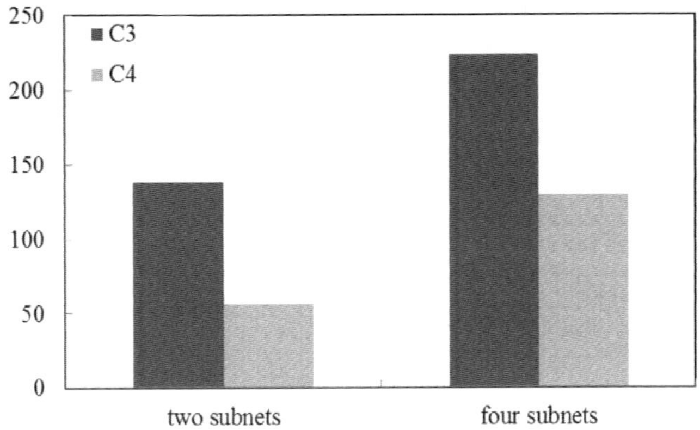

Figure 4. The influence of scanning strategy on scanning time

In order to show the influence of scanning target IP address space size and scanning information granularity on scanning time, we analyze the contents in table 1 in graphical form. Figure 3 shows the influence of IP address space on the network scanning time. It can be found that scanning the four subnets (group 2) takes more time than scanning two subnets (group 1) when using the same scanning strategy. In addition, it can be seen from figure 3 that using these three strategies C1, C2 and C3 to scan the same target consuming more and more time since the three strategies are based on the single information granularity and their information granularity are more and more smaller. Figure 4 shows the influence of scanning strategies on scanning time. In the experiment, scanning strategies C3 and C4 were used to get the service information of the target. C3 is scanning the target directly, and C4 is firstly using coarse-grained granularity scanning to scan the target, removing non-active targets to reduce scanning IP address space, and then using fine-grained granularity scanning to scan the new reduced IP address space. It can be seen from figure 4 that the time consumed by strategy C4 is significantly less than the time consumed by strategy C3, whether scanning two subnets or scanning four subnets. Thus, we can find that designing scanning strategies based on the hybrid information granularity can effectively improve the network scanning efficiency.

5. Conclusion

Network scanning often occurs when cybersecurity personnel conduct a census of network assets, or in the preparatory stage for hackers to conduct network attacks. Since the target network has a large IP address space, various types of equipment, and their ports, services, vulnerabilities vary widely, the network scanning often takes a lot of time when conducting assets census on a large-scale network. Based on the analysis of network scanning principle, in this paper, we defined the information granularity of network scanning, and designed the network scanning strategies based on the information granularity. The experiment proved that using hybrid information granularity design network scanning strategies could effectively improve the efficiency of network scanning.

6. Acknowledgment

In this paper, the research was sponsored by the National Natural Science Foundation of China with Grant No. 61303061 and the Open fund from HPCL with Grant No. 201513-01.

7. References

[1] Li Xiaoyan, Wang Qingxian, Yang Lin, Luo Xiangyang. The Research on Network Security Visualization Key Technology[C]. 2012 fourth International Conference on Multimedia Information Networking and Security (MINES), 2012:983-988.

[2] Fang Lan, Wang Chunlei, Ma Guoqing. A framework for network security situation awareness based on knowledge discovery [C]. 2010 2nd International Conference on Computer Engineering and Technology (ICCET), 2010:226-231.

[3] Weaver N. Potential Strategies for High Speed Active Worms: A Worst Case Analysis [EB/OL]. http://www.cs.berkeley.edu/~nweaver/worms.pdf,2002.
[4] Zou CC, Towsley D, Gong W, et al. Routing worm: A fast selective attack worm based on IP address information [A]. In Proc. of 19th ACM/IEEE/SCS Workshop on Principles of Advanced and Distributed Simulation (PAD/05) [C].2005, 199~206.
[5] Zou CC, Towsley D, Gong W. On the performance of Internet Worm Scanning Strategies [J]. Performance Evaluation, 2006, 63(7):700~723.
[6] Fyodor. The Art of Port Scanning [EB/OL]. http://www.insecure.org/nmap/nmap_doc.html, 2002.
[7] Fyodor. Remote OS Detection Via TCP/IP Stack Fingerprinting [EB/OL]. www.insecure.org, 2002.
[8] Nmap[EB/OL], http://nmap.org/download.html.

The Research of the Parallel Computing Development from the Angle of Cloud Computing

Zhensheng Peng*, Qingge Gong , Yanyu Duan and Yun Wang

Department of Information Engineering, Engineering University of PAP, Xi'an China
Email: *pzsyjsgldd@163.com

Abstract. Cloud computing is the development of parallel computing, distributed computing and grid computing. The development of cloud computing makes parallel computing come into people's lives. Firstly, this paper expounds the concept of cloud computing and introduces two several traditional parallel programming model. Secondly, it analyzes and studies the principles, advantages and disadvantages of OpenMP, MPI and Map Reduce respectively. Finally, it takes MPI, OpenMP models compared to Map Reduce from the angle of cloud computing. The results of this paper are intended to provide a reference for the development of parallel computing.

1. Introduction

Cloud computing is the development of distributed computing, parallel computing and grid computing. These concepts of computer sciences are the implementation of commercial [1]. Therefore, the cloud computing is not just a computing mode while it covers the service, market, technology and many other concepts. The service core of cloud computing is divided into three levels: takes infrastructure as a service—IaaS; takes platform as a service—PaaS and takes software as a service—SaaS (See Figure 1 below).

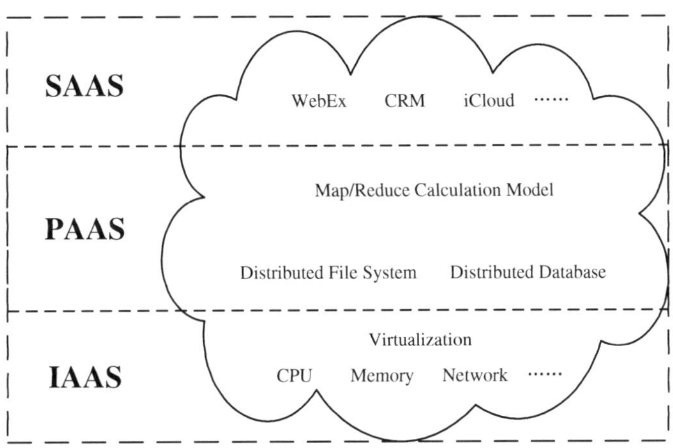

Figure 1 Cloud services

This article mainly analyzes technical from the aspect of cloud computing. To begin with, from technical perspective, after the division of physical host abilities through virtualization technology, the computing power of the host is no longer confined to a single physical host, but a virtual host based on segmentation which makes full use of the computing power of the single host. The second is a distributed computing technology. It makes a single task into many subtasks, and then assigns to each host of the cluster which makes them parallel running coordinately with each other.

We can see that distributed computing is a technique which can aggregate the resources. It is the core concept of cloud computing essence from a macro level. The distributed parallel computing does not mean that there is no non-distributed parallel computing which is the focus of the discussion below.

2. The Overview of Parallel Computing

Parallel computation was appeared in the early 1960s. At that period, the transistor and core memory showed up. The processing unit became smaller and the memory became more compact and cheaper. The development of these technologies gave rise to the occurrence of parallel computer [2]. During this period, the parallel computers are mostly shared memory multiprocessor systems in small scale which called the mainframe.

For a long time, parallel computing has been developing rapidly in the field of high performance computing and the parallel computing architecture has also been in constant changing. The platform relied by parallel computing is called parallel computers which is composed of multiple nodes. The task is decomposed to various nodes and runs in parallel on each node. The early parallel node is not the completely independent host between each other. It more likes various modules of one host. In the 1990s, accompanying with the development of network equipment, MPI/PVM and the release of parallel programming standards, cluster architecture of parallel computer has been emerged. At this time, the node of hardware platform of parallel computing is the independent host. Modern architecture of parallel machine has Symmetrical Multi-Processing (SMP), Distributed Shared Memory (DSM) parallel machine, Massively Parallel Processor (MPP) and Parallel Vector Processor (PVP).The architecture of parallel machine are mainly used for high performance computing in specific areas. This is the foundation of cloud computing—the birth background of PC cluster of workstation [3]. Parallel computing has been brought into people's lives by PC cluster due to the rapid development of the microprocessor performance and network bandwidth.

3. The Implementation of Parallel Computing

The implementation of parallel computing has two levels [4]. First, it is the multi-core parallel computing and the multiple CPU in a single node. Although the parallel computing in single node is not the mainstream of implementation of cloud computing, multi-core is a very important aspect for a single node to improve the performance currently. Second, For cloud computing, the parallel between cluster nodes is emphasized more. At present, the nodes among cluster are usually connected by IP network. On the premise of enough bandwidth, each node is not restricted by geographical and space. So the parallel computing in cloud computing is called distributed parallel computing for many times. However, mutil-CPU and multi-core are the trend of host development. So within a cluster, the parallel requirement of general level is existed such as the parallel between cluster nodes, multiprocessor within node internal and multi-core parallel. The parallel computing between nodes is equivalent to distributed parallel computing as we often said. Different from mature virtualization products, parallel computing has no mature products while only has the relatively mature tools.

The realization of parallel computing relies on the familiar with business by developers and users as well as the correct and skilled use of parallel tool. There are three steps for parallel application software development and use.

(1) On the demand analysis stage, according to business characteristics, the task is divided into multiple tasks which can be executed in parallel as much as possible. This is the basis of the realization of parallel computing and it always be the most effective phase.

(2) During the designing and coding phase, parallel tools are used for program design. The parallel technology used in this stage is the most key technology which is not only difficulty but also not mature. Moreover, most of the technical personnel has not handle the technology applied at this stage at the moment.

(3) Parallel deployment architecture and tools deployment applications are used in deployment stage. Through the deployment, this stage can achieve load balancing of occupied equipment. Load balancing in essence is also belonging to parallel computing. The method used in common is the load

balancing device such as F5 switches, the ESB and etc. The technology involved in this stage is relatively mature which has been widely used in the field.

4. The Overview of Programming model of Parallel Computing

The parallel computing introduced in the second stage is narrow sense of parallel computing. This paper mainly analyzes the OpenMP and MPI in this stage [5], which is the parallel computing technology and Map Reduce which is the typical representative of load balancing technology. It will be introduced in the third phase respectively.

4.1 OpenMP(Open Multi-Processing)

Now from x86 to minicomputer, multi-core is the mainstream of CPU. For single-threaded program, multi-core processors are difficult to improve the process efficiency [6]. For multithreaded program, it can achieve the purpose of improving performance through simultaneous calculation by different cores. Obviously, the execution performance of multi-threaded program depends on the task decomposition. The compilation and maintenance of multi-threaded programs have more complicated than single thread. In general, the execution performance of multithreaded program is still much higher than single thread when the host is multiprocessor, multi-core or has the ability to execute multiple threads at the same time by adopting multithreaded programming technology.

The compilation of traditional multithreaded programming has two ways in general. The first is by calling the operating system function, development tools, multi-threaded control statements or control function. The second is to use P0SIX thread control library functions based on P0SIX's standards and methods. These methods are all through the main thread to generate more sub-threads; then it separate work from main thread to each sub-thread for operation; in the end, it recycle and integrate the results from the main thread.

Since the use of this method, there have been two problems. First of all, the complexity of program development is increased a lot than single-threaded program. Second, the portability of first method also has problems. Third, if you want to have good extensibility, multithreading should not only create a fixed number of threads, but also should increase threads by the growth of CPU cores. Although the above methods can achieve good scalability, the control is rather complicated. The developers need to achieve load balance scheduling of Multithreading on multi-core processors by their own. It is very complex, but using OpenMP could solve the problem.

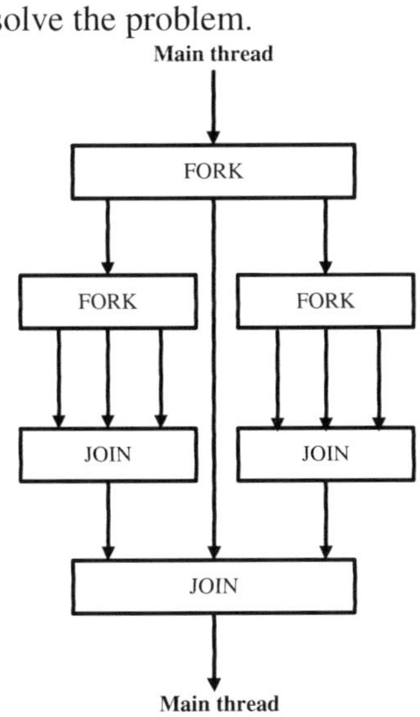

Figure 2 OpenMP programming model of parallel computing

OpenMP is a multi-threaded programming method. Its parallel granularity is in threading level (See Figure 2 below).The range of application is parallel computing within light engine, which can use multi CPU or multi-core better. OpenMP is a set of API and runtime library which can simplify multithreaded development within light engine in order to let the designers and developers focus on the main energy of understanding and breaking up the task. Therefore, the specific details can be accomplished by OpenMP. By using OpenMP, the programming efficiency is much higher than multithreaded programming of general function library or own function library of operating system. It is superior to P0SIX standards and methods. OpenMP is available as shared storage standard. It is an application programming interface (API) designed for compiling parallel programs for multiprocessing machine including a compilation statements and function library used to support it.

OpenMP can multithread the program simply by pseudoinstruction. The simplest case, it can parallel processing the statement within the loop, only through adds one line of compiler directives which used to guide the compile program. OpenMP is originated in C/C++, but now it has Java development kit of OpenMP, which can be used in Java programming.

4.2 MPI (Message Passing Interface)

Although OpenMP is very good, it is only a parallel computing technology in light engine. No matter the parallel computing or cloud computing, the mainstream of the parallel computing is a parallel between the host rather than a single machine. Therefore, MPI is really the representation of parallel computing technology.

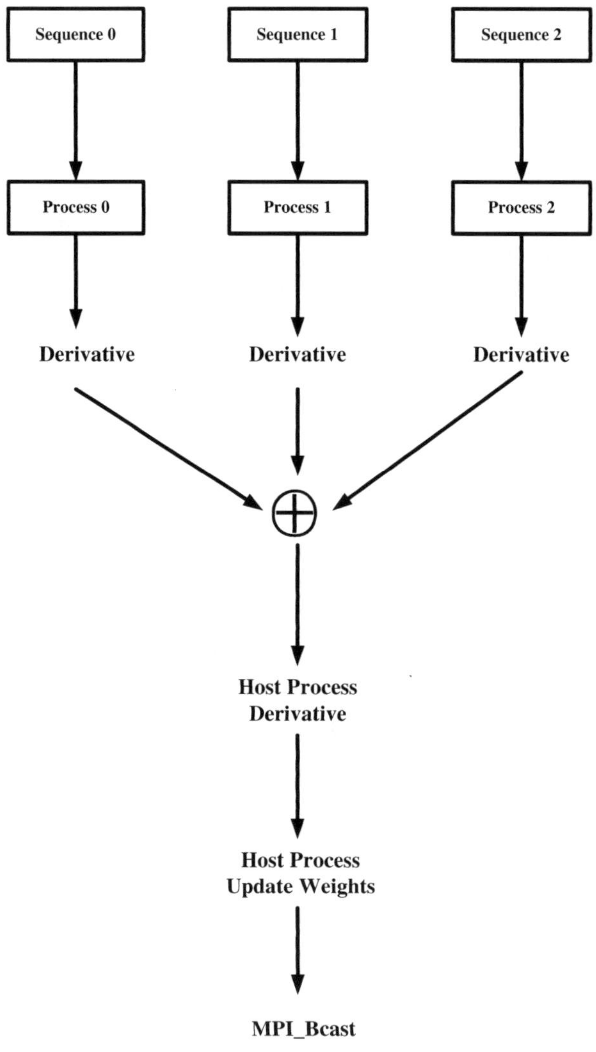

Figure 3 MPICH programming model of parallel computing

MPI standard version 1.0 was born in May 1994. This standard puts forward a kind of function interface description based on message passing. At present, MPI has grown to version 2.0, and it has become an accepted standard of high-performance computing. MPI itself is not a specific implementation but a standard description. The famous concrete implementation and wide use of MPI is the MPICH which is completed by development team of American Argoone National Laboratories. MPICH is free software. It provides binding support for Fortran, C language and so on, which offers to developers in the form of function library (See Figure 3 below). In the program design, using MPI has a larger degree of freedom even it can be used to achieve Map Reduce function of Hadoop.

The way of message passing is a model widely used in parallel computers especially in distributed memory parallel computers. More than a decade, this model has made substantial progress in important calculated applications. In designing MPI, the goal is not to choose someone from existing message passing system but make full use of the most abstract characteristics among these systems.

The main advantage of establishing message passing standard is the portability and easy to use. The standardized benefit is particularly evident in higher level based on low-level messaging program and distributing storage communication environment constructed by abstract program. The definition of message passing standards can offer clearly defined library to manufacturers. Therefore, they can effectively realize these libraries or provide hardware support for library application in some cases. The expansibility is strengthened.

The powerful advantage of MPI is the portability in wide range. The MPI achieved in the upper deck of communication protocol between standard Unix processor will provide the portability for workstation PC system and different kinds of workstation network. This standard includes a point-to-point communication, set operations, process group, communication context, topology structure, binding with Fortran 77 and C language, environmental management and query and interface description.

4.3 Map Reduce

Map Reduce is a programming architecture based on cloud computing system proposed by Google and is a kind of parallel programming model which can deal with huge amounts of data for parallel computing of large-scale data set. The concept and main idea of "Map" and "Reduce" are both originated from functional programming language and vector programming language [7]. Due to that Map Reduce has common characteristics of functional and vector programming language, this programming model is particularly suited for searching, mining, analyzing and machine intelligence learning of unstructured and structured mass data. It can greatly simplify the difficulty of parallel programming.

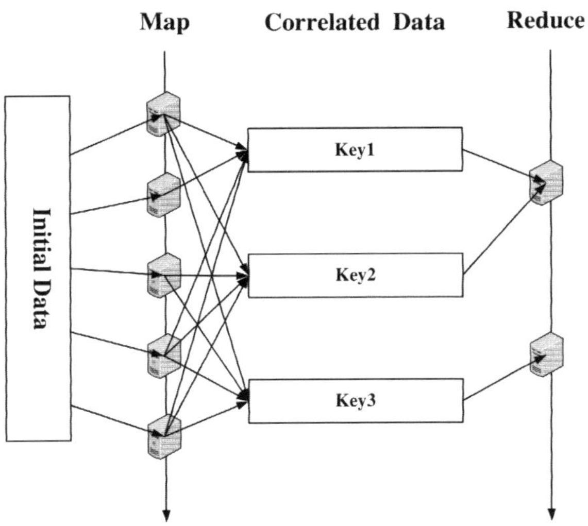

Figure 4 Map Reduce programming model of parallel computing

Originally, some computational problems are very simple but we have to deal with complicated details, fault-tolerant mechanism as well as the data distribution policy, load balance and so many details which led to the complicated and difficult program code. Therefore, all these problems in computing applications can be wrapped in a library and be responsible by a system as a whole. Otherwise, the computing task of data is abstracted as Map and Reduce. The source of this abstraction is based on the Lisp and Map and Reduce concept of many kinds of functional language. Many computing tasks of data are related to Map. The Map is responsible for processing each logical Record among input data and produces a set of intermediate Key/Value pairs, and then performs Reduce operation to all intermediate results which have the same key. The intermediate results from Map should be merged and then generated the final results.

For large-scale data set operation, Map Reduce is managed and controlled by a main node [8]. The concrete computing tasks is accomplished by sending to various sub nodes from the master. In this way, Map Reduce achieves reliable execution and fault-tolerant mechanism of parallel tasks. In each time period, the master node will mark the working condition of various sub notes. Once marked for death, all the tasks of this node will be assigned to other sub nodes for re-performance.

Compared with traditional distributed programming mode, Map Reduce has packaged the parallel processing, fault tolerance, localized calculation, load balance and other details and also provided a simple and powerful interface. By this interface, large data computing tasks can be concurrent and distributed automatically. During operation, system can solve the distribution problem of input data, which can across the program execution and scheduling of cluster nodes and deal with the failure of processing nodes and communications between the management nodes. Therefore, the compilation of parallel programs becomes easier. The programmers can use the resources of large distributed systems without the concurrent processing or distributed programming experience. Meanwhile, it has great economic efficiency by high performance cluster composed by general PC to achieve super high performance. And Map Reduce also has good generality. Many different types of computing applications can be solved simply by Map Reduce programming model.

5. The Comparison of Parallel Computing Programming Model

OpenMp is shared storage in threads level, which is only suitable for shared bus and memory, and SMP machines and DSM with single operating system image [9]. It has poor extensibility but high requirement on the machine. General dual-core machines are suitable for OpenMP because it can improve certain of operating speed. OpenMP has been widely used in scientific computing.

Table 1 The table of the comparison of three parallel computing programming models

Parallel computing programming model	OpenMP	MPI	Map Reduce
Parallel granularity	Threading level	Progressing level	Progressing level
Way of data distribution	Implicit formulation	Explicit formulation	Explicit formulation
Expansibility	Bad	Good	Good
Failure of processing node	No	No	Yes
Data storage and data processing position	Separate	Separate	Transfer from calculation to storage
DFS support	No	No	Yes
Real-time	Good	Good	Bad
Single demand processing	Suit	Suit	Unsuited
Immediately read	Yes	Yes	No
The difficulty of learning	Easy	More difficult	Easy

MPI is parallel granularity in process level with distributed storage and appeared data distribution with good scalability. It is suitable for all kinds of machines but the programming model is complicated: need to analyze program computing tasks, and map it to distributed process collection for calculation. Since MPI is based on message, there is no restriction on task partition and data partition [10]. Powerful flexibility increases the complexity. Since MPI is the process level. It needs to solve the problem of communication delays and unbalanced load to ensure efficiency. MPI program has complicated debugging and bad reliability.

By dispersing the large-scale operation on data set to network nodes, Map Reduce can achieve reliability [11]; each node could periodically report back the finished work and updated status. If a node keeping silence over a predetermined time interval, the master node will record the state as death and then assign the task to other nodes. Therefore, Map Reduce creates parallel computing model based on data partitioning which has powerful fault tolerance and is also very easy to learn. It can be used for distributing, sorting, web log analysis, index construction, document clustering and statistical machine learning.

6. Conclusion

Early parallel computing is mainly used for parallel computers. With the emergence of cluster, cloud computing brings the parallel computing into people's lives. Cloud computing is a kind of distributed parallel computing. Map Reduce is parallel programming model of cloud computing. The main difference between traditional MPI and OpenMP is that Map Reduce is technical delegate of load balancing. OpenMP and MPI are mainly used for concurrent design and coding.

OpenMP is a kind of multithreaded programming method and its parallel granularity is thread level. The range of application is parallel computing in light engine which can make good use of multi CPU or multi-core. And MPI is a message passing standard library and its parallel granularity is process level, which is a parallel between the host rather than a single machine. Therefore, MPI is the representative technology of parallel computing. Map Reduce is parallel programming model for concurrent processing huge amounts of data. It is also a kind of task scheduling model with high efficient at the same time. The greatest advantages of Map Reduce are in shielding the underlying and implementing the details which effectively reduce the parallel programming difficulty and improve programming efficiency.

OpenMP and MPI, as the traditional parallel computation models, OpenMP + MPI which is the hybrid parallel computing model provides a good parallel strategy for computer cluster consisted by multiple processors. As the distributed parallel programming mode of cloud computing, Map Reduce is a hotspot of current researches.

7. References

[1] Hayes, Brian. "Cloud computing." Communications of the Acm 51.7(2008):9-11.
[2] Nogi, Tatsuo. "Parallel computer." US, US4514807. 1985.
[3] Agbaria, Adnan, and R. Friedman. "Starfish: Fault-Tolerant Dynamic MPI Programs on Clusters of Workstations." Cluster Computing 6.3(2003):227-236.
[4] Grama, Ananth, et al. "Introduction to Parallel Computing." 22.2(2003):N12.
[5] Dorta, I., et al. "MPI and OpenMP implementations of branch-and-bound skeletons." Advances in Parallel Computing 13.04(2004):185-192.
[6] Michailidis, Panagiotis D., and K. G. Margaritis. "Open Multi Processing (OpenMP) of Gauss-Jordan Method for Solving System of Linear Equations." IEEE, International Conference on Computer and Information Technology IEEE Xplore, 2011:314-319.
[7] Thusoo, Ashish, et al. "Hive: a warehousing solution over a map reduce framework." Proceedings of the Vldb Endowment 2.2(2009):1626-1629.
[8] Yang, Hung Chih, et al. "Map reduce-merge: simplified relational data processing on large clusters." ACM SIGMOD International Conference on Management of Data ACM, 2007:1029-1040.
[9] Song, Fengguang, S. Moore, and J. Dongarra. "Analytical modeling and optimization for affinity based thread scheduling on multicore systems." (2009):1-10.
[10] Quinn, Michael J. "Parallel Programming in C with MPI and OpenMP." IEEE Distributed Systems Online 5.1(2004):7.1-7.3.

[11] Bersani, Marcello M, et al. "Efficient large-scale trace checking using Map Reduce." International Conference on Software Engineering ACM, 2015:888-898.

Study of Threat Scenario Reconstruction based on Multiple Correlation

Xuejun Yuan, Jing Du, Futong Qin and Yunyan Zhou

Luoyang Electronic Equipment Test Center of China, Luoyang, China
Email: whotrustme@163.com, yuan_boo@163.com, ftqin@163.com, 54zyy@sina.com

Abstract. The emergence of intrusion detection technology has solved many network attack problems, ensuring the safety of computer systems. However, because of the isolated output alarm information, large amount of data, and mixed events, it is difficult for the managers to understand the deep logic relationship between the alarm information, thus they cannot deduce the attacker's true intentions. This paper presents a method of online threat scene reconstruction to handle the alarm information, which reconstructs of the threat scene. For testing, the standard data set is used.

1. Introduction

In response to the security threats faced by the computer systems, the defensive side has deployed a large number of network security products, and achieved positive results. IDS, as a representative of the security products, has good technical indicators and is an effective solution to detect the variety of attacks, ensuring the safety of computer systems. However, the isolated output alarm information, large amounts of data, event errors and omissions, and unorganized alarm flow are presented to the managers, making it difficult to understand the deep logic relationship between the alarm information, and difficult to deduce the attacker's true colors and true intentions. Therefore, how to deal with a large number of alarm data online, mine the deep logical relationship between the alarm information, reconstruct the threat scenario and provide decision support for the managers, become hot spots of the current research.

There are many ways of analyzing and processing the information generated by network security equipment such as IDS. The commonly used methods include: correlation based on attribute similarity [1], correlation based on the attack template [2], correlation based on the causal relationship [3][4], correlation based on timing relationship [5], the cross correlation and so on[6]. These methods have the following problems: low efficiency of the alarm correlation, underuse of the existing data, incompatible with online mode and so on.

To this end, this paper firstly describes the concept of threat scene reconstruction, and analyzes the current typical correlation technologies and methods. Then, a framework of threat scene reconstruction which can be used in the online mode is proposed, and the correlation algorithm is improved. Finally, the experimental test is carried out. The method proposed in this paper can meet the requirements of the online correlation mode, and its resource occupancy is reasonable. With it, it is more convenient for the managers to discover the security threats in time and formulate the corresponding protection strategy.

2. Background

2.1 The Concept of Threat Scene Reconstruction

With the extensive application of the security equipment such as IDS, from the underlying hardware, the basic system, the upper application to the end user, the security of network information system is significantly enhanced. It is difficult for the attackers to just exploit one single vulnerability to achieve its attack attempt. In order to achieve its attack intention, Attackers use a number of vulnerabilities to implement a series of attacks, which gradually expands the attacker's access, and ultimately achieves its intention of attack. The range of the attacks described here is called a threat scenario, also called an attack scenario[7].

In order to detect the attacking intention of the attacker and achieve rapid response, the security managers analyze and associate the alarm information generated by the network security equipment, and then reconstruct a series of attack behaviors implemented by the attacker. This process is called threat scenario reconstruction, also called attack scenario reconstruction.

There are two patterns for threat scene reconstruction. One is online reconstruction, which reconstructs the threat scene while the alarm is generated. In this mode, the traffic of the alarm information is generated constantly, which requires high reconstruction efficiency. The other one is off-line reconstruction, which reconstructs the threat scene after all of the alarms have been generated. The accuracy of threat scenario reconfiguration in this mode is high, but it can only be used in post hoc analysis.

The alarm information, as the data source of the alarm correlation, among which the relationship must be clearly classified. In this paper, the relationship between the alarm information is classified and abstracted into the following three logical relations: redundant relationship, causality relationship, and timing relationship. The core technology of threat scene reconstruction is the alarm correlation technology.

2.2 Alarm correlation technology

1)Correlation based on attribute similarity. The core of the method is to define the similarity function of the important attribute values, determine the threshold and weight, and then correlate the alarm information whose attribute similarity is greater than the threshold.
This method can deal with the alarm information with redundant relationship and greatly reduce the number of alarm information, but it cannot reveal the causal relationship and timing relationship between the alarm information, thus fails to reveal the deep logical relationship of alarm information.

2)Correlation based on the attack template. This method, based on the expert knowledge or empirical data, extracts the characteristic information of the threat scene in a user-defined manner or from the training data set, and then saves it as an attack template which is described by a formal language. By comparing alarm information with the attack templates, the correlation of attacks is achieved.

The advantage of this method is the ability to accurately and quickly detect attack patterns which have been clearly described in the threat scene. The disadvantage is that it is difficult to cope with an unknown attack in the threat scenario or some of the attacks that are not defined in the attack template.

3)Correlation based on the causal relationship. Researchers believe that the attack event always depends on a specific premise, and will have specific consequences. For example, there are two attack events, the time of their occurrences are T_0 and T_1, and $T_0 < T_1$. Comparing the consequences of the attack behavior happened at T_0 and the premise of the attack behavior happened at T_1, if the set *(consequence, premise)* matches, the two are considered as causal and related. The most significant advantage of the causal relationship-based alarm correlation method is that it does not need to capture all the possibilities of the threat scenario, but it can still detect unknown attacks. This method is highly accurate. It need to establish a knowledge database, which contains the premise and the consequences of attacks. Its disadvantage is that if the knowledge database is too large, it cannot adapt to the occasion of online correlation.

IOP Conf. Series: Journal of Physics: Conf. Series **910** (2017) 012003 doi:10.1088/1742-6596/910/1/012003

*4)*Cross *Correlation.* This method is mainly based on the background knowledge, such as network topology information, vulnerability information and host configuration information, to improve the quality of the alarm, and to assess the risk of alarm information. This method is often used in conjunction with other methods.

3. The Design of Threat Scene Reconstruction Method

3.1 The Idea

In order to accurately deduce the attack behavior scene, this paper intends to compress the number of the alarm messages and the size of the causal knowledge, based on the causal relationship. Specific implementation ideas are discussed as follows:

1)Alarm correlation based on the causal relationship. The causal relationship-based correlation method is mostly responsive to the deep logical relationship between the alarm information, and thus it has the highest accuracy. Therefore, this paper chooses the causal correlation method, and uses attack itself as the main body of the correlation.

2)Causal knowledge compression based on cross correlation. The basis of the causal relationship-based correlation is to establish the knowledge of the prerequisite and consequence set of the attack event. If processing the causal knowledge introduce large overhead due to its size, this method cannot adapt to the online mode. For this reason, this paper borrows the idea of cross-correlation, combines the background knowledge of the protected network, such as the network topology information, the vulnerability information and the host information, simplifies the causal knowledge and generates the customized knowledge of the protected network, which in the end successfully reduces the size of the causal knowledge base.

3)Elimination of redundancy alarm information based on attribute similarity. The causal relationship-based correlation method needs to analyze the premise and consequences of the alarm information. If the oversized and redundant alarm information is not preprocessed, information accumulation will inevitably occur in the online mode, and results in the correlation failure. In this paper, the correlation method based on attribute similarity is used to preprocess the alarm information before using the correlation method based on causality, and the redundant relationship between the alarm information is eliminated. In addition, because Distributed Denial of Service (DDoS) attacks can easily generate a large number of alarm information in a short time, this paper will prioritize DDoS attack alarm information.

3.2 Threat Scenario Reconstruction Framework

Based on the above ideas, the threat scene reconstruction method designed in this paper adopts a multi-layer correlation structure. With the preprocessed information, the redundant relation is analyzed by the alarm fusion unit, and the causal relationship and the timing relation are solved by the alarm correlation unit. The information flows through these units in turn, transforming data into knowledge.

The threat scene reconstruction framework is shown in Figure. 1.

At first, the original alarm information is preprocessed by the preprocessing unit. It eliminates the difference among the different types of the alarm information which are generated by different IDS. The main purpose is to convert the original alarm information so that the generated data to meets the relevant format specifications of the alarm information that could be further processed in the following stages.

The alarm fusion unit leverages the correlation method based on attribute similarity, mainly dealing with the redundant relationship of the alarm events. In order to solve the problem of the large number of alarm information generated in a short time, the alarm fusion unit adopts a sliding time window, and preferentially processes the DDoS attack alarm information in batches.

The alarm correlation unit leverages the causal relationship-based correlation algorithm, which is mainly used to exploit the causal relationship and potential timing relationship of the alarm information.

The correlation is based on the causal knowledge, using attack itself as a correlation subject to meet the requirements of the online mode. The input of the alarm correlation unit is the alarm information that has been merged by the alarm fusion unit, and its output is the possible attack sequence.

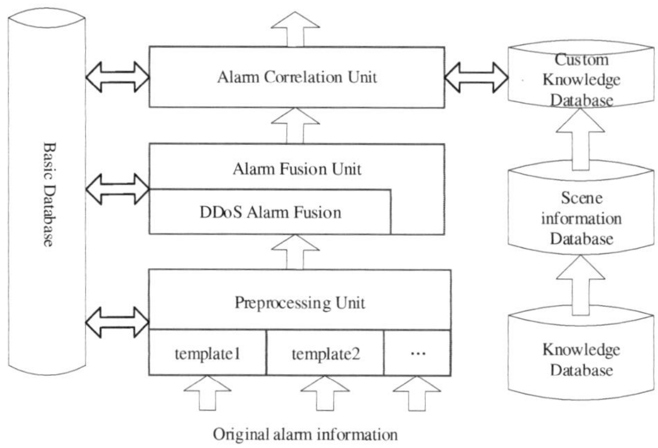

Figure 1 The threat scenario reconstruction framework

The causal knowledge database is the basis of the alarm correlation, and is related to the quality of the correlation effect. It uses predicates to formalize the description of the alarm information, the consequences of information and the super alarm type.

The scene information database stores the static scene information of the network information system. Compared to the business information flow, the scene information of specific network information system is static, which will not be changed frequently. In this paper, we use the idea of cross correlation, simplifies the knowledge with the scene information, generate custom knowledge, thus reduce the correlation matching range.

The basic database, as a storage unit of this framework, has data interaction with each unit. From the original alarm information, the formatted alarm information, the high-level alarm, to the alarm attack sequence, these data are all stored in the basic database.

The threat scene visualization unit displays the threat scene generated in the end in a more intuitive way.

3.3 Design of the Fusion Algorithm based on Attribute Similarity

To meet the requirements of the online processing mode, the algorithm uses a sliding time window to read the preprocessed alarm information. Firstly it checks the sliding time window, then it determines in bulk whether the information within the window matches the DDoS attack alarm characteristics or not. If so, the information will be placed into the DDoS attack alarm queue. Otherwise, the information in the time window will be processed one by one to determine whether it can be merged into the other attack alarm queues.

1)Alarms attribute similarity. In this paper, we select 6 kinds of the alarm attribute for similarity calculation, which include alert time, source IP, target IP, protocol type, source port, and target port. The similarity function is defined as follows:

a) Time attribute similarity function

$$S(time_i, time_j) = \begin{cases} \dfrac{T - |time_i - time_j|}{T}, & |time_i - time_j| \le T \\ 0, & |time_i - time_j| > T \end{cases} \quad (1)$$

Where T denotes the threshold of time, and $time_i$ and $time_j$ are the occurring time of two alarms.

b) IP address attribute similarity function

$$S(ip_i, ip_j) = \frac{n}{32} \tag{2}$$

Where n denotes the number of identical in the IP address bits from high to low, obtained by calculating XOR on ip_i and ip_j in their binary format. In this paper, $n \geq 24$ $(n>23)$, therefore, *0.72* is used as the IP address class attribute similarity threshold.

c) Protocol type attribute similarity function

$$S(type_i, type_j) = \begin{cases} 1, & type_i = type_j \\ 0, & type_i \neq type_j \end{cases} \tag{3}$$

d) IP port attribute similarity function

In this paper, we find that the similarity of the port attributes is not only related to distance, but also has a close relationship with the distribution of port numbers. Consider the characteristic of the port number distribution, expand the definition of the adjacent port, and then determine based on the distance from port to port. Different thresholds are set for different ranges of port numbers. If the difference between ports is less than the given threshold, the pair of ports is considered as adjacent ports. The similarity of the port is defined as follows:

$$S(X_p, Y_p) = \begin{cases} \dfrac{stepL - |X_p - Y_p|}{stepL} & |X_p - Y_p| < stepL \\ 0 & otherwise \end{cases} \tag{4}$$

Where *stepL* denotes the threshold of the difference between the two ports:

$$stepL = \begin{cases} 1 & 0 < \max(\text{port}_i, port_j) \leq 255 \\ 2 & 256 \leq \max(\text{port}_i, port_j) \leq 1024 \\ 4 & 1025 \leq \max(\text{port}_i, port_j) \leq 5000 \\ 8 & 5001 \leq \max(\text{port}_i, port_j) \leq 65535 \end{cases} \tag{5}$$

2)Overall similarity. Whether the alarm information can be integrated depends on the overall similarity of the alarm information. The overall similarity of the alarm information given in this paper is defined as follows:

$$S(X, Y) = \prod_{j=1}^{6} \max\{0, SIM(X_j, Y_j) - V_j\} \tag{6}$$

Where the variable j is the index value of the attribute. V_j denotes the minimum similarity expected for the *j-th* attribute. Because the algorithm selects six important alarm attributes, so j is an integer value from 1 to 6. The variable $SIM(X_j, Y_j)$ is calculated with the Eq.(1) to Eq.(4).

When the overall similarity value $S(X,Y)>0$, the alarm events X and Y are considered to be similar.

3)DDoS attack alarm attribute similarity. According to the characteristics of DDoS alarm information, the fusion method is specially designed to prioritize such information. The alarm information generated by the DDoS attack has the following two characteristics:

a) The alarm information in which the destination IP address equals the target host are the attack packets. In a certain period of time, for these alarm information, the source IP address is not similar but the source port is similar, and the target IP address is the same, and large amount of alarms are generated.

b) If the source IP of the alarm information is the target host, then these are response packet for the attack packets. In a certain period of time, for these alarm information, the source IP address is the same and the destination IP address is different, but the target port is similar, and large amount of alarms are generated.

According to the above characteristics, the proposed batch fusion algorithm is described as follows:

a) In the time slice *t*, $S(sourceIP_i, sourceIP_j)=1$, $S(targetPort_i, \text{tartgetPort}_j)\geq 0.8$; or $S(targetIP_i, targetIP_j)=1$, $S(sourcePort_i, sourcePort_j)\geq 0.8$;

b) If the number of the similar alarms obtained in the previous steps is greater than *T*, which is a threshold for the number of the alarm information, then correlate these alarms to the DDoS attack queue. Among them, the similarity of the IP address is calculated with the formula 2, and the similarity of the port is calculated with the formula 7:

$$S(X_p, Y_p) = \begin{cases} \dfrac{stepL - |X_p - Y_p|}{stepL} & |X_p - Y_p| < stepL \\ 0 & otherwise \end{cases} \tag{7}$$

Where *stepL* has a big threshold within the port range of 5001 to 65535.

$$stepL = \begin{cases} 1 & 0 < \max(\text{port}_i, port_j) \leq 255 \\ 2 & 256 \leq \max(\text{port}_i, port_j) \leq 1024 \\ 4 & 1025 \leq \max(\text{port}_i, port_j) \leq 5000 \\ 200 & 5001 \leq \max(\text{port}_i, port_j) \leq 65535 \end{cases} \tag{8}$$

4)Alert fusion algorithm workflow. The workflow of the alarm fusion algorithm is shown in Figure. 2:

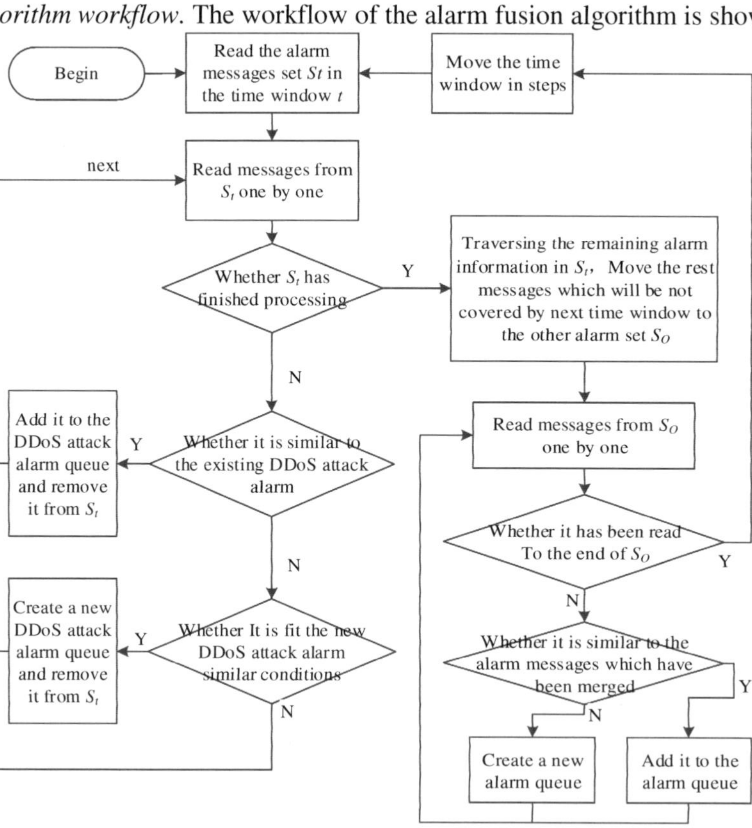

Figure 2 The alert fusion algorithm workflow

3.4 Design of correlation algorithm based on causal relationship

In order to meet the requirement of the online alarm correlation mode, this paper based on custom knowledge database, analyzes the current attack and the subsequent attack list, which can be generated by the current attack itself.

The structure of the custom knowledge database is the same as the knowledge database. They use the predicate to describe the causal relationship between the alarm information, which mainly covers three aspects: firstly, the predicate knowledge, indicating the prerequisite and the consequence information of the alarm information; secondly, the implication of the relationship, indicating the relationship between predicates; and lastly, the super alarm type, indicating the abstraction of each attack corresponding to a type of alarm.

The correlation process is specifically described as follows. When *attack-A* occurs, an attack list *L1* is created based on *attack-A* and the custom knowledge, which lists all the subsequent attacks that may be generated by *attack-A*. When *attack-B* occurs, an attack list *L2* is created based on *attack-B* and custom knowledge database. At the same time, use the previous attack list *L1* match *attack-B*. If the parameters match successfully, then *attack-A* and *attack-B* correlated. Iteratively, whenever a new attack occurs, establish the subsequent attack list and match it with the existing attack lists, to achieve the alarm information correlation.

The workflow of the alarm correlation algorithm is shown in Figure. 3. One thread processes the next alarm message and the other thread outputs the alarm correlation result.

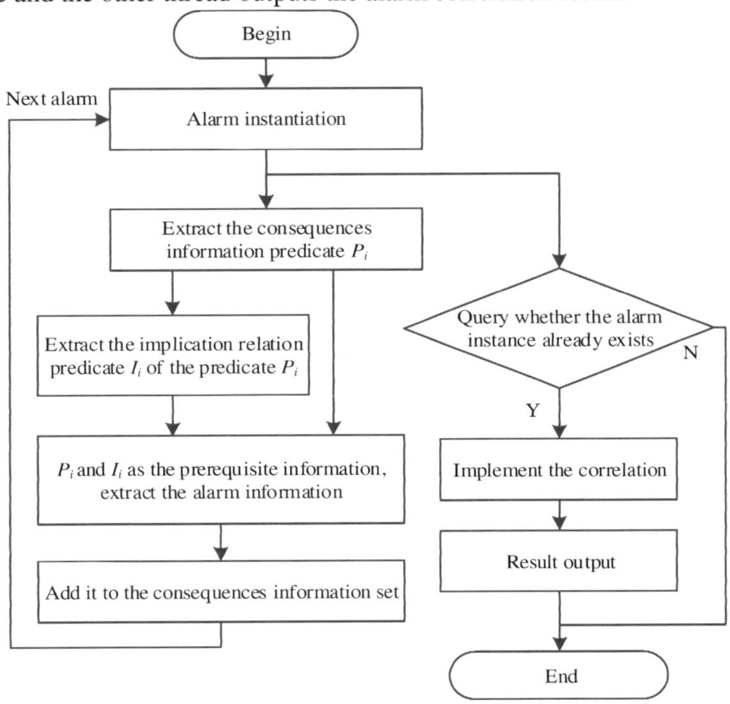

Figure 3 The alarm correlation algorithm workflow

An attack can be divided into several attack stages, and the occurring time of the alarm information generated in the same attack phase, is temporally close to each other. So there is a "blank time window" between different attack phases. This is due to the fact that the consequences of the previous attack phase needs to be confirmed by attackers, and the next attack stage will start only if the previous attack phase works. On the other hand, after the success of the previous stage, both of the system state conversion and the changes on the implementation of the attacking environment require extra time to be paid. These are the two reasons causing the "blank time window" between different attack phases.

In this paper, although the algorithm needs to query all possible follow-up alarm, but on the one hand, follow-up alarm information using the custom knowledge which has been simplified, has reduced the match range. On the other hand, the algorithmic time overhead and resource overhead are introduced in the "blank time window" after the alarm message occurs, so there is little impact on the subsequent alarm information correlation of an attack sequence.

4. Experiment and result analysis

To verify the effectiveness and accuracy of this method, the experiment was tested using the DARPA 2000 LLDoS 1.0 data set [9]. The data set is a commonly evaluated and widely used benchmark data set for intrusion detection.

The software experimental environment consists of the Windows 7 operating system, the Microsoft visual studio 2010 development environment, and the Microsoft SQL server 2008 database. The used hardware experimental environment is an Intel i5 4590 CPU(3.2GHz) with 8G memory. TCPreplay3.4.4 [10]was used for data playback, and the snort intrusion detection system for alarm information collection was on the other side.

Through the alarm fusion processing, the number of alarms is greatly reduced, the details of the results are shown in Table 1:

Table 1 Number of alarms before and after alarm fusion

Attack phase	Inside area			Dmz area		
	input	*output*	*rate*	*input*	*output*	*rate*
1	31	6	19.35%	785	8	1.02%
2	32	12	37.50%	25	8	32.00%
3	35	23	65.71%	80	21	26.25%
4	22	12	54.55%	19	9	47.37%
5	33787	29	0.09%	33943	13	0.04%
total	33907	82	0.24%	34852	59	0.17%

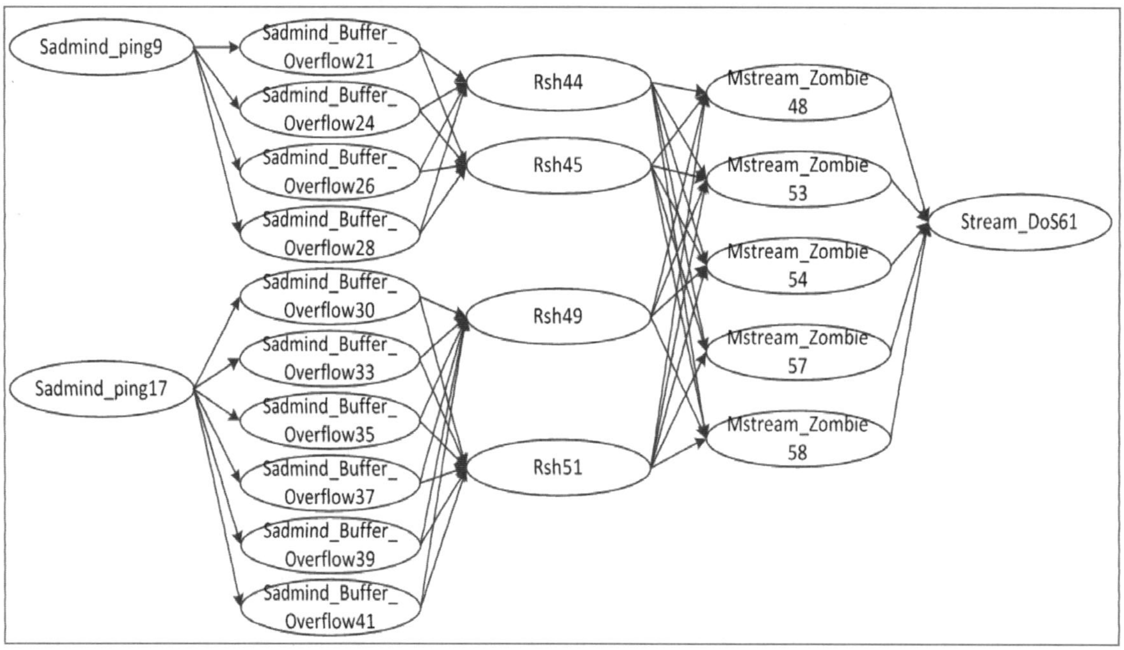

Figure 4 Correlation graph generated by the method described herein

The number of alarms in the *inside area* are reduced to 19%, 37%, 65%, 54% and 0.09% after the alarm fusion. The number of alarms in the *DMZ area* are reduced to 1%, 32%, 26%, 47% and 0.04%. Obviously, the proposed algorithm is effective to eliminate the redundant relationship of the alarm information.

The results of data fusion shows that the fusion efficiency of *phase 1* and *phase 5* is very high. It is found that because the *phase 1* is the pre-detection phase, which results in a large number of redundant information, the integration of the information in this phase is more efficient. And the *phase 5* is a DDoS attack phase, the alarm fusion algorithm detects and fuses the DDoS attack alarms.

Figure. 4 is the correlation graph generated by the method described herein. It can be seen from the figure that the attack process described in this figure is consistent with the DARPA2000 data set and is more clear and intuitive than the literature [3] [8].

5. Conclusion

In this paper, an online threat scene reconstruction method based on multiple correlation is designed, and the related algorithms are improved. The experiment uses the internationally recognized DARPA 2000 data set LLDoS1.0 . The test results show that the threat scene reconstruction method is effective, the alarm fusion effect is ideal, and the correlation graph generated by the alarm correlation method is concise and intuitive. This method can meet the needs of online correlation, and the resource occupancy is reasonable. It is more convenient for the security managers to find security threats in time.

Follow-up work is planned as follows:

A. Improve the causal correlation knowledge database. Design automate mining tools, and update regularly according to the CVE vulnerability library and other related libraries, to enrich the causal knowledge database.

B. Explore further match methods between the "attack - follow-up attack" information and the background knowledge, calculate the probability of the subsequent attacks, reduce the subsequent attack search space, and improve the online matching ability.

6. Acknowlegement

This paper is sponsored by the National Natural Science Foundation of China with Grant No. 61303061, 61402485 and the Open Fund from HPCL with Grant No. 201513-01.

7. References

[1] ValdesA,SkinnerK. Probabilistic alert correlation[C] //Recent advances in intrusion detection. Springer Berlin Heidelberg,2001:54-68.

[2] Mahboubian M, Udzir NI, Subramaniam S,etal. An alert fusion model inspired by artificial immune system[C]//Cyber Security, Cyber Warfare and Digital Forensic(CyberSec),2012 International Conference on.IEEE,2012:317-322.

[3] Ning P,Cui Y, Reeves DS,etal. Techniques and tools for analyzing intrusion alerts[J].ACM Transactions on Information and System Security (TISSEC), 2004,7(2):274-318.

[4] Zhaowen L, Shan L, Yan M. Real-time intrusion alert correlation system based on prerequisites and consequence[C]//Wireless Communications Networking and Mobile Computing (WiCOM), 2010 6th InternationalConference on. IEEE, 2010: 1-5.

[5] Qin X, Lee W. Discovering novel attack strategies from INFOSEC alerts[M]//Data Warehousing and Data Mining Techniques for Cyber Security. Springer US, 2007: 109-157.

[6] Yu Beng L, Ramadass S, Manickam S, et al. A survey of intrusion alert correlation and its design considerations[J]. IETE Technical Review, 2014, 31(3): 233-240.

[7] Cheung S, Lindqvist U, Fong M W. Modeling multistep cyber attacks for scenario recognition[C]//DARPA information survivability conference and exposition, 2003. Proceedings. IEEE, 2003, 1: 284-292.

[8] TIAA project. http://discovery.csc.ncsu.edu/software/correlator /ver1.0/ index.html.

[9] DARPA 2000 dataset. https:// www.ll.mit.edu /ideval/ data/ 2000 data.html.

[10] Tcpreplay tools. http://sourceforge.net/projects/tcpreplay/.

IOP Conf. Series: Journal of Physics: Conf. Series **910** (2017) 012004 doi:10.1088/1742-6596/910/1/012004

Research on Multi-level Electromagnetic Leakage Protection Strategy of Industrial Control System

Yang Li, Feng Li, Xu Wang and Qingrui Guo

State Grid Xinjiang Electric Power Science Research Institute, Changchun road, Xinjiang, Urumqi, China
Email: 282123897@qq.com, sunshine09010208@126.com, liyang20032006@126.com, liyang20062003@126.com

Abstract. The structure of industrial control system (ICS) is different from computer architecture, electromagnetic leakage protection of computer system is no longer suitable for industrial control system. In this paper, a method of electromagnetic leakage multi-level protection for industrial control system is presented, which combines software development, system overhaul and repair management as well as maintainers management. Through the application of the comprehensive and multi-level protection structure, the electromagnetic leakage of the industrial control system is effectively reduced. The defects of traditional protection which focus on local structure and lack of overall protection, are improved.

1. Introduction

With the increasing number of industrial control system, industrial control system security has become one of the cardinal factors of industrial production and socio-economic security. Electromagnetic leakage means that the information system equipment in the work can cause radiation by the ground wires, power lines, signal lines, parasitic electromagnetic signals or harmonics, resulting in electromagnetic leakage. If the electromagnetic signal is received, after extraction and processing, you can restore the original information, resulting in data leakage [1][6]. In the process of industrial control system, digital or analog information interaction is carried out by means of network cable, wireless network and signal line. There are major electromagnetic leakage and anti-control risks.

2. Source of Electromagnetic Leakage of Industrial Control System

Among these components, the control signal, the output signal, the measurement signal and the constant value signal are combined, all of which is combined into a conventional closed loop control system. Manipulating industrial robots as the largest component of the industrial control system, its typical structure usually consists of the control computer, the teaching box, the operating panel, the hard drive storage/the floppy disk storage, the digital and analog input/output, the printer interface, the sensor interface, the axis controller, the auxiliary equipment control, the communication interface, the network interface, and the ethernet interface. There is a risk of exposure to electromagnetic fields if the sensor interface, the digital and analog input/output, the printer interface, the communication interface, the network interface as well as the ethernet interface were unused the professional shielding cables. There is a greater risk of electromagnetic leakage when the main processed mechanism of the industrial control systems controls the computer without any special shielding measures.

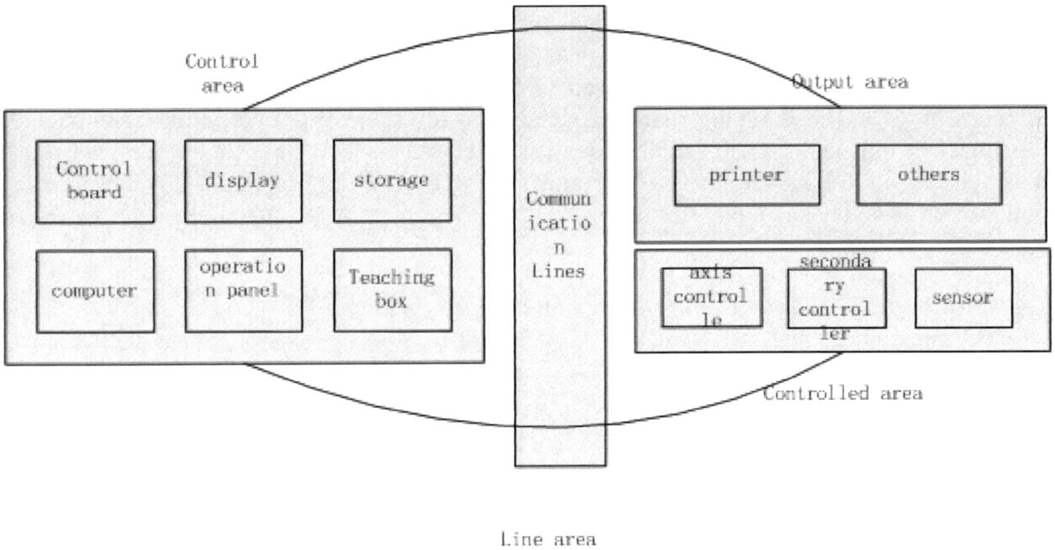

Figure 1 Structure of ICS

3. The electromagnetic leakage prevention of ICS

From the angle of electromagnetic radiation area, the basic structure of industrial control system can be divided into the control terminal area, communication lines area, controlled region and output terminal area [3]. The control terminal area mainly integrates the master control board, the master control display, the memory, the central computer, the operation panel, the display box and other relevant equipment of industrial control system. The communication lines area mainly concentrates the network transmission lines, the communication signal transmission lines and the analog signal transmission lines. The controlled region is mainly integrated with the axis controller, shift and other relevant auxiliary control devices, sensors and other control signal feedback devices, which carries out commands from the control terminal that requires to be issued. The output terminal area mainly integrates the relevant services that debugging the alarm output printers, plotters and other output devices. In virtue of the equipment of industrial control system has the concentricity, the prevention and control of industrial control system and electromagnetic leakage, which can conduct from the perspective of regional control as well as operation and maintenance management. It is conducive to reducing the cost of prevention and control. And it can also enhance the ability of control [4].

3.1 The Prevention and Control of Communication Line Area

The protection of network transmission lines: in the course of the production environment, industrial control system dues to business requirements and the convenience of maintenance & debugging, thus it should be accessed to the wireless network, wired lans, internet network, etc. We can get the data of the industrial control system by the electromagnetic monitoring to the network cables, and after that, we can carry out the replay attack. Therefore, we should restrict the network that industrial control system to run and maintain, and the industrial control system should adopt separate devices to constitute LAN which is independent of any network and its network adopts wire mode to establish a network for avoiding interference and hijack of the wireless signal by other factors. Network structure should be as simple as possible to meet the needs of the business, and should be as far as possible to reduce the device node as well as to avoid increasing uncontrolled electromagnetic leakage of the node and the network cable interface. Therefore the network signal transmission cables should choose the double shielded cables, and avoid transmitting interference and leakage.

The line protection of signal transmission to the telecontrol, telesignalisation and telemetering: it can be divided according to the signal transmission, and transmission lines that are mainly copper wire cables, optical fibre cables and other cables. Some partial cables with the control of robot arm axles are used as a kind of paving method. Some employ standalone weak current routing as a way. This type of

IOP Conf. Series: Journal of Physics: Conf. Series **910** (2017) 012004 doi:10.1088/1742-6596/910/1/012004

signal cable has little influence over electromagnetic leakage and protection, especially electromagnetic can get effective protection after copper with metallic sheath and cabling channel.

The leakage proof outer sleeve can be used for shielding the connection between the equipment, which can achieve a sealed of the radio frequency radiation, electromagnetic interference and the surrounding environment. It has all-around shield and even can achieve 100 dB or else higher shielding effect. The casing is divided into three types: zinc casting outer casing pipe, plastic outer sleeve pipe with inner metal spraying and metal stretched casing. In these three types, the protecting effect of metal stretched casing is the best.

3.2 The Protection of Electromagnetic Leakage on the Control Region

The control site concentrates the core parts of industrial control system, which includes central computers, control panels, main internal memory and so on. It is an instruction issue mechanism and intelligent analysis and processing mechanism of industrial control system.For the reason that the complexity of the structure of the control end area, the shielding technology is proposed to protect the electromagnetic leakage. In accordance with the distribution of the occupied volume of the control area, the shielding room or the shielding cabinet can be used selectively. Shielding room based on " Faraday cage " principle(that is, the potential difference is zero, the electric wave skin effect, the reflection attenuation) is made of a miraculous hexahedral conductive metal materials. Shielding room can suppress and prevent electromagnetic wave propagation in the air. It has two basic functions: to prevent external electromagnetic interference into the shielding room; to prevent the electromagnetic energy in the shielding room leak out. Shielding technology is an important means to prevent leakage of data information, the use of different structures and materials of the shielding room, generally can make the electromagnetic wave attenuation of 60 ~ 140dB. There are many types of shielding chamber by shielding material, such as copper mesh, steel plate, electrolytic copper foil type etc.; according to the structure, there are single plate type, double plate type, multi-layer, etc.; according to the form of the installation of points, there are welding, assembly, etc.. There are several factors that affect the performance of the shielding room: shielding materials, high conductivity metal materials can help to improve the shielding effectiveness of the shielding room, according to the cost and maintenance characteristics of industrial control system, the selection of steel plate;joining and welding technology, welding effect is generally better than other splicing methods, steel plate shielding material is more conducive to welding; the ventilation window and the shielding room door are the key components of the shield room, which directly influence the whole performance of the shielding room; power supply filter, the main reason for the widespread application of the power supply filter is the cost too high, limited by the conditions of the installation site. When the shield cabinet protection, the control side should avoid the use of CRT display, because the CRT display in the shield cabinet to open the hole, it will cause the electromagnetic leakage.

3.3 The Production of Electromagnetic Leakage on Controlled End Region

Controlled end of the industrial control system comprises a mechanical arm, independent small robots, unmanned aerial systems and intelligent sensor acquisition system. Controlled end is mainly responsible for instruction execution, data collection, data monitoring and other contents. The signal reception of controlled end area can be received both wired and wireless, and the advanced controlled end has a local memory. Since most of the controlled ends have a mobile and related mechanical operation function, the electromagnetic protection of the control side is proposed using the overall shielding technology. In addition to the sensor part, the rest of the metal box (shell) should be entire shielded to achieve a certain effect. If necessary, multi-layer or multi shielding materials can be used. Due to the diversity and complexity of the controlled end equipment, the openings or gaps in the mechanical structure are inevitable. But from the point of view of electromagnetic shielding, shielding structure should be seamless. The electromagnetic wave in the controlled end equipment radiates through the gap in the equipment, generating the information electromagnetic leakage. Therefore, it is necessary to use the sealing shield technology to protect the electromagnetic leakage.

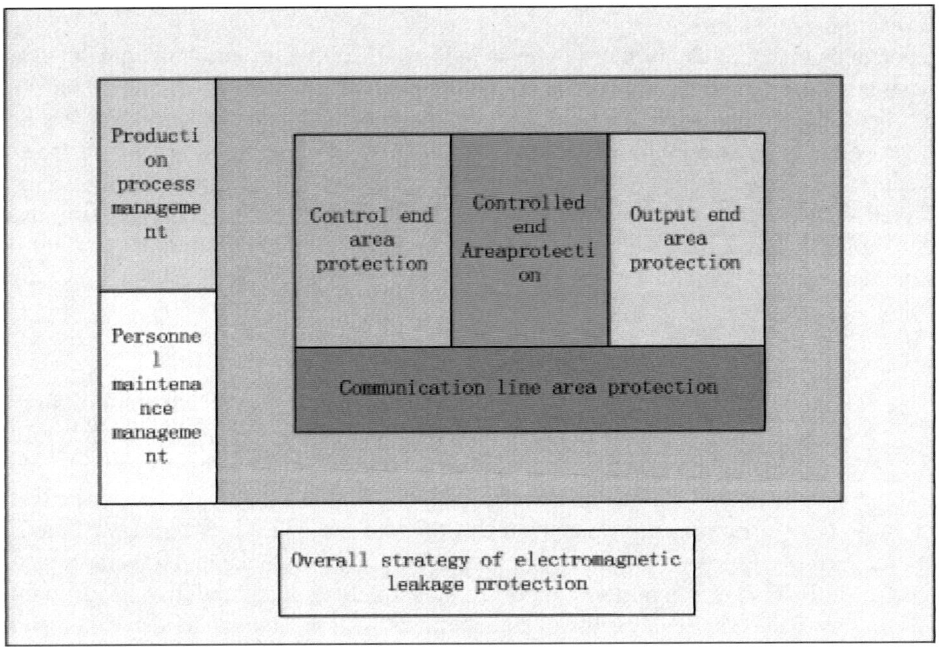

Figure 2 Multi-level protection structure

3.4 The Production of Electromagnetic Leakage on Output Terminal Region

The output of the industrial control system is generally integrated at the control end or the independent unit. The output device is mainly composed of a tape printer or related sound and light alarm equipment. In the process of debugging and running of the industrial control system, the output end provides the necessary information for the operation and maintenance personnel, and it is an integral part of the industrial control system. There is the local data storage in the output end, at the same time, according to the different output signals, the output is prone to cause electromagnetic leakage. The electromagnetic protection of the output end is suggested that the conductive coating comprises a metal conductive layer which is formed by various coating methods, conductive adhesive tape or other special materials. The conductive coating is mainly used to form a complete conductive layer on the surface of nonconductive material (such as plastic) by the method of metallization, in order to achieve the effect of absorbing and shielding electromagnetic waves.

3.5 Software Management of Industrial Control System

The relevant software system of industrial control system is a requisite part, and it can actuate the components of the industrial control system. The internal data flow of industrial control system mainly has four parts: data bus, address bus, control bus and I/O output. Among them, the first three parts of information which has common characteristics. They are all the parallel data streams (8, 16, 32, 64, etc.), provided that these parallel data let out, it was awfully difficult to restore. And it's all because: (a) the parallel multiple data bus with time synchronization and correlative frequency domain. In theory, it is difficult to separate the spectral information that is woven together. (b) These parallel data are binary coded. In different operating systems and applications, the definition of the same code is completely different. Accordingly, it is hard to determine the exact meaning of a group of binary coded. The content of this kind of radiation signal is mostly reflected the operation process of the computer, the radiation frequency is mainly concentrated in 2MHz ~ 450MHz. But the software can read and write storage to produce the continuous or periodic data stream, which can describe the inner data by periodic electromagnetic radiation. Therefore in the industrial control system, we can avoid internal software identifying signal to leak the data by electromagnetic radiation of the data flow.

3.6 The Maintenance Management

The prevention and control of electromagnetism in industrial control system with convenience and high maintenance mainly uses of shielding, coating and other methods. It is mainly based on the components of the cost. Therefore, in the course of the stage of equipment maintenance and debugging, there may be working procedures to disassemble or destroy the integrity of equipment coating on the equipment. These procedures will affect entire electromagnetic leakage protection system. At the same time, the artificial Trojan software implant and other equipment will also affect it. Therefore, the maintenance management should be strengthened. In the case of screening can dismantling or coating damaging, equipment of charged time should be reduced. And equipment maintenance personnel need to carry out the certification management.

3.7 The Production Process Management

Steel structure is mainly used in the factory of industrial control system. Common building materials are used outside the factory building. The distribution terminal of the factory has not been treated with special electromagnetic shielding. After the device is started, the electromagnetic signal can be monitored at the outside of the workshop. Consequently the actual production should be the focus for the production process of optimization management, prevent the concentration of pipelined a process heavy. The fundamental purpose of electromagnetic information leakage protection is to prevent electromagnetic emission by containing important information interception and decoding. So in addition to the above introduced a number of techniques, in the context of the existing equipment without changing the structure, we can use pseudo electromagnetic leakage protection. Pseudo electromagnetic leakage technology is adopted to intentionally designed to make the system produce without information of pseudo noise or pseudo leak, to drown or hide useful equipment in the system, the actual investigation and received basic information is not true information, enhanced the information security[5].

4. Conclusion

This paper analyzes the source of electromagnetic leakage from the angle of industrial control system. It comes up an approach that forms some regional electromagnetic protection to divide up some regions in order to prevent and control the electromagnetic leakage. According to daily production processes, this paper makes some suggestions about software management, operation and maintenance management as well as production flow management. Therefore it forms a set of effective prevention and control system of electromagnetic leakage. Prevention and control system of electromagnetic leakage is proposed in this paper can be extended to the computer control system.

5. References

[1] Guodong Shan, Yingxia Dai, Hang Wang. Research on Computer Electromagnetic Information Leakage and Protection .J. Electronic application technology. 2002 (4): 41.
[2] Jian Wen.The Leakage and Protection of Electromagnetic Energy in the Operation of Information Equipment .J. Information & Communications. 2015 (5): 55~56.
[3] Ting Sun. Electromagnetic Information Leakage .J. Security technology. 2004 (1): 33.
[4] Xiaoshan Wang, An Yang, Zhiqiang Shi, Limin Sun. New Trends of Information Security in Industrial Control Systems .J. Netinfo Security. 2015 (1): 46~47.
[5] Wei Liu, Dong Li, Bo Sun. Safety Analysis of Industrial Control System .J. Netinfo Security. 2012 (8): 57~58.
[6] S. Ghose; D. C. Pande.Identification of electromagnetic leakage from seams and joints and their prevention.C.ICEMIC.2003.1287868:339~342.

IOP Conf. Series: Journal of Physics: Conf. Series **910** (2017) 012005 doi:10.1088/1742-6596/910/1/012005

Network Security Risk Assessment System Based on Attack Graph and Markov Chain

Fuxiong Sun, Juntao Pi, Jin Lv and Tian Cao

School of Information and Safety Engineering , Zhongnan University of Economics and law, Wuhan, China
Email stxsfx@znufe.edu.cn, 1521030213@qq.com, 1530018223@qq.com, tian1163722295@qq.com

Abstract. Network security risk assessment technology can be found in advance of the network problems and related vulnerabilities, it has become an important means to solve the problem of network security. Based on attack graph and Markov chain, this paper provides a Network Security Risk Assessment Model (NSRAM). Based on the network infiltration tests, NSRAM generates the attack graph by the breadth traversal algorithm. Combines with the international standard CVSS, the attack probability of atomic nodes are counted, and then the attack transition probabilities of ones are calculated by Markov chain. NSRAM selects the optimal attack path after comprehensive measurement to assessment network security risk. The simulation results show that NSRAM can reflect the actual situation of network security objectively.

1. Introduction

While network applications have been fully into the social community of all aspects of life, network security issues has been becoming more and more prominent. As of December 2016, the number of PC infected Trojan horses monitored by the 360 security center was 247 million units in the report of China Internet Network Information Center (CNNIC) [1].It is well known that network vulnerabilities lead to network security problems, so network security risk (NSR) assessment as a means of active defense came into being, which can detect potential network vulnerabilities and threats to reduce the enterprise and individual network security risk.

This paper provides a NSR mode to unearth network vulnerabilities and evaluate risk levels. The paper is organized as follows: section 2 introduces associated successes achieved in research; in section 3, the NSRAM is proposed. In Section 4, some simulation experiments are performed to verify the performance of the model. Finally, this paper concludes from this work briefly

2. Related work

In the field of NSR assessment, many great works are accomplished by scholars. According to complex network environment, many intelligent algorithms and theories are introduced into this field. Jinsoo Shin et al. proposed a network security assessment model based on Bayesian network and event tree [2]. Fuzzy, AHP and D-S evidence theories were introduced to evaluate operations of the network services and the overall security situation of the network [3] [4] [5]. An improved immune algorithm was provided into a network security assessment method with risk theory [6]. An algorithm with polynomial complexity was analyzed based on network flow for network security risk assessment [7].

Network attack graph and its probability analysis are frequently used in network risk assessment. Yu Yajun et al. proposed an automatic analysis method of network attack graph based on graph kernel to analyze the attack mode of intruder effectively, and take the corresponding preventive measures according to the possible attack way [8]. For real-time analysis of network risk, Wang Xiao et al presented a real-time risk prediction method based on Markov time-varying model [9]. Qi Yong et al tion probability, which improved the association between vulnerabilities in the network. [10].

CVSS is a universal vulnerability scoring system, as an industry open standard, which measures the severity of vulnerabilities and to help determine the urgency and importance of the required response [11]. Wang Zuoguang et al introduced CVSS into quantitative risk assessment of industrial control systems based on attack-tree [12].

Based on current research findings, this paper proposes a new model NSRAM based on the penetration test, attack graph and CVSS, in which the scalable Markov chain is used to qualitatively calculate the risk value. According to the magnitude of the probability, the risk level of the enterprise's network security can be determined.

3. Model Structure

The structure of NSRAM is shown in Figure.1.

NSRAM is divided into four main modules: information acquisition; penetration test; automatic generation of attack graph; attack graph evaluation.

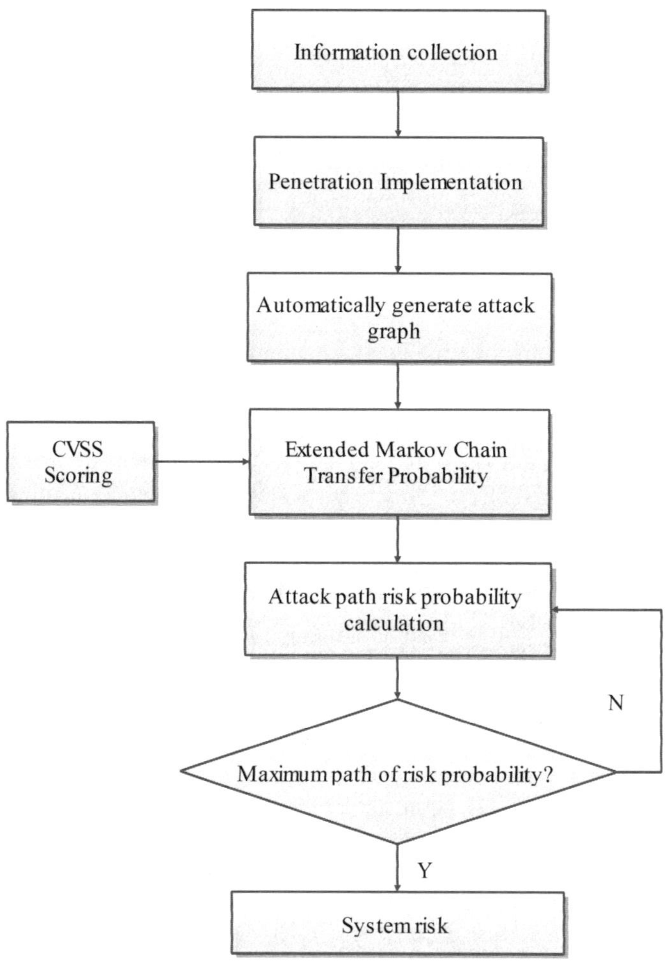

Figure 1 NSRAM Structure

3.1 Information Acquisition

This module is mainly the source of related information, which includes target information, network environment, test steps and so on. The target information consists of a five-tuple pattern (*Host_id*, *Host_vulSet*, *Host_con*, *Host_port*, *Host_data*):

- *Host_id* represents the only host.
- *Host_vulSet* indicates vulnerabilities exist in host, of which value is defined according to CVE (Common Vulnerabilities and Exposures).
- *Host_con* represents network connection between the host and other hosts, which is used to analyze network topology.
- *Host_port* shows what ports are open on the host.
- *Host_data* means host weight which indicates its location and importance in the network.

3.2 Penetration Implementation

The module is an automated test platform which is developed by Ruby script to automatically call security tools to complete network scanning and penetration. Security tools include *burp suite*, *nikto*, *and metasploit* and so on. According to *Host_vulSet*, the module generates the final penetration report.

3.3 Automatic Generation of Attack Graph

In the module, a breadth-first traversal algorithm is designed to generate the attack graph. The attribute of attacker is represented by a five-tuple pattern *Att* (*Att_id*, *Att_start*, *Att_target*, *Att_getAuth*, *Att_other*).

- *Att _id* represents an attacker.
- *Att_start* is the *Host_id* which starts to attack.
- *Att_target* is the *Host_id* of target host.
- *Att_getAuth* is access rights obtained after a successful attack.
- *Att_other* refers to the attack path, attack time, attack ability and other information.

3.4 Attack Graph Evaluation

In the module, the CVSS standard is used to score vulnerabilities on an atomic node. The vulnerability score is related to attack complexity that is the risk probability of an atomic node. Based on extended Markov chain, transition probabilities between nodes can be calculated. According to *Host_con*, all attack paths are drawn, and their attack risk are quantified. Finally, NSRAM takes the largest probability of attack path as the whole risk probability of the system.

4. System Modeling

4.1 Attack Graph Modeling

The attack graph is defined as$A = (A_s \cup A_d, T, R, E)$. Where A_s denotes a set of starting nodes; A_d denotes a set of target reachable nodes; T denotes an attack of an atomic node; $R \subseteq A_s \times A_d$ denotes a transition relation between states; E denotes a set of directed edges. The target result in A is defined as:

$$A(!(Att\_getAuth == root | Att\_getAuth == user)) \tag{1}$$

Equation (1) denotes the root or user permissions obtained on the target host after a successful attack.

The breadth-first algorithm is to build attack graph. From a host *Att_start* of which initial value is *Att_id*, algorithm traverses all hosts that can be connected to it, and tries to attack these connected hosts based on *R*. If an attack succeeds, a five-tuple pattern *Att* is formed. After a loop, the value of *Att_start* is updated by *Att_target* from an *Att*. Above steps are repeated until all nodes have traversed. In order to limit the length of attack paths, the maximum length L is set, which means the length of all traversal paths must not exceed L. Attack graph generation algorithm is shown in Figure.2.

In Figure .2, S is the stack of nodes, and *Pre* (*H*) and *Post* (*H*) are the pre and post node set, which can be obtained by network topology. Function $Exploit(H_i)$ is to determine whether the node is an atomic attack node, and form the atomic ones set. $Exploit(H_i)$ Will run until the traversal length is up to L, which can prevent the reverse circuit. In addition, repeated edge judgments (in steps 12, 13) are used to avoid duplication of paths and edges.

4.2 Markov Chain Risk Assessment

Based on the extended Markov chain, the automatic generation of attack graph is used to evaluate the network security status and realize the quantitative analysis of security risk of the whole system. The Markov Chain is a stochastic process in which the state changes randomly with time. A two-tuple *MC*= (*I*, *P*) is defined to represent the homogeneous Markov Chain, where $I=\{b_1, b_2, ... b_m\}$ is the state space and P is the transition probability matrix.

Input: Attack node *Att_start*, Maximum length of the path *L*, Network connection relationship *R*
Output: $A = (A_s \cup A_d, T, R, E)$
1. $A_s \cup A_d, T, R, E \leftarrow \emptyset$
2. $SetLen(H_i, 0)$ *The initial path node length is set to 0*
3. $A_s \leftarrow A_s \cup \{H_i\}$ *Add the host node to the attack graph*
4. $Push(S, H_i)$ *Add the node to the stack*
5. *While(IsNotEmpty(A))*
6. $H_t \leftarrow Pop(A)$
7. $len \leftarrow GetLen(H_i)$
8. $if(len < L)\{$
9. $T_e \leftarrow \{H_e | H_e \in Exploit(H_i)\}$
10. $T_{te} \leftarrow \{T_{te} | T_{te} \in A_e, < H_i H_{ie} > \in E\}$
11. $for\ each\ H_e \in A_e\{$
12. $\quad if\ (H_e \in A_{et})$
13. $\quad$ *Continue*
14. $\quad H_p \leftarrow Post(H_e)$
15. $\quad SetLen(H_p, len++)$
16. $\quad A_d \leftarrow A_d \cup \{H_p\}$
17. $\quad A_e \leftarrow A_e \cup \{H_e\}$
18. $\quad E \rightarrow E \cup \{< H_e, H_p >\}$
19. $\quad E \rightarrow E \cup \{< H_e, H_p > | H \in Pre(H_e)\}$
20. $\quad Push(S, H_p)$

Figure 2 Attack Graph Generation Algorithm

In the Markov process, the time T and state X are discrete and random. The Markovian property is usually expressed by the conditional distribution rate, and for any positive integer m, q and$0 \leq t_1 < t_2 < \cdots . < t_q < n$, exists:

$$P\{X_{n+m} = b_j | Xt_1 = bi_1, Xt_2 = bi_2, ... Xt_q = bi_q, Xn = bi\} = P\{X_{n+m} = b_j | Xn = b_i\}, b_i \in I \quad (2)$$

In (2), X_{n+m} is the state at time $n+m$ and$n, \mathrm{t_i}, m + n \in T$. The conditional probability P_{ij} of state b_i to b_j is defined as follows:

$$P_{ij}(m, m + n) = P\{X_{n+m} = b_j | X_n = b_i\} \quad (3)$$

P_{ij} Is also called the transition probability of the moment. Based on the extensible Markov chain, after analyzing the attack graph, it is found that the simple Markov chain only indicates the state of the attack, and there is no reason for the migration of the attack state. For this reason, its concept is represented by the triplet *EMC*= (*I*, *P*, *A*), where the set of all available methods is denoted as $A=\{G_1, G_2, G_3, \ldots, G_k\}$. The A is called the set of attack actions.

The nodes in the chain represent the state of the network, and the directed edge represents the weight of the state transition which corresponds to the transition probability. For each node, the sum of the weights is homogeneous. The transition probability matrix is defined as:

$$P = \begin{bmatrix} p_{11} & \cdots & p_{1n} \\ \vdots & \ddots & \vdots \\ p_{n1} & \cdots & p_{nn} \end{bmatrix} \tag{4}$$

In (4), $\forall i = 1,2,3, \ldots \sum_{j=1}^{+\infty} P_{ij}\,(n, n+1) = 1$.

The CVSS defines the complexity of the attack. Most of the researches use it to classify the attack difficulty, as shown in Table 1.

Table 1 The CVSS standard

Id	Weight	Level
E0	0.35	Network risk level is high, the penetration of low degree of difficulty
E1	0.61	Network risk level medium, easy to penetrate the degree of moderate
E2	0.71	Network risk level is low, the penetration of high degree of difficulty
E3	0.71	Network risk level is undefined

But the CVSS standards have few levels of definition which applies to a simple network environment. Therefore, this paper adopts an improved CVSS to calculate the risk probability of nodes, as shown in Table 2.

Table 2 The CVSS standard defines the ease of vulnerability penetration attacks

Id	**Weight**	**Level**
E0	0.1	Public reporting but the attack is only theoretically possible or not publicly reported weaknesses
E1	0.2	Public report but no attack method
E2	0.3	Publicly report and refer to possible attack methods
E3	0.4	Public report and a rough description of the attack method
E4	0.5	No ready attack tool but with more detailed attack steps
E5	0.6	There is no ready attack tool but there are very detailed attack steps
E6	0.7	There is a corresponding attack code and use the method
E7	0.8	Customize the available attack tools with detailed attack steps
E8	0.9	Ready-to-use attack tools with detailed attack steps

After an attack action $G_r \in A$ is selected, the state transfers from node Z_i to the other one Z_j, where $0<i,j<n$, $0<r<k$. If the penetration difficulty of Gr is $E(G_r)$ in CVSS, the transition probability from Zi to Z_j is defined as.

$$P_{ij}\left(Z_i \xrightarrow{G_r} Z_j\right) = \frac{E(G_r)}{\sum_{r=0}^{n} E(G_r)} \tag{5}$$

In the attack graph, I am divided into two sets: target node set Iz and non-target one Ix, where $I_z \cup I_x = I$. The $V(Z_i) \in [0,1]$ is the probability from $Z_i \in I$ to $Z_j \in I_z$ define as follows:

$$V^n(Z_i) = \sum_{Z_i \in STATUS(Z_i)} \sum_{G_i \in REGULATION(Z_i \to Z_j)} P(Z_i \xrightarrow{G_r} Z_j) \times V^{n-1}(Z_j) \tag{6}$$

In (6), $\forall Z_i \in I_z$ then $V^0(Z_i) = 1$, and $\forall Z_i \in I_x$ then $V^0(Z_i) = 0$. $STATUS(Z_i)$ Is the node set that successful attack is accomplished by only one attack action. $REGULATION(Z_i \to Z_r)$ Is optional rule from Z_i to Z_j. After the n iteration, the iterative process is considered convergent if $V^n = V^{n-1}$. For example, there is a directed attack graph with transition probability shown in Figure.3.

It is assumed that the target node is 4 in Figure.3, then $I=\{0,1,2,3,4,5\}$, $I_z=\{4\}$, $I_x=\{0,1,2,3,5\}$, $A=\{G_1,G_2,G_3,G_4, G_5,G_6,G_7\}$. The directed graph is divided into two layers, and to attack the target can be done at most two steps: $Z_0 \to Z_1 \to Z_4$ and $Z_0 \to Z_2 \to Z_4$. Table 3 illustrates the iterative process of attack graph.

Table 3 Iterative Process

Iterative	Z_0	Z_1	Z_2
0	$V^0(0) = 0$	$V^0(1) = 0$	$V^0(2) = 0$
1	$V^1(0) = 0$	$V^1(1) = 1*E(G_4)$	$V^1(2) = P_4*E(G_5)$
2	$V^2(0) = P_1 * 1 * E(G_1) * E(G_4) + P2 * E(G_2) * P_4 * E(G_5)$	$V^2(1) = 1*E(G_4)$	$V^2(2) = P_4*E(G_5)$
3	$V^3(0) = P_1 * 1 * E(G_1) * E(G_4) + P2 * E(G_2) * P_4 * E(G_5)$	$V^3(1) = 1*E(G_4)$	$V^3(2) = P_4*E(G_5)$

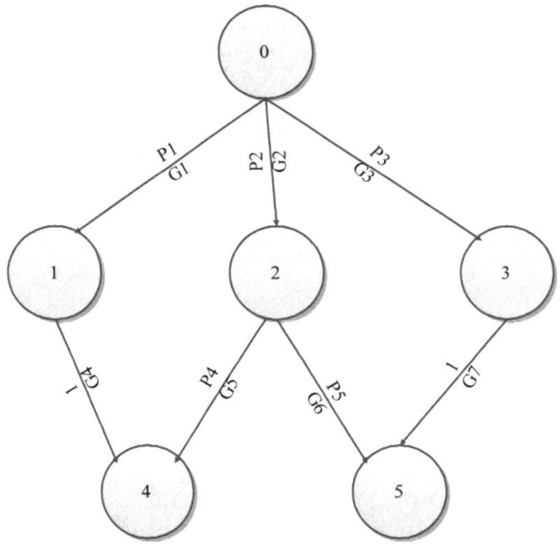

Figure 3 Attack Graph Example

5. Experimental Analysis

5.1 Experimental Environment

In the experimental environment, there are three hosts (*H3*, *H4*, *H5*), one server (*H1*), one router and one firewall (*H2*) shown in Figure.4. The operating systems of three hosts and server are windows XP. The attack host (*A*) is Linux. In *A*, an automated test platform is installed which is developed by Ruby script to automatically call security tools to complete network scanning and penetration. Security tools are installed including *burp suite*, *nikto*, *metasploit* and so on. The three hosts are located in Intranet protected by firewall and server provides some network services to extranet.

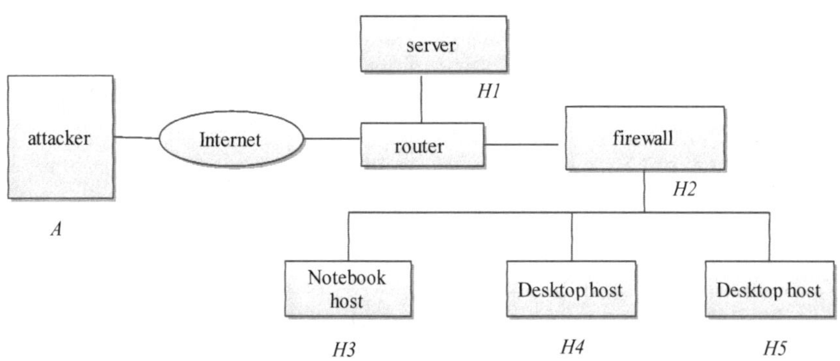

Figure 4 Experimental Environment

In the simulation experiment, there are five typical vulnerabilities exist in network, listed in Table 4.

Table 4 Vulnerability Selection

Host_id	Host type	Host_vulSet	Att_getAuth
H1	server	ServU5.0	Root
H2	Firewall	telnet	Root
H3	Notebook host	SQL	Root or user
H4	Desktop host	RPC	Root
H5	Desktop host	Remote login	Root

5.2 Generate Attack Graphs

By the breadth traversal algorithm proposed in Figure. 2, the attack graph is generated in Figure.5 when the target node is *H3*.

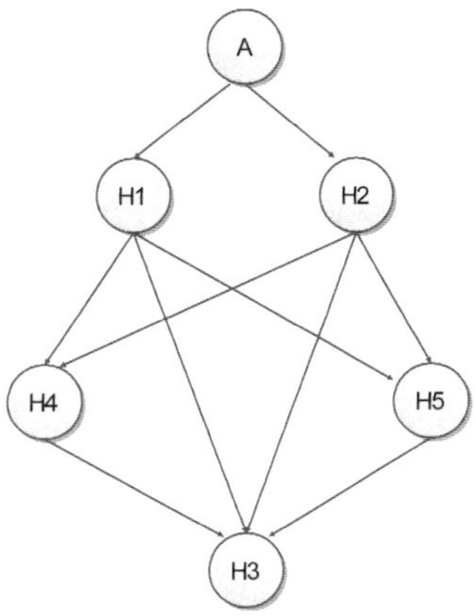

Figure 5 Generated Attack Graph

The probability of a successful attack is determined by the probability that the vulnerability is breached, which can be defined by the penetration attack complexity in CVSS, as shown in table 5.

Table 5 Penetration Attack Complexity Table

Attack action	id	weight
ServU5.0	*E3*	0.4
telnet	*E7*	0.8
SQL	*E5*	0.6
RPC	*E4*	0.5
Remote login	*E8*	0.9

Based on table 5, the corresponding attack graph with transition probability is calculated by (5). Then the transition probability of $A \rightarrow H1$ is 0.4/ (0.8+0.4) =0.33, $A \rightarrow H2$ is 0.8/ (0.4+0.8) =0.67, and so on. By (4), the transfer probability matrix is obtained as shown in Table 6.

Table 6 Transfer probability matrix

End state Start state	*A*	*H1*	*H2*	*H3*	*H4*	*H5*
A	0	0.33	0.67	0	0	0
H1	0	0	0	0.3	0.25	0.45
H2	0	0	0	0.3	0.25	0.45
H3	0	0	0	0	0	0
H4	0	0	0	1	0	0
H5	0	0	0	1	0	0

Attack graph with transition probability is shown in Figure. 6.

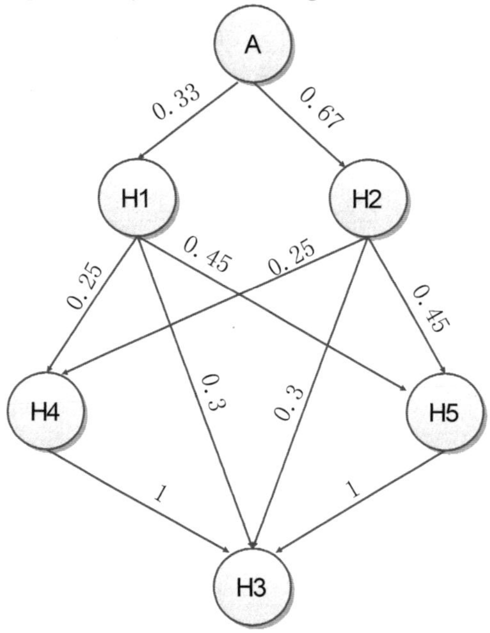

Figure 6 Attack Graph with Transition Probability

5.3 Attack Graph Evaluation

Markov chain is used to calculate the probability of attack of attack graph, and then the complexity of the attack is evaluated. The probability from each node to another node depends mainly on the transition probability *P* shown in Figure.5. There are 6 attack paths to *H3*, for example the *P* of the 1st path is calculated as *P*=0.33*0.4*0.25*0.5*1*0.6 = 0.0099. Table 7 shows the *P* values of all attack paths.

Table 7 Multiple attack paths

Path NO.	Starting node	Target node	Traverse the node path	Attack probability
①	*A*	*H3*	*A→H1→H4→H3*	0.0099
②	*A*	*H3*	*A→H1→H3*	0.02376
③	*A*	*H3*	*A→H1→H5→H3*	0.016038
④	*A*	*H3*	*A→H2→H4→H3*	0.0402
⑤	*A*	*H3*	*A→H2→H3*	0.09684
⑥	*A*	*H3*	*A→H2→H5→H3*	0.130248

According to Figure. 6, the target node *H*3 can be reached by 3 steps at most. The attack path ⑥ is more probably of which probability is 0.130248. By (6), the overall risk value is $V^3(H3)$ =0.0099+0.02376+0.016038+0.0402+0.09684+ 0.130248=0.316986. According to CVSS standard, the network security level is *E0* level which means that the network danger level is high but penetration degree is low.

6. Conclusion

Network vulnerability is one of main security threats of enterprise network. Mining network vulnerability helps to improve the management of network security risk. Network vulnerabilities exist in hosts and various network services, and are closely related to the current network topology. This paper proposed the NSRAM which can automatically and periodically collect network topology and its vulnerability information. Combining the improved CVSS and probability evaluation of attack path, NSRAM determines the risk level of enterprise network security. Simulation experiments verify its feasibility and flexibility.

In large and complex network, the changes of network topology at any time lead to frequent migration of network security risk. Incremental evaluation of attack graph and path will greatly improve the efficiency of the model, which is the focus of the future research.

7. Acknowledgment

This research was financially supported by the education research of Zhongnan University of Economics and law (2016J1408) and Youth project of the National Social Science Foundation (16CXW019).

8. References

[1] China Internet Information Center (ENNIC). The 39th China Internet Development Statistics Report[EB/OL],2017.
[2] Jinsoo Shin,Hanseong Son,Gyunyoung Heo,Cyber Security Risk Evaluation of a Nuclear Instrumentation and Control System Using Bayesian Networks and Event Trees[J]. Nuclear Engineering and Technology,2017(49):517-524.
[3] Sun F X. Danger Theory Based Model for Network Security Risk Assessment[J]. Advanced Materials Research, 2011(187):148-154.
[4] Li Qun Tang,Jian Ping Ma,Hui Qiang Wang. Research on a Method of Service-Oriented Network Layer Network Security Situation Assessment[J]. Applied Mechanics and Materials,2013: 260-264.
[5] Fang Yang, Study on the AHP&Dset Telecom Network Risk Assessment Model and The Application[D]. Beijing University of Posts and Telecommunications, 2010.
[6] Yuan H. A Network Security Risk Assessment Method Based on Immunity Algorithm[J]. Advanced Materials Research, 2010:948-953.
[7] Wu Jinyu.Research on Key Technologies of Network Security Risk Assessment [D]. Beijing University of Posts and Telecommunications, 2013.
[8] Yu Yajun, Liu Zeshen, Pan Zhisong, Hu Guyu, Chang Qing.The Network Attack Graph Analysis Based on Graph Kernel [J]. Military Communications Technology, 2016 (03): 20-25.
[9] WANG Jie, LI Qian-guo, QI Yong. A Real-time Analysis Method of Network Security Risk Based on Markov Model [J] .Computer Science, 2016: 338-341.
[10] Qi Yong, LIU Min, LI Qian-guo.An attack graph model based on extended Markov chain [J]. Computer Engineering and Design, 2014, (12): 4131-4135.
[11] K Scarfone,P Mel. An Analysis of CVSS Version 2 Vulnerability Scoring[C], International Symposium on Empirical Software Engineering and Measuremen. 2009:516-525.
[12] Wang Zuoguang, WEI Qiang, LIU Wen-wen. Quantitative risk assessment of industrial control systems based on attack-tree and CVSS [J] .Application Research of Computers, 2016 (12):3785-3790.

IOP Conf. Series: Journal of Physics: Conf. Series **910** (2017) 012006 doi:10.1088/1742-6596/910/1/012006

Research on Parallelization of GPU-based K-Nearest Neighbor Algorithm

Hao Jiang and Yulin Wu

School of Computer Science and Engineering, Southeast University, Nanjing, China
Email: hjiang@seu.edu.cn, daiboaifadai@126.com

Abstract. Based on the analysis of the K-Nearest Neighbor Algorithm, the feasibility of parallelization is studied from the steps of the algorithm, the operation efficiency and the data structure of each step, and the part of parallel execution is determined. A K-Nearest Neighbor Algorithm parallelization scheme is designed and the parallel G-KNN algorithm is implemented in the CUDA environment. The experimental results show that the K-Nearest Neighbor Algorithm has a significant improvement in efficiency after parallelization, especially on large-scale data.

1. Introduction

In the background of the application of big data, the rapid excavation of useful information in massive data raises a higher demand for computational speed. GPU (Graphic Processing Unit) possesses strong computing power, and its own architecture can well meet the requirements of large-scale parallel computing[1]. Many researches on data mining algorithms have turned to parallel algorithms based on GPU. The K-Nearest Neighbor (KNN) algorithm is a basic classification and regression algorithm method proposed by Cover and Hart in 1968. Because of its good classification performance, KNN has been widely used in many fields. In this paper, the parallelization of KNN algorithm is studied. According to the CUDA programming characteristics and the GPU architecture used in this experiment, the parallel algorithm of KNN algorithm is designed. The tasks that can be parallelized are implemented on GPU, and the correctness of the parallelization algorithm and the improvement in execution efficiency are verified by experiments[2].

2. KNN Algorithm and Its Analysis

The main idea of the KNN algorithm is: Given a training data set, for a newly entered instance (where an instance can be regarded as a point in the feature space), If the majority of the k instances of the nearest neighbor to the instance belongs to a certain class in the training data, the input instance will be sorted to this classification[3]. Specific steps are as Algorithm 1.

The main problem with the implementation of the KNN algorithm is how to quickly search for k nearest neighbors, which is especially necessary when the dimensions of the feature space and the training data sets are large. The simplest method of solving this problem is linear scanning, or called the exhaustive method. It is divided into two steps. First, calculate the distance between the input instance and all instances in the training set based on the given distance metric; second, find the nearest K instances by sorting algorithm. Assuming that the size of the feature set of the training set is n, the training set capacity is N and the test set capacity is M, then the time complexity of the whole KNN algorithm is $O(M \times n \times N) + O(N [\times M \times \log] k)$[4].

Algorithm 1: KNN Algorithm;

Input: Training data set $T=\{(x_1,y_1), (x_2,y_2),\cdots, (x_N,y_N)\}$, wherein, $x_i \in \chi \subseteq R^n$ is the feature vector of the instances, $y_i \in Y = \{c_1, c_2, \dots, c_k\}$s the class of instances, $i = 1,2,\cdots,N$; the feature vector of the instances x;

Output: Class y to which the instance x belongs.

Process:

(1) According to the given distance metric, the distance between the instance X and all data points in the training set T is computed as D;

(2) In the D(x) sort, find the k points with the smallest distance, that is, the nearest k points, the neighborhood of the x that covers the k points is denoted as $N_k(x)$;

(3)In the $N_k(x)$ according to classification decision rules (such as the majority of votes)to determine the class y to which x belongs: $y = \arg\max_{c_j} \sum_{x_i \in N_k(x)} I(y_i = c_i), i = 1,2,\cdots,N; j = 1,2,\cdots,K$,Where I is the indicating function, that is, when $y_i = c_i$, I is 1, otherwise I is 0.

The key part of KNN algorithm is distance calculation, especially when the size of feature space is large. How to optimize the distance calculation is the main problem to improve the KNN algorithm. In practical applications, the training samples are represented by feature vectors. To Euclidean distances, for example, the distance between vectors is actually equivalent to the multiplication of matrices[5].

Assuming that the capacity of the training set T is N, the capacity of the test set P is M, the feature space χ is the n-dimensional real vector space R^n , $x_i, y_j \in \chi, x_i = (x_i^{(1)}, x_i^{(2)}, \cdots, x_i^{(n)}), y_j = (y_j^{(1)}, y_j^{(2)}, \cdots, y_j^{(n)})^T$, then the distance between xi and yj is $L(x_i, y_j) = (\sum_{l=1}^{n} |x_i^{(l)} - y_j^{(l)}|^2)^{\frac{1}{2}}$. T and P can be expressed as follows in Eq. (1) and in Eq. (2), and defines L (T, P) in Eq. (3) as the matrix representation of the Euclidean distance between the instances in the test set P and the training set T; each column vector $L_j = (L(x_1, y_j), L(x_2, y_j), \cdots, L(x_N, y_j))^T$ in L(T,P) is the vector representation of the Euclidean distance between the instance yj in the test set P and the points in the training set T.

$$T = \begin{pmatrix} x_1 \\ \vdots \\ x_i \\ \vdots \\ x_N \end{pmatrix} = \begin{pmatrix} x_1^{(1)} & x_1^{(2)} & \cdots & x_1^{(n)} \\ \vdots & \vdots & \cdots & \vdots \\ x_i^{(1)} & x_i^{(2)} & \cdots & x_i^{(n)} \\ \vdots & \vdots & \cdots & \vdots \\ x_N^{(1)} & x_N^{(2)} & \cdots & x_N^{(n)} \end{pmatrix}_{N\times n} \quad (1)$$

$$P = \begin{pmatrix} y_1 \\ \vdots \\ y_j \\ \vdots \\ y_N \end{pmatrix}^T = \begin{pmatrix} y_1^{(1)} & \cdots & y_j^{(1)} & \cdots & y_M^{(1)} \\ y_1^{(2)} & \cdots & y_j^{(2)} & \cdots & y_M^{(2)} \\ \vdots & \vdots & \vdots & \vdots & \vdots \\ y_1^{(n)} & \cdots & y_j^{(n)} & \cdots & y_M^{(n)} \end{pmatrix}_{n\times M} \quad (2)$$

$$L(T,P) = \begin{pmatrix} L(x_1,y_1) & \cdots & L(x_1,y_j) & \cdots & L(x_1,y_M) \\ \vdots & \vdots & \vdots & \vdots & \vdots \\ L(x_i,y_1) & \ldots & L(x_i,y_j) & \cdots & L(x_i,y_M) \\ \vdots & \vdots & \vdots & \vdots & \vdots \\ L(x_N,y_1) & \cdots & L(x_N,y_j) & \cdots & L(x_N,y_M) \end{pmatrix}_{N\times M} \quad (3)$$

Each instance y_j in the test set P, L_j can be obtained by calculating the distance. Then sort the items in L_j to get the nearest k points to y_j. Finally, the class of y_j is determined according to the classification rules (e.g. majority vote). The foregoing analysis shows that the efficiency of the k nearest neighbor algorithm is the time spent in calculating the distance between an instance and an n-dimensional training set and to sort the nearest k distances, especially the former.

3. Scheme Design of Parallel KNN Algorithm

Through the analysis of section II, in this paper, the parallel strategy of GPU-based KNN algorithm is divided into the following two parts.

3.1 Parallelization of distance calculations

The process of obtaining the distance matrix $L(T,P)$ from the training set T and the test set P is similar to matrix multiplication. The difference is that each item of the result matrix in the matrix multiplication is the dot product of the row vector and the column vector, while each item in $L(T,P)$ is the Euclidean distance computed by row vector and column vector. Therefore the process of obtaining $L(T,P)$ can be completely similar to the process of matrix multiplication, which can be accelerated in the CUDA platform through the GPU parallel computing and improve computing efficiency. In order to maximize the efficiency of matrix multiplication in the CUDA framework and make every thread in the GPU do more work, the matrices T and P can be divided into a number of small matrices. Each block is responsible for calculating the multiplication of the small matrix, making full use of the GPU's computing resources.

3.2 Parallelization using parity sorting algorithm

The idea of parity sorting is repeating scans in arrays twice. The first scan selects data items for each pair of a [j] and a [j + 1], Where j is an odd number (j=1,3,5……). If the value of their keywords is reversed, then exchange them. The second scan performs the same operation on all even data items (j=2,4,6……). Repeat this both sorts until the array is all ordered. Because odd pairs and even pairs are independent of each other, each sort of order, odd pairs or even pairs can be compared and exchanged using the GPU multi-threaded, as to sort quickly. Specifically, assuming that the array size to be sorted is N, the array is divided into N/(1024*2) one-dimensional blocks according to the size of the block (In this experiment, the GPU thread block is assigned up to 1024 thread), making full use of the GPU's thread computing resources. Since every block and every thread within the block are parallel, so in the odds and even pairs of comparison and exchange in each trip, the comparison and exchange of each pair is done by a separate thread, and they are parallel.

To this end, this paper presents a GPU-based parallel algorithm, G-KNN algorithm, to improve the efficiency of the original KNN algorithm. The processes that CPU and GPU collaborate with are shown in Figure. 1.

Specific steps are as follows:

(1) Preprocessing the training set T and the test set P on the CPU and initialize k;
(2) Pass the required calculation data into GPU video memory;
(3) The distance between matrix T and vector P is computed in parallel by GPU multi-core and multithreading;
(4) L is the calculated distance matrix, and sort the column vectors in L corresponding to each test instance;
(5) Returns the sorted distance matrix L back to CPU;
(6) Select the k nearest neighbor distances of the test instance, and use the classification rules to

get the category of the test instance.

In order to maximize the efficiency of the distance calculation in the CUDA framework as well as to make the every thread in GPU to do more work, the matrices T and P are divided into multiple block matrices. Each block is responsible for calculating the multiplication of the block matrix, and Make full use of the computing resources of GPU, as shown in Figure. 2[6].

As in Figure. 2, using the shared memory of the GPU, the matrix data that needs to be calculated is read into the shared memory, and then the data in the shared memory will be calculated.

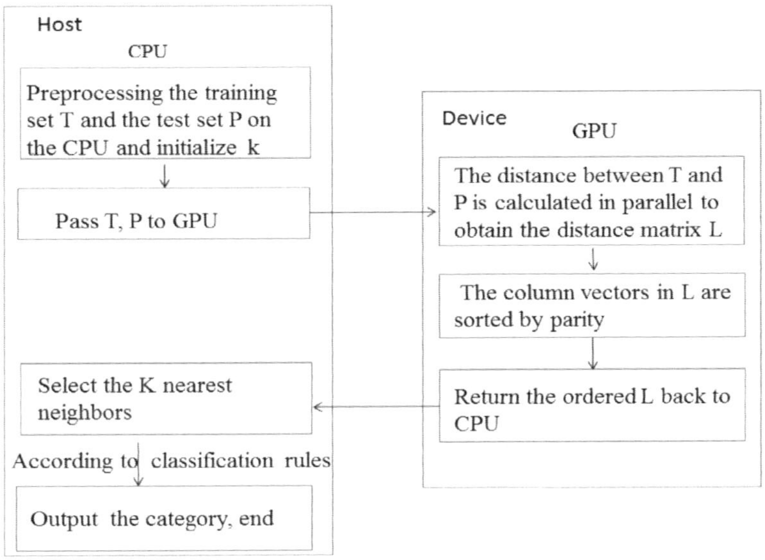

Figure 1. The processing flow of G-KNN algorithm

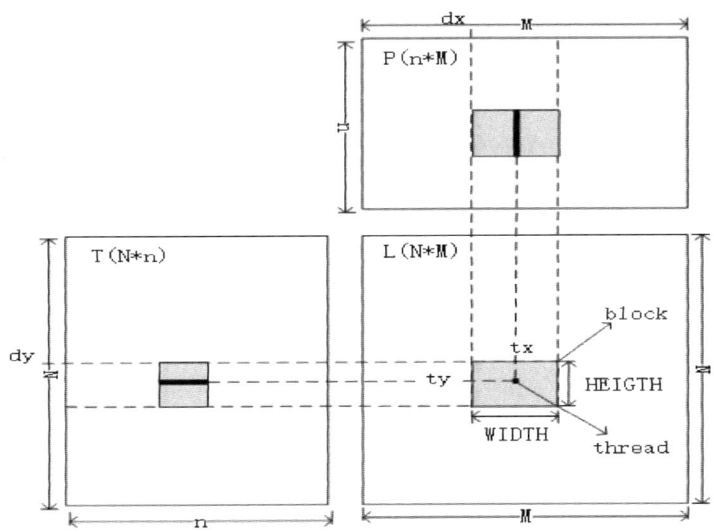

Figure 2 Matrix calculation model

This approach improves performance compared to reading data from a global memory every time. In the memory structure and thread mapping, the grid dimension is set to grid (M/WIDTH, N/HEIGH), the block dimension is set to block (WIDTH, HEIGTH). For example, a block in L in the figure requires data surrounded by the T and P dashed lines, and each block has WIDTH*HEIGHT threads to complete the block calculation. The GPU thread block is allocated up to 1024 threads in the experiments of this paper, thus in the specific implementation, WIDTH and HEIGTH are set to 32, so that GPU thread resources can be fully utilized.

4. Experiment and Analysis

The experimental platform is configured as: Intel(R) Core(TM) i7-4790K and the GTX980Ti of the NVIDIA Maxwell architecture. The experimental data set is a randomly generated N D-dimensional samples, and each sample has a label, which is identified based on the positive and negative of the first dimension data. 90% of the sample data was chosen as training data, and the remaining 10% was used as test data. As the experiment is mainly to verify the efficiency of G-KNN calculation, so the k in the comparison experiment takes 20. The distance measure adopts Euclidean distance, and the classification decision rules adopt majority voting method.

Table 1 The Time Comparison of Two KNN Algorithms When D = 20

	calculation time/s		
N	G-KNN		speedup
1000	0.62	2.11	3.40
2000	0.87	11.97	13.76
4000	1.36	64.28	45.79
8000	2.60	266.88	102.65
16000	6.94	1040.72	149.96

Table 2 The Time Comparison of Two KNN Algorithms When D = 40

	calculation time/s		
N	G-KNN	KNN	speedup
1000	0.65	2.25	3.46
2000	0.97	12.57	12.76
4000	1.48	67.54	45.6
8000	2.80	300.87	107.45
16000	8.55	1277.18	149.37

Table 3 The Time Comparison of Two KNN Algorithms When D = 60

	calculation time/s		
N	G-KNN	KNN	speedup
1000	0.66	2.32	3.51
2000	0.98	12.92	13.18
4000	1.62	78.07	48.19
8000	2.94	312.05	106.14
16000	9.64	1487.62	154.31

Table 4 The Time Comparison of Two KNN Algorithms When D =80

	calculation time/s		
N	G-KNN	KNN	speedup
1000	0.66	2.49	3.77
2000	0.98	14.39	14.68
4000	1.77	82.40	46.55
8000	2.94	319.49	108.67
16000	11.51	1766.87	153.50

Table 1 to 4 respectively record the comparison results of the G-KNN and KNN algorithms when the data dimension D is from 20 to 80. Wherein the data set size N is from 1000 to 16,000.The calculation time and speedup ratio of the G-KNN algorithm and the KNN algorithm are shown in the tables.

Experimental result shows that, the distance calculation in the KNN algorithm occupies the entire algorithm for most of the time, and that especially as the data set and the data object dimension increase, distance calculation accounts for more than 90% of the time. Distance calculation in acceleration algorithm can improve the efficiency of the implementation of the algorithm. As it can be seen from the experimental results acceleration algorithm significantly improved in efficiency.

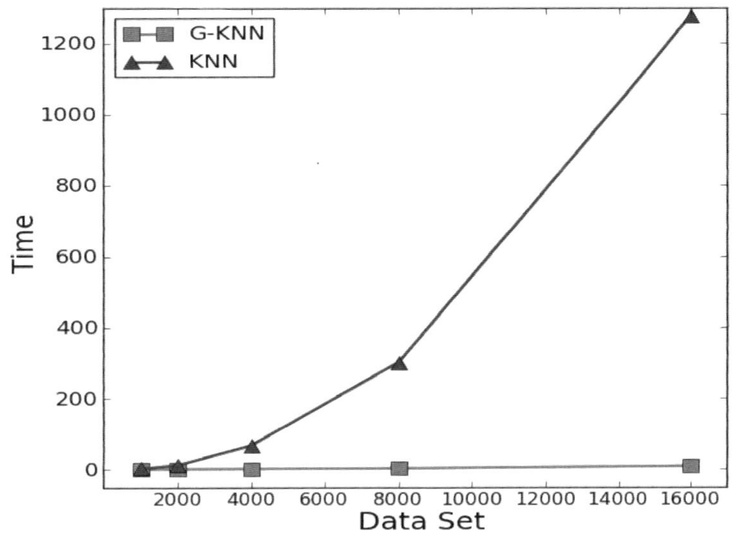

Figure 3 Time contrast at D=40

Figure. 3 illustrates a comparison of the computation time between the KNN algorithm and the G-KNN algorithm when the data set size is N from 1000 to 16000 and the data dimension D=40. It can be seen that the advantages of the G-KNN algorithm in performance are not significant when the data set is small, but with the increasing size of the data set, the advantages of G-KNN algorithm are more and more obvious.

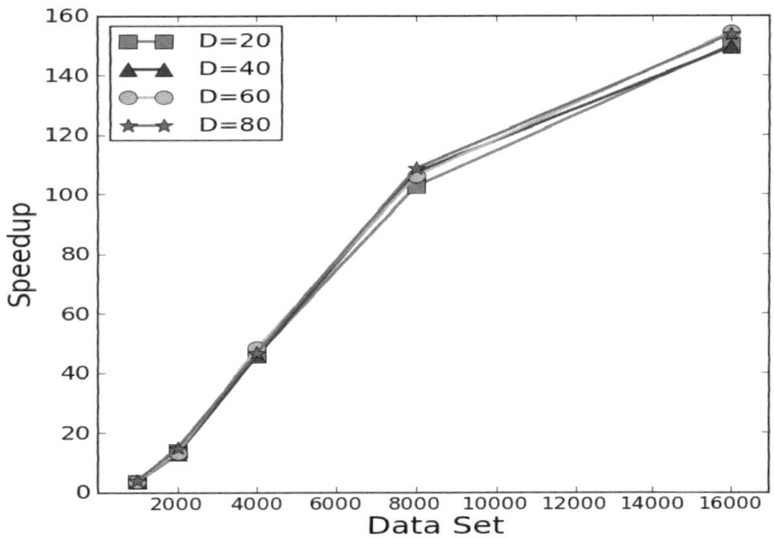

Figure 4 Speedup when D takes different value

Figure. 4 illustrates the speedup ratio of the G-CPU algorithm relative to the KNN algorithm when the data set size is from 1000 to 16000 and the data dimension D is from 20 to 80. It can be seen that the speedup effect of G-KNN algorithm is obvious, and the speedup effect is not related to the value of D. With the increasing size of the data set, the speedup of the algorithm increases, and all exceed 140x when N=16000.

5. Summary

The G-KNN algorithm presented in this paper parallelized the distance calculation and sorting in the KNN algorithm. The performance of the algorithm has been improved significantly, especially in the calculation of the distance of the algorithm. The acceleration effect is more remarkable, and the acceleration ratio of the parallel algorithm increases with the increasing scale of the data set. The KNN algorithm is well suited for parallel execution on the GPU.

6. Acknowledgment

This work was supported by Collaborative Innovation Center of Wireless Communications Technology and the National Natural Science Foundation of China (Grant No. 61672152).

7. References

[1] WU En-Hua. State of the Art and Future Challenge on General Purpose Computation by Graphics Processing Unit[J]. Journal of Software. 2004(10),In Chinese.
[2] Garcia V, Debreuve E, Barlaud M. Fast k nearest neighbor search using GPU[C]. IEEE Computer Society Conference on Computer Vision and Pattern Recognition Workshops. IEEE, 2008: 1-6.
[3] Cover T, Hart P. Nearest neighbor pattern classification. IEEE Transactions on Information Theory, 1967, 13(1):21-27.
[4] Tian Pan,Hua Bei,Lu Li. Implementation of K-nearest Neigbor Algorithm Based on GPU[J].Computer Engineering, 2015,41(2): 189-192,198, In Chinese.
[5] LI Hang. Machine Learning[M].Tsinghua university press,2012. 95-130, In Chinese.
[6] LI Xiao-Wen,CUI Xiang. Performance optimization of matrix multiplication and FFT in GPU. Modern Electronics Technique[J]. 2013, 36(4): 80-84, In Chinese..

Resource Isolation Method for Program'S Performance on CMP

Ti Guan[1,a], Chunxiu Liu[2,b], Zheng Xu[1,c], Huicong Li[1,d] and Qiang Ma[1,e]

[1]State Grid Shandong Electric Power Company, Jinan 250001, Shandong Province, China

[2]State Grid Shandong DeZhou Electric Power Company, Dezhou 253008, Shandong Province

Email: [a]sddl_zdh@163.com, [b]1957626310@qq.com, [c]xuzheng7788@126.com, [d]lhc_111@163.com, [e]xjtusys@126.com

Abstract. Data center and cloud computing are more popular, which make more benefits for customers and the providers. However, in data center or clusters, commonly there is more than one program running on one server, but programs may interference with each other. The interference may take a little effect, however, the interference may cause serious drop down of performance. In order to avoid the performance interference problem, the mechanism of isolate resource for different programs is a better choice. In this paper we propose a light cost resource isolation method to improve program's performance. This method uses Cgroups to set the dedicated CPU and memory resource for a program, aiming at to guarantee the program's performance. There are three engines to realize this method: Program Monitor Engine top program's resource usage of CPU and memory and transfer the information to Resource Assignment Engine; Resource Assignment Engine calculates the size of CPU and memory resource should be applied for the program; Cgroups Control Engine divide resource by Linux tool Cgroups, and drag program in control group for execution. The experiment result show that making use of the resource isolation method proposed by our paper, program's performance can be improved.

1. Introduction

In recent years, data center and cloud computing are more popular, which make more benefits for customers and the providers. However, in data centre or clusters, commonly there is more than one program running on one server, but programs may interference with each other. The interference may take a little effect, however, the interference may cause serious drop down of performance. For programs running in cluster or data centers, serious performance dropdown is not allowed. In order to avoid the performance interference problem, the mechanism of isolate resource for different programs is a better choice.

The common resource isolation method is virtual machine and container. But both the methods are all have shortcomings. For example, the virtual machine often takes a high cost, which consumes a large amount of resource of maintenance the virtual machine. At the same time, the container consumes little resource for resource isolation, but if it running the program processing the database, the database is too large for the container, and can not be used for other programs.

In this paper we propose a light cost resource isolation method to improve program's performance. This method uses Cgroups to set the dedicated CPU and memory resource for a program, aiming at to guarantee the program's performance. There are three engines to realize this method: 1) Program Monitor Engine top program's resource usage of CPU and memory and transfer the information to

Resource Assignment Engine; 2) Resource Assignment Engine calculates the size of CPU and memory resource should be applied for the program; 3) Cgroups Control Engine divide resource by Linux tool Cgroups, and drag program in control group for execution.

The experiment result shows that making use of the resource isolation method proposed by our paper, program's performance can improve 8.65%. The average CPU utilization improves 24.5%, and the average memory utilization improves 21%.

2. Related work

There are many methods for resource isolation of programs.

The first one is the virtual machine. Virtual machine is a fully integrated computer system with a complete hardware system functioning in a fully isolated environment [1, 2, 3]. Virtual machine can be used for workload consolidation and share hosting. The program can be running in virtual machine with an un-interference execution environment. But the shortcoming is the low performance because of the high cost.

The second method is the container. Container is used for consistent development and deployment.For example, Docker and Linux container. Docker [4, 5] is the lightweight container to isolate the programs. It has the advantage of the lower cost, but the shortcoming is the data base commonly not in the container. Similar with Docker, the Linux container also has these characteristic.

Cgroups [6] is a linux resource isolation tools for resource management. Commonly realize resource isolation have to careful allocation and scheduling of physical resources, and Cgroups can realize the function. It can be used easily, and set a control groups for program. For each group, Cgroups can set the size of resource according the operator. Then we can drag the program to the control groups for execution [7].

3. Resource isolation method

Linux resource isolate tool Cgroups is used in our paper for resource isolation. Cgroups is based on the Linux kernel and can be set flexibly. In order to improve the program's performance, the following module should be commit, and the flow chart is in Figure 1.

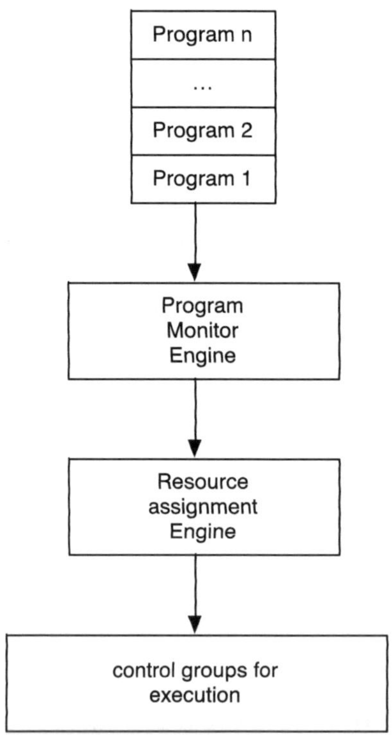

Figure 1 The flow chart of resource isolation method.

Module 1: Program Monitor Engine. This module collects the program's resource usage information about CPU and memory. The general performance profiling tools can be used to understand the program. In our paper we use Linux tool Top. We top program's resource usage of CPU and memory and transfer the information to Module 2.

Module 2: Resource Assignment Engine. The input of the Resource Assignment Engine is the information passed from Module 1, and the output of the Resource Assignment Engine is the size of CPU and memory resource should be applied for the program. The Resource Assignment Engine is used to calculate the size of CPU and memory resource assigning to programs. The calculate method can be found in next section. Then transfer the information to Module 3. This part is very important for our paper. Because the resource assigning to programs have directly effect to the program's performance. If the resource assigning to programs is too large, it is a waste of resource. And if the resource assigning to programs is too small, the program's performance will not good. So the calculate method for assignment resource is the key point.

Module 3: Cgroups Control Engine. The input of the Cgroups Control Engine is the information passed from Module 2, and the output of the Cgroups Control Engine is to set the CPU and memory resource. This part is also important for our paper. The Cgroups Control Engine set the size of resource correctly according to the information passed from module 2. This model divide resource by Linux tool Cgroups, and drag program in control group for execution.

4. Implementation

This section explains the method of calculate the size of CPU and memory resource in detail. In Resource Assignment Engine, the input is the information passed from Program Monitor Engine, and the output of the Resource Assignment Engine is the size of CPU and memory resource should be applied for the program.

Assume that there are n programs have to scheduled to one server. There are p1, p2, …, pn respectively. The CPU resource usage of these program when running alone are Cp1, Cp2, …, Cpn, respectively, and the memory resource usage of these program when running alone are Mp1, Mp2, …, Mpn, respectively.

Then the CPU resource calculated method should obey formula (1).

$$C_{pi-schedule} = \frac{C_{pi}}{C_{p1} + C_{p2} + \ldots + C_{pn}} \times C_{server} \quad (1)$$

In the formula, pi is the number of the program, C_{pi} is the resource usage of program pi when it running alone on a server. C_{server} is the total CPU resource of the server. $C_{pi-schedule}$ is the size of the CPU resource when program pi is scheduled along with other programs.

The memory resource calculated method is different with CPU resource. It is because of our observation in experiment. Commonly when program's memory resource occupies below program's memory resource when running alone, program's performance would drop down dramatically. Based on the belief, the number of program can be scheduled onto one server should obey follow algorithm.

Pseudo-code of caculating the number of program running on one server

```
mem_sum=mem[p1]
for i = p1 to pn do
    mem_sum=mem_sum+mem[i+1];
    if mem_sum >mem_usage
    {
    num_pro=i
    break;
    }
end for
```

Figure 2 Pseudo-code of caculating the number of program running on one server

In Figure 2, we sum the program's usage of memory from p1 until the mem_sum more than the memory size of server. Then we can make sure the number of the program running on one server, which is pi. Then the programs which can not schedule on this server will be scheduled onto the next server.

After fix the number of program, the memory resource calculated method should obey formula (2).

$$M_{pi-schedule} = \frac{M_{pi}}{M_{p1} + M_{p2} + ... + M_{pn}} \times M_{server} \quad (2)$$

In the formula, pi is the number of the program, M_{pi} is the memory resource usage of program pi when it running alone on a server. M_{server} is the total memory resource of the server. $M_{pi-schedule}$ is the size of the memory resource when program pi is scheduled along with other programs.

By the algorithm above, the size of CPU and memory resource should be applied for the program can be calculated reasonably.

5. Experiment

We conduct our experiments on a NUMA server, the server with two sockets and 4 cores on each socket. The detail server information is in TABLE 1. We select programs from NAS Parallel Benchmarks. NAS Parallel Benchmarks is a benchmark suite composed of multithreaded programs. The programs as workloads are in TABLE 2. The CPU and memory event collected by Top are listed in TABLE 3.

We compare three strategies. The first strategy is program running alone on a server. The second strategy is program co-running with other programs. The third strategy is program co-running with other programs, meanwhile with the resource restriction under Cgroups.

Figure 3 shows the program's performance with different strategy, and the baseline is the program's performance when program running alone on a server. There are two workloads.

Table 1 Server configuration

CPU	**Intel Xeon E5620**
core	4 cores@2.13G
L1 caches	32K
L2 caches	256K
L3 caches	4M
Threads per core	1 thread
Sockets	2
Memory	8GB, DDR3

Table 2 Workloads

Workloads	**program**
Workload 1	ft.C.4, mg.C.8
Workload 2	cg.C.4, mg.C.8

Table 3 Collected events.

Item	events
CPU	%CPU
memory	%MEM

The first workload is ft.C.4 and mg.C.8, and the second workload is cg.C.4 and mg.C.8. For worklaod 1 it is shows by the second strategy (program co-running with other programs) almost as 0, it is because program ft.C.4 need almost all of the memory, when ft.C.4 co-running with other programs who use memory, ft.C.4 performance would drop down dramatically. Meanwhile, the co-running program's performance would drop down dramatically too. However, with the third strategy (program co-running with other programs, meanwhile with the resource restriction under Cgroups), because the action of the program monitor engine and resource management engine, the two program's performance are predicted dramatically drop down. Then the resource management engine schedules the two programs to different two servers to avoid performance's serious download. Which is smart than strategy 2.

For workload 2, compare the second strategy, when using the third strategy, the performance of program cg.C.4 improves 20%, and the performance of program mg.C.8 decline 2.7%. And the average performance improves 8.65%. This experiment proves our resource isolation method is good for program's performance.

Figure 4 shows the CPU utilization of the two workload for the three strategy. The CPU utilization when program is running alone is the baseline. It is shows that for workload 1, by the third strategy, the CPU utilization of are always better than the second strategy, and the average CPU utilization improve 24.5%. For workload 2, by the third strategy, the average CPU utilization improves 24.5%.

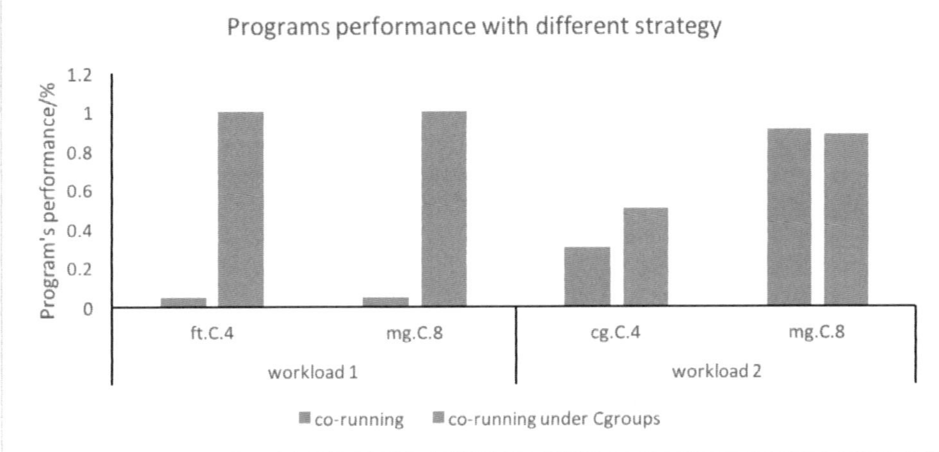

Figure 3 Programs performance with different strategy

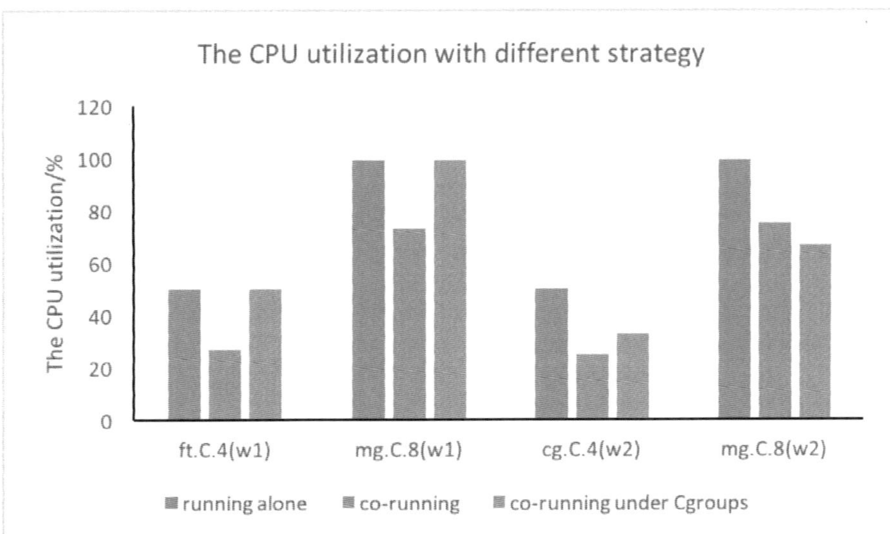

Figure 4 The CPU utilization with different strategy

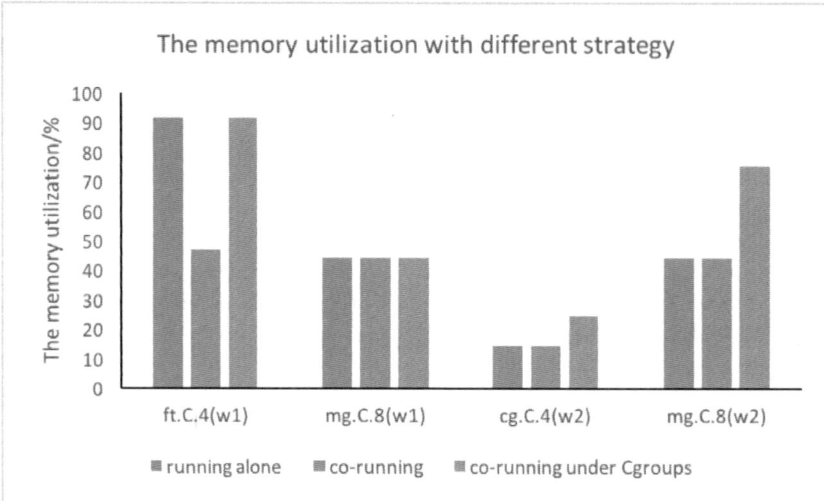

Figure 5 The memory utilization with different strategy

Figure 5 shows the memory utilization of the two workload for the three strategy. The memory utilization when program is running alone is the baseline. It is shows that for workload 1, when use the second strategy, the memory utilization drops down dramatically, resulting the performance's dramatically dropdown. And we get the conclusion that the lack of memory resource would kill the program. For workload 2, compared with the second strategy, by the third strategy the average memory utilization improves 21%.

6. Conclusion

In this paper we propose a light cost resource isolation method to improve program's performance. This method use cgroups to set the dedicated CPU and memory resource for program, aiming at to guarantee the program's performance.

There are three engines to realize this method: 1) Program Monitor Engine top program's resource usage of CPU and memory and transfer the information to Resource Assignment Engine; 2) Resource Assignment Engine calculates the size of CPU and memory resource should be applied for the program; 3) Cgroups Control Engine divide resource by Linux tool Cgroups, and drag program in control group for execution.

The experiment result shows that making use of the resource isolation method proposed by our paper, program's performance can improve 8.65%. The average CPU utilization improves 24.5%, and the average memory utilization improves 21%.

Use footnotes sparingly (or not at all) and place them at the bottom of the column on the page on which they are referenced. Use Times 8-point type, single-spaced. To help your readers, avoid using footnotes altogether and include necessary peripheral observations in the text (within parentheses, if you prefer, as in this sentence).

7. References

[1] H. Jin, A. L. Zhong, S. Wu, and X. H. Shi. Virtual Machine VCPU Scheduling in the Multi-core Environment:Issues and Challenges. Journal of Computer Research and Development, 2011, 48(7): 1216-1224.

[2] M. Liu and T. Li. Optimizing virtual machine consolidation performance on numa server architecture for cloud workloads. In ISCA, 2014.

[3] J. Sugerman, G. Venkitachalam, and B. H. Lim. Virtualizing I/O Devices on VMware Workstation's Hosted Virtual Machine Monitor. Proc. Usenix Annual Technical Conference, June 2001.

[4] C. Boettiger. An introduction to Docker for reproducible research. ACM SIGOPS Operating Systems Review, Special Issue on Repeatability and Sharing of Experimental Artifacts 49(1):71-79
[5] D. Merkel. Docker: Lightweight linux containers for consistent development and deployment. Linux J., 2014(239), March 2014.
[6] Cgroups [EB/OL], https://www.kernel.org/doc/Documentation/cgroup-v1/cgroups.txt
[7] L. Wang, D. Qian, Z. Luan, G. Wei, R. Wang, H. Yang. Speeding up profiling program's runtime characteristics for workload consolidation. PLoS ONE 12(4): e0175861.

A 4-packets Train Measurement Method

Honghua Zhao, Zhimin Miao, Yinjin Fu and Zhuang Miao

College of Command Information Systems, PLA University of Science & Technology, Nanjing, China
Email: zhhahuatian@163.com, Miaozhimin@163.com, yinjinfu@gmail.com, miao_beyond@163.com

Abstract. In order measurement more network performance parameters, a 4-packets trains measurement method was put forward, which could measure the network performance of delay variation and network loss, the method did not need time synchronization. Use the parameters of delay variation and loss, the correlation between destination nodes could be calculated, and the correlation could be used to inference network topology.

1. Introduction

In 1999, Ratnasamy first observed the correlation between nodes when measuring network loss, then the correlation was used in multicast network topology inference [1]. Duffield proved the correctness of topology inference based on correlation and used other performance parameters in multicast network topology inference [2].

In order to improve the correctness of network topology inference and reduce the complex of correlation computing, a multicast network topology inference algorithm based on hamming distance was proposed [3], and the hamming distance could be computed based on link idle percentage. While multicast network topology inference based on hamming distance was only suited for lightly loaded networks and sometimes inferred incorrectly. Li yong-jun[4] improved the topology inference algorithm based on hamming distance, and reduced the probability of wrongness in multicast topology inference.

In unicast networks, the measurement techniques in topology inference are different with common measurement techniques. There are two popular measurement methods in topology inference of unicast network, one is "back to back"packets pair probe [5, 6], and the other is "sandwich" probe [7].

"Back to back"packets pair could probe one way delay and end to end loss; correlation could be calculated based on one way delay or end to end loss. When measuring one way delay, time synchronization was needed between nodes, while the end to end loss is only suit for networks of highly loaded [6]. "Sandwich" could probe the queuing delay, and topology could be inferred based on queuing delay, but it is only suit for networks of lightly loaded [6].

In order to reduce the limits of measurement method in topology inference, a 4-packets train measurement method was proposed, which could probe end to end delay variation[8] and end to end loss. When measuring end to end delay variation, no time synchronization was needed. A topology inference algorithm based on end to end delay variation and end to end loss was put forward and validated theoretically. Compared with "back to back"packets pair probe and "sandwich" probe, topology inference algorithm based on 4-packets train measurement method could suit for networks of lightly loaded, moderately loaded and highly loaded, and it had higher veracity.

IOP Conf. Series: Journal of Physics: Conf. Series **910** (2017) 012008 doi:10.1088/1742-6596/910/1/012008

2. 4-packets Train Measurement

The 4-packets train is defined as below.

a) 4-packets train is composed of double packets pairs. We use (Y_2-X_2,Y_1-X_1) denote one 4-packets train, the packet X_1 and packet Y_1 are packets pair, packet X_2 and packet Y_2 are packets pair,

b) The four packets in one 4-packets train have the same size, for example we set their size 50B;

c) Packet X_1 and packet X_2 have the same destination, packet Y_1 and packet Y_2 have the same destination which is different with destination of X_1;

d) There is stable interval between the double packets pairs in 4-packets train. The interval between packets of one packets pair is so small that the packets in common links have similar situations.

In 4-packets train, every packet has much information including the sequence number and the position number. The sequence number of 4-packets train begins at number 1 and incremented by every 4-packets train. In 4-packets train the position number is only 1or 2, the first packets pair has position number of 1, and the second packets pair has position number of 2. Figure 1 is a sequence of 4-packets train.

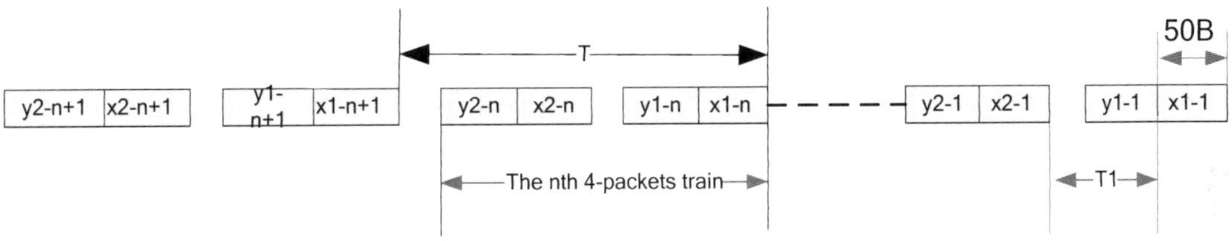

Figure 1 sequence of 4-packets train

When probing end to end delay variation and end to end loss by 4-packets train, a number of 4-packets train were sent from source node to destination node pairs, destination nodes record the receive times of probe packets, then end to end delay variation could be calculated by the intervals between packets in destination nodes and end to end loss could be calculated by the numbers of received packets and probed packets.

Figure 2 is a representative example of end to end delay variation probe by 4-packets train. The 4-packets train in figure 2 is (Y_2-X_2, Y_1-X_1). In the 4-packets train, packets Y_1 and Y_2 have the same destination of node 3; packets X_1 and X_2 have the same destination of node 2. At source node the interval between packets X_1 and X_2 and the interval between packets Y_1 and Y_2 are same and stable. The interval between X_1 and X_2 is recalculated and the received packets number is counted at node 2. The interval between Y_1 and Y_2 is recalculated and the received packets number is counted at node 3 too. End to end delay variation could be calculated by old interval and new interval, which did not need time synchronization. The end to end loss could be calculated by the numbers of sent packets and received packet.

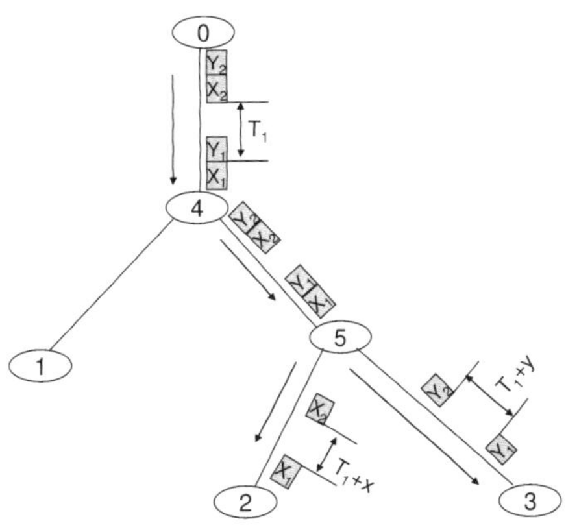

Figure 2 probing by 4-packets train

In order to prevent the packets influenced with each other, we restrict the intervals in one 4-packets train. Suppose the links that 4-packets train passed are $L_1, L_2, \ldots, L_n$, and the bandwidths of these links are $b_1, b_2, \ldots b_n$, and suppose the size of packets in 4-packets train is $s(p)$, then the interval between packets pairs in a 4-packets train must satisfied formula (1).

$$\frac{s(p)}{T_1} < \min(b_i), (1 \le i \le n) \quad (1)$$

3. Correlation Calculation and Topology Inference Algorithm

3.1 Correlation Calculating by Delay Variation

Suppose the 4-packets train is (y_2-x_2, y_1-x_1), the interval between packets pairs is T_1, and the interval between packets in one packets pair is 0. Suppose the interval between 4-packets train is T, the first 4-packets train is sent at time T_0 (T_0 could be 0), then the packets in the first 4-packets train are sent at times $\{T_0, T_0, T_0+T_1, T_0+T_1\}$, the packets in the nth 4-packets train are sent at times $\{T_0, T_0, T_0+T_1, T_0+T_1\}+NT$.

Suppose the destination of packets x_1 and x_2 in 4-packets train is D_x, the destination of packets y_1 and y_2 in 4-packets train is D_y. Suppose the times of packets received at node D_x are T_x, then $T_x=\{T_{x1}, T_{x2}, T_{2x1}, T_{2x2}, \cdots, T_{ix1}, T_{ix2}, \cdots, T_{Nx1}, T_{Nx2}\}$, T_{ix1}, T_{ix2} denotes the received time of packets x_1 and x_2 of the ith 4-packets train. If there is packet lost in a 4-packets train, the received times of packets in the 4-packets train are excluded. End to end delay variation calculated at node D_x of the first 4-packets is T_{dvx1}, $T_{dvx1}=[T_{x2}-(T_0+T_1)]-(T_{x1}-T_0)=(T_{x2}-T_{x1})-T_1$.

The end to end delay variation calculated at node D_x of the nth 4-packets train is T_{dvxn}, T_{dvxn} could be calculated by formula (2).

$$\text{Tdvxn} = (T_{Nx2} - T_{Nx1}) - T_1 \quad (2)$$

The end to end delay variation calculated at node D_y of the nth 4-packets train is T_{dvyn}, T_{dvyn} could be calculated by formula (3).

$$\text{Tdvyn} = (T_{Ny2} - T_{Ny1}) - T_1 \quad (3)$$

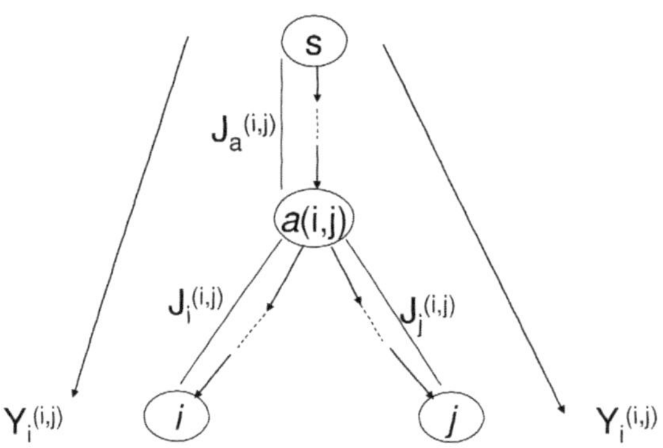

Figure 3 correlation calculation based on delay variation

Figure 3 is an example of correlation calculation based on end to end delay variation. In figure 3, the destinations pair of 4-packets train is (i,j), suppose $a(i,j)$ is father of leaf i and leaf j, suppose $Y_i^{(i,j)}$ denotes the end to end delay variation from source S to leaf i, and $Y_j^{(i,j)}$ denotes the end to end delay variation from source S to leaf j. $J_a^{(i,j)}$ denotes the delay variation of path ($S \rightarrow a(i,j)$), $J_i^{(i,j)}$ denotes the delay variation of path (*a*(*i,j*)→*i*) and $J_j^{(i,j)}$ denotes the delay variation of path (*a*(*i,j*)→*j*). We compute the covariance of $Y_i^{(i,j)}$ and $Y_j^{(i,j)}$.

Because the packets in different links are independent, $Cov(J_i^{(i,j)}, J_a^{(i,j)}) = 0, Cov(J_a^{(i,j)}, J_j^{(i,j)}) = 0, Cov(J_i^{(i,j)}, J_j^{(i,j)}) = 0$. We could get the conclusion of formula (4)

$$Cov(Y_i^{(i,j)}, Y_j^{(i,j)}) = Cov(J_a^{(i,j)} + J_i^{(i,j)}, J_a^{(i,j)} + J_j^{(i,j)}) = Var(J_a^{(i,j)}) \quad (4)$$

Suppose the path (*S*→*a*(*i,j*)) is composed of links $L_1, L_2 \ldots, L_n$, and $M_1^{(i,j)}, M_2^{(i,j)}, \cdots, M_n^{(i,j)}$ denote the delay variations of links $L_1, L_2 \ldots, L_n$,. Because packets in different links are independent, then we could get formula (5).

$$\begin{aligned} Var(J_a^{(i,j)}) &= Var(M_1^{(i,j)} + M_2^{(i,j)} + \ldots M_n^{(i,j)}) \\ &= Var(M_1^{(i,j)}) + Var(M_2^{(i,j)}) + \ldots + Var(M_n^{(i,j)}) \end{aligned} \quad (5)$$

Based on formulas (4) and (5), we know that more common links of the destinations pair, bigger covariance of the delay variation.

We use $J_{i,j}$ denote the correlation between leaves *i* and *j*, where $J_{i,j} = Cov(Y_i^{(i,j)}, Y_j^{(i,j)})$.

3.2 Correlation Calculating by Loss

End to end loss could be probed by 4-packets train measurement method, then the correlations based on end to end loss could be calculated. When calculated the correlations, we use success to replace loss (success=1-loss).

Figure 4 is an example of correlation calculation based on end to end loss. In figure 4, the destinations pair of 4-packets train is (*i,j*), suppose *a*(*i,j*) is father of leaf *i* and leaf *j*, suppose $S_i^{(i,j)}$ denotes the end to end success from source *S* to leaf *i* and $S_j^{(i,j)}$ denotes the end to end success from source *S* to leaf *j*. $P_a^{(i,j)}$ denotes the success of path (*S*→*a*(*i,j*)), $P_i^{(i,j)}$ denotes the success of path (*a*(*i,j*)→*i*) and $P_j^{(i,j)}$ denotes the success of path (*a*(*i,j*)→*j*).

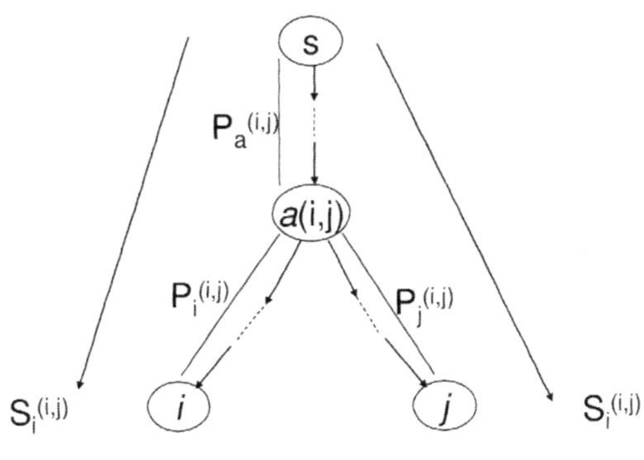

Figure 4 correlation calculation based on loss

Suppose the number of 4-packets train that sent from *S* to (*i,j*) is N_s, then the number of packets pair is $2N_s$. Suppose N_i is the number of probe packets received at leaf *i*, N_j is the number of probe packets received at leaf *j*, and N_{ij} is the number of packets pair received at node *i* and *j*. We get

$s_i^{(i,j)} = \frac{N_i}{2N_s}(N_s \to \infty), s_j^{(i,j)} = \frac{N_j}{2N_s}(N_s \to \infty)$ and $s_{i,j}^{(i,j)} = \frac{N_{ij}}{2N_s}(N_s \to \infty)$, we can know that $s_i^{(i,j)} = P_a^{(i,j)} \times P_i^{(i,j)}$, $s_j^{(i,j)} = P_a^{(i,j)} \times P_i^{(i,j)}$ and $s_{i,j}^{(i,j)} = P_a^{(i,j)} \times P_i^{(i,j)} \times P_j^{(i,j)}$, so we could get formula (6)

$$P_a^{(i,j)} = \frac{s_i^{(i,j)} \times s_j^{(i,j)}}{s_{i,j}^{(i,j)}}(N_s \to \infty) \quad (6)$$

We use $P_a^{(i,j)}$ as correlation between leaves *i* and *j* based on loss. Suppose the path ($S \to a(i,j)$) is composed of links $L_1,L_2...,L_n$, and $K_1^{(i,j)}, K_2^{(i,j)}, \cdots, K_n^{(i,j)}$ denote the success of links $L_1,L_2...,L_n$, we could get formula (7)

$$P_a^{(i,j)} = K_1^{(i,j)} \times K_2^{(i,j)} \times ... \times K_n^{(i,j)} \quad (7)$$

For $K_x^{(i,j)} \leq 1$, based on formulas (6) and (7), we know that more common links between leaves *i* and *j*, less success of $P_a^{(i,j)}$.Topologies could be inferred based on $P_a^{(i,j)}$,which had the same effect with "back to back" packets paris[4].

4. References

[1] N. Duffield, J. Horowitz, L.Presti, D.Towsley, Multicast topology inference from measured end-to-end loss. IEEE Transactions on Information Theory, 2002, 48(1):26-45.
[2] N. Duffield, J. Horowitz, F. Lo Presti, D. Towsley, Multicast topology inference from end-to-end measurements, ITC Seminar on IP traffic, Measurement and Modelling, Monterey, CA, Sep. 2000.
[3] H. Tian, H. Shen. Discover Multicast Network Internal Characteristics Based on Hamming Distance. Proceeding of IEEE International Conference on Communications. Piscataway: IEEE Inc., 2005: 143-147.
[4] L. Yongjun, C. Wandong, W. Wei, et al, Research on network topology identification algorithm based on end-to-end loss performance, Journal on Communications, Vol28, No.10, 2007.10.
[5] M. Shih, A. Hero, Hierarchical Inference of Unicast Network Topologies Based on End-to-End Measurements. IEEE Transactions on Signal Processing, 2007,55(5):1708-1718.
[6] S. Meng-Fu, O. Alfred, Topology Discovery on Unicast Networks: A Hierarchical Approach Based on End-to-End Measurements, http://www.eecs.umich.edu/~msim/Publications/cspl-357.ps.pdf, Mar 2005.
[7] M. Coates, R. Castro, R. Nowak, Maximum likelihood network topology identification from edge-based unicast measurements, ACM Sigmetric, Marina Del Rey, CA., June 2002.
[8] C. Demichelis and P. Chimento, IP Packet Delay Variation Metric for IPPM. htttp://www.ietf.org/internetdrafts/draft-ietf-ippm-ipdv-09.txt, April 2002.

IOP Conf. Series: Journal of Physics: Conf. Series **910** (2017) 012009 doi:10.1088/1742-6596/910/1/012009

Service Discovery Oriented Management System Construction Method

Huawei Li[1,a] and Ying Ren[2,b]

[1]Shandong Business Institute, Yantai, 264001, China
[2]Naval Aeronautical University, Shandong, 264000, China
Email: [a]li_huawei@163.com, [b]ren.ying@163.com

Abstract. In order to solve the problem that there is no uniform method for design service quality management system in large-scale complex service environment, this paper proposes a distributed service-oriented discovery management system construction method. Three measurement functions are proposed to compute nearest neighbor user similarity at different levels. At present in view of the low efficiency of service quality management systems, three solutions are proposed to improve the efficiency of the system. Finally, the key technologies of distributed service quality management system based on service discovery are summarized through the factor addition and subtraction of quantitative experiment.

1. Foreword

The service quality management system is a basic IT facility for service quality and providing data source for business logic. The system covers different core business modules, such as data storage services, data integration services and data access services, and responds to the request of the service sponsor in real time. Because of the continuous development of science and technology, it is inevitable that the quantity, speed and kind of service quality data are expanding constantly, and the quality of service data has entered a chaotic and disordered state. In the worst case, service requesters will be confused and easily frustrated by the mass, erratic, and disorderly quality of service that they face in a short period of time.

This paper provides a unified approach to design a distributed QoS management system. The purpose is to summarize the important factors that affect the performance of the service quality of the system (such as neighborhood effect, synergistic effect and effect way, etc.) and distill lessons to guide the next generation of high performance distributed service management system, so as to enhance the large-scale complex service discovery work performance and efficiency.

2. Total Distributed Architecture

Distributed service quality management system, designed to respond to a massive user QoS query request, through the management of user service interaction data of large-scale QoS information, analysis and prediction of loss, for all types of service discovery, service composition and service recommendation based support. Contains the following key modules.

2.1 Data Preprocessing Module

In distributed QoS management system, because the quality of data collected is a large amount of unstructured data, it is necessary to use stream preprocessing tools to organize in the acquisition phase. Using Cloudera Flume open source tools, Flume uses the distributed architecture of loose coupling, requires only a simple command will be deployed to the global API server, meet the demand of hundreds of MB per second log data acquisition and transmission .

2.2 Persistence Module

Quality of service management data association is high, we use large file write method to optimize the overall performance in the persistence layer. The daemon monitor running on HDFS also supports the processing of large data set.

2.3 Distributed Cache Module

(1)Memcached is a high-performance distributed memory object caching system for dynamic Web applications to alleviate database load. It reduces the number of database reads by caching data and objects in memory, thereby increasing the speed of dynamic, database driven web sites.

(2)The Redis cluster is also a key - value storage system. Similar to Memcached, it supports the storage of value types with more, including string (string), list (linked list), set (set) and hash (hash type)

2.4 Computation Module

(1)Map stage: In accordance with the requirements of different measurement functions, the user of the same clustering cluster is hashed in the Nmapper computing node. Within each mapper computing node, we reconstruct the matrix in a small amount of iterations, in accordance with a predefined matrix decomposition strategy. At this stage, in order to improve efficiency, we do not seek predefined loss function to achieve the global optimum.

(2)Reduce stage: responsible for collecting the processing information of all mapper computing nods, and then re classifying the users according to the requirements of the measurement function, and pushing the classification results back to the mapper in the Map phase.

3. Key index

3.1 Measure factor

In order to study the influence of measure function on the design of distributed QoS system, we set the following different measurement functions:

(1)Random grouping: That is, for any target user u_i, the user is arbitrarily assigned to the nearest neighbor G(i) by using a random measure function. Strategy is as follows:

$$G(i) = \{j|\forall j, rand(j)\%M = m(i), i \neq j\} \tag{1}$$

Among, rand measure function is a hash function with time and random factor. The specific operation is to map the user $u_i$to the 16 bit string space, and then assign it to the Map group belonging to the target user through the redundancy operation. M is the number of mapper in this calculation, $m(i)$ is the mapper index of the target user u_i

(2)PCC: Strategy is as follows:

$$sim(i,j) = \frac{\sum_{s\in S}(r_{is}-r_{\bar{u}_i})(r_{js}-r_{\bar{u}_j})}{\sqrt{\sum_{s\in S}(r_{u_{is}}-r_{\bar{u}_i})^2}\sqrt{\sum_{s\in S}(r_{u_{js}}-r_{\bar{u}_j})^2}} \tag{2}$$

The PCC measure function calculates the similarity of the user's history call record and determines the packet. Therefore, the size and number of mapper generated by the measure function are determined.

(3)PCC*: in the field of service management, we can extend basic PCC with domain knowledge (geographic location information, call failure rate, and call time factor). Strategy is as follows:

$$sim(i,j) = \frac{\sum_{s\in N}(r_{is}-r_{\bar{u}_i})(r_{js}-r_{\bar{u}_j})}{\sqrt{\sum_{s\in N}(r_{u_{is}}-r_{\bar{u}_i})^2}\sqrt{\sum_{s\in N}(r_{u_{js}}-r_{\bar{u}_j})^2}} \tag{3}$$

3.2 Domain Factor
(1)Weighting strategy:The final QoS prediction performance of the target user is affected by multiple neighbors.We can adjust the relative importance of nearest neighbors by weighted and unweighted methods.

We define W(0) as an unweighted policy, and the mathematical representation of the W(0) policy for the target user is as follows:

$$U_i \leftarrow \sum_{j=1}^{G(i)} U_j \tag{4}$$

We define W(1) as a weighted strategy, and the mathematical representation of the W(1) policy for the target user u_i is as follows:

$$U_i \leftarrow \sum_{j=1}^{G(i)} U_j \times w_{ij} \tag{5}$$

Where w_{ij} is defined as follows:

$$w_{ij} = \frac{\mathrm{sim}(i,j)}{\sum_{k=1}^{G(i)} \mathrm{sim}(i,k)} \tag{6}$$

(2) contribution factor: Given a target user, its nearest neighbor user not only plays a role in the decomposition model of the extended matrix, but also can transmit the cooperative information according to its own situation when the prediction result is synthesized. The target prediction $\hat{r_{ij}}$ of the contributing factor Mu is the following formula.

$$\hat{r_{ij}} = \mu \cdot r_{ij}^* + (1-\mu) \cdot \frac{1}{|G(i)|} \sum_{g \in G(i)} r_{gj}^* \tag{7}$$

3.3 Alternative Factor
The calculation method is the key factor that affects the efficiency and accuracy of the matrix decomposition model. In order to extract the key design factors of the distributed QoS management system, the following algorithms are used:

(1)GD: The method of gradient descent is to compute the minimum value in the direction of gradient descent (and also to solve the maximum value along the gradient ascending direction).When the gradient descent algorithm is used to solve the optimization problem, the termination condition of the algorithm iteration is that the magnitude of the gradient vector is close to 0, and a very small constant min value can be set

(2) SGD: The method of stochastic gradient descent takes each sample into a single sample. When the total number of samples is large, the method of the gradient descent iteration is much slower than the gradient descent method.

4. Experimental Results

4.1. Experimental Configure
The dataset used contains two types of QoS attributes: response time, and reliability.The range of response time is 0-20 seconds, and the range of reliability is 0-1. The single service quality management system has been unable to deal with the data of this magnitude, so the key factors that affect the QoS prediction will be quantitatively analyzed through a distributed framework. We randomly remove several QoS values to achieve sparse matrix purposes. By doing so, our QoS matrix is divided into two parts: the training set and the test set. The density matrix introduced in the experiment (density) this concept. Usually, if the density matrix is equal to 5%, we retained 5% QoS data as training set, the remaining 95% as a benchmark to verify the prediction performance of the system. All the experimental results are not separate sampling, we make it by taking the 100 times

mean.

The square root of the mean variance (RMSE) is a generic QoS prediction model testing standards, defined as follows:

$$\mathrm{RMSE} = \sqrt{\frac{1}{N}\sum_{i,j}(r_{ij} - \hat{r}_{ij})^2} \tag{8}$$

Among them, $\hat{r}_{ij}$ is forecasting results, N is the total number of prediction samples.

4.2. Key Index Effect Analysis

In order to measure the effect of key indicators on the final prediction effect, we use factor addition and subtraction to evaluate the predictive performance. Table 1 summarizes the key elements involved in the test.

4.3. Overall Effect

We used the factor addition and subtraction in the quantitative analysis to analyze the key factors. All settings are using the same training set and test set. We set the number of mapper (clustering) to 50.

Table 1 Significance Of Key Indicators

Notation	Definition and Description
rand	Random measure function with time factor
CC	A universal measure of degree of similarity in industry
PCC*	Extended PCC measure function combined with service domain knowledge
W(0)	Matrix decomposition framework without weight
W(1)	A weighted matrix decomposition framework based on user similarity
CI(0)	QoS prediction method without contribution factor
CI(1)	QoS prediction method with contribution factor
GD	Batch gradient descent algorithm
SGD	Stochastic gradient descent algorithm
Adagrad	A randomized gradient descent algorithm with adaptive learning rate

Table 2 Comparison of prediction accuracy of response time

Method	QoS Attribute	D=5%	D=95%
rand+W(0)+GD	Response Time	16.05	16.43
PCC+W(0)+GD		5.13	3.4
PCC*+W(0)+GD		4.36	2.84
rand+W(1)+GD		13.34	14.12
PCC+W(1)+GD		4.3	3.0
PCC*+W(1)+GD		3.85	1.74

Tables I and IIshow the accuracy of the key factor combinations in the two sets of data on QoS response time and reliability. We can observe that:(1) Compared to PCC and extended version PCC, the prediction of random functions is always the worst. Different range of QoS characteristics directly lead to large fluctuations in the RMSE index value.(2) In order to solve the problem, adding the non knowledge measurement function will affect the final prediction effect of the service quality management system. With neighborhood factors as dimensions, both the weighted strategy and the policy with contribution factor are always optimal, regardless of the response time or reliability of the two sets of data.(3) With the method factor as the dimension, the performance is not obvious. In the response time data set, the Adagrad policy produces better prediction results in most cases. However, this difference is not apparent in the reliability data set. Even when the sample density was high, SGD showed greater predictive power.

Table 3 Comparison of reliability prediction accuracy

Method	QoS Attribute	D=5%	D=95%
rand+W(0)+GD	Reliability	3.2	6.2
PCC+W(0)+GD		1.37	2.51
PCC*+W(0)+GD		1.45	3.41
rand+W(1)+GD		2.85	5.6
PCC+W(1)+GD		1.2	2.49
PCC*+W(1)+GD		2.1	2.65

4.4. Distributed Efficiency Analysis

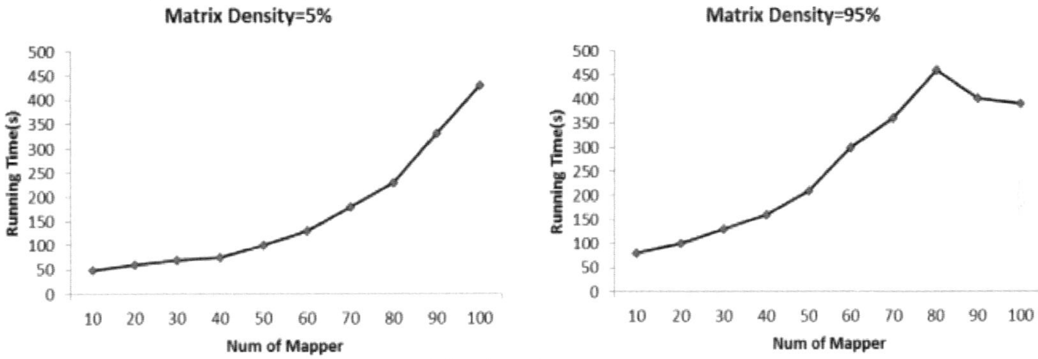

(a)Gradient Descent

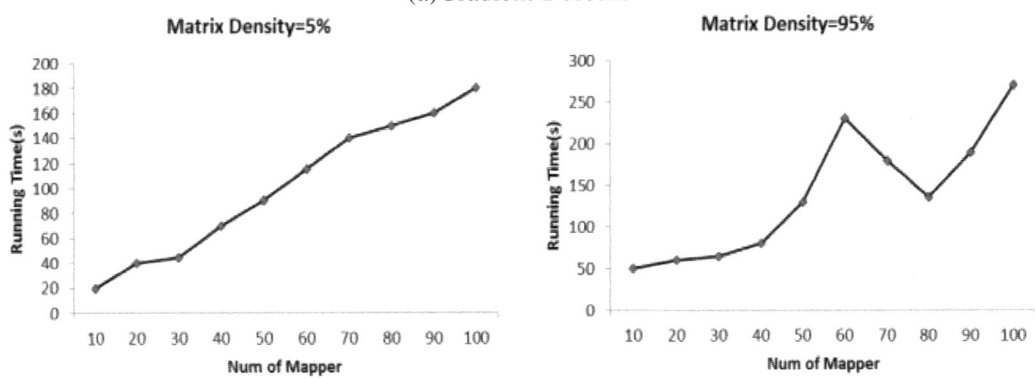

(b)Stochastic Gradient Descent

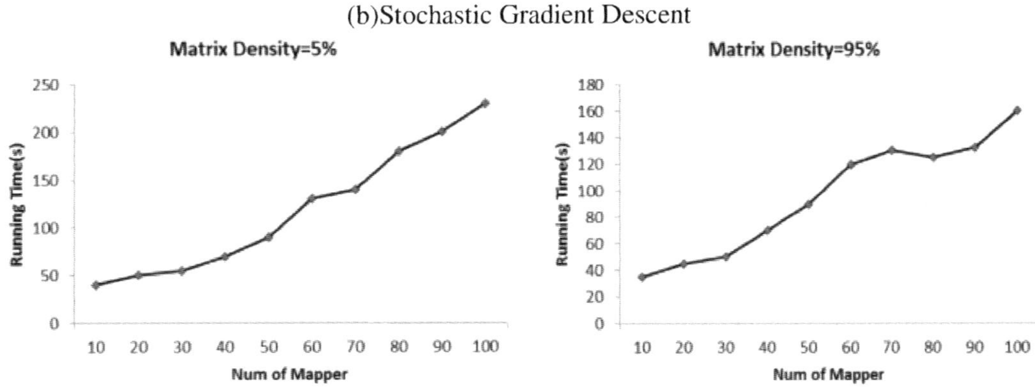

(c)Adaptive Gradient Descent

Figure 1 Comparison of operating efficiencies

We see that the calculation method does not have much effect on the prediction effect, which is in fact expected: The use of the method is to improve the efficiency of the quality of service management system to meet the online query requirements. We focus on the impact of computational approaches on distributed performance. We use the response time data set, the PCC* measure function and the coordination factor in the weighted W(1)and the band contribution factor CI(1) strategy. By adjusting the number of mapper nodes, we observe three sets to obtain globally optimal corresponding time overhead.

Figure I is a comparison of operating efficiency. As can be seen, given a low density learning sample, the actual running time of SGD and Adagrad increase linearly with the number of mapper nodes. At the same time, the actual running of GD which is using the strategy of adding all samples as gradients increases exponentially with the increase of mapper nodes and the operation efficiency is much lower than that of SGD and Adagrad. In the same number of mapper nodes, Adagrad can be better than the original version of SGD because it can determine the learning step size according to the previous historical gradient dynamics. In the case of sufficient sample number (95%), when the number of mapper nodes is relatively small, the inflection point is observed when the number of mapper exceeds the threshold. This phenomenon can be resolved as the time cost of the single wheel gradient will increase when the number of mapper nodes increases. However, when mapper is at a balanced point, the total number of iterations will be reduced (the loss function converges faster), so the overall run time will be lower than the small mapper setting. The stability of the global optimal convergence of the matrix decomposition model is shown in Figure 1.

5. Acknowledgement

Project of scientific research development of universities in Shandong Province in 2017(J17KB175)

6. Reference

[1] Taha Osman,Dhawal Thakker,David A1-Dabass.Semantic-driven matchmaking of web services usingcase-based reasoning[C]//Web Services,206.ICW0S'06.International Conference on.IEEE,2006:29-36.

[2] Luciano Baresi,Matteo Miraz,Pierluigi Plebani.A distributed architecture for efficient web servicediscovery[J].Service Oriented Computing and Applications,2015:1-17.

[3] Zhizhong Liu,YuLan Liu,Yihui He.A two-layered p2p model for semantic service discovery[C]//New Trends in Information Science and Service Science(NISS),2010 4th International Conference on.IEEE,2010:41-46.

[4] Min-Xiou Chen,Fu-Hsing Sung.Integrated service discovery architecture for heterogeneous networks[J].International Journal of CommunicationSystems,2015.

[5] Vinay Kumar Tiwari,Amit Kumar Gautam,Saurabh Diwaker.Web service discovery and integration withqos parameter using soa based repository[C]//Computer Engineering and Applications(ICACEA),2015 International Conference on Advances in.IEEE,2015:1014-1009.

[6] Angelo Furno,Eugenio Zimeo.Efficient cooperative discovery of service compositions in unstructured p2p networks[C]//ParalM,Distributed and Network-Based Processing(PDP),2013 21st Euromicro International Confere.

[7] Wenyu Zhang,Shuai Zhang,Feng Qi,Ming Cai.Self-organized p2p approach to manufacturing service discovery for cross-enterprise collaboration[J].Systems,Man,and Cybernetics:Systems,IEEE Transactions on,2014.44(3):263-276.

[8] Liang Chen,Jian Wu,Zibin Zheng,Michael R Lyu,Zhaohui Wu.Modeling and exploiting tag relevance for web service mining[J].Knowledge and information systems,2014.39(1):153-173.

[9] Jieming Zhu,Pinjia He,Zibin Zheng,Michael R Lyu.Towards online,accurate,and scalable qos prediction for runtime service adaptation[C]//Distributed Computing Systems(ICDCS),2014 IEEE 34th International Conference on.IEEE.2014:318-327.

[10] Pinjia He,Jieming Zhu,Zibin Zheng,Jianlong Xu,Michael R Lyu.Location—based hierarchical matrix factorization for web service recommendation[C]//Web Services(ICWS),2014 IEEE

International Conference on.IEEE,2014:297-304.
[11] Liang Chen,Qi Yu,Philip S Yu,Jian Wu.Ws-hfs:A heterogeneous feature selection framework for web services mining[C]//Web Services(ICWS),2015 IEEE International Conference on.IEEE.2015:193-200.
[12] Shuiguang Deng,Longtao Huang,Wei Tan,Zhaohui Wu.Top-automatic service composition:A parallel method for large-scale service sets[J].Automation Science and Engineering,IEEE Transactions on,2014.11(3):891-905.
[13] Xinhuai Tang,Feilong Tang,Liu Bing,Delai Chen.Dynamic web service composition based on service integration and htn planning[C]//Innovative Mobile and Internet Services in Ubiquitous Computing (IMIS),2013 Seventh International Conference on.IEEE,2013:307-312.
[14] Gopal N Rai,GR Gangadharan,Vineet Padmanabhan.Algebraic modeling and verification of web servicecomposition[J].Procedia Computer Science,2015.52:675-679.
[15] Nikolaos Tsapanos,Anastasios Tefas,Nikolaos Nikolaidis,Ioannis Pitas.A distributed framework fortrimmed kernel k-means clustering[J].Paaern Recognition,2015.48(8):2685-2698.
[16] Aamer Shah,Angshul Majumdar.Accelerating low-rank matrix completion on gpus[C]Hadvances in Computing,Communications and Informatics(ICACCI,2014 International Conference on.IEEE,2014:82-187.

IOP Conf. Series: Journal of Physics: Conf. Series **910** (2017) 012010 doi:10.1088/1742-6596/910/1/012010

An Expressive, Lightweight and Secure Construction of Key Policy Attribute-Based Cloud Data Sharing Access Control

Guofen Lin, Hanshu Hong, Yunhao Xia and Zhixin Sun

The Key Lab of Broadband Wireless Communication and Sensor Network Technology, Nanjing University of Posts and Telecommunications, Nanjing China
Email: 2015070249@njupt.edu.cn, 2014070244@njupt.edu.cn, 2015070108@njupt.edu.cn, sunzx@njupt.edu.cn

Abstract. Attribute-based encryption (ABE) is an interesting cryptographic technique for flexible cloud data sharing access control. However, some open challenges hinder its practical application. In previous schemes, all attributes are considered as in the same status while they are not in most of practical scenarios. Meanwhile, the size of access policy increases dramatically with the raise of its expressiveness complexity. In addition, current research hardly notices that mobile front-end devices, such as smartphones, are poor in computational performance while too much bilinear pairing computation is needed for ABE. In this paper, we propose a key-policy weighted attribute-based encryption without bilinear pairing computation (KP-WABE-WB) for secure cloud data sharing access control. A simple weighted mechanism is presented to describe different importance of each attribute. We introduce a novel construction of ABE without executing any bilinear pairing computation. Compared to previous schemes, our scheme has a better performance in expressiveness of access policy and computational efficiency.

1. Introduction

With improvements in computational technology and large-scale networks, sharing data with others becomes correspondingly more convenient. Additionally, digital resources are more easily obtained via cloud computing and storage. Since cloud data sharing requires off-premises infrastructure that some organizations jointly held, remote storage are somehow threatening privacy of data owners. Therefore, enforcing the protection of personal, confidential and sensitive data stored in the cloud is extremely crucial. The simultaneous participation of large numbers of users requires fine-grained access control for data sharing. Attribute-based encryption (ABE) is a promising cryptographic primitive that offers an interesting solution to secure and flexible data sharing. ABE has an inherent one-to-many property, which means a single key can decrypt different ciphertexts or different keys can decrypt the same ciphertext. There are two types of ABE, called ciphertext policy ABE (CP-ABE) and key policy ABE (KP-ABE). For CP-ABE, the access policy is embedded into a ciphertext and the attribute set is embedded into a private key. For KP-ABE, the access policy is embedded into a private key and the attribute set is embedded into a ciphertext.

However, there are still a lot of open challenges concerning the practical realization of ABEs especially with respect to the limited expressiveness of access policy and heavy computational overhead. In some ABE schemes, the access policy is built based on threshold gate. When the distance between the attribute set and the access policy is shorter than a given threshold, the data receiver can extract the plaintext correctly. For most of current ABE schemes, they adopt access tree as their access policy, which consists of a plenty of "AND" gate, "OR" gate. Although the access tree enriches the form of ABE's access policy, we are still not satisfied with its expressiveness. For one thing, all attributes are considered as in the same status while they are not in most of practical scenarios. For another thing, the size of access policy increases dramatically with the raise of its expressiveness complexity. Together with the growth of mobile applications, mobile cloud services have been introduced as a potential trend in cloud computing. Current research hardly notices that mobile front-end devices, such as smartphones, are poor in computation performance while too much bilinear pairing computation is need for ABE. As a result, the run time may be horribly unacceptable.

1.1. Related Work

In 2005, Sahai and Waters [1] proposed a fuzzy identity-based encryption (FIBE) based on classic identity-based encryption. Since FIBE indicated some many key features of ABE, it laid a theoretical foundation of subsequent research into ABE. In 2006, a formal definition of ABE [2] was proposed, which indicated that its construction based on FIBE is a key policy ABE (KP-ABE). That means each private key is associated with an access policy and each ciphertext is associated with a set of attributes. Inspired by reference, Bethencourt, Sahai and Waters [3] proposed a concrete construction of CP-ABE, in which a data owner can flexibly define the access policy before data is encrypted. However, their construction is more complicated than KP-ABE. Cheung and Newport [4] proposed a novel access tree contain only AND gates with positive and negative attributes to enhance expressiveness of ABE access policy. Waters [5] proposed a high efficient CP-ABE that first makes ciphertext size, encryption time and decryption time increase linearly with the complexity of access policy. In addition, the proposed scheme is proven to be selectively secure with the reduction to Decisional Parallel Bilinear Diffie-Hellman Exponent Assumption (DPBDHE). Green, Hohenberger and Waters [6] proposed a concept of outsourcing decryption ABE (OD-ABE) that delegates a high-performance server to execute part of the decryption while leaving no information with it. Finally, only a small amount of computation is needed for the data receiver to extract a plaintext. In the original OD-ABE, however, the untrusted proxy server risks the confidentiality of ciphertext and transform keys. Furthermore, wrong computation may result in unavailability of the whole system. To solve this problem, an attribute-based encryption with verifiable outsourced decryption (VOD-ABE) [7] was proposed, which demonstrated the necessity of outsourcing verifiability for OD-ABE. Lin, Zhang, Ma and Wang [8] proposed an improved VOD-ABE based on key encapsulation mechanism (KEM). Their experiment demonstrated that ciphertext size, cost of encryption and decryption are nearly half of those in [7]. Based on a concept of hierarchical generalized attributes, a hierarchical multi-authority mechanism for CP-ABE was proposed in [9]. When a user defines an access structure and requests data encryption, each key generation center (KGA) generates corresponding access policy and private key of its level so that security of key management is guaranteed even there exist a compromised KGA engaged in key issuing. In [10], a proxy re-encryption scheme for multi-authority ABE was proposed to realize efficient and fine-grained revocation. In this scheme, access policies are mapped to a weighted access list, which reduces computation overhead during generating and issuing private keys. Hur [11] proposed a secure and efficient attribute-based data sharing system. Without adding any infrastructure, they provided a novel solution to key escrow problem in single-authority system by introducing two-party computation (2PC) between key generation center and data storing center. Oualha and Nguyen [12] considered the limitation of energy and computation of nodes in Internet of Things, and proposed a novel CP-ABE by introducing a pre-computation technique, which computes and stores some essential elements before encryption occurs. Qin *et al.* [13] indicated that although KP-ABE does not allow data owners to define access policies, it helps to build an efficient access control for cloud data storage. Based on this, they proposed a redefinable KP-ABE, which provides flexible access control but also markedly reduces ciphertext size.

As mentioned above, most of current ABE schemes adopt threshold gate or access tree to build their access policy. In these cases, all attributes are considered as in the same status while they are not in most of practical scenarios. Meanwhile, the size of access policy increases dramatically with the raise of its expressiveness complexity. Moreover, current research hardly notices that mobile front-end devices, such as smartphones, are poor in computational performance while heavy computation overhead especially too much bilinear pairing computation is taken over by them. Thus, we are not satisfied with current ABEs in terms of the limited expressiveness of access policy and heavy computation overhead.

1.2. Our Contributions

In this work, we propose a key policy weighted attribute-based encryption without bilinear pairing computation (KP-WABE-WB) aiming to enhance expressiveness and computational efficiency of secure cloud data sharing access control. The main contributions are summarized as follows:

- A simple weighted mechanism is presented to describe different importance of each attribute. As a result, the weighted mechanism not only enhances the expressiveness of access policy but also simplifies its structure.
- We introduce a novel attribute-based encryption without executing any bilinear pairing computation. Therefore, the computational efficiency is markedly improved.
- We propose a reduction of the proposed scheme to Computational Diffie-Hellman (CDH) Assumption. With the help of this reduction, we prove that our KP-WABE-WB is secure under chosen-ciphertext attacks in the random oracle model.

1.3. Organization

The rest of paper is organized as follows: We give essential preliminaries in Section II, propose main construction of KP-WABE-WB in Section III, provide comprehensive analysis in Section IV, and draw conclusion and highlight our future work in Section V.

2. Preliminaries

2.1 Complexity Assumption

Our construction is based on the following complexity assumption:

Computational Diffie-Hellman (CDH) Assumption*:* Given an additive cyclic group $\mathbb{G}$ with prime order q. Let P be a generator of $\mathbb{G}$. There are two random elements $a,b \in Z_q^*$. Given $P_1 = aP$ and $P_2 = bP$, no probabilistic polynomial-time algorithm can compute $P_3 = abP$.

2.2 Modeling Syntax

The model of our construction is shown in Figure. 1. As depicted, there are four entities involved in cloud data sharing, which is the key authority, the data owner, the cloud storage center, and the data receiver. The key authority is responsible for generating and issuing keys. The data owner possesses data to be shared via cloud service. The cloud storage center is responsible for storing and managing encrypted data.

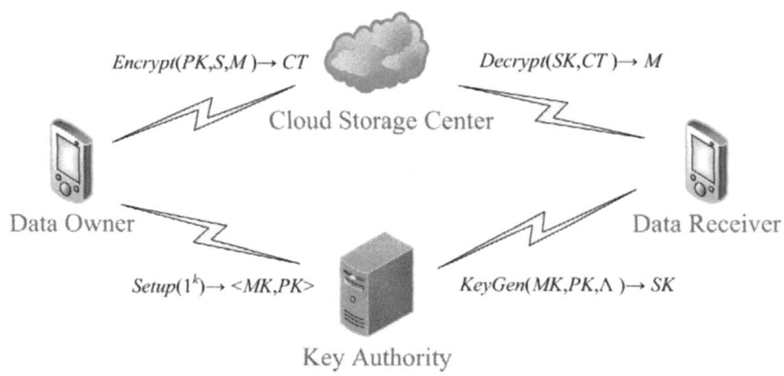

Figure 1 The model of KP-WABE-WB.

The data receiver wants to obtain data stored in the cloud storage center. Precisely, the KP-WABE-WB consists of the following 4 algorithms:

- $Setup(1^k) \to <MK, PK>$: The setup algorithm is run by the key authority. It takes as input a security parameter $k \in Z^+$, then it outputs the master secret MK and the public parameter PK.

Noting that the master secret is kept by the key authority in secret while the public parameter is available for all entities.

- $KeyGen(MK, PK, \Lambda) \rightarrow SK$: The key generation algorithm is run by the key authority when receiving a key generation query from a data receiver. It takes as input the master secret, the public parameter, and an access policy. Then, it outputs a private key SK associated with this access policy.
- $Encrypt(PK, S, M) \rightarrow CT$: The encryption algorithm is run by a data receiver. It takes as input the public parameter PK , an access policy Λ , and a plaintext M . Finally, it returns a ciphertext CT associated with the attribute set S .
- $Decrypt(SK, CT) \rightarrow M$: The decryption algorithm takes as input the private key of a data receiver and a ciphertext. If the attribute set of a ciphertext satisfies the access policy of a private key, it outputs the plaintext correctly. Otherwise, it aborts the decryption.

3. Main Construction

3.1 The Setup Algorithm

Let $\Omega = \{\theta_1, \theta_2, \cdots, \theta_n\}$ be the global attribute space and $W = \{w_1, w_2, \cdots, w_n\}$ be the collection of attributes' weight. Given a security parameter $k \in Z^*$, the setup algorithm chooses a cyclic group $\mathbb{G}$ with prime order q . Let P be a generator of $\mathbb{G}$. For arbitrary $\theta_x \in \Omega$, it selects a unique random element $t_x \in Z_q^*$ and compute $T_x = t_x P$. Choose $y \in Z_q^*$ and compute $Y = yP$. Define a hash function $H : Z_q^* \rightarrow \mathbb{G}$. Then it sets the master secret:

$$MK = < \{t_x\}, y > \tag{1}$$

And outputs the public parameter as follows:

$$PK = < q, \mathbb{G}, P, \Omega, W, \{T_x\}, Y, H > \tag{2}$$

3.2 The Key Generation Algorithm

It takes the public parameter PK , the master secret MK and an access policy Λ as input. It selects a threshold gate (k, n) , in which $n = \sum_{x \in \Lambda} w_x$ and k is selected based on the access policy that satisfies $1 \leq k \leq n$. For arbitrary attribute in access policy, it allocates $|w_x|$ index numbers:

$$\{\forall \theta_x \in \Lambda : index(x,1), index(x,2), \cdots, index(x, w_x)\} \tag{3}$$

The algorithm selects a random element $r \in Z_q^*$ and generate a random polynomial q_{root} with degree $d = k - 1$, which satisfies $q_{root}(0) = ry$. For each index number, set the following polynomial:

$$q_{root}(index(x,i)) = q_{x,i}(0) \tag{4}$$

Define $D = r$ and $D_x = \sum_{i=1}^{w_x} q_{x,i}(0) + t_x$. Finally, it outputs the private key:

$$SK = < D, \{D_x\} > \tag{5}$$

3.3 The Encryption Algorithm

The encryption algorithm takes as input the public parameter PK , an attribute set S and a plaintext M . It first selects a random element $s \in Z_q^*$ and computes $C_1 = sP$ and $C = H(sY) \oplus M$. For each attribute $\theta_x \in S$, it computes $C_x = sT$. Finally, it outputs the ciphertext as follows:

$$CT = < C, C_1, \{C_x\} > \quad (6)$$

3.4 The Decryption Algorithm

The decryption algorithm takes as input a private key SK and a ciphertext CT. Let x be a leaf node in an access policy Λ that represents the attribute θ_x. For each $x \in \Lambda$, define a Lagrange coefficient as follows:

$$\Delta_{idx,\Lambda'}(x) = \prod_{j \in \Lambda', j \neq idx} \frac{x - j}{idx - j} \quad (7)$$

Where $idx = index(x,i)$ and $\Lambda' = \{\forall x \in \Lambda : index(x,i)\}$. If $\theta_x \in S$, it computes:

$$\begin{aligned} & DecryptNode(C_1, C_x, D_x) \\ & = (D_x C_1 - C_x) \cdot \Delta_{idx,\Lambda'}(0) \\ & = ((\sum_{i=1}^{w_x} q_{x,i}(0) + t_x) \cdot sP - sT_x) \cdot \Delta_{idx,\Lambda'}(0) \\ & = \sum_{i=1}^{w_x} q_{x,i}(0) \cdot \Delta_{idx,\Lambda'}(0) \cdot sP \end{aligned} \quad (8)$$

Otherwise, the algorithm returns $\perp$ to indicate error. If the attribute set satisfies the access policy, it extracts the plaintext as follows:

$$M = H(D^{-1} \sum_{\theta_x \in S} DecryptNode(C_1, C_x, D_x)) \oplus C \quad (9)$$

3.5 Correctness Proof

In the decryption algorithm, for any authorized data receiver who launches a decryption query, if attribute set of a ciphertext satisfies the access policy of its private key, it can recover sY as follows:

$$\begin{aligned} & D^{-1} \sum_{\theta_x \in S} DecryptNode(C_1, C_x, D_x) \\ & = r^{-1} \sum_{\theta_x \in S} \sum_{i=1}^{w_x} q_{x,i}(0) \cdot \Delta_{idx,\Lambda'}(0) \cdot sP \\ & = r^{-1} \sum q_{root}(index(x,i)) \cdot \Delta_{idx,\Lambda'}(0) \cdot sP \\ & = sY \end{aligned} \quad (10)$$

4. Comprehensive Performance Analysis

4.1. Security Analysis

It is trivially guaranteed that unauthorized data receivers have no access to the plaintext. Anyone that does not hold authorized attributes cannot extract sY. Accordingly, such a data receiver cannot recover $H(sY)$. As a result, decryption does not work correctly.

Theorem. If CDH assumption is hard, our scheme is secure under chosen ciphertext attacks (CCA-secure) in the random oracle model.

Proof. Due to page limitation, we do not present the proof of the theorem. Interested readers can find the completed proof in the full version of this paper.

4.2. Expressiveness Analysis

4.2.1 Equivalence. Our weighted mechanism is a generalized form of traditional threshold mechanism. So, we can transform the "AND/OR" gate to a generalized form based on the weighted mechanism. For example, there are some data receivers that can decrypt some ciphertexts associated with both attribute "A" and attribute "B". For the threshold mechanism, the access policy of their private key is defined as $(2,2)$ to represent "A and B". For the weighted mechanism, we define weights of "A" and "B" as, respectively, w_A and w_B. In this case, the access policy of their private keys is defined as $(w_A + w_B, w_A + w_B)$ to represent "A and B".

If there are some data receivers that can decrypt some ciphertexts associated with attribute "A" or attribute "B". For the threshold mechanism, the access policy of their private key is defined as $(1,2)$ to represent "A or B". For the weighted mechanism, the access policy of their private keys is defined as $(\min\{w_A, w_B\}, w_A + w_B)$ to represent "A or B". Therefore, the proposed generalization is feasible. We present this expressiveness equivalence in Figure. 2.

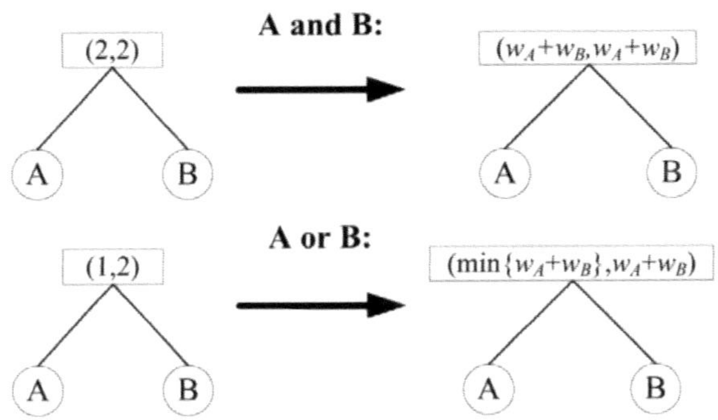

Figure 2 The expressiveness equivalence between the threshold mechanism and the weighted mechanism.

4.2.2 Simplicity. Compared to previous scheme, our weighted mechanism provides superior expressiveness with a simple structure. For example, there are some data receivers that can decrypt some ciphertexts associated with some attribute combinations such as "AB" or "AC" or "BCD". If we use "AND/OR" gates to express this relation, the access policy can be defined as:

$$(\text{A and (B or C)) or (B and C and D)} \tag{11}$$

For the weighted mechanism, we can define weights of A, B, C, and D as, respectively, 4, 3, 2, 1. In this case, the access policy can be defined as $(6,10)$, which is adequate to represent this relation. We show this simplicity of access policy in Figure. 3. Therefore, the proposed mechanism is very efficiency in terms of expressiveness.

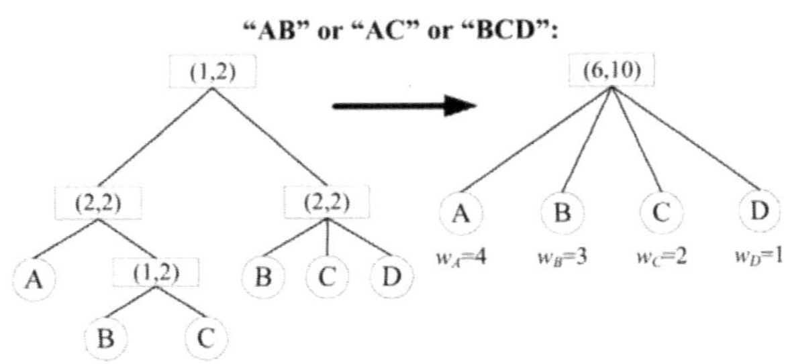

Figure 3 The simplifaction of an access policy by the weighted mechanism

4.3. Computational Efficiency Analysis

In this section, we discuss the computational efficiency of previous schemes [2], [3], [6], [13] and our KP-WABE-WB. Let $\mathbb{G}_*$ be the bit size of an element in $\mathbb{G}_*$, $|\Omega|$ be the total number of attributes in global attribute space, $|Y|$ be the total numbers of attributes in an access policy, $|S|$ be the total numbers of attributes in an attribute set, and $\hat{z}_q^*$ be the bit size of an element in z_q^*. The detailed comparison is listed in Table. I in four aspects, which are the size of public parameter $|PK|$, the size of private key $|SK|$, the size of ciphertext $|CT|$, and the total times of bilinear pairing computation N_{bp}.

Table 1 Computatiomal Efficiency Comparison

Schemes	$\|SK\|$	$\|CT\|$	N_{bp}
[2]	$\|Y\|\mathbb{G}_1$	$\|S\|\mathbb{G}_1+\mathbb{G}_T$	$\|S\|$
[3]	$(2\|S\|+1)\mathbb{G}_1$	$(2\|Y\|+1)\mathbb{G}_1+\mathbb{G}_T$	$2\|S\|+1$
[6]	$(2\|S\|+2)\mathbb{G}_1$	$(2\|Y\|+1)\mathbb{G}_1+\mathbb{G}_T$	$\|S\|+2$
[13]	$(\|Y\|\times L)\mathbb{G}$	$(2\|S\|+1)\mathbb{G}_1+\mathbb{G}_T$	$3\|S\|$
Our scheme	$\|Y\|\mathbb{G}_1+\hat{Z}_q^*$	$(2\|S\|+2)\mathbb{G}_1$	0

As is summarized in Table. I, the size of private key and ciphertext in our scheme is trivial among those schemes. For Qin *et al.*'s [13] construction, an attribute matrix with L rows and D columns is introduced to show hierarchy of different attributes. Meanwhile, their construction is based on composite-order bilinear pairing so the computational efficiency is a little bit unacceptable. Compared to those schemes in Table. I, our KP-WABE-WB is the only one that need no bilinear pairing computation. Therefore, it is efficient to build secure cloud data sharing via our KP-WABE-WB.

5. Conclusion

Attribute-based encryption (ABE) is an interesting cryptographic technique for flexible cloud data sharing access control. In this paper, we propose a key-policy weighted attribute-based encryption without bilinear pairing computation (KP-WABE-WB) for secure cloud data sharing access control. A simple weighted mechanism is presented to describe different importance of each attribute, which not only enhances the expressiveness of access policy but also simplifies its structure. We introduce a novel construction of ABE without executing any bilinear pairing computation. Therefore, the computational efficiency is markedly improved. Compared to previous schemes, our scheme has a better performance in expressiveness of access policy and computational efficiency.

Our future wok will build on the preliminary construction in this work to develop the proposed scheme. Considering some specific industrial scenario including secure data sharing in NB-IoT, we need to design a concrete algorithm of access policy generation based on weighted attribute.

6. Acknowledgment

This work is supported by the National Natural Science Foundation of China under Grant 61672299, Grant 61373135, and Grant 61170276. We thank the anonymous reviewers for their enlightening comments.

7. References

[1] A. Sahai and B. Waters, "Fuzzy Identity-Based Encryption," in *Proc. EUROCRYPT*, 2005, pp. 457-473.
[2] V. Goyal, O. Pandy, A. Sahai, and B. Waters, "Attribute-based encryption for fine-grained access control of encrypted data," in *Proc. ACM CCS*, 2006, pp. 89-98.
[3] J. Bethencourt, A. Sahai, B. Waters, "Ciphertext-Policy Attribute-Based Encryption," in *Proc. IEEE Symp. Secur. Privacy*, 2007,pp. 321-334.
[4] L. Cheung and C. Newport, "Provably secure ciphertext policy ABE," in *Proc. ACM CCS*, 2007, pp. 456-465.
[5] B. Waters, "Ciphertext-Policy Attribute-Based Encryption: An Expressive, Efficient, and Provably Secure Realization," in *Proc. Public Key Cryptogr.*, 2011, pp. 53-70.
[6] M. Green, S. Hohenberger, and B. Waters, "Outsourcing the Decryption of ABE Ciphertexts," in *Proc. USENIX Secur. Symp.*, 2011, pp. 34.

[7] J. Lai, R. H. Deng, C. Guan, and J. Weng, "Attribute-Based Encryption With Verifiable Outsourced Decryption," *IEEE Trans. Forensics Security*, vol. 8, no. 8, pp. 1343-1354, 2013.
[8] S. Q. Lin, R. Zhang, H. Ma, and M. S. Wang, "Revisiting Attribute-Based Encryption With Verifiable Outsourced Decryption," *IEEE Trans. Inf. Forensics Security*, vol. 10, no. 10, pp. 2119-2130, 2015.
[9] D. Pletea, S. Sedghi, M. Veeningen, and M. Petkovic, "Secure distributed key generation in attribute based encryption systems," in *Proc. ICITST*, 2015, pp. 103-107.
[10] X. L. Xu, J. L. Zhou, X. H. Wang, and Y. Zhang, "Multi-authority proxy re-encryption based on CPABE for cloud storage systems," *Journal of Systems Engneering and Electronics*, vol. 27, no. 1, pp. 211-223, 2016.
[11] J. Hur, "Improving Security and Efficiency in Attribute-Based Data Sharing," *IEEE Trans. Knowl. Data. Eng.*, vol. 25, no. 10, pp. 2271-2282, 2013.
[12] N. Oualha and K. T. Nguyen, "Lightweight Attribute-Based Encryption for the Internet of Things," in *Proc. ICCCN*, 2016, pp. 1-6.
[13] B. Qin, H. Deng, Q. H. Wu, J. Domingo-Ferrer, D. Naccache, and Y. Y. Zhou, "Flexible attribute-based encryption applicable to secure e-healthcare records," *International Journal of Information Security*, vol. 14, no. 6, pp. 499-511, 2015.

A Discretization Algorithm for Meteorological Data and its Parallelization Based on Hadoop

Chao Liu[a], Wen Jin, Yuting Yu, Taorong Qiu, Xiaoming Bai and Shuilong Zou[b]

School of Electrical and Information Engineering, Nanchang Institute of Science & Technology, Nanchang, China
Email: [a]mekisstherain@163.com, [b]2392748648@163.com

Abstract. In view of the large amount of meteorological observation data, the property is more and the attribute values are continuous values, the correlation between the elements is the need for the application of meteorological data, this paper is devoted to solving the problem of how to better discretize large meteorological data to more effectively dig out the hidden knowledge in meteorological data and research on the improvement of discretization algorithm for large scale data, in order to achieve data in the large meteorological data discretization for the follow-up to better provide knowledge to provide protection, a discretization algorithm based on information entropy and inconsistency of meteorological attributes is proposed and the algorithm is parallelized under Hadoop platform. Finally, the comparison test validates the effectiveness of the proposed algorithm for discretization in the area of meteorological large data.

1. Introduction

Data discretization as a means of data preprocessing is also increasingly showing its irreplaceable role in the field of data mining. In losts of areas such as rule extraction, rough set and granular calculation, feature clssification that continuous attribute data must be discretized before it can be used. In themeteorological field, the state collects large amounts of data in real time every day. Those data are not only large in volume, but also have many attributes and with continuous values and the data are not complete. Obviously, the discretization of data for such large data needs to face more challenges[1]. Some of the classic data mining methods, such as rough set[2] and other methods to extract the features of the extraction or association rules need to discretization first, and some discretization algorithms based on machine learning, such as clustering[3]. Therefore, it is very important to choose an efficient discretization algorithm. At present, different fields of research scholars have proposed a variety of different discretization algorithm, according to the classification of different standards, data discretization algorithm can be divided into: supervision[4] discrete algorithm and unsupervised[5] discrete algorithm such as equal frequency[6], equal width[7] discrete algorithm. Today's more commonly used discretization algorithms such as chi-square discretization algorithms Chimerge[8]. It is a self-low discrete algorithm. This algorithm introduces the statistical theory into the discretization algorithm. Due to the need to set the threshold, the subjective impact threshold is too high will lead to excessive discretization, too low will lead to insufficient discretization. Later, Liu et al improved the formation of the CHi2[9] algorithm. The algorithm uses a gradual decline in the level of significance, and ultimately can make more intervals merges but Chi2 algorithm processing speed slow in the face of large data, also not good as top-down algorithm. Since then, the algorithm can improve the ability to deal with large samples by changing the way of the breakpoint measurement, but the processing time also increased when the sample is discretized in the face of the large scale data volume. And this algorithm still requires the maximum threshold of the set value inconsistency and the case where the threshold is too high leads to the excessive discretization or too low leads to insufficient discretization.

This paper is based on the discrete algorithm proposed in paper[10]. On the one hand, the setting of the attribute inconsistency threshold is based on the manual method. In this paper, we improve the subjective influence of setting the attribute inconsistency threshold by improving the adjacency interval condition. On the other hand, consider the actual data set, especially in the field of meteorological data sets, the importance of each attribute in the decision-making class is not the same, this paper combines the importance of the attribute to determine whether the need to merge adjacent intervals. The improved method is designed to be able to effectively reduce the number of discrete points to improve the efficiency of the algorithm. At the same time, considering the actual environment, the need for discrete data sets are large data, so this article uses Hadoop platform to improve the method of parallel implementation.

2. Related Theory

2.1 Collection Information Entropy

There is a decision information system[11] $S=(U,AT,V,F), V=C\cup\{d\}$, C is a set of conditional attributes, $\{d\}$ is decision attribute set. The number of objects with the decision property $j(j=1,2,\dots,r(d))$ is k_j in set X, then the information entropy $H(X)$ of the set X is defined as:

$$H(X)=-\sum_{j=1}^{r(d)}p_j\log_2 p_j, p_j=\frac{k_j}{|X|} \tag{1}$$

Among them, $0\le H(X)\le 1$. Information entropy reflects the set X confusion, the smaller the information entropy is , the more object of decision attirbute value in X and the smaller the confusion of the set, and $H(X)=0$ when there is only one decision attribute value in X .

2.2 Breakpoint Importance Definition

According to the formula (1) can be defined as follows: in decision information system $S=(U,AT,V,f)$, P is a set of breakpoint on attribute a, $L=\{Y_1,Y_2,\dots,Y_m\}$ are the number of m subset of U that is divided by P , and the domain U is defined according to the information entropy of the set of breakpoint P :

$$H^P(U)=\frac{|Y_1|}{|U|}H(Y_1)+\dots+\frac{|Y_i|}{|U|}H(Y_i)+\dots+\frac{|Y_m|}{|U|}H(Y_m) \tag{2}$$

After add a new breakpoint $c\notin P$, the information entropy of breakpoint set $P+\{c\}$ can be defined as:

$$H^{P+\{c\}}(U)=\frac{|Y_1|}{|U|}H(Y_1)+\dots+\frac{|Y_i|}{|U|}H(Y_i^{l})+\frac{|Y_i|}{|U|}H(Y_i^{r})+\dots+\frac{|Y_m|}{|U|}H(Y_m)$$

Y_i^{l}, Y_i^{r} are subsets which are devided by breakpoint c . Add a breakpoint $c\notin P$ to P , then the information entropy of the breakpoint c can be defined as:

$$H(c)=H^P(U)-H^{P+\{c\}}(U) \tag{3}$$

The value of the breakpoint information entropy reflects the importance of the breakpoint. The smaller the value of $H(c)$, the smaller the decision attribute value of the equivalence class subset divided by the new breakpoint set, the decrease in the degree of confusion, the greater the $H(c)$ the more important breakpoint c is.

2.3 Inconsistency Introduction

In decision information system $S=(U,AT,V,f)$, P is a set of breakpoint on attribute a, $L=\{X_1,X_2,\ldots,X_m\}$ are the number of m subset of U that is divided by P, and $X_1 \cup X_2 \cup \ldots \cup X_i = U$, the number of objects whose decision attribute value is $j(j=1,2,\ldots,r(d))$ is k_j, M_x is the class label of the largest decision class in the set X, the X inconsistency ξ_x is defined as:

$$\xi_x = \frac{|X|-|M_x|}{|X|} \tag{4}$$

$|M_x| = \max\{k_j\}$ reflects the degree of inconsistency of the decision category in the X set, $0 \le \xi_x \le 1$, ξ_x is smaller, the set X is the more consistent.

2.4 MapReduce Programming Model

MapReduce[12] is a distributed computing framework model was originally introduced by Google engineers Jeffery Dean and Sanjay Ghemawat in 2004. It's divide complex problems running on large clusters into two function, map and reduce function. The map function can be understood as the size of the decomposition problem, which aims to divide a complex large-scale data set into several blocks that can be processed by each individual node in the cluster and analyze the subset of the data into a set of <K1,V2>pairs, then it's processed by framework will be merged, sorted and generated the intermediate results <K2, V2>pairs, and all the same K2 value of the V2 set together as a reduce function input, and after computed by reduced function would got the final result set <K3,V3>pairs, the process shown in Fig 1.

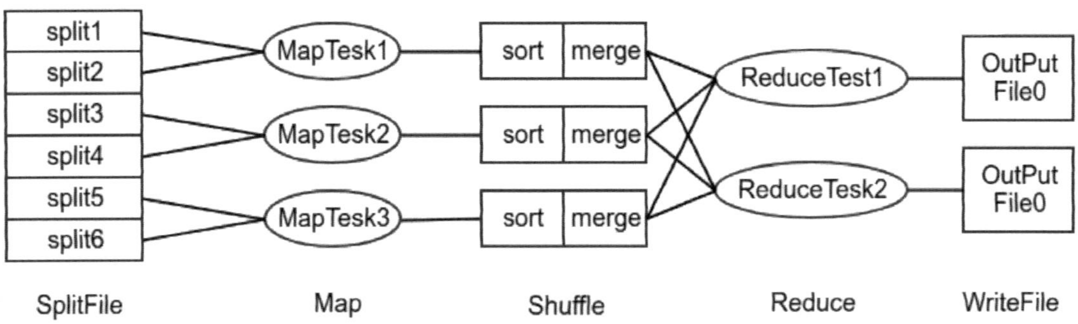

Figure 1 MapReduce calculation mode.

3. Related Theory Algorithm and Improvement Based on Information Entropy and Inconsistency

3.1 An Algorithm Based on Information Entropy and Inconsistency

The algorithm is mainly divided into three steps: the first step is to select candidate discrete points; the second step is to select discrete points according to the information entropy; the third step is to merge the interval according to the inconsistency rate and the importance of the attributes.

3.2 Improved Discretization Algorithm Based on Information Entropy and Inconsistency

The set of breakpoints generated by step2 of the above algorithm divides the continuous condition attribute values into finite value intervals, and merges the adjacent intervals satisfying the given combining condition by step3 to reduce the number of discrete points. However, there are two shortcomings in the interval merging method used in step3 above. One is that it is necessary to set the interval inconsistency threshold manually. This setting is subjective and difficult to determine. By testing in practical meteorological data set, we found that the meteorological data after noise pretreatment is not so by setting the $\xi^{''}_{max}$ to control the artificial noise seem a bit superfluous and the interval combination of discrete points decrease was not obvious; the second is the actual field data from different attributes influence on decision attribute or importance is not entirely consistent, but the algorithm does not consider the merger in the interval. In view of the above shortcomings, this paper improves the interval merging condition by deleting the inconsistency rate threshold and introducing the importance of attributes.

Based on attribute importance to merge interval is under this point: when a condition attribute in the decision table of decision attribute dependence is stronger, the important degree is higher, the number of discretization points or intervals on the properties change little or almost no change;and when a condition attribute in decision table the decision attribute dependence becomes weak, that is the degree of importance becomes smaller ,the number of discretization points or the number of intervals on the attribute changes and more.

In the above algorithm, we have the condition that the discretization interval on the conditional property $a \in C$ can be merged: $\xi Y_i \le \xi^{''}_{max}$ and $\xi Y_{i-1} \le \xi^{''}_{max}$, $\xi Y_i \cup Y_{i+1} \le \xi Y_i$ and $\xi Y_i \cup Y_{i-1} \le \xi Y_{i-1}$, change it to: $\xi Y_i \cup Y_{i-1} \le (\xi Y_i + \xi Y_{i-1}) * (1\text{-}a_i)$, where a_i is the importance of the conditional attribute. Since the study object of this paper is the element in the meteorological field, the importance of the meteorological element is set by the paper[13] when the specific meteorological element is discretized. When the importance of a property is very small, the formula $\xi Y \cup Y_{i-1} \le (\xi Y + \xi Y_{i-1}) * (1\text{-}a_i)$ increase the probability of establishment, then the possibility of merging the two adjacent areas also increased. And when it comes to the importance of a property, $\xi Y \cup Y_{i-1} \le (\xi Y + \xi Y_{i-1}) * (1\text{-}a_i)$,the probability of establishment is reduced, so the possibility of merging between adjacent two sections is also reduced. And the type of conditions for the same attribute in different locations adjacent to the interval can be combined with the above-mentioned algorithm is also dynamic changes. Therefore, the improved interval combination judgment not only takes into account the different importance of attribute, but also meets the different practical requirements. At the same time, the modified merge condition solves the problem of setting the inconsistency threshold $\xi^{''}_{max}$, so that the improved algorithm has less discrete effect caused by the parameter influence. The following is a detailed description of the third step to improve the algorithm:

Step3. Interval merge based on attribute importance and inconsistency

Input: based on information entropy breakpoint set $S^{''}$

Output: based on the inconsistency of the combined breakpoint set $P^{''}$

Algorithm description:

(1) the equivalence class $T^{''}$ is divided according to the breakpoint, and the reference[13] is used to determine the importance of the condition attribute a_i

(2) for p, q $\in T^{''}$, p, q adjacent
(2.1) according to the formula(4), respectively calculate p, q class inconsistency icRateP, icRateQ.
(2.2) merge p, q and assignment set to z
(2.3) according to the formular(4) calculate the inconsistency of z (icRateZ)
(2.4)if cRateZ<=(icRateP+icRateQ)*(1-a_i), then perform the step (2.5), otherwise continue

(2.5) delete the breakpoint between p, q classes from $S^{''}$
(2.6) merge p, q and assignment set top, and get next equivalence class as q
End for

4. Parallel Design Of Discrete Algorithm

Hadoop parallelization of the discrete process is divided into the following two steps correspond to the two jobs; Job1 is mainly on the data set of preprocessing, generate a formatted sample. Job2 is the format of the sample input, using the above 3.1 in step one, step two and improved step three of the algorithm for discretization. Details as follows:

Step1: Input: Complete decision table $S = (U, AT = C \cup D, V, f)$

Output: Candidate formatted sample collection B^{*}

Parallel description:

(1) Map read decision table system (S), attribute name (attId), attribute value (attValue), decision attribute value (convalue)

(2)Map formatted output: key=attId+attValue+conValue,value=1

(3) Reduce read form (2) the Map output

(4) Reduce calculates the number of the same key and assigns it to sumcount

(5)Reduce output: key=attId+attValue+conValue,value=sumcount

Step2: Input: the Step1's Reduce output

Output: The collection of breakpoints for each attribute B^{*}

Parallel description:

(1) Map read from the output of step1 (each line is formatted as Value=attId+attValue+conValue,sumcount)
output after formatted: key=attId (attribute name),
value=attValue+conValue+sumcount

(2) Map output: key=attId,value=attValue+conValue+sumcount

(3) Reduce input: Reduce read (2) the Map output

(4) Reduce calculation: according to the steps of algorithm 3.1,to obtain a candidate breakpoint collection B^{a} ,information entropy breakpoint collection S^{*}, attribute importance degree inconsistency merge breakpoint collection P^{a}.

(5) Reduce output: key=attId, Value= P^{a}

5. Algorithmic Test and Result Analysis

5.1. Test Data and Environment Introduce

The experimental data are derived from the daily value data of chinese climatological ground data provided by the China Meteorological Data Sharing Network, and the data of a ground site in Jinaxi Province from January 1960 to December 2016 are selected. At the same time, according to the paper[13] to provide the importance of meteorological factors on the choice of rainfall and rainfall correlation coefficient of 11 elements as a conditional attribute. And use the rainfall at 20~20 as a decision attribute. Because of the large amount of data, some incomplete data is deleted directly, and finally 1800000 data sets of 1.4G size are formed. The training set of different training sample sizes is extracted from the data set to form the four data sets as shown in Table 1.

Table 1 Experimental use of the data set

dataset	object numbers	size of dataset
D_Set1	225000	179MB
D_Set2	450000	358MB
D_Set3	900000	716MB
D_Set4	1800000	1.4GB

This test environment is Sugon server I620-G20,with 2 Xeon E5-2630CPU,each CPU is 8-core 16 threads, the machine's memory is 128G,hard disk size of 300G,using KVM technology to manage the virtual 8 virtual machines, assigned to each virtual machine resources are 2-core 4 threads CPU,12G memory,20G hard disk, and the operating system adopts Ubuntu16.04 server version, each virtual machine are equipped with Hadoop2.7.2,as a Hadoop nodes, a total of 8 Hadoop nodes, one of which has a separate namenode node responsible for storing the meta information of the files in the distributed system is not the actual computing node, the other seven nodes are the datanode nodes, in the distributed file system datanode node is the actual storage node and also the actual calculate node of algorithm program.

5.2. Test and Result Analysis

The classification algorithm used in the data mining is a classic classification algorithm (KNN classification algorithm), using this algorithm test the data set D_Set1 which processed by discretization. Then we divide D_Set1 into four group data set, each of which is a 1/4-size test set, a 2/4-size test set, a 3/4-size test set, and a test set itself. Each data set randomly selects 20% data as the test set, and selects 80% data as the training set. Using KNN algorithm, in k=5 case, cross validation is used to calculate the prediction accuracy, and the test is taken ten times as the final prediction accuracy data. The test results are shown in table 2.

Table 2 Predictive classification results

indicators	dataset	KNN algorithm	
		DIEIC	Improved DIEIC
Accuracy（%）	D_Set1_1	66.48	67.34
	D_Set1_1	66.33	67.42
	D_Set1_1	66.35	67.30
	D_Set1	66.44	67.21

Compared with the DIEIC algorithm, the improved average prediction accuracy is improved by 0.92%. Although the improved algorithm improve accuracy is not obvious, the main reason is that in the second step algorithm based on information entropy for discrete breakpoint is completed preliminary breakpoint extraction, thus proves the validity of access breakpoint based on information entropy, and the improved DIEIC algorithm is also improved a little discrete effect of instructions on adding attribute importance factor is compared with the discrete demand, it also can effectively reduce the number of discrete points

6. Concluding remarks

The discretization algorithm has a direct impact on the quality of data mining or knowledge discovery. In general, the importance of dependency of different conditional attributes and decision values is different in the decision table. It is one of the keys to improve the discretization effect when considering the importance of the attribute in designing the discretization algorithm. Based on the characteristics of the data set in the meteorological field involved in the research, this paper improves the algorithm based on the information discretization algorithm on information entropy and inconsistency. The improvement of the algorithm is mainly to improve the existing problems of discretization interval merging. The specific improvement includes eliminating the parameters of the original algorithm that need to manually set the inconsistency threshold, and consider different meteorological elements with different decision value dependencies and in the discretization interval combine the dynamic inequality when merging.In addition, for the needs of large meteorological data applications, the improved data discretization algorithm based on attribute importance and inconsistency is processed and implemented in parallel. Improved discretization algorithm and algorithm parallelization in the real meteorological data set under the comparative test and the results of the analysis to verify the effectiveness of the improved algorithm and the feasibility of parallelization. This has a reference value of and knowledge discovery applications for the meteorological large data mining and a certain guiding significance.

IOP Conf. Series: Journal of Physics: Conf. Series **910** (2017) 012011 doi:10.1088/1742-6596/910/1/012011

7. References

[1] Ramírez - Gallego S, García S, Mourio - Talín H, et al. Data discretization: taxonomy and big data challenge[J]. Wiley Interdisciplinary Reviews: Data Mining and Knowledge Discovery, 2016, 6(1): 5-21.
[2] Jiang F, Sui Y. A novel approach for discretization of continuous attributes in rough set theory[J]. Knowledge-Based Systems, 2015, 73: 324-334.
[3] Mincheng C, Jingling Y, Xiaoyan W. A Discrete Random Forest Parallel Classification Algorithm Based on Weakly Correlated Feature Subspace Selection[J]. Computer Science, 2016, 43(6) : 55 – 58.(In Chinese)
[4] Huang W, Pan Y, Wu J. Supervised discretization for optimal prediction[J]. Procedia Computer Science, 2014, 30: 75-80.
[5] Nguyen H V, Müller E, Vreeken J, et al. Unsupervised interaction-preserving discretization of multivariate data[J]. Data Mining and Knowledge Discovery, 2014, 28(5-6): 1366-1397.
[6] Noor Suhana S, Rohani A B. Rough Set Discretization: Equal Frequency Binning, Entropy/MDL and Semi Naives Algorithms of Intrusion Detection System[J]. arXiv, 2016, 282: 77-87.
[7] Jiawei Han, Micheline kamber. data mining concepts and techniques[M] :Beijing: Machinery Industry Press, 2004. (In Chinese)
[8] Kerber R. Chimerge: Discretization of numeric attributes[C]//Proceedings of the tenth national conference on Artificial intelligence. Aaai Press, 1992: 123-128.
[9] Liu H, Setiono R. Feature selection via discretization[J]. IEEE Transactions on knowledge and Data Engineering, 1997, 9(4): 642-645.
[10] Hailiang Y, Deqin Y. A Discretization Algorithm for ontinuous Attributes of Decision Tables Based on Information Theory[J]. Computer Science, 2010, 37(4) : 231 – 233. (In Chinese)
[11] Duolian M, Daoguo L.rough set theory algorithm and application[M].Tsinghua University Press, 2008. (In Chinese)
[12] Li F, Ooi B C, Özsu M T, et al. Distributed data management using MapReduce[J]. ACM Computing Surveys (CSUR), 2014, 46(3): 31. (In Chinese)
[13] Chenyang Z, Zhiqiang M, Limin L. Researchon Meteorological Data Mining Based on Rough Set and Bayesian under Hadoop[J]. computer application and software, 2015, 32(4) : 72 – 76. (In Chinese)

The Survey of Vision-based 3D Modeling Techniques

Mingzhe Ruan

School of Computer Science & Software Engineering, Tianjin Polytechnic University, Tianjin, China

Email: mingzheruan@foxmail.com

Abstract. This paper reviews the vision-based localization and map construction methods from the perspectives of VSLAM, SFM, 3DMax and Unity3D. It focuses on the key technologies and the latest research progress on each aspect, analyzes the advantages and disadvantages of each method, illustrates their implementation process and system framework, and further discusses the way to promote the combination for their complementary strength. Finally, the future opportunity of the combination of the four techniques is expected.

1. Introduction

The three-dimensional model, which can restore the scene of two-dimensional image and the object of the scene, initially belongs to the computer vision research. However, due to its broad application, researchers in the field of computer graphics and the computer vision conducted a lot of related research in this field. The traditional three-dimensional modeling method can be divided into three categories: the first is based on the method of professional modeling software artificially reshaping the geometric modeling, yet because of the high technical requirements of the modeling staff, this kind of time consuming work would have relatively higher cost. The second category is based on three-dimensional scanning technology, using the equipment with distance measurement function to obtain three-dimensional information of objects, in order to obtain high-precision three-dimensional model, but expensive equipment would result in increasing costs. Based on the limitations of the above two methods. The third kind bases on the image-based three-dimensional reconstruction technology, which can efficiently and accurately reconstruct a three-dimensional model with real scene or object by using 3D reconstruction algorithm with a series of two-dimensional images taken in real scene in a short period of time. In this paper, a new VS3D (VSLAM-SFM-3d max) modeling method is proposed to explore and study the 3D reconstruction technology. First, we use the VSLAM technology to explore the unknown scene. Through the use of feedback information, the initial modelling of the scene can be completed with the use of SFM technology and then insert 3D Max into the model, to enrich the details of the scene. Finally, utilize unity3d for its post-production and maintenance in order to achieve the purpose of rapid reconstruction of the three-dimensional model.

2. The VSLAM System

Mobile robot VSLAM technology refers to the robot using the visual system to achieve autonomous positioning and map creation, which is a non-traditional navigation technology. Since the robot position information obtained by the image information estimation usually has a large error, we cannot simply rely on the robot motion [1] to estimate the robot position information. After using the robot motion equation to obtain the robot position estimate, we can use the surrounding environment information obtained by the ranging unit to correct the position of the robot. The above correction process is generally achieved by extracting the environmental characteristics and then re-observing the

IOP Conf. Series: Journal of Physics: Conf. Series **910** (2017) 012012 doi:10.1088/1742-6596/910/1/012012

position of the feature after the robot moves. The core of VSLAM is EKF [2]. The EKF is used to estimate the exact position of the robot in conjunction with the above information. The selected features are commonly referred to as landmarks. EKF will continue to estimate the position of the robot and the location of the surrounding environment. The general procedure for VSLAM is shown in the figure.

When the robot moves, its position will change. At this time, the feature points in the observation information are extracted according to the observation of the robot position sensor. Then the robot would combine the position of the feature points observed with the motion distance, as well as the position of the feature points observed before the robot movement, to estimate the current location and current environmental information by EKF.

Another important source of SLAM data is the robot's own position information estimated by its own motion. The robot's own position data can be estimated by the number of turns of the robot's tire operation, which can be regarded as the initial estimate of the EKF.

Landmarks are features that are easy to be observed and distinguished in the environment. These features are generally used to determine the robot position. We can imagine the above process through the following methods: assuming you are in a strange room, close your eyes, then how do you determine their own position at this time? In general, we will continue to move in the environment, by touching the object or wall to determine their own position. Such touched objects and walls, etc. can be used to estimate their own location landmark.

After determining the characteristics to be extracted according to the above steps, we need to extract the features we need from the information obtained from the ranging unit accurately. There are many ways to extract features, which generally depends on the type of feature that needs to be extracted.

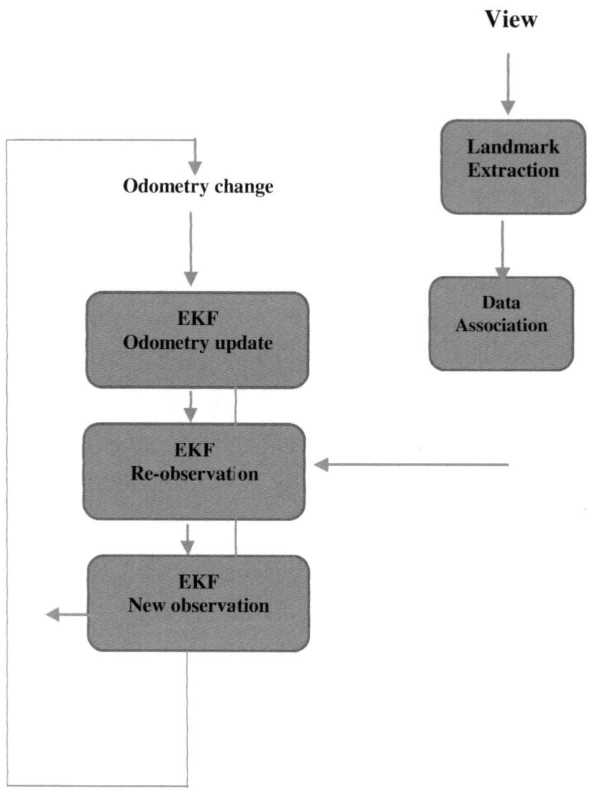

Figure 1 VSLAM positioning mechanism

VSLAM technology obtains two-dimensional image information through the camera, and then it can confirm the two-dimensional image of the landmarks, from which the features can be extracted.

Then by calculating the distance between them to get the location information. In the initial image processing stage, the camera calibration technique corrects the initial visual image and eliminates the effect of visual distortion on the accurate extraction of the image features. Finally, the visual dictionary tree with fast search feature is constructed by visualization of visual feature stratification, which eliminates the single-scale quantization error of visual dictionary. With the progress of VSLAM technology, we can get a lot of detailed scene information, and then use SFM technology to achieve the initial modeling of the scene. SFM technology is the most important part of VS3D modeling method.

3. The SFM System

When a series of detailed scene information is obtained by VSLAM, the scene is modeled using SFM. The SFM uses two-dimensional images to obtain three-dimensional information. Image is two-dimensional data, reflecting the space object or scene characteristics, contours, surface texture and depth information. People can enable the computer to automatically restore the 3D information of the scene by extracting and dealing with one or more images between the feature information and contact relationship, with the use of three-dimensional modeling algorithm. Through this method, it not only can access the space object shape, texture, structure and other external information characteristics, but also store, identify, deal with and analysis them. Three-dimensional modeling algorithm is the bridge of object recovering from the two-dimensional image into a three-dimensional model.

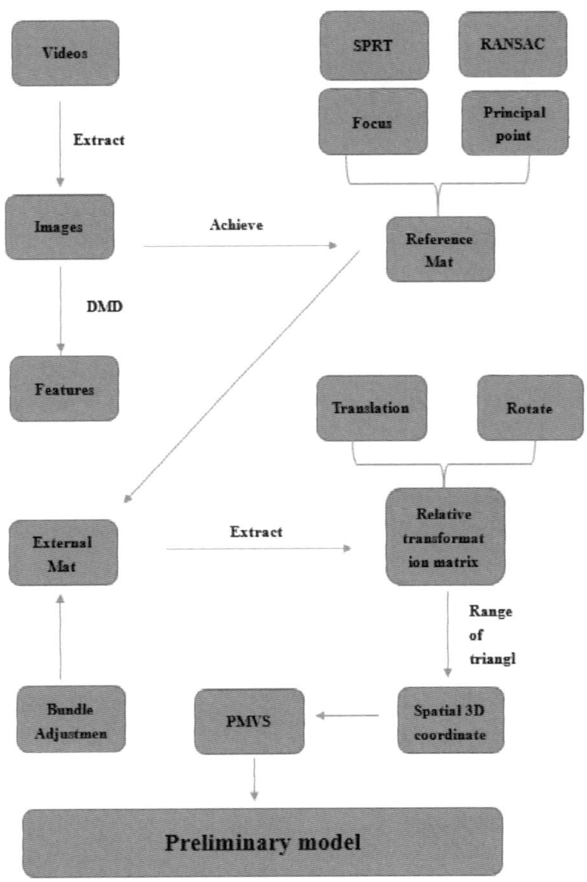

Figure 2 SFM system flow chart

Figure 3 Dense 3D point clouds of bicycles

Therefore, we propose a new Differential Morphological Decomposition (DMD) feature detection and description algorithm. At the same time, the RANSAC [3] algorithm is used to optimize the model checking process for the speed and precision of the model parameters, which saves a lot of data to the model and improves the speed of the RANSAC algorithm. Finally, a method of generating a dense three-dimensional point cloud is proposed. Based on the sparse three-dimensional point cloud, the improved Patch-Based Multiview Stereo (PMVS) [4,5] quasi-dense is proposed by using geometric constraints and adaptive algorithm.

3.1 The Main Steps for SFM Reconstruction Are as Follows:

- Find out the feature points in each picture, and match the two pairs. It is required to be able to accurately identify the local characteristics of the object, and to quickly and accurately match. The commonly used algorithm is the SIFT [6] method proposed by David Lowe.
- According to the matching results, using the projective theorem to calculate the camera position and other scene information, this step also known as Structure from Motion, or Sparse Reconstruction. The results of the standard note are accurate, which is commonly used in the Bundler [8] based on the Lecenberg-Marquardt algorithm.
- Use the scene information and the original photos, we can get photos in the 3D point of the cloud. This step also known as Dense Reconstruction. Using the Multi-View Stereo Reconstruction [7], the 3D point cloud is obtained. The quality of the cloud is affected by the efficiency of the image processing efficiency, reconstruction accuracy and integrity. So far the best algorithm is PMVS.

Build 3D models based on 3D point clouds. Linking the points into the surface, so that it can be used in normal three-dimensional modeling software. The commonly used algorithm is the Poisson Surface Reconstruction [9]. Take a bike as an example, we can get the figure on the left.

As the SFM algorithm is for high-speed processing of two-dimensional information, the cost of modeling the scene can be reduced, resulting in the possibility of a wide range of three-dimensional model reconstruction. After getting the initial modeling scene, import it into 3dmax and complete the details.

Figure 4 Dense 3D point clouds of bicycles

4. 3D Max

With the rapid development of computer technology and the development of powerful three-dimensional modeling software, it is possible to establish a true and reliable three-dimensional shape of three-dimensional objects.

3D Max needs to work in accordance with a certain work process, hence before the formal study, there should be a program flow chart for briefly describing its operational process.

4.1. Create Models

Creating a model is the first step in starting work in 3D Max. If there is no model, the whole work would be an ivory tower. 3D Max provides a wealth of modeling methods. Modeling can start with different 3D basic geometry, or you can use 2D graphics as the basis for lofting or extruding objects. You can also transform objects into a variety of editable surface types and then extend the vertices and use other tools for modeling.

4.2. Design Material

After completing the model creation work, you need to use the Material Editor to design the material. By setting the material for the model to make the model look more real. 3D Max offers a number of material types, materials that can be refracted and reflected, and materials with a rugged surface.

4.3. Set Up Lights and Cameras

Lighting is an indispensable element of the scene, if not having the right light condition, the scene will be greatly eclipsed, and sometimes cannot even show creative intentions.

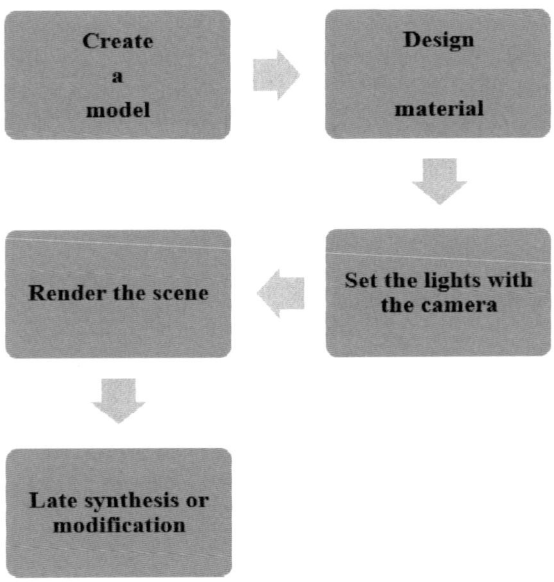

Figure 5 3D MAX flow chart

4.4. Render Scene

After completing the above steps, you also need to render the scene, during which the scene color and environmental effects can be added.

4.5. Post Synthesis or Modification

In most cases, you need to post-modify the rendering, which is, using two-dimensional image editing software, such as Photoshop, to modify, in order to remove the defects caused by model or material, lighting and other problems. In addition, sometimes the rendered image might be used as material applied to the graphic design, film or television post-synthesis work. In either case, you should be aware of the key working points or processes of post-modification or synthesis work so that the two tasks can be better connected.

We found in many 3D modeling software that 3D Max has the following advantages:

Three-dimensional modeling and texture mapping efficiency is high, 3d max Boolean value is double, which can provide more reliable results. Its UV mapping (enhancements) improves performance and increase the speed of UV navigation and editing by up to 5 to 10 times. It resolves the key user requests to improve texture creation performance, visual feedback and workflow efficiency.

3D rendering is realistic, and its new ART is a physics-based fast renderer. ART has a high efficiency, its CPU operation has nothing to do with the display card, and the use of image-based lighting is very excellent, hence it can provide more realistic results for most industries, products and architectural outdoor rendering.

Therefore, the model obtained by the SFM technique is introduced into 3D Max to supplement the scene details, and the obtained model is introduced into Unity for post-production and maintenance.

5. Unity

Unity3D is a comprehensive game development tool released by Unity Technologies that spans multiple platforms and is a full-featured, technically mature professional game engine.

5.1 Through The Analysis of The System, We Can Get Unity3D System Framework Map:

- AI: AI module contains logical units, groups and the objects controlled by AI, where the pathfinding and obstacle-avoidance utilize different turning behaviors to determine the unit states. At the same time, grouped AI manages the behaviors of the group, for instance, the grouped pathfinding. The Artificial Intelligence machine learns to save and load its data using the interface of the persistence data module.
- Persistent data: The persistent data module is responsible for the data in different games, including the data saved and data loaded. Also, this module can be used for saving data for the pathfinding module, including look-up forms and diagrams, as well as for managing the accumulated data for AI machine learning.

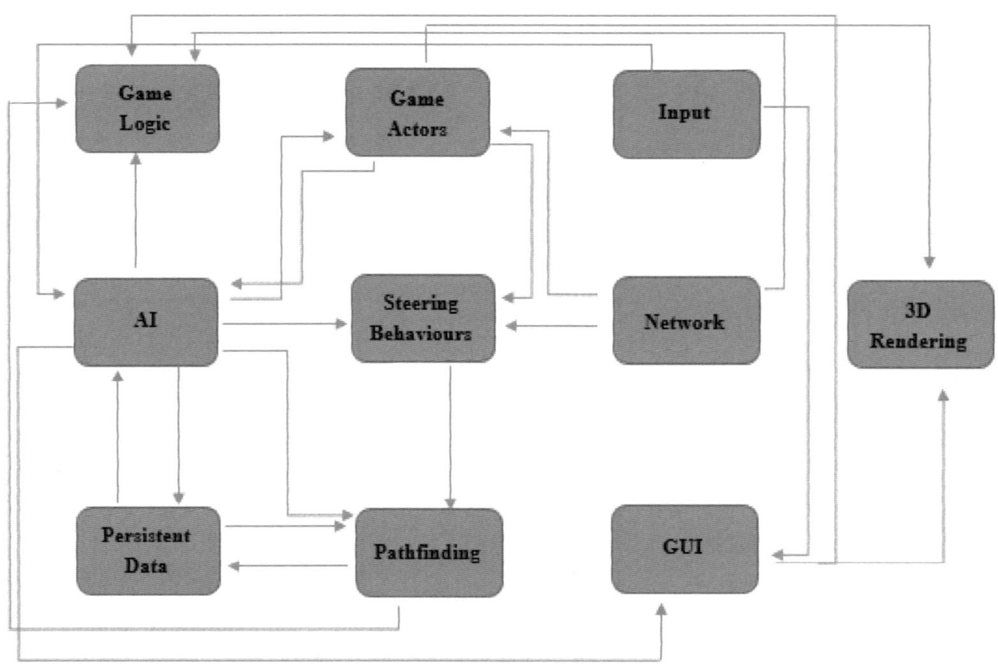

Figure 6 Unity frame structure

- Game actors: The 3D models of landform, units, and buildings can be visualized based on the rendering pipeline in Unity3D. Every game actor has its own AI module for behavior management.
- Steering behaviors: To guide the calculation power of action, which will determine how and how fast and independent game agent active is. It should be used for obstacle avoidance, flow or simple search task.
- Pathfinding: This module is for creating a path grid, obstacle information collection and providing a variety of path seek interface. In order to get better performance, some of the information is saved to load from disk.
- Input: This module tracks the user's input, processes it, and generates feedback.
- Network: Network module is for managing game actors to maintain the game state and keep them consistent on both machines, so that the responsibility of other player can be avoided due to network jittering.
- GUI: The GUI displays all the buttons, menus, small maps and countdown timers. It is responsible for the functions and interactions of these elements and it depends on the logical units used for this game.
- 3D rendering: The module mainly manages Unity3D. The main camera of the scene determines the objects that need to be rendered and sends them through the rendering pipeline. Unity3D provides high pixel and Vertex Shaders.

After the unity produced by unity3d, you can get a more complete scene model.

6. Conclusion

How to obtain the 3D model is a hot research direction of computer vision and computer graphics. Multi-view 3D reconstruction has many unique advantages compared with traditional geometric modeling. At present, through a variety of geometric modeling methods [10-16], it has been able to build some complex three-dimensional scene model, and then use the light, ray tracing algorithm and the radiation algorithm to draw, which can achieve a certain real sense of graphics modeling. However, in the real world, especially for natural scene modeling, since the scene is complicated and the modeling objects are different, if geometric modeling method is used, the modeling process would be complex, slow, and also cannot be in each scene to ensure the realism of the model built. Especially in the virtual reality scene, it requires real-time rendering and interaction. At this time, establishing the real-time model with real complex scene according to the requirements is extremely difficult. In this paper, a new VS3D (VSLAM-SFM-3d max) modeling method is proposed around 3D reconstruction technology. Using VSLAM technology to explore the unknown scene, with the use of feedback information and the use of SFM technology to initial modeling of the scene. Then introducing the model into 3D Max, and adding details of the scene through it. Finally, to use Unity3D for its post-production and maintenance. In the end, we can get three-dimensional reconstruction of the scene.

7. References

[1] E. Royer, M. Lhuillier, M. Dhome, J. M. Lavest. Monocular Vision for Mobile Robot Localization and Autonomous Navigation. International Journal of Computer Vision, 2007,

[2] T. Bailey, J. Guivant, M. Stevens, E. Nebot. Consistency of the EKF-SLAM Algorithm. In Proceedings of the IEEE/RSJ International Conference on Intelligent Robots and Systems.

[3] Zuliani M. RANSAC for Dummies. Santa Barbara, USA:Vision Research Lab, University of California, 2008.

[4] Y. Furukawa, J. Ponce. Accurate, dense, and robust multiview stereopsis. IEEE Transactions on PatternAnalysis and Machine Intelligence, 2010, 32(8): 1362-1376

[5] Y. Furukawa, J. Ponce. Accurate camera calibration from multi-view stereo and bundle adjustment. International Journal of Computer Vision, 2009, 84(3): 257-268

[6] Lowe D G. Distinctive image features from scale-invariantkeypoints. International Journal of Computer Vision, 2004,60(2): 91-110.

[7] Y. Furukawa, J. Ponce. Accurate camera calibration from multi-view stereo and bundle adjustment. International Journal of Computer Vision, 2009, 84(3): 257-268

IOP Conf. Series: Journal of Physics: Conf. Series **910** (2017) 012012 doi:10.1088/1742-6596/910/1/012012

[8] Triggs B, McLauchlan P F, Hartley R I, et al. Bundle adjustment– a modern synthesis//Vision Algorithms: Theory and Practice. Berlin, Germany: Springer, 2000: 298-372.
[9] Kazhdanm, Bolithom, Hoppeh. Poisson surface reconstruction. Procof Eurographics Symposiumon Geometry Processing. 2006.
[10] Ruipeng Gao, Mingmin Zhao, Tao Ye, Fan Ye, Guojie Luo, Yizhou Wang, Kaigui Bian, Tao Wang, Xiaoming Li, "Multi-story Indoor Floor Plan Reconstruction via Mobile Crowdsensing," *IEEE Transactions on Mobile Computing (IEEE TMC)*, vol. 15, no. 6, pp. 1427-1442, 2016.
[11] Ruipeng Gao, Yang Tian, Fan Ye, Guojie Luo, Kaigui Bian, Yizhou Wang, Tao Wang, and Xiaoming Li, "Sextant: Towards Ubiquitous Indoor Localization Service by Photo-taking of the Environment," *IEEE Transactions on Mobile Computing (IEEE TMC)*, vol.15, no.2, pp.460-474, 2016.
[12] Ruipeng Gao, Mingmin Zhao, Tao Ye, Fan Ye, Yizhou Wang, Guojie Luo, "**Smartphone-based Real Time Vehicle Tracking in Indoor Parking Structures,**" *IEEE Transactions on Mobile Computing (IEEE TMC)*, vol. 16, no. 7, pp. 2023-2036, 2017.
[13] Ruipeng Gao, Bing Zhou, Fan Ye, and Yizhou Wang, "Knitter: Fast, Resilient Single-User Indoor Floor Plan Construction," in Proceedings of IEEE INFOCOM, 2017.
[14] Ruipeng Gao, Mingmin Zhao, Tao Ye, Fan Ye, Yizhou Wang, Kaigui Bian, Tao Wang, and Xiaoming Li, "Jigsaw: Indoor Floor Plan Reconstruction via Mobile Crowdsensing," in Proceedings of ACM MobiCom, 2014.
[15] K. Bian, and J. Park. Maximizing Rendezvous Diversity in Rendezvous Protocols for Decentralized Cognitive Radio Networks, IEEE Transactions on Mobile Computing (IEEE TMC), Vol. 12(7):1294--1307, July 2013.
[16] Y Liao, K Bian, L Song, Z Han. Full-duplex MAC protocol design and analysis. IEEE Communications Letters 19 (7): 1185-1188, July 2015.

Secure and Efficient Signature Scheme Based on NTRU for Mobile Payment

Yunhao Xia[1], Lirong You[2], Zhe Sun[2] and Zhixin Sun[1]

[1]Nanjing University of Posts and Telecommunications, Key Lab of Broadband Wireless Communication and Sensor Network Technology, Nanjing, P. R. China
[2]Zhongtian Technology Corporation, Nantong, P. R. China

Email: 2015070108@njupt.edu.com, youlr@chinaztt.com, sunz@chinaztt.com,

sunzx@njupt.edu.cn

Abstract. Mobile payment becomes more and more popular, however the traditional public-key encryption algorithm has higher requirements for hardware which is not suitable for mobile terminals of limited computing resources. In addition, these public-key encryption algorithms do not have the ability of anti-quantum computing. This paper researches public-key encryption algorithm NTRU for quantum computation through analyzing the influence of parameter q and k on the probability of generating reasonable signature value. Two methods are proposed to improve the probability of generating reasonable signature value. Firstly, increase the value of parameter q. Secondly, add the authentication condition that meet the reasonable signature requirements during the signature phase. Experimental results show that the proposed signature scheme can realize the zero leakage of the private key information of the signature value, and increase the probability of generating the reasonable signature value. It also improve rate of the signature, and avoid the invalid signature propagation in the network, but the scheme for parameter selection has certain restrictions.

1. Introduction

With the widespread use of third-party mobile payment, mobile payment has become a serious area of cyber-crime [1]. During the process of mobile payment, the user needs to send the relevant information about the payment, including the order information, payment information, user or merchant information and so on. These data are related to the security of users' funds, which will have serious consequences if they are exploited by lawbreakers. On the other hand, with the advent of the shor quantum algorithm [2] and the rapid development of quantum computers, the threat of RSA and ECC is increasing. With the improvement of computing performance, the low-security public-key encryption algorithm applied to mobile terminals is no longer safe. The security of third-party mobile payment becomes an important obstacle to mobile payment development and popularization [3].

In the face of the threat of quantum computing to the traditional encryption system, this paper researches public-key encryption algorithm NTRU for quantum computation, which is applicable to the terminal of finite computing resources. In order to satisfy the non-repudiation of information, the main method is digital signature. The classic digital signature schemes are RSA [4] and ECC [5]. These two algorithms have the properties of the public key that can be decrypted correctly using the private key, thus can be used directly as the signature algorithm. However, the efficiency of RSA signature algorithm is very low, and the efficiency of ECC signature algorithm is much slower than that of NTRU. The encryption system NTRU [6] is based on the polynomial ring, and it avoids the modular arithmetic of the big integer power, which only involves the addition and subtraction of polynomials and multiplication. The efficiency of encryption algorithm NTRU is better than RSA and ECC, and encryption algorithm NTRU is considered to be the most future public-key encryption algorithm. However, the encryption algorithm NTRU encrypt using private key and this will leak the information of private key. Therefore, this paper redesigns the signature algorithm based on NTRU.

The rest of paper is organized as follows: Section 2 introduces some related work to our research. In section 3, we propose an improved signature algorithm based on NTRU for the defects, and analyze the security and operation performance. Conclude our paper in the section 4.

2. Realted Works

Ajtai provided a ground-breaking conclusion [7]: the security of any cryptographic system based on some of the lattice problems is equivalent to the most difficult password system, and presented the AD encryption system based on the most difficult case of u-SVP (unique Shortest Vector Problem)[8]. The scheme can be proved to be safe, but the system is inefficient and cannot meet the practical requirements. Park in [9] proposed a NFC mobile payment system based on zero-knowledge digital signature of NTRU, instead of using users' private information directly, thus ensuring the security of users' privacy. One innovation of the scheme was the use of NTRU encryption technology, achieving user authentication. However, the shortcomings of proposed method were obvious: the digital signature technology of NTRU was still immature. The NTRU encryption algorithm was applied to PKI in [10], which was found to have a large advantage in speed and the CA certificate was generated faster compared with PKI using RSA.

Hoffstein, Pipher, and Silverman in [11] presented NSS signature scheme based on the difficulty of finding CVP of recent vector problems in the grid. The scheme adopted the structure of NTRU algorithm, so the speed of encryption and decryption was relatively fast. Later, Silverman and Mironov in [12] presented the attack on the signature restoring private key. Subsequently, Hoffstein, Pipher and Silverman in [13] proposed the strengthening version of NSS for the attack of Silverman. However, Gentry and Szydlo researched the characteristics of R-NSS as a single secret key, and in [14] presented a large number of attacks on the signature restoring private key by combining the maximal common factor method. And then, Hoffstein in [15] proposed the NTRUSign algorithm, which has a large structural change compared with NSS and R-NSS algorithm, and its secret key is not the characteristic of a single secret key, using a linear combination of secret keys. However, Nguyen in [16] recovered the private key by a large number of valid signature values according to the characteristics of the uniform distribution of NTRUSign secret-key coefficient. Hu in [17] proposed an improved signature scheme based on R-NSS according to the problem of leaking private key of R-NSS and NTRUSign. Li in [18] proposed a novel identity based undeniable signature scheme, and the paper analyzed the proposed scheme was unfalsified and anonymous under stochastic prediction mode.

Melchor in [19] proposed a method to hide the leaking NTRUSign signature value through gaussian noise. At the same time, a strong parameter setting scheme was proposed according to the latest attack on NTRUSign to realize zero knowledge of signature value and anti-counterfeiting attack. Zhang in [20] presented a new type of NTRU signature algorithm based on CVP problem, and the efficiency was about 1/7 faster than the signature algorithm designed by Hu [17]. However, the proposed method did not disturb the private key directly by signing the private key, so the algorithm still has the problem of leaking private key. Hoffstein J, Pipher J, Schanck J in [21] proposed a new signature scheme based on NTRU, called NTRUMLS. The results showed that the security and efficiency of the scheme were relatively high, but the probability of generating reasonable signature was lower and cannot meet the requirement of mobile payment signature. At the same time Xie in [22] put forward a kind of identity signature scheme based on NTRU lattice, and the security of the scheme was based on γ - SVP difficult problems. An obvious advantage of the proposed method was the efficiency is higher than the same digital signature scheme based on grid.

As a result, the researches on the signature scheme of NTRU were mainly the improvement of NSS, R-NSS and NTRUSign. On one hand, proposed solutions for the classic attack schemes, and on the other, proposed new signature schemes based on NTRU. However, few of the above were suitable for mobile payment. Firstly, NSS, R-NSS and NTRUSign exist the problem of leaking private key. Besides, the probability of reasonable signature value of signature scheme based on NTRU is less than 1, and even some valid probability of signature scheme cannot meet the practical demand. Therefore, it was not suitable for application in a large number of users in mobile payment.

3. Secure and Efficient Signature Scheme For Mobile Payment

Through the research of the proposed signature scheme NTRUMLS [21] by Hoffstein, the analysis showed that the proposed method can realize zero knowledge of signature value. But the proposed scheme had the problem that the effective of signature value was low. For a lot of users of mobile payments, a large number of invalid signature value would greatly aggravate network congestion and the burden of service systems. This paper proposes two ways to improve the signature probability of NTRUMLS by analyzing the influence factors of the probability of reasonable signature value of NTRUMLS. Firstly, select reasonable parameter to narrow down the space to find a valid signature, and ensure that the signature value has a larger probability in the L∩R range. Secondly, during the process of signature, filtrate the conditions that satisfy a valid signature value, and we can ensure that the generated signature values are reasonable.

3.1. Increasing the probability of reasonable signature

According to the related parameters recommended by NTRUMLS [21], the probability of reasonable signature value generated by NTRUMLS is shown in table 3.1.

Table 3.1 The Probability Generated By Recommended Parameters

NTRUMLS parameters(N, p, q, B_s, B_t)	Probability of reasonability signature
(401,3,2^18,240,80)	38%
(439,3,2^19,264,88)	55%
(593,3,2^19,300,100)	41%
(743,3,2^20,336,112)	53%

Table 3.2 The Probability Generated By Another Parameters

NTRUMLS parameters(N, p, q, B_s, B_t)	Probability of reasonability signature
(443,3,2^16,138,46)	8%
(563,3,2^16,174,58)	2%
(743,3,2^17,186,62)	6%
(907,3,2^17,225,75)	2%

In regard to the probability of reasonable signature value generated by NTRUMLS, this paper can be inspired by another parameters.

Compared with table 3.1 and table 3.2, we can find that the change of parameters has a great influence on the probability of reasonable signature value generated by NTRUMLS. Therefore, this paper can be inspired by setting relevant parameters to improve the probability of reasonable signature value generated by NTRUMLS.

Firstly, this paper studies the correlation factors of the probability of reasonable signature value generated by NTRUMLS. For the convenience of research, this paper set $B = \lfloor p^2 N/4 \rfloor$, and $B = B_t = B_s$. In this case, generating a reasonable signature value depends on whether the signature value is in the multidimensional polyhedral formed by lattice $L\left(A - B_{s'}, \frac{q}{2} - B_t\right)$. For valid signature value, the limit of its parameter is less than $\frac{q}{2} - B$. Therefore, during the process of setting parameters, q must be much larger than B, or the probability of generating a reasonable signature will be small. Consider the general form of signature value:

$$(s,t) = (s_0, t_0) + (af + ag) \tag{1}$$

In the formula (1) above, $s_0, t_0 \in R(\frac{q}{2})$, $af \in R(p^2N/4)$, $ag \in R(p^2N/2)$. Therefore:

$$\|(s,t)\| \le \frac{q}{2} + B \tag{2}$$

In the formula (2) above, the coefficients of s and t are less than $\frac{q}{2} - B$, and then we can obtain the probability of generating a reasonable signature value:

$$P((s,t) is valid) \approx \left(\frac{\frac{q}{2} - B}{\frac{q}{2} + B} \right)^{2N} \tag{3}$$

Then make:

$$q \approx \frac{kp^2N^2}{4} \approx kNB \tag{4}$$

We can obtain:

$$P((s,t) is valid) \approx \left(\frac{\frac{q}{2} - B}{\frac{q}{2} + B} \right)^{2N} \approx e^{-9/k} \tag{5}$$

In the formula (5) above k is a small constant. From the above analysis, it can be seen that the probability of generating a reasonable signature is about $e^{-9/k}$. And it is associated with constant k.

$$k \approx \frac{q}{NB} \tag{6}$$

Therefore, the improved method can be realized by raising constants k to improve the accuracy of generating a reasonable signature. The influence of constant k on the probability of generating a reasonable signature is shown in figure 3.1.

According to figure 3.1, constant $k \approx 12$ when the probability of generating a reasonable signature is about 50%. If we raise the constant k to 25, the probability of generating a reasonable signature will be greater than 70%. So it will be a feasible scheme by increasing the constant k. In the formula $k \approx \frac{q}{NB}$, parameter N responds to the security of NTRUMLS, and therefore N cannot be changed. Parameters B_s, B_t are small adjustment of performance and safety, and smaller B_s, B_t can improve performance. However, it brings difficult to find suitable signature value. Therefore, according to the above analysis, it is possible to appropriately improve the value of parameter q. We can know the relation between parameter q and the probability of generating a reasonable signature according to formula (6) $k \approx \frac{q}{NB}$.

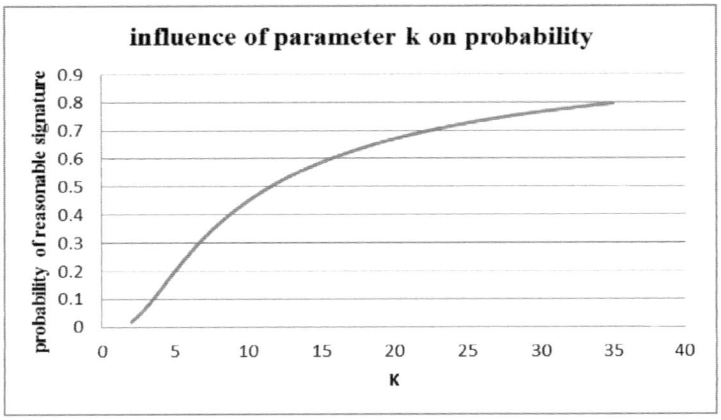

Figure 3.1 The influence of constant k on the probability

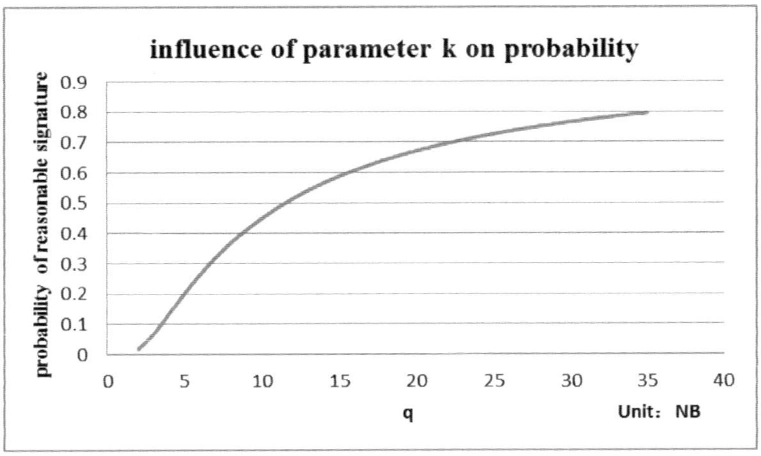

Figure 3.2 The influence of constant q on the probability

According to figure 3.2, it is a feasible scheme to significantly improve the probability of generating a reasonable signature value through change parameter q. In this paper, the value of parameter q in table 3.1 is changed to twice, and it is shown in table 3.3.

Table 3.3 The Probability Generated By Changed Parameter Q

NTRUMLS parameters(N, p, q, B_s, B_t)	Probability of reasonability signature
(401,3,2^19,240,80)	62%
(439,3,2^20,264,88)	74%
(593,3,2^20,300,100)	64%
(743,3,2^21,336,112)	73%

Comparing table 3.1 with table 3.3, it can be seen that the probability of generating a reasonable signature value is obviously improved through double the value of parameter q. However, if you want to make the reasonable signature value close to 100%, the value of parameter q will become very large, and the following will discuss the effect of parameter q on the performance of NTRUMLS. Because key structure of NTRUMLS is similar to the NTRU, so parameter q has little impact on the rate of NTRUMLS through the analysis above. Key sizes of NTRUMLS and cipher text size is related to parameters N, q has no effect on it. Therefore, in order to improve the probability of generating a reasonable signature value, the solution given in this paper is to improve the parameter q appropriately. Table 3.3 is the revised parameter q given in this paper.

3.2. Adding Authentication Conditions for Signature

The template is used to format your paper and style the text. All margins, column widths, line spaces, and text fonts are prescribed; please do not alter them. You may note peculiarities. For example, the head margin in this template measures proportionately more than is customary. This measurement and others are deliberate, using specifications that anticipate your paper as one part of the entire proceedings, and not as an independent document. Please do not revise any of the current designations.

According to literature [20], if $(s,t) \neq (s_p, t_p) \pmod{p}$, then the probability of generating a reasonable signature value is 0. As a result, the first condition that determines whether the signature value is legal is $(s,t) \equiv (s_p, t_p) \pmod{p}$. On the basis of improving parameter q of NTRUMLS, in this paper authentication scheme of the selected signature value is supplemented by the condition $(s,t) \equiv (s_p, t_p) \pmod{p}$ to propose our signature scheme. For the reasonable signature value of NTRUMLS, the following conditions must be satisfied:

$$\|(s,t)\| \le \frac{q}{2} + B \tag{7}$$

$$(s,t) \equiv (s_p, t_p) \pmod{p} \tag{8}$$

It is necessary to add a condition $(s,t) \equiv (s_p, t_p) \pmod{p}$ to determine whether the value of the selected signature value is a reasonable signature value. This can prevent the invalid signature value from spreading in the network, resulting in the waste of network resources. The specific changes are as follows:

Signature:

Describe the proposed signature algorithm here.

Input: (f, g, h, u) u is about the plaintext polynomials that will be signed and the coefficients of that polynomial $\in \{-1,0,1\}$.

Output: (s,t,u)

1: $(s_p, t_p) \leftarrow Hash(h,u)$

2: Repeat

3: Randomly generate $r \in R(\left\lfloor \frac{q}{2p} + \frac{1}{2} \right\rfloor)$, $\lfloor\ \rfloor$ is getting the integers down

4: $s_0 \leftarrow s_p + pr$

5: $t_0 \leftarrow h \otimes s_0 \pmod{q}$ and satisfy $t_0 \in R(\frac{q}{2})$

6: $a \leftarrow g^{-1} \otimes (t_p - t_0) \pmod{p}$

7: $(s,t) \leftarrow (s_0, t_0) + (a \otimes f, a \otimes g)$, determine whether $\| a \otimes f \| \le B_s$且 $\| a \otimes g \| \le B_t$

$\| s \| \le \frac{q}{2} - B_s$ and $\| t \| \le \frac{q}{2} - B_t$ $(s,t) \equiv (s_p, t_p) \pmod{p}$

Output: (s,t,u) , otherwise repeat.

Authentication:

Describe the proposed authentication algorithm here.

Input: (s,t,u,h)

1: $(s_p, t_p) \leftarrow Hash\ (h,u)$

2: If $t \neq h \otimes s (\mathrm{mod}\ \ q)$, output illegal and end, otherwise continue

3: If $\| s \| > \frac{q}{2} - B_s$ or $\| t \| > \frac{q}{2} - B_t$ output illegal and end, otherwise continue

4: If $(s,t) \neq (s_p, t_p)(\mathrm{mod}\ \ p)$ output illegal and end, otherwise output legal

3.3. Performance Analysis of Proposed Signature Scheme

Safety performance:

In this paper the proposed signature scheme is an improvement of NTRUMLS, and there is no significant change in the algorithm structure, so its security depends on NTRUMLS. According to analysis of the literature [20], this signature can guarantee the security of private key information. Therefore, the security of signature algorithm is fully guaranteed.

Efficiency performance:

In this paper, the proposed signature scheme is compared with that of NTRUMLS: firstly, the probability of parameter q is improved greatly, so that the probability of reasonable signature is improved. Second, the authentication conditions are added to avoid the illegal signature value spreading through the network. Table 3.4 reflects experiment contrast of the proposed signature scheme is given in this paper and NTRUMLS. Test environment is: the system is android 5.0, CPU is Intel Z3560 of dominant frequency of 1.8 GHz, memory is 4g. The result is the average of the data of 100.

Table 3.4 Experiment Contrast Of The Proposed Signature Scheme and NTRUMLS

NTRUMLS (N, p, q, B_s, B_t)	Secure level (bits)	Key generat ion time (ms)	Signature Time (ms)	Authenticatio n Time (ms)	probability of reasonable signature	Key size (bytes)
(401,3,2^18,240,80)	112	23.431	5.768	1.001	38%	853
(439,3,2^19,264,88)	128	29.28	4.456	1.113	55%	988
(593,3,2^19,300,100)	192	52.873	10.567	1.897	41%	1335
(743,3,2^20,336,112)	256	79.653	11.912	2.312	53%	1765
(401,3,2^19,240,80)	112	25.727	3.403	1.012	62%	853
(439,3,2^20,264,88)	128	32.149	2.629	1.125	74%	988
(593,3,2^20,300,100)	192	58.054	6.235	1.917	64%	1335
(743,3,2^21,336,112)	256	87.458	7.029	2.337	73%	1765

According to table 3.4, the key generation time and authentication time of the proposed signature scheme are slightly higher than NTRUMLS in the key generation phase. This is because the parameter q is twice larger, which slightly increases the calculation of the key generation, but this is small. The worst time complexity of key generation of the proposed signature scheme in this paper is $O(N\ \hat{}\ 3)$. As can be seen from the above data, the parameter N has great influence on proposed signature scheme, and q has little influence on its speed. The result also verifies the discussion of the key generation and decryption speed of the parameter q for NTRU.

In theory, the parameter q becomes larger and the signature time will become longer in the signature phase. But there will be a significant reduction in the time of generating signature. This is because the signature phase is a process of finding a reasonable signature, and if the selected signature value does not meet the criteria, the signature process will be repeated until it is found to satisfy the given condition. In this paper, the parameter q of the proposed signature is larger than that of

NTRUMLS, and it will greatly increase the probability of generating a reasonable signature value. Then each time the signature process filtering the reasonable signature, the math expectation is smaller, which significantly reduces the signature time. In the meantime, to avoid the illegal signature value which meet the condition $\|(s,t)\| \le \frac{q}{2} + B$ and not meet the condition $(s,t) \equiv (s_p, t_p) \pmod{p}$ to be sent to the validation party, in the proposed signature scheme the authentication conditions are supplemented during the signature phase, in order to ensure that all signature values are valid. The verification condition is the mode of the signature value, so the calculation is very small, and the influence of the signature time is very small, so the signature time can be shortened greatly.

At the same time, under the condition of parameter q of the proposed signature scheme in this paper change for 2 times, the key size and signature size have no change. So the proposed scheme is secure and efficient which can raise the speed and probability of generating reasonable signature value.

4. Conclusion

This paper researches the non-repudiation of data in mobile payment process, considering traditional RSA, ECC and other signature schemes are less efficient and do not have the ability of anti-quantum attack, and we introduce the signature scheme based on NTRU in this paper. However, several classical NTRU signature algorithms, include NSS, R-NSS, NTRUSign, exist the problem of leaking private key information and the classical NTRU signature algorithms NTRUMLS, which has unsatisfactory probability of generating a reasonable signature value. On the basis of these classical signature scheme, this paper proposed an efficient and secure signature scheme which is suitable for mobile payment by increasing the value of parameter q and adding the authentication condition that meet the reasonable signature requirements during the signature phase. Experimental results show that the proposed signature scheme can realize the zero leakage of the private key information of the signature value, and increase the probability of generating the reasonable signature value. It also improve rate of the signature, and avoid the invalid signature propagation in the network, but the scheme for parameter selection has certain restrictions.

5. Acknowledgment

This research is supported by the National Natural Science Foundation of China (NO.61672299, NO61373135). The authors thank the sponsors for their support and the reviewers for helpful comments.

6. References

[1] Emmeline Taylor. “Mobile payment technologies in retail: a review of potential benefits and risks,” International Journal of Retails & Distribution Management, vol. 44, 2016, pp. 159-177, doi:10.1108/IJRDM-05-2015-0065.

[2] Shor P W. “Polynomial-Time Algorithms for Prime Factorization and Discrete Logarithms on a Quantum Computer,” Society for Industrial and Applied Mathematics, vol. 41, 1999, pp. 303-332, doi:10.1137/S0036144598347011.

[3] Yong Wang, Christen Hahn, Kruttika Sutrave. “Mobile payment security, threats, and challenges,” Mobile and Secure Services (MobiSecServ), 2016 Second International Conference on. 26-27 Feb, 2016, pp. 1-5, doi:10.1109/MOBISECSERV.2016.7440226.

[4] B R Ambedkar, Ashwani Gupta and S S Bedi. “Confidentiality of Group Digital Signature Using Proposed Public Keys Encryption Algorithm,” International Journal of Engineering Science and Computing, vol. 6, 2016, pp. 2304-2307, doi:10.4010/2016.550.

[5] Deepti Jyotiyana, Varun P Saxena. “Fault attack for scalar multiplication over finite field (E(Fq)) on Elliptic Curve Digital Signature algorithm,” Recent Advances and Innovations in Engineering (ICRAIE), 2016 International Conference on, 23-25 Dec, 2016, doi:10.1109/ICRAIE.2016.7939539.

[6] Sonika Singh, Sahadeo Padhye. “Generalisations of NTRU cryptosystem,” Security and Communication Networks, vol. 9, 2016, pp. 6315-6334, doi:10.1002/sec.1693.

[7] Ajtai M. "Generating hard instances of lattice problems," Proceedings of the 28th Annual ACM Symposium on Theory of Computing, 1996, pp. 99-108, doi:10.1145/237814.237838.
[8] Ajtai M, Dwork C. "A public-key cryptosystem with worst-case/average-case equivalence," Proceedings of the 29th Annual ACM Symposium on Theory of Computing, 1997, pp. 284-293, doi:10.1145/258533.258604.
[9] Park S W, Lee I Y. "Anonymous authentication scheme based on NTRU for the protection of payment information in NFC mobile environment," Journal of Information Processing Systems, vol. 9, 2013, pp. 461-476, doi:10.3745/JIPS.2013.9.3.461.
[10] Elbasheer M O, Mohammed T. "Signing and Verifying Certificates by NTRU and RSA Algorithms," Cloud Computing (ICCC), 2015 International Conference on. IEEE, 2015, pp. 1-4, doi:10.1109/CLOUDCOMP.2015.7149655.
[11] Hoffstein J, Silverman J H. "NSS: An NTRU lattice-based signature scheme," International Conference on the Theory and Application of Cryptographic Techniques, 2001, pp. 211-228.
[12] Mironov I. "A note on cryptanalysis of the preliminary version of the NTRU Signature scheme," IACR Cryptology ePrint Archive, vol. 5, 2001, pp. 1-6.
[13] Hoffstein J, Pipher J, Silverman J H. "Enhanced encoding and verification methods for the NTRU signature scheme," Previously posted on http://www. ntru. com/technology/tech. technical. html, 2001.
[14] Gertry C, Szydlo M. "Cryptanalysis of the revised NTRU signature scheme," Advances in Cryptology-Eurocrypt 2002, Berlin: Springer-Verlag, 2002, pp. 299-320.
[15] Hoffstein J, Howgrave-Graham N, Pipher J, et al. "NTRUSIGN: Digital signatures using the NTRU lattice," Topics in cryptology—CT-RSA 2003, Springer Berlin Heidelberg, 2003, pp. 122-140.
[16] Nguyen P Q, Regev O. "Learning a parallelepiped: Cryptanalysis of GGH and NTRU signatures," Journal of Cryptology, vol.22, April 2009, pp. 139-160, doi:10.1007/s00145-008-9031-0
[17] Yupu Hu. "A Novel NTRU-Class Digital Signature Scheme," CHINESE JOURNAL OF COMPUTERS, vol.31, Sep. 2008, pp. 1661-1666.
[18] Li S, Wang C, Xu M, et al. "A novel identity based undeniable signature scheme," Conference Anthology, IEEE, 1-8 Jan 2013, pp.1-5, doi:10.1109/ANTHOLOGY.2013.6784885.
[19] Melchor C A, Boyen X, Deneuville J C, et al. "Sealing the leak on classical NTRU signatures," International Workshop on Post-Quantum Cryptography, 2014, pp. 1-21, doi: 10.1007/978-3-319-11659-4_1
[20] Zhang Juanmei, Cao Jie, Liu Nianyi, Yang Yatao, Li Zichen. "A New Design of Digital Signature Scheme Baesd on NTRU," JOURNAL OF SICHUAN UNIVERSITY(ENGINEERING SCIENCE EDITION), vol. 47, 2015, pp. 49-53.
[21] Hoffstein J, Pipher J, Schanck J M, et al. "Transcript secure signatures based on modular lattices," nternational Workshop on Post-Quantum Cryptography, 2014, pp. 142-159, doi:10.1007/978-3-319-11659-4_9
[22] XIE J, HU Y, GAO J, et al. "Efficient identity-based signature over NTRU lattice," Frontiers of Information Technology & Electronic Engineering, vol. 17, 2015, pp. 135-142, doi:10.1631/FITEE.1500197.

SSIM-based Rate Distortion Optimal Differential Energy Watermarking

Guixing Wu and Yudong Xiang

School of Software Engineering, Suzhou Institute for Advanced Study, USTC Suzhou, Jiangsu, P. R. China
Email: gxwu@ustc.edu.cn, sa514005@mail.ustc.edu.cn

Abstract. In this paper, we propose a rate distortion optimal method to maximize the compression performance of differential energy watermarking. First, the compression performance of differential energy watermarking is analyzed, which shows the deterioration of compression performance due to digital marking. Second, we developed an SSIM-based rate distortion optimal method to optimize the compression performance. Experimental results show that the proposed method can achieve better compression performance or big rate reduction while visual quality is not affected.

1. Introduction

Digital watermarking, as a proposed solution to digital rights management, has been studied extensively during the past decades [1], [2], [3], [4]. Digital watermarks are usually embedded into multimedia contents such that they are imperceptible. The embedded watermarks should be robust against various attacks. In addition, the watermark embedding capacity and security are desired to be high.

Since multimedia signals are usually transmitted in com- pressed formats, watermarks are usually embedded in the compressed domain[5]. In [6], [7], the authors investigated watermark embedding within the compressed domain from an information-theoretic perspective. In [8], [9], [10], [11], [12], the authors embedded information into the compressed bitstreams, the tradeoff between the robustness and encoding distortion was investigated. Nevertheless, the resulting dis- tortion or compression rate may increase after information embedding. Hence the embedding of digital information may result in a degradation of compression performance.

In some applications it is desirable to maintain com- pression performance after digital information is embedded into the compressed bit streams. Consider, for example, the application of digital camera. In a digital camera image, date information is often visibly inserted, which degrades the quality of the image. One way to address this problem is to hide the date or similar type of information invisibly into the image. The embedded information is tied to the image, and remains intact during mild distortion and image format conversion. Later, the embedded information can be extracted using a watermark decoder whenever necessary.

Due to the limited storage space of digital camera, the resulting image file size is desired to be maintained or slightly increased after information embedding. However, it is usually difficult to achieve this goal without degrading the quality of the image. In [13], [14], the authors embedded watermarks into the H.264 compressed bit streams while preserving the compression bit rate, but the video quality optimization was not take into consideration.

In this paper, we look into the optimization of compression performance while embedding watermarks into the compressed bit streams. As JPEG is the most widely used image compression format and differential energy water- marking is a well known watermarking method [10], [15], we shall use a JPEG compressed and differentially energy watermarked system as an example to demonstrate how to maximize the compression performance.

The rest of the paper is organized as follows. Section II analyzes the compression performance of differential energy watermarking. In Section III we present an SSIM-based rate distortion optimal algorithm to maximize the compression performance of a JPEG compressed and differentially energy watermarked system. Simulation results are given in Section V.

2. Analysis Of Differential Energy Watermarking

In differential energy watermarking, watermarks are em- bedded in DCT energy differences between DCT blocks of an image. That is, a watermark bit is embedded into $2n$ DCT blocks which are divided into two regions: region A and region B. Each of these two regions contains n DCT blocks. Then the high frequency DCT coefficients in one chosen region are discarded such that the DCT energy difference between these two regions are greater than a desired threshold. The chosen region could be region A or region B depending on whether a bit "1" or a bit "0" is embedded.

Figure 1 compares the compression performance of differential energy watermarking(DEW) with that of standard JPEG compression for the 512×512 image Lena. In Fig. 1 STD-JPG stands for standard JPEG compression. In DEW, 64 bits are embedded, the quality factor of JPEG recompression is set to 25. PSNR means peak signal to noise ratio, it is an image quality metric. From the figure, we can see, the watermark embedding degrades the compression performance. For instance, at PSNR=38dB, the compression bit rate after watermark embedding increases nearly 70%. Similar results were observed for other test images. Therefore, it is necessary to optimize the compression performance to compensate for the rate degradation due to watermark embedding. In this paper, we develop a rate distortion optimal method to maximize the compression performance of JPEG compressed images while DEW is applied to the compression process.

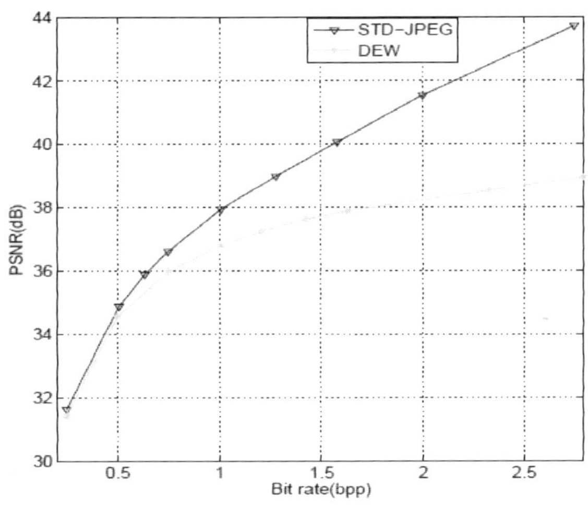

Figure 1 Comparison of performance between the standard JPEG compression and DEW

3. Ssim-Based Rate Distortion Optimization

3.1. Problem Formulation

To maximize the compression performance, we want to solve the following constrained optimization problem

$$\begin{cases} \min\ r \\ subject\ to\ d \le d_{budget}\ \text{and}\ (1-2m)E \ge D \end{cases} \quad (3.1)$$

where r means the compression bit rate, d means the encoded distortion, d_{budget} is the distortion constraint, m is watermark bit 1 or 0, E is the energy difference, D is the energy threshold, and

$$(1-2m)E \geq D \tag{3.2}$$

is the constraint due to watermark embedding

The above constrained optimization problem can be solved by minimizing

$$\min J(\lambda) = \mathrm{d} + \lambda r \tag{3.3}$$

where λ is a fixed parameter that represents the tradeoff of rate for distortion, and $J(\lambda)$ is the Lagrangian cost.

In this paper, we use a widely used quality metric– structural similarity index (SSIM) as the visual quality metric [16]. In the case of high rate quantization, Yeo et al approximate SSIM as [17]

$$SSIM \approx \frac{2\sigma_x^2 + c_2}{2\sigma_x^2 + MSE + c_2} \tag{3.4}$$

where σ_x^2 is the variance of the original image signal x, c_2 is a constant for numerical stability, MSE is the mean square error between an original image and the degraded image.

The distortion can then be approximated as

$$d \approx 1 + \frac{MSE}{2\sigma_x^2 + c_2} \tag{3.5}$$

The Lagrangian cost can be defined as

$$J(\lambda) = SSE + 2(\sigma_x^2 + c_2)\lambda r \tag{3.6}$$

where SSE means the sum of square error between an original image and the degraded image.

3.2. Algorithm Design

Before presenting the developed algorithm, let's first have a brief review of JPEG compression [15]. A JPEG encoder first partitions an input image into 8x8 blocks. Each of these 8x8 blocks is DCT transformed, and the resulting DCT coefficients are then uniformly quantized using an 8x8 quantization table. The coefficient indices from the quantization are finally entropy coded using zero run-length coding and Huffman coding.

The JPEG syntax leaves the selection of the quantization table, the Huffman codewords, and DCT indices to the encoder provided that the encoding process is compatible with the JPEG standard[18]. Since the coefficient indices can be equivalently represented as run-size pairs followed by in-category indices through run-length coding, in [18] Yang and Wang proposed a JPEG-compatible joint optimization algorithm to maximize the compression performance over all possible sequences of run-size pairs (R, S) followed by in category indices $I D$, all possible Huffman tables H, and all possible quantization tables Q.

In this paper, we investigate how to maximize the JPEG compression performance while subjecting to the water- marking constraint (3.2). Therefore, the minimization problem can be converted to

$$\min_{(R,S,ID),H,Q} \{J(\lambda) = SSE + 2(\sigma_x^2 + c_2)\lambda r\} \tag{3.7}$$

where (R, S, ID) and Q are chosen in the set where the watermarking constraint (3.2) is satisfied. Since the Huffman table H is completely determined by the run-size probability distribution P, we use P in the following to replace the Huffman table H in (3.7).

The search for the minimum of $J(\lambda)$ through optimizing (R, S, ID), P and Q simultaneously entails too much computational complexity. To reduce the computational complexity, the following alternating minimization procedure is used to solve the problem (3.7):

1) Fix P and Q. Find an optimal sequence (R, S, ID) that achieves the following minimum

$$\min_{(R,S,ID)} \{J(\lambda) = SSE + 2(\sigma_x^2 + c_2)\lambda r\} \tag{3.8}$$

while satisfying the watermarking constraint (3.2);

2) Fix (R, S, ID). Update P and Q respectively to achieve the following minimum

$$\min\{J(\lambda) = SSE + 2(\sigma_x^2 + c_2)\lambda r\} \tag{3.9}$$

while satisfying the watermarking constraint (3.2).

Note that in (3.9) P can be selected as the empirical runsize distribution of (R, S). The two steps are repeated until the decrease of $J(\lambda)$ is below a prescribed threshold ϵ.

3.3. Optimal DCT Indices Updating

Fix P and Q. The optimal DCT indices (or (R, S, ID) sequence) can be determined independently for each 8×8 image block as $J(\lambda)$ is block-wise additive. Following a similar approach as in [18], we solve the problem using a graph based optimization algorithm while satisfying the watermarking constraint.

Let the cut off index in differential energy watermarking be equal to C. That is, the DCT coefficients after this position C within a DCT block are either discarded or included in the computation of DCT energy. When a DCT block belongs to the region where high frequency DCT coefficients are discarded, the graph utilized in the searching of the optimal result has C + 1 states as shown in Fig. 2. The first C states correspond to the first C DCT coefficient indices of a DCT block in zigzag order. Each state may have incoming connections from its previous 16 states, which correspond to the run R, in an (R, S) pair. The last state is called end state. The end state may have incoming connections from all the other states where the indices are not equal to zeros.

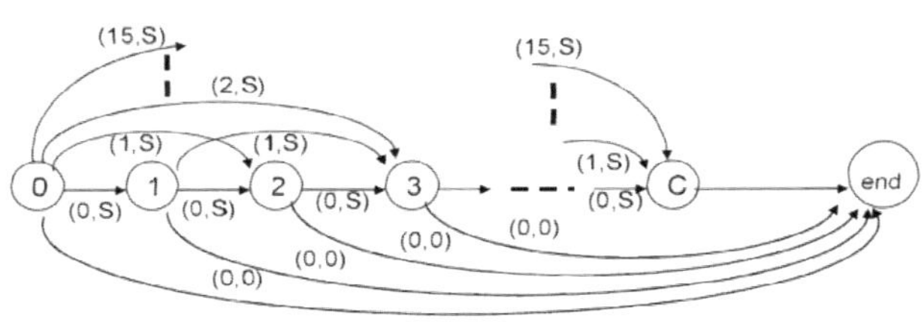

Figure 2 Optimization graph 1

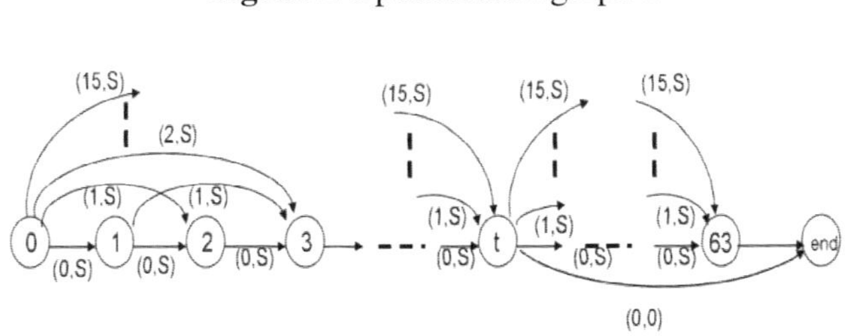

Figure 3 Optimization graph 2

Associated with each transition (r, s) in Fig. 2 is a cost defined as the incremental Lagrangian cost of going from state i − r − 1 to state i when the ith quantized DCT coefficient index eeds s bits to represent its amplitude and all the r DCT coefficients θj before the ith DCT coefficient are quantized to zeros. Specifically, this incremental cost is equal to

$$\sum_{j=i-r}^{i-1} \theta_j^2 + \left|\theta_i - q_i \cdot ID_i\right|^2 + 2(\sigma_x^2 + c_2)\lambda \cdot (-\log_2 P(r,s) + s) \tag{3.10}$$

where ID_i is an index in the size group s that gives rise to the minimum square error to the DCT coefficient θ_i , q_i is the ith quantization step size in the quantization table, and P (r, s) is the probability of the pair (r, s).

Let L_i be the minimum incremental Lagrangian costs from state 0 to state i. Let SSE_i be the cost of dropping all DCT coefficients from i + 1 to 63. Find a certain value k which minimizes the sum of L_i and SSE_i

$$\min_{0 \le i \le C} \{J(i) = Li + SSEi\} \tag{3.11}$$

Then backtrack from state k to find the optimal sequence(R, S, I D) with the minimum Lagrangian cost.

When the DCT block belongs to the region where high frequency DCT coefficients are not discarded, the optimization graph has 65 states as shown in Fig. 3. The first 64 states correspond to the DCT coefficient indices of an image block. The calculation of the minimum incremental Lagrangian cost is similar to that in Fig. 2, but with the following exceptions. To preserve the DCT energy difference, the DCT indices after the cut off index should satisfy the constraint

$$ID_i \ge \frac{[\theta_i]Q_{jpeg}}{q_i} \tag{3.12}$$

where $[\theta_i]Q_{jpeg}$ is the pre-quantized DCT coefficient using standard JPEG recompression. After the cutoff index no transition is allowed across state i where $[\theta_i]Q_{jpeg} \ne 0$. Assume t is the position of last nonzero pre-quantized DCT coefficient within a DCT block, only states from t to 63 are allowed to go to state *end*.

After the calculation of the incremental Lagrangian cost, find a certain value k which minimizes the sum of L_i and SSE_i sequence (R, S, I D) with the minimum Lagrangian cost.

$$\min_{0 \le i \le 63} \{J(i) = Li + SSEi\} \tag{3.13}$$

Then backtrack from state k to state 0 find the optimal sequence (R, S, I D) with the minimum Lagrangian cost.

3.4. Optimal Huffman and Quantization Table Updating

Fix (R, S, I D). The Huffman table H is determined by the empirical run-size distribution P. As the compression rate does not depend on the quantization table Q under this condition, we only need to minimize the sum of square error in the Lagrangian cost J (λ). That is, we need to solve the following minimization problem

$$\min_{Q} SSE \tag{3.14}$$

Among all possible Q, Let $Q = \{q_0, q_1, \cdots q_{63}\}$. Let $\theta_{j,i}$ and $ID_{j,i}$ denote the ith DCT coefficient and the ith DCT index of the jth DCT block in zigzag order respectively.

Then SSE can be expressed as

$$SSE = \sum_{i=1}^{63} \sum_{j=1}^{Num\_Blk} (\theta_{j,i} - ID_{j,i} \cdot q_i) \tag{3.15}$$

Where Num_Blk is the number of 8x8 blocks in an image.

The minimization of SSE can be achieved by minimizing independently the inner summation of (3.15) for each $i = 1, 2 \cdots 63$. Taking the derivative with respect to q_i, we get the minimum of (3.15) with a set of quantization step sizes

4. Experimental Results

Having described the proposed optimization algorithm, in this section, we evaluate the compression performance of the proposed algorithm. In the experiments, the constant $c_2 = (0.03 \times L)^2$, where L is the peak value of an image. We use test images from [17] to make the experiments.

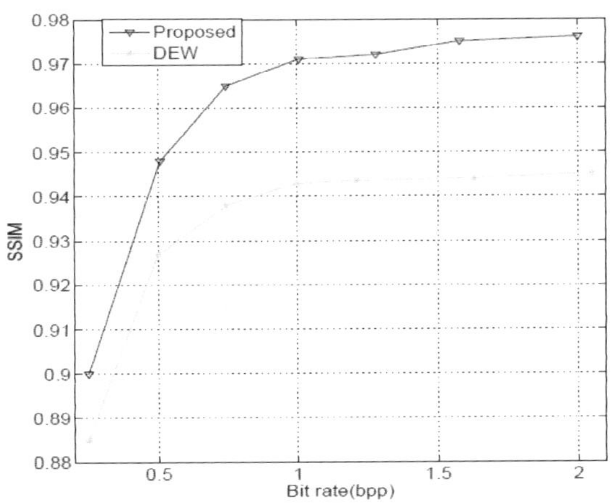

Figure 4 SSIM comparison between the proposed algorithm and DEW

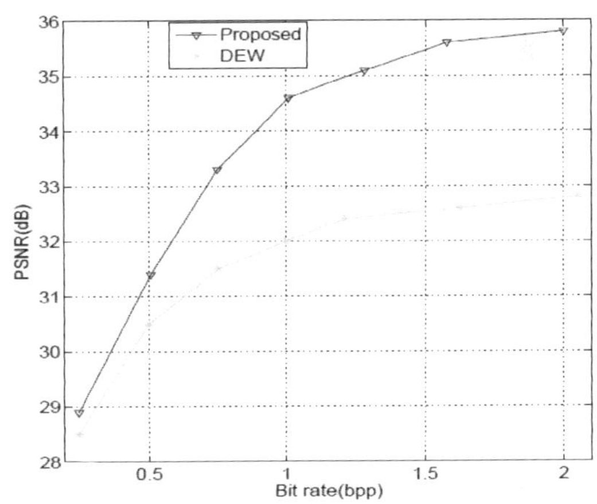

Figure 5 PSNR comparison between the proposed algorithm and DEW

Figure 4 compares the SSIM values of the proposed algorithm with the original DEW algorithm. The test image is 512x512 Lena, The quality factor of JPEG recompression is set to 25. 64 bits of watermark information are embedded. It can be seen from the figure that the proposed algorithm could achieve better compression performance. For example, at *SSI M* =0.94, the proposed algorithm can reduce nearly 50% of the bit rate of a DEW encoder.

Figure 5 compares the PSNR values of the proposed algorithm with the original DEW algorithm. The test image is 512x512 Lena, The quality factor of JPEG recompression is set to 25. We can see that the proposed algorithm could achieve better compression performance. For example, at *PSNR*=32.5*dB*, the proposed algorithm can reduce nearly 50% of the bit rate of a DEW encoder.

Similar results can be observed for other test images and/or other quality factors of JPEG recompression. As the DCT energy difference is preserved in the optimization, the watermark robustness was not affected.

5. Conclusion

In this paper, we developed an SSIM-based rate distortion optimal method to maximize the compression performance of a differential energy watermarking system. We first analyzed the differential energy watermarking, which showed the necessity for the optimization of compression performance. Next we proposed a rate distortion optimal method to optimize the compression performance. From the experimental results we can see that the proposed method can boost the compression performance of a differential energy watermarking system.

6. Acknowledgment
The authors would like to thank General Program of Natural Science Foundation of Jiangsu Province (BK20141209) and Suzhou Fundamental Research Project (SYG201543) for supporting this work.

7. References
[1] A. K. Khanna, N. R. Roy and B. Verma, " Digital image watermarking and its optimization using genetic algorithm," *2016 International Conference on Computing, Communication and Automation,* pp. 1140 - 1144, 2016.
[2] Y. Erfani, R. Pichevar and J. Rouat, " Audio Watermarking Using Spikegram and a Two-Dictionary Approach," *IEEE Transactions on Information Forensics and Security,* Vol.12, No. 4, pp.840 - 852, 2017.
[3] I.J. Cox, M. Miller and J. Bloom, Digital Watermarking, October 2001, 576 pages Cloth.
[4] B. Chen and G. W. Wornell, "Quantization index modulation: A class of provably good methods for digital watermarking and information embedding," *IEEE Trans. Inform. Theory*, Vol. 47, No. 4, pp. 1423–1443, May 2001.
[5] G. Wu and E.-H. Yang, "Joint watermarking and compression using scalar quantization for maximizing robustness in the presence of additive gaussian attacks," *IEEE Trans. Signal Processing,* , Vol. 53, No. 2, pp. 834-844, Feb. 2005.
[6] D. Karakos and A. Papamarcou, "A relationship between quantization and watermarking rates in the presence of additive Gaussian attacks," *IEEE Trans. Inform. Theory*, Vol. 49, No. 8, pp. 1970–1982, Aug. 2003.
[7] A. Maor and N. Merhav, "On joint information embedding and lossy compression in the presense of a stationary memoryless attack channel," *IEEE Trans. Inform. Theory*, Vol. 51, No. 9, pp. 3166–3175, Sept. 2005.
[8] C. -S. Lu, J. -R. Chen, M. Liao, K. -C. Fan, "Real-time MPEG2 video watermarking in the VLC domain," *Proc.16th IAPR Int. Conf. on Pattern Recognition*, Vol. 2, pp.552-555, Aug. 2002.
[9] A. M. Alattar, E. T. Lin and M. U. Celik "Watermarking low bit-rate advanced simple profile MPEG-4 bitstreams," *in Proc. Int. Conf. on Acoustics, Speech and Signal Pro- cessing*, pp. 513-516, 2003.
[10] G.-C. Langelaar and R. L. Lagedijk, "Optimal differential energy watermarking of DCT encoded images and video," *IEEE Trans. Image Processing*, Vol. 10, No. 1 , pp. 148-158, Jan. 2001.
[11] S. A. PARAH,J. A. SHEIKH , N. A. LOAN, "Robust and blind watermarking technique in DCT domain using inter- block coefficient differencing," *in Digital Signal Process- ing*, pp.11-24, Vol. 53, 2016.
[12] L. Wang, H. Ling, F. zou, Z. Lu, "Real-Time Compressed- Domain Video Watermarking Resistance to Geometric Dis- tortions,"*IEEE MultiMedia*, Vol 19, No. 1. pp. 70 - 79,2012.
[13] D. Xu, R. Wang, and Y. Q. Shi, "Data hiding in Encrypted H.264/AVC Video Streams by Codeword Substitution ," *IEEE Trans. on Information Forensics and Security*, Vol.9, No. 4, pp. 596–606, April 2014.
[14] Z. Ma, J. Huang, M. Jiang, X. Niu, "A Video Watermarking DRM Method Based on H.264 Compressed Domain with Low Bit-Rate Increasement,"*Chinese Journal of Electron- ics*, Vol. 25, No. 4, pp. 641-647, April 2016.
[15] W. Pennebaker and J. Mitchell, JPEG Still Image Data Compression Standrad, Kluwer Academic Publishers,1993.
[16] Z. Wang, A. C. Bovik, H. R. Sheikh and E. P. Simoncelli, " Image quality assessment: from error visibility to structural similarity ," *IEEE Transactions on Image Processing*, Vol.13, No. 4, pp. 600–612, 2004.
[17] C. Yeo, H. L. Tan, and Y. H. Tan, "On rate distortion optimization using SSIM ," *IEEE Trans.*

Circuits and Systems for Video Technology, Vol. 23, No. 7, pp. 1170–1181, July 2013.
[18] E.-H. Yang and L. Wang, "Joint optimization of run-length coding, Huffman coding and quantization table with complete baseline JPEG decoder compatibility," *IEEE Trans. Image Processing*, Vol. 18, No.1, pp. 63-74, Jan. 2009.
[19] http://sipi.usc.edu/database/.

IOP Conf. Series: Journal of Physics: Conf. Series **910** (2017) 012015 doi:10.1088/1742-6596/910/1/012015

A Hybrid Multilevel Storage Architecture for Electric Power Dispatching Big Data

Hu Yan[1], Bibin Huang[2], Bowen Hong[3] and Jing Hu[4]

New Energy and Statistics Department, State Grid Energy Research Institute(SGERI), Beijing, China
Email: [1]yanhu916@126.com, [2]huangbibin@sgeri.sgcc.com.cn, [3]hongbowen@sgeri.sgcc.com.cn, [4]hujing@sgeri.sgcc.com.cn

Abstract. Electric power dispatching is the center of the whole power system. In the long run time, the power dispatching center has accumulated a large amount of data. These data are now stored in different power professional systems and form lots of information isolated islands. Integrating these data and do comprehensive analysis can greatly improve the intelligent level of power dispatching. In this paper, a hybrid multilevel storage architecture for electrical power dispatching big data is proposed. It introduces relational database and NoSQL database to establish a power grid panoramic data center, effectively meet power dispatching big data storage needs, including the unified storage of structured and unstructured data fast access of massive real-time data, data version management and so on. It can be solid foundation for follow-up depth analysis of power dispatching big data.

1. Introduction

In 1969, Edgar Frank Codd published an important paper, first proposed the concept of relational data model [1]. After that, many commercial companies such as IBM, ORACLE, Sybase, Microsoft and so on designed many corresponding relational database products based on the model, and later open source relational database software such as MySQL, Postgresql were developed. All relational database design meets the four characteristics of ACID [2], namely atomicity, consistency, isolation, persistence. Relational database, as a broad general-purpose database, has the following advantages, such as 1) data consistency; 2) standardization, data update cost is small; 2) support JOIN and other complex queries; 3) many mature technologies and applications.

With the development of the Internet, data is growing rapidly. Fast writing and high concurrent access have become an urgent need, while the relational database can't well satisfy it for considering more consistency than availability. In order to compensate for the lack of relational database, NoSQL database (Not Only SQL) came into being [3,4,5], it makes a single database distribute to a large number of physical and virtual machines and based on BASE rules, that are basically available, soft state, and ultimately the same. There are mainly two types of NoSQL databases: key-value type and document type. Key-value database organizes data in the Key-Value way and access data through a single key, its representative products are open source HBase, Redis, etc[6,7]. Document database is characterized by a single document to store data, allowing users to access the data through the value, its representative products are Mongdb, CloudDB, and other open source software [8,9].

The application scenarios of Relational database and NoSQL database are different. They are not substitute each other, complement each other. Power dispatching big data includes both structured and unstructured data. Traditionally, the structured power dispatching data is stored by relational database, and the unstructured data is stored in the file [10,11], thus all data are decentralized. Paper[12], presented a distributed key-value pairs based storage method for historical analog data. Paper[13] proposed a version storage method for the grid model. All these methods only solve part storage problems of power dispatching data.

This paper first gives overview of power dispatching data, covering data composition, data structure, data access requirements and so on. Followed a hybrid multilevel data storage architecture for electrical power dispatching big data is proposed, which integrates the advantages of relational database and NoSQL database. Again, the management of the proposed storage system are described, including cluster monitoring, metadata, authority, log. Then, table design examples based on the storage system are given. Finally, the conclusion is given.

2. Power Dispatching Data Overview

There are many kinds of power dispatching systems, for example, Energy Management System(EMS), Operation Management System(OMS), Production Management System(PMS), Geographic Information System(GIS) and so on. Each of them can generate mass of data. The acquisition cycle covers different time scale, from microseconds (such as: high frequency switch), milliseconds (such as: PMU), minutes (such as: power recovery time), day (such as: power generation plan), year (such as: IT asset life), to decade (such as: OT asset life).

The above data can be divided into basic data and derived data according to the source. The basic data includes grid model data, grid analog data, grid management data (such as equipment ledger), location data (such as GIS data), while the derivative data including operational analysis results, statistical data and so on. These data are currently stored in the relational database, accessed by standard SQL but performance requirements are different, including milliseconds, seconds, minutes, hours, etc. Besides, there is another kind of data, that is unstructured data, such as cim / g (<1M), case (30M), grid comprehensive fault reporting and waveform files, teaching plans (10K-30M), these data need to be version management and screen browsing query requires millisecond response.

With the construction of smart grid, the power dispatching data's sources, scale, type expand rapidly and to be more rich. Because these data structures and coding are not unified, the amount of data is huge, the current lack of a unified real-time (historical) data center, resulting in interfaces great variety between systems, and even the emergence of information islands. Data Integration becomes an urgent need.

3. Power Dispatching Big Data Hybrid Multilevel Storage Architecture

3.1. Design Aim

The design aim is to establish a power grid panoramic data center by integrating power grid operation data from different electric professional systems based on the unified grid model, seen in Figure.1.

The designed storage system should have follow features:

- Support storage of PB level data (PB level), such as WAMS data points storage, historical sampling data periodic writing.
- Support efficient query and statistics of PB level data, many applications need real-time response.
- Disaster recovery, multiple copies for fault tolerance and automatic recovery mechanism.
- Support distributed computing.
- Support structured data and unstructured data integration storage and management.
- Support version management of files.
- Support SQL-based access.
- Support complex JOIN queries between different tables.

IOP Conf. Series: Journal of Physics: Conf. Series **910** (2017) 012015 doi:10.1088/1742-6596/910/1/012015

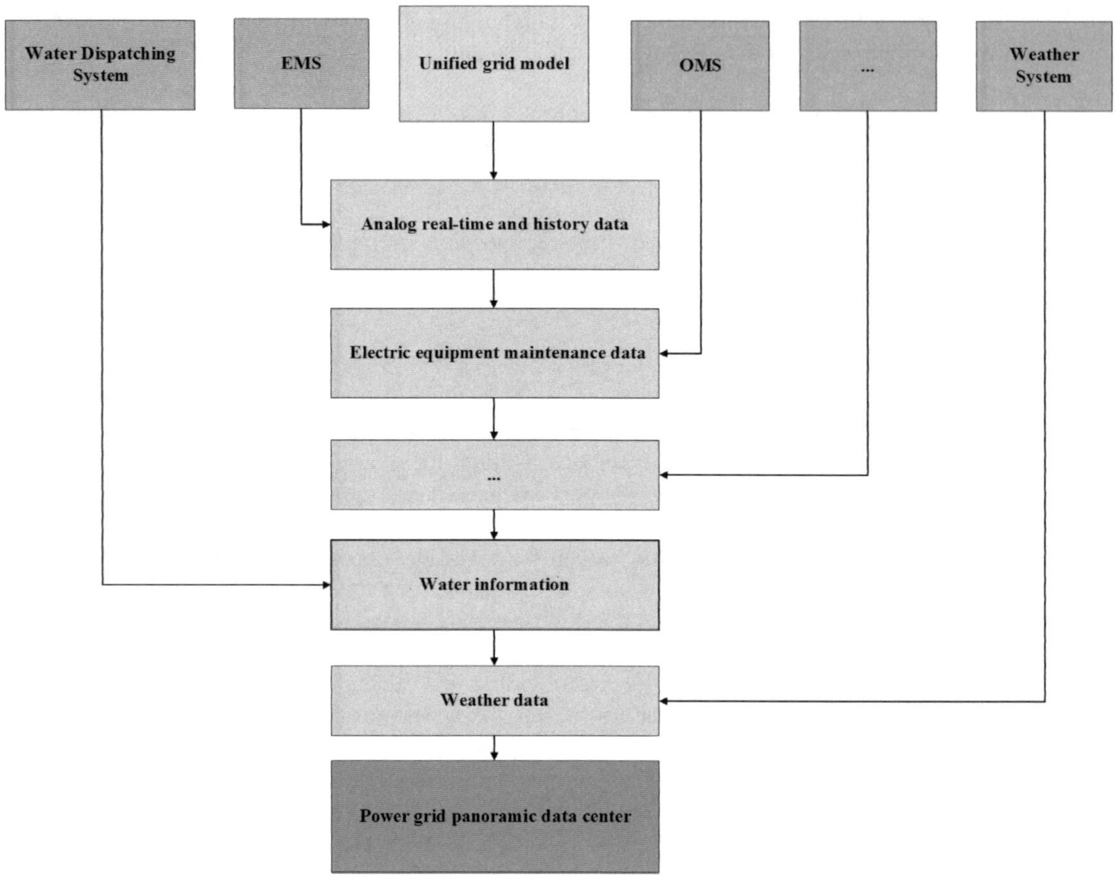

Figure 1. Establishing a power grid panoramic data center

3.2. Archtecture Design

At present, the main database are relational database and NoSQL database, single database may not meet all the different storage and management needs of power dispatching applications. This paper builds a hybrid multilevel storage architecture for power dispatching big data, as shown in Figure 2. The whole architecture is composed of data source acquisition layer, unified data warehouse layer, real-time data storage layer and interface access layer.

3.2.1 Data source acquisition layer. This layer is responsible for transfering data of external power systems, such as OMS, PMS, water dispatching system, etc. into the unified data warehouse. All kinds source of data are transfered by JDBC/ODBC, communication protocol, message queue, message mail, based on data real-time differences.

3.2.2 Unified data warehouse layer. All power dispatching data are stored in the unified data warehouse, bulding *HDFS* and Hive. The warehouse does not consider redundancy, more concerned about the clear data structure, including three data marts:

- Synchronized data mart: all the raw data without ETL will be stored in the synchronization data mart first.
- Unified data chart: application-oriented, stored and organized according to application theme.
- Analysis data chart: statistical analysis results.

The three data charts use ETL to synchronize data.

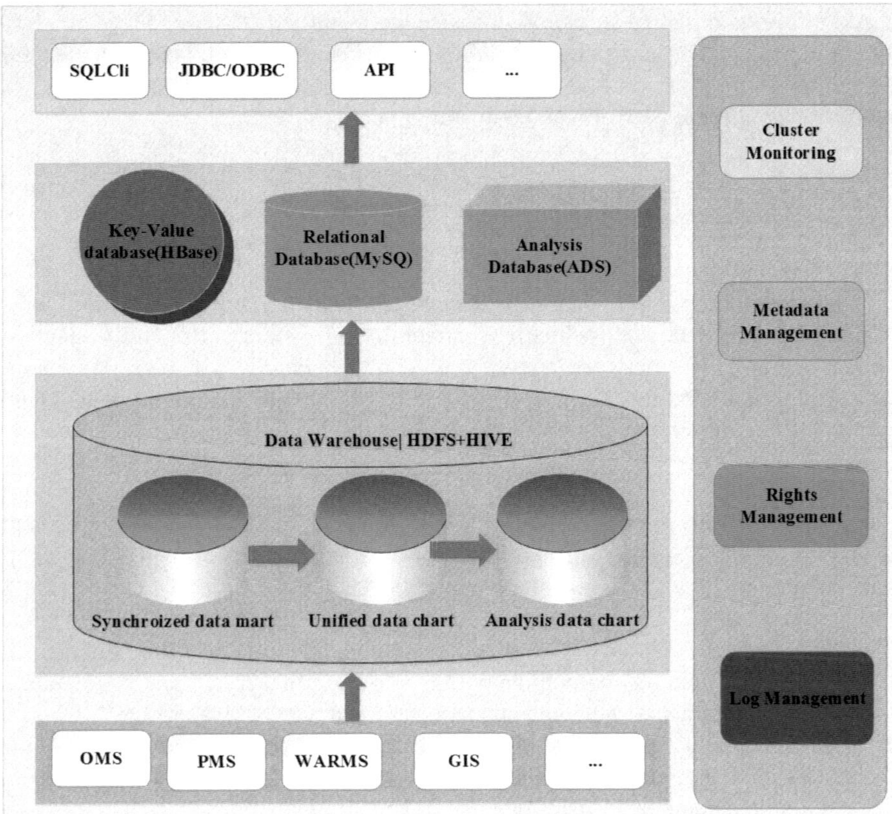

Figure 2. Hybrid Storage Architecture for power dispatching big data

3.2.3 Real-time data storage layer. The layer mainly stores data related to the online applications. In principle, only the daily data is stored, and the storage rules are as follows:

- Grid model data is stored in MySQL database, sampling and other key-value type data is stored in HBase database, and statistical analysis data is stored in the analytical database ADS.
- MySQL database is the prior storage choice, its reliability is relatively high, and response time can reach millisecond level for millions data;
- HBase database only stores data of the most recent period, for example, one week or one month. All the data is stored in Key-Value pattern, querying or writing is by the key and does not support Join queries. The most recent sampling data and the data needed version management are stored here;
- Analytical database ADS is used for large amount of data which needs second level response, complex analysis and son. It updates data in batches, real-time updates are delayed, and does not support the transaction.

3.2.4 Interface access layer. The upper layer application requires a unified interface to access data, blocking interface *differences* of different database. Therefore, in the selection of the database as much as possible to choose database which supports the standard SQL or SQL. Relational database support standard SQL; ADS support class SQL; Some NoSQL database does not support, such as HBase, but the data warehouse tool Hive provides a class SQL access interface, and can be integrated with HBase.The interface access layer provides SQLCli, JDBC, ODBC, API, etc. to access data.

The SQL operation is as follows:

- create /drop: create a new table or drop an existed table.
- select/insert/update/delete with bind: do in batch.
- select/insert/update/delete: do in single row.

- alter: add a column or delete a column or revise a column.

The unstructured data can be access by HDFS command, including put, get, list and so on.

4. Management of the Hybrid Multilevel Mstorage System

4.1. Cluster Monitoring

The proposed data storage architecture involves multiple clusters, requiring centralized monitoring of the cluster status, monitoring cluster load (including CPU, disk capacity, network status, etc.), node status. Ganglia[13,14] is introduced.

Ganglia is an open source cluster surveillance project initiated by UC Berkeley, which includes gmond, gmetad and a Web front end, primarily for monitoring system performance such as cpu, mem, hard disk utilization, I/O load, network traffic, etc. Though Ganglia software, it is easy to see the working status of each node through the curve and can be used to adjust and allocate system resources, playing an important role in improving the overall performance of the system.

4.2. Metadata Management

Metadata management mainly manage table structure information, the relationship among different table structures, and the table version information. All data structure information is maintained by the metadata management module, which can be modified online and the database table structure is traceable.

The unified data warehouse stores all the power dispatching data, so the metadata management module here is only responsible for managing the data in the unified data warehouse. When a database table is created and modified, the metadata management module is automatically triggered, and the metadata management module manages table structure and table structure version information. Metadata is stored in MySQL.

4.3. Rights Management

HBase authority control is introduced. The scope of authority control includes database level, table level, column cluster level and column level, where EXEC is currently only defined in the code but not used; CREATE and ADMIN only relate to table level and database level, column cluster level and column level only relate to READ and WRITE permissions.

4.4. Log Management

All the database operation and access will have logs, on the one hand, it can be used to find a solution to the problem in the future, on the other hand, it can be as evidence data of security control.

5. Table Design

5.1. Table Desgin in HBASE

HBase database is very suitable for storing time-scale data. In the EMS, there are two tables named Manalog_SCADA and Mpoint-SCADA which contains import electrical operation information. Every minute millions of data are wrote into the two tables. These data have the feature of structure simple, large volume, query by id and son. In HBase database, the two table can be designed as follows:

Table 1 Manalog_SCADA Table in HBASE

Column Name	Type	Other
Rowkey	bigint	Id+date
Timestamp	time	Year:month:hour:minute
CF:value	float	
CF:quality	int	Stands for the value status
CF:code	string	Code of electric device
CF:name	string	Name of electric device

Table 2 Mpoint_Scada Table In HBASE

Column Name	Type	Other
Rowkey	bigint	Id+date
Timestamp	time	Year:month:hour:minute:second
CF:pvalue	unsigned int	0 or 1
CF:quality	int	Stands for the value status
CF:code	string	Code of electric device
CF:name	string	Name of electric device

5.2. Table Desgin in HIVE

Grid model has a lot of structured equipment tables. Table information in a period of time will often change, so there is a need for version management. Hive table has a good ability to support for structured data, and can establish partition on the date, achieving version management. For example, the following BUSBARSECTION table, you can add a column to store time, the other column structure unchanged, each line of data on behalf of the device value at some time. As the date partition, you can quickly find a version of the data.

Table 3 Busbarsection Table In HIVE

Column Name	Data format	Other
DEV_ID	117375280623910913	As old column
Time(with date)	2015-11-11 21:11	New column
V	509.561	As old column
PB	20.0	As old column
QB	30	As old column
VB		As old column
ANG	0.0	As old column
...		

6. Conclusion

This paper presents and implements a hybrid multilevel data storage architecture for electrical power dispatching big data, which can not only meet the unified storage of structured data and unstructured data in the field of electric power regulation, solving the information island, but also can meet different access performance requirements of real-time or non-real-time applications, effectively response to large data rapid growth of access performance bottlenecks, in addition, the entire storage architecture has the characteristics of flexibility, high scalability, strong robustness. In the future, based on the power dispatching unified storage, more study of power dispatching data applications will be done, achieving from power dispatching data to value conversion.

7. References

[1] Codd EF."A relational model of data for large shared data banks", Commun ACM[J], vol.13, pp. 377–387,1970.

[2] Carlo Zaniolo."Analysis and design of relational schemata for database systems"[C], international symposium on physical design, University of California, Los Angeles,1976, pp.226-230.

[3] Jaroslav Pokorny. "NoSQL databases: a step to database scalability in web environment"[J], international Journal of Web Information Systems, vol.9, pp. 69–82,2013.

[4] Chang F, Dean J, Ghemawat S, et al. Bigtable： a distributed storage system for structured data[C], ACM Transactions on Computer Systems (TOCS). New York, USA： ACM, 2008:4.

[5] Borthakur D. HDFS architecture guide[J]. Hadoop Apache Project. http://hadoop. apache. org/common/docs/current/hdfs_design. pdf, 2008.

[6] Facebook. Hbase: Bigtable-like structured storage for Hadoop

[7] Baron C. "NoSQL Key-Value DBs Riak and Redis", Database Systems Journal, 2015, vol.5, issue 4.
[8] Harley Vera Olivera, Maristela Holanda, Valeria Guimarâes, et al. "Data Modeling for NoSQL Document-Oriented Databases", CEUR Workshop Proceedings, 2015, vol. 1478, pp.129-135
[9] Kyle Banker."MongoDB in Action",Geography (2011) p. 520.
[10] XIE Shujing, WANG Hao, ZHANG Zhiwei. Discussions about Power Network Modeling Based on CIM [J]. Power System and Clean Energy, 2011, 27(2) :37–41.
[11] YAN Hu, ZHOU Wei, LI Lixin, et al. "Database Platform of EMS System Based on Distributed Key-Value Pair Storage Technology"[J]. Power System Technology, 2012, 36(9) :162–166.
[12] YAN Hu, LI Lixin, YUAN Rongchang, et al. "Study on Multi-dimensional grid model of integrated storage and management"[J], Automation of Electric Power System, 2014, 38(16) :94–98.
[13] M. L. Massie, B. N. Chun, and D. E. Culler. The Ganglia Distributed Monitoring System: Design, Implementation,and Experience. Parallel Computing, 30(7):817–840, 2004.
[14] Ganglia Monitoring System, http://ganglia.info.
[15] Solon Barocas, Andrew D Selbst. "Big Data's Disparate Impact"[J], California Law Review, 2016, 104(3) :671.
[16] Kaile Zhou, Chao Fu, Shanlin Yang. "Big data driven smart energy management: From big data to big insights"[J], Renewable & Sustainable Energy Reviews, 2016, 56:215-225.
[17] Yong Chen, Hong Chen, Anjee GorkhaliKaile Zhou, et al. "Big data analytics and big data science: a survey"[J], Journal of Management Analytics, 2016, 3:1-42.
[18] Paolo Atzeni, Francesca Bugiotti, Luca Cabibbo, et al. "Data Modeling in the NoSQL World"[J], Computer Standards & Interfaces, 2016.
[19] Diego Pasqualin, Giovanni Souza, Eduardo Luis Buratti, et al. "A Case Study of the Aggregation Query Model in Read-Mostly NoSQL Document Stores"[C], international database engineering and applications symposium, Montreal, Canada, 2016, pp.224-229.
[20] Manar Abourezq, Abdellah Idrissi. "Database-as-a-Service for Big Data: An Overview"[J], International Journal of Advanced Computer Science and Applications, 2016, 7(1) :151-157.

Restful API Architecture Based on Laravel Framework

Xianjun Chen[1,a], Zhoupeng Ji[1], Yu Fan[1,b,*] and Yongsong Zhan[2]

[1]Haikou College of Economics, Haikou, China
[2]Guilin University of Electronic Technology, Guilin, China
Email: [a]hingini@126.com, [b]149471869@qq.com

Abstract. Web service has been an industry standard tech for message communication and integration between heterogeneous systems. RESTFUL API has become mainstream web service development paradigm after SOAP, how to effectively construct RESTFUL API remains a research hotspots. This paper presents a development model of RESTFUL API construction based on PHP language and LARAVEL framework. The key technical problems that need to be solved during the construction of RESTFUL API are discussed, and implementation details based on LARAVEL are given.

1. Introduction

In the age of a rapid development software industry, the internet application has got unstoppable growing popularity among which many of them can serve millions of users per day and reach a daily PV of billion level. Monolithic application architecture is not suitable for today's large-scale Internet applications, especially in distributed and heterogeneous computing environments, nor does it satisfied with multi-team collaborative development model. In the foreseeable future, due to the failure of Morgan's theory, the rapid development of Internet applications and Internet of things, the architecture of applications should get more flatten, distributed computing will become the mainstream software development model. In such context, we need a simple and reliable heterogeneous micro-service development model. When we build an application on some micro service infrastructures, the developer can focus on the business process and leave out to the framework all the complex about the inter-process communication.

2. API Technology

REST is a technical specification about heterogeneous inter-communication for Web applications. Adoption of REST can lead to a simple, scalable, effective, safe and reliable architecture. REST is a lightweight RPC protocol built upon HTTP protocol, it's simplicity and web friendliness makes it an unparalleled alternative to SOAP, a popular RPC solution for years. A well-structured Web application can be easily built when using Rest, Many developers successfully created simple and robust API base on Ajax and Restful web service.

2.1 Restful API

Representational state transfer(REST) or RESTful web services is a way of providing interoperability between computer systems on the Internet. REST-compliant Web services allow requesting systems to access and manipulate textual representations of Web resources using a uniform and predefined set of stateless operations[2].

The term representational state transfer was introduced and defined in 2000 by Roy Fielding in his doctoral dissertation[3][4]. Fielding used REST to design HTTP 1.1 and Uniform Resource Identifiers (URI)[3][5][6]. A resource is a kind of information that can be accessed, such as an application

IOP Conf. Series: Journal of Physics: Conf. Series **910** (2017) 012016 doi:10.1088/1742-6596/910/1/012016

object, a database record, an algorithm, and so on. Each resource is identified by a unique URI (Universal Resource Identifier), REST represent URI in the form of "/user/name", and operations on HTTP methods GET, PUT, POST, DELETE, HEADER and OPTIONS, resulting in the next resource being transferred back to the caller. An important characteristic of REST is that the server side keeps stateless between multiple interactions, every server in the clusters can serve the client on every request.

2.2 LNMP Stack

LNMP is an archetypal model of web development stacks, named as an acronym of the names of its original four open-source components: the Linux operating system, the Nginx HTTP Server, the MySQL relational database management system (RDBMS), and the PHP programming language. As a solution stack, LAMP is suitable for building dynamic web sites and web applications.

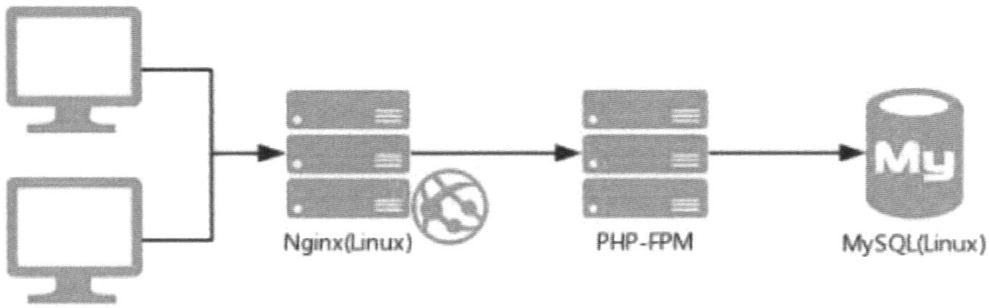

Figure 1 LNMP Stack

2.3 Laravel Framework

Laravel[1] is a free, open-source PHP web framework, created by Taylor Otwell and intended for the development of web applications following the model–view–controller(MVC) architectural pattern. Some of the features of Laravel are a modular packaging system with a dedicated dependency manager.

The laravel framework is easy to understand and powerful, the framework itself provides authentication, routing, session manager, caching, IoC container and tons of most commonly used component, also amazing database migration tools and integrated unit testing support, all these tools give developers the ability to build complex applications.

3. System Architecture design

A sophisticated Laravel based restful web service application should comprise of some key component, such as routing engine, middleware, MVC framework, ORM and authentication component.

When a client sends a request to Laravel restful API through HTTP, the web server first receives the request and passing through to PHP Engine, then the actual execution start from Laravel initialize routines. The Laravel initialize routines finish some configuration, the request is then passed To middleware engine for filtering, after filtering there comes the core routing component responsible for request dispatching according to the route configuration table, finally the MVC controller take over the dispatched request. In general, the controller is responsible for application logic execution and inquire the model for data fetch and persistent, the view component takes over the last piece of work for rendering the page. But when we build Restful API, page rendering is useless, so only controller and model are involved, at the end of the process is only Data encoded with JSON will be returned. Laravel provides a powerful ORM framework for relational database manipulate, the model depends heavily rely on ORM for data processing. The whole process are shown as figure 2.

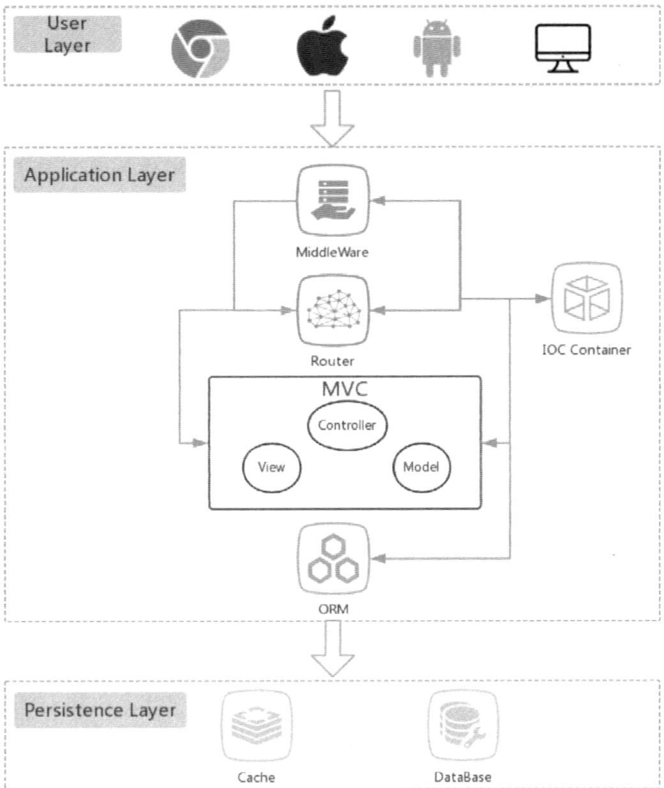

Figure 2 System Architecture

3.1. Middleware

The main purpose of HTTP middleware is to provide a convenient mechanism for filtering HTTP requests before the actual logic process, middlewares chain together to do filter process one by one. These middleware handle reading and writing the HTTP session, determining if the application is in maintenance mode, verifying the CSRF token, and more.

Two kinds of middleware exist in Laravel, one is for HTTP filtering, the other is for routing filtering. The request can hit the controller only after it can pass through all the HTTP middlewares and then routing middleware.

The core component drive Laravel is the class "\ Illuminate \ Foundation \ Http \ Kernel", it acts as a kernel serve in the central location that all requests flow through, the HTTP kernel also defined a list of HTTP middleware that all requests must pass Through all being handled by the application. All middlewares depend on one base class "Illuminate \ Pipeline \ Pipeline", middlewares can form a chainable call stack, benefit from the chainable closure implementation define the base class.

3.2. Routing

All routers[7] of the system are located in the directory named routes, routing definitions are automatically loaded and parsed during initialization. The simplest routing can only accept a URI and a closure, which is executed whenever the URI is requested, but it's a hard coding style. A more common way of define rules of routing is the configuration array in "routes/api.php", each rule is adding when a configuration item is found in the array, the code is as below:

Route::resource('path','PathController');

A single line of "Route::resource" function call define multiple routing actions as shown in the following table. When these routing rules take effect, a request for URL "http://www.yourdomain.com/path" will be dispatched to a controller named "PathController" and the "Index()" function will be called to handle the request.

Table1 URL Pattern

Method	URL Pattern	Action	Router Name
GET	/path	Index()	path.index
GET	/path/create	create()	path.create
POST	/path	store(Request $request)	path.store
GET	/path/{args}	show($args)	path.show
GET	/path/{args}/edit	edit($args)	path.edit
PUT/PATCH	/path/{args}	update(Request $request,$args)	path.update
DELETE	/path/{args}	destroy($args)	path.destroy

3.3. ORM

ORM (Object Relational Mapping) is a programming technique for converting data between database and objects in memory when using object-oriented programming languages. This creates, actually, a "virtual object database" that can be manipulated from within the programming language, and all The modification will be automatically synchronized to the real database in the hard disk.

The Eloquent ORM included with Laravel provides a beautiful, simple ActiveRecord implementation for working with your database. Each database table has a corresponding "Model" which is used to interact with that table. Models allow you to query for data in your tables, as well as insert new records into the table. In Eloquent ORM package, a base class "Model" is defined which contains many database operation methods and utility tools, all application specificated model should inherit from this base class.

The Eloquent model is usually placed in the app directory and can be automatically loaded via composer.json. Usually "Illuminate\Database\Eloquent\Model" come as base， and the subclass definition as：

```
class User extends Model { }
```

Eloquent will map the above "User" model the underline database correspondence of "users" table, so any operations on"User" model will directly affect "users" table.This default behavior can be overridden if a specificated data table name is supplied, as shown in the following code:

```
class User extends Model
{
protected $table = 'user_table';
}
```

4. JWT and Authentication

JSON Web Token (JWT) is a JSON-based open standard for creating access tokens that assert some number of claims.Tokens are signed by the server's key for legitimate verification in both client and server side. The tokens are designed to be compact, URL-safe, and typically be used to pass the identity of authenticated users between an identity provider and a service provider.

4.1. JWT in Laravel

An extension package named "tymon/jwt-auth" developed by the community can be integrated into to Laravel framework which will provide JWT functionality, we can install it from composer on the command below:

```
composer require tymon/jwt-auth
```

After installaiotn, "config/app.php" should be configured for service provider and aliase：

```
'providers' =>
 [
        Tymon\JWTAuth\Providers
        \JWTAuthServiceProvider::class,
 ],
'aliases' =>
 [
```

```
        'JWTAuth'=> Tymon\JWTAuth\Facades
        \JWTAuth::class
 ],
```

And， guards should be set in "config/auth.php" :

```
'api' =>
[
        'driver' => 'token',
        'provider' => 'users',
],
```

Finally， all these should be assembled together in a route group setting to get applied：

```
Route::group(
        ['middleware'=>['api'],'
        prefix'=>'api'],
        function(){
        Route::post(api1, 'ApiController@api1);
        Route::post(api2, 'ApiController@api2);
         }
);
```

5. Cross-domain Solution

A cross-domain request original from same-origin policy, a web application security model of webkit based browser which permits scripts contained in a first web page to access data in a second web page, but only if both web pages have the same origin. The so-called same-origin refers to the domain name, protocol and port are all the same.

Cross-origin resource sharing (CORS) is a mechanism that allows restricted resources on a web page to be requested from another domain outside the domain from which the first resource was served， both browser and server should be supported for CORS to take effect. In server side it is common to set the Access-Control-Allow-Origin attribute in the HTTP header to specify which sites can access。

Since Laravel provides a good middleware extension mechanism, we can implement cross-domain filtering through middleware, the code example is as follows:

```
<?php namespace App\Http\Middleware;
use Closure;
use Response;
class EnableCrossRequestMiddleware
{
    public function handle($request, Closure $next)
    {
        $response = $next($request);
        $response->header('Access-Control-Allow-            Origin', '*');
        $response->header('Access-Control-Allow-            Headers', 'Origin, Content-Type,
Cookie,                  Accept');
        $response->header('Access-Control-Allow-            Methods', 'GET, POST, PATCH,
PUT,            OPTIONS');
    }
}
```

6. Conclusion

Micro service architecture is already a de facto industry standard for web service, and Restful Web Services has become the preferred technology model for micro services application due to its lightweight, scalability, and HTTP protocol compatibility, This paper discusses the conception of how to use Laravel framework to build the basic infrastructures of the Restful service and provides key implementation details for our solution.

IOP Conf. Series: Journal of Physics: Conf. Series **910** (2017) 012016 doi:10.1088/1742-6596/910/1/012016

7. Acknowledgments

This work has been supported by the grant of Hainan Natural Science Foundation(NO:20156239), the grant of Science Research Foundation of Haikou College of Economics(NO: hjkz16-02) . Corresponding author: Fan Yu(149471869@qq.com)

8. Reference

[1] https://laravel.com/
[2] "Web Services Architecture". World Wide Web Consortium. 11 February 2004. 3.1.3 Relationship to the World Wide Web and REST Architectures. Retrieved 29 September 2016.
[3] Fielding, Roy Thomas (2000). "Chapter 5: Representational State Transfer (REST)".
[4] "Fielding discussing the definition of the REST term". groups.yahoo.com. Retrieved 2017-08-08.
[5] RFC 1945
[6] RFC 2616
[7] Zhou Qiao Jun. The Design and Implementation of RESTful Web Services Open Platform[D].Zhejiang University,2016.
[8] Tang Ming Wei. Research and Implementation of Library Management System Based on REST Architecture[J]. New Technology of Library and Information Service,2010,(09):84-89.
[9] Xue Min. Design and Implementation of P2P Network Loan System Based on Laravel Framework[D].JiLin University,2016.
[10] Chen Yan Hui. Shipping Information Interaction Model and Its Application Based on Saa S and Laravel Architecture[D].Zhejiang University of Technology,2016.
[11] Qin Fen. Research And Design Of Mobile Environment For Restful Webservice [D].Beijing industry university,2015.

Direction of Radio Finding via MUSIC (Multiple Signal Classification) Algorithm for Hardware Design System

Zheng Zhang

Greater Atlanta Christian School, Norcross, GA, USA
Email: zhangzhen1708@126.com

Abstract. Concept of radio direction finding systems, which use radio direction finding is based on digital signal processing algorithms. Thus, the radio direction finding system becomes capable to locate and track signals by the both. Performance of radio direction finding significantly depends on effectiveness of digital signal processing algorithms. The algorithm uses the Direction of Arrival (DOA) algorithms to estimate the number of incidents plane waves on the antenna array and their angle of incidence. This manuscript investigates implementation of the DOA algorithms (MUSIC) on the uniform linear array in the presence of white noise. The experiment results exhibit that MUSIC algorithm changed well with the radio direction.

1. Introduction

A radio direction finding (DF) system [1-3] is a receiver arranged in a combination to determine the azimuth angle of a distant emitter. Basically, all DF systems derive the emitter location from an initial determination of the angle-of-arrival (AOA) [4-6]. Radio direction finding techniques have classically been based on multiple-antenna systems employing multiple receivers [7-9]. And MUSIC use simultaneous phase information from each antenna to estimate the angle-of-arrival of the signal of interest. Multiple receivers are impractical. Thus, single channel techniques are of interest, particularly in mobile scenarios. Although the amount of existing research for single channel DF is considerably less than for multi-channel direction finding, single channel direction finding techniques have been previously investigated. DOA (Direction of arrival) estimation is a key research area in array signal processing and many engineering applications, such as wireless communications, radar, radio astronomy, sonar, navigation, tracking of various objects, earthquake, medicine and other emergency assistance devices that need to be supported by direction of arrival estimation. In modern society, DOA (Direction of arrival) estimation is normally researched as a part in the field of array processing, so many works highlight radio direction finding. Over the past ten years, Wireless Local Area Networks (WLANs) have increased quickly because of its flexibility and convenience. In order to satisfy the requirements of advanced services, a high-speed data rate is necessary. Owing to the excessive use of the low end of the spectrum, people begin to search for the higher frequency bands for more applications. With higher user density, higher frequency and higher data rate, multipath fading and cross interference become the main issues. In order to solve these problems and get higher communication capacity, smart antenna systems are proved to be very effective in suppression of the interference and multipath signals

Signal processing in smart antenna systems concentrates on the development of efficient algorithms for Direction of Arrival (DOA) estimation and adaptive beam forming. However, there are many limitations if DOA (Direction of arrival) estimation uses a fixed antenna. Antenna main-lobe beam width is inversely proportional to its physical shape. It is not a practical option to improve the accuracy of angle measurement in accordance with an increase in the physical aperture of the receiving antenna. Some systems such as missile seeker or aircraft antenna have limited physical size, so they are sufficiently wide in beam width of the main lobe to correspond. They do not have a good resolution and if there are multiple signals falling in the antenna's main lobe, it becomes too difficult to distinguish between them.

There are many kinds of super resolution algorithms such as spectral estimation, Bartlett, Capon, ESPRIT [10], and Min-norm [11]. One of the most popular and widely used subspace-based techniques to estimate the DOA (Direction of arrival) of multiple signal sources is the ESPRIT algorithm. Large numbers of computations are needed to search for the spectral angle. When using the ESPRIT algorithm, so in real applications its implementation can be difficult. Compared with spectral ESPRIT algorithm, the MUSIC method has better performance with reduced complexity computation. It can only be used in uniform linear array (ULA) or non-uniform linear array whose arrays are restricted to a uniform grid [12] [13]. MUSIC [14] was the most promising and a leading candidate for further study and actual hardware implementation. However, although the performance advantages of MUSIC are substantial, they are achieved at a cost in computation (searching over parameter space) and storage (of array calibration data) [15]. In this thesis, I will focus on the MUSIC algorithm.

2. MUSIC (Multiple Signal Classification) Algorithm

Multiple Signal Classification (MUSIC) algorithm was proposed by Schmidt and his colleagues in 1979 [16]. It has created a new era for spatial spectrum estimation algorithms. The promotion of the structure algorithm characterized rise and development, and it has become a crucial algorithm for theoretical system of spatial spectrum. The basic idea of MUSIC algorithm is to conduct characteristic decomposition for the covariance matrix of any array output data, resulting in a signal subspace orthogonal with a noise subspace corresponding to the signal components. Then these two orthogonal subspaces are used to constitute a spectrum function, be got though by spectral peak search and detect DOA (Direction of arrival) signals.

MUSIC estimates the frequency substance of a signal or autocorrelation matrix using an Eigen space method. This method assumes that a signal, $x(n)$, consists of p complex exponentials in the presence of Gaussian white noise. Given an $M \times M$ autocorrelation matrix, R_x, if the eigenvalues are sorted in decreasing order, the eigenvectors corresponding to the largest eigenvalues (i.e. directions of largest variability) span the signal subspace. The remaining $M - p$eigenvectors span the orthogonal space, where there is only noise. Note that for $M = p + 1$, MUSIC is identical to Pisarenko harmonic decomposition. The general idea is to use averaging to improve the performance of the Pisarenko estimator. The frequency estimation function for MUSIC is $\widehat{P_{MN}}(e^{j\omega}) = \frac{1}{\sum_{i=p+1}^{M} |e^H v_i|^2}$, where v_i are the noise eigenvectors and $e = [1 \ e^{j\omega} e^{2j\omega} \dots \ e^{(M-1)\omega} \ ]^T$.

3. Numerical and Experimental Results

We estimate implementation of MUSIC algorithms and display that MUSIC algorithm is as precise having appreciably simply structure. Figure 1shows that root locus diagrams produced, for a case when one QPSK emitter was impinging on the 30-element uniform linear arrays (ULA) with λ/2 inter element spacing. Root locus diagram produced by MUSIC algorithm is obviouslyovercrowdedhaving22 pairs of roots in spilt of the fact that only one emitter was impacting on the ULA. Figure 2show mean square error (MSE) in DOA (Direction of arrival) estimation as a function of the SNR value. The MUSIC was estimated when one emitters were impinging on the four-element ULA. The first simulation shows how two signals are recognized by the MUSIC algorithm. There are two independent narrow band signals, the incident angle is 75°, those signal is not correlated, the noise is ideal Gaussian white noise, the SNR is 20dB, the element spacing is half of the input signal wavelength, array element number is 10, the number of snapshots is 200. The sample size was kept constant at T=1000 samples. Emitters were located at 75°.

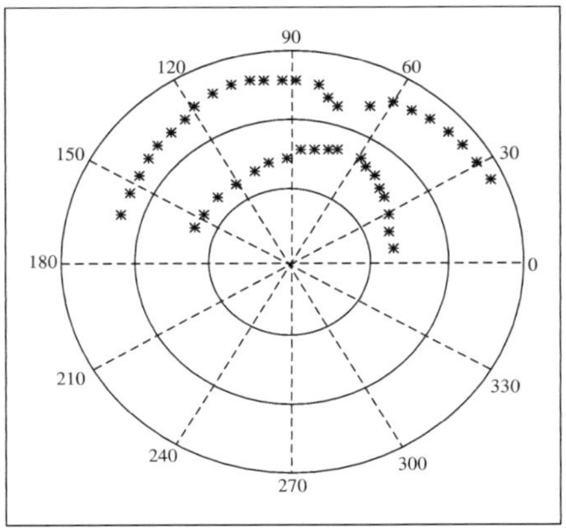

Figure 1 Root locus diagram of the MUSIC algorithm for one source and 30-element ULA.

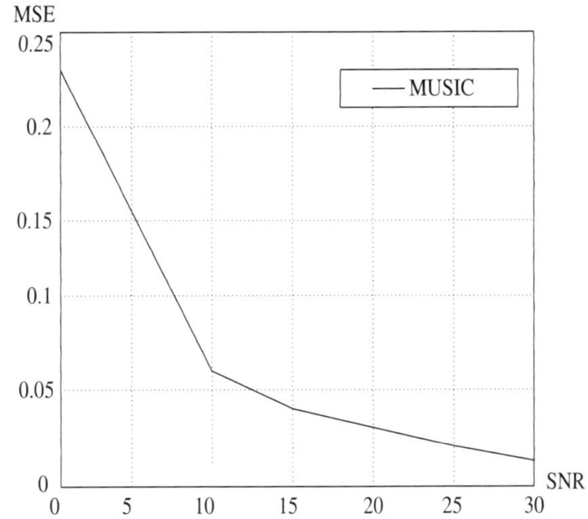

Figure 2 MSE in DOA estimation as a function of the SNR for one source and four-element ULA with sample size T=1000 samples.

In addition, Figures 1-2 shows the polar diagram of the two reciprocal roots provided by the MUSIC algorithm. The MUSIC antenna array as seen in Figure 3. The main board consists of a uniform circular array system with four ADMP504 MEMS microphones (Analog Devices, Norwood, MA, USA), a ADSP21375 (Analog Devices, Norwood, MA, USA) as the core processor, MAXIM MAX11043, 4-Channel, 16-Bit, Simultaneous-Sampling ADCs (Maxim Integrated Products, Sunnyvale, CA, USA) and supplemental hardware circuits. The MAX11043 contains a versatile filter block and programmable-gain amplifier (PGA) per channel. The extended board contains a CSR BC6415 Bluetooth module (Cambridge Silicon Radio, Cambridge, UK), a data acquisition interface and debug interface. The ZYBO is a feature-rich, ready to use, entry level embedded software and digital circuit development platform built around the smallest member of the Z-7101. The Z-7101 is based on the Xilinx All Programmable system on chirp architecture, which tightly integrates a dual- core ARM Cortex-A8 processor with Xilinx 7 Series Field Programmable Gate Array logic. When coupled with the rich set of multimedia and connectivity peripherals available on the ZYBO, the Z-7101 can host a whole system design. The on-board memories, video and audio I/O, dual-role USB, Ethernet, and SD slot will have your design up and ready with no additional hardware needed. Figure 3 illustrates the hardware components that make up the system. Additionally, six connectors are available to put any design on an easy growth path. The SNR error and execution time required for processing uniform linear array with no of elements varying from 7 to 14 is tabulated. we can see that the same SNR value both numerical and experimental approaches gave the same result verifying that with an equivalent physical array aperture of only 1.5 wavelength the DOA (Direction of arrival) in a decoupled array environment can be estimated with errors smaller than 0.01°. The result as shown in Table. 1.

Figure 3 MUSIC antenna arrayfor hardware design system

Table1 Experiment data for searchng accuracy

Test Number	30°	45°	60°	75°
1	29.89°	45.04°	59.60°	74.87°
2	30.05°	45.01°	60.02°	75.12°
3	29.94°	44.57°	60.03°	75.89°
4	30.17°	44.98°	60.01°	75.67°
5	29.95°	44.87°	60.06°	75.95°

4. Conclusion

The MUSIC algorithm was derived. It coined modified root polynomial estimates the DOAs from the polynomial of degree 2L, where L represents number of emitters. The MUSIC algorithm estimates DOAs from the polynomial of degree 2N-2, where N represents number of antennas by selecting L roots that is closest to the unit circle in order to estimate DOA. We have shown by extensive numerical performance evaluation that both algorithms have the same accuracy. Our analysis, based on both numerical and experimental data, shows that with an equivalent physical array aperture of only 1.5 wavelength the DOA in a decoupled array environment can be estimated with errors smaller than 0.01°.

5. References

[1] Luo, Richeng, K. Bai, and X. Gong. "Partial discharges multi-targets localization in power transformers based on spatial-estimation." International Conference on Condition Monitoring and Diagnosis IEEE, 2008:509-512.

[2] Xue, Xiao Feng, et al. "A Study of Decoupling 2D-MUSIC Algorithm and Array Structure." Journal of Air Force Engineering University (2008).

[3] Wen, Fei, et al. "Improved MUSIC Algorithm for Multiple Noncoherent Subarrays." IEEE Signal Processing Letters 21.5(2014):527-530.

[4] Chen, Xiaoming. "Using Akaike Information Criterion for Selecting the Field Distribution in a Reverberation Chamber." IEEE Transactions on Electromagnetic Compatibility55.4 (2013): 664-670.

[5] Schmidt, R. "Multiple emitter location and signal parameter estimation."International Journal of Engineering Research2.3 (1986):276-280.

[6] Kundu, Debasis. "Modified MUSIC algorithm for estimating DOA of signals." Signal Processing 48.1(1996):85-90.

[7] Roy, R., and T. Kailath. "ESPRIT-estimation of signal parameters via rotational invariance techniques." IEEE Transactions on Acoustics Speech & Signal Processing 37.7(2002):984-995.

[8] Hou, Ying Ni, J. G. Huang, and X. A. Feng. "The Study on Beam-space DOA Estimator Based on Beam-forming System." Journal of Projectiles Rockets Missiles & Guidance 27.3(2007):80-82.

[9] Kim, Youngwook, and H. Ling. "Direction of Arrival Estimation of Humans with a Small Sensor Array Using an Artificial Neural Network." Progress in Electromagnetics Research B 27.27(2011):127-149.

[10] Mohan, S, et al. "Localization of multiple acoustic sources with small arrays using a coherence test. " Journal of the Acoustical Society of America 123. 4(2008):2136.

[11] Quazi, A. "An overview on the time delay estimate in active and passive systems for target localization." Acoustics Speech & Signal Processing IEEE Transactions on 29.3 (2003):527-533.

[12] Zhang, Xiaofei, et al. "Blind DOA and polarization estimation for polarization-sensitive array using dimension reduction MUSIC." Multidimensional Systems & Signal Processing 25.1(2014):67-82.

[13] J. Capon. "High-resolution frequency-wavenumber spectrum analysis." Proceedings of the IEEE 57.8(2005):1408-1418.

[14] Scharrer, Roman, and M. Vorlander. Sound Field Classification in Small Microphone Arrays Using Spatial Coherences. IEEE Press, 2013.

[15] Press, W.H; Teukolsky, S.A; Vetterling, W.T; Flannery, B.P. Numerical Recipes: The Art of Scientific Computing, 3rd ed.; Cambridge University Press: New York, NY, USA, 2007; Chapter 11.1.

[16] Pham,T.;Fong,M.F.RealtimeimplementationofMUSICforwidebandacousticdetectionand tracking. In Proceedings of SPIE AeroSense'97: Automatic Target Recognition VII, Orlando, FL, USA, 23 June 1997; pp.250–256.

IOP Conf. Series: Journal of Physics: Conf. Series **910** (2017) 012018 doi:10.1088/1742-6596/910/1/012018

A Method for Digital Chart Quality Evaluation based on Cloud Model

Caixia Yu [1], Jun Xu [1,2], Wenqian Huang [1,3], Luan Jun[4] and Jinzhong Liu [4]

[1]Dalian Navy Academy, Dalian 116018, China
[2]Chinese Academy of Surveying and Mapping, Beijing 100830, China
[3]State Key Laboratory of Geo-information Engineering, Xi'an 710054, China
[4]Navy Publishing House, Tianjin 300450, China
Email: yucaixiaxj@163.com

Abstract. In order to evaluate Digital Chart objectively, quantitively and reasonably, the paper puts forward a method for digital chart quality evaluation based on Cloud Model. Firstly, ssynthesized Cloud Model is improved with power factor, secondly evaluating indexes including one-level and two- level and their powers are given according to the digital chart characteristic, and then calculates all levels of evaluating indexes used the improved synthesized cloud model, lastly, presents the ultimate result in the form of the Cloud Model for quality evaluation and its character figure. It is proved that the method is feasible by a test.

1. Introduction

As the important link in chart quality control, digital chart quality evaluation is of great importance for improving chart quality. At present, national chart quality management and evaluation often depend on experience, stronger subjectivity, less definite magnitude index. Because digital chart quality evaluation, one problem of much objective and much content, whose evaluation indexes are qualitative natural language, takes on much stronger randomicity and fuzziness, traditional methods such as, direct evaluation method, indirect evaluation method, general evaluation method[1], and fuzzy general evaluation method mentioned in document[2-4] are hard to solve the problem rationally and perfectly. Cloud Model can do well in the conversion between qualitative concept and quantitative representation, which is the reason why Cloud Model is the key topic in recent year, but the many researches are mainly on theory [5-9]. In the paper, Cloud Model is applied to digital chart quality evaluation, and then a method for digital chart quality evaluation based on Cloud Model is put forward.

2. Cloud Model Summarization

Objective world and its describing language are full of uncertainty, while the traditional methods often separate randomicity and fuzziness only in each different view, ignoring the stronger correlation and synchronism between randomicity and fuzziness, then Cloud Model is put forward in the situation.

2.1. *Theory of Cloud Model*

U is supposed as one quantitative field showed by exact value, while C is one qualitative concept in U, if $x \in U$, and it is one random value in C, and then $C_T(x) \in [0,1]$ is one random value of stable tendency.

$$\text{If } C_T(x) : U \rightarrow [0,1] \quad \forall x \in U \quad x \rightarrow C_T(x)$$

Then the distribution of x is called Cloud, and each x is called one cloud drop [10, 11].

Numeral characters of Cloud Model, such as Expected value (Ex), Entropy (En), and Hyper entropy (He) reflect the qualitative character of qualitative concept [5, 10]. Three numeral characters of Cloud Model not only integrate randomicity and fuzziness completely, but also draw the cloud shape by thousands of cloud drops, what is the unique of it. It can express one-dimensional normal cloud by these three numeral characters, and record as A=U (Ex, En, He).

2.2. Normal Cloud Generator
Normal Cloud Model is the basic cloud model, including forward cloud generator and backward cloud generator. The specific arithmetic can see the reference [10].

2.3. Synthesized Cloud Model
Synthesized Cloud Model generalizes two or more similar types of child clouds to generate one new high father cloud, and its essence is an advanced concept to synthesize two or more similar types of language to one wider concept language. The specific arithmetic can see the reference [10].

3. Digital Chart Quality Evaluation based on Cloud Model
3.1. Digital Chart Quality Evaluation System
According to reference [1, 2, 10], in the paper, digital chart quality content which is determined by quality character, should include 5 quality indexes as follows, time precision, location precision, attribute precision, content and data, cartography generalization and art, and each quality index includes level 2 indexes. Because different quality indexes play different parts in quality evaluation, then we add weight, seeing table 1.

Table1. Evaluation indexes and their weighting value

Level 1 index	Weight	Level 2 index	Weight
Time precision	0.15	Data adopted	0.7
		Notice correction	0.3
Location precision	0.3	Maths base	0.4
		Plane precision	0.2
		Depth precision	0.2
		Height precision	0.2
Attribute precision	0.15	Features classifications and code correctness	0.5
		Attribute correctness	0.5
Content and data	0.3	Content correctness	0.4
		Logic interconsistency	0.3
		Data perfection	0.3
Cartography generalization and arts	0.10	Cartography generalization	0.6
		Figure quality	0.4

At present, evaluation grade can be divided into the best, better, up to standard, and not up to standard, once there arises one of situations that are mathematical basis is wrong (for instance, coordinate system and its correction, projection, reference parallel or mid-longitude, datum and its correction), or if isolated obstructions, wrecks, shallows, shallow soundings, or the important aids to navigation which may be dangerous for navigation safety are wrong or missed, or if the precision of large- area map don't fit the specifications, or map load and generalization factors don't meet specifications, chart product can be passed[12].

3.2. Quality Evaluation System Steps based on Cloud Model
From table 1, we can see that digital chart quality evaluation which is composed of two grades belongs

to multi-layer synthesized evaluation problem, so synthesized cloud is adopted in the paper. Firstly, we calculate the first synthesized evaluation from the second layer, viz. the values of father cloud are calculated by child clouds, and then we used the new calculated value for synthesized evaluation again to get the final evaluation results.

Digital chart quality evaluation based on cloud model includes three sets-index set, power set, and comment set, which are index set U={ time precision, location precision, attribute precision, content and data, cartography generalization and art}, power set V= {0.15, 0.3, 0.15, 0.3, 0.1} , and comment set={ the best, better, up to standard, and not up to standard}, according to ten-grade rule, the best is [9,10], better is [8, 9), up to standard is [6, 8), not up to standard is [0, 6].

The specific steps are as follows,

(1) We modify the formula in section 1.2.3, and put power W_i to generate cloud model of two layer evaluation index. The specific formula is as follow.

$$\begin{cases} Ex = \dfrac{Ex_1W_1/En_1^2 + Ex_2W_2/En_2^2 + \cdots + Ex_nW_n/En_n^2}{W_1/En_1^2 + W_2/En_2^2 + \cdots + W_n/En_n^2} \\ En = En_1W_1n + En_2W_2n + \cdots + En_nW_nn \\ He = \dfrac{He_1W_1/En_1^2 + He_2W_2/En_2^2 + \cdots + He_nW_n/En_n^2}{W_1/En_1^2 + W_2/En_2^2 + \cdots + W_n/En_n^2} \end{cases} \quad (1)$$

(2) We used the calculated value as the new evaluation value to get the first-grade evaluation cloud model from formula (1). Then the quality of product can be estimated by Ex, and of course can be visualized by character figure.

(3) N cloud drops are got by calculating N time, and by analysis we get the numbers of cloud drops distributing to "the best", "better", "up to standard", and "not up to standard", by which we can further confirm the product quality.

4. Experiment and Analyses

Taking digital chart C12261 for example, fractions of two-layer index by experts are seeing table 2.

According to section 1.3.2, we get the synthesized cloud model from equation (1) as follows,

U1= (8.23, 0.066, 0.02), U2= (8.90, 0.132, 0.02), U3= (8.85, 0.066, 0.02)

U4= (8.62, 0.099, 0.02), U5= (8.31, 0.066, 0.02)

Table2. Evaluation indexes and their weighting value

Level 2 index	Fraction
Data adopted	8.5
Notice correction	7.6
Maths base	9.5
Plane precision	9.0
Depth precision	8.5
Height precision	8.0
Features classifications and code correctness	9.2

IOP Conf. Series: Journal of Physics: Conf. Series **910** (2017) 012018 doi:10.1088/1742-6596/910/1/012018

Table2. Evaluation indexes and their weighting value (Continued)

Attribute correctness	8.5
Content correctness	9.1
Logic interconsistency	8.8
Data perfection	7.8
Cartography generalization	8.5
Figure quality	8.1

By equation (1), we calculate again to get the final synthesized cloud model of the first-layer index-(8.57, 0.48, 0.02), Ex=8.57, from which we see that the quality of digital chart is better. According to normal cloud generator, we calculate 1000 times to get the character figure as follows.

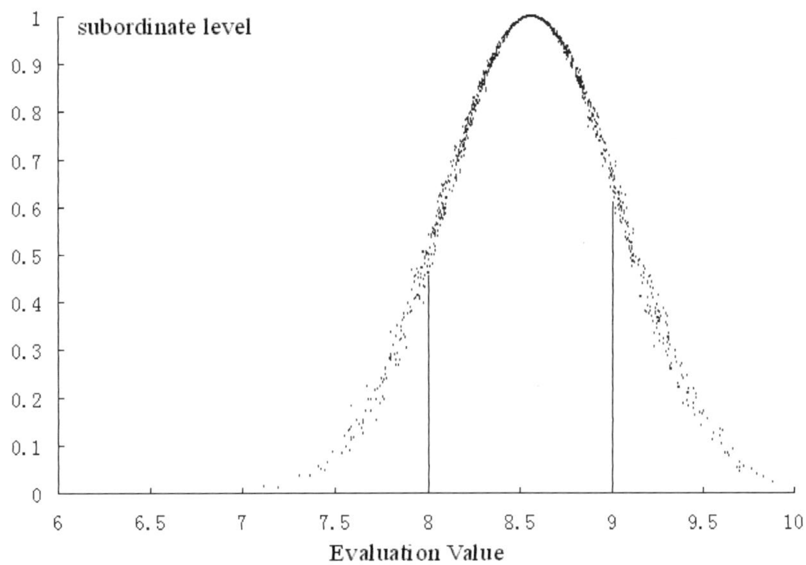

Figure 1. The character figure of C12261quality evaluation

From Figure 1, we can see that most of the cloud drops fall in [8, 9), which again prove the quality of product is better, others fall in [9, 10] and [6, 8), and the cloud drops in [9, 10] are more than that in [6, 8), which prove that the product is more than better. By calculating 1000 times, the number of " the best" drops is 191, "better" is 698, "up to standard" is 111, and "not up to standard" is 0, so subordinate level of comment set is (0.191, 0.698, 0.111, 0), which prove the product is better, and more than better.

5. Conclusions

Cloud Model can preferably solve the qualitative and quantitative problems of digital chart evaluation, taking each evaluation index and power into account, and make the evaluation result to be much practical and reasonable. So we can apply cloud model into other spatial data quality evaluation to get one more perfect, definite, and comprehensive quality evaluation system. Of course, the result depends on some factors, for example, hyper entropy (He).

6. Acknowledgement

National Natural Science Foundation Youth Fund（41501500）; Opening Fund of State Key Laboratory of Geo-information Engineering（SKLGIE2016-M-4-1）

7. Reference

[1] Peng Rencan, Guo Lixin, Chen Huirong. Digital Chart Cartography [M]. Dalian: Dalian Navy Academy, 2009, 177-192

[2] Chen Huirong, Peng Rencan, Li Shujun.The fuzzy synthesis evaluation on digital chart quality [J]. Engineering of Surveying and Mapping, 2010, 19(5): 32-34

[3] Wang Qingguo, Zhu Qing, Wang Xinzhou.The Choice of Evidence for Fuzzy Assessment Model on Map Quality [J]. Science of Surveying and Mapping, 2005, 30(6): 47-48

[4] Zhao Shumao, Zhao Yongjiang, Huang Wanhua. The Fuzzy General Valuation on Map Qualities [J].Journal of Institute of Surveying and Mapping, 2000, 17(1): 70-72

[5] Li Deyi, Liu Changyu. Discussion on Generalization of Normal Model [J].China Engineer Science, 2004, 6(8): 28-34

[6] Fu Bin, Li Daoguo, Wang Mukuai. Review and prospect on research of cloud model [J]. Application Research of Computers, 2011, 28(2): 420-426

[7] Ye Qiong, Li Shaowen, Zhang Youhua, Shu Xingwang, Ni Dongping. Summarize on Cloud Model and its Application [J]. Engineer and Design of Computer, 2011, 32(12): 4198-4201

[8] Fan Dingguo, He Shuo, Duan Fu, Niu Baoning.Synthesis Judgment Method based on Cloud Model [J]. SCI/TECH INFORMATION DEVELOPMENT &ECONOMY, 2003, 13(12): 157-159

[9] Chen Guilin. An uncertain Model of Qualitative and Quantitative Information Transformation—Cloud Model [J]. Research on Computer Application, 2010, 27(6): 2007-2010

[10] Li Deren, Wang Sliang, Li Deyi. Spatial Data Digging Theory and Applicatioon [M]. Beijing: Science Press, 2008, 337-339,342-345,354-356

[11] L1 Deyi, Han Jiawei, Shi Xuemei. Knowledge representation and discovery based on linguistic atoms [J]. Knowledge-based System, 1998(10): 431-440

[12] Li Chunju. Study on Quality Control of Digital Chart [D]. Dalian: Dalian Navy Academy, 2008, 56-60

[13] Luo Sheng, Zhang Baoming, Guo Haitao. A Quality Comprehensive Evaluation Model of Image Map based on Cloud Model [J]. Science of Surveying and Mapping, 2008, 33(3): 44-46

[14] Liu Bingxiang, Li Hailin, Yang libin. Cloud Decision Analysis Methods [J]. Control and Decision, 2009, 24(6): 957-960

The Automatic Recognition of the Abnormal Sky-subtraction Spectra Based on Hadoop

An An and Jingchang Pan*

School of Mechanical Electronic and Information Engineering, Shandong University, Weihai, China
Email: anran0708@163.com, *jingchangpan@163.com

Abstract. The skylines, superimposing on the target spectrum as a main noise, If the spectrum still contains a large number of high strength skylight residuals after sky-subtraction processing, it will not be conducive to the follow-up analysis of the target spectrum. At the same time, the LAMOST can observe a quantity of spectroscopic data in every night. We need an efficient platform to proceed the recognition of the larger numbers of abnormal sky-subtraction spectra quickly. Hadoop, as a distributed parallel data computing platform, can deal with large amounts of data effectively. In this paper, we conduct the continuum normalization firstly and then a simple and effective method will be presented to automatic recognize the abnormal sky-subtraction spectra based on Hadoop platform. Obtain through the experiment, the Hadoop platform can implement the recognition with more speed and efficiency, and the simple method can recognize the abnormal sky-subtraction spectra and find the abnormal skyline positions of different residual strength effectively, can be applied to the automatic detection of abnormal sky-subtraction of large number of spectra.

1. Hadoop Platform

Hadoop[1] is an open source distributed computing platform under the Apache Software Foundation, The core of Hadoop consists of HDFS distributed file system and MapReduce programming model. The HDFS distributed file system is mainly used to solve the problem of large-scale data storage, and the MapReduce programming model is used to solve the problem of distributed parallel computing of large-scale data. The underlying details of the Hadoop are transparent to the user, and we can form a distributed system by making use of a lot of cheap, low - profile minicomputers. The applications are deployed on each node to achieve parallel processing, speed up the processing speed and expand the storage space, so as to be more effective to complete the processing of massive data.

The HDFS cluster is based on the Master/Slaves model, which takes the block as the storage unit, and the default size of each block is 64MB. The figure 1 is the HDFS Structure. From this figure, we can see that Hadoop is mainly composed of Client terminal, NameNode, SecondaryNameNode and DataNode. For Hadoop version 1, a Hadoop cluster has only one NameNode. NameNode is running on the master node of HDFS, and responsible for the management of the namespace about the file system and recording the position in the DataNode of all data blocks and its replication. This information is stored on disks as the HDFS metadata image files and HDFS file change logs. The SecondaryNameNode is used as the slave date node to monitor the HDFS auxiliary background process, and regular merge the HDFS metadata image files and the HDFS file change logs to the disks.

The DataNode is actually used to store the data and also responsible for serving read and write requests about the data block, and through the heartbeat mechanism to communicate with NameNode, to tell NameNode its storage capacity and the current storage of data block information, etc.

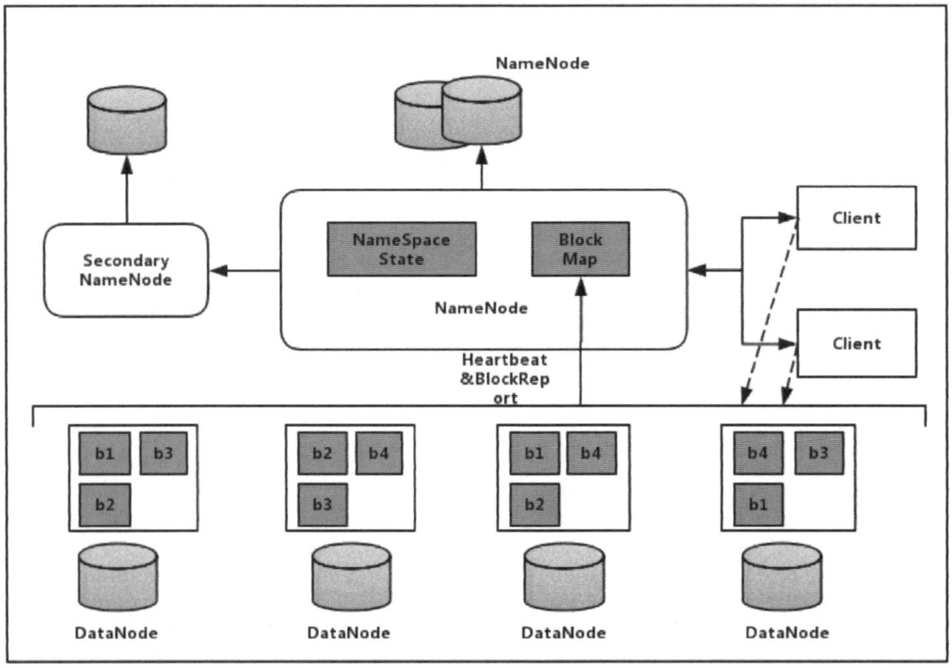

Figure 1 HDFS Structure[1]

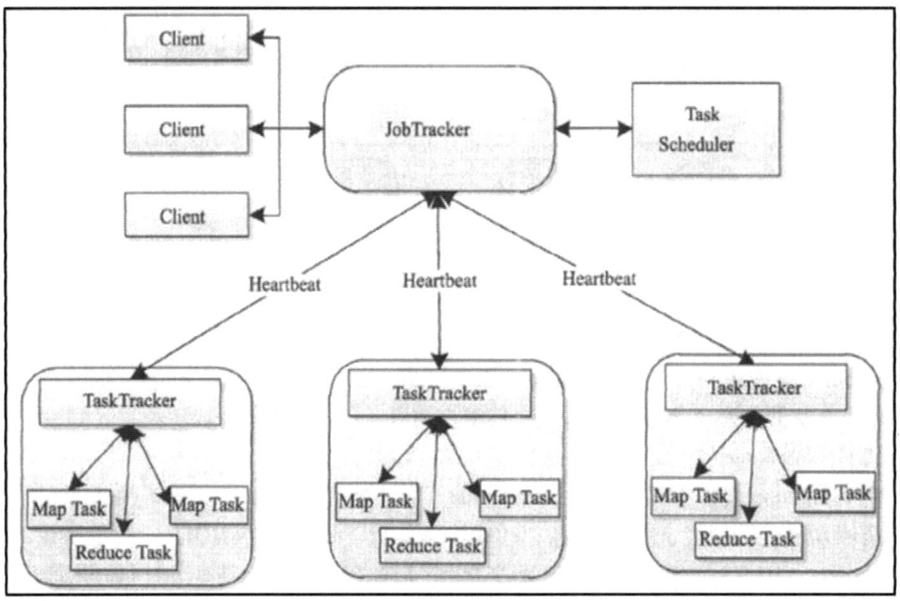

Figure 2 MapReduce Structure[1]

MapReduce is a distributed programming framework, which is mainly composed of JobTracker, TaskTracker and Task. The figure 2 is the MapReduce Structure .The JobTracker, as the MapReduce background process, is running on the master node, and responsible for monitoring resources, scheduling jobs. Such as it will monitor the state of TaskTracker, and transfer the failed tasks to other nodes. The TaskTracker is running on the slave node, for receiving the commands sented by JobTracker, such as starting a new task, killing the failed tasks, etc. In addition, it will regularly report to the JobTracker according to the Heartbeat mechanism about the implementation of tasks and the current storage of data block information, etc. Task is activated by TaskTracker, it is the executor of the task.

One task is generally divided into a number of Map Tasks for the implementation of map () function, a Reduce Task to perform reduce () function.

2. Abnormal Sky-subtraction Recognition algorithm

LAMOST[2,3] is one of the national key scientific projects, which is approved by the state, began in 2001, taking 7 years, costing $235 million to establish, and is China's largest astronomical telescope. Its site located in the National Astronomical Observatory of the Chinese Academy of Sciences. LAMOST contains 16 spectrometers, you can simultaneously observe 4000 target objects, each of the spectrograph will be connected to 250 fibers, and we use 240 fibers to observe the target objects, and the other 10 optical fibers are used to observe the skylight.

The skylight refers to the sky conditions without month and city light interference. When the sun falls 18 degrees below the horizon, the night sky will show a faint diffuse light, we generally call this as sky background or skylight in astronomical observation. The More common skylight emission line are oxygen atomic emission line [OI5577Å] and the OH emission band. The LAMOST process uses this limited skylight fibers to fit a "super skylight" spectrum, and then the spectral data of each observation spectrum is subtracted from the corresponding "super skylight" spectral data to get the sky-subtraction[4] spectral data. On account of the limited technology, the fitted "super skylight" is not a complete representation of the sky background component contained in the target spectrum, this led to the accuracy of the sky is not too high and generated a large number of abnormal sky-subtraction spectra data. Through the analysis of the abnormal sky-subtraction spectra, we can get the anomaly as shown in Figure 3, and the upper line shows the original spectrum, and the middle line represents the fitting of the super skylight spectrum, the lower line represents the spectrum after sky-subtraction processing. It can be seen from the diagram that the spectrum after the sky-subtraction processing is not a relatively smooth curve, but there is uneven band, we can take this phenomenon as some sky-subtraction residual components.

By analyzing the characteristics of the abnormal sky-subtraction spectra, we can know that when identifying the abnormal sky-subtraction spectra, we only need to identify the residuals which are similar to the emission line or the absorption line. So we use continuum normalization method to remove the components of the continuum spectrum, leaving only the characteristic spectral line information and noise information. The continuum normalization is achieved by fitting the spectral continuum, and then the original spectral flow is divided by the continuous spectral flow at the corresponding wavelength, finally, only the spectral lines and noise information remained. As a nonlinear filtering algorithm, median filtering algorithm[5] can effectively remove the transient pulse noise of positive or negative value, and preserve the shape of the continuous spectrum. When the skylight is abnormal, there will be more or less positive and negative pulse noise. In addition, the median filtering algorithm is simple and feasible, we just need to select the appropriate window size to process the algorithm. Therefore, the use of median filtering algorithm will be better suited to carry out the continuous spectrum fitting.

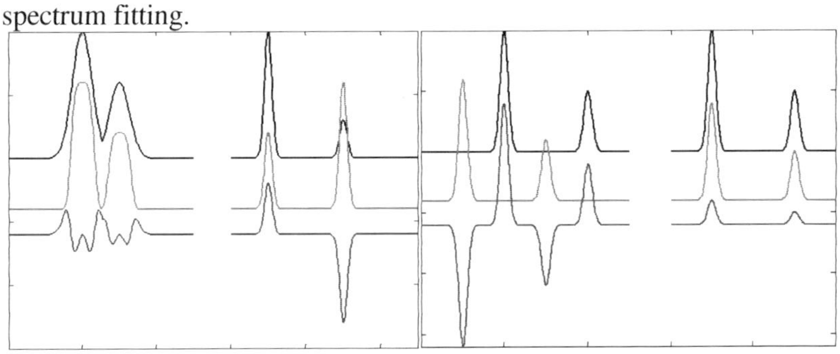

Figure 3 Abnormal sky-subtraction features[6]

In this paper, we first carry out the continuum normalization, and then use the following algorithm to process the recognition of the abnormal sky-subtraction spectra:

Assume that the number of sky emission lines selected is m, the central wavelength of each sky emission lines is $skylines[i], i = 1, 2, 3 \cdots \cdots$, m, the length of the spectrum is n, and the starting pixel position which gained in the FITS files is start, the spacing between adjacent pixels is bata, the spectral flow array is $flux[j], j = 1, 2, 3 \cdots \cdots$, n after the continuous normalization, and the wavelength array is $wave[j], j = 1, 2, 3 \cdots \cdots$, n, now we assume θ_1 as the calculating threshold and θ_2 as the sky line window threshold.

(1) According to the formula $\log_{10}(skylines[i] - start) / bata$, we calculated the subscript position of the skyline wavelength in the wavelength array. Here we assume the position is *skyIndex*.

(2) The average flow μ and standard deviation σ in the wavelength range $wave[skyIndex - \theta_1], wave[skyIndex - \theta_2]$ and the wavelength range $wave[skyIndex + \theta_2], wave[skyIndex + \theta_1]$ of each wavelength are calculated near each skylight emission line after the continuum normalization. As for the selection of the threshold $\theta_1$andθ_2, it is determined by the intensity and the width of the skylight, so as to make the two intervals in the relatively smooth range between the two sides of the skylight.

(3) Between the wavelength range $\{wave[skyIndex - \theta_2], wave[skyIndex + \theta_2]\}$:

a. If the continuum normalization flux on the position of this skyline $flux[skyline[i] > u]$, in the interval to find whether there is a wavelength λ, so that the establishment of $flux[\lambda] > \mu + \theta_2$, If it exists, it is judged that the skylight position is abnormal.

b. If the continuum normalization flux on the position of this skyline $flux[skyline[i] < u$, in the interval to find whether there is a wavelength λ, so that the establishment of $flux[\lambda] < \mu + \theta_2$, If it exists, it is judged that the skylight position is abnormal.

The threshold θ_2 is chosen to reduce the influence of the LAMOST wavelength calibration error, which makes it possible to find the abnormal skyline positions in a certain range near the wavelength of the skylight. In this experiment, we set θ_2 's value to 2. The threshold θ_2 indicates the degree of deviation of the skylight flow relative to the whole flow, In this experiment, we set θ_2 's value to 3.

3. Hadoop Cluster Processing

3.1 Pretreatment

In this paper, the experiment uses 226000 data which are released by the DR1 from LAMOST, and the data is stored in fits format files. The original spectrum file is composed of several smaller Fits files, Hadoop as a large data distributed processing platform, when the size of the file is much smaller than the size of a single block, its performance will be greatly reduced. We need to do the data preprocessing, the small files will be integrated into large files which are more suitable for the Hadoop platform to deal with, and the information needed in this paper will be extracted, irrelevant spectral information will be filtered out at the same time. In this paper, each of the Fits spectral files are treated as follows: First of all, through the Java program to extract MJD, PLANID, SPID, FIBERID data values from the Fits files, and write them to the text files, The information uniquely identifies the different spectral data. Secondly, we need to extract the initial pixel value as CRVAL1, the distance between adjacent pixels as CD1_1 and the Spectral length, and write them to the text files. Finally, the Java program in turn to find the wavelength and flow value of each pixel according to the information, and then write them to the text files, too. These values are separated by spaces, each line represents the data information of a spectral file. In this experiment, this total size of the fits file is 35.2G before the pretreatment, after the pretreatment, this size is compressed to 15.6G, The size of the uploaded file can be greatly reduced by the pretreatment step.

3.2 Upload File to HDFS

Write java programs to upload the text files to HDFS file system of Hadoop as the input file of Hadoop Mapper function.

3.3 Submit Tasks

Submit tasks to Hadoop cluster. Hadoop divides the input files into several pieces based on the size of the block, and each piece starts a Map Task for processing. Each Map Task performs the Map function to read the file information, generates a number of input <Key, Value> pairs according to the contents of the file. The Key is the byte offset of the row in the TXT file which is currently being read, the Value is the content of the current read line, in this paper, the format of the read line is MJD PLANID SPID FIBERID CRVAL1 CD1_1 NAXIS1 Wave[1] Flux[1] Wave[2] Flux[2] … … Wave[NAXIS1] Flux[NAXIS1]. We can find the meaning of each symbol in document 7[7].Then the Map function is used to perform continuum normalization and the recognition of the abnormal sky-subtraction spectra. Finally, we output the result in the form of <Key, Value> pairs, the Key's format is spec-MJD-PLANID-spFIBERID-SPID, it will uniquely identify a spectral data, and the value is the calculation of the abnormal sky-subtraction positions

4. Experiment Analysis

4.1 Performance Analysis of Hadoop

Table1 shows the running time about the automatic identification of the sky-subtraction residual components of the 226000 spectra between the Hadoop cluster model with only three computing nodes and the single model. Table2 shows the performance analysis. In this experiment, 63 Map Tasks and 1 Reduce Task were used to synchronize processing the data in the Hadoop cluster at the same time. So it can be seen from the table1 that the time spent by the Hadoop cluster is less. In addition, the storage capacity of the single mode is limited by the memory and the disk, while the Hadoop cluster can be easily extended to make it easier to store large data.

Table 1 Comparison of running time between single mode and cluster mode

operation mode	runtime
Single model	About 960s
Cluster model	About 120s

Table 2 Performance analysis

Speedup[10]	8
Efficiency[11]	2.6
Cost[12]	360s

4.2 Recognition of Sky-subtraction Residual Components Results

In practical application, we can increase or decrease the threshold value to detect sky-subtraction residual components with different strength according to the actual demand. If we only need to detect some high strength skylight residuals, we can increase the value of the standard deviation multiple correlation threshold in the experiment. In the experiment, we choose 68 high strength and undoped skylines. The following figure 5 is the recognition result of the abnormal sky-subtraction spectra, the black line represents the spectral flux after sky-subtraction process, and the blue line indicates the abnormal sky-subtraction wavelength positions. From the analysis to the experiment we discover that this method can not only effectively reduce the time of identification, but also find out the positions of the skylight residuals. Figure 4 shows the sky-subtraction residuals at the wavelength of 6863 Å and 8344 Å, the position of 8344 Å has the positive and negative feature, probably because the wavelength calibration is not accurate. In the following step, the sky-subtraction precision can be improved by improving the wavelength calibration accuracy. The position of 6863 Å has the anomaly characteristics like a “W” shape, probably because there are contour inconsistency or resolution inconsistency anomalies occurred between the sky spectrum and the original spectrum around the sky emission line position.

Based on the analysis of the experimental results, it is found that the method can find out the residual distribution of abnormal sky-subtraction stellar spectra in varying degrees by adjusting the threshold value. In addition, through the experimental analysis for the abnormal sky-subtraction stellar spectra, the abnormal situation is mainly by fitting the wavelength calibration of imprecise, inconsistent resolution or outline, incorrect "super skylight" caused by background subtraction too much or too little. Because, in reality, the factors that affect the results of the sky-subtraction result is complex, most of which are combined with a variety of factors, so it is difficult to do the repair work about the abnormal sky-subtraction spectra. The most fundamental method to reduce the error of sky-subtraction by increasing the wavelength calibration accuracy, increase the number of skylight fibers, get method to fit the "super light" better agreement with the target spectrum.

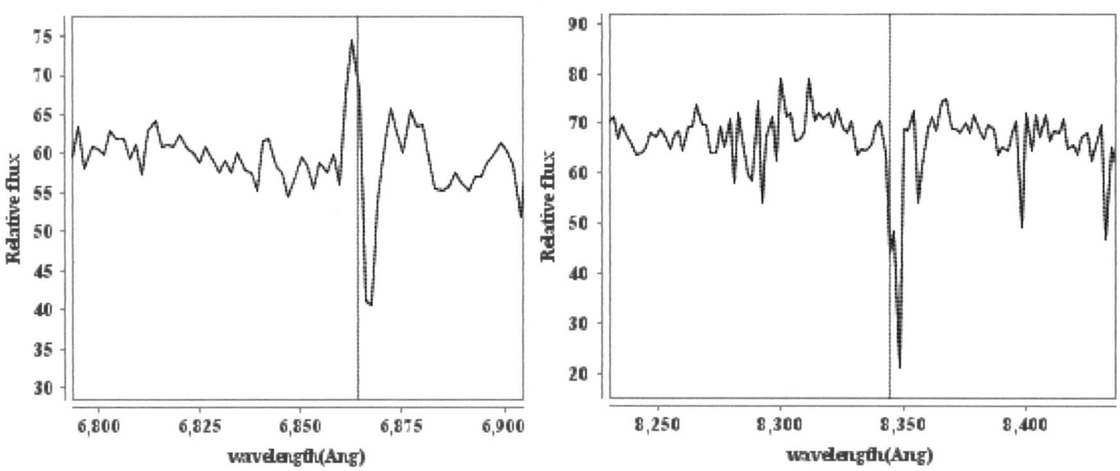

Figure 4 Abnormal sky-subtraction bands

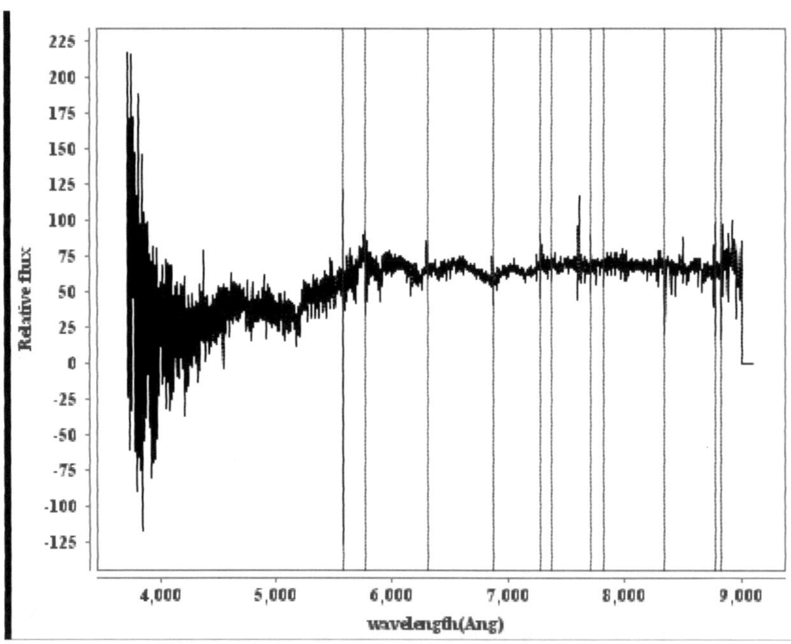

Figure 5 Abnormal sky-subtraction Spectra

IOP Conf. Series: Journal of Physics: Conf. Series **910** (2017) 012019 doi:10.1088/1742-6596/910/1/012019

5. Conclusions

After the completion of the spectral continuum fitting and the continuum normalization steps, the paper proposed a simple and effective method for the automatic detection of different degrees of the residual components based on the Hadoop platform. Through the analysis of the time performance, the experiment proved the efficiency of Hadoop cluster mode compared to single mode in large data processing, such as, the speedup is 8 in this experiment when the Hadoop cluster only have three nodes. In addition, the method proposed in the paper can recognize the abnormal skyline positions of different residual strength effectively and flexibly by adjusting the threshold.

6. Acknowledgment

This work was supported by the National Natural Science Foundation of China (U1431102).

7. References

[1] Dong Xicheng. Hadoop technology insider. BeiJing: China Machine Press.2013.
[2] The official website of LAMOST: http://www.lamost.org/.
[3] Cui, X., et al. The Large Sky Area Multi-Object Fiber Spectroscopic Telescope. Research in Astronomy and Astrophysics, 12(9): 1197-1242, 2012.
[4] Bai Zhongrui. Research on the sky-subtraction method of LAMOST[D]. BeiJing: university of science and technology of china.2007.
[5] S.J. Ko, Y.H. Lee. Center weighted median filter and their applications to image enhancement [J] .IEEE Transactions on Circuits and System, 1991, 38 (1):984-993.
[6] An Ran, Pang Jingchang etc. The Automatic Recognition and Detection of Sky-subtraction Residual Component in The Stellar Spectra. Spectroscopy and Spectral Analysis. 2017, 37(1):273-277.
[7] Fits file format official website of LAMOST: http://dr1.lamost.org/doc/data-production-description.
[8] Basel A. Mahafzah. Parallel multithreaded IDA* heuristic search: Algorithm design and performance evaluation.International Journal of Parallel, Emergent and Distributed Systems, 26(1), 61-82, 2011.
[9] SW Al-Haj Baddar, BA Mahafzah. Bitonic sort on a chained-cubic tree interconnection network. Journal of Parallel and Distributed Computing, 74(1), 1744-1761, 2014.
[10] BA Mahafzah. Performance assessment of multithreaded quicksort algorithm on simultaneous multithreaded architecture. Journal of Parallel and Distributed Computing, 74(1), 1744-1761, 2014.

The Weather Forecast Using Data Mining Research Based on Cloud Computing.

ZhanJie Wang and A. B. M. Mazharul Mujib

Electronics Information & Electrical Engineering, Dalian University of Technology, Liaoning, China

Email: wangzhi@dlut.edu.cn, mazharul.mujib@gmail.com

Abstract. Weather forecasting has been an important application in meteorology and one of the most scientifically and technologically challenging problem around the world. In my study, we have analyzed the use of data mining techniques in forecasting weather. This paper proposes a modern method to develop a service oriented architecture for the weather information systems which forecast weather using these data mining techniques. This can be carried out by using Artificial Neural Network and Decision tree Algorithms and meteorological data collected in Specific time. Algorithm has presented the best results to generate classification rules for the mean weather variables. The results showed that these data mining techniques can be enough for weather forecasting.

1. Introduction

Weather Prediction is the application of science and technology to predict atmospheric conditions ahead of time for a particular region. Prediction is one of the basic goals of Data Mining. Data Mining is to dig out knowledge and rules, which are hidden and unknown. User may be interested in or has potential value for decision-making from the large amounts of data. Such potential knowledge and rules can reveal the laws between the data. There are many kinds of technical methods of data mining, which mainly include: association rule mining algorithm, decision tree classification algorithm, clustering algorithm and time series mining algorithm, etc. [1]. How to store, manage and use these massive meteorological data, discover and understand the law and knowledge of the data, to contribute to weather forecasting completely and effectively has attracted more and more Data Mining researcher's attention[2]. This article constructs the Weather Forecasting platform, using data mining for meteorological forecast and the forecast results are analyzed.

2. Weather Forecasting

Weather forecasting plays a significant role in meteorology [3]. Weather forecasting has remained a formidable challenge because of its data intensive and frenzied nature. Generally, two methods are used to forecast weather: a) the empirical approach and b) the dynamical approach. The first approach is based on the occurrence of analogues and often referred to as analogue forecasting. This approach is useful in predicting local scale weather if recorded cases are plentiful. The second case is based upon equations and forward simulations of the atmosphere and often referred to as computer modeling. Most weather prediction systems use a combination of both of these techniques.

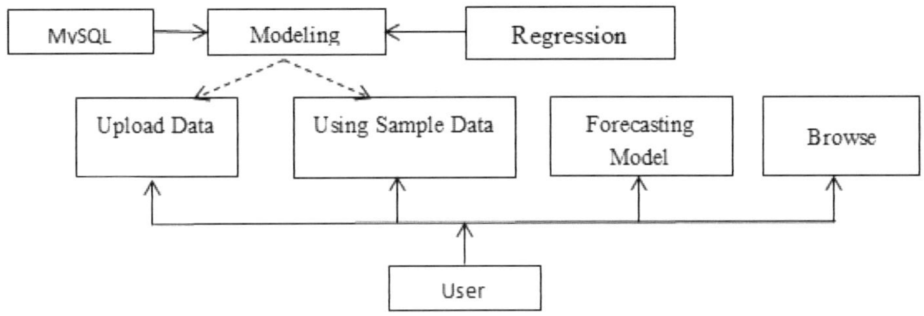

Figure 1 FAAS Models

This framework as a service (FAAS) has selected seven common forecasting methods. These are Regression (R), Logistic Regression, Time Series, Artificial Neural Network, Random Forest, Support Vector Machine and Multivariate Adaptive Regression Splines (MARS). For instance, Regression may encounter them collinearity among variables. Logistic Regression could only deal with the dataset where the dependent variable is nominal.

3. Weather Prediction Architecture

Artificial Neural Networks (ANN) and Decision Trees (DT) were used to analyze meteorological data, gathered in-order to develop classification rules for the Application of Data Mining Techniques in Weather Prediction. Artificial Neural Networks (ANN) has received special attentions among different forecasting methods in recent years [4, 5]. Main reason for the popularity of ANN is its capability of supervised learning from complex relations using non-linear functions [6]. This algorithm combines with both of the time series and regressional approaches. Weather parameters [7] over the study period use available historical data for the prediction of future weather conditions. The targets for the prediction are those weather changes that affect our daily life e.g. changes in minimum and maximum temperature, rainfall, evaporation and wind speed. These techniques are often more powerful, flexible, and efficient for exploratory analysis than the statistical techniques. The most commonly used techniques in data mining are: artificial neural networks, logistic regression, discriminant analysis and decision trees. By this model, temperature (T), rainfall (R) and wind (W) speed can easily be predicted. Now Prediction method that only single parameter for example some researchers [8], [9] used wind speed and other researchers used wind Power for prediction [10], [11]. We have provided more accurate prediction model data and it has more parameters and efficiency. (Figure.2)

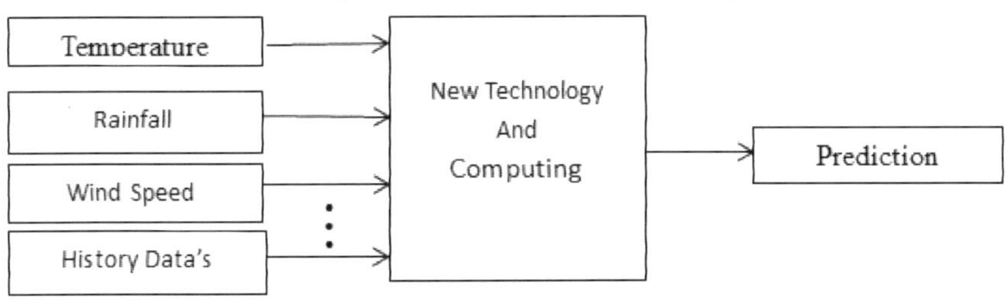

Figure 2 Weather Data Prediction

There are three basic elements of a neuron model. Figure.3 shows the basic elements of neuron model with the help of a perceptron model, which are, (i) a set of synapses, connecting links, each of which is considered by a weight/strength of its own (ii) an adder, for summing the input signals, weighted by respective neuron's synapses (iii) an activation function, for limiting the amplitude of neuron's output. A typical input-output relation can be expressed as shown in Equation 1.

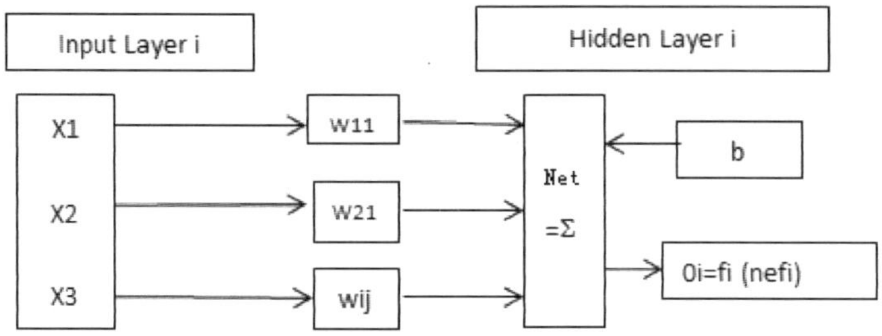

Figure 3 Model of a perceptron

$$netj = \sum_{j=1}^{n} WijXi + bj$$

$$Oi = f_i(neti) \qquad (1)$$

Where = inputs to node in input, = weight between input node and hidden node, b – bias at node, net = adder, f = activation function.

The type of transfer/activation function affects the size of the steps taken in weight and space [12]. ANN's architecture needs determination of number of the connecting weights and the way in which the information flows through this network is carried out via the number of layers, nodes number in each layer, and their connectivity. The output nodes numbers are fixed, according to the estimated quantities. The input nodes numbers are dependent on the existing problem under consideration, and the modeler's choice to utilize knowledge of domain. The neurons in the hidden layer are enhanced gradually, and the network performance in the form an error is examined.

4. Weather Forcast In Cloud Computing

Before Cloud computing has improved the efficiency of data storage, delivery, and dissemination across multiple platforms and applications, allowing easier collaboration and data sharing, including data processing and distribution systems that disseminate key weather forecasting, severe weather warning, and climate information. Data mining techniques and forecasting applications are very much needed in the cloud computing paradigm. In this study, data mining in Cloud Computing allows weather forecasting and data storage, with assurance of efficient, reliable and secure services for their users. The implementation of data mining techniques through Cloud computing will allow the users to retrieve meaningful information from virtually integrated data warehouse that will reduce the costs of infrastructure and storage.

5. Implement And Process

5.1. Forecasting Model

Our prediction model is based on a Multi-Layer Perceptron ANN. The central, innovative point of the method is the construction of the data set and the choice of input data unlike the other works [13][14]. Along with the historical weather data of the site, chosen for the prediction, nearby sites are also taken into account for the evolution of the weather fronts. Therefore, for the identification of the site where to make the wind forecast (called WFS), I have chosen four cardinal directions and the four intermediate directions, two points for each direction, one more far away (1000 km) and second one halfway (500 km) from the place identified. For sites at maximum distance, weather data referred to the instant t-2, for sites at minimum distance, weather data refer to the instant t-1. Because, weather data was referred to instant t0 for the WFS. The implemented MLP, has one hidden layer using transfer function, F(x), the logistic function: $\int F(x)dx = \frac{1}{(1+e^{-x})}$.

The MLP network is trained through the back-propagation learning algorithm. This training algorithm is to minimize the cost function (E) of the network, expressed as mean square error amongst all the outputs of the network and the respective desired values defined in the training set. The mean square error is calculated by using the following expression:

5.2. Input Data

The data used in this study for each point are temperature, rainfall and wind. The data used are the daily average values. The input Dataset variable are (T), (R) and (W) used in this Model and this model is similar to Fig.2. We have collected the data of Dalian weather forecasting and input the data in the model for computing and got the results. Xi (i=1, 2, 3) used are site-dependent. The prediction of weather forecasting, with a resolution of one day, requires the knowledge of the weather conditions in previous time instants to the one chosen for the prediction. Earlier, Mesoscale Meteorological Modeling [15], Numerical Weather Prediction Models [16] and other methods [17] were used for prediction. Unlike these traditional methods, we have improved this model, using different time instants for weather forecasting.

5.3. Simulation Results

In this section we have introduce a case study; the test site is a real weather forecast located in Dalian Liaoning Province, China. In this study of data analysis for Dalian weather forecasting comes from the Dalian Meteorological Bureau in the last five years. In our research, we have set up a new model with more parameters and finished the simulation later. Our provided model has shown more accurate results than the old model. In this prediction model, we have used ANN and FAAS. The figures of merit used for analyzing the performance. The forecasting data are compared with the real data recorded in the site. In this work, we have used last five years data analysis, the variables including in the highest temperature, the lowest temperature and wind speed. Variables of analysis are described in following chart:

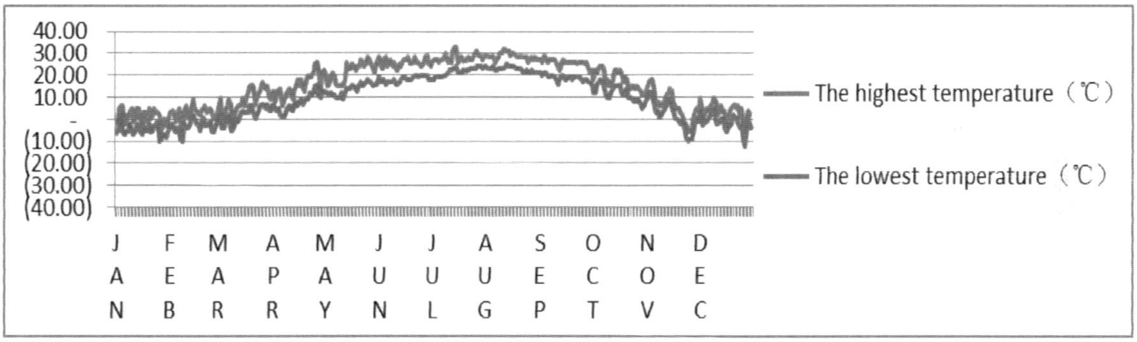

Figure 4 Weather Forecasting Temperature for Last Five Years (2011-2015) Highest and Lowest Value

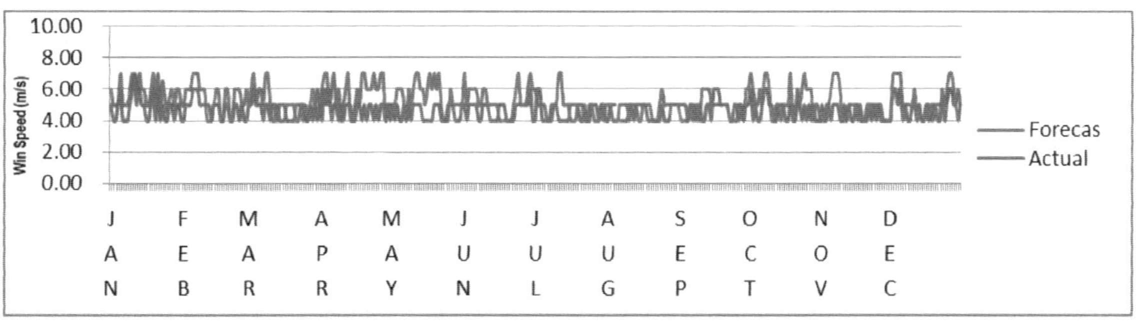

Figure 5 Results of wind speed prediction Value in Last Five Years. (2011-2015)

The results have shown that concerning Wind Speed and Temperature highest and lowest Value displayed the simulation results (figure 4 and 5) in Intervals of last five years. While using more input data values, the execution time can be as longer as expected.

6. Vi. Discussion and Conclusions

In this Study Forecasting Models algorithm are used for classifying weather parameters such as maximum temperature, minimum temperature, and wind speed in terms of the month and years. The implementation of data mining approach to solve the wind forecasting problems for wind farm production, in particular, for predicting wind speed. The data mining prediction algorithm-ARIMA time series prediction algorithm is also integrated into the system. The platform has the ability of mass storage of meteorological data, efficient query and analysis, weather forecasting and other functions. In this study we also adapted the method of Artificial Neural Networks, it can detect the relationships between the input variables and generate outputs based on the observed patterns inherent in the data without any need for programming or developing complex equations to model these relationships. An artificial neural network (multi-layer perceptron) was applied and several simulations have been conducted for comparison purposes. ANN's can detect the relationships between weather parameters and use these for future prediction. Weather conditions are important to climatic change studies because the variation in weather conditions in term of temperature and wind speed can be studied using these data mining techniques. ANNs are implemented, in order to compare their effectiveness in changing the network topology and the training mode. The results obtained from real data are based on time series of meteorological data provided by the Dalian Meteorological Bureau. The test cases pointed out that the proposed approach gives a very interesting performance of the implemented network and shows good performance in term of MSE. For future perspective, there is still significant potential for improvement in weather forecasting by using ANN model, through introducing climate change and global warming variables, in order to forecast more realistic weather parameters.

7. Acknowlegment

This research was supported by the National Natural Science Foundation of China (Grant No. 61672126) and (Grant No. 51579041), We would like to thank the anonymous reviewers for their attentive reading and for their constructive comments that have helped to further strengthen this paper.

8. References

[1] M. Zhang, "Application of Data Mining Technology in Digital Library", Journal of Computers, vol. 6, no. 4, (2011) April, pp. 761-768.

[2] Z. Danping and D. Jin, "The Data Mining of the Human Resources Data Warehouse in University Based on Association Rule", Journal of Computers, vol. 6, no. 1, (2011) January, pp. 139-146.

[3] Introduction to Data Mining and Knowledge Discovery, Third Edition, Two Crowds Corporation, http://www.twocrows.com/introdm.pdf, accessed on 12 April 2009.

[4] L. M. Saini and M. K. Soni, "Artificial neural network-based peak load forecasting using conjugate gradient methods," IEEE Transactions on Power Systems, vol. 12, no. 3, pp. 907–912, . 2002.

[5] S. Fan, C. X. Mao, and L. N. Chen, "Peak load forecasting using the self-organizing map," in Advances in Neural Network-ISNN 2005. New York: Springer-Verlag, 2005, pt. III, pp. 640–649.

[6] Kourentzes, N., "Intermittent demand forecasts with neural networks", International Journal of Production Economics, Volume 143, Number 1, pages 198-206, 2013.

[7] Elia G. P., 2009, "A Decision Tree for Weather Prediction", Universitatea Petrol-Gaze din Ploiesti, Bd. Bucuresti 39, Ploiesti, Catedra de Informatică, Vol. LXI, No. 1.

[8] A. R. Finamore; V. Calderaro; V. Galdi; A. Piccolo; G. Conio; S. Grasso, "A day-ahead wind speed forecasting using data-mining model– a feed-forward NN algorithm", IEEE International Conference on Renewable Energy Research and Applications, 2015, pp. 1230-1235.

[9] E. Erdem, J. Shi, "ARMA based approaches for forecasting the tuple of wind speed and direction", Applied Energy 88, ELSEVIER, 2011, pp. 1405–1414.

[10] N. Chen, Z. Qian, I. T. Nabney, and X. Meng, "Wind Power Forecasts Using Gaussian Processes and Numerical Weather Prediction," IEEE Transaction on Power Systems, vol. 29, no. 2, 2014.
[11] EWEA Wind in power: European Statistics. 2013.
[12] Martin T. H., Howard B. D, Mark B., 2002, Neural Network Design, Shanghai: Thomson Asia PTE LTD and China Machine Press.
[13] N. Chen, Z. Qian, I. T. Nabney, and X. Meng, "Wind Power Forecasts Using Gaussian Processes and Numerical Weather Prediction," IEEE Transaction on Power Systems, vol. 29, no. 2, 2014.
[14] R. R. B. de Aquino, H. T. V. Gouveia, M. M. S. Lira, A. A. Ferreira, O.N. Neto, M. A. Carvalho Jr., "Wind Forecasting and Wind Power Generation: Looking for the Best Model Based on Artificial Intelligence", IEEE World Congress on Computational Intelligence, 2012.
[15] Roger A. Pielke Sr., "Mesoscale Meteorological Modelling", International Geophysics Series, Volume 98, oct, 2013.
[16] David J. Stensrud ,"Parameterization schemes – Keys To Understanding Numerical Weather Prediction Models", Cambridge University Press, 2007.
[17] Renewables – Global Status Report- REN21, 2015.

Data Mining Techniques for Customer Relationship Management

Feng Guo and Huilin Qin

Information School of Beijing Wuzi University, Beijing, China
Email: guofeng80050@sina.com, qinhuilin@bwu.edu.cn

Abstract. Data mining have made customer relationship management (CRM) a new area where firms can gain a competitive advantage, and play a key role in the firms' management decision. In this paper, we first analyze the value and application fields of data mining techniques for CRM, and further explore how data mining applied to Customer churn analysis. A new business culture is developing today. The conventional production centered and sales purposed market strategy is gradually shifting to customer centered and service purposed. Customers' value orientation is increasingly affecting the firms'. And customer resource has become one of the most important strategic resources. Therefore, understanding customers' needs and discriminating the most contributed customers has become the driving force of most modern business.

1. Customer Relationship Management

Customer relationship management (CMR) is a customer centered business strategy. Firms abidingly accumulate, analyze and exploit customer comprehensive information to allocate resources rationally according to customer oriented WORKING mode, and meet customer demand through meaningful communication, understanding and influence customer behavior. Thereby increasing customer satisfaction and loyalty, and to obtain the maximum profit. The general CRM covers customer resource management, marketing management, sales management, customer care and service management, and many other aspects. Its ultimate goal is to provide the basis for business management and decision-making through proactive market research and analysis and processing of customer feedback. Simple from the content of customer relationship management, CMR is customer propaganda, tracking, revisit and classification. Of course, with the development of modern information technology, the pursuit of personalized service has become the new target in the field of after-sales service.

2. Overview of Data Mining Techniques

Firms want to acquaint customer behavior as much as possible, but as the number of customers accumulated in large quantities and customer information has become increasingly complex, more advanced technologies and tools are needed to manage and analyze customer information so as to find out the valuable knowledge of management decision-making. Data mining technology provides a good technical support for CMR to analyze large amounts of complex customer data and explore customers' value.

2.1 Definition

Data mining can extract potentially valuable knowledge, model and rules from mass of data. It is an effective tool to exploit potential associations and patterns. With the help of statistics and artificial intelligence technology, data mining can make further analysis of data and find out knowledge, and build different models in accordance with different business issues, thus provide a technical reference for the firms' decision-making. Data mining involves a considerable number of subject areas, most of them are mature algorithms and techniques in artificial intelligence such as machine learning, statistics, neural network and database techniques. According to the purpose of data mining, it can be divided into association analysis, time series analysis, classification, cluster analysis and data summary.

2.2 Functionality
Through the prediction for future trends and behavior, data mining can help firms make a forward-looking and knowledge-based decision-making. The goal of data mining is to find implied and meaningful knowledge from database. Its functionality is to designate type of pattern that data mining tasks are looking for, and can be divided into two categories: description and prediction. Descriptive mining tasks characterize the general characteristics of the data in the database, while predictive ones extrapolate the existing data to predict. Specifically, firms can discover most valuable customers, make the combined sales more efficient and find fraud with a smaller cost with the help of data mining techniques.

2.3 Methods
As an advanced information processing technology, the essential difference between data mining and traditional data analysis is that it is a process of exploration of data relationships, and in most cases with no assumptions and premises. There are numbers of different ways in data mining for users to embark on a comprehensive analysis of data from different perspectives.

2.4 Correlation and Regression Analysis
Correlation analysis mainly analyzes the closeness of variables, and regression analysis mainly establishes dependencies between variables based on observational data. Both of them can reflect valuable associations and contacts among variables. So they can be collectively referred to as association analysis.

2.5 Time Series Analysis
Similar to association analysis, time series analysis is aimed at tap the intrinsic link between data. The difference is that time series analysis focuses on chronological causal relationship of data, while association analysis focuses on parallel relationship.

2.6 Classification and Prediction Analysis
Classification and prediction analysis are used to extract models describing important data. The model is then applied to determine and classify new observations, or predict future data trends.

2.7 Cluster Analysis
Cluster analysis categorizes data into more than one classes or clusters according to certain characteristics. Objects must have a high degree of similarity if they are from a same cluster, and the difference will be much larger if not. We could see that the cluster analysis is the reverse process of classification and prediction to a certain extent.

3. **Value of Data Mining for CRM**
CMR collect and accumulate a large number of market and customer information for the firms to build a huge data warehouse. The key to data mining is to distinguish truly valuable information and find the association between them from massive data.

Every firm wants to quickly and accurately extract the most important and valuable customer information through the use of certain tools and instruments. After a deep analysis for a huge data warehouse, data mining techniques can find hidden information and knowledge conducive to business operations, thus improve the competitiveness of the firm. Data mining can help firms manage customer life cycle stages, including acquiring new customers, making existing customers create more profits, maintaining valuable customer, etc. It can do in-depth analysis of customer needs, to meet firms' needs of customer relationship in individual market segments. So the firms are able to determine customer characteristics, provide them with targeted services and increase business opportunities.

What's more, a firm can use data mining techniques in CRM to help it discover business trends, reveal the known facts, predict outcomes and analyze needed key factors to complete tasks, so that it could increase revenue, reduce costs, and remain in a more favorable competitive position.

4. Application Fields of Data Mining for CRM

Data mining can filter customer information, and mining implicit, unknown and potentially valuable knowledge on business decisions and rules from a huge number of customer data. To help firms identify new opportunities, predict success in business strategies and make quick decisions. The following aspects of its application are particularly prominent.

4.1 Customer Character Analysis

In addition to customers' address, gender, age, occupation, income, education and other basic information, access, such as hobbies, marriage, spouse, health, home environment and other features can help firms get a more detailed understanding of customers, observe their behavioral regularities and thus better develop customer strategies to improve their campaign response speed. The main idea is to use classification and clustering techniques, divide customers into different group with different demand and transaction practices according to their age, gender, income, trading behavior, etc. And finally arrive at customers' concern, targeted to the development of personalized marketing strategy.

4.2 Customer loyalty analysis

Customer loyalty analysis is to analyze and categorize upscale customers, stable customers, valuable customers, more consumption demand customers, etc. To help firms make customers persistent and stability analysis and respond quickly to customer needs, so that each customer will get a highly personalized service. Statistically, firms must pay much higher cost to get new customers than to keep old customers. The gap is recognized to be more than six to eight times, no matter what business they focus on. Meanwhile, according to 80/20 principles of marketing, that 20% of customers contributed 80% of sales, more strategies should be adopted to develop high consumption customers' loyalty.

4.3 Cross marketing analysis

The previously customer information may contain the key or even critical factors to determine customer behavior. Using data mining techniques, firms can get the key factors affecting customer purchasing behavior from the customer information especially former purchase behavior, and build prediction model to predict customer future purchase behavior. Association analysis is often used to help companies find some implicit, subtle and great commercial value of relationship. This mining process is generally divided into two steps: First, find all the frequent item sets whose frequency should be at least equal to the minimum support frequency, thereby identify all related merchandises that may be purchased together; Then, generate strong association rules according to the frequent item sets. These rules must meet the minimum threshold of confidence; thereby identify all related merchandises that are very likely to be purchased together. To sum up, firms can discover hidden relationship between seemingly independent events; find factors that affect customer behavior through data mining technology. And then target to expand marketing strategy and promote other products in a timely manner.

4.4 Customer acquisition and churn analysis

The growth and expansion of the firm need continuously maintain old customers and draw new customers. Through classification, clustering, decision tree and other techniques, it can extract mass customer information, identify potential customer, and determine what kind of customer are most likely to loss and what features they have. Thus build customer churn prediction model to more accurately identify easily lost customers. This would help firms making plans to take appropriate marketing measures in advance to hold old customers.

4.5 Page Numbering and Running Heads
Through clean and centralize, customers' feedbacks are automatically store in data warehouse. So the firm can track customer behavior, analyze customer satisfaction, credit rating and so on, in order to evaluate and optimize the existing strategy, such as take different credit terms on different credit rating customers. This could maintain customer loyalty and avoid unnecessary risks at the same time. By tracking the evaluation to ensure that the firm customer relationship management to achieve the goals, also establish a good customer relationship.

5. Customer Churn Analysis
Statistically, firms must pay much higher cost to get new customers than to keep old customers. The gap is recognized to be more than six to eight times, no matter what business they focus on. So firms should take strategy to develop high consumption and high loyalty customers' loyalty, and use data mining techniques to identify customers that are likely to loss and take measures to prevent the loss. Commercial enterprises usually record sales information and customer information through transaction records, sweepstakes, membership and other activities. And then dig out customer preferences merchandise mixes, customer habits and characteristics of lost customers from this information. The following will take customer churn analysis as an example to illustrate the process of data mining in CRM.

5.1 Data selection
The target variables are customers' loss states. In another word, by analyzing a specific problem, select a combination of one or a plurality of known amount variables as target variable. What specific consumption characteristics lead to customer loss? Is that income change? Or preference change? Or the demand of shopping environment change?

The input arguments include static and dynamic data. The static data is the data that changes infrequently, such as the customer basic information (gender, marital status, occupation, residential areas, etc.); the dynamic data is the data that changes frequently or regularly, such as the consumption amount in a month, consumption characteristics, etc. There are intrinsic links between the input arguments and the target variables, but cannot be quantified. This provides data mining a play space. If we are unable to determine whether some sort of data is associated with customer churn probability, we could temporarily select it in the model, and examine each variable distribution and correlation in the subsequent step to determine whether to keep it.

Select the modeling data. There two kinds of customer churn: subside and transfer to competitors. The formal is always caused by firms' relocation, or customer service upgrade such as ordinary customers upgrade to the VIP. But the latter is what firms really care about. We must choose the latter loss of customer data to carry out the modeling.

5.2 Data reduction
Take data reduction before building the model, on the one hand, can ensure the correctness and validity of the modeling data, and on the other hand, can make the data more in line with the modeling needs through the adjustment of data format and content. The main tasks include data conversion, integration, sampling, randomization, missing value handling, etc. Such as proportionally extract the information of existing customers and lost customers and merge these two types of data as data source for modeling. The model need large amount of data to test, so the data sample is divided into two part, one for modeling and the other for model checking and correction.

5.3 Model building
Before modeling, we should use correlation analysis to identify the correlation between each input argument and the probability of loss of customers, delete the less correlation variable, thereby reducing the degree of complexity of the model, thus shorten the modeling cycle. There exist many

data mining methods, such as neural networks, regression, decision tree, association analysis, clustering, etc. We can use a variety of methods pre-build some models then compare the pros and cons of these models and select the most suitable for customer churn analysis. There are also a variety of approximation algorithms to simplify the model optimization process. The core part of data mining process is adjustment, which is usually complete by data analysis experts. Different business issues and data distribution properties will affect the modeling and adjust its strategy; therefore business experts are required in order to avoid inappropriate optimization resulting in loss of business information.

5.4 Model evaluation and interpretation

To obtain accurate result, we can use the data not involved in the modeling for model evaluation. The evaluate matric is to use the mode to predict the already known customer state, and compare the predicted result and the actual customer state. The model with the highest prediction accuracy is the best. Through making a reasonable business explanation to the model, firms can identify some potential law to guide businesses behavior. In turn, the explanation can also prove the correctness and effectiveness of the mathematical model.

5.5 Model application

To make the model practical, we can first select a pilot to implement the model, and pay attention to the earnings of the model application during the pilot. If an exception occurs, stop the application and correct the model. After the end of the pilot, if the model is proved to be good, then it can consider a wide range of promotion. The promotion should pay attention to regional differences, and cannot just copy the model. The model bias may increase with the growth of time and significant changes in economic, and then we should consider re-establish a more applicable model.

The churn model can monitor the loss of customer, if the possibility of loss is too high, firms can take measures to increase customer loyalty, and prevent the occurrence of loss of customers, which completely change the situation that firms cannot monitor the loss of customers and cannot effectively achieve customer care even they successfully get many customers.

6. Conclusion

The CRM system using data mining technology has become the key for firms to winning in the increasingly competitive knowledge economy and e-commerce economy mode. Data mining as the premise and foundation of CRM is no longer just confined to customer contact level customer relationship management. On the contrary, it has been deep into customer behavior and customer preferences, and can anticipate their needs and meet their expectations. Thus firms can insight into customer, understand customer value from a more in-depth and comprehensive perspective. Effective use of data mining in CRM, can continually promote a single customer value enhancement and expansion of the customer base, guide high-level decision makers to work out best marketing strategy, reduce operating costs, increase profits, strengthen customer relationships, improve customer satisfaction, improve, improve commodity promotions, enhance information exchange of goods, improve customer loyalty, accelerate the development of firms, and enable firms to gain long-term benefits.

7. References

[1] Guoen Xia. Customer Churn Prediction Based on Simple Support Vector Machine [J]. Research on Application of Computer, 2010, 27(3).
[2] Ying Liu. Application of data mining in customer relationship management [J].Digital Technology and Application, 2015, 1(187).
[3] Kim Sushi K S. Park K,An application of support vector machines for customer chum analysis:credit card case[C]//Proeeoding of the 1 st International Conference on Natural Computation.Berlin, German:Springer,2005:636·647.
[4] Hung S Y,Yen D C,Wang HY.Applying data mining to telecom churn management[J].Export Systems with Appli·cations,2006,31(3):515-524.

[5] Hongxia Wang, Gencai Chen. Data Mining Techniques Applied in Retailing's CRM [J]. Computer and Digital Engineering, 2006(1).
[6] Quinlan J R.C4.5: Programs for machine learning [M]. SanMateo,Cali,f:Morgan Kaufmann, 1993:17—42.
[7] I Chiu, H G Chang, H C Hwang, M C Liaw, S L Hung, D Chen, C Davic. A neural network evaluation model for EKP performance from SCM perspective to enhance enters pries competitive advantage [J]. Expert Systems with Applications, 2008, 35:1809-1816.
[8] Bayard R, Agrawal R Mining the most interesting rules.In: Proc of the ACM S1GKDD Conf on knowledge Discovery and Data rill. Ning San Diego,CA USA,1999.145~154.
[9] MeiZhen Zhu. To Promote Customer Relationship Management Data Mining [J]. Technology and Innovation Management, 2006(4).
[10] Song Zhou. Application of data mining in Journal customer relationship management [J].Science-Technology & Publication, 2014, 3(29-31).
[11] Breman.L, Friedman.J, Olshen.R.et a1.Classification and Regression Trees [J].Monterey, CA.Wadsworth Intimation Group.1994:3-9.
[12] Liuyin Guo, Zhi Fang. Application of Data Mining on Retaining Customers in Telecom Industry [J]. China New Telecommunications, 2005(11).
[13] Zhi Gang Li. Research on Application of association rule data mining in customer relationship management [J]. Chinese information circles, 2012, 10(38-40).
[14] Sheng Tian. The Application of Data Warehouse and Data Mining in CRM [J]. Journal of Information, 2005, 24(9). P18-20.
[15] GU Guifang LI Wenyuan. Research on the application of data mining to customer relationship management in the mobile telecommunication industry[J].Science and Technology Management Research.2007,2(38-41).

Novel Search Schemes for Distributed Cooperative Data Centers

Liang Dong[1,a] and Xiaolu Zhang[2,b]

[1]Yunnan observatory, Chinese academy of science, Kunming, China
[2]School of Information Science and Engineering, Yunnan University, Kunming, China
Email: [a]dongliang@ynao.ac.cn, [b]zxl@ynu.edu.cn

Abstract. Distributed cooperative data centers provide a new data storage and data processing architecture over heterogeneous physical nodes. The major challenge faced by such architecture is to find the right data on some node efficiently. In this paper, we present two alternatives for data indexing based on Chord, a Peer to Peer (P2P) overlay structure, which redefine the routing table structure that underlies the Chord and achieve small search paths. First, we present a regional search algorithm that routes data keys queries by region and super-node information. The search process can be easily implemented via O(log K) hops, while maintaining O(log K) routing information with K regions in the Chord ring at each node. We further propose a Two-hop search based on the regional search scheme which aims to reduce the average search paths to a constant with O(log K) routing states about super-nodes. Results from theoretical analysis and simulations show that our improved routing algorithms can achieve higher search efficiency and the improved membership maintenances can keep routing information sufficiently up-to-date to validate higher search successful rate.

1. Introduction

With rapid increasing of big data, distributed cooperative data centers [1] have been the hotpot of current researched. Heterogeneous and variable data sources from users, companies, sensors and objects are widely distributed among data centers. Owing to the great volume and distribution of data as well as concurrent access of massive users, data indexing in data centers presents many new challenges: (1) the great volume of index data requires large-scale distributed storage for index data. Therefore, an efficient search scheme is needed to accurately locate storage nodes and rapidly find out index data; (2) concurrent access of massive users does not allow performance bottleneck in the search process, so as to guarantee high efficiency and concurrence of the search operation. However, the traditional indexing methods are not efficient for the distributed cooperative data centers due to the high computing complexity. For designing a good data search scheme is necessary. P2P overlay mechanism is usually considered as a technology in which data or programs are stored across a large number of nodes in self-organizing way is more successful than centralized approaches. The basic advantage over other approaches is that each node has equal opportunity to provide services with equal capabilities and roles. Chord protocol [2] as one of the popular structured P2P overlays provides an efficient search speed with less routing information, and has received many attentions to its good scalability, robustness and high search accuracy in distributed system. Yin et al. [3] researched on DHT-based Chord to address the load balancing of distributed file system in cloud computing; Kim et al. [4] presented DChord search scheme which does not depend on the randomness assumption to achieve its performance and deploys it to the Ubiquitous Sensor Network; Ding et al. [5] proposed an efficient quad-tree QT-Chord structure for global index of cloud data management; Zou et al. [6] improved Chord to support a multi-tenant indexing mechanism in clouds.

In a N-node Chord system, each node need maintain $O(\log N)$ other nodes information and $O(\log 2N)$ messages to guarantee the correct routing as nodes join and leave the system, and the average search paths is 1/2 $O(\log N)$. Maintaining small routing tables helps keep the amount of messages required to deal with membership change small. However, this design pays a high price for having low search efficiency since each search needs contacting several nodes in sequence which consumes lots of bandwidth to receive neighbor nodes information. And another deficiency is that the Chord depends on the some certain nodes' routing information without considering the load ability and diversity of nodes.

Chord has been extensively studied over the last several years. In particular, only a small number of algorithms have been proposed to provide efficient routing when the system meets churning and the membership is changing rapidly. Gupta et al. [7-8] describes a one hop routing scheme that every node maintains a full routing table containing information about every other nodes in the overlay. This approach reduces search latency to $O(1)$ hop but increases the bandwidth requirements since the messages of membership change. When nodes join and leave, the scheme need disseminate information about membership changes quickly enough so that nodes maintain accurate routing tables with complete membership information. The Kelips system [9] resolves the data search with $O(1)$ hop which consists of K virtual affinity groups. Each node uses $O(\sqrt{N})$ space to store other nodes information of the same affinity group and K-1 nodes from other K-1 foreign affinity groups. It uses gossip scheme to update routing information. The scheme reduces the search paths by increasing memory usage and complexity of communication overheads. Bidirectional routing scheme [10] focuses on constructing two routing tables in dual directions, clockwise and anti-clockwise. The average search paths can approach 1/3 $O(\log N)$, but the space of routing table is twice than Chord's.

Most obviously, these schemes rely on increasing space for storing other node information to reduce the search paths. In a highly dynamic system, to maintain correct routing process, every node must send information to other nodes to make sure they are still alive, and replace them if not. Above methods make tradeoffs among the amount of storage overhead at each node, the search hops, and the maintenance cost of member change. In this paper, we propose two improved algorithms based on Chord to achieve the high search performance with low maintenance overhead. The regional search algorithm is to design a faster search $O(\log K)$ with $O(\log K)$ routing information, where K is the number of regions in the Chord ring .The Two-hop search is aimed to reduce the search paths to a constant hop, while not increasing routing table size. Routing information per node need be corrected in order for the two algorithms to guarantee correct routing of queries. Thus, the Membership maintenances are designed to maintain accurate routing information when nodes join and leave arbitrarily.

2. Overview of Chord

Chord [2] uses consistent hashing functions such as SHA-I to distribute data keys over uniform nodes responsible for them in a 2^m identifier ring. Every node stores information of its predecessor and successor on the identifier ring. Furthermore, it maintains a routing table called finger table with m other nodes entries. The definition of the finger table is shown in Table I (adapted from [2]).

Table 1 Definition of variables for node n

Notation	Definition
finger[i].start	$(n+2^{i-1})\text{mod}2^m$, $1\leq i\leq m$
.interval	[finger[i].start ,finger[i+1].start)
.node	first node$\geq n$.finger[i].start
successor	the next node in the Chord ring; finger[1].node
predecessor	the previous node in the Chord ring

Chord performs search in $O(\log N)$ hops, where N is the number of nodes in the system. A node forwards a query for a given key to the node in its finger table with the closest identifier less than k. An example is presented in Figure 1(adapted from [1]), whereby node 3 performs a search for key 1. Node

3 invokes the find_successor operation for this key, by visiting node 0, which eventually returns the successor of that key, i.e., node 1.

In order to guarantee the correctness of search and low latency for location, the Chord protocol uses a stabilization scheme running periodically to update finger tables and successor nodes in a dynamic environment.

In Chord, the structure of finger table ensures that node can always forward the query at least half of the remaining identifier space distance to *k*, which accelerates the search process. In fact, each node only maintains a small amount of other nodes, and knows little about far-distant nodes. In this case, Chord node can't find key's successor immediately, thus, routing query is transmitted among some intermediate nodes. On the other hand, nodes are distributed unevenly in the interval. The hash function balances the load, but the different sizes of intervals disrupt it. When variable *i* is small, none or very few nodes distribute in the interval. It will double as variable *i* increases. Note that the search process depends on the finger table; therefore, we introduce two improved search algorithms based on super-node and region information aim to achieve high search efficiency while not increasing the size of routing table.

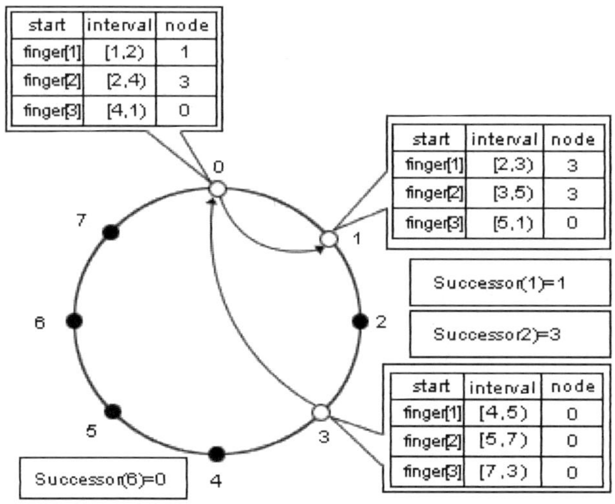

Figure 1 Finger tables and search process for key 1

3. Regional search

This section presents the design of our novel algorithm. The Chord ring is divided into *K* equal-sized regions, and every region selects a super-node based on its capabilities (i.e., lower latency, and higher bandwidth etc.). We will inherit the predecessor and successor list of Chord and add region information to routing table which maintains only about $O(\log K)$ other nodes, called the Rfinger table. The Rfinger table and super-node will help require only $O(\log K)$ search paths. When nodes join and leave, the routing table must be in an up-to-date state, so we outline a membership maintenance mechanism to ensure the correctness of search.

3.1 Reconstructing routing table

We divide a 2^m identifier ring into 2^d equal regions, let *d* be the number of bits of region identifiers. Assuming S_i is the super-node of the ith region, S_i is responsible for maintaining all nodes information in the ith region. The following definitions are applied in reconstructing routing table.

Definition 1: r_i is the region pointed by the ith entry in *n*'s Rfinger table.

$$r_i = (k + 2^{i-1}) \bmod 2^d, \quad 1 \le i \le d$$

where *k* is the region of *n* belonging to.

Definition 2: g[*i*] is the space of region r_i in the ring.

$$g[i] = [r_i \times \frac{2^m}{2^d}, (r_i + 1) \times \frac{2^m}{2^d}) \quad , 0 \le ri \le 2d$$

Definition 3: P_i is the next hop node pointed by the ith entry in n's Rfinger table,

$$P_i = (n + 2^{i-1} \times \frac{2^m}{2^d}) \bmod 2^m \quad , 1 \le i \le d,$$

Let G_i be the set of all nodes in g[i] region, and S_i be the super-node of g[i]. If $P_i \in G_i$, n.Rfinger[i].node=P_i, or $P_i \notin G_i$, n.Rfinger[i].node= S_i.

The Rfinger table of node 1 is shown in Figure 2, which node 1 belongs to region 0. Another case shown in Figure 3, node 33 is not in region 4, so we choose the super-node S35 of region 4 as Rfinger[3].node.

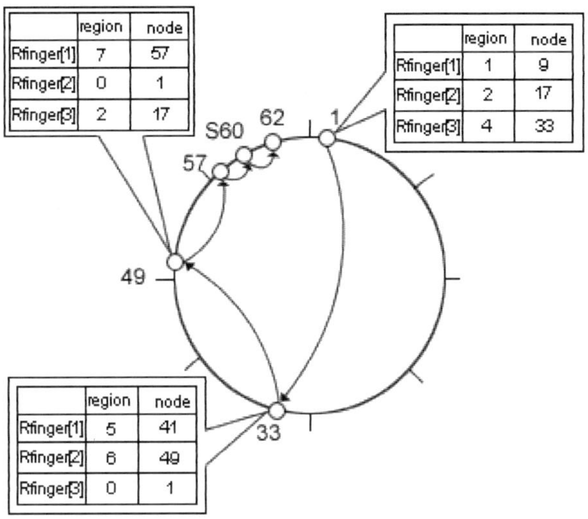

Figure 2 Rfinger tables and search process without Super-node

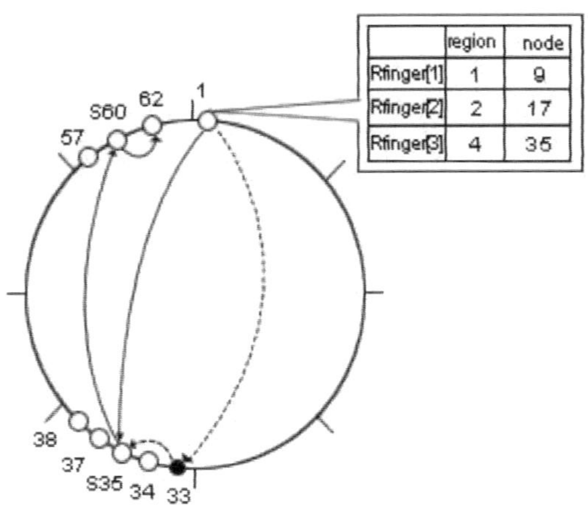

Figure 3 Rfinger tables and search process with Super-node

Apart from the Rfinger table, each super-node also maintains a super table including full super-nodes information of other regions mainly including variable region, ID identifier and IP address, and a region table containing other ordinary nodes of its region including variable ID identifier and IP address. Examples are shown in Table II and III.

Table 2 The super table of S60

Region	ID	IP
0	S2	IP2
1	S10	IP10
2	S17	IP17
3	S30	IP30
4	S35	IP35
5	S41	IP41
6	S50	IP50
7	S60	IP60

Table 3 The region table of S60

ID	IP
57	IP57
62	IP62

3.2 Routing algorithm

We can find that the Rfigner table increases the distance of search paths while not reducing the routing information. Since every super-node contains a few nodes routing information to perform fast routing. The pseudo-code of search process is described as a function *n*.find_nexthop(*id*) in Figure 4, where *n* is the node to invoke a search query, *id* is the target identifier. The parameters we use in our algorithm as follows:

- 2^m is the Chord identifier space.
- 2^d is the number of regions.
- *k* is the region which node *n* belongs to.
- S_k is the super-node of region *k*.
- *r* is the region which *id* belongs to.
- S_r is the super-node of region *r*.
- 2^m is the Chord identifier space.

Table 4 The super table of S35

Region	ID	IP
0	S2	IP2
1	S10	IP10
2	S17	IP17
3	S30	IP30
4	S35	IP35
5	S41	IP41
6	S50	IP50
7	S60	IP60

Figure 2 illustrates an example of search process on an identifier ring with *m*=6, *d*=3. Suppose node 1 wants to locate the identifier 62. It first checks whether the target and itself are in the same region, if not, it will send a request to node 33 whose identifier is closest to the target in its Rfinger table. Similarly, the request is transmitted in the Chord ring until it arrives at node 57 which is in the same region 7 with the target. Then node 57 sends a request to super-node S60. In the end, S60 finds the target in its region table. Figure 3 shows another search condition with the help of super-node. Node 1 sends a request to super-node S35 through its Rfinger table. S35 searches its super table (Table IV) for the super-node S60 of region 7. S60 will check its region table to return the target. This faster search process will be presented as a further improved algorithm in section 5.

```
n.find_nexthop(id);
//if node n is the target id
if n=id then
  return n
end if
//if node n is the super-node of region r
if n= S_r then
  find (id )in n's region table
  return id
end if
else // if id and node n are in the same region r,
// but node n is not the super-node of region r,
if n≠S_r and k=r then
  forward ( query (id), S_r)
  // forward query(id) to S_r
  return S_r
end if
else
 // if node n is the super-node of region k,
 //  but id and node n are not in the same region
if n= S_k and k≠r then
  for i=2^d downto i=1 do
    if i=r then
      forward ( query (id), n.supertable[i].id)
      return n.supertable[i].id
    end if
  end for
end if
else // if above conditions are not satisfied
  for j=d downto j=1 do
    if Rfinger[j].node is closest to id then
        forward ( query (id), n.Rfinger[j].node)
        return n.Rfinger[j].node
     end if
   end for
  end if
```

3.3 Membership maintenance

When nodes join or leave, we must address the question that change information should be disseminated to all nodes in the Chord ring quickly so that nodes can maintain accurate routing tables to ensure the correctness and efficiency of search.

We imitate the process of One-hop [5] to build our membership maintenance mechanism. Every node sends *keep-alive* messages to its successor and predecessor nodes periodically. If either successor or predecessor does not respond, it decides to send a membership change message to the super-node of its region to *notify* the invalid node. After t_{change} time, the super-node sends the change information about invalid nodes which are collected in last t_{change} time to all other super-nodes. Change messages received by a super-node and a notify message are bound and forwarded to all nodes of its region after t_{wait} time. In this hierarchical way, every node in the ring receives the change information within t_{change} + t_{wait} time.

IOP Conf. Series: Journal of Physics: Conf. Series **910** (2017) 012022 doi:10.1088/1742-6596/910/1/012022

3.4 Bandwidth analysis

This improved algorithm performance depends on the number of regions K. Consider above membership maintenance, we analyze the bandwidth use is mostly consumed on the change message data both super-nodes and ordinary nodes. In our system, the upstream bandwidth of super-node is the dominating and limiting factor. We will compute the value for K in a way that reduces the bandwidth utilization of super-node. The m bytes will be described the change event, and the overhead per message (or the size of acknowledged message) will be p. r is the excepted rate of membership changes in this system. The excepted number of events per second in a region is r/K. When a node detects a change event, it sends a notification to its region super-node. Therefore, the downstream bandwidth utilization of super-node is

$$\frac{r \times (m + p)}{K}.$$

Since each message must be acknowledged, the upstream utilization has

$$\frac{r \times p}{K}.$$

In the region, the size of change message sent from one super-node to another batches together events that occurred in the last t_{change} seconds is

$$\frac{r}{K} \times t_{change} \times m + p .$$

During t_{change} period, a super-node sends this message to all other super-nodes (K-1), and receives an acknowledgement from each of them. Therefore, the upstream bandwidth comes out to

$$(\frac{r \times m}{K} + \frac{2 \times p}{t_{change}}) \times (K - 1) .$$

After t_{wait}, a super-node forwards a *notify* message to all ordinary nodes, the aggregate message size becomes

$$(r \times m + \frac{p}{t_{wait}}) \times \frac{n}{K}.$$

The total upstream bandwidth on super-node is

$$\frac{r}{K} \times (m \times n + p) + r \times m + \frac{2 \times p \times K}{t_{change}} + \frac{p \times n}{t_{wait} \times K}.$$

By analyzing this function, we deduce the minimum where K is

$$K = \sqrt{(\frac{r \times m \times n}{2p} + \frac{n}{2t_{wait}} + \frac{r}{2}) \times t_{change}} .$$

This formula allows us to calculate the reasonable number of regions.

4. Algorithm analysis

Chord protocol is a binary search process via $O(\log N)$ hops. Since the distance will halve by finding the closest node to the key from every node's finger table iteratively. We use region information and super-node to locate the target rather than route by node's successor. Our analysis follows a theorem.

Theorem 1. The average search paths of the regional search algorithm based on super-node is

$$\text{Laverage} = \frac{1}{2} \log \text{K} + 2$$

where K is the regions in the ring.
Proof: The regional search algorithm based on super-node uses region information to locate key. The number of forwarding queries between regions distributes in the interval [1, log K]. The search probability of every region is equal. Therefore, the expectation of the average search paths follows:

$$\text{Laverage}=\sum \text{PiCi}=\frac{1}{K}\sum_{i=1}^{\log K} i\times\binom{\log K}{i}.$$

When a routing node finds itself in the same region with the target identifier, it sends a request to the super-node of its region. Given searching local information in region table, we can deduce the average search paths for the following value:

$$\text{Laverage}=\frac{1}{2}\log \text{K}+2.$$

5. Two-hop search

The improved regional search scheme presents a faster search $O(\log K)$ with maintaining an Rfinger table at each node. In this section, we propose a Two-hop search scheme. This scheme routes in two hops on average with $O(\log K)$ routing states on each node whose entry points to the super-node of region k.

5.1 System design

Our design is based on a structure like that used in improved regional search. The ring is split into K equal-sized regions and each region is assigned to one super-node. Every node is required to store $O(\log K)$ routing information. The next hop pointed by the ith entry is redefined to super-node Si belonging to i^{th} region, n.Rfinger[i].node= S_i, $1\leq i\leq d$. Every super-node also keeps routing information for all other super-nodes. Figure 5 gives an example to show the Rfinger tables of node 1 and super-node 35.

5.2 Query routing

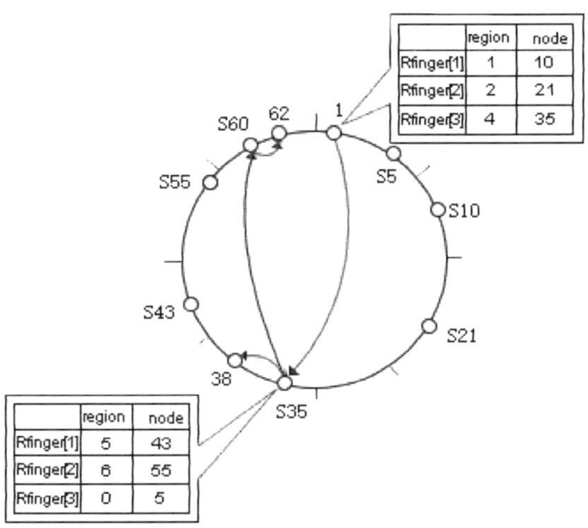

Figure 4 Rfinger tables and Two-hop search process

When a node wants to query the target key, it first checks whether the key and itself are in the same region, then sends this request directly to the super-node of the region. Otherwise, it finds the Rfinger table and forwards the request to next hop super-node whose region table contains the key. In the worst

IOP Conf. Series: Journal of Physics: Conf. Series **910** (2017) 012022 doi:10.1088/1742-6596/910/1/012022

case, it chooses the super-node whose identifier is closest to the target in the Rfinger table. In response, the chosen super-node examines its super table and forwards the request to the super-node responsible for the key. For example, in Figure 5, node 1 requests the target identifier 38(belongs to region 4). It searches its Rfinger table to identify the super-node S35 which belongs to region 4, then, forwards the request to S35, who in turns looks up its region table to find the target.

5.3 Membership maintenance

The mechanism addressing the membership change (i.e., when node joins or leaves) is similar to the improved regional search. The difference occurs in what every node routing table only contains super-nodes information. We optimize the Two-hop membership maintenance for ordinary node and super-node, respectively. The goal is to propagate change information fast and reduce bandwidth consumption. Every node sends *keep-alive* message to its successor and predecessor nodes periodically. If an ordinary node fails, the scheme only sends change message to the super-node of its region and update the region table. In our scheme, every super-node also sends *keep-alive* message to all other super-nodes periodically. If any super-node does not respond, it sends a change message to all other super-nodes. After receiving the message, a super-node will forward it with an attached *notify* information to all nodes in its region.

6. Simulation results

In this section, we simulate the regional search and Two-hop algorithms on P2PSim [21]. P2PSim is a free, multi-threaded and discrete event simulator which supports several P2P protocols, for instance, Chord, One-hop, Kelips, etc. We reconstruct the routing table and modify routing algorithm on Chord protocol to evaluate our improved algorithms on P2PSim with specific topologies, and make a comparison with Chord. We simulate the experiments with 128, 256, 512, 1024 nodes generated by E2Egraph Kingdatas with 8, 16, 32, 64 regions, respectively. Nodes are uniformly distributed in a 2^{16} identifier space we preset. In the simulation, we use the middle node of a region as a special super-node. When a super-node fails, its successor which has a copy of its routing information soon detects and becomes new super-node. During the simulation, requests are generated by nodes in the P2PSim for randomly chosen keys with some nodes joining or leaving the overlay at random time.

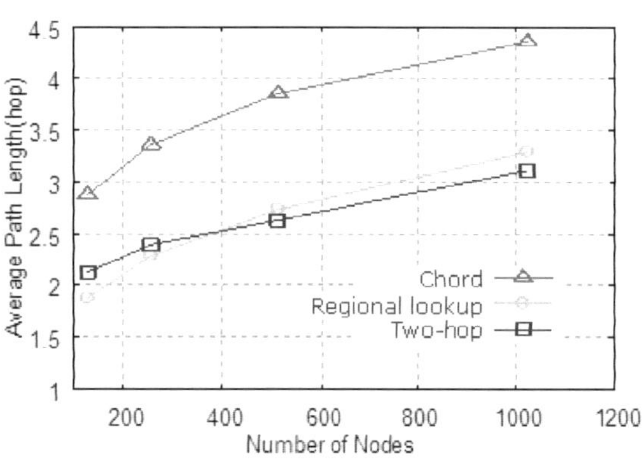

Figure 5 The average search paths

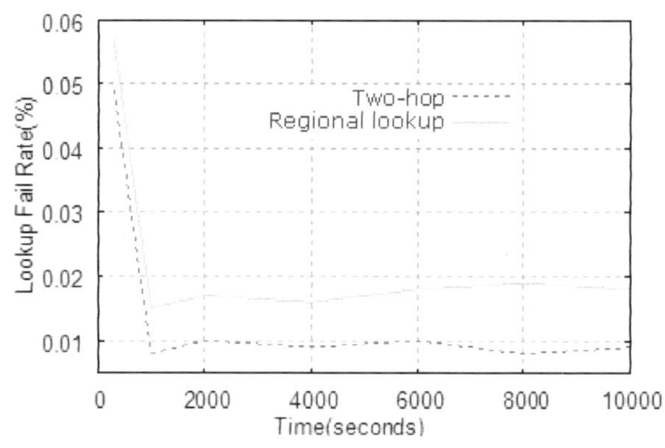

Figure 6 The search failure rate

Figure 6 shows the average search paths. We can see that the two improved algorithms significantly shorten the search paths than chord. And the average search paths in Two-hop are correctly completed by two or three hops. We also simulate the experiment with 1024 nodes to determine the failure rate. The result is shown in Figure 7. In the first 500 seconds, most of new nodes join rapidly. After approximately the first 1000 seconds, the query failure rate stayed steadily. Apparently, the failure rate of Two-hop search is lower than the regional search. This happens because Two-hop search only fails on routing table entries whose information is lost in super-node failure. In our simulation, we further examine the bandwidth consumption of super-node during query and membership maintenance. Figure

8 indicates that the bandwidth use of Two-hop search is lower due to the optimized membership maintenance for two kinds of nodes (ordinary node and super-node).

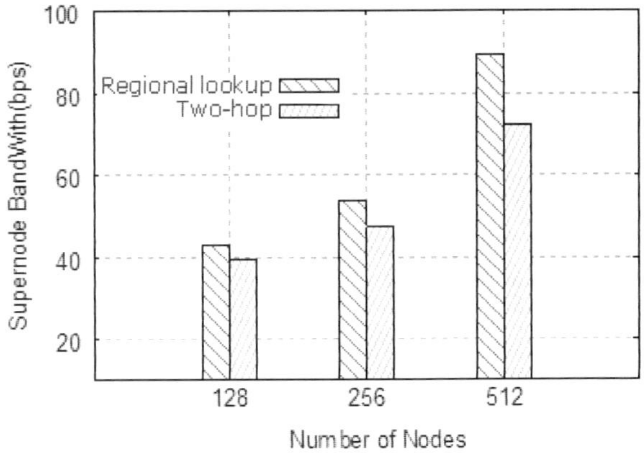

Figure 7 The bandwidth utilization of per super-node

7. Conclusion

This paper presents two novel search algorithms based on Chord for the distributed cooperative data centers. We argue that existing improved algorithms are inefficient and inadequate for reducing search hops by increasing routing table size and maintaining more nodes information. In contrast to existing methods, we design our algorithms based on region information and super-node. In a N-node overlay network and K-region of Chord ring, each node maintains O (log K) other nodes information, and each super-node is responsible for maintaining all other super-nodes routing states. We also present two improved maintenances to disseminate information about membership changes quickly enough so that nodes maintain correct routing tables. Our theoretical analysis and simulation results confirm that these improvements can achieve lower search failure rate and provide more efficient search than Chord, which route O(log K) hops in the regional search algorithm and a constant search in Two-hop, respectively. We believe that the key contribution of this paper is to serve as a reference of the distributed cooperative data centers in which the Chord topological structure and the improved search algorithms can benefit the indexing mechanism.

8. Acknowledgment

This work is sponsored by Youth Fund of National Natural Science Foundation of China (11303094), National Natural Science Foundation Astronomy Joint Fund of China (U1431113), Applied Basic Research Program and Project of Yunnan Province of China (2015FB189), Western Light A Class Talent Program of China, and the Scientific Research Foundation of the Educational Department of Yunnan Province (No. 2015J0007).

9. References

[1] R. Basmadjian, P. Bouvry, G.D. Costa, et al, "Green data centers," Large-scale Distributed Systems and Energy Efficiency: A Holistic View. John Wiley & Sons. P, 2015, pp. 159-196.
[2] I. Stoica, R. Morris, D. Karger, et al, "Chord: A Scalable Peer-to-Peer Search Service for Internet Applications," ACM SIGCOMM Computer Communication Review, vol. 31(4), pp. 149-160, 2001.
[3] X. Yin, J. Yang, C. Qu. "Research on Load Balancing of Distributed File System in Cloud Computing," Computer Science, vol. 41(3), pp. 141-144, 2014.
[4] C.S. Kim, S. Lee, J.I. Han, et al, "DChord: An Efficient and Robust Peer to Peer Search System," Malaysian Journal of Computer Science, vol. 23(1), pp. 37, 2010.
[5] L.L. Ding, B.Y. Qiao, G.R. Wang, et al, "An efficient quad-tree based index structure for cloud data management," In Proceedings of the 12th Conference on Web-Age Information Management, Berlin: Springer-Verlag, pp. 238-250, 2010.

[6] L.D. Zou, Q.Z. Li, L.J. Kong, et al, "Indexing Mechanism of Multi-Tenant Database Based on Chord," Chinese journal of Computers, vol. 38(56), 2015.
[7] A. Gupta, B. Liskov, R. Rodrigues, "One Hop Search for Peer–to-Peer Overlays," In Proceedings of the 9th Workshop on Hot Topics in Operating Systems, pp. 7-12, 2003.
[8] A. Gupta, B. Liskov, R. Rodrigue, " Efficient Routing for Peer-to-Peer Overlays," In Proceedings of the 1st Symposium on Networked Systems Design and Implementation,vol. 4, pp. 9-9, 2004.
[9] I. Gupta, K. Birman, P. Linga, et al, "Kelips: Building An Efficient and Stable P2P DHT through Increased Memory and Background Overhead," Peer-to-Peer Systems II, Springer Berlin Heidelberg, pp. 160-169, 2003.
[10] P. Ganesan, G.S. Manku, "Optimal Routing in Chord," In Proceedings of the 15th Annual ACM-SIAM Symposium on Discrete Algorithms, Society for Industrial and Applied Mathematics, pp. 175-185, 2004.
[11] G. Cordasco, A. Sala, "2-Chord Halved," In Proceedings of the 2nd International Workshop on Hot Topics in Peer-to-Peer Systems, San Diego: IEEE Computer Press, pp. 72-79, 2005.
[12] D. Sun, Y. Wang, "Improvement of Chord routing algorithm," Mechanical & Electrical Engineering Magazine, vol. 12, pp. 30, 2007.
[13] B.Q. Wang, P. He, "H-chord: Routing Model and Algorithm Implementation for Chord Based on Hierarchical Dividing," Computer Engineering and Applications, vol. 42(36), pp. 141-143, 2007.
[14] G. Cordasco, L. Gargano, M. Hammar, et al, "F-Chord: Improved Uniform Routing on Chord," Networks (Wiley InterScience), vol. 52(4), pp. 325-332, 2008.
[15] Y.J. Joung, J.C Wang, "Chord2: A Two-layer Chord for Reducing Maintenance Overhead via Heterogeneity," Computer Networks, vol. 51(3), pp. 712-731, 2007.
[16] E.J.L. Lu, Y.F. Huang, S.C. Lu, "ML-Chord: A Multi-layered P2P Resource Sharing Model," Journal of Network and Computer Applications, vol. 32 (3), pp. 578-588, 2009.
[17] X. Zheng, V. Oleshchuk, "Improving Chord search protocol for P2PSIP-based communication systems," In Proceedings of the 9th International Conference on New Trends in Information and Service Science, pp. 1309-1314, 2009.
[18] R. Cuevas, M. Uruena, A. Banchs, "Routing Fairness in Chord: Analysis and Enhancement," In Proceedings of the IEEE INFOCOM , pp. 1449-1457, 2009.
[19] G. Chiola, M. Ribaudo, "Neighbor-of-neighbor Routing over Deterministically Modulated Chord-like DHTs," In Proceedings of the International Parallel & Distributed processing symposium, pp. 1-8, 2008.
[20] O. Babaoglu, M. Marzolla, M. Tamburini, "Design and Implementation of a P2P Cloud System," In Proceedings of the 27th Annual ACM Symposium on Applied Computing, pp. 412-417, 2012.
[21] P2PSim: A simulator for peer-to-peer (P2P) protocols[EB/OL]. Available at : http://pdos.csail.mit.edu/p2psim/.

IOP Conf. Series: Journal of Physics: Conf. Series **910** (2017) 012023 doi:10.1088/1742-6596/910/1/012023

Research on the Application of Persona in Book Recommendation System

Baozhong Gao[1,a] and Shouyan Du[2,b,*], Xinzhi Li[3,c,*] and Fangai Liu[1,*]

[1]Institute of Information Science and Technology, Shandong Normal University, Jinan, China
[2]Super-computing Center, Computer Network Information Center, Chinese Academy of Science, Beijing, China
[3]School of Information Science and Engineering, Yunnan University, Kunming, China
Email: [a]gaobaozhong@sdnu.edu.cn, [b]dushouyan17@mails.ucas.ac.cn, [c]lixinzhikm@outlook.com

Abstract. Currently, there still exists a host of problems in the book recommendation system, such as low accuracy, weak correlation and poor pertinence. Aiming to unravel these problems, this paper based on the theory of big data and data mining technology, through analyzing internet user behavior and the "5C" model of personal credit evaluation, combined with joint impact weight calculation method, which involves user grade, borrowing credit, book friend recommendation degree, book friend recommended adoption degree, borrowing frequency, borrowing number, and borrowing time interval. User activity and credit are also taken into account in the process of establishing user tagging system so as to build classified book recommendation service. This method is of universal meaning to the book recommendation service of smart campus with user as the core under big data environment.

1 Introduction

Persona is first proposed by the father of interactive design Alan Cooper, he believes that persona is the virtual symbol of real user, which is a target user model based on a series of real data. According to the combination of its user attributes and behavioral data and the combination of dynamic tags and multi-level tags, it can better identify the user's community characteristics and thus customize service for the user. The intelligent recommendation service based on the "persona" has obtained a wide range of applications in the Amazon, Ali, Jingdong and other electricity and information services business [1].

Li Bing explored the characteristics of cigarette retailers based on big data technology and K-means algorithm, realizing the intelligent recommendation for the order of retailers. Wu Mingli and Yang Shuangliang analyze the behavior data of mobile users, combining the two dimensions of time and space and with the use of spark cluster, solving the problems of push service accuracy and large amount of user data [2]. In the space, the book recommendation areas focus on different from other areas of persona, so it's need for their own characteristics of the corresponding design of the persona, in order to achieve accurate recommendation [3]. In this paper, we use spyder integrated development environment and the data of user behavior real-time transmitted to the system data users, through the user multi-dimensional comprehensive analysis to realize the establishment of integral values, and constantly optimize and update the persona of system, in order to achieve the content pushed to the user meet the highest degree of customer demand and interest preferences of the content, and stimulates the user's interest in reading.

2 Big Data Technology and Credit Evaluation

2.1 Big Data Technology

At present, "big data" does not form a recognized definition. National Institute of Standards and Technology NIST's defines big data as data capacity, data acquisition speed or data representation limits the data analysis and processing capabilities of using traditional relational methods, thus need to use the mechanism of horizontal expansion to improve processing efficiency [4]. It is now generally considered that "big data" contains massive structured, semi-structured and unstructured data, when compared with traditional relational databases [5], big data-related techniques have a greater advantage in processing speed and storage capacity when dealing with semi-structured and unstructured data. Common big data processing models usually include data acquisition, data preprocessing, data and text mining and data visualization four modules [6].

2.2 Credit Analysis and Rating

Personal credit investigation, refers to the credit and related institutions based on the agreement to collect, process, and store information scattered in different institutions about the user credit, and the formation of credit data provide a reasonable basis of the user's credit situation for a business activity. Presently, the international financial institutions assess customer credit mainly use the "5C" model expert analysis method to do risk analysis for users, accessing user's ability to repay and willing according to the character, capacity, capital, collateral, and condition five aspects, which can more accurately reflect the user's credit performance in a certain way to determine the degree of risk [7]. In the application of the library book recommendation system based on the persona, we make use of the data mining technology to obtain the user's data from the intelligent campus system and the library system, analyzing the subjective and objective factors that affect the user's credit level, then, the user rating model quantifies the influencing factors with specific scores and synthesizes their performance to form data feedback, thus building a more accurate multi-dimensional persona.

3 Build User Rating System

3.1 Data Collection

Data is the core of the user's interest tag. This project uses data mining technology to collect the relevant data of user from library and intelligent campus system. The data source of persona can be divided into explicit data and hidden data two aspects. As shown in Table 1. The explicit data includes user attributes and library book data, which are relatively stable. For this kind of information, recommendation system does more work as data cleansing without the need of modeling predictions. The hidden data is divided into two categories, namely, dynamic user data and platform data, which are hidden in the user random internet behavior, and need to mine hidden information and extract the information ontology through data analysis and other techniques.

Table 1 System data information

data	type	description
Explicit data	User attributes	Including the user (graduate, teacher, undergraduate) ID number, name, gender code, department code.
	Library Book Data	Including the library book barcode, book name, reader barcode, borrow date, return date, renewal logo.
Hidden data	Dynamic user data	Including the user search book records, collection of book records, borrowed records.
	Platform data	Data obtained from the user and the platform for book interaction.

3.2 User Behavior Image Dimension

According to the principle of presenting similarity of the characteristics within the user group, showing difference between the user groups to achieve the users grouped. This paper analyzes the user behavior, based on the collected and mining data, from different dimensions adding hierarchical organization of the tag for user, in order to make the establishment of the user's tag library, thus achieving the user's behavior, attribute tag. The persona based on our smart campus users mainly include as shown in Table 2.

Table 2 Persona Dimension

type	description
Basic information	Including the users (graduate students, teachers, undergraduates) ID number, name, age, gender, faculties, hobbies, etc.
Borrowing habits	Including the user's borrowing period, borrowing frequency, borrowing time, borrowing time interval, the influencing factors of borrowing books and credit, etc.
Access preferences	According to the number of user access different types of books in the network to determine the user preferences, this persona is portrayed on a monthly basis.
Search Information	Describe the condition of user using the library system search engine to retrieve the content, which can more accurately reflect user's points of interest and needs than the access preferences.
Other features	Including the user active situation, user professional classification, etc.

3.3 User Credit Evaluation Index

This paper is based on the dimension of 5C method when designing the index system of user credit evaluation in colleges, through the basic information, borrowing habits, accessing preferences, search information and other characteristics five dimensions, combining with the service of book recommendation system and obtained data for user credit evaluation, and the credit points determine how much the recommended function and the number and frequency of books the user can obtained. Index selection basis and score description of impact are shown as follows:

- User grade: Low grade users due to have less courses, and more adequate spare time, thus having rather high willingness to borrow. With the increase in user grade, the borrowing willingness becoming lower to some extent, but the system will increase the proportion of recommended professional or competition books.
- Borrow credit: Determined by the user whether is scheduled to return the borrowed books, and the longer the period expires, the lower credit score user has.
- Book friends recommended degree: In book recommendation system, the user with similar interests can recommend books to other users after reading some feeling good, and the higher the recommended number of users, the higher the corresponding score.
- Book friends recommended adoption degree: On the basis of mutual recommendation between book friends, combining with the number of other users who adopted the books to determine, and the higher number of users adopt that book, the higher scores the recommended users will enjoy.
- Borrowing frequency: The higher frequency of borrowing users on behalf of the higher demand for borrowing.
- Borrowing number: The number of borrow book is an indirect reflection of the user's borrowing needs, the more books borrowed, the higher user borrowing needs.
- Borrowing time interval: For some users, they no longer borrow books for a long time after one time, whereas some other users borrow time interval represent as coherent cyclical, showing a higher borrowing demand.

3.4 User Characterization

User characterization is user tagging. The tag is a feature of highly refined hierarchical catalog form which is analyzed by user information. Based on the user's behavior data and a certain model algorithm to get the user's tag weight, combining with different dimensions of tag to form a scoring system to distinguish recommended books priority between users with different characteristics. The static ontology is relatively stable and the weight is set to 1. In this paper, the weight of persona tag is mainly considered by the user grade, borrowing credit, book friend recommendation degree, book friend recommended adoption degree, borrowing frequency, borrowing number, and borrowing time interval seven factors, and each tag weight calculates as following:

3.4.1 User Grade. User grade belongs to static ontology, but is need to combine this factor in the book recommendation system to recommend, and different grades of users book borrowing demand is also different. The user's grade on the weight of the body is represented as G(label), which is the length of time of user until graduation divided by the total length of four years, namely:

$$G(label) = \frac{T_{end}\text{-}T_i}{T_{end}\text{-}T_{start}} \tag{1}$$

3.4.2 Borrowing Credit. The user's borrowing credit mainly about after the user borrow books, whether is return the book on time, and borrowing credit on the weight of the body is represented as C(label), which is the length of time of user borrow books in system divided by the length of time until user return the book, namely:

$$C(label) = \frac{T_{bsystem\text{-}end}\text{-}T_{bsystem\text{-}start}}{T_{b\text{-}end}\text{-}T_{b\text{-}start}} \tag{2}$$

3.4.3 Book friend recommendation degree. The weight of the book friend's recommendation on the ontology R(label) refers to in the mutual recommendation module of the book recommendation system, the ratio of the number of books recommended by the user to other users to the total number of recommended books at the same period, namely:

$$R(label) = \frac{R_{label}}{\sum_{i=0}^{n} R_i} \tag{3}$$

3.4.4 Book Friend Recommended Adoption Degree. The recommendation of the degree of adoption on the ontology A(label) refers to in the mutual recommendation module of the book recommendation system, the proportion of the number of adopted books recommended by the user to other users divided by the total value of adopted recommended books at the same period, namely:

$$A(label) = \frac{A_{label}}{\sum_{i=0}^{n} A_i} \tag{4}$$

3.4.5 Borrowing Frequency. The borrowing frequency of user is used to describe the number of times of borrowing in unit time, which is portrayed on a monthly basis. Borrowing frequency on the weight of the body is represented as F(label), which is the ratio of the number of times the user borrows the book to the total number of borrowing in the system at the same period, namely:

$$F(label) = \frac{F_{label}}{\sum_{i=0}^{n} F_i} \tag{5}$$

3.4.6 Borrowing Number. According to the user's borrowing in the book recommendation system to do the borrowing list, in months as a unit, and the borrowing number of ontology N(label) is the ratio of the number of users borrowed to the total number of borrowed in system at the same period, namely:

$$N(label) = \frac{N_{label}}{\sum_{i=0}^{n} N_i} \tag{6}$$

3.4.7 User Average Borrowing Time Interval. The average borrowing time interval U(label) indicates the sum of time differences between the user's borrowing actions divided by the number of users borrowing except the same borrowing date, which is portrayed on a daily basis. The larger weight value represents longer time interval, m is the number of users borrowing except the same date, and the formula is:

$$U(label) = \frac{\sum_{i=0}^{n}(U_{i\text{-}start} - U_{i\text{-}lastend})}{m} \tag{7}$$

3.5 User Rating Model

In this paper, the multi-dimensional user tagging system based on the data information obtained from intelligent campus and library system can provide detailed data reference for user's rating evaluation. The persona score model is shown as below:

$$people = \sum_{i=1}^{n} label_i \, p(i)$$

$$= \sum_{i=1}^{n} label_i \{[G(label) + C(label) + R(label) +$$

$$A(label) + F(label) + N(label) + U(label)]\} \tag{8}$$

$$= \sum_{i=1}^{n} label_i \{[\frac{T_{end} - T_i}{T_{end} - T_{start}} + \frac{T_{bsystem\text{-}end} - T_{bsystem\text{-}start}}{T_{b\text{-}end} - T_{b\text{-}start}} +$$

$$\frac{R_{label}}{\sum_{i=0}^{n} R_i} + \frac{A_{label}}{\sum_{i=0}^{n} A_i} + \frac{F_{label}}{\sum_{i=0}^{n} F_i} + \frac{N_{label}}{\sum_{i=0}^{n} N_i} + \frac{\sum_{i=0}^{n}(U_{i\text{-}start} - U_{i\text{-}lastend})}{m}]\}$$

$label_i$ indicates the information tag of the user on a dimension, p(i) represents the weight of the tag, G(label)、C(label)、R(label)、A(label)、F(label)、N(label)、U(label) are corresponding to user grade, borrowing credit, book friend recommendation degree, book friend recommended adoption degree, borrowing frequency, borrowing number, borrowing time interval seven factors on the weight of the tag, m is the number of users borrowing except the same date. According to the weighting formula of the user's score model, the weight value of different labels on the ontology of each user is calculated, then the normalized score of the user is obtained and the result is stored in the MySQL database.

3.6 Persona Management

Using K-means clustering method to carry out persona management, combining with the statistical methods of cluster analysis, which can better explain and describe the user's behavioral characteristics. The goal of clustering is to synthesize the data into different classes or clusters according to the similarity of data without any prior knowledge, so that the elements within the cluster have high similarity, while the elements similarity between the cluster is low, so it's also known as non-supervised classification [8]. The clustering process is as follows:

1. The data is normalized to a range of [0,1] through linear function conversion method.
2. Set the cluster center point k = 8 to create the initial partition, that is, selecting 8 elements from unclassified collection U in random, as the center of eight clusters respectively.
3. The similarity of the remaining elements to 8 clusters is calculated separately, and the elements were classified into clusters with the highest similarity.
4. According to the clustering results of the previous round, the center of each eight clusters is recalculated, and the calculation method is to calculate the arithmetic mean of the respective dimensions of all the elements in the cluster.

5. Remove all elements of the new cluster center from the collection, then repeat the third step of the method and re-clustering.

6. Repeat steps 4 and 5 until dissimilarity between the results of this and previous round of clustering are less than the set threshold.

7. Output the final clustering results.

It adopts sliding window filtering algorithm with incremental update mechanism, allowing to achieve in the update operation, only updating the places that needs to be changed and will not repeated. It is required to calculate the user data and store the intermediate values on a monthly basis, and obtain the new persona in the way of association.

4 Experimental results and analysis

4.1. The Presentation of Persona

According to the weight formula, the user's information tag normalized to a range of [0,1], combining with the persona mathematical model for visualization processing, which can intuitively show the weight of user's priority in the recommendation system rated by user's different behavior. Figure 1 shows the corresponding persona weight value for the random selection of 200 users in the borrowing time interval and borrowing number two dimensions.

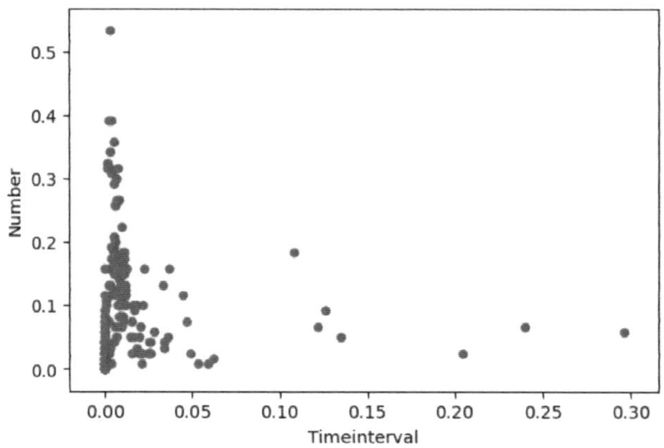

Figure 1 User borrowing time interval and user borrowing number Two-dimensional persona

Table 3 User group classification

User grouping			
Travel preferences class	Application preferences class	Financial preferences class	Anime class
Medical class	Education preferences class	History preferences class	Technology preferences class
Sports preferences class	Photography class	Legal preference class	Social science class
Digital preferences class	Gourmet preferences class	Environmental preferences class	Exam preferences class
Film and television preferences class	Game preferences class	Music preferences class	Car preferences class
Healthy health class	Fashion preferences class	Building preferences class	Electronic machinery class

According to the number of user scores, prioritizing core users, general users, and inactive users, evaluating the potential recommended value for each group. The user is divided into 24 groups as shown in table 3. According to the scoring system, it's divided into one to five level users in the range of [0,1] with a scale interval of 0.2, corresponding to the user's overall demand, activity and credit rating.

4.2. Application Implementation Function

The recommendation system is a collection of software tools and technologies that provides project recommendation information for users [9]. Based on the persona, the book recommendation system according to user's behavior and borrowing records to recommend, and dynamic update content with user's interest changes. The specific application functions are as follows:

Monthly recommendation: According to the user's rating level to determine the recommended number, recommending more books to the active (high demand) level users, while recommending relatively smaller number of books to the lower level (low demand) users. On a monthly basis, the system combining with the user's preferences and real-time book list gives a comprehensive recommendation. For new users who do not borrow books, the system refers to the user's college information and recommends the most popular books recently to them, in order to solve the new user's cold start problem.

Time-based recommendation: According to the user's basic behavior persona, at the beginning of the semester, the end of the semester and before the holiday the recommended system can based on the previous grade professional books borrowing list at the same period, and appropriate increasing in the proportion of recommended books; The system in accordance with the real-time mining of user interest preferences to recommend accordingly.

Book friends recommend: The users in the book recommendation system can recommend books to each other, and the attitude to books of recommended users can be used as one of the evaluation criteria for user interest preferences.

Association recommendation: Based on data mining technology, through the data association and correlation between the data mined from the collection, analyzing the user's borrowing behaviors, so as to achieve the association recommendation.

Others borrowed recommendation: Users with similar preferences can recommend each other's preferred books, making it easier for users to find their own interest books or groups.

5 Conclusion

The traditional library service fails to meet the demands of readers, while resource digitization as well as the establishment of user-based digital library have become development trends. Persona-based recommendation service through making analysis of user behavior, taking the initiative to provide users the recommended information they may need. One of the application is DM-based persona technique in library book recommendation system with bright prospect. It truly realizes user-centered design and creates larger value. In this thesis, the accurate personal credit scoring model established via data mining, requires multi-dimensional information for the sake of giving a scientific weight to each index. Thence the ultimate goal to truly reflect the credit level of the evaluation object will be accomplished. This method provides a universal reference for user-based digital library book recommendation service of the smart campus under big data environment.

6 References

[1] Jiang Jianwu, Li Jingwen, Lu Yanling. Intelligent information push method based on user portrait [J]. Microcomputer & its Applications, 2016, 35 (23): 86-89, 92.
[2] Wu Mingli, Yang Shuangliang. Study and application of user portrait in content push [J]. Computer Knowledge and Technology, 2016, 12 (32): 255-259.
[3] Niu Wenjia, Liu Jiqiang. User network behavior portrait [M]. Beijing: Electronic Industry Press, 2016: 3-51.
[4] Cooper M, Mell P. Tackling Big Data. NIST, 2012.
[5] Li Bing, Wang Yue, Liu Yongxiang. Application of k-means based persona and intelligent recommendation in big data environment [J]. Modern Computer, 2016, 16: 11-15.

[6] Ren Juan. Reader's behavior mining and analysis of big data technology and application [J]. Editor of the Friends, 2017, 1: 31-34.
[7] Wang Guan. Based on the user's Internet behavior data personal credit evaluation system construction analysis - to sesame credit as an example. Beijing: Beijing Jiaotong University [D], 2015.
[8] Chen Li. Research on clustering algorithm in data mining [D]. 2007.
[9] Ricci F, Rokach L, Shapira B. Recommender Systems Handbook[M]. USA: Springer, 2011:5-6.

Guaranteeing Co-running Program'S Performance in Data Center

Ti Guan, Yong Liu, Guanbin Wu, Daning You and Xuekai Zhang

State Grid Shandong Electric Power Company, Jinan 250001, Shandong Province, China

Email: sddl_zdh@163.com, 15963922979@163.com, wuguanbin@sd.sgcc.com.cn, youdaning@sd.sgcc.com, 84190464@qq.com

Abstract. The data centers are more popular nowadays. In order to improve the utilization of the servers, programs commonly co-running with each other on one server, and programs may interference with each other. Otherwise, for programs who running in cluster or data centers, commonly they can be classified to different kind of program, for example, the batch program, the latency-sensitive program and so on. Different kind of programs has different kind of duty for performance. In order to improve the utilization of the servers while concurrently guarantee the performance of latency-sensitive program, we propose performance guaranteed program co-running algorithm. We design three engines to realize the performance guaranteed program co-running algorithm, they are Program Monitor Engine, Resource Division Engine and Resource Partition Engine respectively. The Program Monitor Engine collects the program's resource usage information about CPU and memory, and then label the programs. The Resource Division Engine calculate the size of CPU and memory resource should be divided for latency-sensitive program. The Resource Division Engine set the resource control group with the CPU and memory value. All the latency-sensitive programs are running in this resource control group. Meanwhile, out side the resource control group, multiple batch programs can be executed. The experiment result show that by making use of the performance guaranteed program co-running algorithm, most of the latency-sensitive program's performance sustain with in 95%, at the same time, the batch programs are used to increase the throughput as more as possible.

1. Introduction

The data centers are build based on the Internet network infrastructure to accelerate the calculation, display and data storage, which are more popular nowadays [1]. It provides a serious of value-added service on the internet, including the specific application domain name, virtual hosting space, hosting and other business. Data center is a complex set of facilities, not only including computer systems and other related equipment (such as communication and storage system), also contains redundant data communication connection environmental control equipment, monitoring equipment, and various safety devices.

In data center or cluster, in order to improve the utilization of the servers [2], programs commonly co-running with each other on one server, and programs may interference with each other. Some times the interference is small, which do little effect on programs; but some times the interference is big, which drop down program's performance dramatically.

Otherwise, for programs who running in cluster or data centers, commonly they can be classified to different kind of program, for example, the batch program, the latency-sensitive program and so on. Different kind of programs has different kind of duty for performance. Take an example, commonly people have little expect on batch program, but the latency-sensitive program should guarantee QoS agreement. The QoS requirement can be appointed by the custom, for instance the 85% of the program's performance when running alone.

In order to improve the utilization of the servers while concurrently guarantee the performance of latency-sensitive program, a smart algorithm should be proposed to solve the trade-off problem. In this paper we propose performance guaranteed program co-running algorithm to guarantee the latency-sensitive program's performance. This algorithm can guarantee the latency-sensitive program's performance and improve the utilization of the servers as much as possible. We design three engines to realize the performance guaranteed program co-running algorithm, they are Program Monitor Engine, Resource Division Engine and Resource Partition Engine respectively.

The Program Monitor Engine collects the program's resource usage information about CPU and memory, and then label the programs. We use Linux tool Top to collect the CPU and memory usage information off-line. Then this information are transferred to Resource Division Engine.

The input of the Resource Division Engine is the program queue and the information transferred from the Program Monitor Engine. The output of the Resource Division Engine is the size of CPU and memory resource should be divided for latency-sensitive program, and the rest of the resource is applied for the batch program. Then the value of the CPU and memory resource for resource guarantee group are transferred to the Resource Partition Engine.

The input of the Resource Partition Engine is the information passed from Resource Division Engine, and the output of the Resource Partition is the size to set the CPU and memory resource After processed by the Resource Partition Engine, A resource control group would be set up for latency-sensitive programs, and all the latency-sensitive programs are running in this resource control group. Meanwhile, out side the resource control group, multiple batch programs can be executed.

The experiment result show that by making use of the performance guaranteed program co-running algorithm, most of the latency-sensitive program's performance sustain with in 95%, at the same time, the batch programs are used to increase the throughput as more as possible.

2. Related work

There are a lot of related work on co-running program's performance in CMP, cluster and data centers.

Some work scheduling programs on CMP to avoid serious interference [3, 4]. They make use of on-chip event to predict program's performance. Some work scheduling programs on cluster to increase the utilization [5]. They use pressure test programs to predict program's performance when co-running with other programs.

Otherwise we have to introduce the implementation tool Cgroups. Cgroups [6] is a Linux resource isolation tools for resource management. It is provided by Linux kernel and isolate the process by making use of material resource, such as CPU, memory and network. Cgroups provide a virtual file system as the group management subsystem and interface setting. For each group, Cgroups can set the size of resource according the operator. Then we can drag the program to the control groups for execution [7].

3. Resource isolation method

We use Linux resource isolate tool Cgroups to realize the performance guaranteed program co-running algorithm. One one hand, this algorithm can ensure program's performance by divide enough resource for program, one the other hand, this algorithm can improve the server's resource utilization by make the most use of the surplus resource. In order to realize the plan, the following engine should be designed, and the flow chart is in Figure 1.

Module 1: Program Monitor Engine. This work should be done off-line. For new programs come into cluster, this engine collects the program's resource usage information about CPU and memory, and then label the programs. The label (resource usage information) can be used for Resource Division Engine to decide how to divide the resource. The general performance profiling tools can be used to help us to collects the program's resource usage information about CPU and memory. In this paper we use Linux tool Top. Every program's (latency-sensitive program and batch program) resource usage of CPU and memory information are topped and passed to Module 2.

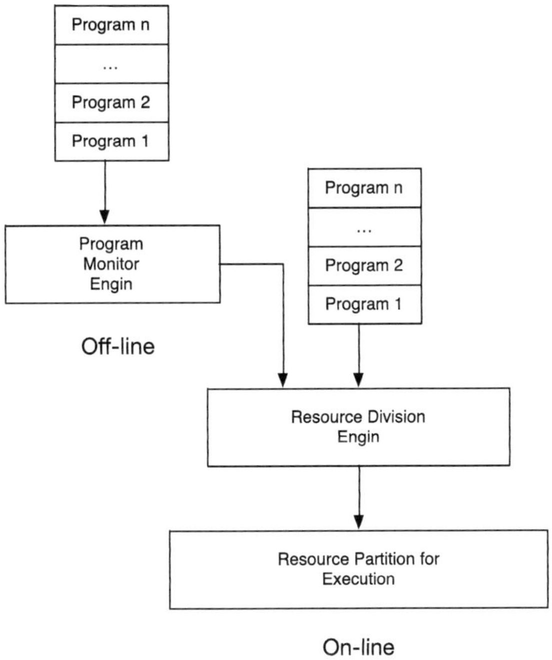

Figure 1 The flow chart of performance guaranteed program co-running algorithm.

Module 2: Resource Division Engine. This part is executing on-line. In order to ensure the efficiency of on-line process. This engine should be simple and efficient. The input of the Resource Division Engine is the program queue and the information transferred from Module 1, which are the resource usage information on CPU and memory of the latency-sensitive program and batch program. The output of the Resource Division Engine is the size of CPU and memory resource should be divided for latency-sensitive program, and the rest of the resource is applied for the batch program. Furthermore, we set a resource guarantee group on the server, and the resource size of the resource guarantee group is the sum of the latency-sensitive program scheduling on this server, which are in formula (1) and formula (2).

$$C_{guarantee-group} = C_{lat-sen1} + C_{lat-sen2} + ... + C_{lat-senn} \tag{1}$$

$$M_{guarantee-group} = M_{lat-sen1} + M_{lat-sen2} + ... + M_{lat-senn} \tag{2}$$

In the above two formula, $C_{guarantee\ group}$ is the size of the CPU resource of the resource guarantee group for latency-sensitive program, and $M_{guarantee\ group}$ is the size of the memory resource of the resource guarantee group for latency-sensitive program. $lat-sen1$, $lat-sen2$, ..., $lat-senn$ are the latency-sensitive programs scheduled to the server.

The the resource size for batch program is the size of server resource minus the size of the resource guarantee group, which is in formula (3) and formula (4).

$$C_{batch-group} = C_{server} - C_{guarantee-group} \tag{3}$$

$$M_{batch-group} = M_{server} - M_{guarantee-group} \tag{4}$$

IOP Conf. Series: Journal of Physics: Conf. Series **910** (2017) 012024 doi:10.1088/1742-6596/910/1/012024

In the above two formula, $C_{batch-group}$ is the size of the CPU resource for the batch program, and $M_{batch-group}$ is the size of the memory resource for batch program. C_{server} is the CPU resource of the server and M_{server} is the memory resource of the server.

Then the value of $C_{guarantee\ group}$ and $M_{guarantee\ group}$ are transferred to Module 3.

Module 3: Resource Partition Engine. The input of the Resource Partition Engine is the information passed from Module 2, and the output of the Resource Partition is the size to set the CPU and memory resource. After processed by the Resource Partition Engine, A resource control group would be set up for latency-sensitive programs, the size of the resource control is the value calculated by formula (1) and formula (2), and all the latency-sensitive programs are running in this resource control group. Meanwhile, out side the resource control group, multiple batch programs can be executed

By this method, the latency-sensitive programs can be executed in resource control group with enough resource, the sufficient resource would guarantee the latency-sensitive program's performance. Meanwhile, the batch programs can be executed outside the resource control group, in order to improve the resource utilization, we can start multiple batch programs to fulfill the resource outside the resource control group, with the price of scarifies every single program's performance.

However, the number of co-running programs outside the resource control group is not unlimited. Too many co-running programs contending the limited resource will result in killing of the server. In the next section we will discuss the number of the co-running batch programs.

4. the discussion of batch programs

This section explores the number of batch programs co-running on the server. As we know, when too many co-running programs contending the limited resource, the server will be killed.

On one hand we want to run as many as possible programs to improve the resource utilization and the throughput of the server. One the other hand we are afraid the server down because of the server can not deal with too many programs. So we have to found the trade-off point.

Assume there are $C_{batch-group}$ CPU resource and $M_{batch-group}$ memory resource for the batch program. We have to set the maximum number of batch program. We start abtch programs one by one, and accumulate the program's resource usage on CPU and memory, onec the sum of the CPU and memory are outstrip the 200% of the resource for batch programs, then wen stop start the new program until some batch program finish. The maximum number of batch program should obey the algorithm in Figure 2.

The maximum number of batch program co-running on server

```
cpu_sum=cpu[p1]
mem_sum=mem[p1]
for i = p1 to pn do
   cpu_sum=cpu_sum+cpu[i+1];
   mem_sum=mem_sum+mem[i+1];
   if cpu_sum >cpu_batch*2 or mem_sum >mem_batch*2
   {
   num_pro=i
   break;
   }
end for
```

Figure 2 Pseudo-code of caculating the maximum number of co-running program running on one server

5. Experiment

The experiment servers are two socket, 4 cores are on one socket, and the detailed server information is in TABLE 1. We select programs from NAS Parallel Benchmarks and take them as latency-sensitive programs. We select programs from The Princeton Application Repository for Shared-Memory Computers (PARSEC) and take them as batch programs. The programs used by the experiment are in TABLE 2.

Table 1 Server configuration

CPU	Intel Xeon E5620
core	4 cores@2.13G
L1 caches	32K
L2 caches	256K
L3 caches	4M
Threads per core	1 thread
Sockets	2
Memory	8GB, DDR3

Table 2 Workloads

Benchmark	program	feature
NAS Parallel Benchmarks	ft.C.4, cg.C.4, mg.C.8	latency-sensitive
PARSEC 3.0	blackscholes, facesim, bodytrack	batch

We compare co-running program's performance with the program's performance when running alone. Then check the QoS of latency-sensitive program's performance. We select three workloads. In every workload there is one program selected from latency-sensitive programs and one program selected from batch programs. We run as more as possible batch programs according our algorithm.

Figure 3 shows the latency-sensitive program's performance and the batch program's performance of workload1 when co-running on one server. The workload is ft.C.4 and blackscholes. When ft.C.4 co-running with blackscholes, there are two batch programs can co-running at the same time. It is because ft.C.4 consume almost all the memory of the server, there is no memory space for the third batch program, although there is idle CPU resource. It is shows that the latency-sensitive program can sustain it's performance with 95%, and the batch program executing with 62%. The low performance is the trade off of throughput.

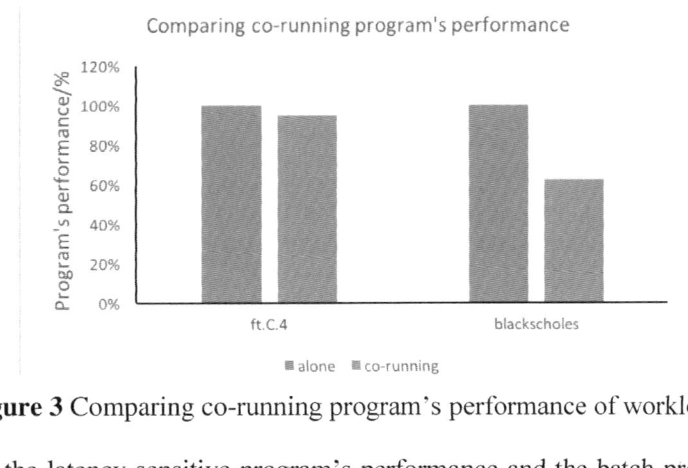

Figure 3 Comparing co-running program's performance of workload 1

Figure 4 shows the latency-sensitive program's performance and the batch program's performance of workload2 when co-running on one server. The workload is cg.C.4 and facesim. When cg.C.4 co-running with facesim, there are three batch programs can co-running at the same time. It is because there is no more CPU space for the forth batch program. It is shows that the latency-sensitive program can sustain it's performance with 97%, and the batch program executing with 45%.

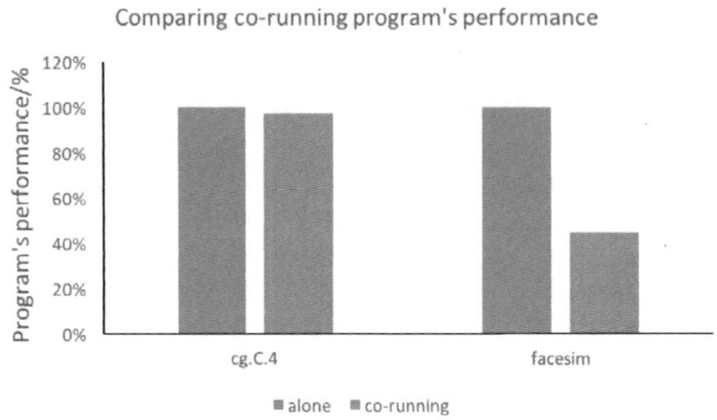

Figure 4 Comparing co-running program's performance of workload 2

Figure 5 shows the latency-sensitive program's performance and the batch program's performance of workload3 when co-running on one server. The workload is mg.C.4 and bodytrack. When mg.C.4 co-running with facesim, there is only one batch programs can co-running. It is because there is no more CPU space for the other batch program. It is shows that the latency-sensitive program can sustain it's performance with 98%, and the batch program executing with 92%.

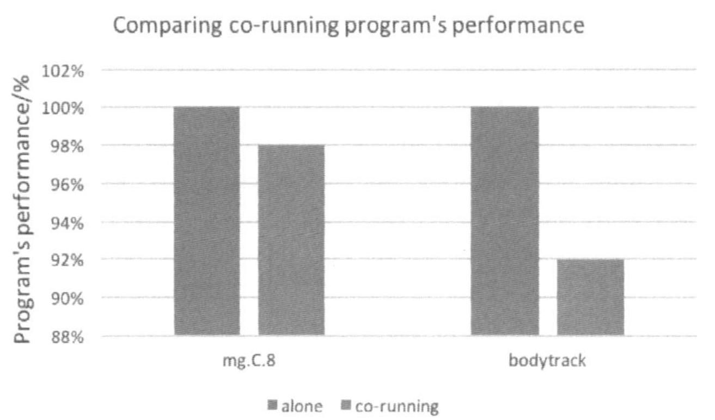

Figure 5 Comparing co-running program's performance of workload 3

6. Conclusion

In order to improve the utilization of the servers while concurrently guarantee the performance of latency-sensitive program. In this paper we propose performance guaranteed program co-running algorithm to guarantee the latency-sensitive program's performance. There are three component to realize the algorithm, there are Program Monitor Engine, Resource Division Engine and Resource Partition Engine respectively.

The experiment result show that by making use of the performance guaranteed program co-running algorithm, most of the latency-sensitive program's performance can be guaranteed by the expect value, and the batch programs are used to increase the throughput of the server.

7. References

[1] A. Mohammad, L. Alexander. A scalable, commodity data center network architecture[J]. ACM SIGCOMM Computer Communication Review, 2008, 38(4): 63-74

[2] X. Li, Z. Qian, S Lu. Energy efficient virtual machine placement algorithm with balanced and improved resource utilization in a data center[J]. Mathematical and Computer Modelling, 2013, 58(5): 1222-1235

[3] S. Zhuravlev, S. Blagodurov, A. Fedorova. Addressing Shared Resource Contention in Multicore Processors via Scheduling[C]. Fifteenth International Conference on Architectural Support for Programming Languages and Operating Systems, 2010: 129-141

[4] J. Mars, N. Vachharajani, R. Hundt. Contention Aware Execution: Online Contention Detection and Response[C]. The 8th annual IEEE/ACM international symposium on Code generation and optimization (CGO '10), 2010: 257-265

[5] J. Mars, L. Tang, R. Hundt. Bubble-Up: increasing utilization in modern warehouse scale computers via sensible co-locations[C]. The 44th Annual IEEE/ACM International Symposium on Microarchitecture (MICRO-44), 2011: 248-259

[6] Cgroups [EB/OL], https://www.kernel.org/doc/Documentation/cgroup-v1/cgroups.txt

[7] L. Wang, D. Qian, Z. Luan, G. Wei, R. Wang, H. Yang. Speeding up profiling program's runtime characteristics for workload consolidation. PLoS ONE 12(4): e0175861.

A Hadoop-based Molecular Docking System

Yueli Dong, Quan Guo and Bin Sun

Department of Software Engineering, Dalian Neusoft Institute of Information, Dalian, China
Email: dongyueli@neusoft.edu.cn

Abstract. Molecular docking always faces the challenge of managing tens of TB datasets. It is necessary to improve the efficiency of the storage and docking. We proposed the molecular docking platform based on Hadoop for virtual screening, it provides the preprocessing of ligand datasets and the analysis function of the docking results. A molecular cloud database that supports mass data management is constructed. Through this platform, the docking time is reduced, the data storage is efficient, and the management of the ligand datasets is convenient.

1. Introduction

Traditionally, lab chemical experiments are needed in the field of drug design, which cost much money and manpower. Computer-aided drug design can be used to focus on the docking objects set, and the period time of making drug can be shortened. With the completion of human genome project and the rapid development of structure biology and protein purification technologies, the suitable targets of receptor molecules have increased dramatically, at the same time the commercial small molecule database has been updated continuously. It is necessary to use computing technology to optimize the process. Computer aided drug molecular design[1] focuses on the molecular docking object sets. However, the search space of molecule docking is tremendous. A rough estimate of docking covered in the search space includes at least 10^{30} solutions which need a large amount of computation time.

Virtual screening procedures search collections of small molecules, seeking those members that contain a set of features that matches a defined search goal. With growth of structural and non-structural data of the ligands, it is an important issue to store and manage a large amount of data. Cloud computing technology completes storing and computing of massive data by distributing data to each computing node of cluster through the network. Tsai[2] constructed a cloud-computing system for traditional Chinese medicine(TCM) intelligent screening system (iScreen). Capuccini, M.[3] developed a method to run existing docking-based screening software on distributed cloud resources, Hongjian Li[4] have developed a publicly-accessible web platform called istar.

In this paper, we proposed a method to construct a molecular database for ligand preprocessing. We implemented a Hadoop -based virtual screening platform using AutoDock Vina[5]. Hadoop[6] is an open-source framework, originally introduced by Google for parallel processing of many small data items, is widely used in massive data processing. Hive is an open source data warehouse tool which is based on the Hadoop. Hive can map the file into a data table, and provide HQL statements. AutoDock Vina is a well-known tool for protein-ligand docking built in the same research lab as the popular tool AutoDock4[7]. Our system can reduce a timeline and cost of drug discovery.

2. Data Model

The non-drug-like molecules (according to user-defined filters) can be selected before docking. Some ligands datasets which cannot be the targets should be eliminated from the datasets for docking process, so preprocessing of data is needed to improve the efficiency of docking. During the process of ligand preprocessing, users search for the datasets according to the properties of the compound, and the selected ligands are submitted for docking.

Two tables are necessary for handling properties and the mol2 file of ligand. Table 1 shows the properties of ligand. Users can retrieve data fields for virtual screening and the retrieved data automatically compose a file in mol2 format. Table2 stores the data of the ligands. The results of docking are stored in table 3.

Table 1 The Properties Table of Ligand

Attribute	Description	Type	comment
Ligand_ID	Ligand ID	string	Primary key
Ligand_name	Ligand name	string	Unique
XlogP	Log of the fat water distribution coefficient	double	-
Rotable_bond	number of rotable bonds	int	-
H_bond_donors	number of H bond donors	int	-
H_bond_acceptors	number of H bond acceptors	int	-
Molecular_weight	sum of the weights of the molecule	double	-
Apolar_desolvation	Apolar desolvation	double	-
atom_count	Atoms of which it is made	int	-
bond_count	bonds count	int	-

Table 2 The file table of ligand

Attribute	Description	Type	comment
Ligand_ID	Ligand ID	string	Primary key
Ligand_name	Ligand name	string	Unique
Content	Whole contents of Mol2 file	string	-
Ligand_name	Ligand name	string	Unique

Table 3 The result table of docking

Attribute	Description	Type	comment
Ligand_ID	Ligand ID	string	Primary key
Ligand_name	Ligand name	string	Unique
score	the minimum value of the scores	string	-

Our platform provides the service for query and the service for preprocessing which inquires information of these ligand data tables.

3. Architecture of Hadoop-Based Molecular Docking Platform

The architecture of molecular docking platform based on Hadoop is shown in Figure 1.

The molecular docking platform includes four layers, they are User Layer, Hive Layer, MapReduce layer and HDFS(Hadoop Distributed File System) layer. User Layer provides services for users, include preprocessing of ligands and operating of docking results. After preprocessing, the drug-like datasets of ligand will be selected as the input file of AutoDock Vina. Hive Layer provides data support. MapReduce is a high performance computing model of Hadoop. The application is executed in many small fragments of works by different nodes in the clusters. Map/Reduce framework divides the work in two phases: the map phase and the reduce phase separated by data transfer between nodes in the cluster. The Reduce stage produces another sets of key-value pair, as final output based on user defined reduce function. HDFS is a Hadoop distributed file system running on commodity hardware.

In HDFS, data is organized into files and directories. Files are divided into uniform sized blocks and distributed across cluster nodes. After docking, the minimum value of the score in the out file will be stored in the result table. Users can retrieve and analysis the information after docking by using the services of this platform.

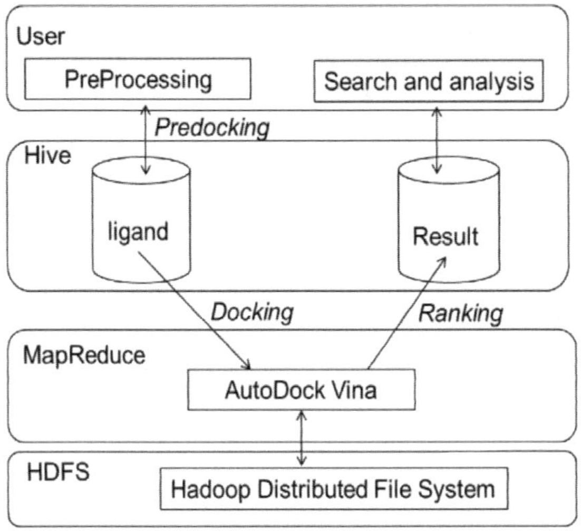

Figure 1. The architecture of molecular docking platform based on Hadoop

4. AN Algorithm of Docking Process

The essential enhancement is the data management from the main jobs in virtual screening, such as docking and ranking. Users not only can control and inspect the computing process but also attain consistent and logically organize data while computing is in progress or finished.

4.1. Preprocessing

All structures and properties of the molecules required for docking and ranking were calculated and stored in the database(Table1 and Table2). For each molecule in the ZINC library25 CHARMm26 atom types were assigned. Then preprocessing system was applied to calculate the atomic and chemical properties of each molecule. Even though not all of these properties were used in docking and ranking, they were prepared for different kinds of filters. Besides these properties, the mol2 file of each molecule was also stored in the database.

4.2. Docking

In order to improve the efficiency of docking, molecules from the library ligand table were selected according the drug-like rules. The poses of each molecule in the PDB format, with their interaction energies with the receptor and efficiencies (electrostatics and VDW), were stored in the table Pose of the database. During the docking, the computing clients acquired the 3D structure of the molecules directly from the database and stored poses and energies in the database after each docking process finished.

4.3. Ranking

We used a developed approach based on calculations of quantum mechanics to efficiently rank the poses. Algorithm used of docking process is as follows.

1) Connect to Mol2File database;

2) Lock the data in table Ligand_file according to the properties of the ligands and set 'unfinished' to ligands;

3) While there are still unfinished ligands exist do

4) Select a ligand within unfinished molecules in table Ligand_file;

5) Generate the pose records randomly and store them into table pose in Pose database;

IOP Conf. Series: Journal of Physics: Conf. Series **910** (2017) 012025 doi:10.1088/1742-6596/910/1/012025

6) Compute each pose in clients parallel;
7) Store the interaction energy w in Ranking table;
8) Update the tag of ligand to 'finished';
9) Endwhile.

5. Experiments and performance evaluation

We implemented a Hadoop-based molecular docking platform in order to validate the performance of the platform based on Hadoop. This section describes the hardware components in the platform and the testing scenarios. Figure2 shows the deployment of the experimental system.

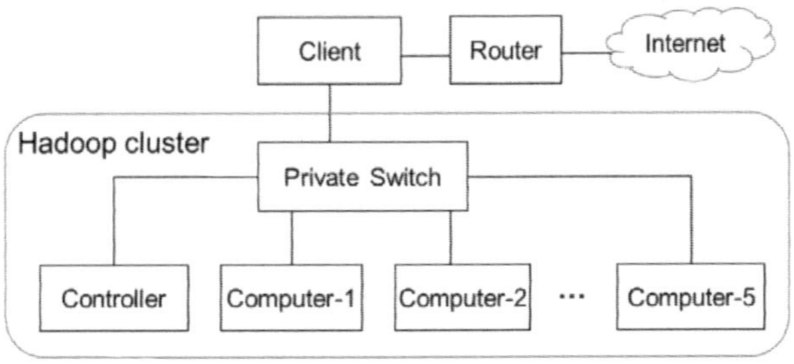

Figure 2. The deployment of the experimental system based on Hadoop

This experimental system assigns one node as controller for the management and five nodes as slaves of the Hadoop cluster. The CPU of controller and each computer is Inter(R) Core™ 2Quad, 2.66Hz processor, and the memory is 8GB, the hard drive of 160GB. We construct a database that store existing chemical databases using MySQL.

We performed docking jobs with 2000 ligands on a target receptor. Figure 3 shows the comparison of execution time as the number of docking jobs increases. We compared three different approaches to execute docking jobs. The first approach is to execute docking jobs on only single node which has the best computing performance. The second approach is to execute docking jobs on Grid-based molecular docking system[8] with 5 computation nodes. The third approach is to execute docking jobs using our Hadoop -based molecular docking system. In Figure3, we can see that the performance of our molecular docking system is better than other approaches.

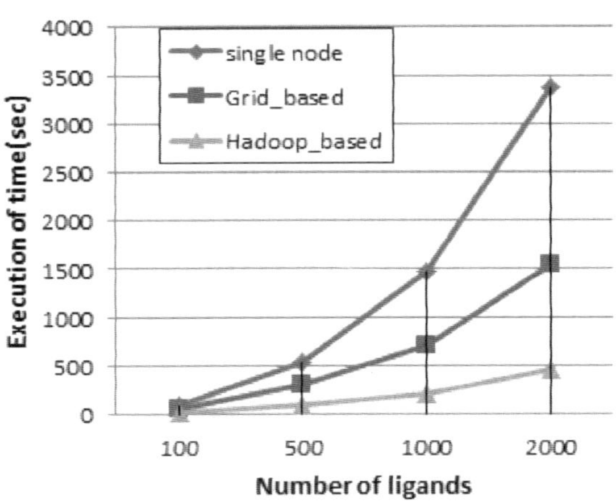

Figure 3. The comparison of execution time

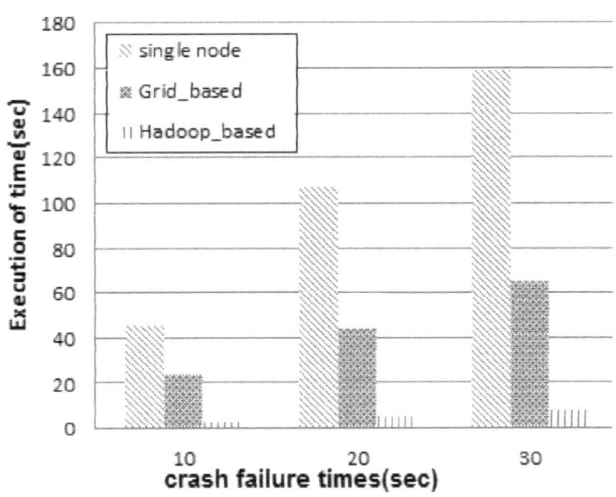

Figure 4. The comparison of crash failure time

When 2000 docking jobs are executed, the execution time of first approach is 455.3s, the execution time of second approach is 1547.9s and third approach is 3368.9s.

When a crash failure happens, in the first case, the docking jobs will be restarted, that means more time will be cost. In the second case, resource manager will migrate the remaining jobs to restart docking in the other computing node. In the third case, resource manager performs rescheduling and job migration when a crash failure happens. Figure 4 shows the comparison of the extra time for three cases to deal with the failure when crash failure happens during the docking process of 2000 ligands.

According to the performance testing on the experiment platform, it shows that the Hadoop-based molecular docking system has a good performance, scalability and fault tolerance. Meanwhile, this platform provides preprocessing function and result processing function for users to access.

6. Conclusion

In this paper, we design and develop a Hadoop-based molecular docking system, which reduces an unmanageable number of compounds to a limited number of compounds for the target of interest. We constructed a ligand database for virtual screening and developed the preprocessing service and ranking service for uses. Through performance testing on the experiment platform, our system can reduce the timeline and cost of drug discovery. We plan to improve the algorithm to improve efficiency, and make various experiments with large datasets to measure the platform in the future.

7. Acknowledgment

These works are mainly supported by National Science Foundation under Grant No.61170168 and No.61170169.

8. References

[1] Chen Kai-xian, Jiang Hua-liang, Ji Ru-yun, Computer aided drug design -- The principle, method and application, Shanghai: Shanghai scientific & Technical Publishers, 2000.
[2] Tsai, Tsung Ying, K. W. Chang, and Y. C. Chen, "iScreen: world's first cloud-computing web server for virtual screening and de novo drug design based on TCM database@Taiwan," Journal of computer-aided molecular design, vol.25, 2011, pp.525-531.
[3] Capuccini, M, et al, "Large-scale virtual screening on public cloud resources with Apache Spark," Journal of Cheminformatics, in press.
[4] Li H, Kwong-Sak L, Ballester P J, et al, "istar: A Web Platform for Large-Scale Protein-Ligand Docking," Plos One, 2014, vol.9, pp.1-12.
[5] Trott O, Olson A J, "AutoDock Vina: improving the speed and accuracy of docking with a new scoring function, efficient optimization, and multithreading," Journal of Computational Chemistry, vol.31, 2010, pp.455-461.
[6] Apache Hadoop framework, Accessed: June 2017, http://hadoop. apache.org/.
[7] Ewing T J, Makino S, Skillman A G, et al, "DOCK 4.0: search strategies for automated molecular docking of flexible molecule databases," J Comput Aided Mol Des, vol. 15, 2001, pp.411-428.
[8] Dong Yueli, Guo Quan, Sun Bin, et al, "Dynamic task migration optimizing for molecule docking," Journal of Jilin University (Engineering and Technology Edition), vol.45, 2015, pp.1253-1259.

Digital Topographic Mapping in Urban Obstructed Environment Based on Multi-GNSS Network RTK Technology

Qiuying Guo[1,a], Tonglong Zhao[1], Chao Zhang[2] and Xuxiang Wu[1]

[1] College of Surveying and Geo-informatics, Shandong Jianzhu University, Fengming Road, Jinan 250101,P.R. China
[2]School of Environment Science and Spatial Informatics,China University of Mining and Technology,Jiangsu Xuzhou 221116, P.R. China
Email: [a]qyguo@sdjzu.edu.cn

Abstract. Digital topographic mapping experiments were carried out based on network RTK technology using GPS/BEIDOU/GLONASS multi-constellation compatible GNSS receivers in urban obstructed environment. Operation scheme and technique flow were discussed. Experimental results show that the horizontal position and elevation of the points measured by RTK can reach 2cm and 3cm precision level respectively in open environment. RTK initialization time needs about 3-5s. While in obstructed environment, such as high building and tree shanding, the RTK initialization time needs about dozens of seconds or tens of seconds, and sometimes floating solutions or even differential solutions were obtained. The impact of dense and tall building on RTK measurement is more seriously. It is more likely to get RTK fixed solution in the south side of high building than the north side of the building.

1. Introduction

Digital City is the development trend of modern cities. Obtaining and updating urban spatial data is the foundation of Digital City. GNSS Continuously Operating Reference Station (CORS) is one of the urban spatial data infrastructure and is the foundation of building a modern digital city. CORS system has been developed into a network integrated service system and will play an important role in the construction of Digital City. With the development and improvement of multi-constellation GNSS, especially rapid development of BEIDOU, the horizontal position and elevation of the points can be measured quickly by network RTK technology with 2-3cm precision level using only one GNSS receiver. In order to test the performance of network RTK in urban obstructed region, the experiment and analysis of the digital topographic mapping using SDCORS network RTK service were carried out in various environment.

2. Experiment environment of network RTK

In order to test the performance of digital topographic mapping based on network RTK technology in urban obstructed environment. We selected two representative experimental areas in our university, which are Zone 1, Zone 2. Figure 1 gives the plan of each zone. Zone 1 is around the Yifu Building. Zone 2 is around the Yingxue Lake. From Figure.1, we can see that: There are buildings, trees and roads in the Zone 1. There is a large area of calm water, woods, grass, roads in the Zone 2.

IOP Conf. Series: Journal of Physics: Conf. Series **910** (2017) 012026 doi:10.1088/1742-6596/910/1/012026

Zone 1: Around Yifu Building Zone 2: Around Yingxue Lake

Figure.1. Plan of the experimental areas

3. Network RTK experiments and result analysis

The instrument used in the digital topographic mapping experiment is GPS/BEIDOU/GLONASS three-constellation compatible receiver. The use of network RTK technology in mapping needs to meet the corresponding technical specifications and follows certain operating methods. According to RTK surveying technical specifications, each detail point needs to be observed one round and each round needs to collect five epochs for mapping. The final results of the observation point is the average value of each epoch. In RTK surveying, we should also follow the following operation procedure to avoid obtaining wrong integer ambiguity. First, all the observations are performed after the RTK fixed solution is stable and convergent. Secondly, the operation scheme "first fixed not recorded, second fixed recorded" is performed.

After the RTK point correction, the topographic detail points can be collected. One person is responsible for holding the RTK mobile station for data acquisition, and another person follows to draw the sketch and record the attribute information of detail points.

In the field operation process, the RTK initialization time needs about 3-5seconds and the horizontal position and elevation of the topographic points can reach 2cm and 3cm precision level respectively in the open environment or in not seriously sheltered environment(see Figure.2). While in the seriously obstructed environment (see Figure.3), the RTK initialization time needs about dozens of seconds or even a few minutes and sometimes ambiguity fixed solutions can not be obtained.

Figure.2. Observation environment where RTK can obtain ambiguity fixed solutions

Figure.3. Some seriously obstructed environment where RTK cannot obtain ambiguity fixed solutions

For those points which cannot obtain ambiguity fixed solutions, we use RTK technology combining with distance measurement method to obtain positions of seriously obstructed detail points. As shown in Figure 4, P is a point in seriously obstructed environment (see Figure.5) where RTK cannot obtain ambiguity fixed solutions. A and B are points in not seriously obstructed environment where RTK can obtain ambiguity fixed solutions. The distance of AP(S1) and BP(S2) can be measured by a electromagnetic distance measuring instrument or a tape. The coordinates of point P can be calculated according to the linear intersection method. The calculation method is as follows:

$$S_{AB} = \sqrt{(x_B - x_A)^2 (y_B - y_A)^2} \tag{1}$$

$$\alpha_{AB} = \arctan\left(\frac{y_B - y_A}{x_B - x_A}\right) \tag{2}$$

$$\angle A = \arccos[(S_1^2 + S_{AB}^2 - S_2^2)/(2S_1 * S_{AB}) \tag{3}$$

$$\alpha_{AP} = \alpha_{AB} - \angle A \tag{4}$$

$$\left.\begin{aligned} x_P &= x_A + S_1 \cos\alpha_{AP} \\ y_P &= y_A + S_1 \sin\alpha_{AP} \end{aligned}\right\} \tag{5}$$

In order to validate the performance of the above-mentioned linear intersection method, we compared the coordinates of point P calculated by fomula (1)-(5) with the coordinates of point P measured by traditional method using a total station. The differences of x and y component are 0.5cm and 1.3cm, respectively. The bias of horizontal position is 1.4cm. Therefore this method can meet the needs of mapping in seriously obstructed environment.

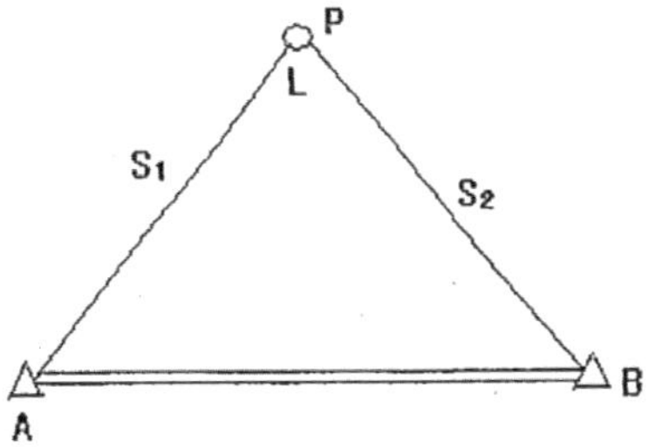

Figure.4. Linear intersection method

Figure.5. The corner point of Building

After the completion of RTK field data acquisition, the exported data files from RTK field electric handbook are stored in the computer. Finally, the topographic mapping software CASS 7.0 is used to draw the maps. Figure.6-7 give the digital topographic maps corresponding to Zone 1 and Zone 2 as shown in Figure.1.

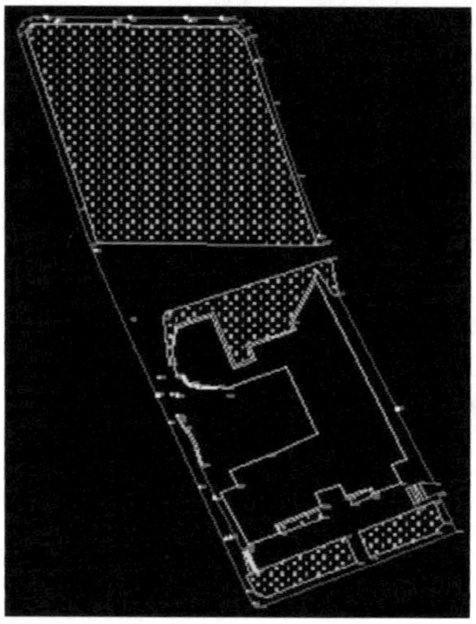

Figure.6 Digital topographic map of Zone 1

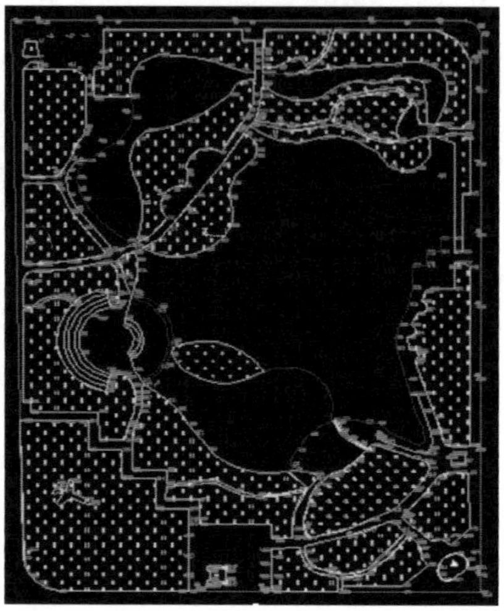

Figure.7 Digital topographic map of Zone 2

4. Conclusions

The experimental results of digital topographic mapping in various urban environment based on GPS/BEIDOU/GLONASS Multi-GNSS RTK technology showed that: (1) In open environment or in not seriously sheltered environment, the horizontal position and elevation of the topographic points can reach 2cm and 3cm precision level respectively using RTK technology. The observed number of satellites is about 20(the cut-off elevation angle is 10 degree) and PDOP is less than 2. The RTK initialization time is only 3-5 seconds. (2)While in obstructed environment, the RTK initialization time needs about dozens of seconds or even a few minutes and sometimes ambiguity fixed solutions cannot be obtained. For the seriously obstructed topographic points, using RTK technology combining with distance measurement method can also obtain the positions of detail points with the 2-3cm precision level.

5. Acknowledgments

This work is supported by the Shandong Provincial Natural Science Foundation, China (ZR2017MD029), the Technology Project Plan of Ministry of Housing and Urban-Rural Development of the People's Republic of China(2014-K8-070) and Shandong Provincial Department of Housing and Urban Rural Construction Technology Project Plan(2017-R1-004). The authors greatly appreciate the editor and the anonymous reviewers on this manuscript.

6. References

[1] CH/T 2009-2010. Specifications for global positioning system real-time kinematic (RTK) surveys[S].
[2] Li Yongning, Liu Limin, Cui Liyan, etc.. Effects of Canopy Openness on Positioning Availability and Initialization Time of GNSS RTK in Forests [J]. Scientia Silvae Sinicae, 2014, 50(2): 78-84.
[3] Guo Qiuying, Mi Xinggang, Zhang Jixiang, etc.. Experiment and Analysis of Efficiency and Precision on SDCORS Network RTK in Urban Area [J]. GNSS World of China, 2013, 38(6): 36-39.
[4] Li Zhiming. The application of Network RTK in Large Scale Mapping Control Survey [J]. Geospatial Information, 2012, 10(3):134-135.
[5] Huang Jinyu. Research on the Application of Network RTK in Digital Topographic Survey [J]. Beijing Surveying and Mapping, 2015,(2):86-88.
[6] Zhou Guowen. Principle and application of network RTK technology based on CORS system [J]. Geomatics & Spatial Information Technology, 2014, 37(3):147-148.
[7] Ahmed El-Mowafy. Performance analysis of the RTK technique in an urban environment [J]. Australian Surveyor, 2000, 45:47-54.

[8] Pirti, MA. Yucel, K. Gumus. Testing real time Kinematic GNSS methods in obstructed and unobstructed sites [J]. Geodetski Vestnik , 2013, 57(3): 498-512.

Comparison of Various Turbulence Models for Unsteady Flow around a Finite Circular Cylinder at *Re*=20000

Di Zhang

School of Civil, Environmental & Mining Engineering, The University of Western Australia, 35 Stirling Highway, Crawley, WA 6009, Australia.
Email: di.zhang@uwa.edu.au.

Abstract. This paper compares the performance of eight Reynolds-Averaged Navier–Stokes (*RANS*) two-equation turbulence models and two sub-grid scale (*SGS*) large eddy simulation (*LES*) models in the scenario of unsteady flow around a finite circular cylinder at an aspect ratio (*AR*) of 1.0 and a Reynolds number of *Re*=20000. It is found that, among all the eight *RANS* turbulence models considered, the *K-Omega-SST* model (*viz. SST-V2003*) developed by Menter *et al.*[1, 2] possesses the best overall performance (being closest to the numerical results of the two *LES* models considered, which can be deemed as the quasi-exact solution in view of the very fine computational mesh employed by the two *LES* models in this study) in terms of the mean surface pressure coefficient distribution (*i.e.* C_p), the mean drag coefficient (*i.e.* C_d), the mean streamline profiles in some characteristic planes (such as the mid-height plane and the symmetry plane of the cylinder) and the distribution of mean bed-shear-stress amplification on the bottom wall.

1. Introduction

The experimental and numerical study on the complex three-dimensional flow structures around a bluff body remains one of the most active areas of research in fundamental fluid dynamics over the past decades, mainly due to the extensive presence of such flows in nature and engineering applications, such as the wind field around high-rise buildings, the pollutant transport around chimney stacks, the aerodynamics force on cooling towers, the flow field around offshore structures, the heat exchange on electronic circuit boards, and so on [3-6]. Although many earlier studies focused on the analysis of the flow past a nominally infinite two-dimensional (2D) circular or square cylinder, recently most attentions have been paid to the unsteady flow around finite-height cylinders [7-12], with one end immersed in the free stream (*viz.* the free end) and the other end mounted on a flat wall (*viz.* the base end), which are more consistent with the structures in reality. Correspondingly, due to the combined influence of the downwash flow from the free end and the boundary layer near the bottom wall, the three-dimensional (3D) flow structure around a finite cylinder is usually much more complicated than that behind an infinite one.

It can be concluded from the existing literature that all the following six factors can have some effects on the flow structure around a finite-height cylinder [6-8]: 1). the turbulence intensity of the approaching flow; 2). the cross-section shape of the cylinder; 3). the *Re* number (*viz.* $Re=UD/\nu$, where D is the characteristic width of the cylinder, U is the free stream velocity, and ν is the fluid's kinematic viscosity.); 4). the boundary-layer thickness on the bottom wall relative to the cylinder height (*viz.* δ/h); 5). the ratio of the cylinder height to the characteristic width of the cylinder (*viz.* $AR=h/D$); 6). the blockage ratio of the channel (*viz.* $\beta_1=h/H$ and $\beta_2=D/B$, where H and B are the height and width of the channel or the computational domain, respectively.). On one hand, the respective effect extent of the aforementioned factors may vary significantly from each other when it comes to a specific condition. On the other hand, in view of multiple influencing factors and complex flow field under this circumstance, in the past, researchers often only discuss one or two factors' influence in each article for simplifying the problem, therefore, the integrated effect of simultaneously changing several parameters remain to be investigated in the future.

Considering that high Reynolds numbers always result in very complicated vortex structures in the wake of a finite circular cylinder (especially when mounted on a non-slip bottom wall), this paper presents a detailed discussion on the flow around a finite circular cylinder with AR=1 at a relatively high Re numbers (*viz.* 20000). The purpose of this study is to quantitatively and qualitatively make a detailed comparison of different turbulence models when it comes to the mean pressure coefficient profile C_p in the mid-height plane of the cylinder, the mean drag coefficient C_d of the cylinder, the time-averaged velocity and pressure fields, and the distribution of mean bed-shear-stress amplification on the bottom wall.

2. Configuration and Numerical Model

2.1 Test Configuration

A similar geometric configuration as that used by Zhang *et al.* [6, 8] is employed in this study. As illustrated in *Fig.* 1, a finite circular cylinder with a non-dimensional width of D=1 and a non-dimensional height of h=1 is vertically mounted on a plane boundary, and the (streamwise) length, (transverse) width and (spanwise) height of the computational domain are, respectively, L=30D, W=22D and H=2D (which gives an area blockage ratio of 2.27%). Further, the junction section between the cylinder and the bottom wall is centered at the origin of the coordinate system, which means that the inlet boundary is located at 10D upstream of the cylinder and the outlet boundary is situated at 20D downstream of the cylinder (*i.e.* L_1=10D, L_2=20D).

2.2 Governing Equations

2.2.1. RANS models. The unsteady 3D Reynolds-averaged Navier-Stokes (RANS) equations employed in this study can be obtained by taking ensemble average of instantaneous mass and momentum conservation equations for incompressible and isothermal flows, as shown in the following [6-8]:

$$\frac{\partial \overline{u}_i}{\partial x_i} = 0 \tag{1}$$

$$\frac{\partial \overline{u}_i}{\partial t} + \frac{\partial}{\partial x_j}(\overline{u}_i\overline{u}_j + \overline{u_i'u_j'}) = -\frac{1}{\rho}\cdot\frac{\partial \overline{p}}{\partial x_i} + \frac{\partial \overline{\tau}_{ij}}{\partial x_j}. \tag{2}$$

where the subscripts i and j indicate the ith and j^{th} components of the Cartesian coordinate respectively, $\overline{u}$ and $\overline{p}$ represent the time-averaged velocity and pressure fields respectively, t is the time, ρ is the density of the fluid (defined as a constant in this study, *viz.* ρ=1000), $\overline{\tau}_{ij} = 2\nu S_{ij} = \nu\left(\partial\overline{u}_i/\partial x_j + \partial\overline{u}_j/\partial x_i\right)$ is the mean viscous stress tensor, $S_{ij} = \left(\partial\overline{u}_i/\partial x_j + \partial\overline{u}_j/\partial x_i\right)/2$ is the mean strain-rate tensor, and ν is the kinematic viscosity of the fluid (*i.e.* ν=5×10^{-5}, which leads to a Reynolds number of Re=2×10^4). Obviously, the Reynolds stress tensor term (*i.e.* $\overline{T_{ij}} = \overline{u_i'u_j'}$) can be obtained by using the Boussinesq eddy-viscosity hypothesis:

$$\overline{T}_{ij} = \overline{u_i'u_j'} = \frac{2}{3}k\delta_{ij} - \nu_t\left(\frac{\partial \overline{u}_i}{\partial x_j} + \frac{\partial \overline{u}_j}{\partial x_i}\right) \tag{3}$$

$$k = \frac{1}{2}\overline{u_i'u_i'} = \frac{1}{2}\left(\overline{u_x'u_x'} + \overline{u_y'u_y'} + \overline{u_z'u_z'}\right) \tag{4}$$

where k is the turbulent kinetic energy, δ_{ij} is the Kronecker delta symbol and v_t represents the turbulent viscosity of the flow. In order to obtain the turbulent viscosity v_t, different RANS turbulent models are adopted and compared with each other here, including the standard *k*-epsilon ($k-\varepsilon$) model [13], the renormalization group *k*-epsilon (RNG $k-\varepsilon$) model [14], the realizable $k-\varepsilon$ turbulence model [15], Lien's cubic non-linear low-Reynolds $k-\varepsilon$ model [16], Launder and Sharma low-Reynolds $k-\varepsilon$ turbulence model [17], Shih's quadratic non-linear $k-\varepsilon$ turbulence model [18], the standard high Reynolds-number *k-omega* ($k-\omega$) model [19] and the *K-Omega–SST* ($k-\omega-SST$) turbulence model [1, 2]. For the conciseness of the presentation, the details of these models are not presented here, readers can refer to *Ref.* [1, 2, 13-19] for further details.

2.2.2. LES models. The spatial filtering of unsteady 3D instantaneous mass and momentum equations in a Newtonian incompressible flow results in the following LES equations [20-23]:

$$\frac{\partial \tilde{u}_i}{\partial x_i}=0\cdot \tag{5}$$

$$\frac{\partial \tilde{u}_i}{\partial t}+\frac{\partial}{\partial x_j}(\tilde{u}_i\tilde{u}_j)=-\frac{1}{\rho}\cdot\frac{\partial \tilde{p}}{\partial x_i}+2\partial_j\left(v\tilde{S}_{ij}\right)-\partial_j\left(\tilde{\tau}_{ij}\right)\cdot \tag{6}$$

$$\tilde{S}_{ij}=\frac{1}{2}\left(\frac{\partial \tilde{u}_j}{\partial x_i}+\frac{\partial \tilde{u}_i}{\partial x_j}\right),\quad \tilde{\tau}_{ij}=(\widetilde{u_iu_j}-\tilde{u}_i\tilde{u}_j). \tag{7}$$

where $\tilde{u}$ and $\tilde{p}$ are the filtered velocity and pressure fields respectively, $\tilde{S}_{ij}$ is the filtered or resolved scale strain-rate tensor, and $\tilde{\tau}_{ij}$ is the unknown SGS stress tensor, representing the effects of the SGS motions on the resolved fields of LES, which needs to be modelled by using a so-called SGS model, for the purpose of closing the above governing equations. By employing the Boussinesq eddy-viscosity hypothesis, the SGS stress tensor can be expressed as:

$$\tilde{\tau}_{ij}=v_{sgs}\left(\frac{\partial \tilde{u}_i}{\partial x_j}+\frac{\partial \tilde{u}_j}{\partial x_i}\right)+\frac{1}{3}\delta_{ij}\tilde{\tau}_{ll}\cdot \tag{8}$$

For the Smagorinsky SGS model [20, 21]:

$$v_{sgs}=\left(C_S\tilde{\Delta}\right)^2 S,\quad S=\left(2\tilde{S}_{ij}\tilde{S}_{ij}\right)^{\frac{1}{2}},\quad \tilde{\Delta}=(\Delta x\Delta y\Delta z)^{\frac{1}{3}}. \tag{9}$$

For the One-Equation eddy viscosity model [22, 23]:

$$\frac{\delta k_{sgs}}{\delta t}+\nabla\cdot\left(k_{sgs}\tilde{U}\right)-\nabla\cdot\left(\left(v_{sgs}+v\right)\cdot\nabla k\right)=-D_{ij}:B_{ij}-C_E\cdot k^{\frac{3}{2}}/\tilde{\Delta}$$

$$v_t=C_k\tilde{\Delta}\sqrt{k_{sgs}},\quad D_{ij}=symm\left(\nabla\tilde{U}\right),\quad B_{ij}=\frac{2}{3}K_{sgs}I_{ij}-2v_{sgs}dev(D_{ij}). \tag{10}$$

where v_{sgs} and k_{sgs} are the sub-grid turbulent viscosity and the sub-grid turbulent kinetic energy respectively, C_E=1.048, C_k=0.094 and C_S=0.1 are the model parameters, and $\tilde{\Delta}$ represents the local grid-resolution quality.

2.3. Boundary Conditions and Numerical Schemes

Four kinds of boundary conditions are involved:

1. Inlet: For the velocity field, a fixed uniform velocity is prescribed (*i.e.* u_I=1, $v_I=w_I$=0), and, for the pressure field, the zero-gradient condition is imposed (*i.e.* $\partial p_I/\partial n=0$).

2. Outlet: For the velocity field, the convective outflow boundary condition is adopted:

$$\frac{\partial \phi}{\partial t} + \vec{u}_c \frac{\partial \phi}{\partial n} = 0 . \quad (11)$$

where ϕ denotes all the three velocity components (*viz. u, v* and *w*), and $\vec{u}_c$ is the advective velocity at the outlet boundary. For the pressure field, the homogeneous Dirichlet condition is utilized at the outlet boundary (*i.e. p*=0).

3. Bottom wall and the surface of the obstacle: No-slip impermeable boundary condition is prescribed for the velocity field (*i.e. u*=*v*=*w*=0), and the zero-gradient condition is employed for the pressure field.

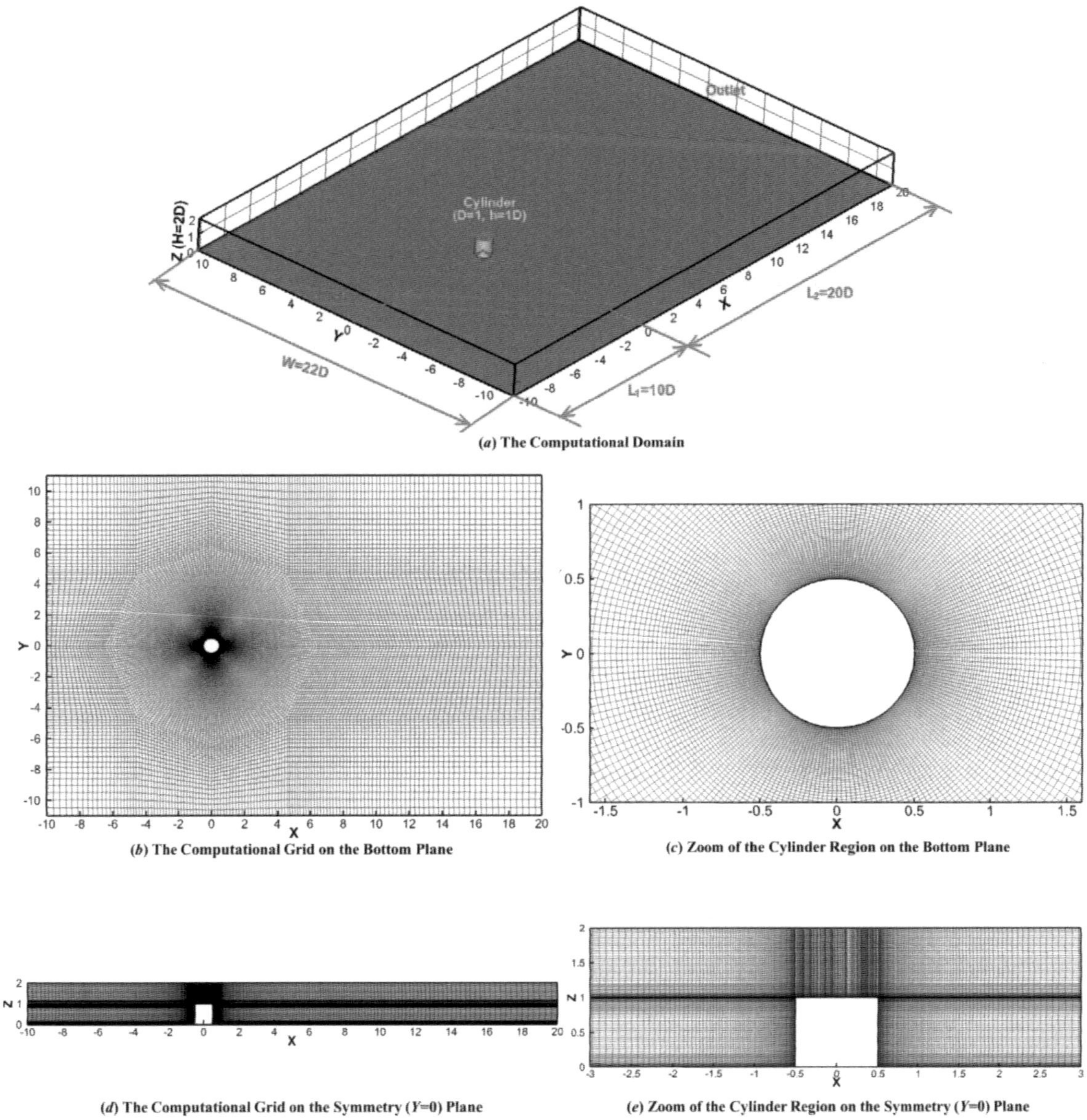

Figure.1. Schematic of the computational domain and the *Level-1* mesh employed by different RANS models.

4. Top and lateral boundaries of the computational domain: Free-slip condition is prescribed, which means that the velocity component normal to the boundary is zero (*i.e.* $u_{b\perp} = 0$) and the normal

gradients of both the pressure and the tangential velocity component are zero (*i.e.* $\partial u_{b//}/\partial n = \partial p_b/\partial n = 0$, where $u_{b//}$ is the tangential velocity component).

The numerical simulations were carried out by using the Open Source Field Operation and Manipulation (OpenFOAM) C++ libraries. Specifically, PisoFoam, one of the standard solvers provided by OpenFOAM, is selected to solve the aforementioned governing equations. The Pressure Implicit with Splitting of Operators (PISO) [24] algorithm is employed to deal with the pressure-velocity coupling on a collocated grid system in the context of finite volume method (FVM). In order to satisfy the boundedness property, the *limitedLinear* TVD/NVD scheme is employed to discretize the convection terms of the scalar transport equations, and the *limitedLinearV* scheme, which stands for the improved version of the *limitedLinear* TVD/NVD scheme for vector fields and whose limiter is formulated to take into account the direction of the field, is adopted to discretize the convection terms of the vector transport equations [25-28]. In addition, the Gauss linear scheme is chosen to discretize the diffusion term and the pressure gradient term when discretizing the governing equations, and the second-order Crank–Nicolson method is adopted for the temporal discretization.

3. Results and Discussion

3.1. Mean Drag Coefficient and Surface-Pressure Profile

Considering that in normal conditions LES turbulence models have a higher demand on the resolution of the computational mesh relative to various RANS turbulence models, two levels of grids are used in this study, namely *Level-1* for eight RANS models and *Level-2* for two LES models. As shown in the *Table-1*, for the *Level-1* mesh, it totally consists of approximately 3.26 million grid points, the circumference of the cylinder is evenly divided into 200 parts, and the near-wall grid size (defined as the distance between the centroid of the first cell and the non-slip boundary) is about 0.0016, which leads to a distance in wall units of less than 1.0 (*viz.* $(\Delta y)^{+}_{max}<1.0$). However, for the finer *Level-2* mesh, it totally consists of approximately 10.3 million grid points, the circumference of the cylinder is evenly divided into 248 parts, and the near-wall grid size is about 0.0005, which leads to a distance in wall units of less than 0.5 (*viz.* $(\Delta y)^{+}_{max}<0.5$). In addition, the time step is fixed as Δt=0.0002, which can ensure that the maximum Courant–Friedrichs–Lewy (CFL) number in all cases is not larger than 0.3 in order to improve both the temporal accuracy and the numerical stability.

In view of a much finer grid resolution is employed by the two LES models and the C_d values of two LES models are well consistent with each other (*i.e.* C_d=0.802 for the one-equation eddy viscosity LES model and C_d=0.820 for the Smagorinsky LES model), it is reasonable to assume that the C_d values obtained by adopting the two LES models can be (at least to some degree) treated as the exact value of the time-averaged drag coefficient (*i.e.* $C_{d\text{-}exact}\approx$0.80). From *Table-1*, it can be concluded that the Standard K-Omega Model (1998) and the K-Omega–SST Model (2003) possess the best performance when evaluating the C_d value (*i.e.* C_d=0.766 and C_d=0.761, respectively), when compared with the other six RANS turbulence models, namely Shih's non-linear RSA *k-ε* Model (C_d=0.703), Lien-Cubic Low-*Re* *k-ε* Model (C_d=0.678), Realizable *k-ε* Model (C_d=0.668), Launder-Sharma Low-Re *k-ε* Model (C_d=0.631), Standard *k-ε* Model (C_d=0.625) and RNG *k-ε* Model (C_d=0.612).

Fig. 2 presents a quantitative comparison of the distribution of the mean surface-pressure coefficient (C_p) along the circumference at the mid-height of the cylinder (*i.e.* Z/D=0.5). Actually, the numerical results of two existing experimental studies have also been added for comparison, namely Okamoto and Sunabashiri[7] (AR=1, δ/D<0.11, Re_D=2.5×10^4~4.7×10^4, $H/D\approx$7) and Kawamura *et al.*[5] (AR=1, δ/D=0.10, Re_D=3.2×10^4, H/D=15). It is clear from *Fig.* 2 that, even for the time-averaged surface-pressure profile at the mid-span of the cylinder, relative to the remaining six RANS turbulence models, the Standard *K-Omega* Model and the *K-Omega–SST* Model possess the best prediction accuracy as well (sharing the best consistency with the prediction results of the two LES models). Concretely, in the range of θ=[50°, 120°], the mean pressure profile of the *K-Omega–SST* Model is closest to that of the two LES model, but in the range of θ=[120°, 180°], the mean pressure profile of the Standard *K-Omega* Model is closest to that of the two LES model.

From *Table-1* and *Fig. 2*, it can be concluded that, in terms of both the mean drag coefficient and the mean surface-pressure distribution at the mid-height of the cylinder, the Standard *K-Omega* Model (1998) and the *K-Omega–SST* Model (2003) can result in a better prediction accuracy relative to the other six RANS turbulence models considered (at least for the simulation case carried out in this study). Therefore, in the following two sections when analyzing the other two aspects of the low field, only the numerical results of the aforementioned two RANS models are presented and compared with that of the two LES models considered.

3.2. Mean Velocity and Pressure Fields

In this section, the mean velocity and pressure fields are investigated by examining the time-averaged streamlines and pressure contours in two characteristic planes (*i.e.* the mid-height plane of the cylinder (Z/D=0.5) and the symmetry plane (Y/D=0)).

Fig. 3 indicates that, in the mid-height plane of the cylinder, a strong positive/negative pressure region can be generated before/behind the cylinder because of the obstruction effect of the circular cylinder. Moreover, it is clear that the same topology can be observed for the two LES model and the *K-Omega–SST* model, in terms of the mean streamlines in this plane, containing two symmetrically distributed spiral centers (*i.e. C* and *D*) and one saddle node (*i.e. R* in *Fig.* 3). Nevertheless, when it comes to the standard *K-Omega* model, differences exist in terms of the overall topology of the time-averaged streamlines in the Z/D=0.5 plane when compared with that of the other three turbulence models, as verified by *Fig.* 3(*d*). The *X*-location of *R* is found to be with only slight difference for all the aforementioned four turbulence models, in view of that $X(R)$=1.524 for the Smagorinsky LES model, $X(R)$= 1.504 for the one-equation eddy viscosity LES model, $X(R)$=1.22 for the *K-Omega–SST* model and $X(R)$= 1.56 for the standard *K-Omega* model.

Fig. 4 presents the time-averaged streamlines and pressure contours in the symmetry plane, and the same color scale is used for all the cases for the convenience of comparison. Obviously, the overall topology is almost the same for all the four turbulence models presented in this figure, except for the fact that the standard $k-\omega$ model fails to predict the time-averaged horse-shoe vortices and the corresponding separation points in front of the cylinder, as confirmed by *Fig.* 4(*d*). In the near wake behind the cylinder, one clockwise vortex core *A* (caused by the downwash flow and located in the upper region) and another anti-clockwise vortex core *B* (induced by the very weak upwash flow close to the bottom wall and located in the lower region) coexist in the symmetry plane for all the four turbulence models. Besides, the flow separating from the leading edge of the cylinder tip reattaches onto the free-end surface, and consequently an isolated secondary recirculation region (its center is denoted by the capital letter *E* in *Fig.* 4) is formed above the tip end for all the four turbulence models. It should be emphasized that the flow field above the free end is of significant importance since the curvature of the streamlines within this region is directly related to the angle of the downwash flow behind the cylinder. Additionally, *Fig.* 4(*a*, *b*, *c*) illustrates that two mean horseshoe-vortex cores can be identified in the symmetry plane for the two LES models and the $k-\omega-SST$ model, marked by *H1* and *H2*. But, no such vortex phenomena can be captured in *Fig.* 4(*d*) for the standard $k-\omega$ model.

From *Figs*. 3-4, it can be concluded that, in terms of the mean streamlines and pressure contours in both the mid-height plane of the cylinder and the symmetry plane, the *K-Omega-SST* model (employing a relative coarse mesh) will have a better consistency with that of the two LES turbulence models considered (employing a much finer mesh) and therefore is superior to the standard *K-Omega* model (using the same coarse mesh as the *K-Omega-SST* model).

Table.1. Grid resolution and the mean drag coefficient C_d.

Turbulence Model	Grid Resolution	Number of Nodes	Node-Number Along the Circumference	Near-wall Grid Dimension	Near-wall Grid Y^+	Mean Drag Coefficient C_d
Standard *K-Omega* Model (1998)	Level-1	3.26 Million	200	0.0016	1	0.766
K-Omega–SST Model (2003)	Level-1	3.26 Million	200	0.0016	1	0.761
Shih's non-linear RSA *k-ε* Model (1993)	Level-1	3.26 Million	200	0.0016	1	0.703
Lien-Cubic Low-Re *k-ε* Model (1996)	Level-1	3.26 Million	200	0.0016	1	0.678
Realizable *k-ε* Model (1995)	Level-1	3.26 Million	200	0.0016	1	0.668
Launder-Sharma Low-Re *k-ε* Model (1974)	Level-1	3.26 Million	200	0.0016	1	0.631
Standard *k-ε* Model (1972)	Level-1	3.26 Million	200	0.0016	1	0.625
RNG *k-ε* Model (1992)	Level-1	3.26 Million	200	0.0016	1	0.612
One-Equation LES Model (1986)	Level-2	10.3 Million	248	0.0008	0.5	0.802
Smagorinsky LES Model (1963)	Level-2	10.3 Million	248	0.0008	0.5	0.820

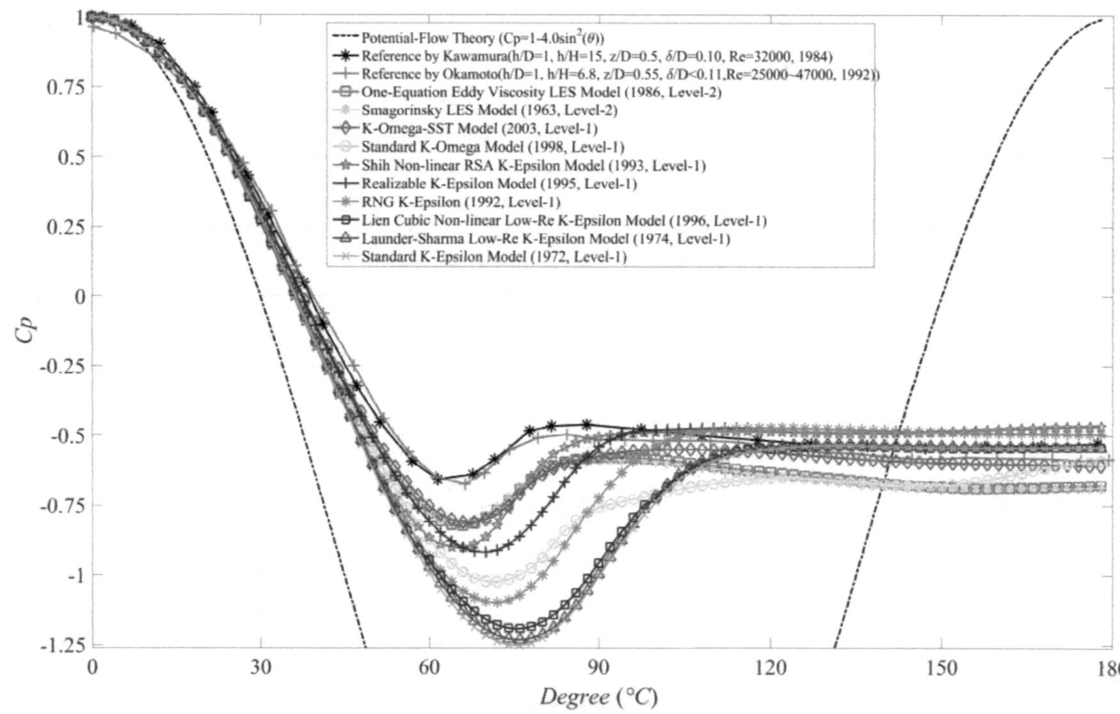

Figure. 2. Comparison of the mean surface-pressure coefficient along the circumference at Z/D=0.5.

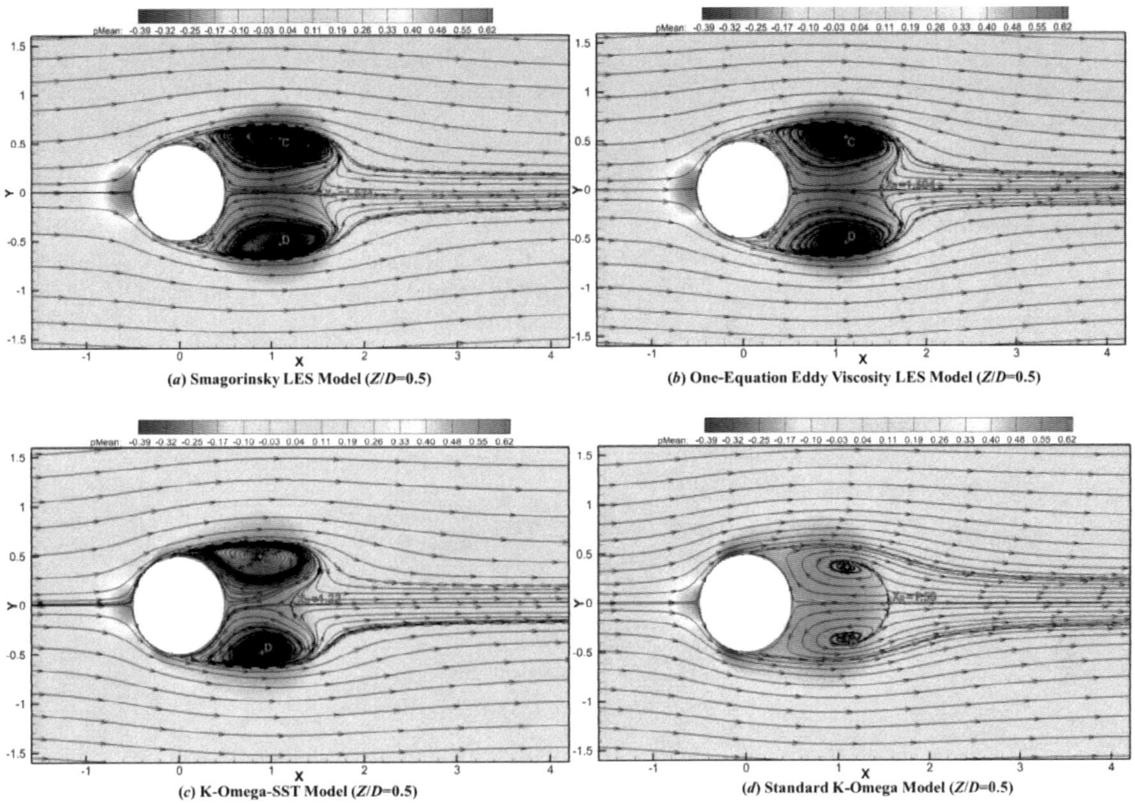

Figure. 3. Comparison of the mean streamlines and pressure contours on the *Z*/*D*=0.5 plane.

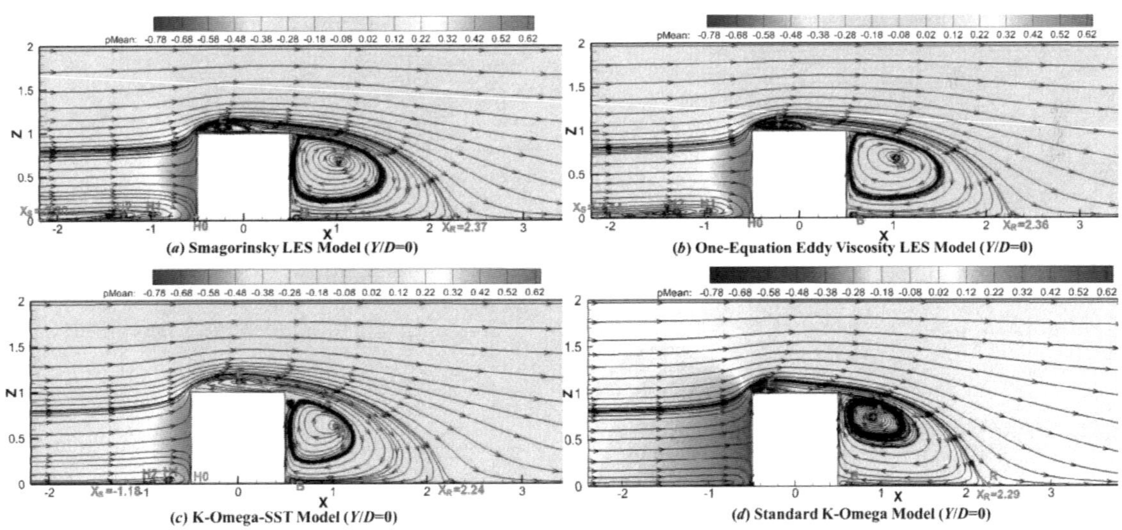

Figure. 4. Comparison of the mean streamlines and pressure contours on the *Y*/*D*=0 plane.

3.3. Bed-Shear-Stress Distribution

This section aims to investigate the distribution of the time-averaged bed shear stress and further give the full picture of the bed-shear-stress amplification (based on the velocity gradient at the height of *Z*/*D*=0.0001) around the cylinder for different turbulence models. *Fig.* 5 demonstrates a comparison of the time-averaged bed-shear-stress amplification along the symmetry line at the upstream side of the cylinder. It is evident that the both the *K-Omega-SST* model and the standard *K-Omega* model possess the same main features (namely the overall shape) as the experimental and numerical results of Roulund *et al.* [9] ($|\tau_{max}|\approx 2.0$, AR=1, δ/D=1, δ/H=1, $Re_D=1.7\times10^5$) and *Ming et al.* [8] ($|\tau_{max}|\approx 1.45$, AR=1, δ/D=1, δ/H=1, $Re_D=1.7\times10^5$). However, when it comes to the location and value of the most negative amplification factor, the *K-Omega-SST* model results in a value of $|\tau_{max}|\approx 1.3$, which is slightly more accurate than that of the standard *K-Omega* model ($|\tau_{max}|\approx 1.1$). Furthermore, the Smagorinsky LES model and the one-equation eddy viscosity LES model will lead to a value of $|\tau_{max}|\approx 1.8$ and $|\tau_{max}|\approx 2.2$ respectively, which are almost identical to the experimental results of Roulund *et al.* [9] ($|\tau_{max}|\approx 2.0$) due to the much finer computational mesh adopted by the two LES models. Additionally, *Fig.* 6 compares the full picture of the mean bed-shear-stress amplification around the cylinder for different turbulence models. Two symmetrically distributed points with the largest bed shear stress can be identified at an angle of θ=[55°, 60°] (θ represents the angle measured from the negative X axis) for each test case, being consistent with the conclusions of Roulund *et al.*[9] θ=[45°, 70°]. Obviously, no significant difference can be obtained in terms of the maximum bed-shear-stress amplification at these locations

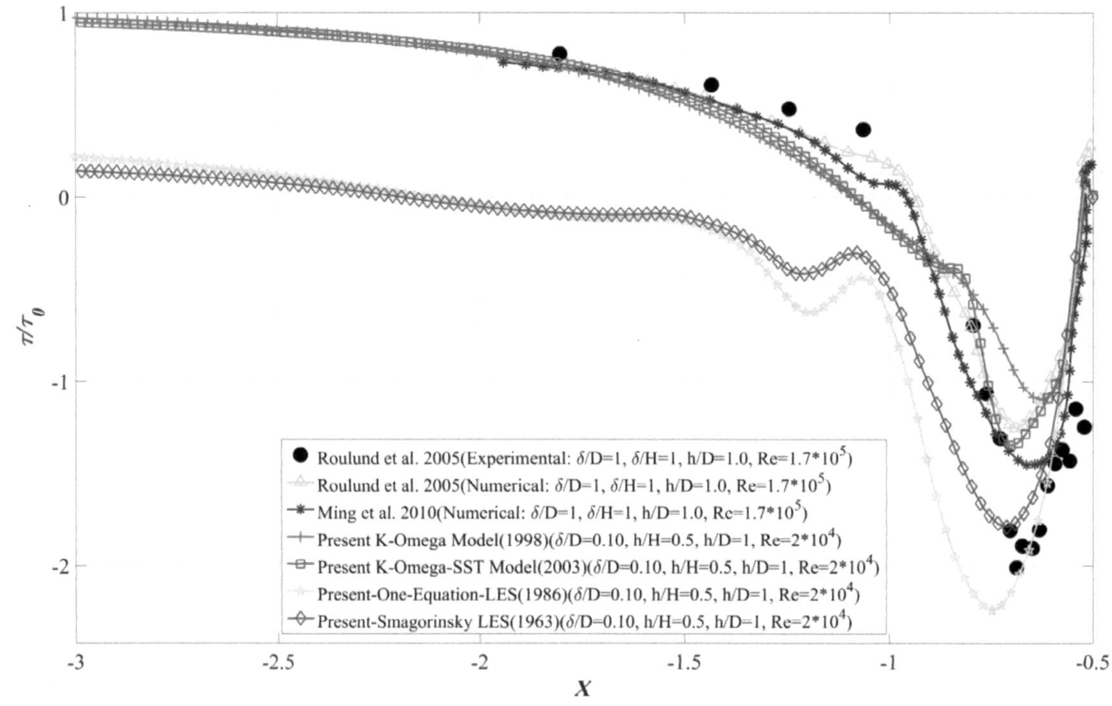

Figure. 5. Comparison of the bed-shear-stress along the symmetry line in front of the cylinder.

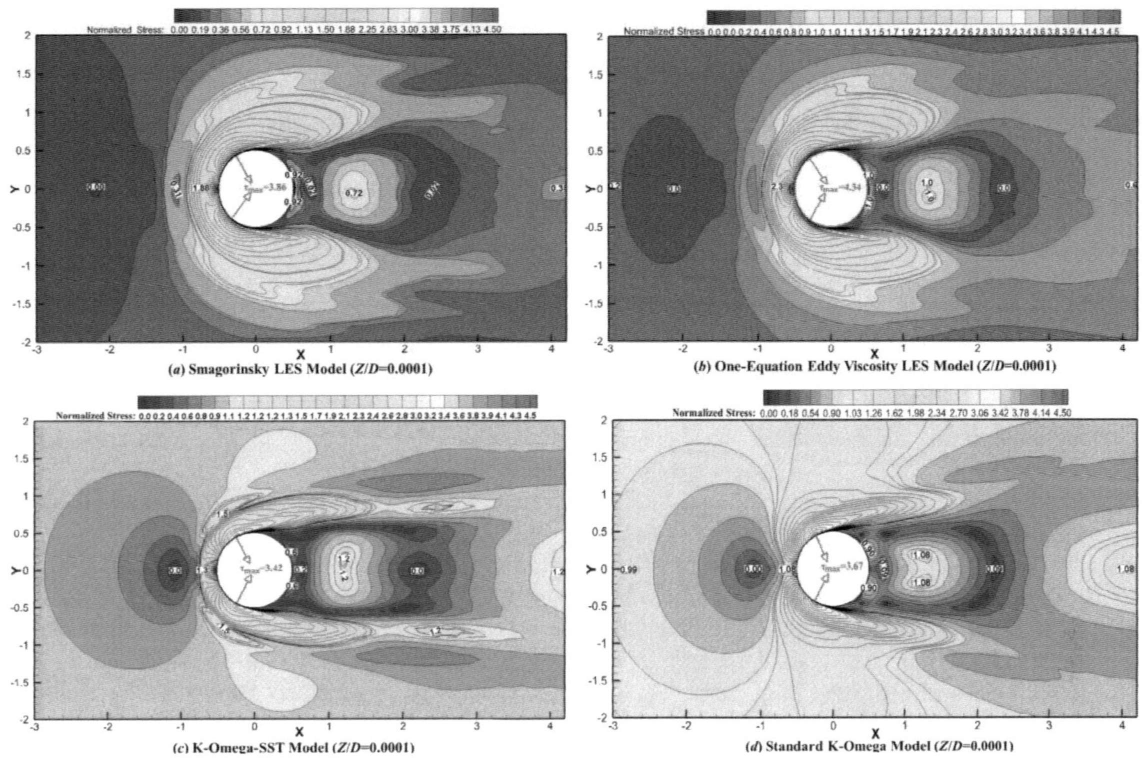

Figure. 6. Comparison of the full picture of the mean bed-shear-stress on the bottom wall.

for different turbulence models, in consideration of that $|\tau_{\text{max-lateral}}|\approx 3.86$ for the Smagorinsky LES model, $|\tau_{\text{max-lateral}}| \approx 4.34$ for the One-Equation LES model, $|\tau_{\text{max-lateral}}| \approx 3.42$ for the *K-Omega-SST* model and $|\tau_{\text{max-lateral}}| \approx 3.67$ for the standard *K-Omega* model.

4. Conclusions

Turbulent flow past a finite circular cylinder (*AR*=1) is simulated at a relatively large Reynolds number (*Re*=20000). The focus is to examine the influence of different turbulence models on several aspects of the flow, namely the mean drag coefficient, the mean surface-pressure coefficient, the time-averaged velocity and pressure fields, and the mean bed-shear-stress amplification. It can be concluded that, in terms of all the aforementioned aspects, the numerical results of the K-Omega-SST model (employing a relative coarse mesh) will have a better consistency with that of the two LES turbulence models considered (employing a much finer mesh) and therefore is more accurate than the standard K-Omega model and the other six remaining RANS turbulence models (using the same coarse mesh as the K-Omega-SST model). Therefore, the superiority of the K-Omega-SST model (*viz.* SST-V2003, developed by Menter *et al.* [1]) has been effectively validated for the present numerical case and this turbulence model is highly recommended in the scenario of the adverse pressure gradients and separating flow around a finite cylinder.

5. Acknowledgments

The author wish to acknowledge the financial support provided by Australian Research Council Discovery Grant (Project ID: DP150104644) and the computational resources provided by the Pawsey Supercomputing Centre with funding from the Australian Government and the Government of Western Australia.

6. References

[1] F.R. Menter, M. Kuntz, R. Langtry. "Ten years of industrial experience with the SST turbulence model". *Turbulence, Heat and Mass Transfer*, **4**(1), 625-632 (2003).

[2] F.R. Menter. "Two-equation eddy-viscosity turbulence models for engineering applications". *AIAA Journal*, **32**(8), 1598-1605 (1994).

[3] C.H.K. Williamson, "Vortex dynamics in the cylinder wake," *Annual Review of Fluid Mechanics*. **28**(1), 477-539 (1996a).

[4] D. Sumner, J.L. Heseltine. "Tip vortex structure for a circular cylinder with a free end". *Journal of Wind Engineering and Industrial Aerodynamics*, **96**(6), 1185-1196 (2008).

[5] T. Kawamura, M. Hiwada, T. Hibino, I. Mabuchi, M. Kumada. "Flow around a finite circular cylinder on a flat plate: Cylinder height greater than turbulent boundary layer thickness". *Bulletin of JSME*, **27**(232), 2142-2151 (1984).

[6] D. Zhang, L. Cheng, H. An, M. Zhao. "Direct numerical simulation of flow around a surface-mounted finite square cylinder at low Reynolds numbers". *Physics of Fluids*, **29**(4), 045101 (2017).

[7] S. Okamoto, Y. Sunabashiri. "Vortex shedding from a circular cylinder of finite length placed on a ground plane". *Journal of Fluids Engineering*, **114**(4), 512-521 (1992).

[8] D. Zhang, L. Cheng, H. An, S. Draper. "Flow around a surface-mounted finite circular cylinder completely submerged within the bottom boundary layer". *Submitted to Physics of Fluids*.

[9] A. Roulund, B.M. Sumer, J. Fredsøe, J. Michelsen. "Numerical and experimental investigation of flow and scour around a circular pile". *Journal of Fluid Mechanics*, **534**, 351-401 (2005).

[10] O. Lehmkuhl, I. Rodríguez, R. Borrell, J. Chiva, A. Oliva. "Unsteady forces on a circular cylinder at critical Reynolds numbers". *Physics of Fluids*, **26**(12), 125110 (2014).

[11] J.A. Bourgeois, P. Sattari, R.J. Martinuzzi. "Alternating half-loop shedding in the turbulent wake of a finite surface-mounted square cylinder with a thin boundary layer". *Physics of Fluids*, **23**(9), 095101 (2011).

[12] D. Sumner, N. Rostamy, D.J. Bergstrom, J.D. Bugg. "Influence of aspect ratio on the mean flow field of a surface-mounted finite-height square prism". *International Journal of Heat and Fluid Flow*, **65**, 1-20 (2017).

[13] B.E. Launder, D.B. Spalding. "The numerical computation of turbulent flows". *Computer Methods in Applied Mechanics and Engineering*, **3**(2), 269-289 (1974).

[14] V. Yakhot, S.A. Orszag, S. Thangam, T.B. Gatski, C.G. Speziale. "Development of turbulence models for shear flows by a double expansion technique". *Physics of Fluids A: Fluid Dynamics*, **4**(7), 1510-1520 (1992).

[15] T.H. Shih, W.W. Liou, A. Shabbir, Z. Yang, J. Zhu. "A new k-ϵ eddy viscosity model for high reynolds number turbulent flows". *Computers & Fluids*, **24**(3), 227-238 (1995).

[16] F.S. Lien. "Low-Reynolds-number eddy-viscosity modelling based on non-linear stress-strain/vorticity relations". *Engineering Turbulence Modelling and Measurements*, **3**, 91-100 (1996).

[17] B.E. Launder, B.I. Sharma. "Application of the energy-dissipation model of turbulence to the calculation of flow near a spinning disc". *Letters in Heat and Mass Transfer*, **1**(2), 131-137 (1974).

[18] T.H. Shih, J. Zhu, J. Lumley. "A Realizable Reynolds Stress Algebraic Equation Model". *NASA Technical Memorandum*, 105993, (1993).

[19] D.C. Wilcox. "Turbulence modeling for CFD (Vol. 2, pp. 103-217)". *La Canada, CA: DCW Industries* (1998).

[20] Y. Zhiyin. "Large-eddy simulation: Past, present and the future". *Chinese Journal of Aeronautics*, **28**(1), 11-24 (2015).

[21] J. Smagorinsky. "General circulation experiments with the primitive equations: I. The basic experiment". *Monthly Weather Review*, **91**(3), 99-164 (1963).

[22] A. Yoshizawa. "Statistical theory for compressible turbulent shear flows, with the application to subgrid modeling". *The Physics of fluids*, **29**(7), 2152-2164 (1986).

[23] W. Rodi. "Comparison of LES and RANS calculations of the flow around bluff bodies". *Journal of Wind Engineering and Industrial Aerodynamics*, 69, 55-75 (1997).
[24] O. Ubbink. "Numerical prediction of two fluid systems with sharp interfaces". Doctoral dissertation, University of London, (1997).
[25] D. Zhang, C. Jiang, D. Liang, Z. Chen, Y. Yang, Y. Shi. "A refined volume-of-fluid algorithm for capturing sharp fluid interfaces on arbitrary meshes". *Journal of Computational Physics*, **274**, 709-736 (2014).
[26] D. Zhang, C. Jiang, D. Liang, L. Cheng. "A review on TVD schemes and a refined flux-limiter for steady-state calculations". *Journal of Computational Physics*, **302**, 114-154 (2015).
[27] D. Zhang, C. Jiang, C. Yang, Y. Yang. "Assessment of different reconstruction techniques for implementing the NVSF schemes on unstructured meshes". *International Journal for Numerical Methods in Fluids*, **74**(3), 189-221 (2014).
[28] D. Zhang, C. Jiang, L. Cheng, D. Liang. "A refined *r*-factor algorithm for TVD schemes on arbitrary unstructured meshes". *International Journal for Numerical Methods in Fluids*, **80**(2), 105-139 (2016).

Research on Recent GPS Crustal Deformation Characteristics in the Northeastern Edge of Qinghai-Tibet Plateau

Haiping Ma[1], Yanqiang Wu[2], Jian'gang Feng[1], Rong Xu[1], Shayi Wu[1] and Qian Wang[1]

[1] Lanzhou Institute of Seismology, China Earthquake Administration (CEA), Lanzhou, P. R. China.
[2] The First Monitoring and Application Center, CEA, Tianjin, P. R. China.
Email: lutmhp@163.com

Abstract. Based on the GPS data in the northeastern edge of Qinghai-Tibet Plateau, the velocity field and the extension change rate of baseline between stations are analyzed. The crustal movement characteristics and their dynamic changes in this region are discussed. The result indicates that the baseline in the regions demonstrates an overall characteristic of shrinking in the NE direction and extending in the NW direction; The extrusion of Bayan Har Block in SE direction since 2009 has amplified obviously, which leads to the enhancement of correlated movement in SE direction in the NW side of Longmenshan Fault. In addition, Tianzhu empty regions and the area near the epicenter of Menyuan Ms6.4 earthquake in 2016 are right in the boundary of this differentiated movement. Recently, a certain amount of strain energy might have been accumulated in the nearby area of the northern margin of Alkin Fault Zone, west boundary of Erdos Block and west to105°N on Northern Xiqinling Fault. The seismic risk in the west and east section of the intersecting area between Bayan Har Block and Qaidam Block is high. The west section of Northern Xiqinling Fault belongs to the inner ring of clockwise vortex and possesses medium-strong seismic risk.

1. Introduction

The northeastern edge of Qinghai-Tibet Plateau is where the Tibet Block, Alxa Block and Erdos Block converge [1]. Considering that it is located in the drastically-uplifting frontier part in the northeastern Qinghai-Tibet Plateau, this region becomes one of the zones where the structural activities are the most active and the seismic activities are the most frequent in Qinghai-Tibet Plateau. In this region, active fault zones, such as Qilian Mountain, Haiyuan Fault, Liupanshan, northern fault zone of West Qinling Mountain, have been developed from north to south since the Neotectonic era. Driven by the main active force by northward extrusion and impact onto the Eurasia Continent by the Indian Plate, and blocked by the rigid Alxa Block and Erdos Block, the characteristics of geological structural activities in the active fault zone in the northeastern edge of Qinghai-Tibet Plateau are mainly demonstrated as NE extrusion-thrust and sinistral strike slip-rotation [2]. Huining *M*s7.0 earthquake in 1352, Zhongwei *M*s7.2 earthquake in 1561, Guyuan *M*s7.0 earthquake in 1622, Tianshui South *M*s8.0 earthquake in 1654, Zhongwei *M*s7.5 earthquake in 1709, Yinchuan-Pingluo *M*s8 earthquake in 1739, Wudu *M*s8 earthquake in 1879, Haiyuan *M*s8.5 earthquake in 1920, Gulang *M*s8 earthquake in 1927 and over ten M6~6.9 earthquakes occurred in this region [3, 4], and thus this region is equipped with strong seismic development background.

Earthquake results from the sudden release of all the long-term accumulated strain during crustal movement [5]. The energy for earthquake preparation comes from the accumulation of strain energy generated in crustal movement process under the impact of structural boundary dynamic force [6]. Considering that GPS observation can provide the relative movement information in multiple spatial dimensions, applying these observation results into the research of relation between temporal-spatial distribution dynamics of crustal movement and strong earthquakes as well as the seismic deformation process and mechanism might become a vital aspect for seismic forecasting research to expand into physical forecasting. Generally, strong earthquakes occur in the regions where the crustal movement differs greatly. In other words, strong earthquakes take place in the regions where the magnitude and direction of movement velocity vector differ greatly. Strain can only be accumulated when there is difference in crustal movement. When the regional crustal movement is consistent, no strain will be accumulated, the risk of strong earthquakes is not high [7]. In this paper, based on the GPS mobile observation data in the northeastern edge of Qinghai-Tibet Plateau between 1999 and 2015, the velocity field in the research region is analyzed; then, the GPS continuous observation data since 2010 are utilized to analyze the extension change rate of baseline between stations; next, the current crustal movement characteristics and their dynamic changes in this region are studied. The research result can offer reference basis for the forecasting of strong seismic risks in this region.

2. Velocity Field Characteristics Based on GPS Mobile Observation

The GPS observation data collected in multiple periods since 1999 are processed by GAMIT/GLOBK software and QOCA software [8, 9] to obtain the GPS velocity field 1999-2007 year, 2009-2013 year and 2013-2015 year. The specific processing strategy can refer to literature [10]. On the basis of ITRF2005 reference framework velocity field result, the stable South China Block is adopted as the reference datum and then transformed. Fig.1 presents the comparison result of the velocity field in above three periods.

Based on Figure 1, under the extrusion impact of Indian Plate and certain blockade impact of relatively stable Alxa Block and Erdos Block, the fault zone extending from the north boundary of Qilian Block to Haiyuan the NWW direction demonstrates obvious sinistral strike-slip and extrusion differentiated movement characteristics. In other words, the north of this fault zone demonstrates W-NWW movement while the south demonstrates NE-E-ES movement. The velocity field in the region ranging between 36°N and 42°N, between 94°E and 100°E, forms a counterclockwise vortex across different geological structural elements, and the direction of its movement vector is shown as NE-near NS-NW direction on the whole from its south to its north, and demonstrates counterclockwise rotation trend. The region surrounding by north to 34°N, east to 100°E, the north boundary of Qilian Block, Haiyuan Fault and Liupanshan Fault forms a clockwise vortex across different geological structural elements, and the direction of its movement vector is shown as NE-near EW-SE-S direction on the whole from its west to its east, and demonstrates clockwise rotation trend. In the area near the block boundary and active fault controlled by vortex movement, the abnormal crustal deformation zone is where the crustal movement velocity is very high or the magnitude and direction of two adjacent observation stations differ obviously [11]. The direction of velocity vector on both sides of the north boundary of Qilian Block-Haiyuan Fault-Liupanshan Fault differs greatly. These are the actively-moving faults in Neotectonic ear and the major seismic activity zones. Considering that the north boundary of Qilian Block and the west boundary of Erdos Block are extruded, the velocity on their NE side is small. The velocity of Bayan Har Block is obviously larger than that of other areas in the research region.

In Fig.1, the black arrow represents the velocity field result 1999-2007 year in the northeastern edge of Qinghai-Tibet Plateau (observed before Wenchuan *M*s8.0 earthquake). This period is not influenced by the Wenchuan *M*s8.0 earthquake. Therefore, the velocity field in this period can be adopted as the basic movement deformation background field in a relative sense in the northeastern edge of Qinghai-Tibet Plateau while the velocity fields in the later periods serve as the dynamic fields. It can be known from Fig.1 that the direction of the movement vector in the east of Bayan Har Block turns from NEE and EW directions 1999-2007 year to EW and SE directions 2009-2013 year and 2013-2015 year, and the velocity increases obviously, which indicates that the extrusion of Bayan Har Block in SE direction since 2009 has amplified obviously and reflects that the Wenchuan *M*s8.0 earthquake leads to the Longmenshan Fault unlocking in a wide range, causing the enhancement of correlated movement in SE direction in the NW side of Longmenshan Fault. In the three velocity fields, the magnitude and direction of velocity vector on both sides of the west section of the intersecting area between Bayan Har Block and Qaidam Block differ, and the vector direction is NS in the north and NEE in the south, and the velocity in the south is obviously larger than that in the north; compared to the velocity field 1999-2007 year, the difference between the velocity magnitudes on both sides of the east section of the intersecting area between Bayan Har Block and Qaidam Block 2009-2013 year and 2013-2015 year is enlarged, and the velocity increase in the south is obviously larger than that in the north; in addition, the movement vectors on both sides differ, NE direction in the north but EW and SE directions in the south.

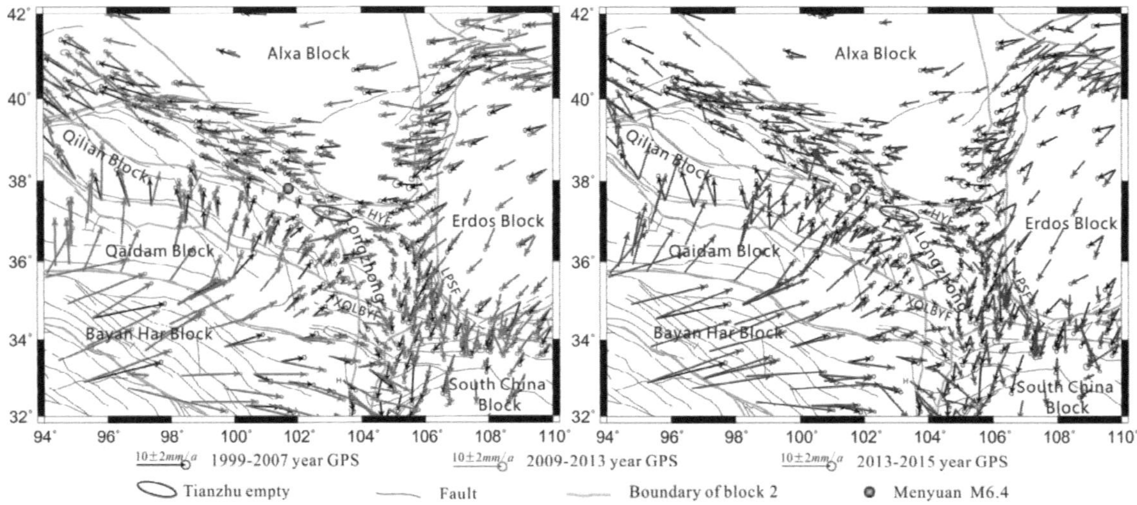

Figure.1. Results of GPS velocity field in the study area (South China Datum), (Black arrow represents 1999-2007year, red arrow represents 2009-2013 year, blue arrow represents 2013-2015 year, HYF represents Haiyuan Fault, LPSF represents Liupanshan Fault, XQLBYF represents Northern Xiqinling Fault)

In the mid-west Qilian Block, the direction of velocity vector is demonstrated as near NS direction on the whole 1999-2007 year and 2009-2013 year. 2013-2015 year, the direction of velocity vector changes obviously and is demonstrated as NNE direction on the whole, which enlarges the difference between the movement directions of mid-west Qilian Block and its northern regions. In the three velocity fields, the movement direction of east Qilian Block is NE direction and the movement direction on its NE side is NWW direction. In addition, in the velocity field 2013-2015 year, the velocity magnitude in the east Qilian Block obviously increases, which indicates that the magnitude and direction of velocity vector on both sides of the east section in the north boundary of Qilian Block differ greatly. It is shown by research that strong earthquakes occur in the regions where the magnitude and direction of movement velocity vector differ greatly. Strain can only be accumulated when there is difference in crustal movement. Tianzhu empty and the area near the epicenter of Menyuan *M*s6.4 earthquake on January 21st, 2016 are right in the outer ring of the clockwise vortex, and the movements on both sides differ greatly, and thus they are where the strain is accumulated.

The movement directions on both sides of the west section of Haiyuan Fault Zone are opposite, namely W direction in the north side and E direction in the south side. The difference in the directions of velocity fields on both sides of the east section of Haiyuan Fault Zone obviously attenuates 2013-2015 year, and the velocity direction on the north side turns from W direction in the previous two periods to SW direction 2013-2015 year. Haiyuan Fault Zone is the sinistral strike-skip fault zone. In the velocity field 2013-2015 year, the velocity direction in the large area on NE side changes into extrusion movement relatively perpendicular to the fault zone. This might be the cause for the velocity increase on the SW side. When the regional crustal movement is consistent, no strain will be accumulated and the risk of strong earthquakes is not high, which indicates that the current risk of strong earthquake in the east section of Haiyuan Fault Zone is low.

Compared to the previous two periods, the 2013-2015 year movement velocity on both sides of the section west to 105°N on Northern Xiqinling Fault differs greatly, larger on the south side and smaller on the north side. This area is in the inner ring of the clockwise vortex, and the direction turns from near EW direction to SE direction. This area is where the strain is accumulated and has the possibility of earthquakes.

It is shown by the velocity field in three periods that the direction of velocity vector on both sides of the north boundary of Qilian Block-Haiyuan Fault-Liupanshan Fault differs greatly, and these are the actively-moving faults in Neotectonic ear; the velocity field in the region ranging between 36°N and 42°N, between 94°E and 100°E, forms a counterclockwise vortex across different geological structural elements; the region surrounding by north to 34°N, east to 100°E, the north boundary of Qilian Block, Haiyuan Fault and Liupanshan Fault forms a clockwise vortex across different geological structural elements. The extrusion of Bayan Har Block in SE direction since 2009 has amplified obviously, which reflects that the Wenchuan *M*s8.0 earthquake leads to the Longmenshan Fault unlocking in a wide range, causing the enhancement of correlated movement in SE direction in the NW side of Longmenshan Fault. The magnitude and direction of velocity vector on both sides of the west section of the intersecting area between Bayan Har Block and Qaidam Block differ to a certain extent, and the difference between the movement velocity magnitudes on both sides of the east section enlarges, and the movement vectors differ as well. The difference between the movement directions of mid-west Qilian Block and its northern regions enlarges, and the magnitude and direction of velocity vector on both sides of the east section in the north boundary of Qilian Block differ greatly. In addition, Tianzhu empty and the area near the epicenter of Menyuan *M*s6.4 earthquake on January 21st, 2016 are right in the boundary of this differentiated movement. The movement directions on both sides of the west section of Haiyuan Fault Zone are opposite. In recent periods, the movement velocity magnitude on both sides of Northern Xiqinling Fault ranging between 103.5°N-105°N differs greatly.

3. Time sequential movement characteristics of GPS baseline

GPS observation is divided into mobile observation and continuous observation. The mobile observation can guarantee the spatial resolution of the crustal deformation research while the continuous observation compensates for the insufficient time resolution in mobile measurement network and can objectively reflect the time dynamic process of crustal deformation. GPS baseline time-series between stations is basically not influenced by the reference datum and can directly reflect the extension of crustal structure on the line between two stations, namely the dynamic change in relative movement between two stations. In this paper, the data from GPS continuous observation stations collected in the construction of China Continental Structural Environmental Monitoring Network Project since 2010 are utilized to analyze the characteristics of baseline movement velocity of 48 GPS baselines in the northeastern edge of Qinghai-Tibet Plateau in recent seven years. Fig.2 shows the calculation results of extension change rate of baselines between GPS stations in the northeastern edge of Qinghai-Tibet Plateau, in which the red line represents that the baseline is shorten, the blue line represents that the baseline is extends, and the line thickness represents the magnitude of extension deformation. Table 1 presents the statistical results of GPS baseline movement velocity in the research region corresponding to Figure 2, in which the positive velocity represents that the baseline is extended and the negative velocity represents that the baseline is shorten.

It can be known from Figure 2 and Table 1 that the baseline extension change rate in the northeastern edge of Qinghai-Tibet Plateau demonstrates the overall characteristic of shortening in NE direction and extending in NW direction. The baselines NMEJ-NMAY, YANC-GSPL and GSWD-SNTB are the inner baselines of Alxa Block, Erdos Block and South China Block respectively, and the movement velocities are respectively 0.24±0.01 mm/a, -0.07±0.02 mm/a and -0.69±0.02 mm/a, which indicates that the crustal deformation rate inside the three rigid blocks is very small.

The baselines QHGE-QHMD, QHMD-QHMQ and QHMQ-GSMA are respectively located in the west, middle and east section of the intersecting area between Bayan Har Block and Qaidam Block, and their movement velocities are respectively 8.43±0.02 mm/a, -2.84±0.01 mm/a and -0.13±0.01

mm/a. Specifically, the baseline QHGE-QHMD in the west section is extended. In combination of Figure 1, the directions of velocity vector on both sides in the west section differ, N direction on the north side and NEE direction on the south side, which further indicates that the deformation characteristic in this region is extension deformation, and the movement velocity on both sides differ greatly. In addition, the extension rate of this baseline is the largest among the 48 baselines, which indicates that its sinistral strike-slip is the largest and thus the strain accumulation effect in this region is the most obvious. In the middle section, the baseline QHMD-QHMQ is shortening. In combination of Figure 1, the magnitude and direction of velocity vector on both sides in the middle section differ slightly. In the east section, the baseline QHMQ-GSMA has a relatively smaller shortening rate. In combination of the velocity field results in the last two periods, the velocity magnitude on both sides of the east section differs obviously, which indicates that the strain accumulation rate in the east section in recent years has slowed down.

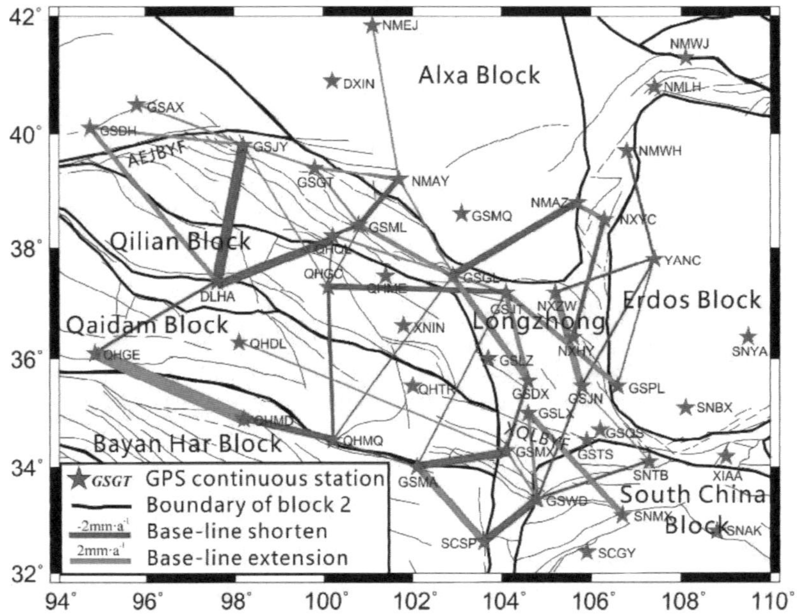

Figure 2. Shrinkage ratio of GPS station baseline in the study area.(AEJBYF represents the northern margin of Alkin Fault, XQLBYF represents Northern Xiqinling Fault)

The movement velocities of baselines GSAX-GSJY and GSDH-GSJY crossing the northern margin of Alkin Fault Zone are respectively 0.4±0.01 mm/a and 0.8±0.01mm/a, which indicates that the sinistral strike-slip movement in the northern margin of Alkin Fault Zone is relatively weak, and a certain amount of strain might exist. The movement velocities of baselines YANC-NMWH, YANC-NXZW and YANC-GSJN in the west boundary of Erdos Block are respectively -0.66±0.02 mm/a, -0.61±0.01 mm/a and -0.73±0.02 mm/a, which indicates that the shortening movement on both sides of the west boundary of Erdos Block is relatively small and a certain amount of strain might have been accumulated. In addition, under the blockade of Alxa Block and Erdos Block, the movement in NE direction in the northeastern edge of Qinghai-Tibet Plateau turns to S direction (the baselines QHGC-QHQL and QHGC-QHMQ in NS direction west to the Longzhong Basin Structure are shortening). Therefore, the baselines GSDX-GSJT, NXZW-GSJN and GSWD-GSLX in NS direction inside the Longzhong Basin Structure close to the west boundary of Erdos Block are extending.

The movement velocities of baselines GSMX-GSDX, GSWD-GSJN and GSWD-GSLX crossing Northern Xiqinling Fault are respectively -1.09±0.01 mm/a, -0.54±0.02 mm/a and 1.03±0.02mm/a, which indicates that the mutual movement on both sides of this fault zone in recent years has attenuated, and a certain amount of strain energy might have been accumulated in nearby regions. In the meanwhile, the discussion result on velocity field in Fig.1 further illustrates that the movement velocity on both sides of the west section of Northern Xiqinling Fault differs greatly. In addition, this region belongs to the inner ring of clockwise vortex, which confirms that this region is where the strain is accumulated and possesses medium-strong seismic risk.

Table1. Annual variation ratio of GPS baselines

Baseline	Trend	State	Rate/mm·a^{-1}	Error/mm
GSAX-GSJY	NW	Extension	0.40	0.01
GSDH-DLHA	NW	Extension	1.71	0.01
GSDH-GSJY	NWW	Extension	0.80	0.01
GSDX-GSGL	NW	Extension	2.41	0.01
GSGT-GSML	NW	Extension	0.49	0.01
GSJT-GSGL	NWW	Extension	0.54	0.02
GSJT-GSPL	NW	Extension	1.83	0.02
GSJY-GSGT	NWW	Extension	0.39	0.01
GSMA-SCSP	NW	Extension	4.08	0.02
GSML-GSGL	NW	Extension	1.60	0.01
GSWD-GSMX	NWW	Extension	0.51	0.02
NMAZ-NXYC	NW	Extension	1.49	0.03
QHGE-QHMD	NW	Extension	8.43	0.02
QHQL-GSGL	NW	Extension	0.40	0.03
NXZW-GSJN	NW	Extension	2.98	0.01
GSGL-NMAY	NWW	Extension	0.33	0.03
SNMX-GSLX	NW	Extension	2.23	0.03
SNTB-GSJN	NW	Extension	0.001	0.02
QHGC-GSJY	NW	Extension	0.13	0.02
QHDL-GSMX	NW	Extension	0.15	0.01
GSGT-NMAY	NNW	Extension	0.73	0.02
GSDX-GSJT	NS	Extension	1.61	0.01
GSWD-GSLX	NS	Extension	1.03	0.02
NMEJ-NMAY	NS	Extension	0.24	0.01
DLHA-GSJY	NE	Shorten	-5.06	0.01
DLHA-QHQL	NEE	Shorten	-3.88	0.01
GSML-QHQL	NE	Shorten	-1.12	0.01
GSMX-GSDX	NNE	Shorten	-1.09	0.01
GSWD-SNTB	NE	Shorten	-0.69	0.02
NMAY-GSML	NE	Shorten	-1.95	0.01
NXYC-NXHY	NNE	Shorten	-1.44	0.02
SCSP-GSWD	NE	Shorten	-3.29	0.02
QHGE-DLHA	NE	Shorten	-0.85	0.01
YANC-GSPL	NNE	Shorten	-0.07	0.02
YANC-NXZW	NEE	Shorten	-0.61	0.01
YANC-GSJN	NE	Shorten	-0.73	0.02
GSGL-NMAZ	NE	Shorten	-3.00	0.02
GSWD-GSJN	NE	Shorten	-0.54	0.02
GSMA-GSJT	NE	Shorten	-0.004	0.02
QHGC-GSML	NE	Shorten	-0.03	0.02
QHMQ GSGL	NE	Shorten	-0.003	0.02
GSMA-GSMX	NEE	Shorten	-3.57	0.01
QHGC-GSJT	EW	Shorten	-2.25	0.02
QHGC-QHMQ	NS	Shorten	-0.93	0.02
NMWH-YANC	NS	Shorten	-0.66	0.01
QHGC-QHQL	NS	Shorten	-0.06	0.02
QHMD-QHMQ	NWW	Shorten	-2.84	0.01
QHMQ-GSMA	NWW	Shorten	-0.13	0.01

In summary, it is shown by the GPS baseline extension change rate result that the baselines in the northeastern edge of Qinghai-Tibet Plateau demonstrate an overall characteristic of shortening in NE direction and extending in NW direction. A certain amount of strain energy might have been accumulated in the areas near the northern margin of Alkin Fault and the west boundary of Erdos Block. It can be judged based on the velocity field analysis result that the seismic risk in the west and east section of the intersecting area between Bayan Har Block and Qaidam Block is high; the west section of Northern Xiqinling Fault belongs to the inner ring of clockwise vortex and possesses medium-strong seismic risk.

4. Cognition and discussion

(1) The velocity field in the region ranging between 36°N and 42°N, between 94°E and 100°E, forms a counterclockwise vortex across different geological structural elements; The region surrounding by north to 34°N, east to 100°E, the north boundary of Qilian Block, Haiyuan Fault and Liupanshan Fault forms a clockwise vortex across different geological structural elements. The extrusion of Bayan Har Block in SE direction since 2009 has amplified obviously, which reflects that the Wenchuan *M*s8.0 earthquake leads to the Longmenshan Fault unlocking in a wide range, causing the enhancement of correlated movement in SE direction in the NW side of Longmenshan Fault.

(2) The magnitude and direction of velocity vector on both sides of the west and east section of the intersecting area between Bayan Har Block and Qaidam Block differ, and the difference between the movement velocity magnitudes on both sides of the east section has enlarged in recent years. Besides, the directions of movement vectors on both sides of the northern boundary of Qilian Block differ greatly. In addition, the movement directions on both sides of the west section of Haiyuan Fault Zone are opposite. In recent periods, the movement velocity magnitude on both sides west to 105°N on Northern Xiqinling Fault differs greatly.

(3) The baselines in the northeastern edge of Qinghai-Tibet Plateau demonstrate an overall characteristic of shortening in NE direction and extending in NW direction. A certain amount of strain energy might have been accumulated in the areas near the northern margin of Alkin Fault and the west boundary of Erdos Block. The seismic risk in the west and east section of the intersecting area between Bayan Har Block and Qaidam Block is high; the west section of Northern Xiqinling Fault belongs to the inner ring of clockwise vortex and possesses medium-strong seismic risk.

5. Acknowledgments

This research was funded by the Earthquake Science and Technology Development Fund of GEA (Grant No.2016M02, 2016Y02, 2017IESLZ07) ,the Earthquake Tracking Task of CEA (2017010221) ,Grant of NSFC (51778590 ,51408567, 41304048).The authors are grateful to the First Crust Deformation Monitoring and Application Center of the China Earthquake Administration for providing the time series data of GPS reference stations.

Ma Haiping, born in 1986, is an assistant professorat Lanzhou Institute of Seismology, Earthquake Administration of Gansu Province. She is engaged in crustal deformation analysis. E-mail: lutmhp@163.com

Corresponding author: Feng Jian'gang, born in 1977, is associate research professor at Lanzhou Institute of Seismology, CEA. He works on seismology and the application of digital seismic data. E-mail: fjgspring@163.com.

6. References

[1] Zhang Dongju. Research on Stress and Strain Field Analyze and Tectonic Activity of Northeast Regions of Qinghai- Tibet Plateau [D]. Chang' An University, 2006. (in Chinese with English abstract)

[2] Zhang Qiang, Zhu Wen-Yao, Xiong Yong-Qing. Net Rotation of the ITRF96 [J]. Chinese Journal of Geophysics, 2000, 43(5):598-606 (in Chinese with English abstract)

[3] Department of Earthquake Disaster Prevention, State Seismological Bureau.1995.Catalogue of Chinese Historical Strong Earthquakes (23th Century B.C.--1911 A.D.)[Z].Seismological Press ,Bering(in Chinese)

[4] Department of Earthquake Disaster Prevention, China Earthquake

Administration.1999.Catalogue of Chinese Modern Earthquakes[z].China Science and Technology Press, BeiJing. 1637(in Chinese)

[5] WANG Shuang-xu, ZHANG Xi, ZHU Yi-qin, et al. Study on the Characteristics of Strong Earthquake Deformation Anomalies and For ecasting Index in the North Segment of the South—north Seismic Belt of China[J]. Northwestern Seismological Journal, 2012,34(2):105~113(in Chinese with English abstract)

[6] WANG Shuang-xu. Consideration on both the geodetic deformation monitoring of the north-south seismic belts and the big earthquake predictions[J]. Earthquake Research in Sichuan,2011,36(4):13~15. (in Chinese with English abstract)

[7] Jiang Zaisen, Yang Guohua, Fang Ying, et al. Study on the Relations between Distribution of Crustal Movement and Strong Earthquakes Using GPS Survey [J]. Recent Developments in World Seismology, 2007, 7:32-42 (in Chinese with English abstract)

[8] Herring, T. A., King, R. W. and McClusky, S. C. GAMIT Reference Manual. GPS Analysis at MIT. Release 10.4. Massachussetts Institute Technology, http://www-gpsg.mit.edu/~simon/gtgk/index.htm(last accessed 2011 October 5), 2010a

[9] Herring, T. A., King, R. W. and McClusky, S. C. GLOBK Reference Manual. Global Kalman filter VLBI and GPS analysis program. Release 10.4. Massachussetts Institute Technology, http://www-gpsg.mit.edu/~simon/gtgk/index.htm (last accessed 2011 October 5), 2010b

[10] Dong, D., Hering, T. A. and King, R. W., Estimating regional deformation from a combination of space and terrestrial geodetic data. J. Geophys. Res. 1998, 72, 200~214

[11] Chen Zhiliang, Liu Yuping, Zhang Xuanyang, et al. GPS Monitoring, Crustal Deformation and Geodynamics in the East Qinghal-tibet Plateau[M].Guangdong Science & Technology Press,2015(in Chinese with English abstract)

Different Monitoring Methods for Building Deformation of Practical Exploration

Xinhua Zhang

College of Civil Engineering, Shanghai Normal University, Shanghai, 201418, China
Email: zhxh152@shnu.edu.cn

Abstract. During the whole life of the building, the common inclination and displacement of the buildings affect the use and safety of the buildings. These problems must be observed by tilt and displacement observation. This paper mainly introduces the building slant measurement method, and the measurement data is prepared to deal with the measurement data in order to obtain the correct data performance of the construction and the stability of the construction and use.

1. Introduction

Today, building, due to the survey, design, construction, operation and maintenance management and the causes of natural disasters arising tilt, uneven settlement in especially in high-rise building, the inclination deformation of content than the subsidence deformation, even a slight tilt will also affect the structure of the beautiful and normal use, but excessive tilt lowers build, safety, and even collapse. Therefore, the construction of this kind of construction is the main content of oblique deformation observation. When a building is seriously deformed, effective engineering must be carried out, such as the high rise of the high rise due to uneven subsidence, must be correct, before it can be used. With the rapid development of social economy, engineering survey services become more and more wide, the structure of the deformation measurement is one of the, deformation measurement, the observation object and object in addition to the building itself, it is including construction sites, the base field, top and bottom structures and internal, etc. And the means of observation are not limited to conventional plane coordinates, elevation and variation, but also modern means, such as close-up photography, non-visible light perspective and so on. In addition to the amount of deformation directly provided by the external observation, the deformation law can be found through data processing and geometric analysis. The corresponding relationship can be analyzed by means of the measurement of the slant volume and the settlement amount of the high-rises, and the whole process of rectifying work is provided.

2. Monitoring purpose

As the basic building of daily life work, the stability and security of people are always concerned. In recent years, due to a variety of reasons, some buildings have been inclined, severely affected normal use, and even endanger user safety. In accordance with relevant regulations, when the subject tilts value above a certain number, the building is considered unsafe. When the inclination of the building has reached or even exceeded the allowable inclination value (1/20), the building has the proper use and economic value, and the proper corrective measures are adopted to support the building. Therefore, accurately grasp the structure of the actual situation is particularly important, the measurement accuracy, building tilt data for relevant units to provide real and effective results of the measurement data needs to be correct. The measures include the settlement observation of the building and the oblique monitoring of the inner and outer tilt of the building. Through data analysis, judgment and foundation settlement calculation of building tilting tilt consistency, subsidence and tilt rate exceed the national standard requirements, for the relevant departments or units of experts and scholars on the deformation structures provide powerful basis rectifying make decisions.

3. Monitoring programmer

The tilt of buildings is calculated by measuring the horizontal displacement of the top of the building relative to the bottom or the upper layer of the building relative to the lower layer. The common methods of measuring building slanting have the instrument observation method, the observation method of the lead hammer, the measurement method of the inclinometer, the basic subsidence observation method, and the near sighting photogrammetry. The overall tilt of the rigid building can be determined by measuring the relative subsidence of the top or base. Tilt was measured, generally using the direct projection method, is the extension of vertical intersection of two sides wall line, measuring 1.5 times to 1.5 times the height of building in the distance, with two precision theodolite or total station set up line in the north and south direction of the point will be projected onto the ground, corner sometimes when leave the building a close observation location and bend at the eyepiece end ocular observation and measuring its tilt. External observation is something near perpendicular to the settlement of a contour line on the direction of the precision theodolite, total station or obtain houses near that point contour line on what direction with the highest position of the top floor at the bottom of the slope displacement; The north and south direction at the same time in close to a settlement observation point perpendicular to the line set up total station or precision theodolite observation, get that point line near the building periphery upstairs in the north-south direction base slope displacement and the roof of their building. According to the height of the building, calculate the inclination rate of buildings in the direction of north and south. This is the most common observation of the sloping value of the building in the construction, but due to the uneven formation of the building in the process of masonry painting. The direct projection method need to be in vertical intersection of two side's wall, on the extension of 1.5 - 2.0 times the height of high-rise buildings in the distance, the corner with two precision theodolites is projected onto the ground, and then take its tilt amount. However, in high-density buildings, it is difficult to implement the gap between them. Therefore, this method cannot be implemented in big cities.

Besides external observation and internal observation method, it is inside the building floors, in close to a settlement observation point perpendicular to the corresponding contour direction with total station or precision theodolite observation, get the settlement points corresponds to the layer to its highest position at the bottom of the slope displacement; At the same time, in the south and north of this layer, which is perpendicular to a settlement point, the sloping displacement of the contour line of the house near the point in the south and north of the site is higher than that of the highest position. Four observation points are taken on each floor, the lowest point is the base of the lowest point, the floor surface is selected 0.5 m high, and the height of the second point is 1m; To the third point height of 1.5 m, the fourth height is 2 m, according to the result of measurement can also calculate the tilt rate of the house, calculating formula for the parties to the upward sloping rate: $\sum$(each layer between the north and the south (something) in the direction of the average tilt rate)/4

4. Monitoring content

Deformation of the building has two kinds of static deformation and dynamic deformation, the static deformation is usually refers to the deformation observation results only show how without under the action of external force during a certain period of deformation value, the deformation value function only change over time. Refers to the external forces under the influence of dynamic deformation and the deformation, dynamic system represented by outside force function for the change of time, the observation is the result of the said buildings at some point of the instantaneous deformation. The slope observation of the observation of deformation observation of buildings, crack observation of buildings and deflection observation of buildings.

IOP Conf. Series: Journal of Physics: Conf. Series **910** (2017) 012029 doi:10.1088/1742-6596/910/1/012029

4.1. Settlement observation

The settlement of buildings is the result of the combination of foundation and superstructure. This is to use level, according to the standard, regular observation of building settlement observation point elevation difference between each point and level, and calculate the observation point of elevation and subsidence and calculate according to the historical data analysis of subsidence, subsidence of buildings is the difference between observation points and leveling point's deformation values. As a result, the stability of the leveling points determine the reliability of the observed values, standard points should be not less than three basis points, the stationing position is not affected by the surrounding environment, leave the building proper distance, but also about observation condition requirement. The number and location of observation point, should according to the structure of the covering surface size, importance, basic form, form, load distribution and geological conditions, can reflect the status of building settlement. The settlement observation technology and data are the basis of the research to solve the settlement of complex foundation, and help to improve the foundation design. At the same time, the relative settlement difference of the settlement observation point is analyzed, and the security condition of the building is monitored.

4.2. Displacement observation

The displacement observation of a building refers to its horizontal displacement observation, which mainly measures the movement of the overall plane position of buildings over time. The foundation is affected by horizontal stress, which makes the plane of the building move, such as the foundation is in the landslide zone or affected by the earthquake. Displacement observation is the measurement of the displacement of the building in the plane position with time. Due to different building structures, different horizontal displacement observation requirements are required. Different methods are needed. The commonly used displacement observation method is base line method, traverse measurement method, triangulation method, and forward intersection method.

4.3. Oblique observation

The oblique observation of buildings, because the upper structure of high-rise buildings and the overall stiffness of the base area, the inclination of the foundation, such as uneven settlement, reflects the tilt of the upper main body. The tilt observation of buildings is the unevenness of the foundation settlement, and the safety of buildings is monitored by oblique observation.

4.4. Fracture observation

Fracture observation, when the foundation of the building is caused by uneven settlement, the walls appear in different sizes of cracks, and the external erosion media infiltrates the buildings and destroys the buildings. Fracture observation can be informed of the status quo and development trend of crack, so that according to the observation data analysis of the causes of cracks, measure the influence degree of the safety of the building, in order to take effective measures to deal with. According to fracture observation and settlement observation, the laws, characteristics and causes of deformation are analyzed to take measures to ensure the safety of buildings. According to the specific situation of the building, the crack observation means, the plaster mark, the thin iron mark or the metal bar sign.

4.5. The deflection observation

Deflection observation, the structure member of the building will generate the deflection with the increase of the load; the size of the flexure will have a great influence on the stress state of the structural member of the building. Therefore, the flexure deformation of building structural components must not exceed a certain value; otherwise the building will be endangered. The common building foundation deflection observation, the deflection observation of the building truss, the deflection observation of the bridge cantilever beam, the deflection observation of the building main body. The deflection observation of the building is the degree of deflection of the building member. For vertical high-rise buildings, three displacement observation points should be set up, medium and lower, and displacement observation: the deflection of the component can be calculated by using the displacement of three points.

4.6. The method of building deformation observation
In the observation of deformation of buildings, besides the above five deformation observations, there is also torsional observation, vibration observation, bending observation, offset observation, etc. The above five kinds of deformation observation of buildings, and not every building, must be applied to the different USES of buildings, the above five kinds of deformation observation of buildings, and not every building, must be applied to the different USES of buildings, the deformation observation point is also different. The method of building deformation observation should be determined according to the nature of the building, the size of the building, the use function, the observation precision and the surrounding environment. In the general case, the settlement deformation of buildings is measured by precision level measurement, liquid static level measurement or micro level measurement. For horizontal displacement, if it is a straight line building, it is usually observed by reference line method. If it is a curved structure, it is usually observed by traverse method. For the deformation of cracks or expansion joints of buildings, special instruments or other methods for measuring gap can be used to observe.

5. Observational data analysis
To deformation observation data monitoring buildings safe use and develop its project benefit, in the engineering field observation data, the exact analysis, still need to do to deformation observation data to make the right process. It should be careful, comprehensive and scientific to deal with deformation observation data and conform to the principle of experimental data processing. The deformation observation data of the actual project is processed and sorted into a diagram for practical application. To discuss the reason of building deformation, the value function of the relationship between load and deformation are given, so as to make the right judgment for building operation state and deformations forecast, and provide the basis for modifying design parameters. The content of the deformation observation data analysis is the geometric analysis of the deformation of the building, which is the geometric description of the spatial change of the building. The second is the physical explanation of the deformation of the building, which explains the deformation of the building.

6. Conclusion
The practice shows that the measurement of the inclination of buildings is one of the important means to provide scientific basis for the safe and effective means of building emergency. Only by building subsidence observation, oblique observation, such as scientific observations, a combination of observation data, can grasp the building safety, which will provide a scientific basis for experts to make scientific decision.

7. Acknowledgement
In this paper, the research was sponsored by Shanghai municipal education commission key course construction project 2015 - civil engineering surveying (A-0131-17-003026).

8. References
[1] Jiye Li, Jie dong. Measurement engineering practical technology and case. Beijing: China building materials industry press, 2012.11:159-172.
[2] Wusheng Hu, Civil engineering constructional measurement handbook [M]. Beijing: people's transport press. 2011.
[3] Qiuhong Geng, Qingzi Liu. The application of subsidence observation technology in high-rise building construction [J]. Shanxi architecture. 2008, 34 (22) : 336-337.
[4] Yeqing Tang. New technology of building correction [J]. Architecture technology, 1995 (6) : 323-327.
[5] Zhengjun Za, Feng Bao, a method of building tilt measurement [J]. Sichuan surveying and mapping, 2006 (4) : 175-177.
[6] Xianghong Hua, Xinzhou Wang, etc., the ground subsidence has a negative effect on the safety of buildings. Engineering survey, 2007 (4) : 58-60.

[7] Bing Wang, You Pan, Several problems of settlement monitoring of construction engineering [J]. Construction technology, 2005 (4) : 276-277.
[8] Fuchang Li. Analysis of construction surveying in building survey [J]. The gold card engineering (economics and law), 2010 (12) : 169-170.
[9] Tilong Gu, Jingqian Qiu. Multi-storey residential buildings tilt accident analysis and its rectifying treatment [J]. Jiangsu building, 2006 (109) : 5-12.

Study on the Algorithm of Judgment Matrix in Analytic Hierarchy Process

Zhiyong Lu, Futong Qin and Yican Jin

Luoyang Electronic Equipment Test Center of China, Luoyang, China
Email: luzhiyong2002@sohu.com, ftqin@163.com, solarflare@yeah.net

Abstract. A new algorithm is proposed for the non-consistent judgment matrix in AHP. A primary judgment matrix is generated firstly through pre-ordering the targeted factor set, and a compared matrix is built through the top integral function. Then a relative error matrix is created by comparing the compared matrix with the primary judgment matrix which is regulated under the control of the relative error matrix and the dissimilar degree of the matrix step by step. Lastly, the targeted judgment matrix is generated to satisfy the requirement of consistence and the least dissimilar degree. The feasibility and validity of the proposed method are verified by simulation results.

1. Introduction

Analytic Hierarchy Process (AHP) is a systematic analysis method which proposed by Professor Thomas L. Saaty who was in the University of Pittsburgh in the mid-1970s. Its basic principle is to make decision by qualitative and quantitative analysis on the basis of goals, guidelines, programs and other levels which were decomposed by complex system. It has been widely used in auxiliary decision-making, pattern recognition, effectiveness evaluation and many other aspects in many areas such as social, economic, military, management, etc.

The judgment matrices are needed in AHP analysis, but the judgment matrices given by experts are often inconsistent and need to be adjusted. Research on the consistency adjustment of judgment matrix has been the difficulty and hotspot of AHP method. The traditional methods[1][2] of consistency adjustment of judgment matrix is that the inconsistent matrix can be obtained by properly perturbing a completely consistent judgment matrix. By constructing such a perturbation matrix, we can find the elements that have the greatest disturbance to the original judgment matrix. Through the adjustment of the element we can adjust the consistency of the judgment matrix. However, this method does not take into account the expert judgment information provided by the judgment matrix, nor does it guarantee that the elements of the resulting judgment matrix must be in the range of 1 to 9 and its reciprocal, which is contrary to the principle of constructing judgment matrix proposed by Professor Saaty. In this paper, an AHP judgment matrix generation algorithm based on pre-ordering and round-up function was proposed, and it can solve this problem.

2. Traditional Method of Constructing AHP Judgment Matrix

Professor Saaty described the method of constructing AHP judgment matrices as follows:

2.1 Establish Target Hierarchical Hierarchy Model

Suppose that the problem can be divided into a hierarchical model as shown in Figure.1.

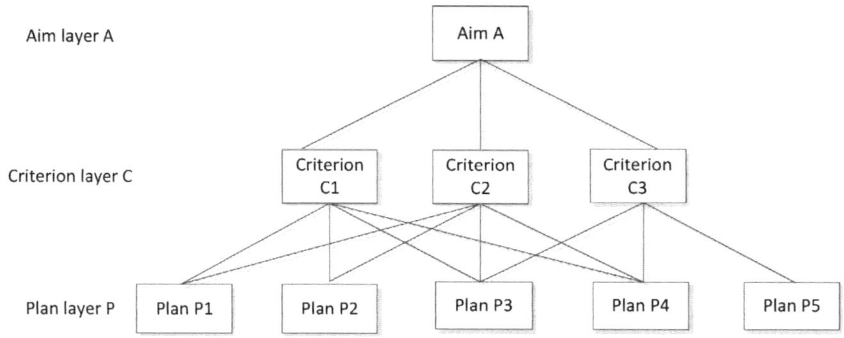

Figure 1 A Hierarchical Hierarchy Model

2.2 Construct The Judgment Matrix

The values of the elements in the judgment matrix are generally determined by 1 ~ 9 scale method[3][4], mainly through expert assessments or from historical (empirical) data. Table 1 lists the meaning of the scale 1 ~ 9.

Table 1 The meaning of scale in judgment matrix

Scale	Meaning
1	Indicating that the two factors are of equal importance
3	Indicating that the former factor is slightly more important than the latter
5	Indicating that the former factor is significantly more important than the latter
7	Indicating that the former factor is strongly more important than the latter
9	Indicating that the former factor is extremely more important than the latter
2,4,6,8	Indicating an intermediate value of the above-described adjacency determination
reciprocal	If the ratio of the importance of factor i and j is aij, then the ratio of the importance of factor j and i is 1/aij.

2.3 Consistency Check

In order to avoid the interference of other factors to the judgment matrix, it is required that the judgment matrix satisfy the general consistency in practice and the consistency check should be carried out. Only through the check, it can be determined that the judgment matrix is logically reasonable, and can continue to analyze the results. The consistency check of the judgment matrix according to the formula (1).

$$CR = CI / RI \tag{1}$$

In the formula, CR is the consistency ratio. When $CR < 0.10$, it can be considered that the consistency of the judgment matrix is acceptable, otherwise the judgment matrix should be modified appropriately. CI is the consistency index, which can be calculated by the formula(2)[5].

$$CI = (\lambda_{max} - n)/(n - 1) \tag{2}$$

In the formula, λ_{max} is the maximum eigenvalue of judgment matrix, n is the number of factors, RI is the random consistency index, which can be found in table 2[4].

Table 2 Average Random Consistency Index of 3 ~ 11th Order Matrix

Order	3	4	5	6	7	8	9	10	11
RI	0.58	0.90	1.12	1.24	1.32	1.41	1.45	1.49	1.52

The maximum eigenvalue λ_{max} is the relative weight of each factor of the judgment matrix against its criterion. For the calculation of λ_{max}, first we can be use sum vector method[4][6] to solve the eigenvectors W, then do as follows:

First, each column of the judgment matrix is normalized.

$$\overline{a_{ij}} = \frac{a_{ij}}{\sum_{k=1}^{n} a_{kj}}, i, j = 1, 2, \dots, n \tag{3}$$

Next, each column of the normalized judgment matrix is added by rows

$$\overline{W_i} = \sum_{j=1}^{n} \overline{a_{ij}}, j = 1, 2, \dots, n \tag{4}$$

Then, the vector $\overline{W} = \left[\overline{W_1}, \overline{W_2}, \dots, \overline{W_n}\right]^T$ is normalized

$$W = \frac{\overline{W_i}}{\sum_{k=1}^{n} \overline{W_k}}, i = 1, 2, \dots, n \tag{5}$$

The result $W = \left[W_1, W_2, \cdots, W_n\right]^T$ is the desired eigenvector.

Finally, the maximum eigenvalue of the matrix is as follows.

$$\lambda_{max} = \sum_{i=1}^{n} \frac{(AW)_i}{nW_i} \tag{6}$$

3. AHP Judgment Matrix Generation Algorithm Based on Pre-sort and Top-rounding Function

3.1 Definitions and Propositions

Definition 1: Initial Judgment Matrix: The matrix $A = (a_{ij})_{n \times n}$ generated by the following rules is called the initial judgment matrix.

Step1 Compare the importance of the factors belonging to the same goal, pre-sort the factor set, the factor set in descending order is $r_1, r_2, \cdots, r_n$, in which $n \geq 3$.

Step2 According to expert experience, construct a matrix $A = (a_{ij})_{n \times n}$. Because the factor set has been pre-sorted, the first row of elements are all integers, and $1 = a_{11} \leq a_{12} \leq \cdots \leq a_{1n}$.

Definition 2: Comparison Matrix: The matrix $B = (b_{ij})_{n \times n}$ generated by the following rules is called the comparison matrix.

Step 1 The first row of the matrix B has a relationship $b_{11} \leq b_{12} \leq \cdots \leq b_{1n}$, and $b_{1j} \in \{1, 2, 3, 4, 5, 6, 7, 8, 9\}$, while the first column element's value of B is $b_{j1} = 1 / b_{1j}$, in which $j = 1, 2, \cdots, n$.

Step 2 The kth row element value of the matrix B is $b_{kj} = \left\lceil \frac{b_{(k-1)j}}{b_{(k-1)k}} \right\rceil$, correspondingly, the kth column element value is $b_{jk} = 1 / b_{kj}$,in which $j = k, k+1, \cdots, n$, $2 \leq k \leq n$.

Step 3 Until the element value b_{nn} is calculated, the algorithm ends, the comparison matrix is generated.

Definition 3: Relative Error Matrix: The matrix $E = (e_{ij})_{n \times n}$ generated by the following rules is called the relative error matrix

The matrix E is obtained by subtracting the comparison matrix from the initial judgment matrix and comparing it with the comparison matrix for the corresponding position, and has a relationship $e_{ij} = \left|a_{ij} - b_{ij}\right| / b_{ij}$, $i = 1, 2, \cdots, n$, $j = 1, 2, \cdots, n$.

Definition 4: Transition Judgment Matrix: The generated matrix transformed from the initial judgment matrix is called the transition judgment matrix $G = (g_{ij})_{n\times n}$.

Definition 5: Matrix Dissimilarity: The Euclidean distance between the principal eigenvectors of two matrices is called the matrix dissimilarity. The dissimilarity of matrices $A = (a_{ij})_{n\times n}$ and $B = (b_{ij})_{n\times n}$ counted as $d(A, B)$. There is

$$d(A, B) = \sqrt{\sum_{k=1}^{n} \left| w_A^{\ k} - w_B^{\ k} \right|^2}$$

w_A and w_B are the main eigenvectors of matrices A and B. The smaller value indicates that the matrices A and B are more similar.

The smaller the value of $d(A, B)$ indicates that the matrices A and B are more similar.

Definition 6: Target Judgment Matrix: Traverse the comparison matrix that has the same order with the initial judgment matrix, if the generated transition judgment matrix not only satisfies the consistency requirement but also has the smallest matrix dissimilarity with the initial judgment matrix, the transition judgment matrix is the target judgment matrix $D = (d_{ij})_{n\times n}$.

Proposition 1: All comparison matrices meet the consistency requirements. The correctness of the proposition is verified by the simulation in this paper 3.2.

3.2 The Simulation Verification of Proposition 1

Whether the comparison matrix can meet the requirement of consistency, it's need to check. Computer simulation is used to verify the consistency of the results. The following is the verification matrix consistency verification traversal nesting algorithm:

Step1 Initialize.

Ⅰ. Enter the order of the matrix n, $n \geq 3$;

Ⅱ. Use variable $m = n - 1$ controle the nesting level of process.

Step2 For $n_1 = 1$ to 9;

Step3 if($m -- \geq 1$) For $n_2 = n_1$ to 9;

Step4 if($m -- \geq 1$) For $n_3 = n_2$ to 9;

⋮

Stepn-1 if($m -- \geq 1$) For $n_{n-2} = n_{n-3}$ to 9;

Stepn if($m -- \geq 1$) For $n_{n-1} = n_{n-2}$ to 9;

Ⅰ. suppose $b_{11} = 1, b_{12} = n_1, \cdots, b_{1n} = n_{n-1}$, vector($b_{11}, b_{12}, \cdots, b_{1n}$)as the first row of the comparison matrix, call the pre-sort and the rounding function algorithm to construct the comparison matrix B;

Ⅱ. Normalize the comparison matrix according to equation (3)

Ⅲ. The comparison matrix added by rows according to equation (4) , get the vector $\overline{w}$.

Ⅳ. According to equation (5) get the normalized w, which is the desired feature vector.

Ⅴ. Calculate the largest eigenvalue $\lambda_{\max}$ of the comparison matrix according to equation (6).

Ⅵ. If $CR = (\lambda_{\max} - n)/(n-1)/RI \geq 0.10$, it does not satisfy the consistency requirement.

Stepn+2 The algorithm end.

Consistency checking of all possible AHP judgment matrices of the 3~30th order comparison matrix by the consistency verification traversal nesting algorithm, simulation results show that all meet the requirements. Table 3 lists the number of all possible AHP judgment matrices of 3~29th order comparison matrices.

If first row vector of the matrix given by experts is (1, 1, 2, 4, 6, 6, 7, 8, 9, 9), the resulting comparison matrix is B, which is as follows.

IOP Conf. Series: Journal of Physics: Conf. Series **910** (2017) 012030 doi:10.1088/1742-6596/910/1/012030

$$B=\begin{pmatrix} 1 & 1 & 2 & 4 & 6 & 6 & 7 & 8 & 9 & 9 \\ 1 & 1 & 2 & 4 & 6 & 6 & 7 & 8 & 9 & 9 \\ 1/2 & 1/2 & 1 & 2 & 3 & 3 & 4 & 4 & 5 & 5 \\ 1/4 & 1/4 & 1/2 & 1 & 2 & 2 & 2 & 2 & 3 & 3 \\ 1/6 & 1/6 & 1/3 & 1/2 & 1 & 1 & 1 & 1 & 2 & 2 \\ 1/6 & 1/6 & 1/3 & 1/2 & 1 & 1 & 1 & 1 & 2 & 2 \\ 1/7 & 1/7 & 1/4 & 1/2 & 1 & 1 & 1 & 1 & 2 & 2 \\ 1/8 & 1/8 & 1/4 & 1/2 & 1 & 1 & 1 & 1 & 2 & 2 \\ 1/9 & 1/9 & 1/5 & 1/3 & 1/2 & 1/2 & 1/2 & 1/2 & 1 & 1 \\ 1/9 & 1/9 & 1/5 & 1/3 & 1/2 & 1/2 & 1/2 & 1/2 & 1 & 1 \end{pmatrix}$$

After the consistency check, λ_{max} =10.0629, CR =0.0047<0.1, which is satisfied the consistency requirements.

Table 3 The number of all possible AHP judgment matrices of 3~29th order comparison matrices

Order	Number of matrices	Order	Number of matrices	Order	Number of matrices
3	45	12	75582	21	3108105
4	165	13	125970	22	4292145
5	495	14	203490	23	5852925
6	1287	15	319770	24	7888725
7	3003	16	490314	25	10518300
8	6435	17	735471	26	13884156
9	12870	18	1081575	27	18156204
10	24310	19	1562275	28	23535820
11	43758	20	2220075	29	30260340

3.3 Algorithm Flow

First, the initial judgment matrix is generated and the transition judgment matrix is equivalent to the initial judgment matrix. Then all the comparison matrices of the same order as the initial judgment matrix are traversed. According to the descending order of the elements of the error matrix, which is generated by the initial judgment matrix and the comparison judgment matrix, replacing the value of the corresponding place of the transition judgment matrix one by one with the value of the corresponding place of the comparison matrix, and replacing the element values of the corresponding symmetric place until the consistency requirement is satisfied. Finally compare and find out all the comparison matrices and the corresponding transition judgment matrix which satisfies the consistency requirement and the matrix dissimilarity is the smallest, the transition judgment matrix is the target judgment matrix. The specific algorithm flow is as follows.

Step1 Initialize

Ⅰ. Input the order of the matrix n , $n \ge 3$.

Ⅱ. Use variable $m = n - 1$ control the nesting level of process.

Ⅲ. The smallest matrix dissimilarity is $d = 2000000.0$.

Step2 Generating the n order initial judgment matrix $A=(a_{ij})_{n\times n}$, the transition judgment matrix $G=(g_{ij})_{n\times n}=A$, the target judgment matrix $D=(d_{ij})_{n\times n}=A$.

Step3 For $n_1 = 1$ to 9

Step4 if($m -- \ge 1$) For $n_2 = n_1$ to 9

Step5 if($m -- \ge 1$) For $n_3 = n_2$ to 9

⋮

Stepn if($m -- \ge 1$) For $n_{n-2} = n_{n-3}$ to 9

Stepn+1 if($m -- \ge 1$) For $n_{n-1} = n_{n-2}$ to 9

Ⅰ. Make $b_{11}=1, b_{12}=n_1, \cdots, b_{1n}=n_{n-1}$, vector($b_{11}, b_{12}, \cdots, b_{1n}$) as the first row of the comparison matrix, generating the comparison matrix B.

Ⅱ. Generating the relative error matrix $E=(e_{ij})_{n\times n}$ from matrices A and B.

Ⅲ. According to the descending order of the elements of the error matrix E, replacing the value of the corresponding place of the transition judgment matrix one by one with the value of the corresponding place of the comparison matrix, and replacing the element values of the corresponding symmetric place until the consistency requirement is satisfied. If the matrix dissimilarity is smaller than the previous one, correct the dissimilarity, and the target judgment matrix is the transition judgment matrix.(The verification consistency algorithm is the same as the judgment matrix consistency verification traversal nesting algorithm).

Stepn+2 In this case, the target judgment matrix corresponding to the matrix dissimilarity value is the solution.

4. Example Analysis

4.1 Example

The above algorithm can be used to achieve by matlab programming, the following is an example of matrix adjustment.

Initial judgment matrix $A=\begin{pmatrix} 1 & 2 & 5 & 9 \\ 1/2 & 1 & 2 & 5 \\ 1/5 & 1/2 & 1 & 1/3 \\ 1/9 & 1/5 & 3 & 1 \end{pmatrix}$

The dominant eigenvector of matrix A is w_A=[0.5335, 0.2636, 0.0896, 0.1133], $CR=0.1874>0.10$, so it does not satisfy the consistency requirements, it needs to be adjusted.

According to the generation algorithm in this paper, until the success of the adjustment, then $B=\begin{pmatrix} 1 & 2 & 5 & 9 \\ 1/2 & 1 & 3 & 5 \\ 1/5 & 1/3 & 1 & 2 \\ 1/9 & 1/5 & 1/2 & 1 \end{pmatrix}$, $E=\begin{pmatrix} 0 & 0 & 0 & 0 \\ 0 & 0 & 1/3 & 0 \\ 0 & 1/2 & 0 & 5/6 \\ 0 & 0 & 5 & 0 \end{pmatrix}$, $D=\begin{pmatrix} 1 & 2 & 5 & 9 \\ 1/2 & 1 & 2 & 5 \\ 1/5 & 1/2 & 1 & 2 \\ 1/9 & 1/5 & 1/2 & 1 \end{pmatrix}$

The dominant eigenvector of matrix D is w_D=[0.5526, 0.2689, 0.1202, 0.0583], $CR=0.0028<0.10$, it satisfies the consistency requirements.

According to the adjustment method which is described by Saaty, The final judgment matrix is $D_s=\begin{pmatrix} 1 & 2 & 5 & 9 \\ 1/2 & 1 & 2 & 5 \\ 1/5 & 1/2 & 1 & 2.1264 \\ 1/9 & 1/5 & 0.4703 & 1 \end{pmatrix}$

The dominant eigenvector of matrix D_s is w_{Ds}=[0.5521, 0.2686, 0.1219, 0.0573], $CR=0.0027<0.10$, it satisfies the consistency requirements.

4.2 Validity Analysis

The validity of the generating algorithm can be measured in terms of the reachability of the algorithm, matrix dissimilarity, and the maximum adjustment of the matrix elements.

The AHP judgment matrix generation algorithm based on pre-sort and round-up function adjusts the initial judgment matrix based on the comparison matrix which satisfies the consistency requirement, for the n-order initial judgment matrix which does not satisfy the consistency, the target judgment matrix can be solved by adjustment in at most n(n-1)/2 times. Therefore, the algorithm is convergent.

In the part 4.1, the dissimilarity between the target judgment matrix generated by the algorithm proposed in this paper and the initial judgment matrix is $d(A,D)=\sqrt{\sum_{k=1}^{n}\left|w_A^{\,k}-w_D^{\,k}\right|^2}=0.0660$. However, the dissimilarity of the matrix generated by the method described by Saaty is $d(A,Ds)=\sqrt{\sum_{k=1}^{n}\left|w_A^k - w_{Ds}^k\right|^2}=0.0674$, thereby $d(A,D)<d(A,Ds)$.

In the part 4.1, the algorithm proposed in this paper adjusts the element of the initial judgment matrix $a_{43}=3$ to $d_{43}=0.5$, the adjustment rate is 83.33%. But the method described by Saaty adjusts the element of the initial judgment matrix $a_{43}=3$ to $ds_{43}=0.4703$, the adjustment rate is 84.32%.

5. Conclusion

Compared with the traditional Adjustment Algorithm for AHP Judgment Matrix, On the basis of making full use of the information of the initial judgment matrix, the AHP judgment matrix generation algorithm based on pre-sort and round-up function finds out a target judgment matrix which satisfies the consistency requirement, and the matrix dissimilarity and the adjusted ratio is smaller than the traditional AHP judgment matrix adjustment method, and can ensure the elements of the generated target adjustment matrix are in the range 1 to 9 and their reciprocal. Therefore, the AHP judgment matrix generation algorithm based on pre-sort and round-up function is reasonable and effective, and has strong operability and practicability.

6. References

[1] Saaty T L. Decision-making with the AHP: Why is the principal eigenvector necessary[J]. European Journal of Operational Research, 2003, 145:85-91.
[2] Satoshi Matsuda. A Neural Network Model for the Decision-Making Process Based on AHP, Proceedings of International Joint Conference on Neural Networks, 2005:821-826
[3] Saaty T L. Principles of the analytical hierarchy process. Expert judgment and expert systems, 1987.
[4] Saaty T L. What is the analytical hierarchy process. Mathematical models for decision support, 1988.
[5] Saaty T L. Decision-making - The Analytic Hierarchy and Network Processes (AHP/ANP). Journal of Systems Science and Systems Engineering, 2004, 13(1):1-35
[6] Saaty T L. A Scaling Method for Priorities in Hierarchical Structures. Journal of Math. Psychology, 1977, 15(3):234-281.

An Efficient Algorithm for Server Thermal Fault Diagnosis Based on Infrared Image

Hang Liu[1,a], Ting Xie[1,b], Jian Ran[1,c] and Shan Gao[2,d]

[1]School of Electronic Science and Technology, Dalian University of Technology, Dalian, China
[2]Network and Information Center, Dalian University of Technology, Dalian, China
Email: [a]liuhang@dlut.edu.cn, [b]tingxie9116@mail.dlut.edu.cn, [c]ran554300274@mail.dlut.edu.cn, [d]gshan@dlut.edu.cn

Abstract. It is essential for a data center to maintain server security and stability. Long-time overload operation or high room temperature may cause service disruption even a server crash, which would result in great economic loss for business. Currently, the methods to avoid server outages are monitoring and forecasting. Thermal camera can provide fine texture information for monitoring and intelligent thermal management in large data center. This paper presents an efficient method for server thermal fault monitoring and diagnosis based on infrared image. Initially thermal distribution of server is standardized and the interest regions of the image are segmented manually. Then the texture feature, Hu moments feature as well as modified entropy feature are extracted from the segmented regions. These characteristics are applied to analyze and classify thermal faults, and then make efficient energy-saving thermal management decisions such as job migration. For the larger feature space, the principal component analysis is employed to reduce the feature dimensions, and guarantee high processing speed without losing the fault feature information. Finally, different feature vectors are taken as input for SVM training, and do the thermal fault diagnosis after getting the optimized SVM classifier. This method supports suggestions for optimizing data center management, it can improve air conditioning efficiency and reduce the energy consumption of the data center. The experimental results show that the maximum detection accuracy is 81.5%.

1. Introduction

Recently there has been a boom in data center usage, it is prompted by the great developments in new information technologies and applications (e-commerce, Big Data, and cloud computing) and other services [1]. Servers are complicated to maintain, and require a special environment in which to operate. In order to make sure the server operates at a temperature below its overheating threshold and prevent abnormal shutdown due to excessive temperature or overload operation, many data centers have to improve the power of air conditioning, resulting in high energy consumption and low efficiency of the data center. A report by Koomey in 2010 reveals that the use of electricity in data centers during 2010 was about 1.1%-1.5% of the worldwide electricity and this proportion is growing [2]. Remarkably, the computer room air conditioner (CRAC) units account for nearly 40% of the total power consumption in a data center [3]. One of the main factors that affect the cooling efficiency is the uneven distribution of temperature inside the data center [4]. The super hot servers with hotspots in data center account for about 5%. We need to solve the problem of heat dissipation of both these 5% super high heat cabinet and the vast majority servers. However, we cannot handle these problems in a same way. Research shows if the temperature variation inside data center changes from 10°C to 2°C, it can reduce about 25% energy consumption of the CRAC [5]. Therefore it is important to monitor the temperature within data centers, and then find the cause of the hotspots. Thus we can exclude the local high temperature and balance the internal temperature of data center by adjusting the load distribution or the refrigeration resource distribution. It improves the efficiency of air conditioning and realizes the energy saving of the data center.

At present, for monitoring the temperature inside a large data center, the traditional solution is to place temperature sensors in the key position of a data center, then all measurement data is collected by a specific method. The application of wired sensors is not widespread due to the expensive installation and configuration costs. Although wireless sensors have the advantages of low cost and non-invasive measurement, reference [6] shows that the wireless electromagnetic environment in large-scale data center is not conducive to the sensor network. Moreover, the spatial resolution of temperature sensors is not enough to get detailed information. If the sensor is not deployed in right position, we cannot accurately obtain the size and temperature of hot spots without sufficient information. Infrared thermography (IRT) has been applied widely in the field of militaries, medical diagnostics and security monitoring with the advantage of non-contact detection, no electro-magnetic interference, safety and reliability, and wide monitoring range [7,8,9]. Compared with temperature sensors, the thermal camera can obtain more intuitive two-dimensional thermal images, fault associated with abnormal temperature distribution can be easily detected by IRT and the image processing technology can further analyze the cause of hot spots.

In order to obtain a rapid and accurate fault diagnosis result according to the inspecting data of electric equipments, some intelligent diagnosis systems are constructed based on image processing and artificial intelligence. The application of intelligent fault diagnosis system can assess the abnormal degree of electronic equipment, even if the experts or experienced personnel are not at scene [10]. Generally, the intelligent diagnosis system based on infrared image consists of four steps, as shown in Figure 1. Firstly, shot the infrared images of electronic equipment under different fault conditions by infrared camera. Secondly, find the region of interest (ROI) and extract the feature information of segmented regions, these features have enough discrimination. Finally use artificial intelligence algorithm and decision to determine the equipment state.

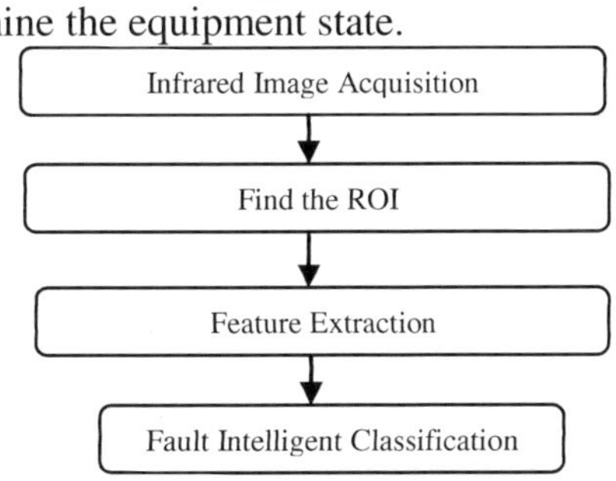

Figure 1 General steps of intelligent diagnosis system

This paper proposes an efficient algorithm for server thermal fault diagnosis based on infrared images. The intent is to optimize the management of data center, improve the efficiency of CRAC and reduce the energy consumption of data center. The innovation of this article is to propose an infrared image processing technology based on modified entropy features, it is superior to texture and Hu moments features, and it produces higher fault diagnosis rate. The rest of this paper is as follows. In section 2, the realization of the intelligent diagnosis method based on infrared image is proposed, and the feature extraction algorithms are mainly discussed. Section 3 explains the implement of the proposed method. Experimental results and classification performance of SVM classifier are presented in Section 4. Finally, summarize and conclude our work.

2. Methodology

The classification and diagnosis system of the server thermal failure is presented in this article. Firstly, the image distortion caused by different shooting positions is eliminated by image standardization, and the ROIs of the infrared image are segmented manually. We use three algorithms for feature extraction.

For the larger feature space, a principal component analysis (PCA) is employed to reduce the feature space dimensions, and guarantee high processing speed without losing the fault feature information. Finally, different feature vectors are taken as the input of support vector machine (SVM) respectively to train classification model, and then make thermal fault diagnosis after getting the optimized SVM classifier. The key issues of deciding a server condition depend on the extracted information which can be distinctive enough to be classified and the pattern recognition method in intelligent diagnosis.

We extract the features from target images which reflect the characteristics of important and primitive. In this research, three feature extraction algorithms are applied to the segmented images, namely the texture feature extraction algorithm, Hu moments feature extraction algorithm and modified entropy feature extraction algorithm.

2.1 Texture Feature

Texture as an important characteristic of the image reflects the spatial structural information, and it is also an important factor that cannot be ignored during classification. Traditionally, they are calculated considering the neighborhood pixels on the image. A considerable number of approaches have been reported to extract texture features. The most widely used is the grey level co-occurrence matrix (GLCM) proposed by Haralick et al. (1973). GLCM of image reflects the integrative information of pixels with regard to direction, adjacency spacing relationship and the range of variance change. It is the basis to analyze image textures. Seven texture characteristic parameters are extracted using GLCM, which are used as the basis of classification. They are contrast, uniformity, entropy, variance, covariance or product moment, inverse difference moment, and correlation.

The texture features vector is used to train SVM model, then thermal fault conditions of the server are diagnosed by the optimized SVM classifier.

2.2 Hu Moments Feature

Hu moments are used in pattern recognition to provide a scale, orientation and position-invariant representation of an object [11]. The practice shows that the characteristics of the image cannot be guaranteed by the use of the origin moment or the central moment. To this end, m.k.hu first proposed the invariant moments in 1961, He gave the definition of moments of continuous functions and the basic properties of moments, and proved the properties of translation invariance, rotation invariance and scale invariance.

They are suitable for describing the overall shape of the object, so it has a wide applications in edge extraction, image matching and object recognition. The amount information included in seven Hu moments are different because of different calculations. The useful information of image generally focused on lower order moments which are relatively small in calculation, the calculation of the high order moments are large that contain some details easily affected by noise, the difference between each target of the high order moments is not easy to distinguish. In this research, seven Hu moments feature vectors of the image are adopted as the input data. The feature space is M1 M2, M3, M4, M5, M6 and M7.

2.3 Modified Entropy Feature

The relationship between gray value of infrared image and its temperature is shown in formula (1)

$$T = \frac{T_{max} - T_{min}}{255} \cdot i + T_{min} \tag{1}$$

T represents the temperature value of a point in gray image, i stands for the gray value of the point, T_{max} and T_{min} respectively represent the maximum and minimum temperature of the image area. Entropy is an uncertainty measurement of events occurrence in information theory according to Shannon entropy. For an infrared image of the server with normal operation condition, the heat distribution is determined. When there is a hot spot, the heat distribution of the image is destroyed, and the entropy changes as well. The global entropy of an image represents the amount of information contained the aggregation feature in gray level distribution. Gray value of each pixel is a certain level of the 0~255 gray-level for an image I, let P_i represents the proportion of pixel i in the gray image I,

The global entropy of image *I* is defined by *H* (*I*).

$$H(I) = -\sum_{i=0}^{255} P_i \log P_i \tag{2}$$

Although the global entropy can represent statistical properties of the image, it cannot reflect the spatial characteristics of the gray distribution. For this problem, a feature extraction algorithm of local entropy is applied in this paper. The mean value of neighborhood gray are taken as the spatial feature in the image, and combined with the gray value of pixels, a two- tuple characteristic is constituted, denoted as (*i*, *j*). *i* and *j* respectively represent the gray value of pixels and the mean gray value of neighborhood.

$$P_{ij} = \frac{f(i, j)}{M \times N} \tag{3}$$

Formula (3) reflects the proportion of the two-tuple in the image, where $f(i, j)$ is the frequency of the corresponding two-tuple. $M \times N$ is the scale of the local window. In formula (1), the variable P_i replaced with P_{ij}, thus the local entropy with $M \times N$ local window is acquired.

Local entropy of the image can be used to reflect comprehensive features of the gray value with position information and the gray level distribution of the surrounding pixels. It is the result of the interaction of all pixels in the local window, which is not sensitive to the single point noise, so the local entropy feature has certain anti-noise filtering ability. In order to extract the statistical characteristics of local entropy, this article proposes the concept of modified entropy feature algorithm. For an image with size of X×Y, calculate the mean value of local entropy for each row and column pixels, therefore we obtain X and Y feature vectors, which are corresponding to the row and column number.

The modified local entropy reflects the inhomogeneous gray distribution of the row (column) pixels and the relationship between neighborhood pixels. It is helpful for the thermal fault diagnosis of server by taking the global and modified local entropy as the input features.

3. Implementation of the Proposed Method

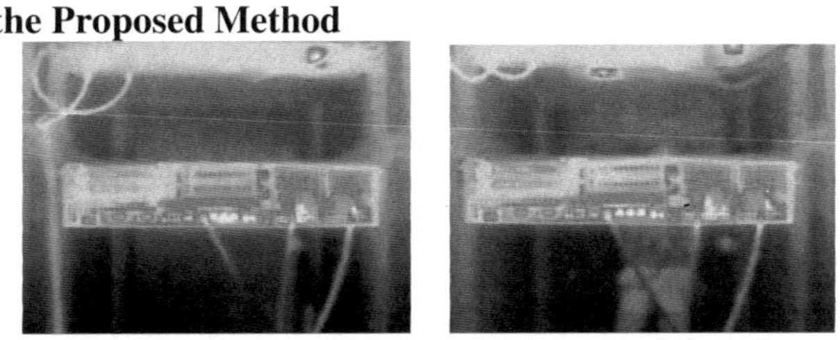

(a) Main Fan Failure (b) Inlet Air Blocking.

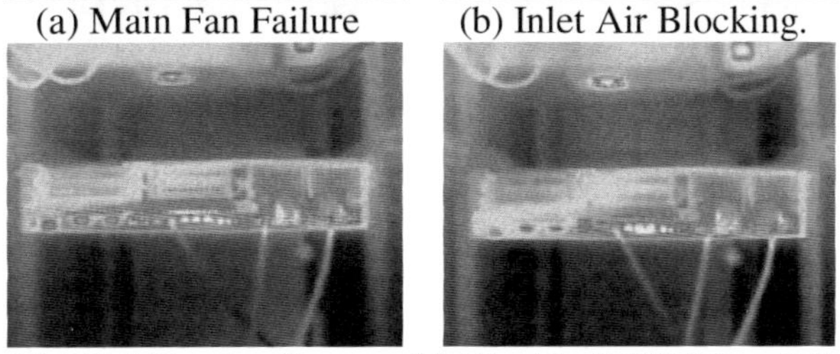

(c) CPU Overloading (d) The No.2 CPU Fan Failure.

Figure 2 Infrared Images of Four Thermal Fault States of The Server

Our approach was applied in a laboratory at Dalian University. The FLIR E8 thermal camera with fusion technology is set up against the rack that equipped with a server of the DELL710 model. The distance between target server and the thermal camera is between 0.5 and 1 meter. During monitoring process, the ambient temperature is between 20°C and 30°C, other information related to the accuracy of the results is recorded, such as ambient humidity is between 16% and 25%. In this research, the thermal faults are classified into 4 classes, namely main fan failure, inlet air blocking, CPU overloading and the No.2 CPU fan failure. The infrared thermal images of server operating under four fault states (as shown in Figure 2) are taken as the research objects. Sample number of each running state is 255, and a total of 1275 infrared images are obtained.

3.1 Image Standardization

Because of the different viewing angle and distance of infrared camera, the geometrical distortion of the infrared image appears in the process of image acquired, which can introduce errors into the extracted features and cause inaccurate recognition. Hence, this paper adopts an image standardization method to obtain the thermal distribution features of the server on a uniform model and provide information for evaluating the operating state [12]. Firstly, prepare a template, which is scaled to a unified rectangle with an aspect ratio equal to the server facade. Then the thermal distribution of each sever is standardized by image registration. Image registration technology can establish one to one pixel relationship between infrared images and visible images, so the temperature of any position on the visible image can be known, and it is easy to feature extraction.

The server region is segmented manually into a template which is mapped to a standard rectangle that is proportional to the size of the actual server. In this research it will be mapped to a server image of 87 ×424 pixels (shown in Fig. 3) for subsequent image processing.

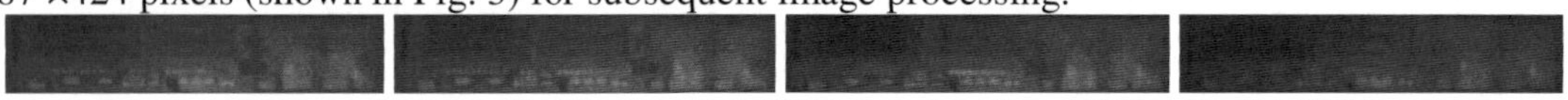

(a) Main Fan Failure (b)Inlet Air Blocking(c) CPU Overloading (d)The No.2 CPU Fan Failure

Figure 3 Standardized Image of The Corresponding States

3.2 Parameter Optimization

The application of SVM with radial basis kernel function (RBF) obtains preferable performance than the MLP artificial neural network in thermal faults detecting of electrical installations [13]. In this paper, support vector machine (SVM) is used as a classifier which has good generalization ability and it is better able to solve the small practical problem [14]. Gauss function is usually used as the radial basis function (RBF) kernel. The classification performance of SVM is affected by parameters selection of penalty parameter C and kernel function parameterγ. Parameter C controls the tradeoff between the maximum margin and the minimum error rate of the two classes. The parameter$\gamma$is the parameter of Gauss radial basis kernel function, which determines the complexity of the distribution of sample data in high dimensional space. Therefore it is necessary to search the parameters, the goal is to determine a pair of good parameter combination (C, γ) for the classifier predict correctly with sample data.

Cross validation is a method can avoid over fitting of SVM model [15]. For a K-fold cross validation, initially the training data is divided into k subsets of equal size, then each subset is used as the test data set, the rest of the k-1 subset is used as the training data set. Every subset of the whole training set is predicted once, thus the problem of over fitting can be avoided. According to the lack of theoretical guidance, it requires the machine learning engine for more training. Herein we use the grid search algorithm to select the best combination. Some parameters are shown in TABLED I.

Table 1 Parameter Optimization Of Svm

SVM kernel function	RBF
K-Cross Validation	K=5
Range of loss parameters C	[-5,5]
Range ofγvalue	[-5,5]

Additionally, considering the time required in the practical application, we adopt the principal component analysis (PCA) algorithm to reduce the dimensions of the extracted features for large feature space. PCA can not only eliminate redundant information which reduces the correlation among the feature vectors, especially decreases data quantity, but also make the sample space more compact and reasonable for SVM [16].

4. Experimental Result and Classification Performance Analysis

Data set contains totally 1275 infrared images of the server under different thermal fault conditions captured by thermal camera, and the number of infrared images of each thermal failure state is equal. In this experiment, 60% of the data are selected randomly as training data and the remaining 40% of the samples chosen for testing.

For each infrared image, the Hu moments method extracts 7 characteristics contained in the feature vectors. Another 7 features are extracted by the texture feature extraction algorithm. However, the feature space of the global entropy and modified local entropy features are too huge for SVM. Specifically, firstly extract the global entropy feature of each image which is only one feature, then use 9 ×9 local window to extract the row and column mean value of local entropy features, they are 87 features and 424 features respectively, and totally 512 image features. It necessitates the use of PCA, which can speed up the data processing speed without loss of the original data and make the feature information more representative. The new set of features of the modified entropy compressed by PCA contains 49 dimensions. The combination input features are mapped into a new feature space with 50 dimensions by PCA.

In this thesis, we use the SVM classifier combined with different input characteristics to analyze the thermal faults of server. The accuracy of thermal faults diagnosis with different input features are shown in TABLEII. The results showed that the classification accuracy of both the CPU overloading condition and the No.2 CPU fan failure is the highest. It is obvious that the all kinds of characteristics algorithm introduced in this article can identify the causes of the above two hotspots completely. Particularly, the features of global entropy and modified local entropy are superior to the Hu moments and texture features for the remaining thermal fault cases. Different feature vectors are used to train SVM classifier. In experiment, we also use 5-fold cross validation and grid search method to optimize the combination of penalty parameter C and kernel function parameterγ. According to the results in TABLEIII, we noticed that when all features are combined as input data, the maximum correct rate of the thermal fault diagnosis reaches up to 81%, and the parameter C=1 andγ=0.0313. In this paper, the classification performance of SVM with the global entropy and the modified local entropy features is better than that of Hu moments and texture features. The results of this study can reliably predict server thermal faults, it provides the suggestions for thermal management, and improves the cooling efficiency in data centers. This study has a certain practical value.

Table 2 Fault Diagnosis With Different Inputs

States of Server	**Fan Failure (%)**	**Inlet Air Blocking (%)**	**CPU Overloading (%)**	**The No.2 CPU Fan Failure (%)**
Hu Moment + SVM	28.4314	14.7059	100	50
Textrue + SVM	51.9608	43.1373	100	100
Modified Entropy Feature + PCA + SVM	57.8431	55.8824	99.0196	100
All Features Above + PCA + SVM	64.7059	57.8431	99.0196	100

Table 3 Classification Performance of SVM

Number	**Feature As Input**	**PCA**	**C**	**γ**	**Diagnosis Accuracy (%)**
1	Hu moment feature	No	32	4	48.2927
2	Texture Feature	No	32	0.1768	73.6585
3	Modified Entropy Feature	Yes	1.4142	0.0313	77.3171
4	All Featuers Above	Yes	1	0.0313	81.4634

5. Conclusion

In this paper, an efficient algorithm of thermal fault diagnosis based on infrared image is presented. In the intelligent fault diagnosis system, we extract the Hu moments feature, texture feature and global entropy and the modified local entropy feature as different inputs of SVM classifier. 5-fold cross validation method and grid search method are used to adjust the penalty parameter C and kernel function parameter γ. The comparison results demonstrate that the global entropy and modified local entropy features developed in this thesis obtain a higher accuracy of the server thermal fault diagnosis than that of Hu moments feature and texture feature. The experimental results show that SVM trained by all feature vectors produced better performance with 81% of accuracy. The proposed method has been validated through actual thermal faults diagnosis.

6. References

[1] Alkharabsheh S, Fernandes J, Gebrehiwot B, et al. A Brief Overview of Recent Developments in Thermal Management in Data Centers[J]. Journal of Electronic Packaging Transactions of the Asme, 2015, 137(4).

[2] Koomey, J., 2011, Growth in Data Center Electricity Use 2005 to 2010, Analytics Press, Oakland, CA.

[3] He Z, He Z, Zhang X, et al. Study of hot air recirculation and thermal management in data centers by using temperature rise distribution[J]. Building Simulation, 2016, 9(5):1-10.

[4] Liu H, Lee E K, Pompili D, et al. Thermal camera networks for large datacenters using real-time thermal monitoring mechanism[J]. The Journal of Supercomputing, 2013, 64(2):383-408.

[5] Moore J, Sharma R, Shih R, et al. Going beyond CPUs: The potential of Temperature-Aware Solutions for the Data Center[C] 2004.

[6] Liang, Chieh Jan Mike, et al. RACNet: a high-fidelity data center sensing network. International Conference on Embedded Networked Sensor Systems, SENSYS 2009, Berkeley, California, USA, 2009, pp.15-28.

[7] Jadin M S, Taib S. Recent progress in diagnosing the reliability of electrical equipment by using infrared thermography[J]. Infrared Physics & Technology, 2012, 55(4):236-245.

[8] Bagavathiappan S, Lahiri B B, Saravanan T, et al. Infrared thermography for condition monitoring – A review[J]. Infrared Physics & Technology, 2013, 60(5):35-55.

[9] Huda A S N, Taib S. Application of infrared thermography for predictive/preventive maintenance of thermal defect in electrical equipment[J]. Applied Thermal Engineering, 2013, 61(2):220-227.

[10] Huda A S N, Taib S, Jadin M S, et al. A semi-automatic approach for thermographic inspection of electrical installations within buildings[J]. Energy & Buildings, 2012, 55(55):585-591.

[11] Hu J J, Li H C, Tai H M. Thermal distribution monitoring of the container data center by a fast infrared image fusion technique[J]. Computers & Mathematics with Applications, 2012, 64(5):1484–1494.

[12] Liu H, Xie T, Ran J, et al. A modified approach for thermal distribution monitoring of the container data center by thermal image registration[C] Image and Signal Processing, BioMedical Engineering and Informatics (CISP-BMEI), International Congress on. IEEE, 2016: 259-263.

[13] Jadin M S, Taib S, Ghazali K H. Feature extraction and classification for detecting the thermal faults in electrical installations[J]. Measurement, 2014, 57(57):15–24.

[14] Zou H, Huang F. A novel intelligent fault diagnosis method for electrical equipment using infrared thermography[J]. Infrared Physics & Technology, 2015, 73:29-35.

[15] K. Ron, A study of cross-validation and bootstrap for accuracy estimation and model selection, in: Proceeding of the International Joint Conference on Artificial Intelligence, 1995, pp. 1137–1145.

[16] Balaguer A, Ruiz L A, Hermosilla T, et al. Definition of a comprehensive set of texture semivariogram features and their evaluation for object-oriented image classification[J]. Computers & Geosciences, 2010, 36(2):231-240.

An ASI Suppression Scheme Based on Array Synthesis

Jinhua Zhu, Peixi Liu and Wu Luo

The State Key Laboratory of Advanced Optical Communication Systems and Networks, Peking University, Beijing, China

Email: zhujinhua@pku.edu.cn, liupeixi@pku.edu.cn, Luow@pku.edu.cn

Abstract. An adjacent satellite interference (ASI) suppression scheme based on transmitting antennas array synthesis is proposed in this paper. The scheme ensures the ASI under the upper bound specified by international regulations, and simultaneously maximize the main-axis radiated power. Specifically, the interference suppression problem is converted into a uniform linear array (ULA) synthesis problem, and then solved by two algorithms. Numerical results are presented to illustrate the performance of the scheme.

1. Introduction

Due to the wide beam, the very small aperture terminal (VSAT) and the ultra small aperture terminal (USAT) usually lead to strong ASI. It may cause serious deterioration to the performance of the adjacent satellite system. Various international regulations have been issued to restrict the ASI. ITU-R S.728-1 [1] and FCC 47 C.F.R.§25.222 [2] specified the upper bound of the off-axis effective isotropic radiated power (EIRP) spectrum density. And ITU-R S.738 [3] presented a criterion for deciding whether coordination is needed between adjacent satellite systems. If the noise temperature increment caused by the ASI is greater than the specified threshold (6%), then interference coordination is necessary. By comparing the regulations, it can be deduced that ITU-R S.728-1 has a strict limit on USAT and can be an alternative for the noise temperature increment criterion [4].

In ITU-R S.728-1, it is recommended that the off-axis EIRP spectrum density of the VSAT earth stations operating with geosynchronous orbit (GEO) satellites in the 14GHz frequency band should not exceed the following value [1] in the systems where the satellite spacing is near 2°. We need to keep the off-axis EIRP spectrum density no greater than $ITU(\theta)$ and, at the same time, maximize the main-axis radiated power to improve the communication quality.

$$ITU(\theta)[dBW/40kHz] = \begin{cases} 25-25\log\theta & 2^\circ < \theta \le 7^\circ \\ -6 & 7^\circ < \theta \le 9.2^\circ \\ 18-25\log\theta & 9.2^\circ < \theta \le 48^\circ \\ -14 & 48^\circ < \theta \le 180^\circ \end{cases} \tag{1}$$

2. Problem Description

For a given envelope shape of the antenna beam pattern, the main-axis radiated power cannot increase any more, as long as the ASI in a certain off-axis angle reaches the specified upper bound. In order to avoid the bottleneck effect caused by the ASI in a certain direction, the envelope of the antenna beam pattern is desired to match the $ITU(\theta)$ curve. We use an antenna array to achieve this, because the envelope of its beam pattern can be easily controlled by adjusting the element weights. The structure of an array antenna is shown as Figure. 1.

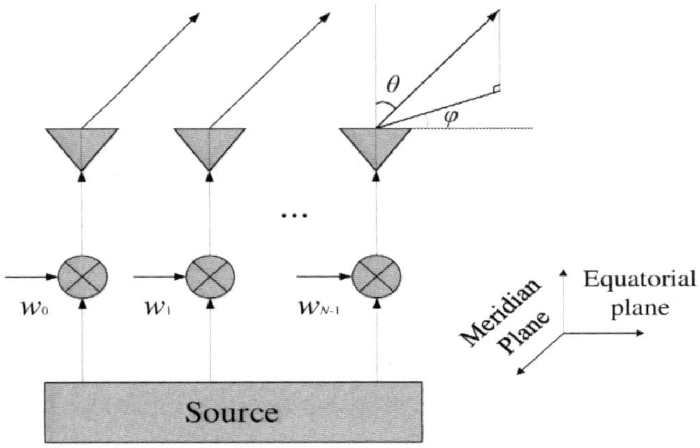

Figure 1 Structure of Array Antenna

Define the weight vector

$$\mathbf{w} = \begin{bmatrix} w_0 & w_1 & \cdots & w_{N-1} \end{bmatrix}^H \tag{2}$$

and the steering vector [5]

$$\mathbf{a}(\theta,\varphi) = \begin{bmatrix} e^{j\phi_0(\theta,\varphi)} & e^{j\phi_1(\theta,\varphi)} & \cdots & e^{j\phi_{N-1}(\theta,\varphi)} \end{bmatrix}^T, \tag{3}$$

where $\phi_n(\theta,\varphi)$ is the phase difference between the signals on element 0 and element n. Then the beam pattern of the array can be expressed as

$$G(\theta,\varphi) = \mathbf{w}^H \mathbf{a}(\theta,\varphi). \tag{4}$$

Uniform rectangular array (URA) is a widely used type of antenna. For URA, $\phi_{m,n}(\theta,\varphi) = kd\sin\theta\,(m\sin\varphi + n\cos\varphi)$, where k is the wave number and d is the element spacing. Hence, the beam pattern of an $M \times N$ array is given by

$$G(\theta,\varphi) = \sum_{m=0}^{M-1}\sum_{n=0}^{N-1} w_{mn} e^{jkd\sin\theta(m\sin\varphi + n\cos\varphi)}. \tag{5}$$

In the equatorial plane ($\varphi = 0$), (5) can be simplified to

$$G(\theta) = \sum_{n=0}^{N-1}\left(\sum_{m=0}^{M-1} w_{mn}\right) e^{jnkd\sin\theta}. \tag{6}$$

By Comparing (6) with the beam pattern of a ULA

$$G(\theta) = \sum_{n=0}^{N-1} w_n e^{jnkd\sin\theta}, \tag{7}$$

we obtain the following formula

$$w_n = \sum_{m=0}^{M-1} w_{mn}. \tag{8}$$

Therefore, the equatorial beam pattern of a URA is equivalent to the pattern of a ULA with the element weights given by (8).

Based on this, the original problem can be converted into a ULA synthesis problem. In order to optimize the equatorial beam pattern of an $M \times N$ URA, we just need to perform array synthesis on a N -element ULA, and distribute its n -th element weight to the n -th column of the rectangular array, as shown in Figure. 2. The method of distribution will influence the beam pattern in the meridian plane; however, that is not concerned in this problem.

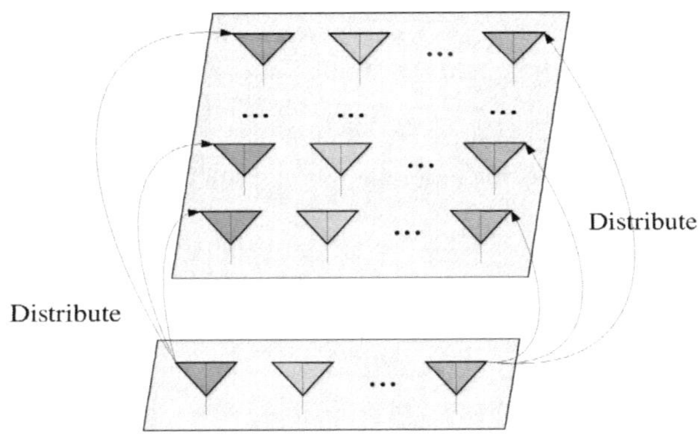

Figure 2 Weight Distribution

3. Array Synthesis

In this part, we discuss the array synthesis algorithms for the ULA.

3.1 Virtual Interference Algorithm

To show the idea of the proposed virtual interference algorithm, we consider the antenna array operating on receiving mode. The weights that maximize the output signal-to-interference-plus-noise ratio (SINR) often form nullings in the incident directions of the interferences. Moreover, an interference with larger power usually leads to a deeper nulling. Inspired by this, we place a large number of virtual interferences in the sidelobe region. By adjusting the power of the interferences, the envelope of the beam pattern can be arbitrarily controlled [6].

Assume that one desired signal and M uncorrelated virtual interferences are incident on the N -element array from angles $\theta_0, \theta_1, \theta_2, \ldots \theta_M$. The power of the interferences are $P_1, P_2, \ldots, P_M$. Additive white Gaussian noise with variance σ^2 is introduced by each element. The upper bound of the sidelobe is set as $\mathrm{ITU}(\theta)\varepsilon$, where ε determines the sidelobe height. If ε is too small, the sidelobe height constraint will be impossible to satisfy; if ε is too big, the main-axis radiated power will not be sufficiently large. In practice, the value of ε is usually decided by trial.

The steps of the virtual interference algorithm are described as follows:

1) Initialize P_m to σ^2;
2) According to the maximum SINR criterion, fix the gain in angle θ_0 and minimize the output power of the interferences and the noise. The weight vector is expressed as [7][8]

$$\boldsymbol{w} = \frac{\boldsymbol{R}^{-1}\boldsymbol{a}(\theta_0)}{\boldsymbol{a}^H(\theta_0)\boldsymbol{R}^{-1}\boldsymbol{a}(\theta_0)}, \tag{9}$$

where

$$\boldsymbol{R} = \sum_{m=1}^{M} P_m \boldsymbol{a}(\theta_m)\boldsymbol{a}^H(\theta_m) + \sigma^2 \boldsymbol{I}. \tag{10}$$

3) If the beam pattern does not exceed the specified upper bound $\mathrm{ITU}(\theta)\varepsilon$, the algorithm terminates. Otherwise, adjust the power of the virtual interferences to [9][10]

$$P_m^{(i+1)} = P_m^{(i)} \left[\frac{\boldsymbol{w}^H \boldsymbol{a}(\theta_m)\boldsymbol{a}^H(\theta_m)\boldsymbol{w}}{\mathrm{ITU}(\theta_m)\varepsilon} \right]^{\eta}, \tag{11}$$

where η is the speed factor and ε is the sidelobe height factor;
4) Go to step 2).

3.2 LP Algorithm

In the LP algorithm, we first compute the auto-correlation sequence of the element weights, and then obtain the weights by performing spectrum factorization.

Let $\{x_m\}$ be the auto-correlation sequence of $\{w_n\}$ [11], then $\{x_m\}$ is given by

$$x_m = \sum_{n=0}^{N-1-m} w_n w_{n+m}. \tag{12}$$

For a ULA, the beam pattern can be expressed as

$$\begin{aligned} G(\theta) &= \boldsymbol{w}^H \boldsymbol{a}(\theta)\boldsymbol{a}^H(\theta)\boldsymbol{w} \\ &= \left(\sum_{n=0}^{N-1} w_n e^{-jnkd\sin\theta}\right)\left(\sum_{n=0}^{N-1} w_n e^{jnkd\sin\theta}\right) \\ &= \sum_{n=0}^{N-1} w_n w_n + 2\sum_{n=0}^{N-2} w_n w_{n+1}\cos(kd\sin\theta) \\ &+ 2\sum_{n=0}^{N-3} w_n w_{n+2}\cos(2kd\sin\theta) + \ldots \\ &= x_0 + 2\sum_{m=1}^{N-1} x_m \cos(mkd\sin\theta). \end{aligned} \tag{13}$$

Our aim is to limit the height of the sidelobe and maximize the main-axis gain.

$$\begin{aligned} &\max\ x_0 + 2\sum_{m=1}^{N-1} x_m \\ &\text{s.t. } x_0 + 2\sum_{m=1}^{N-1} x_m \cos(mkd\sin\theta) \le \mathrm{ITU}(\theta), \\ &\theta \in \Theta_{\text{side}}, \end{aligned} \tag{14}$$

where Θ_{side} is the sidelobe region. Equation (14) is a LP problem of $\{x_m\}$. It can be solved quite efficiently by MATLAB functions or toolboxes (e.g. linprog).

However, the auto-correlation sequence $\{x_m\}$ cannot be directly applied to the array, and the weight sequence $\{w_n\}$ has to be computed. Here we use a technique called spectrum factorization. Its steps are described as follows [12]:

1) Interpolation

$$\underbrace{\{x_0, x_1, \ldots, x_{N-1}, 0, \ldots, 0, x_{N-1}, \ldots, x_1\}}_{2R+1}, \tag{15}$$

where R is 5 to 10 times larger than N ;

2) Perform fast fourier transformation (FFT), logarithmization and inverse FFT (IFFT) on (15) and form the vector y

$$\boldsymbol{y} = \text{IFFT}\left(\ln\left(\text{FFT}\left(\boldsymbol{x}\right)\right)\right). \tag{16}$$

Due to the symmetry of (15), y can be expressed as

$$\boldsymbol{y} = \left\{y_0, y_1, \ldots, y_R, y_R, \ldots, y_1\right\}. \tag{17}$$

3) Let the vector v be

$$\boldsymbol{v} = \left\{y_0 / 2, y_1, \ldots, y_R\right\}. \tag{18}$$

4) Perform FFT, exponent and IFFT on v and form the weight vector

$$\boldsymbol{w} = \text{IFFT}\left(\exp\left(\text{FFT}\left(\boldsymbol{v}\right)\right)\right). \tag{19}$$

From the derivation of the array synthesis algorithms, it can be deduced that the LP algorithm can only be applied to ULA or URA, and the virtual interference algorithm is suitable for arbitrary arrays. According to the commutation relation between the transmitting and receiving antennas, the scheme discussed above can also be applied to the array synthesis problem of the receiving antenna.

4. Numerical Results

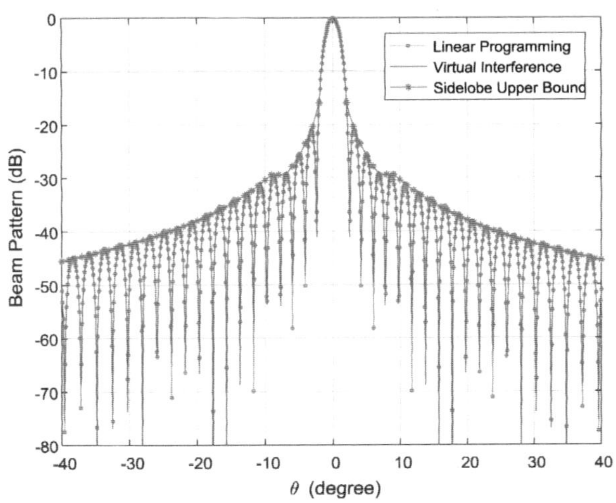

Figure 3 Synthesis Result

The beam pattern of a 60×60 rectangular array computed by the virtual interference algorithm ($\varepsilon = 0.278$, $\eta = 0.0005$) and the LP algorithm ($R = 6000$) are shown as Figure. 3. The element spacing is half-wavelength. We can see that the envelope of the beam pattern matches the $\mathrm{ITU}(\theta)$ curve, and the synthesis results of the two algorithms are similar.

When the sidelobe height factor ε is set as a proper value, the time cost of the two algorithms are close to each other. In the example above, it takes 2.49s and 2.83s respectively for two algorithms to solve the array synthesis problem (Test environment: Intel(R) Core(TM) i5-5200U CPU 2.20GHz$\times$ 2, 8GB RAM, MATLAB 2016b). The virtual interference algorithm is even a little bit faster. However, as we mentioned before, the factor ε in this algorithm needs to be determined by trial. This will lead to some inconvenience.

5. Conclusion

An ASI suppression scheme for transmitting antennas is proposed in this paper. In order to suppress the ASI and maximize the main-axis radiated power, we need to solve a ULA synthesis problem, and distribute the weights of the ULA to the elements of the original array. Two array synthesis algorithms are studied, and numerical results show that the synthesis result and the time cost of the algorithms are similar, but the LP algorithm is more convenient.

6. References

[1] International Telecommunication Union, "ITU-R Recommendation S.728-1, Maximum Permissible Level of Off-Axis EIRP Spectrum density from Very Small Aperture Terminals (VSATs)," 1995.

[2] Federal Communications Commission, "FCC 47 CFR Ch.1 25.222, Blanket Licensing for provisions for Earth Stations on Vessels (ESVs) receiving in the 10.95-11.2 GHz(Space-to-Earth), 11.45-11.7GHz((Space-to-Earth)), 11.7-12.2GHz((Space-to-Earth)) frequency bands, operating with Geostationary Satellites in the Fixed-Satellite Service," 2010.

[3] International Telecommunication Union, "ITU-R Recommendation S.738, Procedure for determining if coordination is required between geostationary-satellite networks sharing the same frequency bands," 1992.

[4] Y Ma, J Zhu, W Luo, "ASI regulations comparison for GSO satellite communication system at Ku-band," International Conference on Control, pp. 171-175, 2015.

[5] SN Shahab, AR Azinun, NH Noordin, SS Balasim, "Assessment of MVDR Adaptive Beamforming Algorithm in Uniform Linear Arrays, Uniform Rectangular Arrays and Uniform Circular Arrays Configurations," APRN Journal of Engineering and Applied Sciences, vol. 11, pp. 3911-1917, 2016.

[6] Y Ding, C Liu, Y Zhao, H, Chen, "Robust Wideband Beamforming Algorithm to Suppress Multipath Interference," Journal of Signal Processing, vol. 30, pp. 1104-1111, 2014.

[7] SR Tuladhar, JR Buck, "Unit circle MVDR beamformer," IEEE International Conference on Acoustics, Speech and Signal Processing, pp.2484-2488, 2015.

[8] L Zhang, W Liu, B Peng, "Generalized eigenvector problem for Hermitian Toeplitz matrices and its application to beamforming," Signal Programming, vol. 92, pp. 374-380, 2012.

[9] X Liu, C Liu, G Liao, "Improved pattern synthesis method with linearly constraint minimum variance criterion," IEEE International Conference on Wireless Communication, pp. 63-66, 2010.

[10] C Liu. "Robust Beamforming and DOA Estimation," Fourier Transform Applications, Dr Salih Salih, ISBN: 978-953-51-0518-3, InTech, 2012.

[11] K Huang, YC Eldar, ND Sidiropoulos, "Phase Retrieval from 1D Fourier Measurements: Convexity, Uniqueness, and Algorithms," IEEE Transactions on Signal Processing, vol. 64, pp. 6105-6117, 2016.

[12] J Jezek, M Hromcik, M Sebek, "New algorithm for spectral factorization and its practical application," Control Conference, pp. 3104-3109, 2001.

A Novel Signal Similarity Evaluation Algorithm and its Application in Irregular Shape Recognition

Yi Wei, Chenhui Yang, Hang Yang and Yaming Xu

School of Automation, Wuhan University of Technology, 122 Luoshi Road, Wuhan, PR China
Email: weiyi@whut.edu.cn, 270917365@qq.com, 648719145@qq.com, xmxu@sgg.whu.edu.cn

Abstract. Irregular shape recognition has been a tough problem in image understanding field. One type of feature extraction techniques involves transforming shapes into signals through certain process, and thus the shape recognition problem becomes comparing similarity between signals. This paper proposes a novel signal similarity evaluation scheme that may classify signals into different categories to fulfill the shape recognition task. Experiments on pavement guide marks recognition are conducted to study the potential of the proposed method. Results demonstrate its ability in irregular shape recognition applications with higher accuracy and faster speed.

1. Introduction

Pavement guide mark recognition has been playing an important role in the Intelligent Driving System. However due to the convexity-concavity characteristics of the boundaries, the recognition of guide marks becomes a big challenge and existing methods cannot meet the requirement. With the combination of data processing theory, transformations that convert shapes into signals have been applied and therefore corresponding signal similarity evaluation techniques are keys to solve the problem.

Pavement guide marks are one of the most commonly seen transportation signs for drivers. Their shapes are well-defined but with both convex and concave points along their borders. Therefore the automatic recognition of guide marks is a difficult task faced by many researchers. The paper chooses guide marks recognition as its goal and leads to the study of a novel approach of signal similarity evaluation scheme. Beginning with the projection to a guide mark boundary to complete the shape transformation into signals, the signal centroid is then calculated and fused into the original signal to form a new signal for the final similarity measurement. Conclusions are made based on the similarity of whether the unknown shape is one of the guide marks.

All experiments in the study are run in natural scenes with different sorts of real pavement status. The results demonstrate the potential of this new similarity comparison algorithm in irregular shape recognition applications.

2. Signal Centroid And Similarity Evaluation

Centroid is a notation originally derived from Physics. It is introduced in this work to describe one of characteristics of a signal and believed to be able to distinguish different signals under certain circumstances. If a signal is considered as a mass point system, the definition of the signal centroid can be written as follows:

$$\begin{cases} x_c = \dfrac{\sum_{i=1}^{N}(i \times S^2(i))}{\sum_{i=1}^{N} S^2(i)} \\ y_c = \dfrac{\sum_{i=1}^{N} S^3(i)}{\sum_{i=1}^{N} S^2(i)} \end{cases} \tag{1}$$

Where $\{S(i)\}, i = 1,2,\cdots,N$ is a random signal, and (x_c, y_c) is its centroid. To make a decision on whether two signals are similar, the distance between two signal centroids are calculated. The smaller the distance is, the closer two signals are.

$$\text{Dis} = \sqrt{(x_{tc}\text{-}x_{rc})^2 + (y_{tc}\text{-}y_{rc})^2} \tag{2}$$

Where (x_{tc}, y_{tc}) and (x_{rc}, y_{rc}) are centroids of the unknown signal and the reference signal respectively.

3. Algorithm Design

The centroid based signal similarity evaluation method demonstrates its good performance in random time series signals. However in the pavement guide marks recognition applications, guide marks are both irregular and well defined. On one hand, they have convex and concave points along their boundaries, and on the other hand guide marks themselves show similarities to some extent. Below is an example.

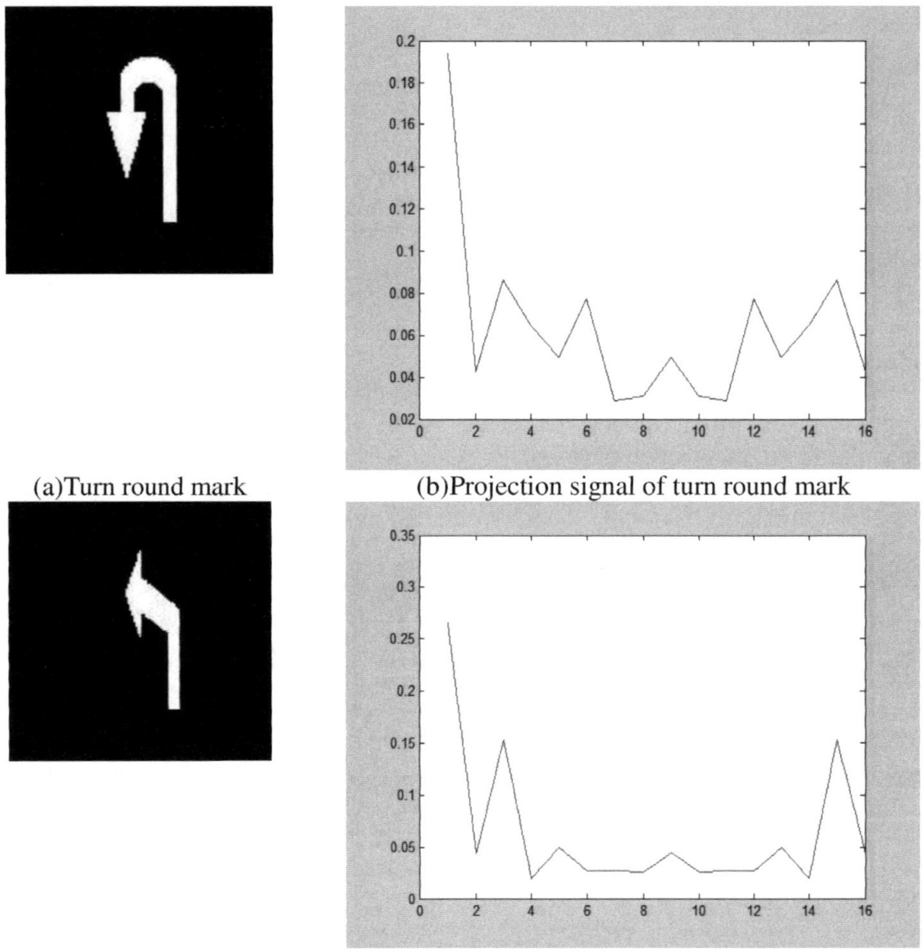

(a)Turn round mark (b)Projection signal of turn round mark

(c)Left turn mark (d)Projection signal of left turn mark

Figure 1 Two guide marks with similar projection signals

Given facts above, a new signal comparison scheme must be presented to distinguish shapes or signals as shown in Figure.1. By introducing centroid, the paper proposes a novel approach to reform the projection signal by adding its centroid at the end. Suppose $\{S|S(1), S(2), \cdots, S(N)\}$ is the original projection signal of a guide mark and (x_c, y_c) is its centroid. According to the new scheme, firstly a new signal $\{NS|S(1), S(2), \cdots, S(N), x_c, y_c\}$ is formed. Then the similarity between an unknown signal $NS_t$and the reference signal NS_r is calculation based on the linear combination of the Euclidean distance and Chessboard distance:

$$\mathrm{Sim} = \alpha \mathrm{Sim}_E + \beta \mathrm{Sim}_C \tag{3}$$

Where α and β and two weights,

$$\mathrm{Sim}_E = \sqrt{\sum_{i=1}^{N+2}(NS_t(i)\text{-}NS_r(i))^2} \tag{4}$$

is the Euclidean distances between NS_t and NS_r, and

$$\mathrm{Sim}_C = \max_{1 \le i \le N+2} |NS_t(i)\text{-}NS_r(i)| \tag{5}$$

is the Chessboard distance between the two.

4. Experimental Results

Pavement guide marks under tested in this paper are all obtained from the camera mounted on the moving vehicle in real scene. In addition to the complexity in shape, negative effects such as weather, light and other natural factors are unavoidable and therefore make the guide marks recognition even more difficult. The study adopts area projection in both vertical and horizontal directions as two signals representing a guide mark to conduct proposed signal similarity evaluation.

(a)Original image of a guide mark (b)Binary image of the guide mark

Figure 2 Straight and right turn mark

Given a $h \times w$ binary image contains a straight and right turn guide mark as shown in Figure 2(b), it can be represented as

$$B(x,y) = \begin{cases} 1, & foreground \\ 0, & background \end{cases} \tag{6}$$

The projection of $B(x,y)$ onto the X axis is defined as $hp(x)$, and the projection of $B(x,y)$ onto the Y axis is defined as $vp(y)$

$$\begin{cases} hp(x) = \sum_{y=1}^{h} B(x,y) \\ vp(y) = \sum_{x=1}^{w} B(x,y) \end{cases} \tag{7}$$

From (7), $hp(x)$ is the accumulation of all foreground in each column and $vp(y)$ is the sum of white pixels in each row. In order to calculate the similarity between a tested signal and the reference, normalizations in both axes of signals must be finished firstly. Both $hp(x)$ and $vp(y)$ have to be rescale into the same length k, and signal magnitude have to be limited within $0 \sim 1$ through the following procedures:

$$\begin{cases} Hp_k(x) = \dfrac{hp_k(x) - \min\limits_{1\le i\le k} hp_k(i)}{\max\limits_{1\le i\le k} hp_k(i) - \min\limits_{1\le i\le k} hp_k(i)} \\ Vp_k(y) = \dfrac{vp_k(y) - \min\limits_{1\le i\le k} vp_k(i)}{\max\limits_{1\le i\le k} vp_k(i) - \min\limits_{1\le i\le k} vp_k(i)} \end{cases} \tag{8}$$

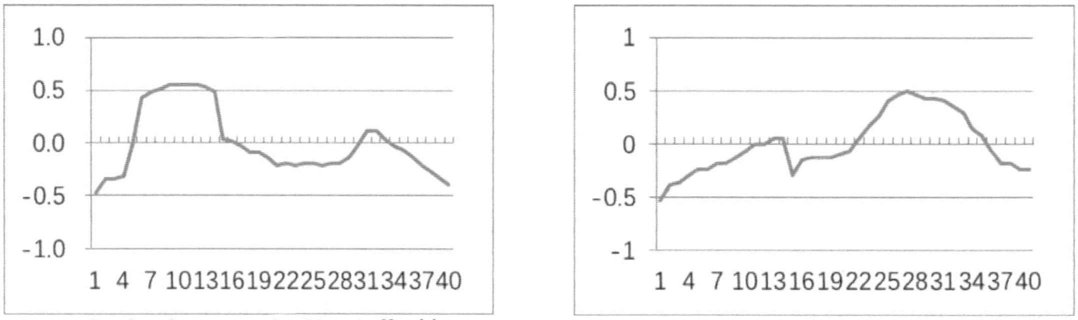

(a) Projection onto the X axis $Hp_k(x)$ (b) Projection onto the Y axis $Vp_k(y)$

Figure 3 normalized projection signals of the straight and right turn mark

Applying the new signal comparison scheme, two new signals with $k+2$ as their lengths are formed for both Hp_k and Vp_k respectively and are written as NHp_{k+2} and NVp_{k+2}. According to (3),(4) and (5), similarities between unknown shapes and a reference guide mark are calculated separately as Sim_{Hp} and Sim_{Vp}. The final decision on whether the unknown shape is the reference guide mark is made based on

$$Sim = \sqrt{{Sim_{Hp}}^2 + {Sim_{Vp}}^2} \tag{9}$$

Four tested images are given in Figure.4 and Table.1 is the similarity evaluation of those 4 with respect to 9 references.

(a)Unknown target 1

(b) Unknown target 2

(c) Unknown target 3

(d) Unknown target 4

Figure 4 A group of unknown targets

It can be seen from Table.1, all unknown targets can be classified into the right group. Large experiments have been run and results have demonstrated the proposed method shows promising irregular shape recognition potential in both accuracy and speed.

IOP Conf. Series: Journal of Physics: Conf. Series **910** (2017) 012033 doi:10.1088/1742-6596/910/1/012033

Table 1 Similarity Evaluation of 4 Unknown Targets With Respect To References

Target \ references	Target 1	Target 2	Target 3	Target 4
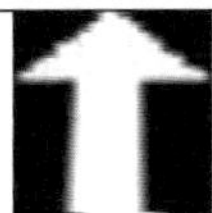	2.61	12.38	19.61	21.92
	9.43	7.51	4.4	6.35
	8.99	7.08	5.1	4.69
	8.21	4.83	8.29	8.01
	8.43	2.16	11.29	11.29
	6.21	8.43	10.83	11.78
	9.51	7.31	5.32	2.08
	9.19	4.82	9.51	10.24
	8.10	7.45	7.74	6.82

5. Conclusions

Irregular shape recognition has been one of the most difficult tasks in the image understanding area. It becomes tougher when dealing with convex-concave shapes. This work transforms shapes into signals firstly and then introduces a novel signal similarity evaluation scheme to fulfill the task. Experimental results have proven its advantages in pavement guide marks recognition and it may be used in other shape recognition applications as well

6. Acknowledgment

This research was financially supported by the National Science Foundation 41474005.

7. References

[1] Wu Xiaojun, Zou Guanghua. High performance template matching algorithm based on edge geometric features [J]. Chinese Journal of scientific instrument, 2013, 34 (7): 1462-1469.

[2] Tan Feigang, Liu Weiming, Huang Ling, et al. Target re-recognition algorithm based on weighted Euclidean distance [J]. Journal of South China University of Technology (NATURAL SCIENCE EDITION). 2015 (9): 88-94.

[3] Zhang Chuang, Wang Tingting, et al., et al. Image edge detection based on Euclidean distance map [J]. Journal of image and graphics, 2013, 18 (2): 176-183.

[4] Cao qingjuan and Wang yongjun, Similarity search of time series based on centroid, Journal of Chongqing Institute of Science and Technology, 2010,12(5):142-145

[5] Ismail M F, Jaafar T R, Pin N C, et al. Sobel operator for edges detection in surface texture analysis[J]. 2015.
[6] Wan Jine, Yuan Baoshe, Gu Chao, et al. Recognition algorithm based on character normalization and double projection correlation [J]. computer applications. 2013, 33 (3): 645-647.
[7] Shi S, Mingb Y U, Honglic Y U, et al. SIFT Image Matching Algorithm Based on Similarity Measure Optimization[J]. Video Engineering. 2013.

IOP Conf. Series: Journal of Physics: Conf. Series **910** (2017) 012034 doi:10.1088/1742-6596/910/1/012034

Feature Extraction Method based on AIC-FisherApplication in UltrasonicTesting

Banteng Liu, Mengjie Li and Lisha Wang

Zhejiang Shuren University, Hangzhou, China
Email: hupo3@sina.com, 842695193@qq.com, 1821760599 @qq.com

Abstract. In ultrasonic testing, the feature extraction of ultrasonic signal is equivalent to defect feature extraction. How to extract the effective signal features is the key to the defect identification. Traditional of ultrasonic signal features (echo time and peak value) insufficient to reflected defects characteristics. Commonly use features extraction and has supervision of discriminant method to optimization signal features. Fisher linear discriminant is to reduce the high-dimensional features to one-dimensional space, and to calculate the feature projection matrix according to the sample point distribution in one dimensional space. But the sample point distribution misclassification rate of one-dimensional feature is not optimal. So paper proposed a feature extraction method based on AIC-Fisher. The dimension of dimensionality reduction and dimensionality reduction are determined by Fisher criterion, sample point distribution error rate and AIC criterion. Then, the optimal feature set is used to quantify the defect depth. Finally, the experimental results show that this method can obtain the effective features, and has a high accuracy in the identification of defect depth.

1. Introduction

Ultrasonic testing is a common method of non-destructive testing. It has the advantages of accurate positioning, high sensitivity and fast response. And it is widely used in industrial, medical, military, agriculture, biotechnology and other fields. Defect detection is one of the main functions of ultrasonic testing. Due to the complex relationship between the ultrasonic signal and the defect reflection surface, it is difficult to find the effective feature in the practical application. Therefore, the feature extraction of the ultrasonic echo is the difficulty of the defect identification [5]. The traditional feature extraction is to extract the signal waveform or energy distribution in time-frequency domain. It cannot fully describe the characteristics of defects, especially it is not aware of the defect characteristics, and cannot directly select the effective signal features. So, many scholars often use multi feature framework to select more common features and combine supervised feature extraction methods, In order to extract the effective features adaptively. Linear discriminant analysis (Fisher) is the most commonly used supervised feature extraction method, but it has the problem of high false recognition rate after dimensionality reduction.

This paper proposed a feature extraction method based on AIC-Fisher, considering the distribution of the sample points of the Fisher linear discriminant, raises each kind of fault sample point error rate as dimensionality reduction index were constructed to determine the number of dimensionality reduction based on dimension reduction model AIC criterion. Finally, the defect depth recognition experiment is designed to verify the effectiveness of feature extraction. This method can obtain the effective defect feature, and has high accuracy in defect recognition. Compared with traditional feature extraction methods such as PCA and ICA, it has a great advantage.

IOP Conf. Series: Journal of Physics: Conf. Series **910** (2017) 012034 doi:10.1088/1742-6596/910/1/012034

2. feature Extraction method based on AIC-FISHER

Feature extraction based on AIC-Fisher method proposed in this paper, the diagram is as follows:

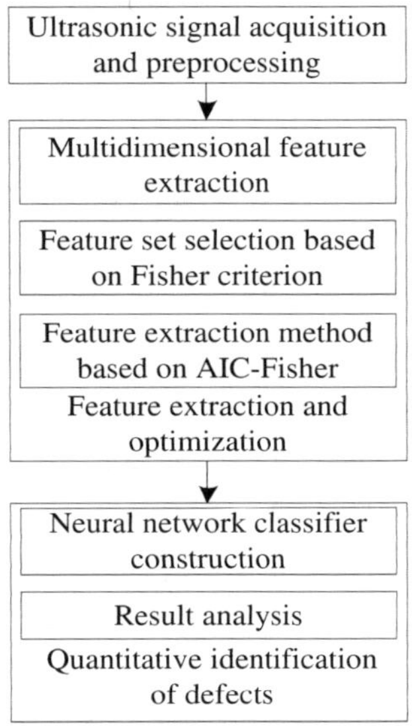

Figure 1.Ultrasonic testing of defects depth quantitative block diagram

As shown in Figure 1, proposed a feature extraction method based on AIC-Fisher, which is used in ultrasonic detection of defects depth recognition. Firstly, collecting the ultrasonic flaw signals and the ultrasonic signal is divided and de-noised. Then, the feature extraction and optimization of the processed signal, including three steps: multi feature extraction, feature vector selection based on Fisher and feature dimension optimization based on AIC-Fisher. Multi feature extraction includes time domain envelope spectrum feature extraction, frequency domain amplitude spectrum feature extraction and wavelet decomposition energy distribution feature extraction. Feature vector selection based on Fisher is to select the dimension reduction feature according to the change of Fisher criterion value. Feature extraction method based on AIC-Fisher is based on the sample point error rate as the index to calculate the AIC value, AIC value to determine the minimum dimension is used to obtain the optimal feature subset. Finally, neural network is used to identify the defect depth.

2.1 Fisher-Based Feature Selection

This paper proposed a feature vector selection method based on Fisher, the specific method is: Searching the Fisher criterion function value of the largest combination of different dimensions, and drawing the *J* curve under different dimensions. The Fisher criterion function is the trace of the discrete matrix ratio, the expression is as follows:

$$J(a) = \frac{a^T S_B a}{a^T S_w a} \tag{1}$$

$J(a)$ is the criterion value. S_W is the within-class scatter matrix. S_B is the between-class scatter matrix. a is a vector map.

$$S_W = \sum_{i=1}^{C} S_i = \sum_{i=1}^{C} \sum_{n_i \in \omega_i} (n_i - \mu_i)(n_i - \mu_i)^T \tag{2}$$

$$S_B = \sum_{i=1}^{C} P_i(\mu_i - \mu)(\mu_i - \mu)^T \tag{3}$$

X is a feature set, a total of N samples belong to the C category. Each class contains *n* samples. μ_i is the mean value of each category; μ is class center; P_i is a priori probability of class *i*. The feature of large contribution to the growth of the curve in the *J* value growth curve is selected as the feature set used to reduce the dimension.

2.2 Feature Extraction Method based on AIC-Fisher

In this paper, feature extraction method based on AIC-Fisher is proposed, which combines the AIC criterion and the Fisher criterion to further eliminate the redundant vectors in the feature set:

Step1: According to the distribution of mapping matrix of sample points by Fisher pre reduction, if the category of the sample point and the center distance is greater than other types of center distance, the point is considered as a miscarriage of justice, the calculation of different dimensions of the error rate.

$$P = \frac{D_W}{Q} \tag{4}$$

Step2: According to the AIC criterion to calculate the AIC value of the dimension reduction matrix of different dimensions:

For the *K* dimension of the sample data includes *N* categories, each class contains a *Q* sample, *D* is the sample point error of different categories.

$$AIC = -2\sum_{m=1}^{N} \ln \frac{N}{K} \cdot \frac{Q(m)}{D_{max} - D_{min}} + 2K \tag{5}$$

Step3: Drawing the change curve of AIC value, the lowest point of the curve corresponds to the best dimension reduction.

3. Experimental design and analysis

3.1 Experimental Design

In the experiment, ten kinds of defects with the same shape but different depth were identified. In a uniform thickness of the steel plate, there are ten kinds of defects of aperture 1.00mm, depth of 0.5mm~5mm, each defect difference between 0.5mm. Each defect was scanned 30 times, and a total of 30*10 ultrasonic echo signals were collected. The schematic diagram of the test piece and the defect is shown in figure 2:

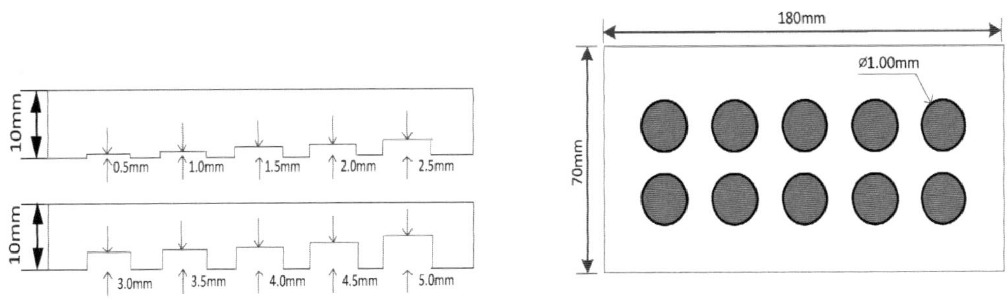

Figure 2 The specimen and sketch of defects

3.2 Result analysis

According to the above, the experiment according to the different dimensions of the Fisher criterion value change, make the *J* value curve as shown in figure 3:

It can be seen from Figure 3 that the contribution of the first 8 dimensions to the growth of the J value is large, so choose the feature of the 8 dimension as the feature set for AIC-Fisher optimization. AIC value curves are obtained by the AIC-Fisher method, as shown in Figure 4:

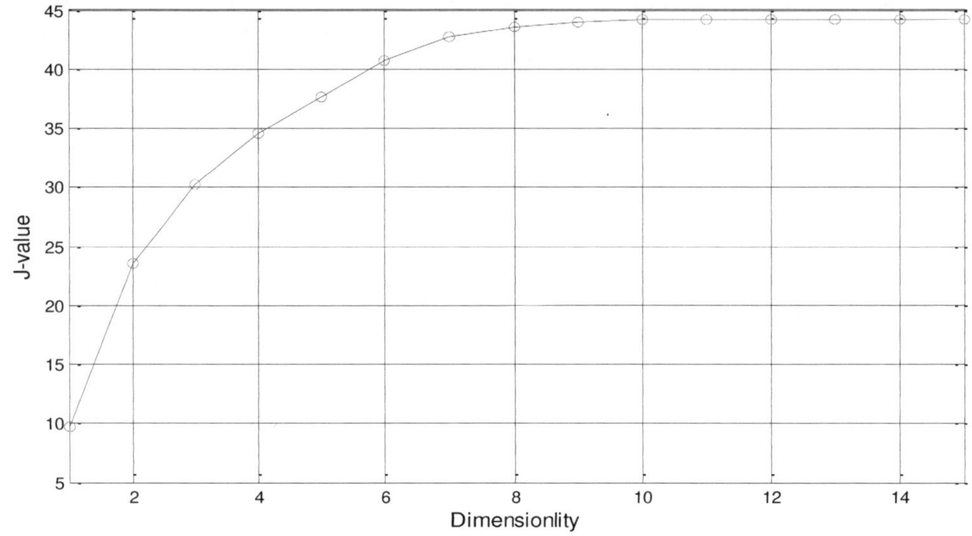

Figure 3.J value change curves

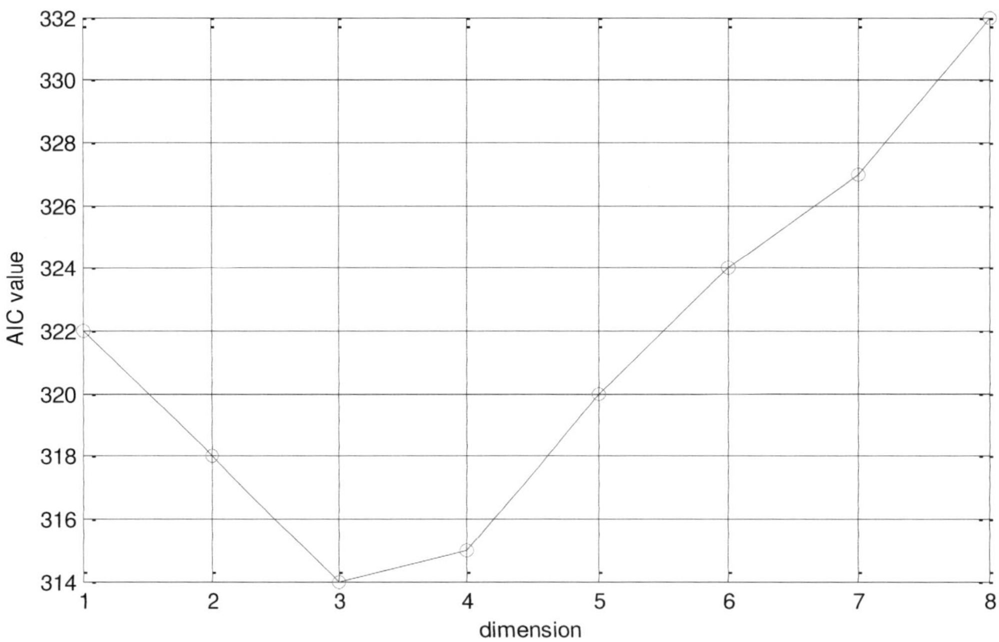

Figure 4.AIC Value Change curves

From the above diagram, reduced to 3 dimensional minimum AIC value can be obtained. In order to compare the superiority of this method, the 7 dimension, the 8 dimension, the 9 dimension and the 24 dimension are taken as examples to reduce the dimension distribution map, as shown in figure 5~8:

Finally, the 3D feature set is used as the sample set of neural network recognition. There are ten kinds of defects in the sample set, each group has 30 sets of data, and the 20 sets of data is used as the training set, and the 10 sets of data is used as the test set for the quantitative identification of neural network. Specific results are shown in table 1:

It can be seen from the above table that the accuracy of Fisher method is higher than that of PCA and ICA. And in the multidimensional Fisher method, the recognition effect of the sample to the 1 dimension is not good and the false alarm rate is high. According to the overall data, the recognition rate of the Fisher criterion is improved, and the optimal feature obtained by the AIC-Fisher method has the highest recognition rate.

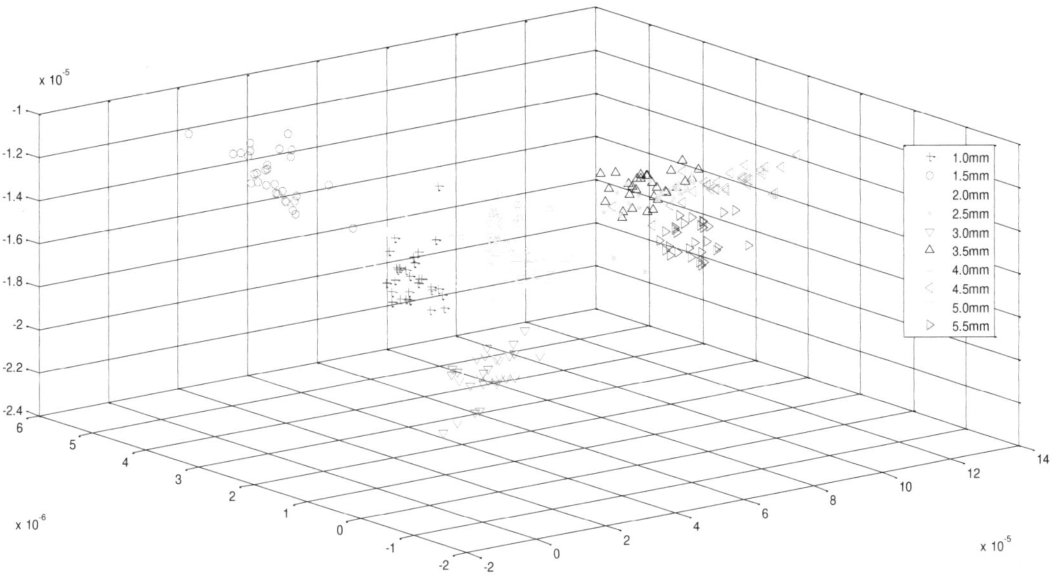

Figure 5 .24D-Fisher map

Table 1 Results of neural network analysis

Optimization method	false positive(%)	Correct recognition rate(%)
PCA	44.0	68.3
ICA	42.3	69.8
Fisher	40.0	70.3
AIC-Fisher	12.3	93.6
5D-Fisher	36.7	73.0
6D-Fisher	35.0	75.6
7D-Fisher	34.7	74.9
8D-Fisher(1D)	35.3	74.9
8D-Fisher(2D)	26.3	80.2
8D-Fisher(4D)	14.0	91.3
8D-Fisher(5D)	21.6	87.4
8D-Fisher(6D)	22.3	85.3
8D-Fisher(7D)	24.7	82.7
9D-Fisher	36.3	75.1

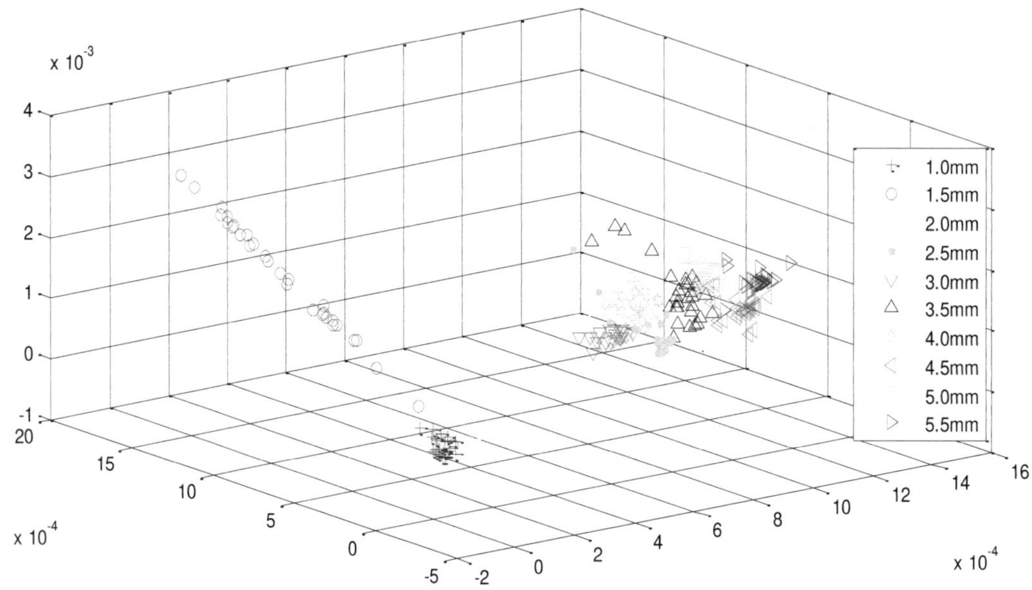

Figure 6 7D-Fisher map

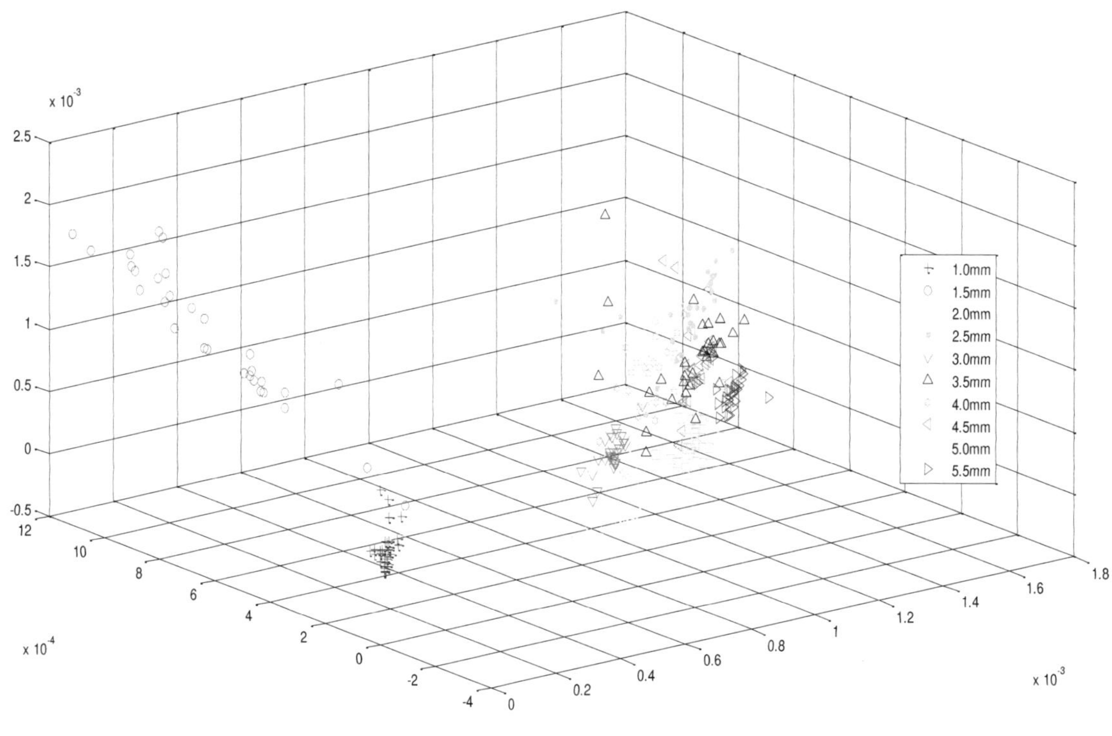

Figure 7 9D-Fisher map

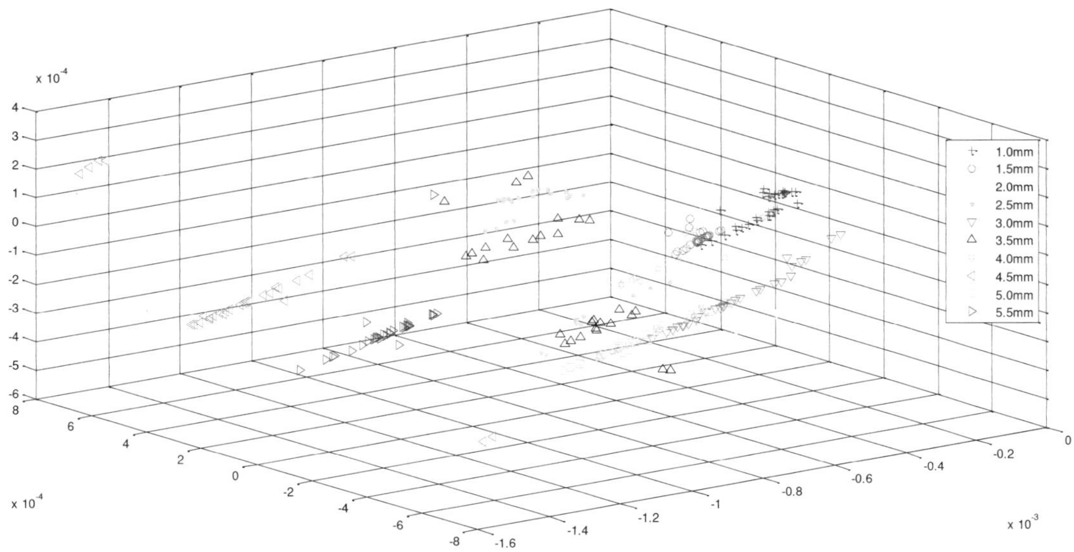

Figure 8 .AIC-Fisher map

4. Conclusion

In this paper, according to the limitation of the feature extraction of ultrasonic signal and the limitation of Fisher dimension reduction method, a method of ultrasonic signal feature extraction based on AIC-Fisher is proposed. In this method, according to the change curve of the Fisher criterion value in different dimensions to select a reasonable feature set. According to the distribution of known sample points, the paper points out the error rate index of the sample points, and combine the AIC criterion to determine the dimension of the dimension reduction of the feature set. Finally, the calculation of RBF neural network identification is reduced, and the accuracy of quantitative recognition of defect depth is improved.

5. Acknowledgement

Foundation item: National innovation and entrepreneurship training program for college Students in 2016 (201611842002); In 2017, students in Zhejiang province science and technology innovation plan and Xinmiao Talents Program（2017R421028）

6. References

[1] Meng Qingfeng. Study of Signal Feature Extraction Methods and Applications[D]. Xidian University, 2006.
[2] Wang Lihua, Zhang Yonghong. Application of Feature Extraction and Selection of Ultrasonic Signal in Ultrasonic Flaw Echo Identification [J]. Machinery manufacturing and automation, 2011, 40 (6): 114-117.
[3] Zhang Xuegong. Introduction to Statistical Leaning Theory and Support Vector Machines[D]. Journal of Automation, 2000, 26 (1): 32-42.
[4] Shigeo Abe, Support Vector Machines for Pattern Classification, Springer, London, 2005.
[5] Che Hongkun. Study on Key Technologies of Non-imaging Ultrasonic Defect Identification and its Application [D]. Zhejiang University, 2011.
[6] Qin Xuanyun. Nearest neighbor cluster algorithm based on AIC criterion[J]. Systems Engineering and electronic technology, 2005, 27 (2): 257-259.
[7] Yang Ping, Xia Ping, Zhang Bin. Study on feature extraction and optimum feature subset selection of defect ultrasonic echo Signal [J]. Machinery manufacturing, 2013, 51 (6): 30-32.
[8] Dai Wenzhan, Zhou Changliang. Linear discriminant analysis based on improved Fisher criterion. Computer Engi- neering and Applications, 2013, 49(3): 210-21.

Dynamic Simulation Analysis of Key Components of Special Transport Vehicle for Blowout Prevention Equipment Based on ADAMS

Huixiang Yang, Junpeng Zhao, Dawei Jiang*, Guowa Song and Bangcheng Zhang

Changchun University of Technology, Changchun 130012, China
Email: yanghuixiang@mail.ccut.edu.cn, zhaojunpeng163@163.com, jiangdawei@mail.ccut.edu.cn, edu.cn, sgymrr@163.com, zhangbangcheng@mail.ccut.edu.cn

Abstract. The mounted structure is a part of the whole vehicle, and determines the rationality of the vehicle design. First of all, through the ADAMS and CATIA data transmission interface, the use of CATIA to create a special model of the blowout prevention equipment, the model is imported into ADAMS; Secondly, ADAMS related settings in the definition of force, boundary conditions, and calculate the time step and other related parameters. Finally, the simulation results of the flipping mechanism are obtained by dynamic simulation analysis. The results show that the design of the truck loading structure meets the actual performance requirements.

1. Introduction

In general, the equipment used for the transport of the jetting equipment is very large in size and weight, usually around 25t[1], and the loading structure as the carrying base of the carrier, the full force exerted by the upper equipment and Moments are acting on it. At the same time, due to the rugged road of the transport vehicle, and the operating conditions and road conditions are more complex, carrying equipment volume and quality is large, and the loading structure is complex, through its dynamic correlation analysis, can fully understand the movement characteristics of the loading structure , In order to avoid defects in the design process, in which case the frame itself must have sufficient bending rigidity and strength to ensure that the relative position between the load-bearing upper assembly is kept constant during operation and minimized Small body of the deformation; and sufficient strength is to ensure the reliability of the frame and the premise of life[2]. Spraying workover equipment for the use of special transport vehicles used in the chassis and its sub-frame shown in Figure 1, the test deformation and fracture of the situation shown in Figure 2.

Figure 1 Two types of chassis and subframe

Figure 2 Subframe deformation and fracture diagram

The loading structure is a key component of the jet repair work over vehicle, so it needs to carry on the dynamic simulation analysis

2. Overview of Virtual Prototyping Technology

Virtual prototyping technology is to solve the shortcomings of traditional product design process, the use of computer technology developed from high-tech. Traditional product design and manufacturing process, the general process is the concept of design, product design, prototype experiments, product improvement [3-5]. In the prototype experiment process, need to spend a lot of manpower and material resources, increase the cost of product design, extend the design cycle, the traditional design means cannot meet the needs of market competition. Through the analysis of different mechanical systems can get the optimal performance of the work, but also can reduce the complex mechanical system production costs, the number of tests, research and development cycle, improve the mechanical system design quality and R & D efficiency [6].

3. Dynamic Simulation and Analysis Process of Special Transport Vehicle for Blowout Prevention Equipment

The process of dynamic simulation analysis of special transport vehicle with ADAMS software is mainly divided into four steps. The analysis flow is shown in Figure. 3

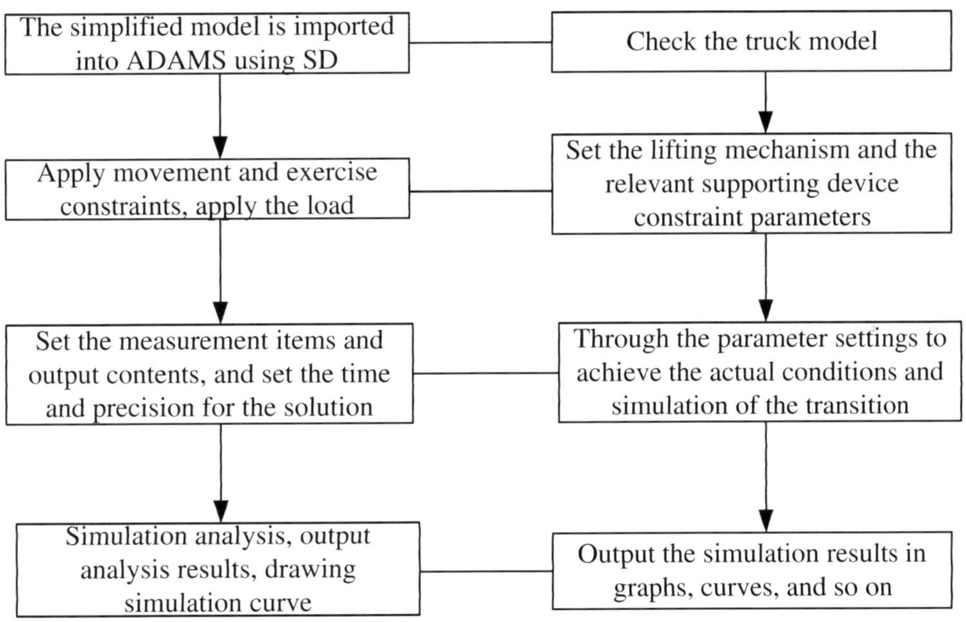

Figure 3 Analysis of dynamic simulation of special vehicle

3.1 The Basic Theory of Dynamics Analysis

When solving the dynamic problem of the lifting mechanism of the special transport vehicle, it is possible to simplify the motion analysis of the mechanical system if the coordinate system can be set reasonably. In the process of mechanical system motion analysis, three coordinate systems are commonly used: ground coordinate system, local component reference coordinate system, frame coordinate system. The speed of solving the dynamic equation depends largely on the choice of generalized coordinates. Through the study of the motion of the rigid body in the inertial space, this paper adopts the Euler parameter as the generalized coordinate of rotation, and its variables are not too many, Euler angle when there is no singularity.

The generalized coordinates used in Admas are determined by the Cartesian coordinate system of the component center and the Euler angle of the component position, $q=[x,y,z,\psi,\theta,\varphi]^T$, set $R=[x,y,z]^T, \gamma=[\psi,\theta,\varphi]^T$, $q=[R^T,\gamma^T]^T$. The coordinate transformation matrix between the component centroid reference coordinate system and the ground coordinate system is[7]:

$$A^{gi}=\begin{bmatrix}\cos\psi\cos\phi-\sin\psi\cos\theta\sin\phi & -\cos\psi\sin\phi-\sin\psi\cos\theta\cos\phi & \sin\psi\sin\theta \\ \sin\psi\cos\phi+\cos\psi\cos\theta\sin\phi & -\sin\psi\sin\phi+\cos\psi\cos\theta\cos\phi & -\cos\psi\sin\theta \\ \sin\theta\sin\phi & \sin\theta\cos\phi & \cos\theta\end{bmatrix} \quad (1)$$

The above three Euler shafts as a unit vector can be used to establish a new coordinate system - Euler axis coordinate system. The coordinate transformation relationship between the Euler axis coordinate system and the component centroid Cartesian coordinate system is given by the following matrix:

$$B=\begin{bmatrix}\sin\theta\sin\phi & 0 & \cos\theta \\ \sin\theta\cos\phi & 0 & -\sin\theta \\ \cos\theta & 1 & 0\end{bmatrix} \quad (2)$$

The angular velocity of the component movement is expressed as:

$$\omega=B\dot{\gamma} \quad (3)$$

When using Adams for kinetic analysis, the component of the angular velocity in the Euler axis coordinate system is set to:

$$\omega_e=\dot{\gamma} \quad (4)$$

Combined with the constraint equation, the following expression is deduced by the energy form of Lagrange first type equation:

$$\frac{d}{dt}(\frac{\partial T}{\partial \dot{q}_j})-\frac{\partial T}{\partial q_j}=Q_j+\sum_{i=1}^{n}\lambda_i\frac{\partial \phi}{\partial q_j} \quad (5)$$

In the above formula: T -the kinetic energy of the system's generalized coordinates; q_j -generalized coordinates; Q_j - q_j direction broadness; $\sum_{i=1}^{n}\lambda_i\frac{\partial \varphi}{\partial q_j}$ - q_j directional constraint force. By further introducing generalized momentum:

$$P_j=\partial T/\partial \dot{q}_j \quad (6)$$

The simplified expression of the reaction force is:

$$C_j=\sum_{i=1}^{n}\lambda_i\frac{\partial \phi}{\partial q_j} \quad (7)$$

This equation (4-5) can be simplified as:

$$\dot{P}_j-\frac{\partial T}{\partial q_i}=Q_i-C_j \quad (8)$$

Kinetic energy can be further expressed as:

$$T=\frac{1}{2}\dot{R}^T M\dot{R}+\frac{1}{2}\dot{\gamma}^T B^T JB\dot{\gamma} \quad (9)$$

Where M is the mass of the component and J is the inertia of the component in the centroid coordinate system.

3.2 Establishment of Dynamic Model for Blowout Prevention Equipment

Although ADAMS / View comes with modeling tools, but only a few relatively simple structure of the model, and for some of the complex structure, more model of the operation of the model cumbersome, longer time, greatly reducing the efficiency, ADAMS The CAD model data interface can be used to import the model data from other professional solid modeling software (CATIA, Pro/E, etc.) through the corresponding data interface to obtain the required dynamic model.

The SimDesigner for CATIA V5 products product series developed specifically for CATIA enables MSC to quickly exchange data between ADMAS and CATIA and seamlessly link 3D models drawn with CATIA V5 to ADAMS. Therefore, the three-dimensional model of the flip mechanism is presented in CATIA, which fully exploits the three-dimensional solid modeling function of CATIA and the dynamic analysis function of ADAMS [8-10]. SimDesigner Generative Products can be used to output the required cmd format for ADAMS in the CATIA model, import it into the ADAMS / View function component, and then define the relevant parameters or component attribute values, such as grid model and size, component constraints, force size and direction, component name and color, etc. The ADAMS analysis model is shown in Figure 4.

Figure 4 ADAMS dynamics model

In ADAMS, there are six relative degrees of freedom in the space, including the degree of freedom of translation (3) and the degree of freedom of rotation (3), there are seven translational degrees of freedom in the mobile subcomponent, and there are five degrees of freedom in the rotation subcomponent.

After simplification of the model, remove the screws, nuts and locating pins and other components. The rest of the dynamic part of the road model by the 218 rigid body, seven mobile vice, five rotation vice, two fixed sub-composition. There are 12 degrees of freedom in the whole flipping mechanism, 6 on the body of the special vehicle and the flipping mechanism, where the turning mechanism is the degree of freedom of rotation.

The vehicle is in the flip work is to control the lift hydraulic cylinder through the hydraulic drive to achieve, so that the turning mechanism in the set time to do a uniform rollover movement, when the flip mechanism to flip to a predetermined position (flip to 95° position) to achieve anti- The workover equipment is docked up and down with the wellhead during the set time. In addition to the corresponding mobile vice and rotating vice, the rest of the components are fixed with the auxiliary, are fixed in the special transport vehicle pre-set on the ground.

3.3 Dynamic Simulation Analysis of Blowout Prevention Equipment

When using ADAMS/View to calculate the contact force, according to the different calculation model can be divided into impact function and compensation method two. It is worth noting that the contact force calculated by the impact function method not only includes the damping force caused by the relative motion between the members, but also the elastic force caused by the mutual cutting of the members.

The dialog box for defining the contact force in ADAMS is shown in Figure 4-3. The values and units of each parameter are shown in Table 1.

Table 1 Contact Force Parameter Table

Parameter name	unit	Value
Stiffness	newton/mm	200
Force Exponent	newton·sec/m	2.2
Damping		10.0
Penetration Depth	mm	0.1
Static Coefficient		0.3
Dynamic Coefficient		0.2
Stiction Transition Velocity	mm/sec	100
Friction Transition Velocity	mm/sec	1000

The finite element model of the special transport vehicle for jetting and repairing wells is added in ADAMS, and the effect of driving force is shown in Figure.5.

First define the coordinate system, set the length of the direction of the transport vehicle for the x-axis, the front to the rear of the direction is positive, and the special transport vehicle placed on the ground perpendicular to the direction of the z axis, pointing to the sky direction is positive, y axis is also parallel to the The direction of the truck body, the acceleration of gravity Z = -9806.65mm / s2, the direction of the negative direction along the z axis.

In the simulation experiment of this process, because the parts of the special transport vehicle are rigid, the contact form can be defined as the contact between the entities, and the contact force can be obtained by the Impact function, set the special vehicle and the ground were solid 1, 2, the relative slip between the two when the friction coefficient shown in Figure 6 curve, so the friction can be obtained by the Coulomb method.

Figure 5 ADAMS simulation software to add motion pairs, constraints, driving force renderings

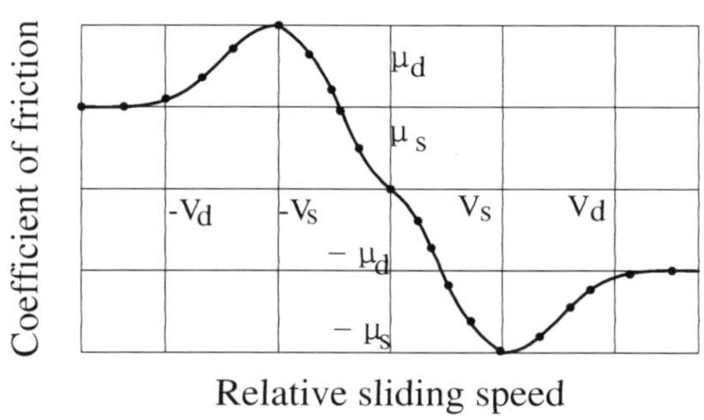

Figure 6 Friction coefficient and sliding velocity curve

In the ADAMS software dynamics simulation, the four auxiliary hydraulic cylinders of the special vehicle are set to the corresponding contact constraints (Solid-Solid between the support cylinder block and the piston rod) and the friction coefficient (μ = 1) The fixed cylinder and the piston rod of the truck are fixed with a downward freedom, and the start time and the starting position of the driving force function are set to zero, and the end time and the ending position are set to 10 s and 303 mm, respectively. Lifting hydraulic cylinders Cylinders and special transport vehicles Fixed lifting joints (contact-constrained Solid-Solid) with a certain angle of rotation, lift the hydraulic cylinder of the piston rod and the special transport vehicle rollover mechanism of the lifting support Contact-bound Solid-Solid) Rotary pair with an angle, lift cylinder block and piston rod (contact constraint Solid-Solid) with upward freedom of the movement of the vice. Set the start dialog box to set the start time and start position of the drive force function dialog box to 10s and 0mm respectively, and set the end time and the ending position to 30s and 2300mm respectively. (Contact constraint Solid-Solid) has an upward movement pair, set the movement pair dialog box, set the start time and starting position of the driving force function dialog box to 30s and 0mm respectively, and set the movement counter dialog box, Set the end time and the end position to 60s and 700mm, respectively. And then change the cylinder and the piston rod of the lift cylinder to the movement of the downward freedom, and set the additional end time and the ending position of the mobile sub dialog box to 90s and 0mm respectively. The support cylinder and the piston rod are changed upward The degree of freedom of the movement of the vice, set the mobile sub-dialog box additional end time and termination position were set to 100s and 0mm.

Through the simulation analysis of the actual working condition, the simulation curve is obtained under the condition of lifting and loading a complete operation. Figure 7, 8, 9, 10 are the curves of the lift cylinders x, y, z and the resultant moment, Figure 11, 12, 13,14 is the x, y, z direction and the resultant torque change curve of the articulated position of the hydraulic cylinder and the flip frame.

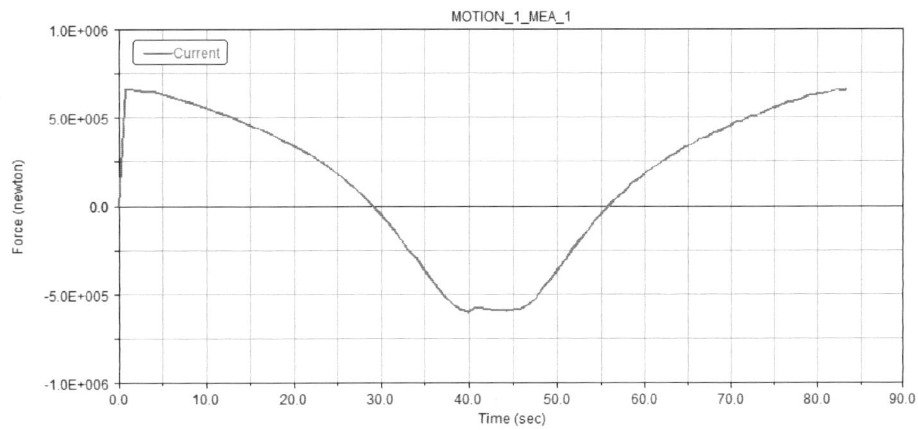

Figure 7 Lift the hydraulic cylinder x direction torque change map

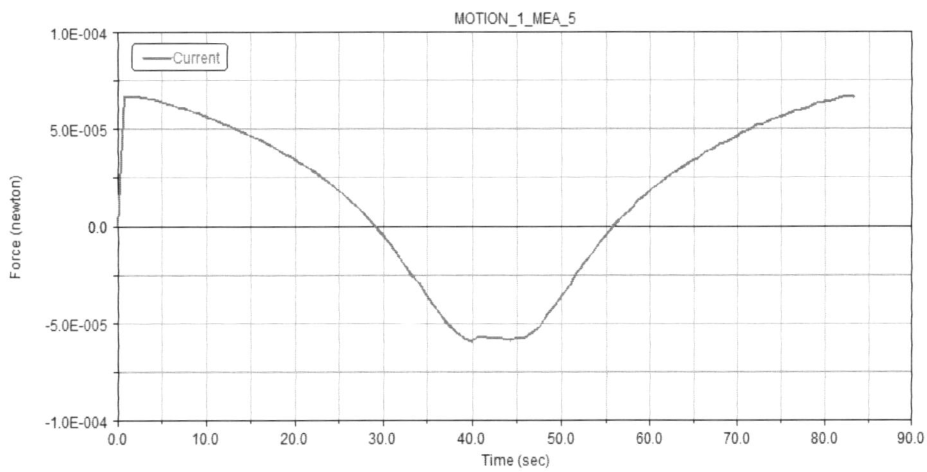

Figure 8 Lift the hydraulic cylinder y direction torque change map

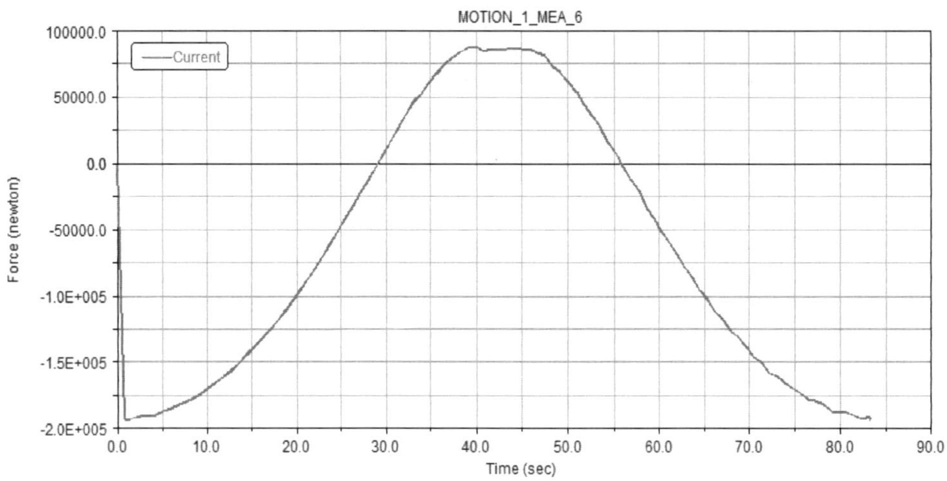

Figure 9 Lift the hydraulic cylinder z direction torque change map

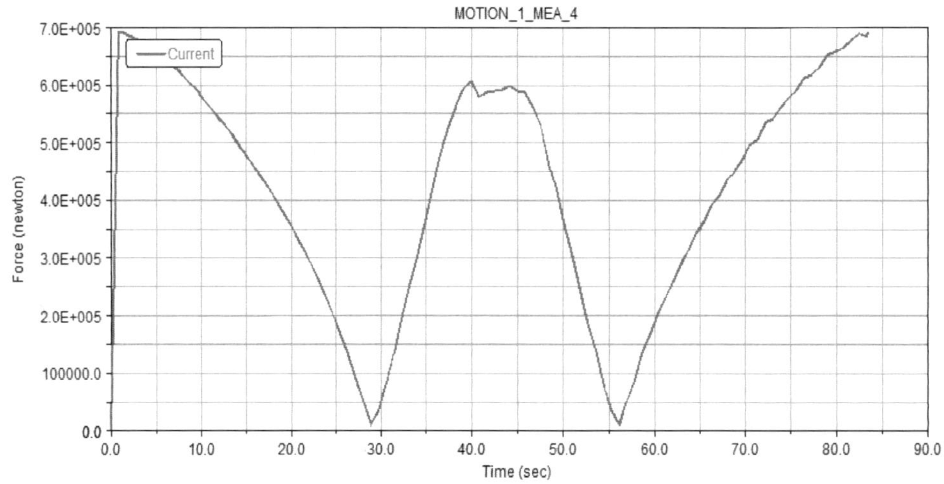

Figure 10 Lifting of hydraulic cylinder

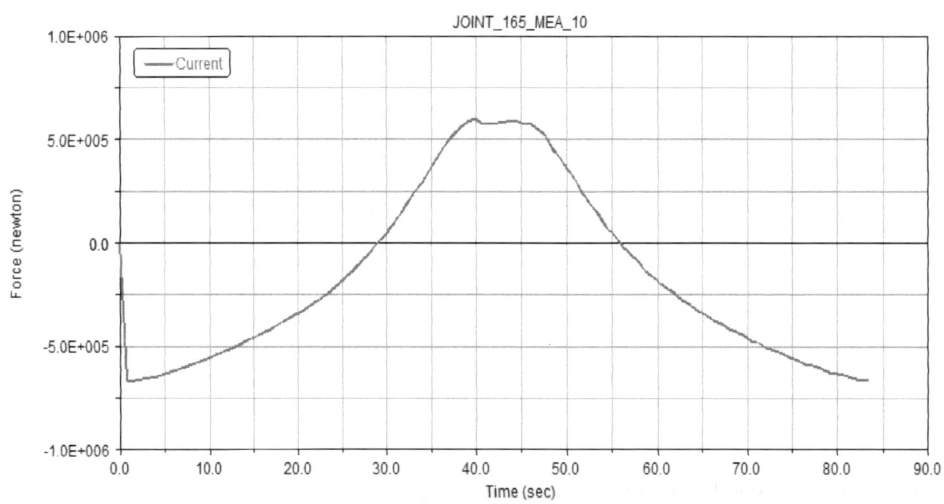

Figure 11 Hinge position x direction torque change map

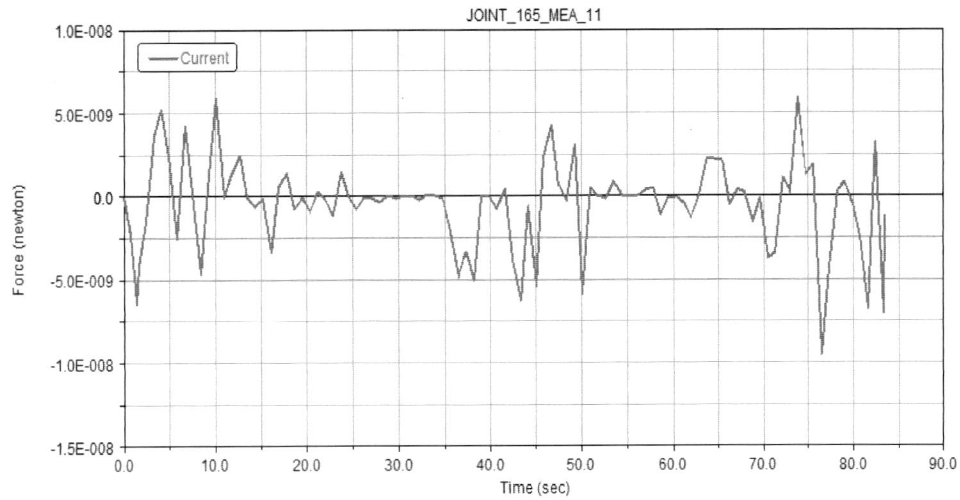

Figure 12 Hinge position y direction torque change map

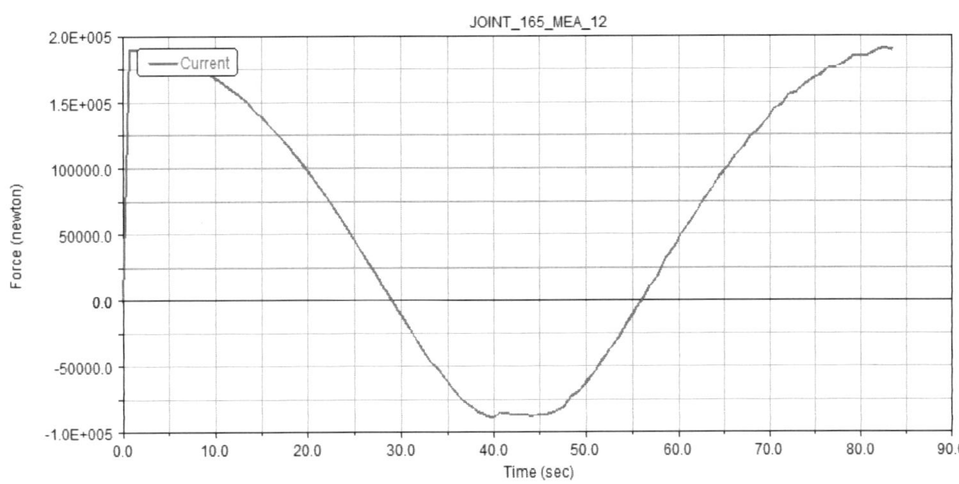

Figure 13 Hinge position z direction torque change map

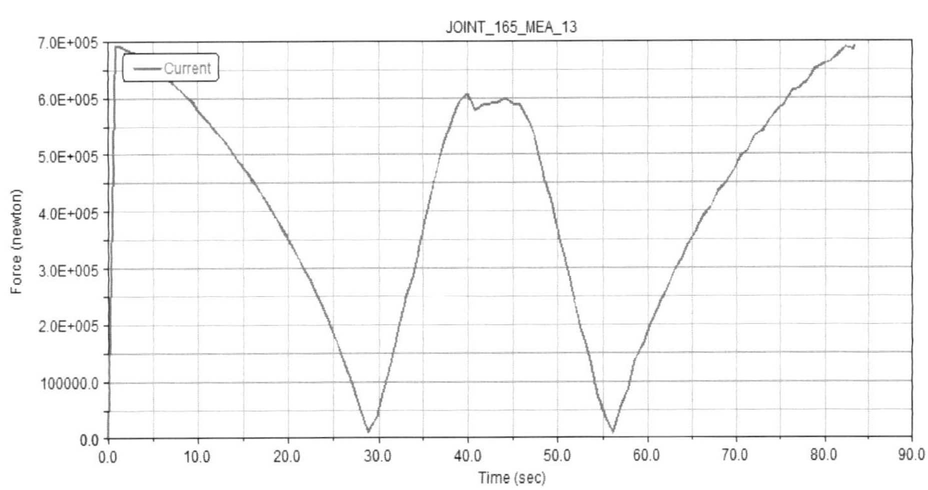

Figure 14 Rotation position of the joint

From the simulation results can be seen throughout the operating cycle of the hinge and lift the hydraulic cylinder of the force changes. When the lifting mechanism is in the initial position, lift the hydraulic cylinder thrust is the largest, hinge at the force is also the largest, the maximum contact force of 352KN. When the time is 30s, that is, when the flip frame is turned to 71°, the center of gravity of the flip frame and the anti-scratch workover device will be turned to the vertical direction. The contact force between the lifting hinge and the hydraulic cylinder is the smallest, close to 0N. When the flip frame from 71 ° gradually increased to 95°, due to the center of gravity to move back, lift the hydraulic cylinder from the thrust into the tension, then lift the triangular block hinge A contact force and gradually increased to 258KN. 45s after the flip for the process of return, is the reverse phase of the lifting process.

4. Summary

After the establishment of the three-dimensional simulation model for the special maintenance vehicle, the ADAMS simulation software is used to limit the contact and the contact load, and finally the torque curve of the lifting cylinder and its articulated position is obtained. Through the simulation of the actual work, the analysis of the force curve of the lifting mechanism is carried out to verify the rationality of the loading structure and the stability of the special transport vehicle, thus ensuring the safety and reliability of the working process of the overturning mechanism. The actual application of the structure, design and further improve the reference provided.

5. Acknowledgement

Jilin Provincial Development and Reform Commission Item Number (No. [2015] 114).

6. References

[1] CAI Bin, PENG Yong, YAN Wen-hui, etal. Study on the development of workover equipment for non-chilling wells [J]. Drilling & Production Technology, 2008 (06): 106-109.
[2] ZHANG Yan-ling. Stress Analysis and Structural Optimization of Sub-frame of Wheel Loader [J]. Construction Machinery, 2013, 44 (7): 1-3.
[3] Li Zenggang .ADAMS entry details and examples [M]. Beijing: Defense Industry Press, 2012.4.
[4] YOU Yu-zhuo. Simulation and Experimental Research on Reliability of Heavy Duty Truck Based on ADAMS. Jilin University .2010: 26-27
[5] Zheng Jianrong. ADAMS-virtual prototype technology entry and improve. Beijing: Machinery Industry Press .2012.
[6] TAKANG S, NAGAI M, TANIGUCHI T, et al. Study on a vehicle dynamics model for improving roll stability. JSAE Review, 2013, 24(2):149-156.
[7] DAHLBERG E, STENSSON A. The dynamic rollover threshold- a heavy track sensitivity study [J].International Journal of Vehicle Design,2010,40(1/2/3):228-250.
[8] Gao Qinhe, Guo Xiaosong. Modeling Modeling of Multi - stage Hydraulic Cylinder System Based on ADAMS. Machine Tool & Hydraulics. 2013,1: 93-97.
[9] DAI Wen-yue, LIANG Hao. Cynamic Simulation and Integrated Optimization Design of Loader Working Device [J]. Journal of Jilin University (Engineering Science), 2004,04: 602-605.
[10] YIN Hui-jun, WEI Zhi-lin, HUANG Chang-chun, HUANG Gui-dong, SHEN Guang-li.Optimization Design of Lifting Triangular Arm of Dump Truck [J]. Journal of Agricultural Mechanization, 2007,09: 186-188.

Mobile Robot Designed with Autonomous Navigation System

Feng An, Qiang Chen, Yanfang Zha and Wenyin Tao

Suzhou Industrial Park Institute of Services Outsourcing, Suzhou, China
Email: anf@siso.edu.cn, chenq@siso.edu.cn, zhayf@siso.edu.cn, taowy@siso.edu.cn

Abstract. With the rapid development of robot technology, robots appear more and more in all aspects of life and social production, people also ask more requirements for the robot, one is that robot capable of autonomous navigation, can recognize the road. Take the common household sweeping robot as an example, which could avoid obstacles, clean the ground and automatically find the charging place; Another example is AGV tracking car, which can following the route and reach the destination successfully. This paper introduces a new type of robot navigation scheme: SLAM, which can build the environment map in a totally strange environment, and at the same time, locate its own position, so as to achieve autonomous navigation function.

1. Introduction

The story “old horse is a good guider" is about a king with his army, lost in mountains, but with the help of several old horses they finally found the way out. In the story, the old horse could know the way out because they had been passed through this mountain before. But, for robots, it seems that it’s much more complicated than ever. First of all, most robots moving in an unknown environment, such as exploring the kind of car, sweeping robot, this is not experienced "old horse" to rely on; secondly, GPS communication mode is used in the traditional navigation robot is no longer applicable, because most of the work on the interior, even underground river environment; finally, the robot is designed for specific purposes, so they are unable to withstand the pressure of navigation algorithm. Based on the above reasons, we introduce the new mobile robot navigation and positioning scheme - SLAM.

2. Design Objectives

When the robot is in an unknown environment, it needs to know the following three things: where am I? What is this place? How to get out of this place?

In SLAM theory, the first problem is called Localization, the second is called Mapping, and the third is the subsequent Path Planning. To achieve the three goals we need to give the robot "armed", which includes the installation of the eyes, the robot legs, and most important: the brain.

3. Robot's “eyes"

The "eye" of a robot is a camera or a laser sensor. Through these sensors, the robot can get the understanding of the surrounding environment, so as to lay the foundation for building. In recent years, more and more engineers use the following two sensors as the robot's eyes.

Kinect is released in June 2010 by Microsoft, which was initially widely used in the field of games! It has three lenses; the middle of the lens is a RGB color camera, used to capture color images. The left and right sides of the lens is infrared transmitter and infrared CMOS camera composed of 3D structure light depth sensor, used to collect the depth of data (objects in the scene to the camera distance).

Kinect and laser sensor are shown in figure I.

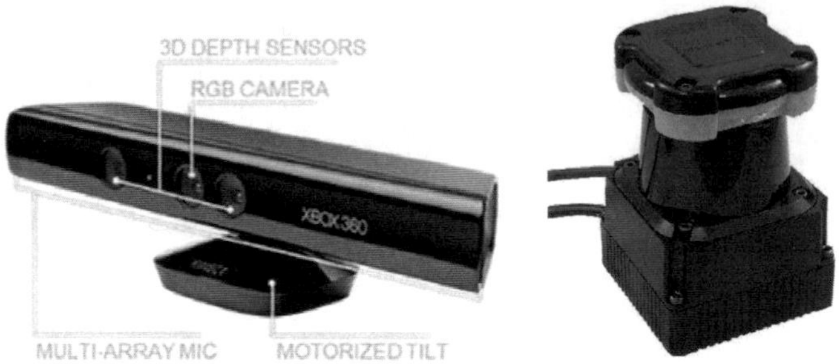

Figure 1 Kinect & laser sensor

The collected and processed images are shown in figure II.

When the robot is equipped with a Kinect or a laser sensor, in her eyes, a strange external environment is recorded, especially with the distance between obstacles. When the robot moves, the eyes of the external environment is certainly changing, the robot will collect all of these image information, by the brain for analysis and processing. Smart brain according to this image can draw a picture of the external environment of the real map, and in the process of continuous incremental, find out their location (positioning). If there is a set of goals, the robot will further path planning, in order to achieve the fastest and safest way to reach the destination.

Figure 2 Kinect capture pictures(RGB, depth and pcl)

4. Robot's "brain"

The brain of the robot, which is the control center of the whole system, is usually held by a PC (or embedded control board). PC runs on the software, to control the robot to walk, collect information, avoid obstacles, to build plans and positioning, etc. This is a complex and difficult work, because all of these work are interrelated, interaction, fortunately, the emergence of ROS.

Figure 3 ROS robot

SLAM exists in the form of ROS development kit, we install the ROS system for the robot's brain, when receiving sensors to collect environmental information, the SLAM package will analyse and process the complete and build the environment map and its positioning etc.. The specific composition of the SLAM package as shown in figure III:

Of course, the development of the present, SLAM is no longer a simple algorithm or software, but includes hardware and software platform, a number of processes, a variety of solutions and mapping system. Generally speaking, the robot to complete the SLAM process generally is: the robot continuously through the depth of the camera or the laser sensor, the odometer information collection robot path of their own, and then analyze the information, draw their own position and the environment map. The robot is to infer the location information (EKF) by the way of probability, because the acquisition environment, the path information will have errors, noise. Thus, in all of these steps, the EKF is the core, which keeps the (probabilistic) update of the robot's position.

When the odometer changes because the robot moves the uncertainty pertaining to the robots new position is updated in the EKF using Odometer update. Landmarks are then extracted from the environment from the robots new position. The robot then attempts to associate these landmarks to observations of landmarks it previously has seen. Re-observed landmarks are then used to update the robots position in the EKF. Landmarks which have not previously been seen are added to the EKF as new observations so they can be re-observed later. All these steps will be explained in the next chapters in a very practical fashion relative to how our ER1 robot was implemented. It should be noted that at any point in these steps the EKF will have an estimate of the robots current position.

The general framework for SLAM is shown in figure IV.

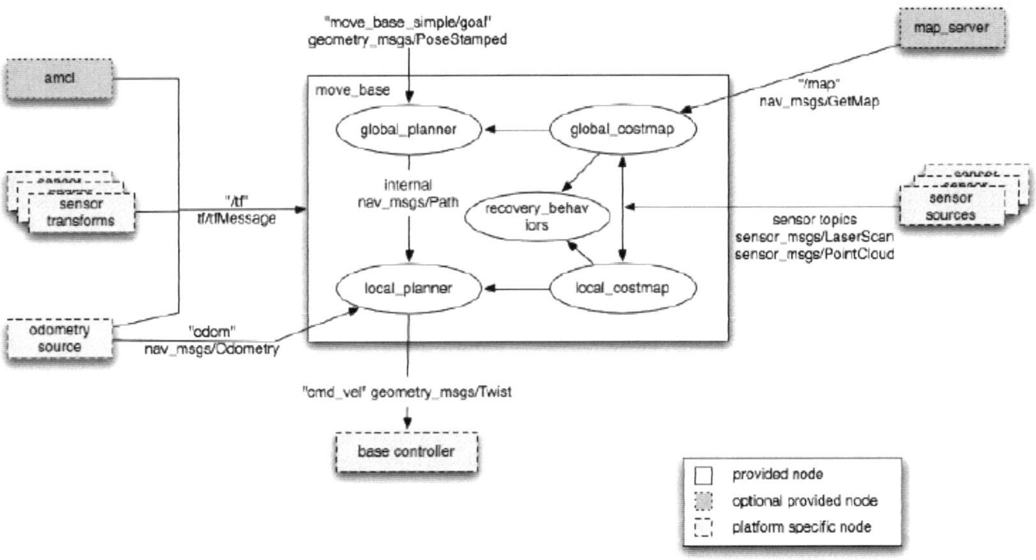

Figure 4 SLAM components

5. Acknowledgment

The mechanism of SLAM is discussed in this article, which can make the robot recognize the road, autonomous navigation even in an absolutely new environment, just like the "old horse". Especially with the increasing demand of service robots, SLAM will have more and more applications. However, due to the complexity of the actual environment, such as indoor light will change, the obstacles move, let a robot at a slower speed (2cm/s), this paper looks easy to realize the function, in the actual environment are often stretched, run everywhere. The good news is that companies such as Google have released a SLAM development kit, marking the industry's research and development in this field into a new stage. Challenges to the actual environment, is the main direction of development of SLAM technology, which includes the dynamic scene, semantic maps, multi robot cooperation, etc. It's believed that with the continuous progress of SLAM technology, it will soon come into our life.

6. References

[1] G. Eason, B. Noble, and I. N. Sneddon, "On certain integrals of Lipschitz-Hankel type involving products of Bessel functions," Phil. Trans. Roy. Soc. London, vol. A247, pp. 529–551, April 1955.
[2] Laurens Valk, A Beginner's Guide to Building and Programming Robots, starch press, 2010.
[3] Carol Carol Fairchild, ROS Robotics By Example.
[4] Aaron Martinez, Learning ROS for Robotics Programming, Packt Publishing, 2013
[5] J. Clerk Maxwell, A Treatise on Electricity and Magnetism, 3rd ed., vol. 2. Oxford: Clarendon, 1892, pp.68–73.
[6] I. S. Jacobs and C. P. Bean, "Fine particles, thin films and exchange anisotropy," in Magnetism, vol. III, G. T. Rado and H. Suhl, Eds. New York: Academic, 1963, pp. 271–350.
[7] K. Elissa, "Title of paper if known," unpublished.
[8] R. Nicole, "Title of paper with only first word capitalized," J. Name Stand. Abbrev., in press.
[9] Y. Yorozu, M. Hirano, K. Oka, and Y. Tagawa, "Electron spectroscopy studies on magneto optical media and plastic substrate interface," IEEE Transl. J. Magn. Japan, vol. 2, pp. 740–741, August 1987 [Digests 9th Annual Conf. Magnetics Japan, p. 301, 1982].
[10] M. Young, The Technical Writer's Handbook. Mill Valley, CA: University Science, 1989.

Dynamic Gesture Recognition based on LeapMotion and HMM-CART Model

Qixiang Zhang and Fang Deng

College of Computer Science and Technology, Beijing University of Posts and Telecommunications, Beijing, 100876, China
Email: manatea@bupt.edu.cn, dengfang@bupt.edu.cn

Abstract. This paper focuses on improving the recognition accuracy of dynamic gesture learning, and proposes dynamic gesture track recognition strategy based on HMM-CART model structure. It integrates the modeling ability of the HMM (Hidden-Markov-Model) for time series data and the interpretability of CART (Classification-And-Regression-Tree) for its ability of fast classification and regression, and use it for dynamic gesture recognition. Time series data of movement gestures collected by LeapMotion sensor will be first divided into four channels: finger shape, palm normal vector, palm ball radius and palm displacement vector, and then HMM in HMM Layer will be built for each channel, finally the likelihood probability of model calculated for each sub-sequence of observation should be classified as the input of the CART model in CART Layer to identify the gesture. Experiments show that the accuracy of dynamic gesture recognition method using HMM-CART model is higher than that of traditional single channel HMM.

1. Introduction

In recent years, gesture recognition in many different areas attracted the interest of a growing number of studies, such as human-computer interaction, robotics, computer games, and automatic sign language interpretation and so on. In the development of human-computer interaction which has gone through the interaction from the initial two-dimensional image by simple operator interface to interact more intuitively and naturally, and gesture recognition has also been improved from a wearable device detection to more convenient camera-based visual gesture recognition. Currently more sophisticated visual identification products with high efficiency depth feature extraction technique is widely used in the world: Microsoft's Kinect using both sides of the infrared sensor and the middle of the RGB camera through the optical coding technology to obtain depth information and human skeletal information[1], Intel's Real Sense is similar to Kinect, and its black box model of the package allows users to store features in accordance with established models[2], LeapMotion is innovative in terms of gesture recognition software, unlike Kinect, LeapMotion device is clear for gesture recognition and accurate calculation of finger and hand structure. Compared to the depth of the camera like Kinect and other similar devices, it produces more limited gesture local information and interactive area, the advantage lays in the local data recognition more accurate [3].

Gesture recognition is the process of eigenvalue extraction and gesture model parameter estimation of the image after segmentation of the gesture area. It is a process of mapping a point or track in a parameter space to a subset of the space. It includes static gesture recognition and dynamic gesture recognition, and dynamic gesture recognition can eventually be converted to static gesture recognition. Lathe et al. [4] achieved gesture recognition by calculating the similarity of the hausdorff distance calculation between the target data and the gesture templates obtained from sample gesture training, which is simple and fast, but the degree of division between different gestures depends on the choice of eigenvalues, so the recognizable gestures are more limited. Tusor at al. [5] used a scheme based on artificial neural networks and fuzzy theory to classify different gestures by setting a priori fuzzy rule. Its recognition accuracy based on a large number of training data, although the use of fuzzy network to improve the training iteration speed, but the training process is still too long, which does not apply to real-time recognition. X Wang et al.[6] Used Baum-Welch algorithm to train the HMM (Hidden-Markov-Model), which can handle time-based information of different lengths and has high recognition efficiency for dynamic gestures. However, with the development of interactive applications and the improvement of recognition accuracy requirements, a single HMM has been unable to meet the needs of the application, therefore, XY Wang et al.[7] Proposed a HMM-FNN-based model structure that combines the HMM with a fuzzy neural network to establish a more complex gesture by setting up fuzzy rules. Indeed, it has improved the accuracy, but still has the shortcomings such as the dependence of fuzzy rules on prior experience and insufficient convergence rate due to more hidden layer nodes.

This paper focuses on improving the recognition accuracy of dynamic gesture learning, and proposes dynamic gesture track recognition strategy based on HMM-CART model structure. It integrates the modeling ability of HMM for time series data and the interpretability of CART (Classification-And-Regression-Tree) for its ability of fast classification and regression, and uses it for dynamic gesture recognition. Time series data of movement gestures collected by LeapMotion sensor will be first divided into four channels: finger shape, palm normal vector, palm ball radius and palm displacement vector, and then HMM in HMM Layer will be built for each channel, finally the likelihood probability of model calculated for each sub-sequence of observation should be classified as the input of the CART model in CART Layer to identify the gesture. Experiments show that the accuracy of dynamic gesture recognition method using HMM-CART model is higher than that of traditional single channel HMM.

The rest of the paper is organized as follows: Section 2 introduces the collection and pre-processing of gesture data. Section 3 describes the structure of dynamic gesture model based on HMM-CART, including the structure and training principle of related learning model. Section 4 describes the experiment that introducing the classification and training process of new model and compare it with traditional model according to the experimental results. Finally, Section 5 ends the paper with a summary.

2. Data Acquisition and Preprocessing

Compared with kinect, LeapMotion uses a similar infrared camera technology, and its exquisite design is specifically used to capture the trace of the hand. It can be through the infrared light reflection of the LED tracking about 1 meter in diameter within the coordinates of the coordinates of the hand. Figure. 1 shows that the LeapMotion device captures 215 frames per second and the detection accuracy can reach 0.01 mm. API support cross-platform, allowing access to the original data such as coordinates of five fingers, the strength of grip hand, the hand rotation angle, which help us to build a data set to achieve and analyze the algorithm of gesture recognition. We will have theses tracking data collected and extracted through the information acquirement of the LeapMotion device.

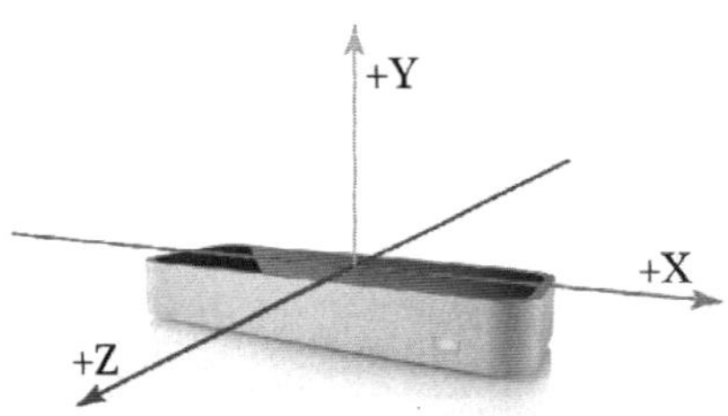

Figure 1 LeapMotion device

The acquisition of gesture data needs to define the start and end of gesture position. Therefore, this paper summarizes a threshold judgment scheme based on hand shape template and gesture movement velocity and direction change rate. We determine starting and ending points of the gesture through whether the gesture movement speed exceeds the threshold, and set the direction change rate to prevent the gesture of the inflection point when the speed may be lower than the threshold. Like gesture V, in acquisition of gesture V, the velocity in bottom inflection point of V may be lower than the threshold.

The definition and extraction of features are related to the accuracy of the overall model training. The LeapMotion device contains a wide variety of features in its original dataset that it allows access to. In order to avoid the high-dimensional observation sequence learning, it is necessary to do dimensionality reduction first. According to the characteristics of hand shape, the dynamic change of the gesture in space could be divided into several channels in three-dimensional space such as trace of palm position, the trace of normal vector of the palm, the change of ball radius holding in palm and the stretch of five fingers. Wherein the variation of each gesture information is represented by a time series of successive characteristics, the Gaussian HMM training by continuous observation of sequence can retain more original information than that by Discrete HMM, and

The Gaussian HMM modeling is carried out for each one-dimensional feature of the above division.

Each gesture detection process may have a range of size, shape distortion and other issues, and each gesture to achieve the shape and size of the space is different. Even the same set of gestures training data its eigenvalues will vary greatly if not get data preprocessed, so we need to normalize the gesture data into the same format under the data. This paper realizes the same specification data mapping by normalizing the eigenvalues of coordinates.

The experiment will be normalized using the L^2 norm, and the norm $x(x_1, x_2 ..., x_n)$ for the vector L^2 is defined as follow:

$$norm(x) = \sqrt{x_1^2 + x_2^2 + ... + x_n^2} \tag{1}$$

To normalize x to the unit L^2 norm, as create a mapping from x to x' so that the norm x' of L^2 is 1, as shown in (2):

$$1 = norm(x') = \frac{\sqrt{x_1^2 + x_2^2 + ... + x_n^2}}{norm(x)} = \sqrt{x'^2_1 + x'^2_2 + ... + x'^2_n} \tag{2}$$

Which is:

$$x'_i = \frac{x_i}{norm(x)}$$

The normalized gestational feature observation sequence can be expressed as: $m = \{x_i \mid i \in [1, n], x_i \in [-1, 1]\}$

3. Model Design and Training

The dynamic gesture modeling method of HMM-CART model proposed in this section preserves the ability of time model modeling based on HMM, and also makes full use of the rule interpretability of CART model and fast classification and regression ability.

3.1 HMM-CART Model

As shown in Figure. 2, the HMM-CART model is divided into four layers: Input Layer, HMM Layer, Mid Layer, and CART Layer. In Input Layer, the data is reduced dimension and normalized by the feature, and then divided into four channels: finger shape, palm normal vector, palm ball radius and palm displacement vector, where each channel contains Part of the information of a gesture that will be modeled as a separate observation sequence into HMM Layer according to the design of the different gestures.

The HMM is similar to the traditional HMM, but the difference is that the HMM in the HMM Layer only covers a certain kind of information of the gesture, and the traditional HMM does not carry out the multi-channel division of the feature.

Each channel in HMM Layer is concerned only with which models are associated with the sequence, for example, the gesture sequence feature is based on the four observed sequences of s_v , s_r , s_n , s_e , Where s_v is the displacement vector sequence, s_r is the palm ball radius sequence, s_n is the palm normal vector sequence and s_e is the finger extension degree sequence, each data sequence has a corresponding HMM model. We can evaluate the observed sequence values by entering the HMM and then enter the score results into the next layer. At HMM Layer, each node represents the HMM of a gesture under that channel. The total number of HMM nodes is $n_g \cdot n_c$, where n_g represents the number of design gestures and n_c represents the number of channels.

The middle layer evaluates the likelihood probability of each model as a set of likelihood probability features and classifies the eigenvector into CART Layer after performing the normalization algorithm.

As the input root node of CART Layer, the decision tree as a binary tree structure contains several decision points and the result nodes, which can be easily, expressed as if-then rules. Starting from the root decision point, enter the data for two sub-program decision-making calculation, and ultimately reach the output of a result node. The output layer contains all possible classification results and integrates the results as the final output of the model.

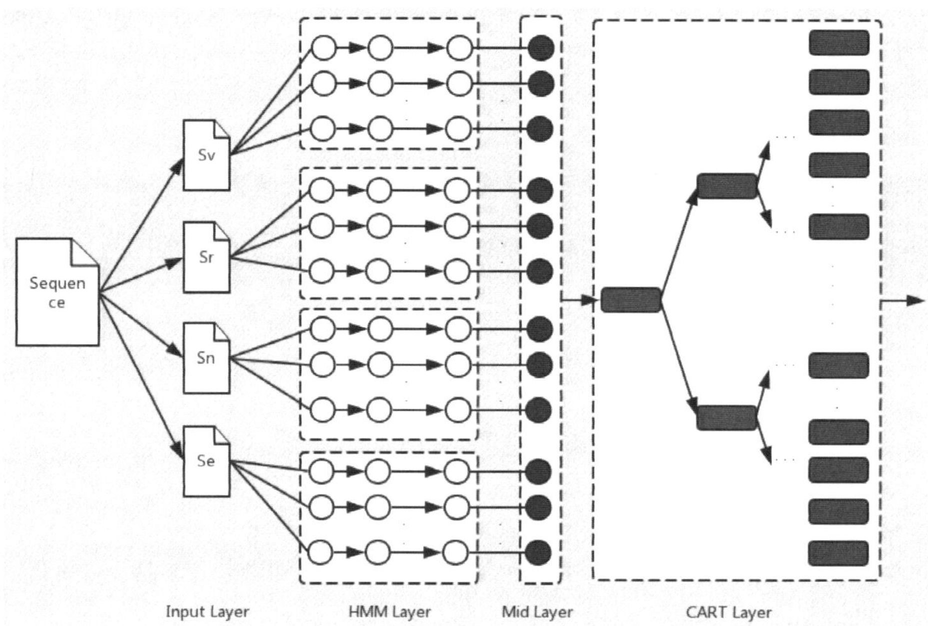

Figure 2 HMM-CART model

3.2 Model Training

Hidden-Markov-Model is distinguished in timing-based pattern recognition applications such as gestures, voice, handwriting, and semantics. The HMM can map the transition probability between the observed and implicit states whose implicit state is invisible and the output of the observed state that is dependent on the implicit state is visible. There is a transition probability between each implied state, and the transition probability of the implied state to the observed state is obeying the distribution, thus a HMM can be represented by the implicit state transition probability matrix and the confusion matrix as well as the initial state probability matrix.

Training an HMM model is the process of giving an observation state, finding the optimal state transition matrix and confusing the matrix. In the training process, by continuously updating the HMM parameters the $P(O \mid \lambda)$ value reaches a local optimum. The training algorithm uses the Baum-Welch algorithm, which will iterate over the model parameters according to the maximum likelihood estimate.

We assume that the initial HMM parameter is $\lambda = (A, B, \pi)$, first calculate the forward variable α and the backward variable β as shown in follow formula:

$$\alpha_t(i) = P(O_1 O_2 \dots O_t, \lambda) \tag{3}$$

$$\beta_t(j) = P(O_{t+1} O_{t+2} \dots O_T \mid \lambda) \tag{4}$$

Where the forward probability α is the sum of all the probabilities reaching the state i at time t, and the backward probability β is the probability of partially observing the sequence $O_{t+1} O_{t+2} \dots O_T$ when the state i is reached, so $P(O \mid \lambda)$ can be further expressed as:

$$P(O \mid \lambda) = \sum_{i=1}^{N} \alpha_t(i) \beta_t(i), 1 \le t \le T \tag{5}$$

According to (5), we define the expectation ξ and γ, where ξ is the probability that the state q_t is i at time t and the state q_{t+1} is j when time is t + 1:

$$\xi_t(i, j) = \frac{P(q_t = i, q_{t+1} = j \mid O, \lambda)}{P(O \mid \lambda)} \tag{6}$$

We define γ as the probability of the state i at t when given the observed state sequence and the HMM parameter:

$$\gamma_t(i, j) = P(q_t = i \mid O, \lambda) \tag{7}$$

Finally, we use (6) and (7) to re-iterate to calculate $\lambda = (A, B, \pi)$, the formula is as follows:

$$\overline{\pi_i} = \gamma_1(i), 1 \le i \le N \tag{8}$$

$$\overline{a_{ij}} = \frac{\sum_{t=1}^{T-1} \xi_t(i, j)}{\sum_{t=1}^{T-1} \gamma_t(i)}, 1 \le i \le N, 1 \le j \le N \tag{9}$$

$$\overline{b_{ik}} = \frac{\sum_{t=1, \alpha_t = k}^{T} \gamma_t(i)}{\sum_{t=1}^{T} \gamma(j)}, 1 \le j \le N, 1 \le k \le M \tag{10}$$

It is possible to obtain a maximum likelihood estimate of the HMM model after a number of iterations. The maximum likelihood estimation of the HMM model can be obtained by iteratively calculating the above three equations.

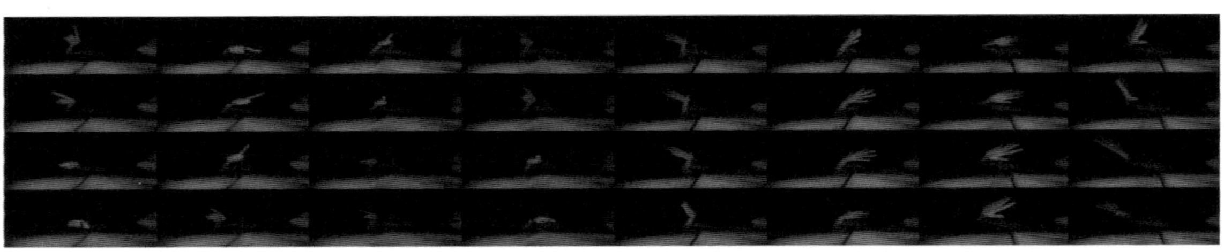

Figure 3 Four kinds of custom dynamic gestures (8 for a group, from top to bottom)

The best segmentation of the classification regression tree is based on the feature realization of the model dataset. It constructs the prediction criterion by constructing the binary tree, which can deal with large amount of data and high data effectively, and can also deal with the continuous variable data in this paper. CART is based on the Gini impurity to make the segmentation decision. For continuous target variable, the minimum variance is used as the splitting rule to generate the binary tree. For a set of data $D = \{(x_1, y_1), (x_2, y_2), ..., (x_n, y_n)\}$, we define two sets of regions to be segmented:

$$\begin{cases} R_1(j,s) = \{x \mid x_j \leq s\} \\ R_2(j,s) = \{x \mid x_j > s\} \end{cases} \tag{11}$$

The selection rule for the partitioned region must satisfy the sum of the weights of the two regions. So that each partition of area data have to ensure relative polymerization to achieve the purpose of training.

Over fitting may be presented during the binary state of the training process, since the characteristics of the model training samples contains too precise and noise information, it is necessary to prune the tree generated by the method of cross-validated with the validation data set, and by adding a penalty factor to select the high error rate for branch reduction to solve the problem of fitting.

4. Experiments

In this paper, the CPU for the Intel Core i5-3470 3.20GHz, 8G memory for the ordinary PC deployment of the gesture recognition experiment, as shown in Figure. 3 gesture definition is divided into four kinds of fingers: draw round, draw fork, left slide, and right slide. Each gesture collected 100 data, which divided into four channels as follows: displacement vector sequence, palm ball radius sequence, palm normal vector sequence and finger extension degree sequence. Total 1600 data that division of the four different channels for a group of 400 groups, will be in accordance with the ratio of 8: 2 cross validation.

In addition to the definition of simple gestures in the experiment, such as round, fork, but also defines a more complex left slide, right slide gestures. As is shown in Figure. 3, the first two kinds of gestures are the process of the extension of index finger. The latter two gestures are the level of the process of placing from the left or from the right in the opposite direction with five fingers stretch.

First, during the training the sequence of displacement vector, palm ball radius, palm normal vector and finger extension degree will be extracted for HMM training. And then the HMM-CART classification regression of the likelihood of each HMM will have completed.

The training process of the experiment is described in the previous section, and the training is divided into two phases: Firstly, 80×4 sets of data are input into the HMM model for parameter training. Then, 320 sets of likelihood probability vectors are obtained by inputting the training data of each training set into the HMM as training input train of CART model.

Figure. 4 shows a change in the palm displacement characteristics of a left slide and a right slide gesture. It can be seen in Figure 4 that although the Y and Z axes have a slight change, the X axis completely covers the gesture information.

Finally, the HMM-CART model and the traditional HMM model were predicted by 80 sets of test data. The test results are shown in Table 1.

It can be seen from the experiment that the multi-channel decomposition of the gesture feature can make the recognition effect of the HMM for the single channel feature significantly improve, and still maintain a high recognition rate after classification and aggregation.

Table 1 Comparison of HMM - CART Model and Traditional HMM Model Recognition Rate

Hand Gesture Recognition Model	Circle	Cross	Left Slic	Right Slide
Recognition rate of Conventional HMM (%)	91.3	76.6	92.6	80.8
Recognition rate of HMM-CART (%)	95.7	93.3	98.1	95.6

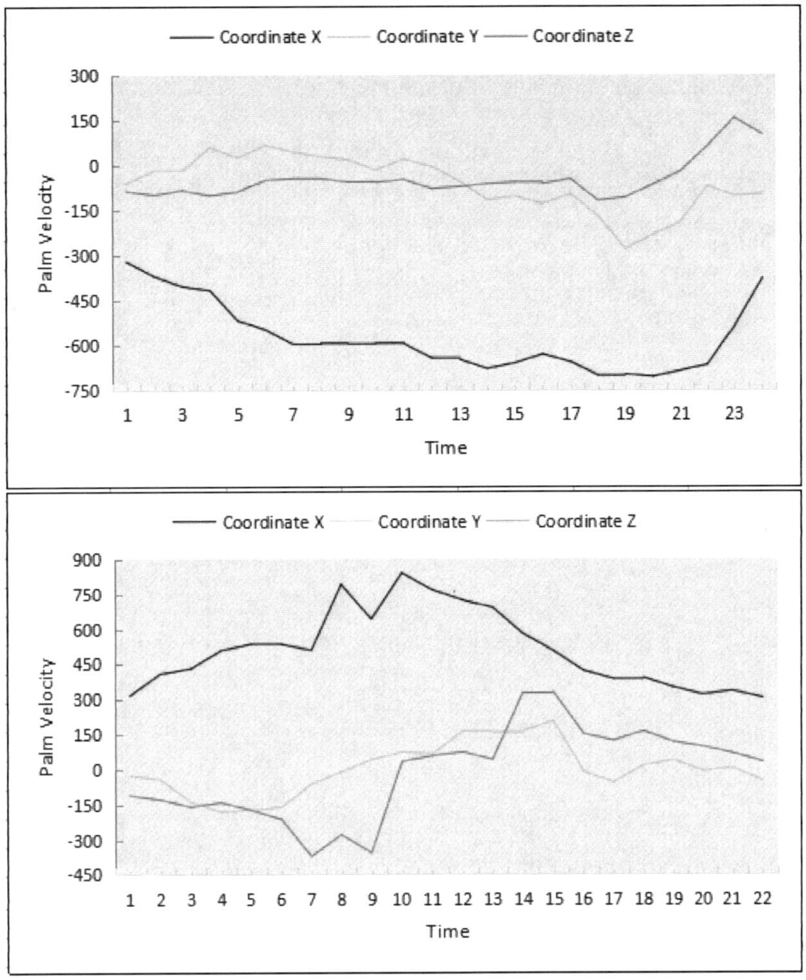

Figure 4 Left slide (top) right slide (bottom) palm displacement characteristics

5. Conlusion

In this paper, HMM-CART model is proposed for modeling and recognizing dynamic gestures. First, it uses the threshold acquisition method to collect the gesture information through the LeapMotion sensor, and then split it into four component channels and normalize the timing data as the input data of the HMM-CART model.

It integrates the modeling ability of the Hidden-Markov-Model for time series data and the interpretability of Classification-And-Regression-Tree for its ability of fast classification and regression, and uses it for dynamic gesture recognition.

In the future, we will collect data through multiple LeapMotion devices and integrate gesture implicit features to improve the recognition of gestures.

6. References

[1] Zhang Z. Microsoft Kinect Sensor and Its Effect[M]. IEEE Computer Society Press, 2012.
[2] Lin C P, Wang C Y, Chen H R, et al. RealSense: directional interaction for proximate mobile sharing using built-in orientation sensors[C]// ACM International Conference on Multimedia. ACM, 2013:777-780.
[3] Marin G, Dominio F, Zanuttigh P. Hand gesture recognition with leap motion and kinect devices[C]// IEEE International Conference on Image Processing. IEEE, 2015:1565-1569.
[4] Latha P S, Babu M R, Siva K, et al. Hand Gesture Recognition using skeleton of Hand and Hausdroff Distance[J]. International Journal of Advanced Research in Computer Science, 2011,

2(2):346-354.
[5] Tudor B, Várkonyi-Kóczy a R. Circular fuzzy neural network based hand gesture and posture modelling[C]// Instrumentation and Measurement Technology Conference. IEEE, 2010:815-820.
[6] Wang X, Xia M, Cai H, et al. Hidden-Markov-Models-Based Dynamic Hand Gesture Recognition[J]. Mathematical Problems in Engineering,2012, (2012-04-24), 2012, 2012(1):137-149.
[7] Wang X Y, Dai G Z, Zhang X W, et al. Recognition of Complex Dynamic Gesture Based on HMM-FNN Model[J]. Journal of Software, 2008, 19(9):2302-2312.

Stabilized Stepwise Orthogonal Matching Pursuit for Sparse Signal Approximation

Mingjiang Wang*, Guanghong Liu, De Zhang, Kuoye Han and Yanmin Chen

Academy of Information and Science, China Electronics Technology Group Corporation, Beijing, China

*Wangmingjiang.1987@163.com

Abstract. Orthogonal Matching Pursuit (OMP) algorithm is equipped with the capability to decompose any signal into a linear expansion of waveforms, which are selected from a redundant functional dictionary. Nevertheless, classical OMP algorithm suffers a heavy computational burden due to its single element selection strategy in each repetition. Recently, an accelerated implementation called stage wise orthogonal matching pursuit (StOMP) algorithm has been proposed through exploiting a multiple elements selection scheme based on an iterative threshold. However, as the defined threshold is a function of an empirical and undetermined parameter, such a reconstruction scheme is not optimal and the algorithm may get obstructed in some specific conditions. This manuscript presents an adaptive threshold selection strategy which takes signal structure into consideration and furthermore, a regularized iterative framework for sparse signal approximation is suggested. Compared with classical StOMP approach, these efforts can provide robust and more attractive approximation performance for sparse signal recoveries. Experimental results present the substantial improvements of these optimizations.

1. Introduction

Sparse signal decompositions[1] have received considerable attentions in recent years and applications are found in a wide range of field such as source coding[2] and signal acquisition[3]. Typically, a signal decomposition model approximates observed signal using a linear combination of elementary waveforms which are selected from a large collection. Nowadays, a variety of algorithms can be utilized for sparse signal approximation such as convex optimization tools[4-5], and various iterative methods[6-8]. While convex optimization techniques offer valuable procedures for computing sparse representations and L1-minimzation solutions, which can be commonly found in the literature of Compressed Sensing(CS)[9-11], some more powerful solvers for such problems can be accomplished by a great number of greedy algorithms[12-16]. Greedy methods rely on gradual approximation progresses, either by iteratively identifying the support of signal until a convergence criterion is met, or alternatively by obtaining an improved estimate of sparse signal in each iteration. Among these competitive greedy algorithms, a fundamental implementation is Orthogonal Matching Pursuit (OMP)[17], which repeats by finding the column of measurement matrix most correlated with current residual and then updating estimated coefficients through seeking a least square solution (LS)[18-19] until the algorithm reaches certain stop criterion.

Nevertheless, while OMP can provide good approximation performances and stable recoveries, one crucial problem when applying the algorithm to large-scale data is that it suffers a slow convergence rate as the algorithm only selects one atom each time. D.L.Donoho in [20] proposed an innovative elements selection strategy in order to speed up the pursuit progress of OMP, called Stagewise Orthogonal Matching Pursuit (StOMP) algorithm. The implementation intelligently selects multiple elements through introducing a threshold-based elements selection strategy into each iterative step and can substantially accelerate the convergence rate of pursuit progress. However, as the threshold suggested in[20] depends on an undetermined parameter and the selection of the unresolved parameter just relies on some empirical values, implementation in[20] is not robust and recovery progresses may get obstructed under certain conditions. The absence of guidance to robust parameter selection and unstable recovery capacity make the implementation rather inconvenient, thus dim the capability of StOMP algorithm in some practical applications.

This manuscript presents an optimized stepwise orthogonal matching pursuit scheme with the aim of providing stable and more appealing recovery performances for high dimensional sparse signal approximations. Similarly, we mainly concentrate on two aspects of conventional greedy algorithms, elements selection and coefficient update. On the one hand, this literature suggests an adaptive threshold concerning the internal structure of original signal, which can constantly provide robust recoveries with favorable convergence under general circumstance. On the other hand, a regularized framework for selected coefficients estimation is further developed, which carries greedier and more touching approximation performance compared with current StOMP algorithm. The brief review of this paper is organized as follows. The main part starts with the framework of classical greedy pursuit strategies, and restrictions of current OMP and StOMP algorithm are also investigated in Section II. In Section III, a robust elements selection scheme based on an "adaptive" threshold is firstly suggested. Then we turn to selected coefficients update step and a regularized coefficients update framework is further devised in this part. Section IV provides numerical results to demonstrate the attractive improvements of proposed algorithm and finally, discussion and conclusions are presented in Section V.

2. Greedy Pursuit Framework

Greedy pursuit methods build up an approximation through making locally optimal choices at each step. Nowadays, there is a large and growing family of greedy pursuit techniques for signal approximation with a long and multi-rooted history. The main idea of these approaches can be stated as projecting the data in a given direction and then testifying the derivation of estimated signal from measurements. A typical candidate for these techniques is OMP algorithm, which was first taken up by Mallat and Zhang [17] and used for signal approximation. This section will briefly cover the pursuit procedures of OMP and then touch upon the kernel of StOMP algorithm in the following part.

2.1 Overview of Orthogonal Matching Pursuit

OMP aims at estimating original signal by iteratively projecting data onto different directions. The algorithm begins with finding the column of measurement dictionary most correlated with the measurements, and then repeats this step by correlating the columns with evaluated residual, which is obtained by subtracting the contribution of a partial estimate from original measurement vector.

$$y = Ax + e \tag{1}$$

Consider a general noisy measurement model in CS, where $A \in \mathbb{R}^{M \times N}$ is the measurement matrix ($M < N$). For a given $y \in \mathbb{R}^M$, we want to recover an K - sparse vector $x \in \mathbb{R}^N$ under the assumption that the error $e \in \mathbb{R}^M$ is bounded and the measurement matrix A satisfies the restricted isometric property (RIP)[21]:

$$(1 - \delta_K) \| x \|_2^2 \le \| Ax \|_2^2 \le (1 + \delta_K) \| x \|_2^2 \tag{2}$$

Then the solution of Eq.(1) generally leads to a l_1 minimization [22-24] problem and the evaluation of x can be effectively estimated with the assistance of OMP algorithm when the measurement matrix A satisfies RIP condition [25].For a given vector y and a matrix A, OMP devotes to identifying a small support set Λ and an estimated vector $\hat{x}$ to approximate y using.

$$\hat{y} = A_{\Lambda}\hat{x}_{\Lambda} \tag{3}$$

The algorithm is initialized with setting the first residual $r^{[0]} = y$, setting $x^{[0]} = 0$ and $\Lambda^{[0]} = \varnothing$. Each repetition it updates these three quantities. The iterative pursuit progress can be briefly summarized as the following steps:

a) Initialize the residual $r^{[0]} = y$, the sparse estimate $x^{[0]} = 0$ and initialize the index set of selected columns $\Lambda^{[0]} = \varnothing$. Set the iteration counter $i = 1$.

b) Calculate the correlation of column vectors of A with the current residual $r^{[i]}$, $h_j^i = \langle A_j, r^{[i]} \rangle$ and find the column best correlated with current residual $j^{[i]} = \arg\max_j \langle A_j, r^{[i]} \rangle$. Update the index set $\Lambda^{[i]} = \Lambda^{[i-1]} \cup j^{[i]}$.

c) Compute new estimation by projecting y onto the linear space spanned by the columns indexed by $\Lambda^{[i]}$, i.e., $\hat{\boldsymbol{x}}_{\Lambda^{[i]}} = \boldsymbol{A}^{\dagger}_{\Lambda^{[i]}}\boldsymbol{y}$. Update the residual $r^{[i]} = y - A_{\Lambda^{[i]}}\hat{x}_{\Lambda^{[i]}}$.

d) If certain stopping criterion is satisfied, then stop. Otherwise, set $i = i + 1$ and turn to step 2.

While OMP enjoys superior approximation capability in the context of CS, one crucial problem the algorithm encounters is that it suffers a slow convergent performance for large scale application. The pursuit algorithm has to operate as many cycles as the number of atoms needed to be selected to approximate measurement y, as it selects only one atom at each repetition. When the sparsity K gets large, the pursuit progress becomes impractically slow and inefficient.

2.2 Technique of Stagewise Orthogonal Matching Pursuit

Paper [20] first proposed an inventive elements selection strategy based on a defined threshold with the motivation of keeping the algorithm a low computational time consuming. Instead of choosing the element that is maximally correlated with the residual $j^{[i]} = \arg\max_j \langle A_j, r^{[i]} \rangle$ in OMP, the stagewise strategy replaces the maximum with a threshold criterion and selects all the indexes satisfying the following condition in each iterative step:

$$T^{[i]} = \{ j : \langle A_j, r^{[i]} \rangle \geq \lambda^{[i]} \} \tag{4}$$

where $\lambda^{[i]}$ is the defined threshold at iteration i. While some choices for $\lambda^{[i]}$ are suggested in several literatures, most of these approaches remain to be dissatisfactory and undesirable. For example, a simple non-iterative threshold was presented in paper [26]. Although it's by far the most computationally simple procedure, this suggestion has limited recovery guarantees. Another technique employing a dynamic threshold in the iterative framework is presented in StOMP. In this approach the threshold is calculated concerning current residual r and an undetermined parameter t:

$$\lambda^{[i]}_{stomp} = \frac{t \, \| r^{[i-1]} \|_2}{\sqrt{M}} \tag{5}$$

While the threshold provides performance guarantees for some specific coefficient distributions, similar guidance is not available for other general cases. From a practical point of view, the selection of parameter t required in this method is critical for its performance. Despite the suggestion considering a value $2 \leq t \leq 3$ in [20], there do not seem to be any other guidelines available. The indeterminacy and fluctuation of parameter selection make the implementation rather inconvenient and unstable. Actually, specific cases may be encountered when using this improper threshold is that the algorithm will terminate and get obstructed permanently when all inner products fall below the casual threshold, thus lead to an inaccurate recovery.

3. Stabilized Stepwise Orthogonal Matching Pursuit Scenario

This section presents a stable elements selection program through introducing an adaptive threshold, which takes signal structure into consideration. Furthermore, we employ a regularized framework at the coefficient update step to provide a more appealing performance for sparse signal approximation applications. The removal of undeterminable factor and the introduction of regularized coefficients update scheme maximally stabilize the proposed algorithm, and thus can constantly provide an accurate even better recovery compared with classical StOMP approach.

Consider a signal to be recovered is K sparse, theoretically, there are at most K columns of A contributing to the residual at i th iteration (for simplicity, we assume the first K columns here), that is:

$$r^{[i]} = \gamma_1^{[i]} A_1 + \gamma_2^{[i]} A_2 + \cdots + \gamma_K^{[i]} A_K \tag{6}$$

Then the l_2 norm of the residual $r^{[i]}$ at current iteration can be calculated as:

$$\begin{aligned} \| r^{[i-1]} \|_2^2 &= \| \gamma_1^{[i]} A_1 + \gamma_2^{[i]} A_2 + \cdots + \gamma_K^{[i]} A_K \|_2^2 \\ &\le \sum_{k=1}^{K} (\gamma_k^{[i]})^2 \| A_k^{[i]} \|_2^2 \\ &\le K (\gamma_{j_{\max}}^{[i]})^2 \| A_{j_{\max}}^{[i]} \|_2^2 \end{aligned} \tag{7}$$

where $\gamma_j^{[i]}$ is the projection coefficient of $r^{[i]}$ on the direction of j-th column vector A_j, and $j_{\max}$ is the column index which remains $\left\| \gamma_j \cdot A_j \right\|_2$ to be largest at current step. Then, a preferred reference for threshold can be acquired from Eq. (7):

$$\lambda^{[i]} = \frac{\| r^{[i-1]} \|_2}{\sqrt{K} \cdot \| A_{j_{\max}} \|_2} \le | \gamma_{j_{\max}}^{[i]} | \tag{8}$$

Setting Eq. (8) as the criterion of iterative threshold, we suggest selecting all the elements that satisfy the following condition at i the iteration:

$$T^{[i]} = \{ j : | \gamma_j^{[i]} | \ge \lambda^{[i]} \} \tag{9}$$

Although this referential value operates similarly to that in [20], essential difference lies in the fact that this factor is adaptive and based on the correlation of the current residual and the intrinsic information of signal structure. The elimination of indeterminate parameter t makes this choice to be constantly robust and preferable. Thus obstruction hazards concealed in classical StOMP pursuit progress will vanish and never occur in this scheme as there are always competitive candidates coming into the subset at each iterative step.

Once qualified elements are selected, coefficients update follows in subsequent procedure. StOMP provides an approximate estimation by taking orthogonal projections. The estimation for x is acquired by projecting the measurement orthogonally onto the selected columns of A :

$$\hat{x}^{i}_{T^{[i]}} = A^{\dagger}_{T^{[i]}} y \tag{10}$$

where † represents pseudo-inverse operator. Then the algorithm subtracts off the contribution of the estimated vector $\hat{x}^{i}_{T^{[i]}}$ and iterate on the residual. In contrast with StOMP, this manuscript suggests a regularization framework for the coefficient update step with the aim of offering a more fascinating approximation capacity for sparse signals. Similarly, at the beginning of each iterative step, an intermediate estimation $\hat{x}^{i}_{T^{[i]}}$ is accomplished by finding a least-square solution:

$$\hat{x}^{i}_{T^{[i]}} = \arg\min_{\tilde{x}_{T^{[i]}}} \left\| y - A_{T^{[i]}} \tilde{x}_{T^{[i]}} \right\|_2^2 \tag{11}$$

Then, the regularized coefficients update scheme differs in last step. Instead of keeping all the calculated coefficients as the current estimate, the regularized scheme only retains the largest K elements of intermediate estimation to define a new support $T^{[i]}_{opt}$ and then decrease the new estimate contribution to measurement:

$$r^{[i]} = \| y - A_{T^{[i]}_{opt}} \hat{x}_{T^{[i]}_{opt}} \|_2^2 \tag{12}$$

The regularization framework not only carries profits for the stability of pursuit progress, but also provides sufficient insurance for the qualification of selected coefficients at each step, thus benefits the accuracy of sparse signal recovery

Algorithm 1.

Stabilized Stepwise Orthogonal Matching Pursuit (Stabilized StOMP)

Input: y, A, K

Initialization: $r^{[0]} = y$, $\hat{x}^{[0]} = 0$, $T^{0}_{opt} = \phi$, $i = 0$

For $i = 1, i := i + 1$ until $\| r^{[i]} \| < threshold$ is met do

Step1: $g^{[i]} = A^{T} r^{[i-1]}$

Step2: $j^{[i]}_{max} = \arg\max_{j} | g^{[i]}_{j} | \left\| A_{j} \right\|_2$

$$\lambda^{[i]} = \frac{\| r^{[i-1]} \|_2}{\sqrt{K} \cdot \| A_{j^{[i]}_{max}} \|_2}$$

$$T^{[i]} = T^{[i-1]}_{opt} \cup supp\{ j : | g^{[i]}_{j} | \geq \lambda \}$$

Step3: $\hat{x}_{T^{[i]}} = A^{\dagger}_{T^{[i]}} y$

$\hat{x}^{[i]} = \hat{x}^{T^{[i]}}_{K}$, $T^{[i]}_{opt} = supp(\hat{x}^{[i]})$

Step4: $r^{[i]} = r^{[i-1]} - A_{T^{[i]}_{opt}} \hat{x}^{[i]}$

end

$r = r^{[i]}, \hat{x} = \hat{x}^{[i]}$

Output: r and $\hat{x}$

The full scenario of the stabilized stepwise Orthogonal Matching Pursuit algorithm (Stabilized StOMP) can be listed as the pseudo code in Algorithm.1 below:

4. Experimental Results Analysis

In order to evaluate the approximation capacity of the stabilized StOMP scheme, numerical experiments are presented in this section through comparisons with classical OMP and StOMP algorithms.

4.1 Robustness Analysis

This experiment is conducted on a sparse dataset with the aim of investigating the stability improvement for proposed method. Experiment setup is depicted as follows. A sparse signal x is generated with a length of $n=1000$, and the sparsity of the signal is $s = ceil(1000 \times 0.075) = 75$ (blue plot in Figure.1). The magnitudes of all nonzero elements are normalized and their positions follow random distribution. After Fourier transform operation on signal x (multiplication with a $n \times n$ Fourier transform matrix F), measurements are acquired through two time random down-sampling of the transformed signal. Then, subsequent signal recoveries are performed separately through OMP, StOMP and the proposed algorithm. The threshold parameter for StOMP is empirically chosen as $t_s = 2.5$. The relative recovery error at the i- th iteration is defined as $e = \| x - \hat{x}^{[i]} \|_2 / \| x \|_2$ and the iterative convergence criterion is chosen as $\| r \|_2 \leq 0.1$. The recoveries and convergence rates for all three methods are presented in Figure.1 and Figure 2 below. It can be observed from Figure.1 that while OMP algorithm can reconstruct the sparse spectrum exactly with a relaxed convergence rate (red plot in Figure.1 and Figure.2), classical StOMP just fails to provide an accurate recovery (purple plot in Figure.1) due to sharp obstruction after 8 iterations. This is led by the fact that the empirical selection of parameter t_s is inappropriate and an overlarge threshold assignment just blocks the update of support set in the elements selection step, thus induce a suspension with a large recovery error (pink curve in Figure 2). Nevertheless, through the introduction of an adaptive threshold concerning the signal structure, the proposed scenario can remain constantly robust and prevent the recovery progress from getting obstructed.

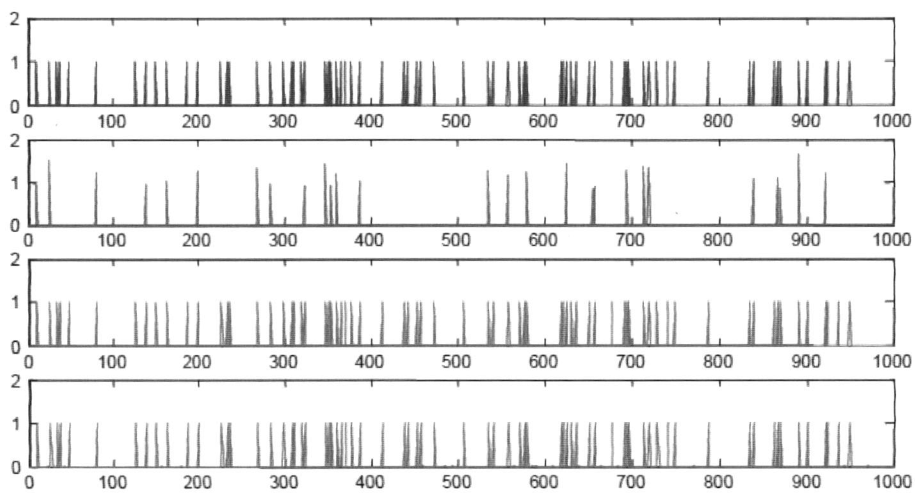

Figure 1. Recoveries for OMP, StOMP and proposed method.

It can be found from Figure.1 that the proposed algorithm can stably provide an accurate recovery (green profile) within a reduced computational time consuming and reach to convergence within only 20 iterations (green curve in Figure.2),while that for OMP need to perform at least 75 iterations. Note that the final approximation residual for proposed method is 0.009242 while those for OMP and StOMP turn to be 0.009177 and 0.799, respectively.

4.2 Investigation of Approximation Capacity on Real Dataset

General approximation performance for proposed algorithm has been also explored on real image dataset. In the following experiment, the size of tested imagery data is 256*256 and it can be demonstrated that the image vector f has a sparse representation on DCT basis. Explicit settings are specified in illustration of Figure.3 and characters of running results for all three algorithms are listed in Table I. The parameter ***psnr*** is defined as $psnr = 20 * \log 255 \cdot \sqrt{N} / \|f - \hat{f}\|_2$. It can be observed from Figure.3 that while the implementation can accomplish a blurred approximation with $psnr = 26.2136dB$ when t_s is assigned to 2.0, the recovery for assignment $t_s = 3.0$ just reveals disordered, and the output $psnr$ for this running has decreased to 19.5093dB due to an unexpected obstruction. In fact, a proper choice for t_s is rather difficult in practical applications. Although we can persistently alternate to a lower factor, these attempts seem unwise and inefficient. Figure.3 (e) shows the result of proposed algorithm. It can be found from this picture that the stabilized algorithm can provide a comparative approximation ($psnr = 28.289dB$) concerning with classical OMP algorithm ($psnr = 28.6253dB$) while consumes a lower computation time. Statistics in Table I reveals that the introduction of regularized strategy in element update procedure can substantially benefit the approximation performance for sparse signals, as the $psnr$ evaluation for proposed algorithm is better than classical StOMP($t_s = 2.0$).

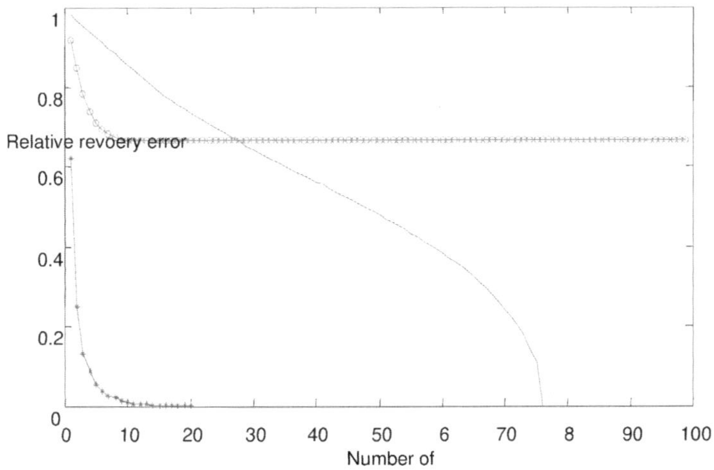

Figure 2. Convergence for OMP, StOMP and proposed algorithm.

Table 1 Specifications For Three Recovery Methods

	OMP	*StOMP(2.5)*	*StOMP(3.0)*	*Stabilized StOMP*
Time consuming (sec)	8.6547	5.6142	4.7839	5.6707
Psnr(dB)	28.6253	26.2136	19.5093	28.2890

The original image vector is acquired from 'Lena' image datasets through DCT transformation, quantization and multiplication with an inverse DCT matrix. The sparsity of original image is 9340/(256*256) =0.1579. (a) Original sparse image vector. (b) Approximation by OMP algorithm. (c) Recovery by StOMP algorithm ($t_s = 2.0$). (d) Reconstruction by StOMP algorithm ($t_s = 3.0$). (e) Recovery by stabilized StOMP algorithm of running results for all three algorithms are listed in Table

I. The parameter psnr is defined as $psnr = 20 * \log 255 \cdot \sqrt{N} / \left\| f - \hat{f} \right\|_2$.

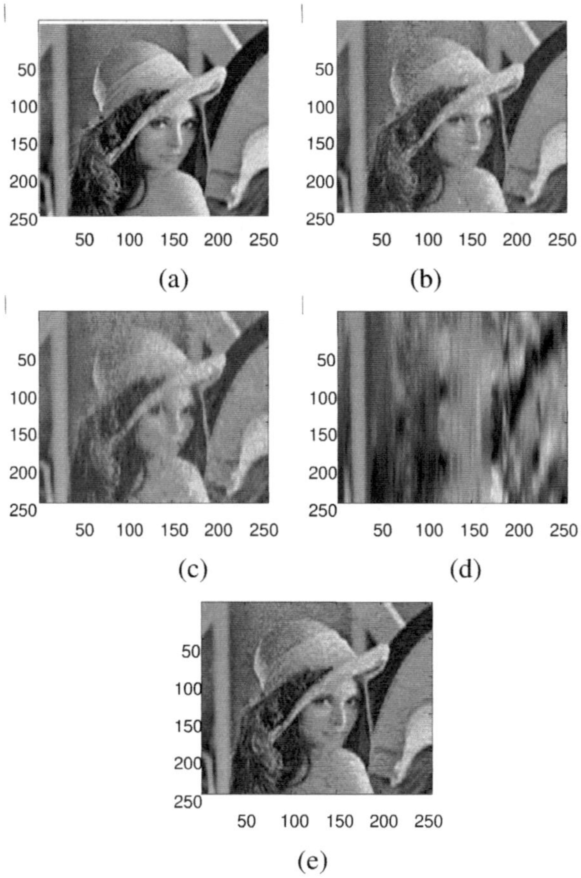

Figure 3. Image approximation under DCT bases for three methods with 50% samples.

5. Discussion And Conclusions

Sparse signal approximation can be found in a variety of areas in modern signal processing. Classical StOMP algorithm can significantly decrease the iteration number of pursuit progress but suffers an unstable recovery performance due to the indeterminacy of parameter. This manuscript suggests a new guidance through the introduction of an adaptive threshold concerning original signal structure. Compared with classical StOMP algorithm, the elimination of undefined parameter can substantially stabilize the pursuit progress and keep the algorithm from getting obstructed. Furthermore, in order to improve the recovery accuracy of pursuit progress, this paper presents an additional regularization framework for coefficients update step. Compared with conventional StOMP, the regularized scenario can provide more fascinating approximation performances for sparse signal approximation applications. Experimental tests validate the benefits and efficiency of the proposed algorithm. Numerical results show that the new algorithm outperforms classical StOMP approach in both the robustness and the approximation accuracy when applied to high dimensional signals.

6. References

[1] Remi Gribonval and Emmanuel Bacry, "Harmonic Decomposition of Audio Signals With Matching Pursuit," IEEE Transactions on signal processing, vol.51, no.1, Jan. 2003.

[2] V.K.Goyal, M.Vetterli and N.T.Thao, "Quantized overcomplete expansions in RN :Analysis, synthesis and algorithms," IEEE Transactions on Information Theory,vol.44,no.1,pp.16-31, Jan.1998.
[3] E.Candes, J.Rombergand T.Tao, "Robust uncertainty principles: Exact signal reconstruction from highly incomplete frequency information," IEEE Transactions on Information Theory, vol.52, pp.489-509, Feb.2006.
[4] S.Becker, J.Bobin, and E.Candes, "NESTA:A fast and accurate first-order method for sparse recovery," SIAM J Imag Sci,vol.4,no.1,pp.1-39,2011.
[5] S.Becker,E.Candes, and M.Grant, "Templates for convex cone problems with applications to sparse signal recovery," Math Pro Comp, vol.3, no.3,pp.165-218,2011.
[6] T.Blumensath and M.Davies, "Iterative thresholding for sparse approximations," J Fourier Anal Appl, vol.14, no.5, pp.629-654,2008.
[7] J.Portilla, "Image restoration through l0 analysis-based sparse optimization in tight frames," presented at the Proc of IEEE Int Conf Image Proc, 2009.
[8] R.Garg and R.Khandekar, "Gradient Descent with Sparsification: an iterative algorithm for sparse recovery with restricted isometry property," presented at the Proc of Int Conf Machine Learning, Montreal, Canada, 2009.
[9] D.L. Donoho, "Compressed sensing," IEEE Transactions on Information Theory, vol.52, no.4, pp.1289–1306, 2006.
[10] Lawrence Carin, Dehong Liu, and Bin Guo, "Coherence, Compressive Sensing, And Random Sensor Arrays," IEEE Antennas and Propagation Magazine, vol.53, no.4, pp.28-39, Aug.2011.
[11] R.Baraniuk, "Compressive sensing," IEEE Signal Processing Magazine, vol.24, no.4, pp.118-120, 2007.
[12] T.Blumensath and M.Davies, "Gradient pursuits," IEEE Trans Sig Proc, vol.56, no.6, pp.2370-2382, 2008.
[13] T.Bulmensath and M.Davies, "Iterative hard thresholding for compressive sensing," Appl Comput Harmon Anal, vol.27, no.3, pp.265-274, 2009.
[14] W.Dai and O.Milenkovic, "Subspace pursuit for compressive sensing signal reconstruction," IEEE Trans Inform Theory, vol.55, no.5, pp. 2230-2249,2009.
[15] I.Daubechies, M.Defrise and C.De Mol, "An iterative thresholding algorithm for linear inverse problems with a sparsity constraint," Comm Pure Appl Math, vol.57, no.11, pp.1413-1457, 2004.
[16] MDavenport and M.Wakin, "Analysis of orthogonal matching pursuit using the restricted isometry property," IEEE Transactions on Information Theory, vol.56, no.9, pp.4395-4401,2010.
[17] S.Mallat and Z.Zhang, "Matching pursuits with time-frequency dictionaries," IEEE Transactions on Signal Processing, vol.41, no.12, pp.3397-3415, Dec.1993.
[18] S.Mallat, G.Davis and Z.Zhang, "Adaptive time-frequency decompositions," SPIE J Opt Eng,vol.33,no.7,pp.2183-2191,1994.
[19] S.F.Cotter,J.Adler,B.D.Rao and K.Kreutz-Delgado, "Forward sequential algorithm for best basis selection," presented at the Proc IEEE Vision, Image Signal Proc,pp.235-244,1999.
[20] Donoho, D.L, Tsaig, Y, Drori, I, and Starck,J-L, "Sparse solutions of underdetermined linear equations by stagewise orthogonal matching pursuit," IEEE Transactions on Information Theory, vol.58, pp.1094-1121, Feb.2012.
[21] E.candes and T.Tao, "Decoding by linear programming," IEEE Trans Inform Theory, vol.51, no.12, pp.4203-4215,2005.
[22] J.Friedman,T.Hastie and R.Tibshirani, "Regularization paths for generalized linear models via coordinate descent," J Stats Software, vol.33,no.1,pp.1-22,2010.
[23] I.Loris, "On the performance of algorithms for the minimization of l1-penalized functions," Inverse Problems, vol.25, no.3, 2009.
[24] S.Osher,Y.Mao,B.Dong and W.Yin, "Fast linearized Bregman iterations for compressive sensing and sparse denoising," Commun Math Sci,vol.8, no.1,pp.93-111,2010.

[25] M.A.Davenport and M.B.Wakin. "Analysis of orthogonal matching Pursuit using restricted Isometry property," IEEE Transactions on Information Theory, vol.56, no.9, pp.4395-4401, Jun.2012.
[26] K.Schnass and P.Vandergheynst, "Average performance analysis for thresholding," IEEE Signal Processing Letters, vol.14, pp.431-433, Nov.2007.

IOP Conf. Series: Journal of Physics: Conf. Series **910** (2017) 012039 doi:10.1088/1742-6596/910/1/012039

An Adaptive Backlight Dimming with the Edge Enhancement Technique

Yeuhorng Shiau, Chenghan Li, Zhaoyuen Huang and Yungtai Chang

Dept. of Electrical Engineering, National Yunlin University of Science and Technology, Yunlin, Taiwan
Email: shiauyh@yuntech.edu.tw, d10110004@yuntech.edu.tw

Abstract. Backlight dimming, which decreases the backlight luminance level and performs pixel compensation, is known as the most effective way to reduce the power consumption of portable electronic devices with LCD. In this paper, we propose an adaptive backlight dimming technique that preserves the quality of details in images even when the backlight luminance of LCD devices is greatly reduced. The proposed backlight dimming method consists of the following two steps: clipped point selection and image compensation. In the first step, to reduce power consumption, the proposed approach selects a suitable clipped point according to the chosen level for a given image based on the maximum and average value. In the second step, several parameters are adaptively chosen according to the characteristic of a given image. These parameters are then used to compensate pixels with an edge enhancement technique. The simulation results show that at the same clipping point the proposed method has higher detailed information than others.

1. Introduction

Liquid crystal displays (LCDs) are widely used in many kinds of portable electronic devices, such as wearable devices, smart phones, tablet personal computers, and laptop computers. With the increasing demand for portable handled devices, reducing their power consumption has become a very important issue. In terms of the overall system power consumption on those portable devices, LCD consumes 20%–45% of total system power due to different applications [1]. Furthermore, the backlight of an LCD accounts for the largest portion of the total power consumption. Hence, decreasing the backlight luminance level can significantly reduce the power consumption of LCD devices. However, this may cause significant changes in visual perception. To eliminate the visual changes, we can adjust the image brightness to the desired level by performing pixel compensation, which is a technique used for compensating gray levels of pixels to increase the transmittance level of liquid crystals (LCs). Backlight dimming, which decreases the backlight luminance level and performs pixel compensation, is known as the most effective way to reduce the power consumption of LCD devices.

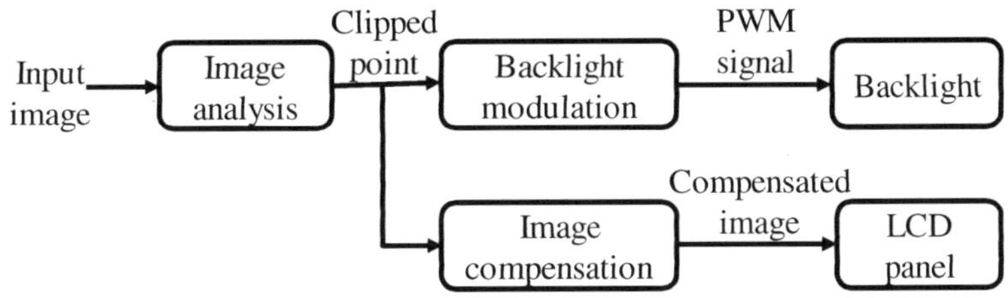

Figure 1 Overall block diagram of the global backlight dimming

Generally, backlight dimming methods are divided into two kinds: global dimming algorithms [2]-[12] and local dimming algorithms [13]-[18]. Global dimming controls the backlight based on the luminance of entire image simultaneously, while local dimming splits the image area into several blocks and controls the brightness of each block separately. Local dimming has the advantage in regard to power management efficiency. However, the local dimming technique must be equipped with the appropriate LCD device which can adjust the backlight of each block individually, while the small size LCD usually does not contain this function. The local dimming algorithm also involves higher computational complexity. On the contrary, global dimming is more common for any kind of LCD than local dimming, and offers lower computational complexity.

In this paper, an adaptive backlight dimming method based on global dimming is presented. The rest of this paper is organized as below. In Section II, the related works are presented and the features of the proposed method is emphasized. In Section III, the proposed method is described in detail. In Section IV, the simulation results to evaluate image quality for the previous methods and the proposed method are presented. Finally, conclusions are given in Section V.

2. Related Works

Global dimming usually contains three procedures, as shown in Fig. 1. The main task of image analysis is to find a clipped point, which is the gray level of the reduced backlight luminance level, based on the characteristic of image. Then, the backlight luminance is modulated in the backlight modulation via the dimming rate which can be determined by the clipped point. The technique of the backlight modulation is pulse width modulation (PWM) in general. Finally, the luminance reduction resulting from backlight modulation is then compensated by image compensation. Therefore, the power consumption of the backlight can be saved while the brightness of image is still preserved in visual.

Although backlight dimming has been successfully applied, the backlight luminance level may be decreased below the maximum luminance level of a given image while the great reduction of power consumption is required. That means the gray levels above the clipped point cannot be restored. The distortion due to this problem is so called a clipping artifact. Clipping artifacts can degrade image quality severely. Hence, many researches focused on the two tasks, the optimal clipping point selection [6], [8]-[12] and pixel compensation [2]-[5], [7], [9], to obtain high image quality in visual. The optimal clipping point selection is an excellent idea to obtain high image quality according to the target peak signal-to-noise ratio (PSNR). However, the process of selection is an exhausted iteration with high computational complexity. The pixel compensation methods except [9] suffered from a problem that while in the low clipping point, the edge information of image is greatly reduced.

In this paper, avoiding high complexity for clipping point selection, a low complexity method, which selects the adaptive clipping point based on the maximum and average value of image, is presented. Then, the image is compensated by an edge enhancement method to preserve the detail information of image even in the low backlight luminance. The simulation results show that at the same clipping point the proposed methods has higher detailed information than others.

3. Proposed Method

The proposed backlight dimming method consists of the following two steps: clipped point selection and image compensation, which are illustrated in Fig. 2. In the first step, to reduce power consumption, the proposed approach selects a suitable clipped point according to the chosen level for a given image based on the maximum and average value. In the second step, several parameters are adaptively chosen according to the characteristic of a given image. These parameters are then used to compensate pixels with an edge enhancement technique. The operation of each step is described in detail as follows.

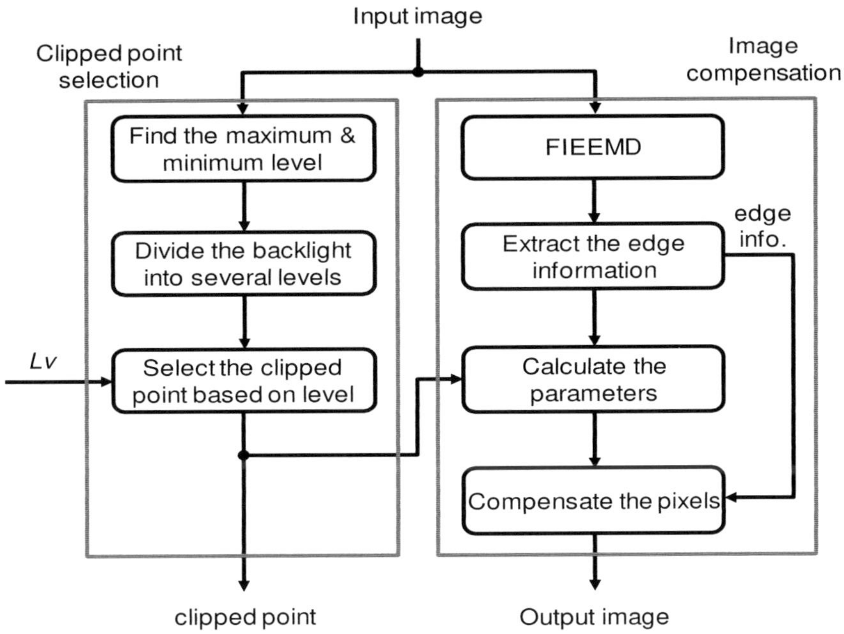

Figure 2 Flow diagram of the proposed algorithm.

3.1 Clipped Point Selection

Although the optimal clipping point selection is an excellent method to obtain high image quality according to the target peak signal-to-noise ratio (PSNR), the process of selection is an exhausted iteration with high computational complexity. The higher the computational complexity, the higher power consumption and longer processing time the algorithm requires during processing. It is always a tradeoff. In our method, avoiding high computational complexity, a low complexity clipping point selection is applied. First, the gray level data of input image are obtained from the *value* component (V) of *HSV* domain (*hue*, *saturation*, and *value*), which are calculated as follows:

$$V(x, y) = \max(R(x, y), G(x, y), B(x, y)), \tag{1}$$

where $V(x, y)$ is the gray level of the pixel of input image located at the (x, y) coordinate, $R(x, y)$, $G(x, y)$, and $B(x, y)$ are the three component of the *RGB* domain (*red*, *green* and *blue*) of input image, and max is the operation to find the maximum value among input elements.

Then, the maximum and average value of $\boldsymbol{V}$ are calculated as below:

$$Avg\_V = \frac{\sum_{j=0}^{H-1}\sum_{i=0}^{W-1} V(i, j)}{H \times W}, \tag{2}$$

$$Max\_V = \max(V(i, j) \text{ for } i = 0,1,\ldots, W-1;\ j = 0,1,\ldots, H-1), \tag{3}$$

where W and H are the width and height of the input image, respectively. While Max_V is used as the clipping point, the compensated image yields almost no artifacts. While Avg_V is applied as the clipping point, there may be about half the pixels suffering from distortion. Thus, Max_V and Avg_V are chosen as the maximum and minimum value of the clipping point respectively in our method. Besides, an adjustable level of output image quality is established for users that can reconstruct different image quality according to a given parameter Lv [5], [7]. The mechanism of adjustment is show as follows:

$$Diff = \frac{(Max\_V - Avg\_V)}{8}, \tag{4}$$

$$BL = Avg\_V + (Lv \times Diff),\ 0 \le L \le 8, \tag{5}$$

where *BL* is the clipping point according to different *Lv*. The greater the value of *Lv*, the higher the quality is, but the less reduction of power consumption the device can achieve.

3.2 Image compsensation

Many pixel compensation methods suffered from a problem that while in the low clipping point, the edge information of image is greatly reduced. In our compensation method, the edge information is first extracted from the given image. After the pixels are compensated, the edge information is used to enhance the detail of image. The Retinex theory [19] is applied here to extract the edge information. The theory assumed that the intensity of visible light reaching a camera depends on the product of the illumination and the reflectance. Thus, an observed image can be expressed as below:

$$I(x, y) = L(x, y) \times D(x, y), \tag{6}$$

where $I(x, y)$ is the pixel of observed image located at (x, y) coordinate, $L(x, y)$ is the illumination of the scene and $D(x, y)$ is the reflectance of the objects. In general, the illumination represents the ray of light source from the sun or sky, and the reflectance represents the detail information of the object surfaces.

To extract the reflectance ***D***, the fast illumination estimation by using the concept of bi-dimensional empirical mode decomposition (FIEEMD) [20] is adopted here to estimate the illumination ***L***. The estimation of ***L*** is shown as follows:

$$E'_U(x, y) = local\_\max_{(i,j)\in\Omega(x,y)} (V(i, j)), \tag{7}$$

$$E'_L(x, y) = local\_\min_{(i,j)\in\Omega(x,y)} (V(i, j)), \tag{8}$$

$$L(x, y) = \frac{E'_U(x, y) + E'_U(x, y)}{2}, \tag{9}$$

where $\Omega(x, y)$ is a square region of size 3×3 centered at the (x, y) coordinate, *local*_max and *local*_min are operations to find the local maximum and minimum value among the square region $\Omega(x, y)$, respectively. Once the illumination ***L*** is obtained, the reflectance ***D*** can be extracted as below:

$$D(x, y) = \frac{I(x, y)}{L(x, y)}, \tag{10}$$

The reflectance ***D*** is then saved as the edge information of input image.

The second step of image compensation is to prepare the parameters for the pixel compensation. Figure 3 shows the common idea of the pixel compensation. The black line shows the original pixels without backlight dimming. The dot line represents the pixels after backlight dimming without pixel compensation. After pixel compensation, the red line shows the pixels with clipping artifacts where the pixels whose gray levels beyond the clipped point are saturated at the clipped point. To smooth clipping artifacts, a method with two points is adopted in our method as show in Fig. 4. The advantage of two-point method is that the gradient among pixels whose original gray levels are around the clipped point can be reduced. The values of A and B are determined as follows:

$$A_1 = BL \times 0.5, \tag{11}$$

$$A_2 = BL + (255 - BL) \times 0.25, \tag{12}$$

In the pixel compensation step, a gamma correction is adopted in our method which is combined with gamma correction method described above. Figure 5 presents the transformation curve of two-point method with gamma correction. If $V(x,y) \le A_1$, the parameter γ_1 is chosen. If $A_1 < V(x,y) \le A_2$, the parameter γ_2 is chosen. If $A_2<V(x,y) \leq 255$, the parameter γ_3 is chosen. In this paper, γ_1, γ_2, and γ_3, are selected as 0.9, 0.3, and 1, respectively. After the compensated image $\boldsymbol{V'}$ is obtained, the reflectance $\boldsymbol{D}$ is then combined with the compensated image $\boldsymbol{V'}$ to produce edge enhancement image $\boldsymbol{V''}$ by using the formula as below:

$$V''(x,y) = V'(x,y) \times D(x,y). \tag{13}$$

According to [7], the reconstruction of the output image with RGB domain can be derived by using $\boldsymbol{V''}$ as the following equations.

$$\begin{bmatrix} R_o(x,y) \\ G_o(x,y) \\ B_o(x,y) \end{bmatrix} = k \times \begin{bmatrix} R(x,y) \\ G(x,y) \\ B(x,y) \end{bmatrix}, \tag{14}$$

where

$$k = \frac{V''(x,y)}{V(x,y)}. \tag{15}$$

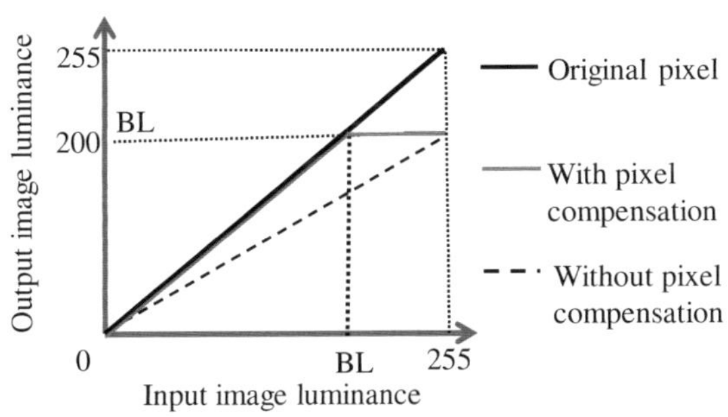

Figure 3 Common idea of pixel compensation

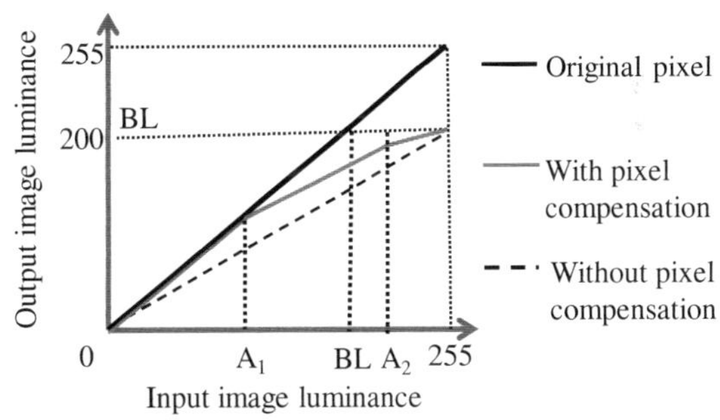

Figure 4 Two-point method of pixel compensation.

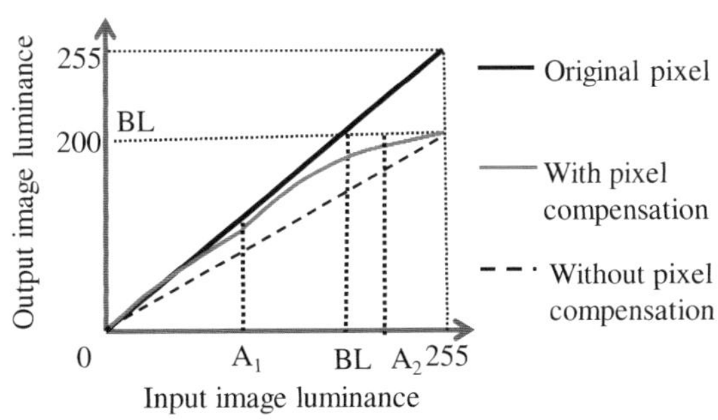

Figure 5 Two-point method with gamma correction.

4. Simulation Results

Several experiments were performed to evaluate the objective and subjective performance of the proposed method. For objective assessment, the PSNR, discrete entropy (DE) [21], and edge-based contrast measure (EBCM) [22] are applied as quantitative measures. The PSNR can show the distortion between the original image and the compensated image. The higher PSNR means the less distortion. The DE of image $\boldsymbol{I}$ measures the content of image, in which a higher value indicates that an image has more detail information. The EBCM is based on the observation that human perception mechanisms are very sensitive to edges. Thus, it is suitable for evaluating the contrast of images. The gray level corresponding to object contours is obtained by computing the average value of the pixel gray levels weighted by their edge values.

We use eight testing images to explore the performance of objective testing for the five image compensation methods. The compared results of PSNR, DE, and EBCM for the five methods are shown in Table I, Table II, and Table III, respectively. Note that the clipped points (BL) of all five methods are the same. It can be seen that the proposed methods obtains the comparable image quality as compared with other methods in PSNR. In DE and EBCM, our method is the best one among these five methods. That means our method can really enhance the edge information. Figure 6-7 show the close-up views of image results under the lower clipped point to illustrate the performance of visual quality with the five methods. It can be seen that our method can efficiently enhance the details of the image to obtain the better results as compared with other methods.

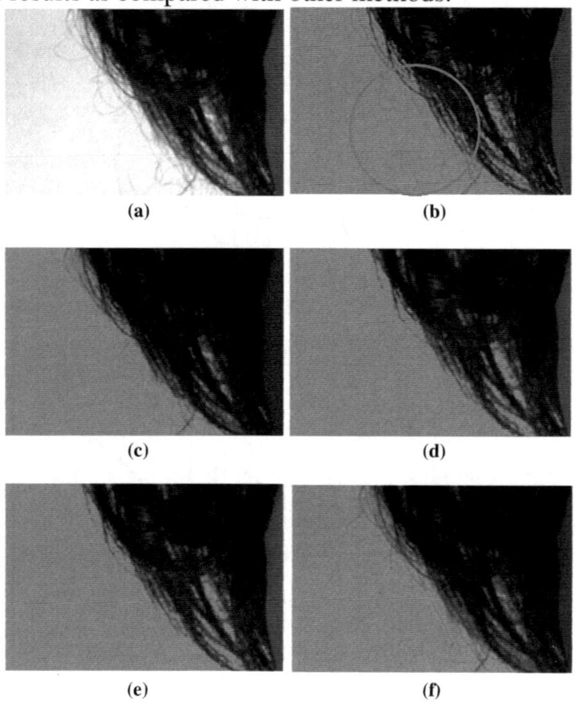

Figure 6 The close-up view of hair with five methods. (a) the original image, (b) the proposed method, (c) [4], (d) [5], (e) [6], (f) [9].

Table 1 Comparison of psnr for image compensation with five methods.

Image	BL	[4]	[5]	[6]	[9]	Proposed
House	178	15.282	19.636	18.223	17.069	18.5799
Motor	156	16.269	23.196	20.343	19.504	22.1741
Girl	179	16.010	17.839	17.398	16.571	17.1146
Bird	178	16.127	22.798	20.374	19.676	21.9667
Hat	169	15.782	26.771	20.939	20.913	24.6183
Barn	174	15.318	25.178	20.360	19.615	23.1582
Lighthouse	178	15.184	26.708	20.672	20.509	23.2433
Mountain	168	15.554	21.716	19.194	17.782	19.9809

IOP Conf. Series: Journal of Physics: Conf. Series **910** (2017) 012039 doi:10.1088/1742-6596/910/1/012039

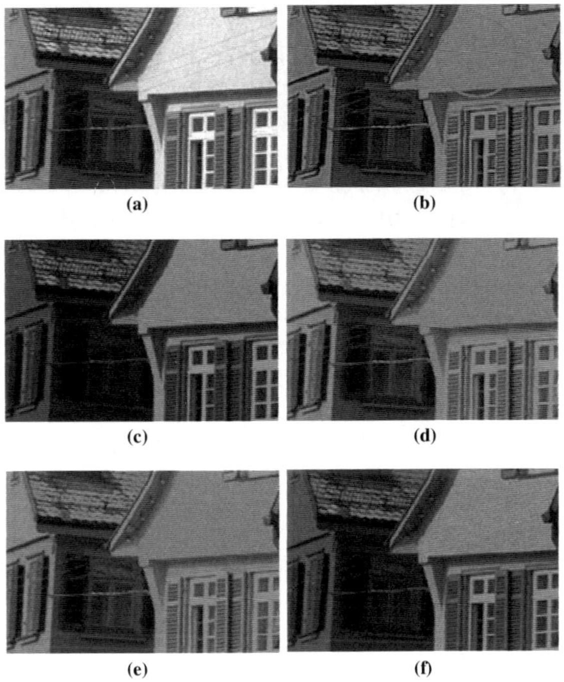

Figure 7 The close-up view of house with five methods. (a) the original image, (b) the proposed method, (c) [4], (d) [5], (e) [6], (f) [9].

Table 2 Comparison of DE for image compensation with five methods.

Image	BL	[4]	[5]	[6]	[9]	Proposed
House	178	4.4834	4.2645	4.5106	4.6698	4.6322
Motor	156	4.5814	4.5646	4.7229	4.7156	4.8341
Girl	179	4.2887	4.0131	4.2754	4.4984	4.5179
Bird	178	4.7032	4.2674	4.5761	4.6406	4.7099
Hat	169	4.5177	4.4740	4.6347	4.6361	4.7228
Barn	174	4.4747	4.5168	4.6266	4.6072	4.6903
Lighthouse	178	4.2264	4.4516	4.4229	4.4185	4.5189
Mountain	168	4.5612	4.4430	4.6526	4.6878	4.7185

Table 3 Comparison of EBCM for image compensation with five methods.

Image	BL	[4]	[5]	[6]	[9]	Proposed
House	178	0.0846	0.0674	0.0709	0.0795	0.1157
Motor	156	0.0852	0.0771	0.0819	0.0822	0.1237
Girl	179	0.0454	0.0402	0.0440	0.0445	0.0624
Bird	178	0.0325	0.0285	0.0306	0.0308	0.0383
Hat	169	0.0345	0.0316	0.0335	0.0355	0.0436
Barn	174	0.0441	0.0399	0.0428	0.0429	0.0588
Lighthouse	178	0.0531	0.0473	0.0521	0.0521	0.0755
Mountain	168	0.0843	0.0700	0.0782	0.0792	0.1183

5. Conclusion

In this paper, an adaptive backlight dimming technique is proposed. This technique preserves the quality of details in images even when the backlight luminance of LCD devices is greatly reduced. The proposed backlight dimming method consists of the following two steps: clipped point selection and image compensation. In the first step, to reduce power consumption, the proposed approach selects a suitable clipped point according to the chosen level for a given image based on the maximum and average value. In the second step, several parameters are adaptively chosen according to the characteristic of a given image. These parameters are then used to compensate pixels with an edge enhancement technique. The simulation results show that at the same clipping point the proposed method can efficiently enhance the details of the image to obtain the better results as compared with other methods.

6. Acknowledgment

This work was supported by Ministry of Science and Technology, ROC, Taiwan, under Grant MOST 105-2221-E-224-063 -.

7. References

[1] I. Choi, H. Shim, and N. Chang, “Low-power color TFT LCD display for handheld embedded systems,” in *Proc. Int. Symp. Low-Power Electron. Des.*, Aug. 2002, pp. 112–117.

[2] P. Tsai, C. Liang, T. Huang, and H. H. Chen, “Image enhancement for backlight-scaled TFT-LCD displays,” *IEEE Trans. Circuits Syst. Video Technol.*, vol. 19, no. 4, pp. 574–583, Apr. 2009.

[3] N. Chang, I. Choi, and H. Shim, “DLS: Dynamic backlight luminance scaling of liquid crystal display,” *IEEE Trans. Very Large Scale Integr. Syst.*, vol. 12, no. 8, pp. 837–846, Aug. 2004.

[4] C. Lai and C. Tsai, “Backlight Power Reduction and Image Contrast Enhancement Using Adaptive Dimming for Global Backlight Applications,” *IEEE Trans. Consumer Electron.*, vol. 54, no. 2, pp. 669-674, May 2008.

[5] H. Cho and O. Kwon, “A backlight dimming algorithm for low power and high image quality LCD applications,” *IEEE Trans. Consumer Electron.*, vol. 55, no. 2, pp. 839–844, May 2009.

[6] S. Kang and Y. Kim, “Image integrity-based gray-level error control for low power liquid crystal displays,” *IEEE Trans. Consumer Electron.*, vol. 55, no. 4, pp. 2401–2406, Nov. 2009.

[7] Y. Lai, Y. Lai, and P. Chen, “Content-based LCD backlight power reduction with image contrast enhancement using histogram analysis,” *J. Display Technol.*, vol. 7, no. 10, pp. 550–555, Oct. 2011.

[8] S.-J. Kang and Y. H. Kim, “Multi-histogram-based backlight dimming for low power liquid crystal displays,” *J. Display Technol.*, vol. 7, no. 10, pp. 544–549, Oct. 2011.

[9] S. I. Cho, S.-J. Kang, and Y. H. Kim, “Image quality-aware backlight dimming with color and detail enhancement techniques,” *J. Display Technol.*, vol. 9, no. 2, pp. 112–121, Feb. 2013.

[10] S.-J. Kang and Y. H. Kim, “Segmentation-based clipped error control algorithm for global backlight dimming,” *J. Display Technol.*, vol. 10, no. 7, pp. 568–573, Jul. 2014.

[11] S.-J. Kang, S. Bae, J.-J. Yun, and M.-Y. Lee, “Color Distortion-Aware Error Control for Backlight Dimming,” *J. Display Technol.*, vol. 11, no. 1, pp. 79–85, Jun. 2015.

[12] S.-J. Kang, and S. Bae, “Fast Segmentation-Based Backlight Dimming,” *J. Display Technol.*, vol. 11, no. 5, pp. 399–402, May. 2015.

[13] H. Chen, J. Sung, T. Ha, Y. Park, and C. Hong, “Backlight local dimming algorithm for high contrast LCD-TV,” in *Proc. ASID*, Oct. 2006, pp. 168–171.

[14] H. Chen, J. Sung, T. Ha, and Y. Park, “Locally pixel-compensated backlight dimming for improving static contrast on LED backlight LCDs,” in *SID 07 Dig.*, May 2007, pp. 1339–1342.

[15] J. Hong, S. Kim, and W. Song, “A clipping reduction algorithm using backlight luminance compensation for local dimming liquid crystal displays,” *IEEE Trans. Consumer Electron.*, vol. 56, no. 1, pp. 240–246, Feb. 2010.

[16] X. Zhang, R. Wang, D. Dong, J. Han, and H. Wu, “Dynamic backlight adaptation based on the details of image for liquid crystal displays,” *J. Display Technol.*, vol. 8, no. 2, pp. 108–111, Feb. 2012.

[17] H. Cho, B. Cho, H. Hong, E. Oh, and O. Kwon, "A color local dimming algorithm for liquid crystals displays using color light emitting diode backlight systems," *Opt. Laser Technol.*, vol. 47, pp. 80–87, Apr. 2013.
[18] S.-C. Hsia, M.-H. Sheu, J.-R. Chang Chien, and S.-K. Wang, "High-Performance Local Dimming Algorithm and Its Hardware Implementation for LCD Backlight," *J. Display Technol.*, vol. 9, no. 7, pp. 527–535, July 2013.
[19] E. H. Land and J. J. Mccann, "Lightness and retinex theory," *J. Opt. Soc. Am.*, vol. 61, no. 1, pp. 1–11, Jan. 1971.
[20] Y.-H. Shiau, P.-Y. Chen, H.-Y. Yang and S.-Y. Li, "A Low-Cost Hardware Architecture for Illumination Adjustment in Real-Time Applications," *IEEE Trans. on Intelligent Transportation Systems*, vol. 16, pp. 934-946, April 2015.
[21] C. E. Shannon, "A mathematical theory of communication," *Bell Syst. Tech. J.*, vol. 27, pp. 379–423, Jul. Oct., 1948.
[22] A. Beghdadi and A. L. Negrate, "Contrast enhancement technique based on local detection of edges," *Comput. Vis. Graph. Image Process.*, vol. 46, no. 2, pp. 162–174, May 1989.

IOP Conf. Series: Journal of Physics: Conf. Series **910** (2017) 012040 doi:10.1088/1742-6596/910/1/012040

WCNNM Algorithm for Wideband DOA Estimation

Ruiyan Du[1,a], Fulai Liu[1,b] and Yajian Cheng[2,c]

[1]Engineering Optimization & Smart Antenna Institute, Northeastern University at Qinhuangdao, Qinhuangdao, China

[2]School of Computer Science and Engineering, Northeastern University, Shenyang, China

Email: [a]ruiyandu@126.com, [b]fulailiu@126.com, [c]chengyajian@126.com

Abstract. This paper presents a novel direction-of-arrival (DOA) estimation method for multiple wideband (WB) signals impinging on a nested array, named as WCNNM algorithm. Making use of the proposed method, the DOA estimation problem can be effectively transformed into a convex nuclear norm minimization problem, by solving it to surrogate to recover the signal component from the observed matrix so that the number of incoming signals and signal subspace can be obtained. By the signal subspace, we can obtain the DOA estimation. Compared with the related works, the presented method can not only improve the DOA estimation performances, such as higher resolution, smaller root-mean-square-error (RMSE), but also increase the maximal number of detectable signals. Simulation results are presented to verify the efficiency of the pro-posed method.

1. Introduction

The direction-of-arrival (DOA) estimation problem has been intensively studied in the past few decades. It has been widely applied in radar, sonar and wireless communication, as it is a crucial parameter for sorting and recognition of sources, directing jamming and passive location, etc. Under the key assumption that the number of active sources is less than the number of sensors, several high-resolution algorithms, such as multiple signal classification (MUSIC) [1] and estimation of signal parameter via rotation invariance techniques (ESPRIT) [2] are proposed to resolve the uncorrelated signals. However, when the number of sources is larger than the number of sensors, often referred to as the underdetermined case, the above high-resolution algorithms will fail in accurately estimating the DOAs. Several methods are presented to resolve the above difficulty. For examples, the DOA estimation algorithms are based on higher-order cumulants [3-5]. A Khatri-Rao (KR) subspace based algorithm is proposed for nonstationary sources in underdetermined case [6].

Recently, many sparse signal recovery (SSR) based DOA estimation methods are proposed [7-9]. For examples, a covariance sparsity-aware DOA estimation method is presented for the nonuniform noise [7]. The presented approach can handle more sources and obtain better performance. A co-prime array is use to estimate DOA [8]. To overcome the grid mismatch problem, a new method is proposed for the co-prime array [9]. Thus, it is of considerable interest to determine the DOAs for the WB signals [10-11]. An effective DOA estimation method is proposed for multiple WB signals [10]. This approach exploits the interpolation technique to form a set of virtual arrays. A subspace-based method is presented [11]. It can achieve higher-resolution by formulating a combined signal subspace. The KR-DOA method [6] is extended to the case of WB non-stationary signals by using the incoherent signal-subspace (ISS) processing. However, its estimation effectiveness is degraded for low signal-to-noise ratio (SNR) scenarios [11].

In this paper, an efficient DOA estimation approach is developed for the WB signals impinging on a nested array. The WB DOA estimation problem is transformed into a convex nuclear norm minimization problem. And then, the DOAs are estimated via a one-dimensional search. The rest of the paper is organized as follows. The data model is described in Section 2. Section 3 introduces the proposed algorithm. Section 4 shows some simulation. Finally, the conclusion is given in Section 5.

2. Data Model

Consider a two-level linear nested array with M sensors, including two concatenated uniform linear arrays (ULAs)-inner and outer, where the inner ULA has M_1 sensors with spacing d_1 and the outer ULA has M_2 sensors with spacing $d_o = (M_1+1)d_1$, $M = M_1 + M_2$. More precisely, it is a linear array with sensors locations given by the union of the sets $S_{inner} = \{md_1, m = 1, 2, \cdots M_1\}$ and $S_{outer} = \{n(M_1+1), n = 1, 2, \cdots M_2\}$, where the first sensor is set to be the reference, as shown in figure 1.

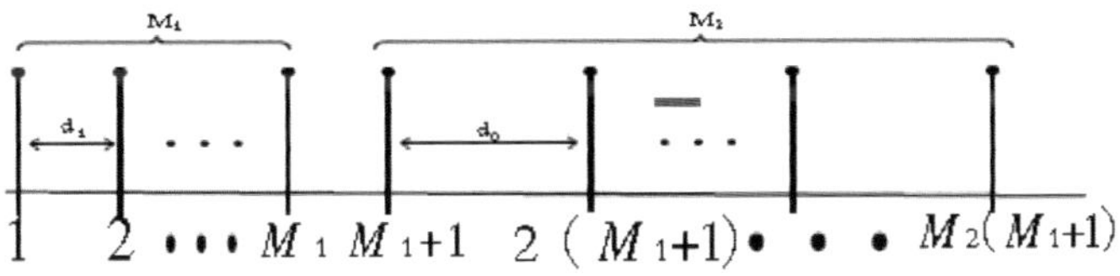

Figure 1. Nested array configuration

Assume that the Q WB signals have a common bandwidth B (Hz) with center frequency f_0 (Hz). The observed q th bandpass signal $\bar{x}_q(t)$ at a reference point can be given by :

$$\bar{x}_q(t) = s_q(t)e^{j2\pi f_0 t} \tag{1}$$

where $s_q(t)$ denotes the q th baseband signal. If the signal is observed over a duration $[t_0, t_0 + T_0]$, the baseband signal can be written as follows

$$s_q(t) = \sum_{i=1}^{I} S_q(f_i)e^{j2\pi f_i t}, \; t_0 \le t \le t_0 + T_0 \tag{2}$$

where the Fourier coefficients $S_q(f_i) = \frac{1}{T_0}\int_{t_0}^{t_0+T_0} s_q(t)e^{-j2\pi f_i t}dt$ with $f_i = (i - (I+1)/2)/T_0$, $i = 1, \cdots, I$. I stands for the number of frequency components. Consider the propagation delay $\tau_{q,m}$ of the q th signal at the m th sensor, the modulated bandpass signal at the reference point can be expressed as

$$\bar{x}_q(t + \tau_{q,m}) = \sum_{i=1}^{I} S_q(f_i)e^{j2\pi(f_0+f_i)(t+\tau_{q,m})} \tag{3}$$

where $\tau_{q,m} = md_1\sin(\theta_q)/c$, c denotes the propagation speed. The demodulated signal can be expressed as

$$x_q(t, \tau_{q,m}) = \bar{x}_q(t + \tau_{q,m})e^{-j2\pi f_0 t} \tag{4}$$

Let $X_q(t) = [x_q(t, \tau_{q,1}), x_q(t, \tau_{q,2}), \cdots x_q(t, \tau_{q,M})]^T$. Then, we have the following representation

$$X_q(t) = \sum_{i=1}^{I} a(\theta_q, f_0 + f_i)S_q(f_i)e^{j2\pi f_i t} \tag{5}$$

where $a(\theta_q, f_0 + f_i)$ denotes the $M \times 1$ steering vector of the q th signal, it has the following form

$$a(\theta_q, f_0 + f_i) = \left[e^{j2\pi(f_0+f_i)\tau_{q,1}}, \cdots, e^{j2\pi(f_0+f_i)\tau_{q,M}}\right]^T \tag{6}$$

The received data vector has the form

$$X(t) = \sum_{i=1}^{I} [A(\theta, f_i) S(f_i) + N(f_i)] e^{j2\pi f_i t} \tag{7}$$

where $A(\theta, f_i) = [a(\theta_1, f_0 + f_i), \cdots, a(\theta_Q, f_0 + f_i)]$ stands for $M \times Q$ array steering matrix, $S(f_i) = [S_1(f_i), \cdots, S_Q(f_i)]^T$ denotes the $Q \times 1$ signal vector, $N(f_i) = [N_1(f_i), \cdots, N_M(f_i)]^T$ represents the $M \times 1$ noise Fourier coefficient vector, and define $y(i)$ as the $M \times 1$ Fourier coefficient vectors of $X(t)$:

$$y(i) = A(\theta, f_i) S(f_i) + N(f_i),\ i = 1, \cdots, I \tag{8}$$

The common assumptions are listed as follows:

A1) The source DOAs $\theta_q (q = 1, \ldots, Q)$, are distinct from one another.

A2) The source signals $s_q(t) (q = 1, \ldots, Q)$, are statistically independent. This guarantees that the covariance matrix of the source signals is diagonal.

A3) The noise is spatially and temporally white, and uncorrelated with the source signals.

Based on the above assumptions, the covariance matrix can be represented as

$$R_y = E(y(i) y^H(i)) = A_i R_{s_i} A_i^H + \sigma_N^2 I \tag{9}$$

where $E(\cdot)$ denotes the expectation operator. $R_{s_i} = \mathrm{diag}(\rho_{1,i}^2, \rho_{2,i}^2, \cdots, \rho_{Q,i}^2)$, $\rho_{q,i}^2$ and σ_N^2 represent the signal power and the noise power, respectively. I is an $M \times M$ identity matrix, for brevity, we substitute A_i for $A(\theta, f_i)$.

$$z = \mathrm{vec}(R_y) = (A_i^* \odot A_i) p_i + \sigma_N^2 I_e \tag{10}$$

Let where $\mathrm{vec}(\cdot)$ denotes the vectorization operation. We can see that z in (10) behaves like the received signal at an array whose manifold is given by $A_i^* \odot A_i$, $p_i = [\rho_{1,i}^2, \rho_{2,i}^2, \cdots, \rho_{Q,i}^2]^T$ denotes the equivalent source signal vector, it consists of the powers $\rho_{q,i}^2$ of the actual sources, and these powers behave like fully coherent sources, $I_e = [e_1^T, e_2^T, \cdots, e_M^T]^T$ with e_i being a column vector of all zeros except a 1 at the i th position. In addition, $(\cdot)^*$ stands for complex conjugation without transposition and symbol $\odot$ denotes the KR product.

3. Algorithm Formulation

3.1 Spatial Smoothing

Since the difference co-array of a two level nested array has sensors located from $(-M^2/4 - M/2 + 1)d_1$ to $(M^2/4 + M/2 - 1)d_1$. We firstly construct a new matrix $\mathbf{A}_i$ of size $((M^2 - 2)/2 + M) \times Q$ from $A_i^* \odot A_i$ where we have removed the repeated rows and sort them so that the jth row corresponds to the sensor location $(-M^2/4 - M/2 + j)d_1$ in the difference co-array of the two level nested array. This is equivalent to removing the corresponding rows from the observation vector z and sorting them to get a new vector z_1 given by

$$z_1 = \overline{A}_i p_i + \sigma_N^2 \overline{e} \tag{11}$$

where $\overline{e} \in C^{((M^2-2)/2+M)\times 1}$ is a vector of all zeros except a 1 at the $(M^2/4+M/2)th$ position. We now devide these $(M^2/2+M-1)$ sensors into $(M^2/4+M/2)$ overlapping subarrays, in which case the jth subarray has sensors located at $\{(-j+1+m)d_1, m=0,1\cdots, M^2/4+M/2-1\}$. The jth subarray corresponds to the $(M^2/4+M/2-j+1)th$ to $((M^2-2)/2+M-j+1)th$ rows of z_1 which we denote as

$$\overline{z}_1 = \overline{A}_{ij} p_i + \sigma_N^2 e_j \tag{12}$$

We can check that $\overline{z}_1 = \overline{A}_{i1}\Phi^{j-1} p_i + \sigma_N^2 e_j$ where $\Phi^{j-1} = diag(e^{-j(2\pi/\lambda)d_1 \sin\theta_1}, \cdots, e^{-j(2\pi/\lambda)d_1 \sin\theta_Q})$. Viewing $\overline{z}_1$ as a newly received vector, the corresponding covariance matrix $\overline{R} = \overline{z}_1\overline{z}_1^H$ can be obtained. Taking the average of $\overline{R}$ over all j. We get

$$\overline{R}^{ave} = \frac{1}{(M^2/4+M/2)} \sum_{j=1}^{M^2/4+M/2} \overline{R} \tag{13}$$

According to [12], $\overline{R}^{ave}$ can also be expressed as $\overline{R}^{ave} = R^2$, where R has the form

$$\mathbf{R} = \frac{1}{\sqrt{M^2/4+M/2}}(\overline{\mathbf{A}}_{i1}\mathbf{R}_{S_i}\overline{\mathbf{A}}_{i1}^H + \sigma_N^2 I) \tag{14}$$

The matrix R in (14) has the same form as the conventional covariance matrix used in subspace based DOA estimation techniques when applied on a ULA with $M^2/4+M/2$ sensors whose array manifold is represented by $\overline{A}_{i1} \in C^{(M^2/4+M/2)} \times Q$. As long as $Q \le M^2/4+M/2$, R has a low rank component $\overline{A}_{i1} R_{S_i} \overline{A}_{i1}^H$.In practice, since the covariance matrix R_y of the received signal is usually unavailable, we will use the sample-average estimated array auto-covariance matrix given by

$$R_y = \frac{1}{I}\sum_{i=1}^{I} y(i) y^H(i) \tag{15}$$

In this case, the smoothed matrix has the form

$$R = \frac{1}{\sqrt{M^2/4+M/2}}(\overline{A}_{i1} R_{S_i} \overline{A}_{i1}^H + E) \tag{16}$$

where R_{S_i} is a diagonal matrix containing the diagonal of $\frac{1}{I}\Sigma_{i=1}^{I} S(f_i)S^H(f_i)$ and E represents the noise, due to finite samples effect, signal and noise cross correlation terms are not 0. Here the matrix $\overline{A}_{i1} R_{S_i} \overline{A}_{i1}^H$ is still low rank, there is no easy way to find its rank.

3.2 Low Rank Matrix Recovery and DOA Estimation

In order to address the above problem, low rank matrix recovery is performed by nuclear norm minimization, moreover, it can also avoid spurious peaks for DOA estimation. Assume that the matrix $B = A + C$ in which A is low rank and C has bounded entries, for some sufficiently small ε ,the following in equation $\| C \|_F \le \varepsilon$ is satisfied, where $\| \cdot \|_F$ denotes Frobenius norm. For reasonably large samples, the noise E is considered to have a small Frobenius norm since the cross correlation terms become smaller with the number of samples increasing.

Due to the matrix $\bar{A}_{i1} R_{S_i} \bar{A}_{i1}^H$ is inherently low rank, thus it can be estimated by the following convex nuclear norm minimization given by

$$\min_{R} \| R \|_* \\ s.t. \| R - R \|_F \le \varepsilon \tag{17}$$

where $\| \cdot \|_*$ denotes the nuclear norm (sum of singular values). The optimal solution R^* is an estimation for the matrix $\bar{A}_{i1} R_{S_i} \bar{A}_{i1}^H$, and it is a positive semidefinite matrix. Then we can perform the singular value decomposition (SVD) of R^*, namely, SVD$(R^*) = U \Lambda V^H$. If the rank$(R^*) = r$, we can order the singular values such that the last $M^2/4 + M/2 - r$ singular values are equal to 0. Denote the corresponding columns of V as $V_n = \left[v_1, \cdots, v_{M^2/4+M/2-r} \right]$, and $\bar{a}_{i1}$ as the steering vector of the co-array corresponding to the direction θ_q. Combining the resulting measurements for all the different frequencies, we construct the new combined DOA estimation spectrum:

$$M(\theta) = \frac{1}{\frac{1}{I} \sum_{i=1}^{I} \bar{a}_{i1}^H V_n V_n^H \bar{a}_{i1}} \tag{18}$$

the estimated DOAs are corresponding to the largest values of the spectrum $M(\theta)$.

It is clear that the low rank recovery stage succeeds in recovering $\bar{A}_{i1} R_{S_i} \bar{A}_{i1}^H$, true DOAs will be recovered. When we choose $\varepsilon = \sqrt{M^2/4 + M/2} \sigma_N^2$ the estimated covariance matrix R can be replaced by the exact covariance matrix R_y. Suppose that total observation time is T and divide it into L segments, there are I samples within each segment. Therefore, we have $I \times L$ samples $X = [x, \cdots, x(I \times L)]$, for each segment l, we employ DFT to get the $M \times I$ corresponding frequency coefficient matrix:

$$y = [y_l(1), \cdots, y_l(I)],\ l = 1, \cdots, L \tag{19}$$

Consider all the segments, we can get the $M \times L$ coefficient matrix for each frequency index i:

$$Y^i = [y_l(1), \cdots, y_L(i)],\ i = 1, \cdots, I \tag{20}$$

The resulting sample covariance matrix for frequency index i can be written as $R_y = \frac{1}{I} \sum_{i=1}^{I} Y^i (Y^i)^H$, following the spatial smoothing technique and low rank matrix recovery in the previous process, the correspondingly optimal solution R^* can be obtained, we can conduct DOA estimation based on the optimal solution.

4. Identifiability

To obtain better DOA estimation and derive the maximal number of identifiability signals, the identifiability of the proposed method is presented below. Since the rank of the matrix $(A_i^* \odot A_i)_{smooth}$ affects the identifiability, it is necessary to further investigate the rank of the matrix $(A_i^* \odot A_i)_{smooth}$. We use the following lemma (see property 2 in [13]).

Lemma 1. For two matrices $A \in c^{n \times k}$ and $B \in c^{m \times k}$, with $\mathrm{krank}(A) \ge 1$ and $\mathrm{krank}(B) \ge 1$, it holds true that $\mathrm{rank}(A \odot B) \ge \mathrm{krank}(A \odot B) \ge \min\{\mathrm{krank}(A) + \mathrm{krank}(B) - 1\}$.

The rank properties of KR product have a strong connection to a concept called Kruskal rank, or k-rank for short. The k-rank of a matrix A, denoted by $\mathrm{krank}(A)$, is supposed to be equal to r when every collection of r columns of A is linearly independent, but there exists a collection of $r+1$ linearly dependent columns. However, the rank of A is the maximal number of linearly independent columns, the relationship between them is $\mathrm{rank}(A) \ge \mathrm{krank}(A)$.

Theorem 1. Let A be a vandemonde matrix as follows:

$$A = \begin{bmatrix} 1 & 1 & \cdots & 1 \\ a_1 & a_2 & \cdots & a_q \\ \vdots & \vdots & \ddots & \vdots \\ a_1^{m-1} & a_2^{m-1} & \cdots & a_q^{m-1} \end{bmatrix} \in c^{m \times q} \tag{21}$$

where $a_n = e^{-j\frac{2\pi}{\lambda} nd\sin(\theta_i)}$, $a_q \ne a_k$ for all $q \ne k$.

The KR product $A^* \odot A$ via $m^2/4 + m/2$ factors spatial smoothing in a two level nested array with m sensors, $(A_i^* \odot A_i)_{smooth}$ is of full column rank if $q \le m^2/4 + m/2$.

Proof : for a vandemonde matrix A, it is easy to know that $\mathrm{krank}(A^*) = \mathrm{krank}(A) = \min\{m, q\}$. And for distinct $a_1, \cdots, a_q$, it must hold true that $\mathrm{rank}(A) = \min\{m, q\}$. Moreover, $\mathrm{krank}(A^*) = \mathrm{krank}(A) = \min\{m, q\}$. Since the dimension of the matrix $A^* \odot A$ is $m^2 \times q$, but it has precisely as many distinct rows as the number of degree of freedoms (DOFs) of the difference co-array which is $(m^2-2)/2 + m$ for two level nested array. After $m^2/4 + m/2$ factors spatial smoothing, each subarray has $m^2/4 + m/2$ distinct rows. As a consequence, we have $\mathrm{rank}\{(A^* \odot A)_{smooth}\} \ge \mathrm{krank}\{(A^* \odot A)_{smooth}\} \ge \min\{q, m^2/4 + m/2\}$, where the last inequality is due to Lemma 1 and spatial smoothing. For $1 \le q \le m^2/4 + m/2$, $\mathrm{rank}\{(A^* \odot A)_{smooth}\} \ge q$, so $(A^* \odot A)_{smooth}$ has full column rank.

This concludes the proof.

Consider $\bar{R}^{ave} = \hat{R}^2$, where $\hat{R}$ has the form $\hat{R} = \frac{1}{\sqrt{M^2/4 + M/2}}(\bar{A}_{i1} R_{s_i} \bar{A}_{i1}^H + \sigma_N^2 I)$. Moreover, $\bar{A}_{i1} = (A_i^* \odot A_i)_{smooth}$,

$$\bar{A}_{i1} = \begin{bmatrix} 1 & 1 & \cdots & 1 \\ a_1 & a_2 & \cdots & a_Q \\ \vdots & \vdots & \ddots & \vdots \\ a_1^{\frac{M^2}{4}+\frac{M}{2}-1} & a_2^{\frac{M^2}{4}+\frac{M}{2}-1} & \cdots & a_Q^{\frac{M^2}{4}+\frac{M}{2}-1} \end{bmatrix} \tag{22}$$

where $\bar{A}_{i1} \in c^{(M^2/4+M/2) \times Q}$. A direct consequence of theorem 1 is that the virtual array response matrix $\bar{A}_{i1} = (A^* \odot A)_{smooth}$ has full column rank if $Q \le M^2/4 + M/2$.

Remark:

1) To correctly estimate the DOAs of the WB signals, $Q \le M^2/4 + M/2$ must be satisfied, $\bar{A}_{i1} \hat{R}_{s_i} \bar{A}_{i1}^H$ is inherently low rank matrix with rank Q and we can correctly use the convex programming method to estimate it. Thus the proposed method can identify the maximal number of the WB signals is $M^2/4 + M/2 - 1$. However, the augmented matrix approach cannot identify more than

$1/2(M^2/4+M/2)-1$ sources, because it uses only one half of co-array [14], and the conventional subspace-based methods can only handle $M-1$ sources.

2) From difference co-array perspective of the nested array, it is a filled ULA with $2M_2(M_1+1)-1$ elements whose positions are given by the set $S_{ca}=\{nd_1, n=-N,\cdots,N, N=M_2(M_1+1)-1\}$. In this paper, the spatial smoothing technique only uses $M^2/4+M/2(M=M_1+M_2)$ DOFs offered by the difference co-array.

5. Simulation Results

In this section, we construct several simulations to evaluate the proposed method for WB sources DOA estimation in underdetermined case by using a linear nested array. In the examples, we consider a nested array with $N=6$ sensors and $Q=8$ WB sources impinging on it from different directions $[-40^\circ,-30^\circ,-20^\circ,-10^\circ,5^\circ,20^\circ,30^\circ,60^\circ]$. Suppose the WB sources have the same center frequency $f_0=100\text{Hz}$ and the same bandwidth $B=40\text{Hz}$. By using the singular value thresholding approach [14] to solve (17) and recover the low rank matrix, we use the routine singular-value-rpca [15] under the svt package available online. The routine uses a parameter λ, which is related to the parameter ε in (17), in the simulation experiments, we set $\lambda=5$. The demodulated data is sampled at a frequency of 300Hz.

The array output is divided into $I=33$ narrowband components via DFT and we choose the segment number to be $L=100$. Therefore we use a total of $I\times L=3300$ samples. In the following simulations, K Monte Carlo runs are performed to calculate the average result for each experiment and the input signal-to-noise ratio (SNR) is defined as $10\log 10(\rho_{q,i}^2/\sigma_n^2)$, where $\rho_{q,i}^2$ denotes the power of signal source and σ_n^2 stands for the noise power.

Figure 2 shows the representative DOA estimation spectrum obtained using the proposed method, with respect to various angles at a SNR of 0dB. We can see that the proposed method can use 6 sensors to resolve the 8 sources well.

We consider the performance of our proposed method by studying the root mean square error (RMSE) of the angle estimates as a function of the SNR and the snapshot number. By choosing the reasonable value for which the RMSE is minimized, the RMSE is defined as follows [16]

$$\text{RMSE}=\sqrt{\frac{1}{K\times Q}\left\{\sum_{k=1}^{K}\sum_{q=1}^{Q}\left(\theta_{qk}-\theta_q\right)^2\right\}} \tag{23}$$

where θ_q and θ_{qk} denote the true and estimation DOAs, respectively.

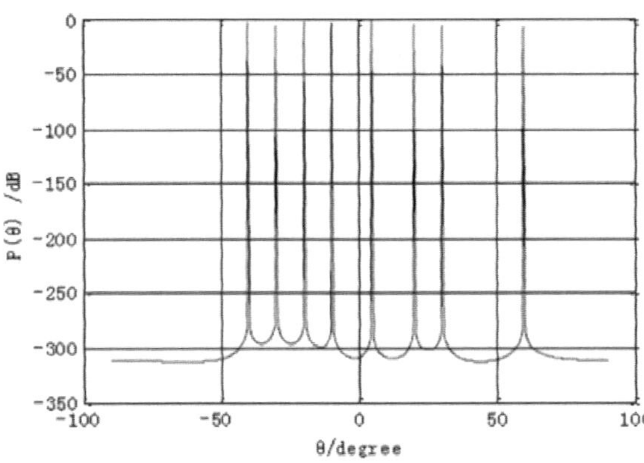

Figure 2. DOA estimation spectrum.

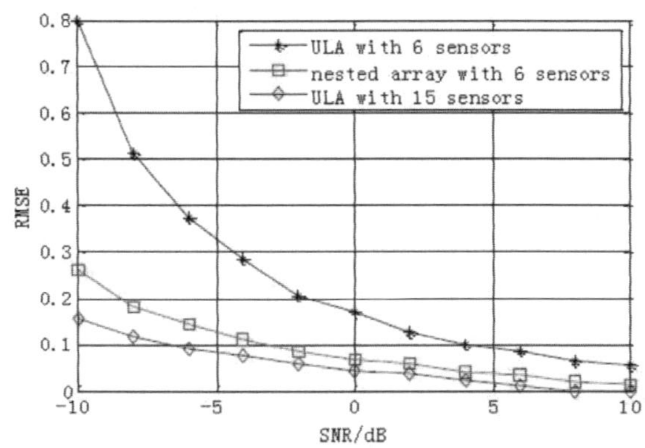

Figure 3. RMSE curves versus the SNR, using nested array,traditional ULA with 6 sensors and 15 sensors

In the first simulation, since the nested array has 6 sensors and 11 DOFs, we also consider the corresponding RMSE for conventional MUSIC applied to 6-sensor and 15-sensor ULAs. We plot the RMSE for the source at 30°, the performance is similar for the other sources, and we fix the number of snapshots is 256. Figure 3 shows the RMSE of the three methods as a function of SNR, we can see that the performance of all three methods improve with the SNR increasing. In addition, the proposed method performs reasonably better than the corresponding ULA with same number of sensors, and performs close to the much longer ULA with 15 sensors. For the purpose of comparison, we also plot the Cramer-Rao Bound (CRB) for the WB signals, which can be applied to the underdetermined case. The RootCRB (N) represents the square root of CRB bound with N sensors, we compare the proposed method with the RootCRB (6) and RootCRB (15). Figure 4 gives RMSE curves with SNR from -10dB to 10dB, we fix the number of snapshots is 256. The proposed method shows a more satisfactory performance and the RMSE is reasonably close to RootCRB (15) at the same SNR. When SNR is set to 10dB and the snapshot number varies from 50 to 500 with a step of 50, the RMSE curves of the proposed method, RootCRB (6) and RootCRB (15) are shown in Figure 5, respectively. The Figure 5 shows that the performance of the proposed method is better than the ULA with 6 sensors, but still not good as the ULA with 15 sensors at the same snapshot number.

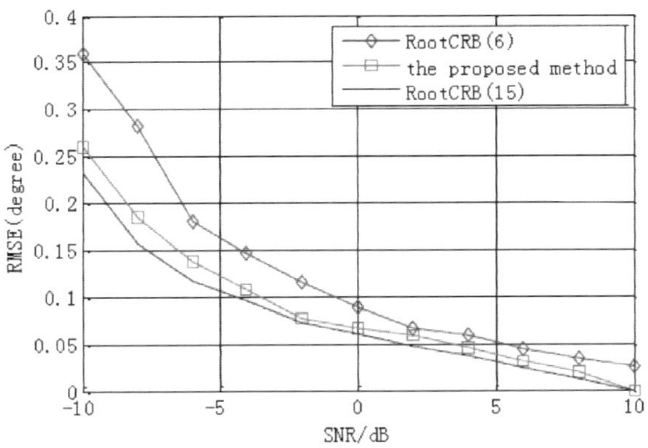

Figure 4. RMSE curves versus SNR.

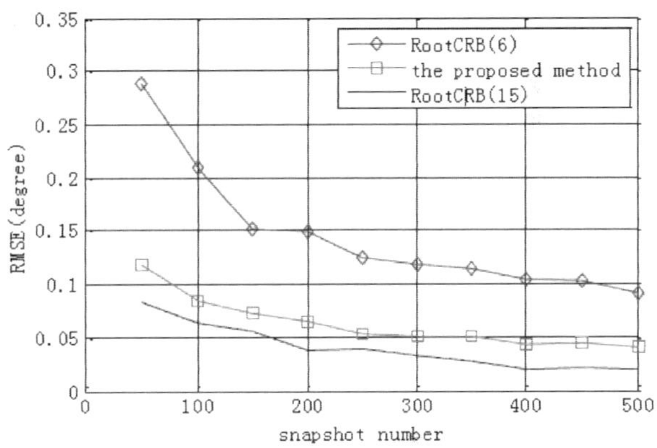

Figure 5. RMSE curves versus snapshot.

6. Conclusions

In this paper, a WCNNM algorithm is proposed for multiple WB signals impinging on a nested array. The proposed method can improve the DOA estimation performances, as well as increase the maximal number of detectable signals. Simulation results validate the effectiveness of the presented approach and it still has better performance even when the total number of incident sources exceeds that of array elements.

7. Acknowledgment

This work was supported by the Program for New Century Excellent Talents in University (NCET-13-0105), and by the Support Program for Hundreds of Outstanding Innovative Talents in Higher Education Institutions of Hebei Province, under Grant No.BR2-259, and by Natural Science Foundation of Hebei Province (No. F2016501139), and by the Specialized Research Fund for the Doctoral Program of Higher Education of China (No.20130042110003), and by the Fundamental Research Funds for the Central Universities under Grant No. N142302001, and by the Fundamental Research Funds for the Central Universities under Grant No.N162304002.

8. References

[1] R.O. Schmidt, Mutiple emitter location and signal parameter estimation, IEEE Transactions on Antennas and Propagation, vol.35,no.3, pp.243-258, 1979.

[2] R. Roy, T. Kailath, ESPRIT-Estimation of signal parameters via rotational invariance techniques, IEEE Transactions on Acoustics, Speech and Signal Processing, vol.37, no.7, pp.984-995, 1989.
[3] P. Chevalier, L. Albera and A. Ferreol, On the virtual array concept for higher order array processing, IEEE Transactions on Signal Processing, vol.53, no.4, pp.1254-1271, 2005.
[4] P. Chevalier, A. Ferreol and L. Albera, High-resolution direction finding from higher order statistics: The 2q-MUSIC algorithm, IEEE Transactions on Signal Processing, vol.54, no.8, pp.2986-2997, 2006.
[5] P. Pal, P.P. Vaidyanathan, Multiple level nested array: An efficient geometry for 2qth order cumulant based array processing, IEEE Transactions on Signal Processing, vol.60, no.3, pp.1253-1269, 2012.
[6] W.K. Ma, T.H. Hsieh and C.Y. Chi,Direction-of-arrival estimation of quasi-stationary signals with less sensors than sources and unknown spatial noise covariance: A Khatri-Rao subspace approach, IEEE Transactions on Signal Processing, vol.58, no.4, pp.2168-2180, 2010.
[7] Z.Q. He, Z.P. Shi and L. Huang, Covariance sparsity-aware DOA estimation for nonuniform noise, Digital Signal Processing, vol.28, pp.75-81, 2014.
[8] Y.D. Zhang, M.G. Amin and B. Himed, Sparsity-based DOA estimation using co-prime arrays, IEEE International Conference on Acoustics, Speech and Signal Processing, pp.3967-3971, 2013.
[9] Z. Tan, Y.C. Eldar and A. Nehorai, Direction of arrival estimation using co-prime arrays: A super resolution viewpoint, IEEE Transactions on Signal Processing, vol.62, pp.5565-5576, 2014.
[10] B. Friedlander, A.J. Weiss, Direction finding for wide-band signals using an interpolated array, IEEE Transactions on Signal Processing, vol.41, no.4, pp.1618-1634, 1993.
[11] H. Wang, M. Kaveh, Coherent signal-subspace processing for the detection and estimation of angles of arrival of multiple wideband sources, IEEE Transactions on Acoustics, Speech and Signal Processing, vol.33, no.4, pp.823-831, 1985.
[12] P. Pal, P.P. Vaidyanathan, Nested arrays: A novel approach to array processing with enhanced degrees of freedom,IEEE Transactions on Signal Processing, vol.58, no.8, pp.4167–4181, 2010.
[13] W.K. Ma, T.H. Hsieh and C.Y. Chi, Direction-of-arrival estimation of quasi-stationary signals via khatri-rao subspace,IEEE International Conference on Acoustics, Speech and Signal Processing, pp.2165-2168, 2009.
[14] J.F. Cai, E.J. Candes and Z. Shen, A singular value thresholding algorithm for matrix completion,SIAM Journal of Optimal, vol.20, no.4, pp.1956-1982, 2010.
[15] [Online]. Available: http://perception.csl.illinois.edu/matrix-rank/sample-code.html.
[16] F.L. Liu, J.K. Wang, C.Y. Sun and R.Y. Du,Spatial differencing method for DOA estimation under the coexistence of both uncorrelated and coherent signals,IEEE Transactions on Antennas and Propagation, vol.60, no.4, pp.2052-2062, 2012.

IOP Conf. Series: Journal of Physics: Conf. Series **910** (2017) 012041 doi:10.1088/1742-6596/910/1/012041

Traffic Video Image Segmentation Model Based on Bayesian and Spatio-Temporal Markov Random Field

Jun Zhou[1,a], Xu Bao[1,b], Dawei Li[2,c] and Yongwen Yin[3]

[1]Key Laboratory for Traffic and Transportation Security of Jiangsu Province, Huaiyin Institute of Technology , Huai'an, China
[2]Jiangsu Key Laboratory of Urban ITS, Jiangsu Province Collaborative Innovation Center of Modern Urban Traffic School of Transportation, Southeast University, Nanjing, China
[3]Transportation Bureau, Huai'an, China
Email: [a]joujou1980@163.com, [b]112026245@qq.com, [c]328650320@qq.com

Abstract. Traffic video image is a kind of dynamic image and its background and foreground is changed at any time, which results in the occlusion. In this case, using the general method is more difficult to get accurate image segmentation. A segmentation algorithm based on Bayesian and Spatio-Temporal Markov Random Field is put forward, which respectively build the energy function model of observation field and label field to motion sequence image with Markov property, then according to Bayesian' rule, use the interaction of label field and observation field, that is the relationship of label field's prior probability and observation field's likelihood probability, get the maximum posterior probability of label field 's estimation parameter, use the ICM model to extract the motion object, consequently the process of segmentation is finished. Finally, the segmentation methods of ST - MRF and the Bayesian combined with ST - MRF were analyzed. Experimental results: the segmentation time in Bayesian combined with ST-MRF algorithm is shorter than in ST-MRF, and the computing workload is small, especially in the heavy traffic dynamic scenes the method also can achieve better segmentation effect.

1. Introduction

Segmentation of moving objects in complex scenes is one of the hotspots in computer vision research in recent years. Markov random field model (MRF) defined by Gibbs distribution is widely used in image segmentation. Such as: Pappas[1] used MRF model to segment the still image, Liu[2] used the method of MRF model combined with fuzzy clustering for color image segmentation. Andrey [3] used MRF for segmentation on video texture image, although these studies are very successful, and they all have adopted the model of 2D-MRF in the space images (such as still images). But for traffic videos, this kind of dynamic image sequences, segmentation effect of above-mentioned method is not as good as still images. This paper will be based on the 2D-MRF model, and extend the 2D-MRF to the ST-MRF model. The model not only can optimize the image distribution on the space but also optimize the image on the time axis. On the time axis, the image sequence of successive frames is associated on each pixel. ST-MRF takes this association into consideration, which will benefit the segmentation of dynamic images. For the lack of research on dynamic images of above-mentioned algorithm, ST-MRF combines still information and dynamic information, and this algorithm is less sensitive to light, so there is no need for images to denoise to detect moving targets.

Recently segmentation technology of the moving object in complex scene is one of the hot spots of computer vision.

2. ST-MRF Model

2.1 ST-MRF Theoretical Knowledge

Aimed at a series of points $S=\{i_1,i_2,...,i_s\}$ on a set of 2D images, for these points, there is a corresponding prior observation field $D=\{d_1,d_2,...,d_s\}$ (also called time field) and a corresponding label field $L=\{l_1,l_2,...,l_s\}$, also known as the space field. Among them, the prior observation field D of the image, which is the frame difference image, recording the relationship between pixels and its frame can be obtained according to the subtraction between two continuous frames of of continuous time:

$$D(t+1)=|S(t+1)-S(t)| \tag{1}$$

The label field records the relationship between each pixel and background as well as the neighborhood. In order to simplify the model, two mutually adjacent groups are only considered here. For the S on the two-dimensional images, there are 8 adjacent points on the spatial domain and 18 adjacent points on the time domain [4]. The spatio-temporal neighborhood system on the image S can be described by the Gibbs random field with the ST-MRF distribution. In this paper, we assume that each frame of gray image on the video image is a ST-MRF. At this time, random variable x_i has become the label value at s_i of the pixel. (As the gray value), the definition of the energy function is: the sum of the potential of all groups in a neighborhood system of a pixel point i in a certain frame image. The energy function consists of two parts, the formula is:

$$U(x)=U_{space}(x)+\alpha U_{time}(x) \tag{2}$$

Among them: $U_{space}(x)$ is the energy function of label field (spatial field). It records the relationship information between each pixel gray value and the background as well as the neighborhood on a frame image; $U_{time}(x)$ is the energy function of observation field (time field).It records the relationship information between the gray value of each pixel in the current frame and intercepted frame images among continuous frames. α is the weight coefficient for the two parts. In the image segmentation, single pixel cannot be used as a scene in ST-MRF (site), a larger group of pixel is needed. Therefore, we need to divide a 720 * 640 image into 90 * 80 blocks, as each block consists of 8 x 8 pixels; blocks between adjacent images are linked through their vector.

2.2 The Construction of the Energy function of ST-MRF

$$U(y)=\sum_k U(y_k) \tag{3}$$

$$\begin{aligned} U(y_k) &\equiv U_N(N_{yk})+U_{pre}(D_{xyk},M_{xyk}) \\ &= U_N(N_{yk})+U_D(D_{xyk})+U_M(M_{xyk}) \\ &= a(N_{yk}-\mu N_y)^2+b(M_{xyk}-\mu M_{xy})^2+cD_{xyk}^2 \end{aligned} \tag{4}$$

Among them, G (t-1) =g, G (T) =h: the value of image G at the time of t-1 is g, at the time of t, the value is h. For each pixel, it can be expressed as: $G(t-1;\ i,\ j)=g(i,\ j),\ G(t;\ i,\ j)=h(i,\ j)$;

X (*t*-1) =*x*, *X* (*t*) =*y*: the target map X was detected that its label distribution is X at the time of t-1, at the time of t, the label distribution was detected is y. For each block, it can be expressed as: $X_k(t-1)=x_k,\ X_k(t)=y_k$, K for block number;

U: energy function;

N_{yk} : The number of the block with the same number in adjacent blocks of a block, if you consider using the 8- neighborhood groups, then μN_y =8 is the maximum value;

D_{xyk} : Texture correlation between representative image G (t-1) and G (T);

M_{xyk} : In the current target map X (T) =y, the effective value of the previous map X (t-1) =x, that is, the number of pixels in the occlusion part in the partial occlusion two blocks, obviously $\mu M_{xy} = 64$ is the maximum value;

$U(y_k)$: The energy functions of ST-MRF.

3. The Segmentation Method of the Bayesian Combined with ST - MRF

3.1 The Bayesian Rules

Assuming that the random fields described by the image data is Y=y, Y represents a random field known as the observation field and time field, y represents a reality of the random fields; Assuming that the random fields described by the image data is X=x,, X represents a random field known as label field or space field. X is a reality of the random fields that is a result of segmentation.

According to Bias theorem, the image segmentation problem is expressed as:

$$P(X = x \mid Y = y) = \frac{P(Y = y \mid X = x) P(X = x)}{P(Y = y)} \tag{5}$$

Among them, $P(X = x \mid Y = y)$ is the posterior probability of the label field under the condition of given observation data $Y = y$; $P(Y = y \mid X = x)$ indicates the joint distribution of the observed field $Y = y$ under the condition of a realistic condition of the given label field $X = x$. $P(Y = y)$ Is the joint distribution of the observed field $Y = y$.

According to the formula (5) for image segmentation directly, the calculation is very complex and it is not the correct result. In order to carry out the MAP (A Posterior Maximum) image segmentation under the Bayesian framework, two hypotheses are needed.

Hypothesis 1: Assume that each component of the observation field $Y = y$ is independent under a realistic condition $X = x$ in a given label field.

If the label field has a total of K markers, that is, each position of the label field has a total of K species, that is, all the observed field data will be divided into K species, which can be expressed as:

$$P(X = x \mid Y = y) = \frac{\prod_{k=1}^{K} [P(y^k \mid X = x) P(X = x)]}{P(Y = y)} \tag{6}$$

Among them, $P(y^k \mid X = x)$ represents the probability distribution of the K species composition y^k of the observation field under the condition of $X = x$.

Hypothesis 2: Assume that the same type of pixels in the image is subject to the same distribution, such as the Gauss distribution.

Under the assumption of 1 and 2, if the distribution function is expressed as $P(y_s^k \mid X_s = k)$, then the image segmentation problem is expressed as:

$$\hat{x}_s = \arg\max_{x_s} \{ P(y_s^k \mid X_s = k) P(X_s = k \mid X_{N_s}) \} \tag{7}$$

Among them, $P(X_s = k \mid X_{N_s})$ represents the local probability of label field, N_s is the neighborhood location set for location s . If the same type of distribution of pixels' observation field is represented by the Gauss distribution, is expressed as:

$$P(y_s^k \mid X_s = k) = \frac{1}{\sqrt{2\pi\sigma_k^2}} \exp[-\frac{(y_s^k - \mu_k)^2}{2\sigma_k^2}] \tag{8}$$

3.2 Algorithm Estimation of Bias Combined with ST-MRF Model

For the target map X, it is detected that label distribution is X at the time of t-1, it is detected that the label distribution is y at the time of t, image G has the value of g at the time of t-1 and has the value of h at the time of t. A posterior probability model can be established, and maximize it. A maximum possible realization of the label field can be obtained. The posterior probability can be expressed as Bias: $P(X(t)=y \mid G(t-1)=g, X(t-1)=x, G(t)=h)$

$$\frac{P(G(t-1)=g, X(t-1)=x, G(t)=h \mid X(t)=y) P(X(t)=y)}{P(G(t-1)=g, X(t\text{-}1)=x, G(t)=h)} \tag{9}$$

In the formula, $P(G(t-1)=g, X(t-1)=x, G(t)=h)$ is constant, so the maximum value of the posterior probability is determined by the fractions of the formula (9).

$P(X(t)=y)$ is the probability of the block C_k ,according with $X_k(t-1)=y_k$. y_k is the Vehicle label in traffic video. For each C_k , the probability of each block regards as Gauss distribution, blolzmann distribution can be used.

$$P(X(t)=y)=\prod_k \exp[-U_N(N_{yk})]/Z_{NK}=\prod_k \exp[-\frac{1}{2\sigma_{N_y}^2}(N_{yk}-\mu N_y)^2]/Z_{NK} \tag{10}$$

In the formula, N_{yk} is the adjacent block of the block C_k , the blocks all have the same vehicle label,

When N_y is 8, the value of the energy function $U_N(N_{yk})$ is the minimum, when N_y =0, the value of the energy function $U_N(N_{yk})$ is the maximum. So the maximum value of the posterior probability is:

$$P(G(t-1)=g, X(t-1)=x, G(t)=h \mid X(t)=y)$$

$$\prod_k \exp[-U_{pre}(M_{xyk}, D_{xyk})]/Z_{DMK}$$

$$\prod_k \exp[-U_M(M_{xyk})]/Z_{MK} \bullet \prod_k \exp[-U_D(D_{xyk})]/Z_{DK} \tag{11}$$

$$\prod_k \exp[-\frac{1}{2\sigma_{M_{xy}}^2}(M_{xyk}-\mu M_{xy})^2]/Z_{MK} \bullet \prod_k \exp[-\frac{1}{2\sigma_{D_{xy}}^2}(D_{xyk}-\mu D_{xy})^2]/Z_{DK}$$

The image segmentation based on maximum a posteriori probability (MAP) criterion is to find the mark set X , which makes the maximum of the posterior probability distribution about X . Considering the problem of computational efficiency, we can use conditional iterative method (ICM). ICM algorithm is an iterative algorithm; maximizing the conditional probability point by point through the realization of the pixel value is updated.

3.3 The Realization of Maximum a Posteriori Probability

3.1.1 ICM iterative algorithm. ICM (condition model iterative) [6] is a deterministic algorithm based on local conditional probability which was proposed by Besag in 1986, it complete the image segmentation through updating the image mask point by point.

Assuming that each pixel of the image data $y=\{y_1, y_2, ..., y_n\}$ is independent on the condition of giving segmentation result x , and the conditional distribution of y_i on x depends only on its mark x_i :

$$f(y_i \mid x) = f(y_i \mid x_i) \tag{12}$$

Therefore, the conditional distribution of y on x represents:

$$f(y \mid x) = \prod_{i \in s} f(y_i \mid x_i) \tag{13}$$

Given the image data and the mark of the neighborhood of i, according to the Bayesian formula:

$$P(x_i \mid y, x_{N_i}) \propto f(y_i \mid x_i) P(x_i \mid x_{N_i}) \tag{14}$$

Maximize the formula (14); we can get the classification mark of the pixel:

$$\hat{x}_i = \arg\max_{x_i} P(x_i \mid y, x_{N_i})(i = 1, 2, \ldots, n) \tag{15}$$

3.1.2 The basic steps of ICM algorithm. If the Gauss mixed model is used to model the observation field data and Potts model is used to model the label field data, and the fixed potential function β is used, so the basic steps of the ICM algorithm can be expressed as:

Step1 set categories K, the potential function B (β) of the image;

Step2 uses the K- mean value algorithm to calculate the initial segmentation results

Step3 estimation of field parameters μ and σ^2;

Step4 calculated the observed field energy;

Step5 calculate the label field energy;

Step6 according to the principle of minimum energy, estimate the new segmentation results;

Step7 according to current segmentation results of parameters and the iteration of the image, according to the formula (15), calculate the maximum possible category of each point;

Step8 determine the convergence; determine whether the$|\Delta U|$ is less than the set t. If the satisfaction is out; otherwise return to Step3 to the next iteration.

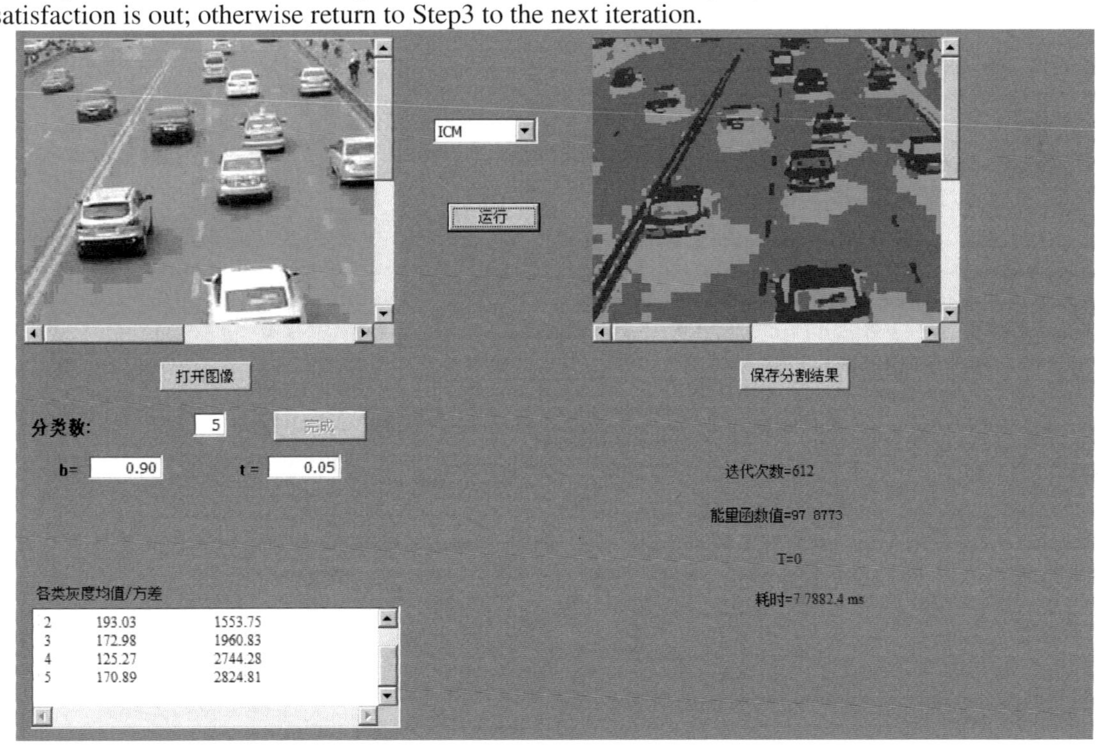

Figure 1 The Segmentation Interface of ICM

3.1.3 Segmentation Result. In Figure 1, the left image is a video image, and the right image is the segmentation result. In the segmentation process, set the image classification number K=5, classification number is the number of initialization classification, that is, the foreground category which is relatively easy to separate from background, the more the category, the segmentation is more detailed. Potential function or β = 0.9 and t which is threshold for the convergence of the energy function, that is the difference among two adjacent iterations of energy levels $|\Delta U|$, its value is set to 0.05, the number of iterations of the algorithm is 612, the value of the minimization of the energy function is 978773, consuming time is 77882.4ms. Fig. 2 is the change curve of energy function.

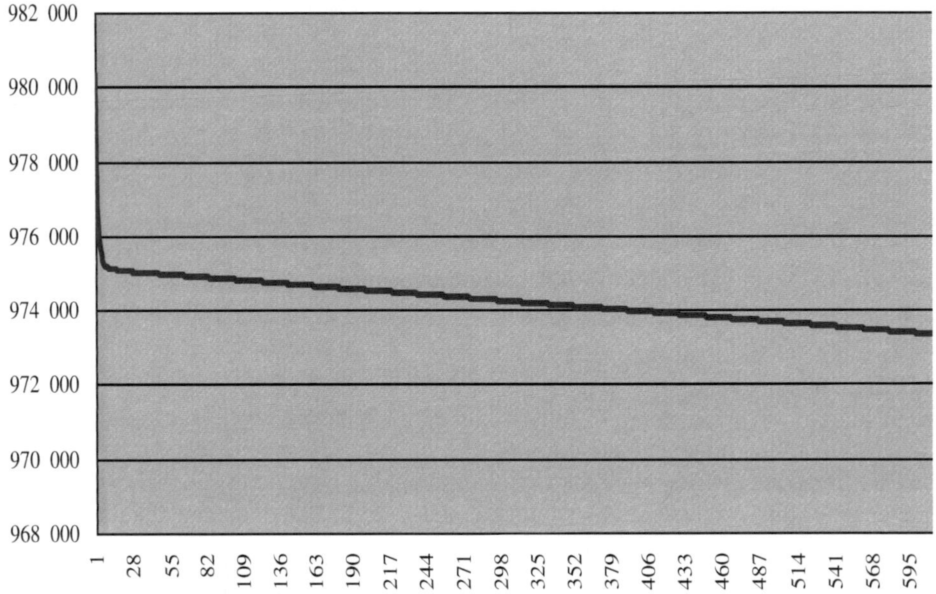

Figure 2 The Energy Value Change Curve of ICM Algorithm

4. Experimental results and analysis

The two Figure shown in the 3 picture ,their pixel size are 627 x 461and 643 x 466 respectively. The segmentation results are shown in figure 4.We can conclude that the segmentation effect in ST-MRF methods and ST-MRF combined with Bias is almost on the static image. But for the segmentation data,we can conclude that the segmentation time in Bayesian combined with ST-MRF algorithm is shorter than in ST-MRF, and the computing workload is small, especially in the heavy traffic dynamic scenes the method also can achieve better segmentation effect.

(a) The 263rd frame (b) the 287rd frame

Figure 3 The image to be segmented

(a)ST-MRF segmentation (b) ST-MRF segmentation

(a)the method in this paper for segmentation (b)the method in this paper for segmentation

Figure 4 Segmentation Results of Two Methods

The above-mentioned simulation results shows that the segmentation method based on Bayesian combined with ST-MRF can better segment the moving objects from the complex background. But there are still many issues that need to continue to study. First of all, the most important problem is that the computing time is too long, because the parameter estimation of the model is complex, leading to the parameter estimation need to spend more time computing. The results will affect the entire image segmentation time. In order to improve the speed of operation, according to the experimental experience, join the artificial selection in the ICM iterative algorithm, that is , adjust it according to the characteristics of the image, it is conducive to the segmentation of dynamic video image.

5. Conclusion

Reliable image segmentation is an important prerequisite to achieve automatic detection of the video image, and it is significant for the development of automatic incident detection and traffic flow monitoring system. This paper is aimed at the problem of occlusion in image segmentation.,establish the image segmentation algorithm of the Bayesian combined with ST-MRF model, The algorithm can not only optimize the spatial image distribution but also optimize the image distribution on the time axis.It can solve the problem of occlusion between targets effectively. The research results of this dissertation are: 1) establish energy function aimed at video image with Markov; 2) establish image segmentation model based on Bayesian and spatio temporal Markov random field ; 3) Use the iterative conditional model (ICM) algorithm to achieve the maximum a posteriori (MAP) estimation problem to achieve the extraction of moving objects;4) ST-MRF segmentation method and Bayesian combined with ST-MRF segmentation method are compared and analyzed.

This work was supported in part by National Natural Science Foundation of China(no. 51608115), Natural Science Foundation of the Jiangsu Higher Education Institutions of China(no.14KJB580002), Six Talent Peaks Project in Jiangsu Province (no. 2015-XXRJ-017).Natural Science Foundation of JiangsuProvince(no.BK20150613), Jiangsu province natural science in colleges and universities research major projects, Projects of International Cooperation and Exchange of the National Natural Science Foundation of China (No. 51561135003), Philosophy and Social Science of Universities in Jiangsu(Grant no. 2016SJB790057), Department of housing and urban rural development(Grant no. 2015-R2-063), Social Science Foundation of Jiangsu

Province (Grant no. 14EYD010), Social Development of Huai'an(Grant no.HASZ201635).

6. References

[1]. XIANG R H, WANG R S. A Range Image Segmentation Algorithm Based on Gaussian Mixture Model[J]. Journal of Software, 2003,14(7):1250-1257.
[2]. Cressie N. Verzelen N. Conditional-Mean Least-Squares Fitting of Gaussian Markov Random Fields to Gaussian Fields[J]. Computational Statistics & Data Analysis, 2008, 52(5):2794-2807.
[3]. Andrey P. Tarroux P. Unsupervised Segmentation of Markov Random Field Modeled Textured Images Using Selectionist Relaxation[J]. Pattern Analysis and Machine Intelligence, IEEE Transactions on, 1998, 20(3):252-262.
[4]. LIU G Y, MA G R, WANG L G. Image Modeling and Segmentation in the Wavelet Domain Based on Markov Random Field[M]. BEIJING: SCIENCE PRESS,2010.
[5]. LI X C, ZHU S A. A Survey of the Markov Random Field Method for Image Segmenttion[J]. Journal of Image and Graphics, 2007, 12(5):789-798.
[6]. CAO J Z, SONG A G. Research on the texture image segmentation method based on markov random field[J]. Chinese Journal of Scientific Instrument,2015,36(4):776-786.
[7]. HUANG X W, ZHU L, ZHONG X R. A Novel Moving Object Segmentation Technology Based on Spatiotemporal Markov Random Field[J]. Journal of Electronics &Information Technology, 2006, 28(2):367-371.

Diffusion Tensor Tractography Reveals Disrupted Structural Connectivity during Brain Aging

Lan Lin, Miao Tian, Qi Wang and Shuicai Wu

Biomedical Engineering Department, Beijing University of Technology, Beijing, China,

Email: lanlin@bjut.edu.cn, tianm@emails.bjut.edu.cn, qiwang@bjut.edu.cn, shuicaiwu@bjut.edu.cn

Abstract. Brain aging is one of the most crucial biological processes that entail many physical, biological, chemical, and psychological changes, and also a major risk factor for most common neurodegenerative diseases. To improve the quality of life for the elderly, it is important to understand how the brain is changed during the normal aging process. We compared diffusion tensor imaging (DTI)-based brain networks in a cohort of 75 healthy old subjects by using graph theory metrics to describe the anatomical networks and connectivity patterns, and network-based statistic (NBS) analysis was used to identify pairs of regions with altered structural connectivity. The NBS analysis revealed a significant network comprising nine distinct fiber bundles linking 10 different brain regions showed altered white matter structures in young-old group compare with middle-aged group ($p < .05$, family-wise error-corrected). Our results might guide future studies and help to gain a better understanding of brain aging.

1. Introduction

With the rapid growth of people aged 60 or older in China, it has been increasingly important to understand the characteristics of age-related brain and cognitive changes. Brain aging is a complicated biological process and changes in a myriad of ways, often related to the brain shrinkage, replicating cells' damage, and decline in cognitive functions. Much of the cognitive changes were the result of aging: poorer vision and hearing, increased forgetfulness, slower information processing, etc. Brain aging itself does not constitute a disease, but those aged related changes are amplified in most common neurodegenerative diseases, such as Alzheimer's disease, Parkinson's disease, stroke, etc [1-3].

Human brain, as the center of the nervous system, is one of the most remarkable aspects of the adult nervous system. The average human brain is estimated to contain approximately 100 billion neurons [4], each connected by synapses to several thousand other neurons. An increasing number of theoretical and empirical studies study the organization of the human brain as a synthesis of the sciences of complex networks. By mapping brain, as a set of nodes and edges, connectome analysis [5] could help us to understand the brain from the perspective of the network. The application of graph theoretical analysis to human neuroimaging data has uncovered topological features of the connectome [6, 7]. Tractography based on diffusion tensor imaging (DTI) [8], has been widely adopted as a tool by researchers to investigate structural connectome, due to its advantage of in-vivo white matter (WM) mapping. Although some research has examined the organization of the brain network during brain aging at the nodal-level [9, 10], relatively few studies have specifically performed for network level.

In the graph theoretical analysis of structural brain network, a large number of univariate statistical tests are undertaken to test the local level hypothesis related to network nodes or edges. The more inferences are made, the more likely erroneous inferences are to occur. There is no universally accepted

IOP Conf. Series: Journal of Physics: Conf. Series **910** (2017) 012042 doi:10.1088/1742-6596/910/1/012042

approach for the problem of multiple comparisons; most approaches require a stricter significance threshold for individual comparison, which may not offer sufficient statistical power. The aims of this study were to explore the DTI to investigate the topological WM changes during the normal aging by using network-based statistic (NBS) approach, to translate conventional cluster statistics to a graph during brain aging.

2. Subjects and Method

2.1 Subjects and Data Acquisitions

The study included seventy-five right-handed healthy elderly subjects (age range, 50-70 years; 39 females, 36 males). Subject inclusion criteria: No history of psychiatric disorders and other diseases which can affect the central nervous system; the Mini-Mental State Examination (MMSE) scores ≥ 26; Hamilton Depression Rating Scale ≤ 10. The subjects were divided into two groups: middle-aged group (MAG) (aged 50 to 60) and young-old group (YOG) (aged 60 to70). Subject characteristics are shown in Table 1.

Table 1 Subject characteristics

Group	Age	Gender(Male/Female)	Education
MAG:(50-60)	56.1±2.7	17/20	16.0±4.0
YOG:(60-70)	65.0±2.7	21/18	18.0±1.6

The MRI data were collected with GE Signa II 3.0T. DTI scans were collected by using a single-shot spin echo diffusion sensitized echo-planar imaging (EPI) sequence. Scanning parameters: eight directions without proliferation-sensitive gradient, b=0 s/mm^2; 51 directions are applied to the proliferation of sensitive gradient, b=1000 s/mm^2; TR=12500ms, TE=71ms, matrix=128 × 128, FOV = 250 × 250 mm^2, slice thickness= 2.6 mm. The T1-weighted data were acquired using 3D Spoiled Gradient-echo(3DSPGR) sequence (TR = 5.3 ms, TE = 2.0 ms,TI = 500; flip angle = 15°; matrix = 256 × 256; FOV = 256 × 256 mm^2) with 204 contiguous1.0 mm thick coronal slices.

2.2 Image Processing

DTI images were processed by using PANDA (a Pipeline for Analysing braiN Diffusion images, http://www.nitrc.org/projects/panda/) [11] (Figure 1). DICOM images were first converted to NIFTI format. In order to eliminate motion artifact and image distortions, all images have been processed by eddy correction and head movement correction. Voxel-wise diffusion tensor matrix was calculated by weighted linear least squares fitting to the log-transformed diffusion data and registering images to a standardized template in the MNI space.

The procedure used for getting connectivity matrixes was similar to described in previous studies [12]. A graph of a network is determined by a set of nodes, and a collection of edges describing the interactions between the nodes. The standard Automated Anatomical Label (AAL) [13] atlas, which is the most commonly used in structural network investigations, was used to divide the brain into 90 cortical and sub-cortical parcellation regions. The inverse of the estimated deformation field is applied to the AAL labels, producing masks for 90 regions in DTI native space. Reconstruction of white matter tracts was conducted by continuous tracking (FACT) [14] algorithm to generate network edges. If the FA value was lower than 0.2 or the angle between the current and the previous path segment was higher than 45 degrees, the tracking would be stopped. Based on the nodes and edges, we got the fiber number between two regions (which is called FN). A symmetric 90×90 FN connectivity matrix was generated. If at least three fibers are located between two regions, the weight of the edge would be kept; otherwise, the weight would be placed at zero. Weighted adjacency matrices were used to express the FN for each pair of nodes in the whole-brain tractography.

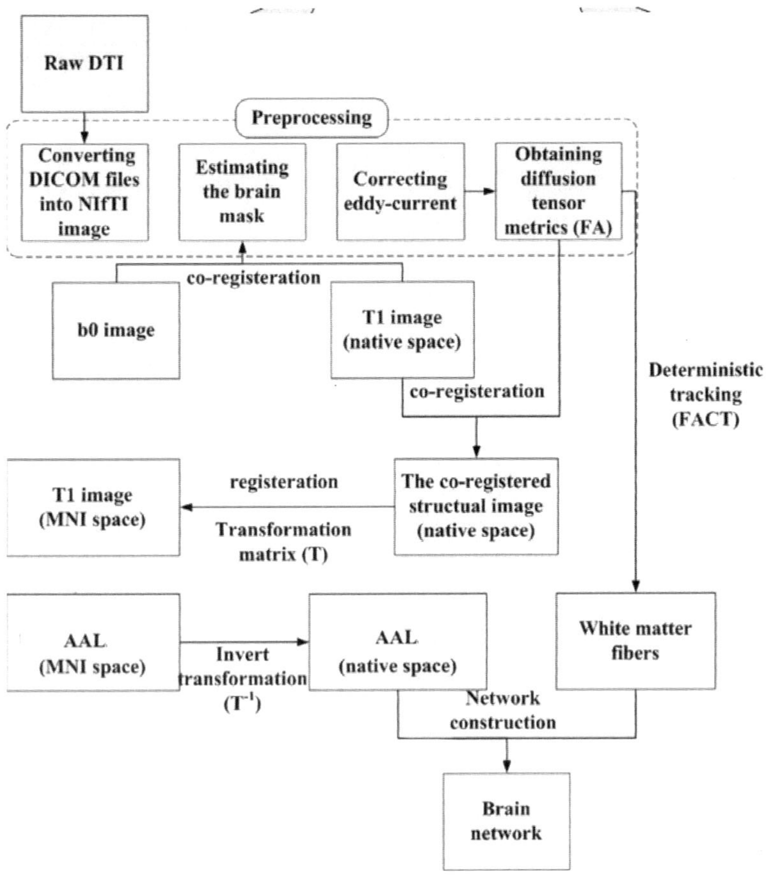

Figure 1 Pipeline of brain network construction

2.3 NBS Analysis

NBS introduced by Zalesky et al. [15, 16], is a nonparametric statistical test to avoid the multiple comparison problems encountered when used to isolate regional brain networks that differ significantly between two distinct groups. It identifies differences between groups, isolates sets of highly interconnected regions, thus giving information on the structural organization of the whole-brain. The framework of NBS (Melbourne Neuropsychiatry Center, the University of Melbourne and Melbourne Health, Australia, http://www.nitrc.org/projects/nbs/) was used in this study. General linear models (GLM) were formulated to quantify the inter-group differences in connectivity measures.

A 90× 90 connectivity matrix stores a totals of 90(89 − 1)/2=4005 unique edges. For individual edge of these associations, a two sample t-test contrasting the MAG and YOG is calculated independently for each pair of regions in the AAL template, to test the null hypothesis of equality based on the differences in structural connectivity values (i.e., FN) between MAG and YOG groups. Pairs of regions with a *t*-statistic exceeding a predefined *t* threshold of 2.2 were identified. The basis of the network-based statistic is to correct for multiple comparisons by testing for evidence against the null at the level of graph components, rather than at the level of individual connections. All subthreshold edges are assessed for mutual connections to form clusters in a topological space. Permutation testing is then applied to compute p-values for every component previously identified. Finally, previous steps are repeated for each of the 5000 random permutations where participants are randomly assigned to one of the two groups with the same size as the origin groups of MAG and YOG, with noting the maximum component size resulting in a null distribution for the largest component size. The final hypothesis test is then performed for the empirically determined components by comparing their sizes with the proportion of permutations yielding a component with equal or greater size. The final result controls a family-wise error (FWE) rate at cluster level with $p<0.05$. Two hypotheses (MAG >YOG and MAG <

YOG) were evaluated independently. This method effectively controls for false-positives due to multiple comparisons

3. Result

NBS is used to isolate the components of a 90 × 90 undirected connectivity matrix that differs significantly ($p<0.05$, corrected) between two groups. A pair of nodes showing a weaker association in the YOG is hereby designated as a disconnection and the set of all such disconnections is referred to as the disconnected subnetwork. The alteration of WM architecture was hallmarked by reduced strengths in nine connections, using the FN model and a statistical threshold of t = 2.2 (Figure 2). The NBS identified a single disconnected sub-network that showed significantly (p = 0.030±0.005, corrected) decreased inter-regional connectivity in the YOG compared with the MAG. A subnetwork of nine different nodes interconnected by 10 edges linking frontal lobe and central structures was identified as showing decreased connectivity. We did not identify any network with significantly increased connectivity in the YOG compared with the MAG.

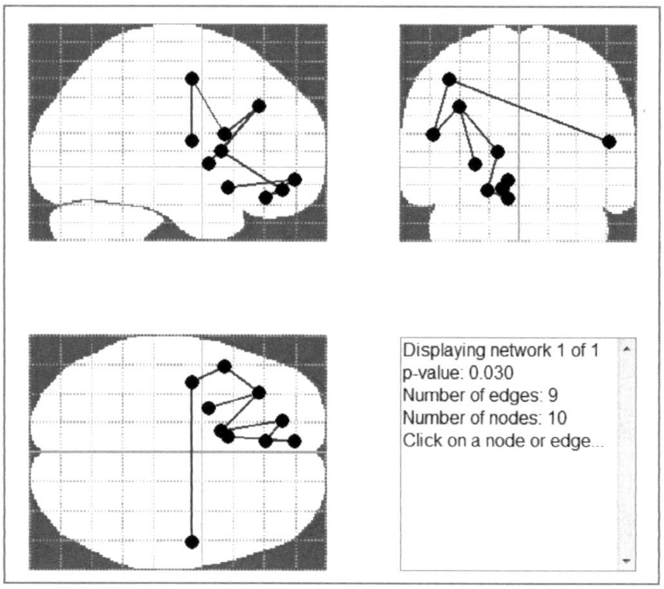

Figure 2 Significantly decreased white matter connectivity in the YOG compared with the MAG

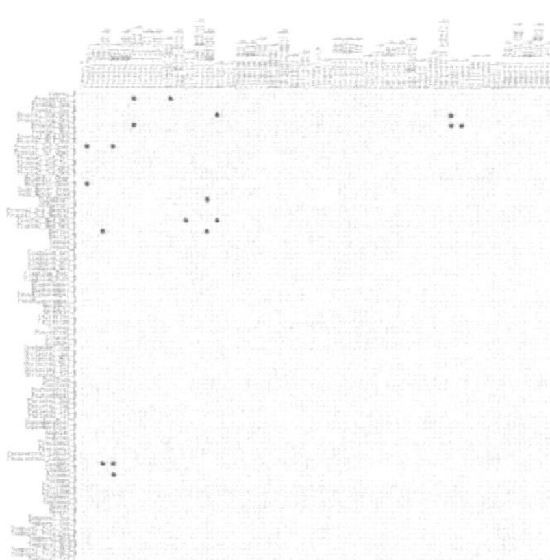

Figure 3 The connection matrix of disconnected subnetwork(From left to right and from top to bottom and regions in AAL atlas)

Table II lists the nodes comprising the disconnected subnetwork as well as their degrees. The connection matrix of disconnected network is given in Figure 3. No individual edge can be declared significant, only the disconnected subnetwork can be declared significant.

Table 2 Node Name and Degree

Node Name	degree
Caudate_L	2
Frontal_Inf_Oper_L	2
Frontal_Mid_Orb_L	2
Frontal_Mid_L	3
Frontal_Sup_Orb_L	2
Olfactory_L	1
Precentral_L	2
Putamen_L	1
Rectus_L	2
Rolandic_Oper_R	1

Table 3 Network-Based Statistics Reveals Group Difference between YOG and MAG

Region 1	**Region 2**	**T value**
Precentral_L	Frontal_Inf_Oper_L	2.49
Frontal_Mid_L	Frontal_Inf_Oper_L	2.27
Precentral_L	Rolandic_Oper_R	2.26
Olfactory_L	Frontal_Mid_Orb_L	2.48
Frontal_ Mid_ Orb _L	Rectus_L	2.44
Frontal_Sup_Orb_L	Rectus_L	2.31
Frontal_Sup_Orb_L	Caudate_L	2.43
Frontal_Mid_L	Caudate_L	2.95
Frontal_Mid_L	Putamen_L	2.26

Pairwise connectivity that was significantly reduced in YOG is summarized in Table III. The disconnected subnetwork mainly affected the left hemisphere. All connections between nodes were impaired in old subjects, that is, for each edge within the cluster, the FN was consistently reduced in YOG

4. Discussion

Age-related changes in brain network vary considerably across individuals and across brain regions, with some brain regions appearing more susceptible than others to the effects of aging. Much of the basic research in brain network has focused on nodal level, and indeed analysis of brain network from network level can help the understanding of the brain organization of both normal subjects and neurodegenerative disease subjects. In this study, we found that the impaired structural connectivity affecting subnetworks during brain aging. The disconnected subnetwork identified in this study involved the frontal lobe to central structures disconnectivity of the left hemisphere. This finding is in line with structural imaging study in brain aging [17], which suggests that the most grey-matter changes occur in frontal, temporal, regions. Further studies of subjects through fMRI are needed to evaluate the relationship among affected subnetworks and brain functional network.

5. Acknowledgment

This work was supported by grants from Natural Science Foundation of Beijing (7143171).

6. References

[1] R . Peters, “Ageing and the brain,” Postgrad Med J, vol. 82, No. 964 , pp. 84–88, Feb 2006.

[2] L. Lin, C. Jin, Z. Fu, B. Zhang, G. Bin and S. Wu, "Predicting healthy older adult's brain age based on structural connectivity networks using artificial neural networks," Comput Methods Programs Biomed, vol. 125, pp. 8–17, Mar 2016.

[3] N. Raz, U. Lindenberger, K . Rodrigue, K . Kennedy, D. Head, A. Williamson, C. Dahle, D. Gerstor and J. Acker, "Regional brain changes in aging healthy adults: general trends, individual differences and modifiers," Cerebral cortex, vol. 15, No. 11, pp. 1676–1689, 2005.

[4] D. Pelvig, H. Pakkenberg, AK. Stark and B. Pakkenberg, "Neocortical glial cell numbers in human brains," Neurobiology of aging, vol. 29, pp. 1754–1762,2008.

[5] O. Sporns, G. Tononi and R. Kötter, "The human connectome: a structural description of the human brain," PLoS computational biology, vol. 1, pp. e42,2005.

[6] A. Fornito, A. Zalesky and M. Breakspear,"Graph analysis of the human connectome: promise, progress, and pitfalls," NeuroImage, vol. 80, pp. 426-444, 2013.

[7] A. Fornito, A. Zalesky and M. Breakspear,"Graph analysis of the human connectome: promise, progress, and pitfalls," NeuroImage, vol. 80, pp. 426-444, 2013.

[8] L. Lin, Z. Fu, B. Zhang, G. Bin and S. Wu, "Recent Advances in DTI Connectomes of Brain Diseases," China Medical Devices, No. 6 , pp. 1–6,2015.

[9] A. Perry, W. Wen, A. Lord, A. Thalamuthu, G. Roberts, P. Mitchell, P. Sachdev and M. Breakspear,"The organisation of the elderly connectome," Neuroimage, vol. 114, pp. 414-426, 2015.

[10] L. Lin, C. Jin, X . Xu and S. Wu ,"Changes in cortical network topology with brain aging," Information Technology and Applications: Proceedings of the 2014 International Conference on Information technology and Applications, Xian, China, 8-9 August 2014. CRC Press, vol. 59, pp. 313, 2005.

[11] Z. Cui, S. Zhong, P. Xu,Y. He and G. Gong, "PANDA: a pipeline toolbox for analyzing brain diffusion images," Frontiers in human neuroscience, July 2013.

[12] C. Jin, Y. Chao, L. Lin, Z. Fu, B. Zhang and S. Wu, "The Study of Graph Measurements for Hub Identification in Multiple Parcellated Brain Networks of Healthy Older Adult," Journal of Medical and Biological Engineering, pp. 1-13, 2017, In Press.

[13] N. Tzourio-Mazoyer, B. Landeau, D. Papathanassiou, F. Crivello, O. Etard, N. Delcroix, B. Mazoyer and M. Joliot, "Automated anatomical labeling of activations in SPM using a macroscopic anatomical parcellation of the MNI MRI single-subject brain," Neuroimage, vol. 15, pp. 273-289, 2002.

[14] S. Mori and P. Barker, "Diffusion magnetic resonance imaging: its principle and applications," The Anatomical Record, vol. 257, pp. 102-109, 1999.

[15] A. Zalesky, A. Fornito and E. Bullmore, "Network-based statistic: identifying differences in brain networks," Neuroimage, vol. 53, No. 4, pp. 1197-1207, 2010,

[16] A. Zalesky, L. Cocchi, A. Fornito, M. Murray and E. Bullmore,"Connectivity differences in brain networks," Neuroimage, vol. 60, No. 2, pp. 1055-1062, 2012,

[17] G. Bartzokis, M. Beckson, P. Lu, K. Nuechterlein, N. Edwards and J. Mintz, "Age-related changes in frontal and temporal lobe volumes in men: a magnetic resonance imaging study," Archives of general psychiatry, vol. 58, No.5, pp. 461-465, 2001,

IOP Conf. Series: Journal of Physics: Conf. Series **910** (2017) 012043 doi:10.1088/1742-6596/910/1/012043

Enhanced Security and Pairing-free Handover Authentication Scheme for Mobile Wireless Networks

Rui Chen[1,2,a], Guangqiang Shu[2,b], Peng Chen[1,c] and Lijun Zhang[3,d]

[1]College of Computer Science, SiChuan University, Chengdu, China
[2]College of Computer Science, SiChuan Normal University, Chengdu, China
[3]Chengdu Ruibei Yingte Information Technology Limited Company, Chengdu, China
Email: [a]crs1934@hotmail.com, [b]254625466@qq.com, [c]7199611@qq.com, [d]630452696@qq.com

Abstract. With the widely deployment of mobile wireless networks, we aim to propose a secure and seamless handover authentication scheme that allows users to roam freely in wireless networks without worrying about security and privacy issues. Given the open characteristic of wireless networks, safety and efficiency should be considered seriously. Several previous protocols are designed based on a bilinear pairing mapping, which is time-consuming and inefficient work, as well as unsuitable for practical situations. To address these issues, we designed a new pairing-free handover authentication scheme for mobile wireless networks. This scheme is an effective improvement of the protocol by Xu et al., which is suffer from the mobile node impersonation attack. Security analysis and simulation experiment indicate that the proposed protocol has many excellent security properties when compared with other recent similar handover schemes, such as mutual authentication and resistance to known network threats, as well as requiring lower computation and communication cost.

1. Introduction

Mobile intelligent terminal technology and wireless communication with mobile wireless network (MWN) are currently being developed and used in several areas along with the rapid development of wireless networks. All types of mobile applications and mobile internet services in intelligent mobile nodes (MNs) (i.e., laptops, smartphones, tablets, and PDA) enhance the variation and convenience of our lives. However, network and information security are increasingly gaining attention with the development of various network attack technologies.

An anonymous mobile node handover authentication (AMNHA) protocol allows MNs to roam seamlessly over multiple access points (APs) in MWNs while providing mutual authentication for two sides of communication and protecting the privacy of MNs. Typical MWNs often comprise three types of entities, including MNs, APs, and an authentication server (AS)[1]. After registering to an AS, the MN can access the network and acquire network service from an associated AP. When an MN moves from one AP to another, it should perform a handover authentication to the new AP to protect them from unauthorized user access or unauthorized access by legitimate users. Meanwhile, a shared session key will be generated after the AMNHA protocol is executed to ensure communication data consistency, integrity, and accuracy.

Several protocols have been recently proposed by employing various methods. However, a secure and efficient roaming AMNHA protocol is difficult to design given the broadcast nature of wireless networks and the limited power and resources of mobile devices.

Based on bilinear pairings and identity-based signature, He *et al.*[2] presented a new AMNHA protocol, which uses a pseudo-ID to provide mutual authentication and privacy protection between communication parties. However, this protocol is insecure and prone to key compromised attack. Subsequently, a number of improved approaches have been proposed [3, 6, 7, 8]. He *et al.*[4] and He *et al.*[5] outlined some characteristics that a secure and efficient AMNHA protocol should satisfy and proposed different solutions based on bilinear pairings and elliptic curve cryptosystem (ECC).

However, these protocols [2-8] are unsuitable for use in a mobile environment, where a number of wireless mobile devices with limited power and processing capability are deployed, due to complexities and time-consuming operation in model cryptography that results from adopting a bilinear pairing operation [16]. To address the above mentioned problems, a number of efficient pairing-free methods and its corresponding protocols have been proposed [9-15]. In these protocols, the bilinear pairing operation is completely removed.

Cao *et al.*[9] presented a new identity-based AMNHA protocol without bilinear pairing. Their scheme can support strong security while minimizing the message exchange time. However, the scheme in [9] is not secure and vulnerable to two types of attacks as pointed out by [10]. Thereafter, Cao *et al.*[11] presented a new uniform handover authentication scheme based on their previous method[9]. Their improved approach provides strong security and robust efficiency while supporting user mutual authentication.

However, Li *et al.*[12] noted that the improved scheme of Cao *et al.* cannot provide user anonymity and user untraceability given that the real identity of the MN is transmitted in clear text form over the wireless networks. They then presented a new privacy-aware AMNHA protocol for MWNs, which is proven secure under an extended Canetti–Krawczyk model. By adopting the identity-based cryptography and pseudo-ID for the MN, their protocol can achieve user anonymity and good performance.

Unfortunately, Chaudhry *et al.*[13] and Xie *et al.*[14] indicated that Li *et al.*'s protocol still has a number of drawbacks, and an adversary may impersonate a legal AP to communicate with an innocent MN. Thus, the protocol is prone to potential fraud attacks and cannot provide mutual authentication between the MN and AP. They then proposed different solutions.

Xu *et al.*[15] recently presented a novel privacy preserving AMNHA scheme without pairing operation for MWNs. They claimed that their scheme has excellent security and efficiency while supporting user anonymity and untraceability. However, we found that this scheme remains vulnerable to MN impersonation attack, which will result in the failure of mutual authentication. Furthermore, we propose a new pairing-free AMNHA protocol that overcomes the weaknesses of Xu *et al.*'s protocol.

2. Review of Xu Yang et al.'s Scheme

This section gives a brief review of Xu *et al.*'s scheme [15]. Their protocol is divided into following two phases:

2.1 Pre-deployment Phase

The pre-deployment phase includes the following two processes:

1) System initialization

In this process, the AS needs to complete a few steps:

a) Let q be a k-bit prime. Selects a tuple $\{F_q, E/F_q, G, P\}$, where F_q is a finite field, E/F_q is an elliptic curve E over F_q, G is a cyclic additive group of order q, and P is a base point of G over E/F_q.

b) Randomly selects two integers $s,r \in Z_q^*$ as the system master keys. Then, calculates $PK_1 = s \cdot P$, $PK_2 = r \cdot P$ is the corresponding public keys.

c) Selects four cryptographic hash functions H_1, H_2, H_3, and H_4, where $H_1 : \{0, 1\}^* \times G \rightarrow Z_q^*$, H_2: $\{0, 1\}^* \times \{0, 1\}^* \times G \rightarrow \{0, 1\}^k$, $H_3 : \{0, 1\}^k \times G \rightarrow \{0, 1\}^k$, $H_4 : \{0, 1\}^* \times \{0, 1\}^* \times \{0, 1\}^* \rightarrow \{0, 1\}^k$

d) Publishes the public parameters (i.e., F_q, E/F_q, G, P, PK_1, PK_2, H_1, H_2, H_3, H_4) and keeps s,r secret.

2) Key pre-distribution

a) For an MN, the AS selects a set of independent pseudo-IDs ($PID = pid_1, pid_2, \ldots$) to replace the MN's real identity(*i.e.* ID_{MN}). Then, the AS generates the private key ($S_{pidi} = r + H_1(pid_i \| PK_2)s$) for each pseudo-ID ($pid_i \in PID$) and transfers all tuples (pid_i, S_{pidi}) to the MN via a security tunnel. The MN can calculate the corresponding public key ($PK_{pidi} = S_{pidi} \cdot P = PK_2 + H_1(pid_i \| PK_2)PK_1$).

b) For an AP with real identity ID_{AP}, the AS computes its private key ($S_{AP} = r + H_1(ID_{AP} \| PK_2)s$) and the AP can then calculate the corresponding public key ($PK_{AP} = S_{AP} \cdot P = PK_2 + H_1(ID_{AP} \| PK_2)PK_1$). At last the AS securely transfers the private key S_{AP} to the AP.

2.2 Handover Authentication Phase

The MN should perform the handover authentication process to access the network service when it leaves the current AP and moves into a new AP. The details are as follows.

1)MN → AP: {Mi, RMN, H3(PTK1 || RAP)}

The MN picks a random integer $a \in Z_q^*$, generates a timestamp ts, and computes $R_{MN} = a \cdot P$. Then, the MN selects an unused pseudo-ID pid_i and computes $b = H_4(M_i)$ and $R_{AP} = b \cdot P$, where $M_i = \{pid_i \| ID_{AP} \| ts\}$. Next, the MN calculates the AP's public key ($PK_{AP} = PK_2 + H_1(ID_{AP} \| PK_2)PK_1$), shared secret ($K_1 = S_{pidi} \cdot R_{AP} + a \cdot PK_{AP}$), and session key ($PTK_1 = H_2(pid_i \| ID_{AP} \| K_1)$).

The MN computes the authentication messages $H_3(PTK_1 \| R_{AP})$ and sends $\{M_i, R_{MN}, H_3(PTK_1 \| R_{AP})\}$ to the AP.

2)AP → MN: $H_3(PTK_1 \| R_{MN})$

Upon receiving $\{M_i, R_{MN}, H_3(PTK_1 \| R_{AP})\}$, the AP initially validates the timestamp ts and terminates the session if the ts is invalid. Otherwise, the AP computes $b = H_4(M_i)$ and $R_{AP} = b \cdot P$. The AP then calculates MN's public key ($PK_{MN} = PK_2 + H_1(pid_i \| PK_2)PK_1$), shared secret ($K_2 = S_{AP} \cdot R_{MN} + b \cdot PK_{MN}$), and session key ($PTK_2 = H_2(pid_i \| ID_{AP} \| K_2)$. The AP then verifies the following equation: $H_3(PTK_2 \| R_{AP}) = H_3(PTK_1 \| R_{AP})$. If the result is negative, then this message is not sent. Otherwise, the AP computes the authentication message $H_3(PTK_2 \| R_{MN})$ and sends it to the MN.

3)MN: Verification

After receiving $\{H_3(PTK_2 \| R_{MN})\}$ from the AP, the MN generates $H_3(PTK_1 \| R_{MN})$ and compares it with the received $H_3(PTK_2 \| R_{MN})$. The authentication fails if they are unequal. Otherwise, the MN believes that the AP is a legal access point and a secure communication channel is erected between them.

3. Impersonation Attack on Xu et al.'s Protocol

In this section, we prove that the Xu *et al.*'s scheme cannot support the mutual authentication of communication parties and is suffer from the mobile node impersonation attack. A malicious attacker can gain access to the network service from an AP by pretending as a legal user. Furthermore, a legal MN can use a randomly selected identity to deceive an AP, which results in completely untraceability and will become unable to meet the security requirement of conditional privacy preservation [4].

The details are as follows.

Step1: Assume A be the adversary. When A roams into the coverage range of an AP, A randomly selects an identity ID_A, generates a timestamp ts, and computes the following:

$$M_i = \{ID_A \| ID_{AP} \| ts\},\ b = H_4(M_i)$$
$$R_A = R_{AP} = b \cdot P$$
$$PK_{AP} = PK_2 + H_1(ID_{AP} \| PK_2)PK_1$$
$$PK_A = PK_2 + H_1(ID_A \| PK_2)PK_1$$
$$K_1 = b(PK_A + PK_{AP})$$
$$PTK_1 = H_2(ID_A \| ID_{AP} \| K_1)$$

Step2: A sends $\{M_i, R_A, H_3(PTK_1 \| R_{AP})\}$ to AP.

Step3: AP verifies the message as follows:

$$b = H_4(M_i)$$

$$R_{AP} = b \cdot P$$
$$PK_A = PK_2 + H_1(ID_A \parallel PK_2)PK_1$$
$$K_2 = S_{AP} \cdot R_A + b \cdot PK_A$$
$$PTK_2 = H_2(ID_A \parallel ID_{AP} \parallel K_2)$$

Step4: AP checks whether $H_3(PTK_2 \parallel R_{AP})$ is equal to $H_3(PTK_1 \parallel R_{AP})$. If equal, A is authenticated by the AP. Otherwise, the message is dropped.

The login request message $\{M_i, R_A, H_3(PTK_1 \parallel R_{AP})\}$ of A can successfully pass the identity verification because:

$$\begin{aligned} K_2 &= S_{AP} \cdot R_A + b \cdot PK_A \\ &= S_{AP} \cdot b \cdot P + b \cdot PK_A \\ &= b(PK_{AP} + PK_A) \\ &= K_1 \end{aligned}$$

The AP believes A is a legal MN and provides it network service.

However, A is an adversary.

Therefore, the mutual authentication and resistance of impersonation attack cannot be achieved in Xu *et al.*'s protocol.

4. Proposed handover authentication protocol

In this section, we design a new improvement to AMNHA protocol over Xu *et al.*'s approach.

The proposed protocol also contains two phases:

4.1. Pre-deployment phase

The pre-deployment phase has following two processes:

1)System initialization

The AS initially performs these four steps:

a) Selects a k-bit prime q and chooses the tuple $\{F_q, E/F_q, G, P\}$, where F_q is a finite field, E/F_q is an elliptic curve E over F_q, G is a cyclic additive group with order q, and P is a base point of G over E/F_q.

b) Selects two random values $s, r \in Z_q^*$ and uses them as system private keys. Then, calculates the corresponding public keys as $PK_1 = s \cdot P$, $PK_2 = r \cdot P$.

c) Selects four cryptographic hash functions, H_1: $\{0, 1\}^* \times G \rightarrow Z_q^*$, H_2: $\{0, 1\}^* \times \{0, 1\}^* \times G \rightarrow \{0, 1\}^k$, H_3 : $\{0, 1\}^k \times G \rightarrow \{0, 1\}^k$, H_4 : $\{0, 1\}^* \times \{0, 1\}^* \times G \times \{0, 1\}^* \rightarrow \{0, 1\}^k$

d) Publishes the tuple (F_q, E/F_q, G, P, PK_1 PK_2, H_1, H_2, H_3, H_4) and keeps the system private keys secret.

2)Key pre-distribution

a) For MN, the AS selects a set of independent pseudo-IDs (PID = pid_1, pid_2, ...) to replace the MN's real identity(i.e. ID_{MN}). Then, the AS calculates the private key ($S_{pidi} = r + H_1(pid_i \parallel PK_2)s$) for each pseudo-ID ($pid_i \in PID$) and transfers all tuples (pid_i, S_{pidi}) to the MN via a security tunnel. The MN can generate the corresponding public key ($PK_{pidi} = S_{pidi} \cdot P = PK_2 + H_1(pid_i \parallel PK_2)PK_1$). To achieve user anonymity, each key pair and the corresponding pseudo-ID are only used once in the handover authentication phase.

b)For AP with real identity ID_{AP}, the AS computes its private key as $S_{AP} = r + H_1(ID_{AP} \parallel PK_2)s$; then the AP's public key PK_{AP} is $PK_{AP} = S_{AP} \cdot P = PK_2 + H_1(ID_{AP} \parallel PK_2)PK_1$. Finally, the AS sends the secret key S_{AP} to the AP through a secure channel.

4.2. Handover Authentication Phase

This phase is performed when the MN leaves the current associated AP coverage and roams into another AP. Suppose each AP broadcasts a beacon message periodically, which contains its identity and other relevant network information. The MN initially selects a target AP and extracts the identity from the beacon message, and a handover authentication phase is then executed with AP. The steps of the handover authentication phase are described in Fig. 1.

The details are as follows.

1)MN → AP: {M_i, $H_3(PTK_1 || R_{AP})$}

The MN generates a random integer $a \in Z_q^*$, a timestamp ts, and computes $R_{MN} = a \cdot P$. Then, the MN selects an unused pseudo-ID pid_i and computes $M_i = \{pid_i \| ID_{AP} \| R_{MN} \| ts\}$, $b = H_4(M_i)$, and $R_{AP} = b \cdot P$. Subsequently, the MN calculates the AP's public key ($PK_{AP} = PK_2 + H_1(ID_{AP} \| PK_2)PK_1$), shared secret ($K_1 = (S_{pidi} + a)(R_{AP} + PK_{AP})$), and session key ($PTK_1 = H_2(pid_i \| ID_{AP} \| K_1)$).

Finally, the MN computes the authentication messages ($H_3(PTK_1 \| R_{AP})$) and sends the login request message $\{M_i, H_3(PTK_1 \| R_{AP})\}$ to AP.

2)AP → MN: $H_3(PTK_2 || R_{MN})$

Upon receiving $\{M_i, H_3(PTK_1 \| R_{AP})\}$, the AP initially checks whether ts is fresh and computes $b = H_4(M_i)$ and $R_{AP} = b \cdot P$ or otherwise terminates the session. Next, the AP generates the MN's public key ($PK_{MN} = PK_2 + H_1(pid_i \| PK_2)PK_1$), shared secret ($K_2 = (S_{AP} + b)(PK_{MN} + R_{MN})$), and session key ($PTK_2 = H_2(pid_i \| ID_{AP} \| K_2)$).

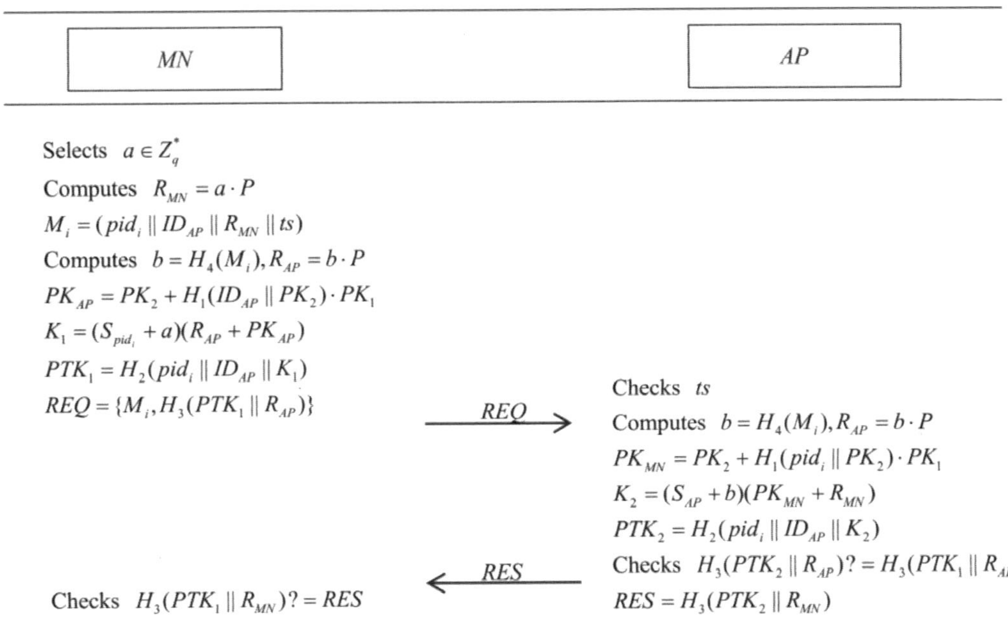

Figure 1 Handover authentication phase

The AP then checks whether $H_3(PTK_2 \| R_{AP})$ is the same as the received $H_3(PTK_1 \| R_{AP})$. If the two values are not the same, then the authentication will fail. Otherwise, the AP computes the authentication message $H_3(PTK_2 \| R_{MN})$ and sends it to the MN as a response message.

3)MN: Verification

Upon receiving the message $H_3(PTK_2 \| R_{MN})$, the MN initially computes $H_3(PTK_1 \| R_{MN})$ and compares it with $H_3(PTK_2 \| R_{MN})$. If the result is negative, then the session is terminated. Otherwise, the MN believes that AP is a legal service provider.

Once the mutual authentication is finished, a secure communication channel is established between two communicating parties using the shared session key PTK_1 or PTK_2.

5. Security Analysis

5.1. Mutual Authentication and key Agreement

In our protocol, the two communicating parties (i.e., MN and AP) can authenticate each other because they share the same secret:

$$\begin{aligned} K_1 &= (S_{pidi} + a)(PK_{AP} + R_{AP}) \\ &= (S_{pidi} + a)(R_{AP} + H_1(pid_i \parallel PK_2)PK_1 + R_{AP}) \\ &= (S_{pidi} + a) \cdot P \cdot (S_{AP} + b) \\ &= (PK_{MN} + R_{MN})(S_{AP} + b) \\ &= K_2 \end{aligned}$$

Hence, MN and AP can calculate the same session key

$$\begin{aligned} PTK_1 &= H_2(pid_i \parallel ID_{AP} \parallel K_1) \\ &= H_2(pid_i \parallel ID_{AP} \parallel K_2) \\ &= PTK_2 \end{aligned}$$

Based on the Computational Diffie-Hellman(CDH) problem[17] and Decisional Diffie-Hellman (DDH) problem[18], no one can forge the private key of MN and AP and generate the shared K_1 or K_2 secret, as well as the session key. Only the legitimate MN and AP can pass the identity authentication through the valid $H_3(PTK_1 \parallel R_{AP})$ and $H_3(PTK_2 \parallel R_{MN})$ and establish the shared session key for both parties.

Consequently, the proposed protocol supports mutual authentication for both parties of communication, as well as the key establishment between MN and AP.

5.2. User Anonymity and Untraceability

To protect privacy, the AS generates a family of unlinked pseudo-IDs for each MN, which is used to replace the real identity during the handover authentication stage. Hence, only the AS knows the correspondence between the real identity of the MN and the pseudo-IDs. Furthermore, given that the login request message $\{M_i, H_3(PTK_1 \parallel R_{AP})\}$ will constantly change because the number a is randomly selected by the MN during each session run, an attacker can obtain the pseudo-ID through the M_i. However, the attacker cannot extract the real identity of the MN from the pseudo-ID and trace the behavior of the MN given that no linkage between them is present.

Hence, our scheme can provide user anonymity and untraceability.

5.3. Perfect Forward Secrecy

In the proposed protocol, a temporal Diffie–Hellman parameter (i.e., a), which is selected randomly by the MN during each session run, is used to generate the shared secret and session keys and further establish the security communication channel between the MNs and APs. Hence, no relation exists between the session key and other session keys or the long-term system private key. Furthermore, even though the latter are compromised by an attacker, the preceding session keys cannot be obtained and the transmitted information between the MNs and APs cannot be extracted.

Therefore, the new protocol can provide perfect forward secrecy.

5.4. Resist Impersonation Attacks

The private keys of the MN and the AP are computed by the AS based on the pseudo-IDs of the MN and the identity of the AP, respectively, and then delivered through a secure channel. Hence, without the system private key, an attacker cannot generate a valid public/private key pairs, and launch impersonation attacks by forging the legitimate authentication message.

In addition, any changes in M_i will result in a series of values change, such as b, R_{AP}, K_2, PTK_2 and $H_3(PTK_2 \parallel R_{AP})$. Thus, the AP can easily detect the impersonation attacks by checking $H_3(PTK_1 \parallel R_{AP})$.

Without the private key of the AP, an attacker cannot generate a valid K_2 and PTK_2; thus, the MN can easily find this type of attack by comparing the received $H_3(PTK_2 \parallel R_{MN})$.

Thus, our protocol can effectively prevent the impersonation attack.

5.5. Resist Replay Attacks

Our protocol can efficiently withstand message replaying attacks given that each login request message includes a time stamp ts. Thus, the AP can easily detect this type of attack by checking whether the ts is fresh.

Furthermore, an adversary, including malicious APs, cannot replay this message to other APs even the time *ts* is fresh due to the login request message that contains the identities of two communication parties.

6. Performance evaluation

In this section, we briefly compare the security properties and performance of our protocol with three recent protocols.

6.1 Comparison of security properties

Table 1 provides a summary and comparison of the security properties of our improved AMNHA protocol and other relevant works [2,14,15].

The data presented in Table 1 indicate that He *et al.*'s scheme[2] cannot prevent key compromise attacks, whereas Xu *et al.*'s scheme[15] cannot resist impersonation attacks and provide mutual authentication. The proposed AMNHA protocol and Xie *et al.*'s protocol satisfies all the security requirements, thereby making them more secure and efficient than other two schemes and suitable for wireless networks.

6.2 Comparison of computation costs

To illustrate the performance comparisons, some of the symbols are defined below:

• T_{BP}: Time cost for a bilinear pairing operation

Table 1 Comparison of security properties

	He[2]	**Xu[15]**	**Xie[14]**	**Ours**
Mutual authentication	YES	NO	YES	YES
User anonymity and untraceability	YES	YES	YES	YES
Perfect forward secrecy	YES	YES	YES	YES
Resistimpersonation attack	YES	YES	YES	YES
Resist replay attack	YES	NO	YES	YES
Resist key compromised attack	NO	YES	YES	YES

Table 2 Running time of related operations (ms)

	T_{BP}	T_{PM}	T_{PA}	T_{MTP}	T_{EPM}	T_{EPA}
AP	1.212	0.001	0.005	0.135	0.458	0.004
MN	3.729	0.002	0.015	0.375	1.451	0.019

Table 3 Comparisons of computation cost (ms)

	AP	**MN**	**Total Time**
He[2]	$3T_{BP} + T_{PM} = 3.637$	$T_{BP} + T_{PM} = 3.731$	7.368
Xu[15]	$4T_{EPM} + 2T_{EPA}= 1.84$	$5T_{EPM} + 2T_{EPA} = 7.293$	9.133
Xie[14]	$6T_{EPM} + 3T_{EPA}= 2.76$	$5T_{EPM} + 2T_{EPA} = 7.293$	10.053
Ours	$3T_{EPM} + 2T_{EPA} = 1.382$	$4T_{EPM} + 2T_{EPA}= 5.842$	7.224

• T_{PM}: Time cost for a scalar point multiplication based on pairing
• T_{PA}: Time cost for a point addition based on pairing
• T_{MTP}: Time cost for a map-to-point hash operation
• T_{EPM}: Time cost for an elliptic curve scalar point multiplication
• T_{EPA}: Time cost for an elliptic curve point addition

Table 2 shows the execution time of the bilinear pairing and elliptic curve operations. We assume that an AP runs on an Intel Core i5 2.4 GHz processor and an MN runs on 1 GHz processor, and the operating system is Ubuntu 16.04 with 2 GB of RAM. The running time of the bilinear pairing operations and elliptic curve operations is based on the PBC[19] library and MIRACL[20] library. We use the MNT family of curves[21] with an order of 160 bits and the embedding degree $k = 6$ for ECC operation.

A comparison of the computation costs between the new protocol and other three relevant protocols are listed in Table 3. Table 3 shows that the total computation cost of our protocol is 7.224 ms, and, which is a slightly less than the total time of He *et al.*'s protocol[2].

In comparison with the schemes proposed by Xu *et al.* and Xie *et al.*'s, the total execution time of MN and AP in our protocol has decreased by 20.9% and 28.14%, respectively.

6.3 Comparison of communication cost

In the simulation, we assume that the security level of ECC and the bilinear pairing is 80 bits. Thus, the size of p is 64 bytes, and the element in G_1 (additive group in bilinear pairing) is 128 bytes; the size of q is 20 bytes, so the element in G (ECC) is 40 bytes. We suppose that the length of the output of all hash functions and all identities are 20 bytes, and the length of the timestamp is 4 bytes. The transmitted messages between MN and AP in our protocol are $\{M_i, H_3(PTK_1 \| R_{AP})\}$ and $\{H_3(PTK_1 \| R_{MN})\}$, where $M_i = (pid_i \| ID_{AP} \| R_{MN} \| ts)$; thus, the total size is 124 bytes. We can similarly compute the communication cost of the scheme of He *et al.*, Xu *et al.*, and Xie *et al.* as 232 bytes, 124 bytes, and 268 bytes, respectively.

Our proposed protocol and Xu *et al.*'s have the same value, which are reduced by 8% and 53.7% when compared with the protocol of He *et al.* and Xie *et al.*, respectively.

7. Conclusion

In this paper, we reviewed the recent studies on MWNs, and found that these AMNHA protocols can be divided into two types; that is, one that is based on bilinear pairing operations and ECC operations, and the other is only based on ECC operations. We then analyzed the pairing-free protocol of Li *et al.* and found that their scheme suffers from impersonation attacks. Through a randomly selected identity, an adversary (or legitimate MNs) can access the network service by cheating the AP. To address the above problem, we present an improved AMNHA protocol for MWNs. The proposed protocol not only satisfies the entire security properties and is robust against network threats but also has a lower computation and communication cost when compared to other recent protocols.

8. Acknowledgment

This work was supported in part by the National Natural Science Foundation of China under grants U1435213 and 61172180, the Specialized Research Fund for the Doctoral Program of Higher Education under Grant 20120181110053, and the Youth Science and Technology Foundation of Sichuan Province under Grant 2013JQ0002.

9. References

[1] D. He, C. Chen, S. Chan and J. Bu. "Strong roaming authentication technique for wireless and mobile networks," International Journal of Communication Systems, vol. 26, no. 8, 2013, pp. 1028-1037.

[2] D. He , C. Chen, S. Chan and J. Bu, "Secure and Efficient Handover Authentication Based on Bilinear Pairing Functions," IEEE Transactions on Wireless Communications, vol. 11, no. 1, Jan. 2012, pp. 48-53.

[3] D. He , C. Chen, S. Chan and J. Bu, "Analysis and Improvement of a Secure and Efficient Handover Authentication for Wireless Networks," IEEE Communications Letters, vol. 16, no. 8, 2012, pp. 1270-1273.

[4] D. He, S. Chan and M. Guizani. "Handover authentication for mobile networks: security and efficiency aspects," IEEE Network, vol. 29, no. 3, 2015, pp. 96-103.

[5] D. He, D. Wang, Q. Xie and K. Chen. "Anonymous handover authentication protocol for mobile wireless networks with conditional privacy preservation," Science China Information Sciences, vol. 60, no. 5, 2017, pp. 052104.

[6] D. He, M. K. Khan and N. Kumar. "A new handover authentication protocol based on bilinear pairing functions for wireless networks," International Journal of Ad Hoc & Ubiquitous Computing, vol. 18, no.1/2, 2015, pp. 67-74.
[7] W. Wang, L. Hu. "A secure and efficient handover authentication protocol for wireless networks," Sensors, vol. 14, no. 7, 2014, pp. 11379-11394.
[8] J. L. Tsai, N. W. Lo and T. C. Wu. "Secure handover authentication protocol based on bilinear pairings," Wireless Personal Communications, vol. 73, no. 3, 2013, pp. 1037-1047.
[9] X. Cao, W. Kou and X. Du. "A pairing-free identity-based authenticated key agreement protocol with minimal message exchanges". Information Sciences, vol. 180, n. 15, 2010, pp. 2895-2903.
[10] S. H. Islam, G. P. Biswas. "An improved pairing-free identity-based authenticated key agreement protocol based on ECC". Procedia Engineering, vol. 30, no.4, 2012, pp. 499-507.
[11] J. Cao, M. Ma and H. Li. "An Uniform Handover Authentication between E-UTRAN and Non-3GPP Access Networks," IEEE Transactions on Wireless Communications, vol. 11, no. 10, 2012, pp. 3644-3650.
[12] G. Li, Q Jiang and F Wei and C. Ma "A New Privacy-Aware Handover Authentication Scheme for Wireless Networks," Wireless Personal Communications, vol.80, no. 2, 2015, pp. 581-589.
[13] S. A. Chaudhry, M. S. Farash, H. Naqvi, S. H. Islam and T. Shon. "A Robust and Efficient Privacy Aware Handover Authentication Scheme for Wireless Networks". Wireless Personal Communications, vol.93, no. 2, 2017, pp. 1-25.
[14] Y. Xie, L. Wu, N. Kumar and J. Shen. "Analysis and Improvement of a Privacy-Aware Handover Authentication Scheme for Wireless Network". Wireless Personal Communications, vol. 93, no.2, 2017, pp. 523-541.
[15] Y. Xu, Y. Zhang, J. K. Liu and Y. Zeng. "A Trust and Privacy Preserving Handover Authentication Protocol for Wireless Networks," Trustcom/BigDataSE/ISPA, 2016, pp. 138-143, DOI: 10.1109/TrustCom.2016.0056
[16] X. Cao, X. Zeng, W. Kou, L. Hu, "Identity-based anonymous remote authentication for value-added services in mobile networks," IEEE Transactions on Vehicular Technology, vol. 58, no. 7, 2009, pp. 3508-3517.
[17] I. Shparlinski. "Computational Diffie-Hellman Problem," Encyclopedia of Cryptography and Security, 2011, pp. 240-244, DOI 10.1007/978-1-4419-5906-5_882.
[18] C. Ran. "Decisional Diffie-Hellman Assumption," Encyclopedia of Cryptography & Security, 2005, pp. 140-142, DOI 10.1007/0-387-23483-7_99.
[19] Pairing based cryptography benchmarks. (Available from: http://crypto.stanford.edu/pbc/times.html).
[20] S. Michael. "Multiprecision Integer and Rational Arithmetic C/C++ Library(MIRACL)," Published by Shamus Software Ltd., http://www.shamus.ie/.
[21] A. Miyaji, M. Nakabayashi and S. Takano. "New explicit conditions of elliptic curve traces for FR-reduction," IEICE Transactions on Fundamentals, vol. 84, no. 5, 2001, pp. 1234-1243.

IOP Conf. Series: Journal of Physics: Conf. Series **910** (2017) 012044 doi:10.1088/1742-6596/910/1/012044

The Design and Realization of Multimedia Textbook The Equipment of Ship's Command and Control(C^2) System

Hao Zhang[1,a] **and Hui Yang**[2,b]

[1]Information Warfare Department, Dalian Naval Academy, Dalian, China
[2]International Programs Department, Naval Command College, Nanjing, China
Email: [a]zhanghao_zhh_81@126.com, [b]yanghui011001@126.com

Abstract. Based on the powerful multimedia interactive integration platform of Director, the paper designs to make a multimedia textbook The Equipment of Ship's Command and Control(C2) System with reference to the characteristics of the equipment, while making use of various image and sound editing software like Photoshop and Video Studio who are rich in functions. It attempts to solve problems in the teaching of the course, such as unitary teaching mode, outmoded course content, unsatisfactory teaching result, the lack of fitted textbook, etc.

1. Introduction

Currently, the teaching of the equipment of ship's C2 system takes the traditional form of instructing and lecturing, which is outdated in teaching theory and unsatisfactory in teaching result. The low efficiency of talent training can barely meet the requirements for the cultivation of highly qualified talents in command in the new era. In order to solve these problems, reduce the span of talent training and enhance the efficiency of teaching, it attempts to design and realize a multimedia textbook The Equipment of Ship's C2 System by using multimedia software like Director, Photoshop, Cool Edit Pro, Video Studio, etc.

2. Basic Principles Of Design

In consideration of the characteristics, problems and requirements, the design of the multimedia textbook abides by the following principles:

2.1 Pedagogic Principle

As a new form of teaching, the development of the multimedia textbook aims to enhance the overall comprehension of the equipment of ship's C^2 system by officers and soldiers, while at the same time, reducing the time of learning. Therefore, the content and structure of the multimedia textbook should conform to teaching rules and cognition rules, making it easily acceptable and convenient for self-study.

2.2 Scientific Principle

As a carrier of knowledge, the content of the multimedia textbook should be accurate, reliable and scientific. For example, the definitions for basic concepts and statement for combat theories should be absolutely correct. The explanation of important ideas should be in an easily understandable way, like through visualized images, simplified sentences, etc.

2.3 Systematic Principle

A set of complete and systematic structure is required. With relevant theories on the equipment of ship's C^2 system as the core, the multimedia textbook should be constructed in a hierarchical way to meet the demand of officers and soldiers' different levels of learning. The logic in the structure is emphasized.

2.4 The Principle of Advancement

Making full use of multimedia technology, the textbook should follow the trend of equipment development and consist with the demand of information warfare, reflecting the application of scientific development in military.

2.5 The Principle of Combination of Theory and Practice

Theories should be always combined with practices in the statement of concepts, the explanation of operating methods, and the selection of exercises. Specifically, relevant data and typical cases in actual maneuvers can be brought into the textbook, as multimedia technology makes it easy and informative. In this way, the use of C^2 system can be mastered by officers and soldiers. Meanwhile, their analysis and problem-solving ability can be better trained.

3. Overall Design Ideas

3.1 Structure Design

The design of structure, as the core and basis of multimedia textbook design, is to set the framework for the whole program. The structure design of *The Equipment of Ship's C^2 System* covers three parts: the content, the supplement, and the extension.

3.1.1 The Content. The content is fundamental to structure design. This textbook expounds the equipment of type 056 ships' C^2 system in five chapters, namely, general introduction to the system, basic components, working mechanism, position setting and personnel responsibility, operating principles. The details of design are shown in figure1.

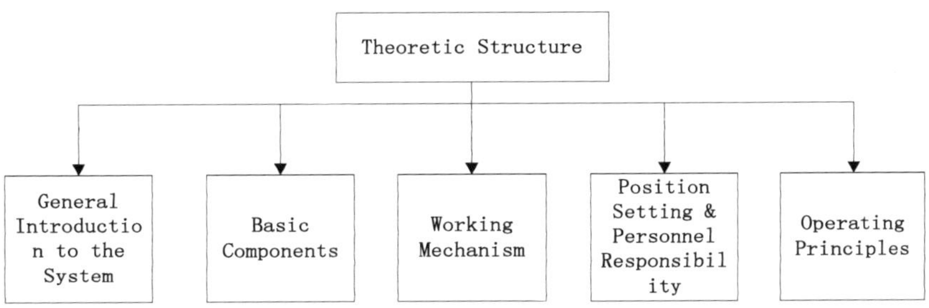

Figure 1 The Design of Content

3.1.2 The Supplement. The supplement is important to structure design. It assists the running of the content and *shows* other useful information to consummate the structure design. The supplement of *The Equipment of Ship's C^2 System* consists of seven parts. The details are shown in figure 2.

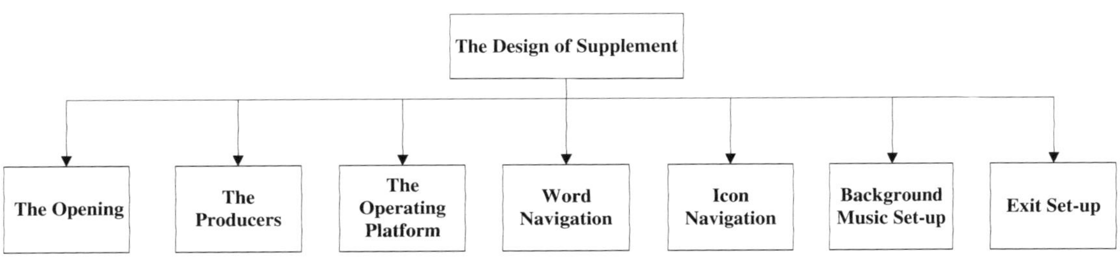

Figure 2 The Design of Supplement

3.1.3 The Extension. Question bank is the extension part of this textbook. The question bank focuses on the basic knowledge of the equipment of ship's C^2 system, and sets questions on relevant concepts and theories, equipment composition, technique and tactic norms, working mechanism, application scope, etc. It takes the form of true or false questions, choice questions, and essay questions. The users can check the answer after finishing each question. It is possible for them to mark the difficult question and come back to it until they finish the whole test. The users can concentrate on the difficult questions, taking notes and practicing again and again. In this way, the study of the officers and men can be better measured and planned.

3.2 Interface Design

Interface design is to determine the components and style of the interface. The interface design of this textbook includes: the design of dominative interface, the design of subsidiary interface and the design of the question bank.

3.2.1 The Design of Dominative Interface. As the catalogue of the main page, the dominative interface contains all information about the multimedia textbook, which makes it the first level of design in interface design. The dominative interface of *The Equipment of Ship's C^2 System* is made up of the background, 7 interactive buttons, and navigational icons of 5 chapters, which is shown in figure 3.

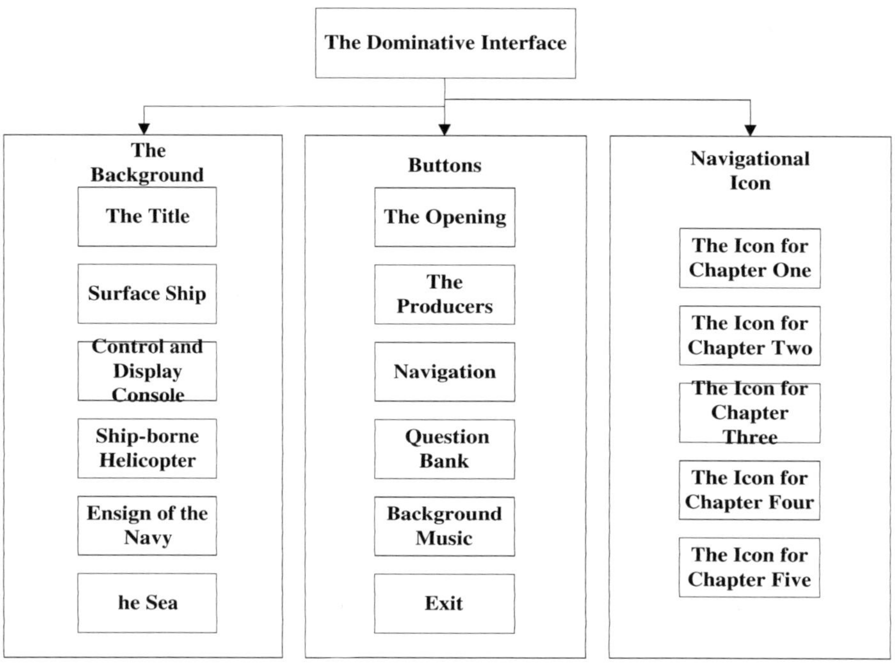

Figure 3 The Structure for Dominative Interface

3.2.2 The Design of Subsidiary Interface. The subsidiary interface contains information about the chapters, whose design is similar to the dominative interface, containing the background and 8 controlling buttons. The detailed composition is shown in figure 4.

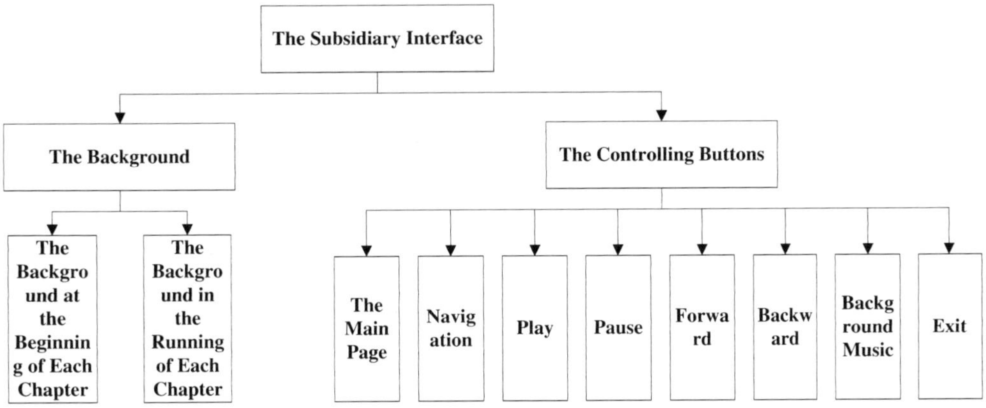

Figure 4 The Structure of the Subsidiary Interface

3.2.3 The Design of Question Bank Interface. The question bank interface includes the first page interface and the questions interface, whose design is for user friendly. Its structure is shown in figure 5.

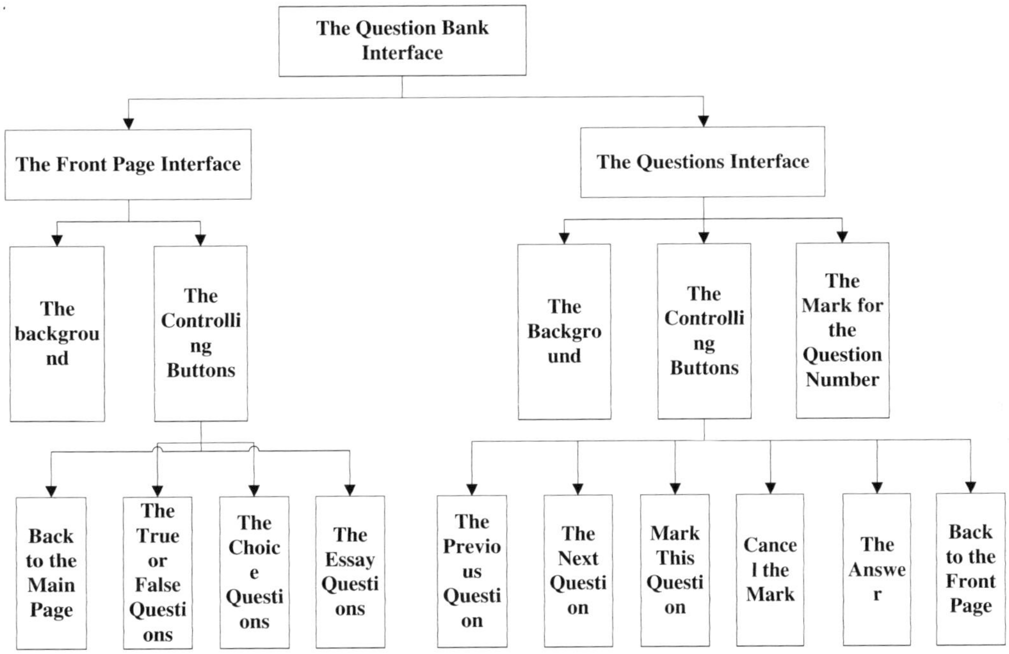

Figure 5 The Structure of the Question Bank Interface

3.3 Function Design

The function design is to clarify the functional components and its applications, which determines the running effects and evaluation level of the multimedia textbook in practice. It is a key part in design. Following the concept of "diversification and enrichment", the function design mainly makes use of Lingo language to fulfill the navigational function design and controlling function design.

Noticeably, the design of the textbook should be more convenient and efficient for the benefits of users' self-study at different levels.

4. The Realization Of The System

4.1. The Overall Procedure

The realization of the multimedia textbook *The Equipment of Ship's C^2 System* depends on the audios, images and texts produced by the main platform Director and other software like Video Studio, Photoshop, Cool Edit Pro, etc.Among them, the initialization is the premise; the core module is the main part; and the publish is a key step.

4.2. The Initialization

As the preparation work, the initialization of the multimedia textbook includes interface settings and actor import.

4.2.1 Interface Settings. In Director, the interface settings are to set the size, color, and background environment of videos and images. The detailed steps for setting are:

- Entering into the operating interface of Director.
- Set the layout of the operating interface.

Among them, the stage toolbar, attributes inspector, procedures working group, design working group can be concealed so that the size of three windows can be larger and more convenient to operate.

- Click the "Attributes Inspector" in icon toolbar and a pop-up dialog will show on the right side of the operating interface.
- Click the column of "Film" in the pop-up dialog and set the size of the stage 1024×768, number of channels 150, the color of background environment #000000(black), format of color RGB, the palette System-Win.
- Click "Save All" in icon toolbar to save the settings of the interface.

4.2.2 Actor Import. Actors are basic elements for film-making. In Director, all images, videos and texts are called actors. Various types of actors are imported in the following ways:

a) Images and Videos

- Choose a blank cell in "Actors List Window". Click the right key and the shortcut menu pops up.
- Click "Import" in the shortcut menu and it comes the pop-up dialog "Import Files into 'internal'".
- Choose the targeted images and videos, i.e. "main interface. png", and click "Import" to complete the action.

b) Texts

- Choose a blank cell in "Actors List Window". Click the "Text Window" in icon toolbar and the text window will pop up.
- Input text in the text window. Set the font and close the window.

4.3. The Realization of the Core Module

As the most important part in the realization of the multimedia textbook, the core module determines the running of textbook in each step. It includes the module of buttons, the module of interface switchover, the module of sound, the module of transitional effect, the module of script.

4.3.1 The Module of Buttons. Buttons directly operate and control the textbook, whose function determines how convenient the textbook can be. In Director, the looks of the buttons are imported images edited by Photoshop, while the functions of the buttons are realized by Lingo language within Director. Take the button of "Play" in subsidiary interface for example, the steps for realizing its function are as follows:

- Click "Play" in the stage window with the right button, and the shortcut menu pops up.
- Click "Scrip…", and the Scrip window pops up.
- Write code in Lingo language in the Script window:

```
on mouseup me
    go to the frame+0
end
```

The function of this sentence is: when clicking it with the left button of the mouse, the film will start to play from the current frame.

- Close the Script window and finish the setting of its function.

Likewise, change the second line of the code "go to the frame+0" to "pause, go to the frame+1, go to the frame－30", the functions of pause, forward and backward can be realized.

4.3.2 The Module of Interface Switchover. The switchover and skip between interfaces are the embodiment of *multimedia* textbook's powerful interaction and flexibility. In Director, the control of interfaces and the switchover between them can be realized by making use of the mark channel in mirror window to mark the first frame of each interface and establishing connection between the controller and the marker through Lingo code. Take "Chapter One General Introduction to the System" for example, the steps for switching to an interface include:

- Click the first frame of "The Interface of Chapter One" in marker channel. The word "New Marker" will pop up.
- Change the word "New Marker" into "Chapter One General Introduction to the System" to finish marking.
- Click the pull-down button on the left side of the marker channel to check all the markers being set.
- Enter into the first chapter's navigation icon in the main interface and the Script window under the navigational words "General Introduction to the System". Write the sentence in Lingo code:

```
on mouseup me
    go to Chapter One General Introduction to the System
end
```

The function of this sentence is: when clicking it, the film will skip to the marked frame in chapter one and start to play from that frame on.

Likewise, "Chapter One General Introduction to the System" can be changed into "Chapter Two Basic Components" "First Working Patterns" etc. Meanwhile, the marker should be set at the beginning of the chapter. In this way, the switchover between interfaces of other chapters can be realized. Therefore, the function of switchover between interfaces and the control of navigation of the multimedia textbook are achieved through the marker channel.

4.3.3 The Module of Sound. Sound is the most direct and effective way to convey information. The demonstration *effect* of the multimedia textbook depends largely on the setting of the sound. In Director, the sound is played through the sound channel in mirror window. Noticeably, there are 8 sound channels in Director, with two revealed ones (sound 1 and sound 2) and 6 concealed ones. What's more, in Director, two or more sound records cannot be played simultaneously in one sound channel. There are three types of sound in this textbook: audio of the content, background music, and click sound. In this way, the audio of the content will be played in Sound 1, background music in Sound 2, click sound in Sound 3 and Sound 4. The setting of different types of sound is as follows:

a) Audio of the Content

Audio of the content is the sound document recorded by the text of each chapter. Take Chapter One Section One's sound record "1.2" for example, the steps for its setting are:

- Drag the audio 1.2 in Actors List Windows to Sound 1.
- Adjust the length of the audio to approximately 30 seconds to cover the corresponding text part.

b) Background Music

The setting of the background music of dominative interface and subsidiary interface is similar to that of the audio of the content. Drag the two sound records to Sound 2 and adjust the length to cover all frames of the dominative interface and subsidiary interface.

c) Click Sound

Click sound is usually short. They are set on navigational icon and buttons in interfaces. Take the navigational icon of chapter one as an example. Enter into the Script window of chapter one's navigational icon and write the sentence in Lingo code:

```
on mouseenter me
sound (4). play (member ("icon sound 1"))
end
on mouseleave me
stop sound (4)
end
```

The function of this sentence is: when the cursor touches "navigational icon of chapter one, play icon sound 1 in sound channel 4; when the cursor removes from the icon, the sound will stop in channel 4.

Take the navigational key of the dominative interface as an example. Enter the Scrip window of navigational key and write the sentence in Lingo code:

```
on mouseenter me
    sound (3). play (member ("touch sound"))
end
```

The function of this sentence is: when the cursor touches navigational key, the click sound will be played in sound channel 3.

The settings of various types of sound are basically handled through these operations. The sound is multifunctionalized in this way.

4.3.4 The Module of Transitional Effect. Transitional effect is the visual effect of the smooth and natural cohesion among interfaces of each chapter. There are several transitional effects in Director, which are placed in the transitional effect channel in mirror window. Besides, the transitional effect can also be realized by inserting key frame. In this textbook, the transitional effect of each chapter is set by transitional effect channel, while the transitional effect of different sections in each chapter is set by inserting key frame. The details are as follows:

a) Transitional Effect Channel

As transition takes place at the end of the previous interface and the beginning of the next interface, the transitional effect is put in the first frame of the next interface in Director. Take the transition of chapter one and chapter two as an example, the steps for setting are:

- In the transitional effect channel, double click the first frame of chapter two's interface, it comes the pop-up dialog "Frame Attribute: Transitional Effect".
- Choose "Strips" in the column of "Catalog" in the pop-up dialog and "Strips on Top, Build Right" in the column of "Effect". Adjust the duration to 1.20 seconds and the smoothness to the minimum.
- Click "Yes" to finish the action.

b) Insert Key Frame

In mirror window, all "actors" are called "elf". The key frame is to define the character of any frame in elf. If two frames are inserted simultaneously at the frames shortly prior to and after an elf, then, when the film is running, the character of the previous frame will be transited to the next frame, achieving the transitional effect. Take the transition of the first and second sections of chapter one for example. The steps for setting are:

- Click Frame 361 of elf "Text 111" with right-hand button, the shortcut menu will pop up.
- Click "Insert Key Frame".
- Set the transparency to 100% in Attributes Inspector.
- Repeat the previous steps on Frame 370 and set the transparency to 0%. The set of the key frame is completed and the transitional effect of gradual disappearance is achieved.

Through these two methods, the function of the module of transitional effect is achieved, which will ensure the smooth running of the textbook.

4.3.5 The Module of Script. The module of script is used for the automatic control of the multimedia textbook, which mainly assists the buttons in the textbook. In Director, the running process is usually controlled by the settings of the script channel. For this textbook, it is required that at the completion of one audio, it will stop at its last frame and then automatically skip to the next audio. The action will repeat. Take the audio of content "1.3" for example, the steps for fulfilling this function through script channel are:

- Double click the last frame of "1.3" at the script channel, and the Script window will pop up.
- Write Lingo code in Script window:

```
on exitframe me
    if soundbusy (1) then
        go to the frame
    else
        go to the frame+1
    end if
end
```

The function of this sentence is: at the time of leaving the frame, if the sound is still playing in Sound 1, the film will stay at this frame; if the play is finished in Sound 1, the film will automatically skip to the next frame. The sound control is fulfilled in this way.

Besides, if the process should be controlled so that at the completion of one audio, it will pause and not skip to the next audio, the sentence should only be changed into "go to the frame". Through the script channel, the functions of the multimedia textbook will be more abundant.

4.4. The Publish of the System

The last step is the publish of the system, which means the production and export of a document "the multimedia textbook *The Equipment of Ship's C^2 System.* exe". The detailed steps are:

- Click "File/Publish settings" in the menu bar, and the "Publish Settings" dialog will pop up.
- Check the needed options in the columns of "Formats" and "Projector" in "Publish Settings" dialog. Other columns will stay at the default settings.
- Click "..." at the right lower side of "Program File's Customized Icon" in "Projector", and choose "download the icon for national flag in the world 0.ico" in the pop-up dialog. Click "open", and the setting of the icon for exe file will be completed.
- Click "OK" to finish the setting.
- Click "Publish" in the icon toolbar, and export the document "the multimedia textbook *The Equipment of Ship's C^2 System.* exe". The multimedia textbook is published at last.

5. Conclusion

Based on relevant theories and knowledge on the equipment of ship's C^2 system, the paper adopts hierarchical design model and uses diversified multimedia technology in the fine design and reasonable arrangement of the textbook. It meets the demand of the forces for study with content of different levels of difficulty, clear cohesion among different parts, and detailed explanation for key points. What's more, the platform of Director is multi-functional, interface friendly, interactive, practical and operable, which is beneficial for the study of officers and men. In a word, the multimedia textbook is meaningful and helpful for the cultivation of personnel who will use the ships' C^2 system in military forces.

6. Reference

[1] Xiu Zhihong, and Zhanghao. Ship's C2 System and Its Operational Usage. Haichao Press, 2014.in Chinese
[2] Zhou Dexing, and Wang Lin. Director MX and the Development of Lingo. Electronic Industrial Press, 2005. in Chinese
[3] Yan Zhihan, and Li Zhiya. Mastering Director MX 2004. Tsinghua University Press, 2005. in Chinese
[4] Yan Chen. The Art of Multimedia: the Course for Director MX Series. Chemistry Industrial Press, 2006. in Chinese
[5] Zhang Shengli, and Gen Zheqiao. A Practical Course for Director MX 2004. Tsinghua

University Press, 2005. in Chinese
[6] Peng Peng. A Course on Director's Interactive Artistic Design. Tsinghua University Press. 2007. in Chinese

A Study on Spectral Signature Analysis of Wetland Vegetation Based on Ground Imaging Spectrum Data

Chengxing Ling[1,a], Hua Liu[1], Hongbo Ju[1], Huaiqing Zhang[1], Jia You[1] and Weina Li[1]

Institute of Forest Resource Information Techniques, CAF, 100091, Beijing, China
Email: [a]lingcx@ifrit.ac.cn

Abstract. The objective of this study was to verify the application of imaging spectrometer in wetland vegetation remote sensing monitoring, based on analysis of wetland vegetation spectral features. Spectral information of Carex vegetation spectral data under different water environment was collected bySOC710VP and ASD FieldSpec 3; Meanwhile, the chlorophyll contents of wheat leaves were tested in the lab. A total 9 typical vegetation indices were calculated by using two instruments' data which were spectral values from 400nm to 1000 nm. Then features between the same vegetation indices and soil water contents for two applications were analyzed and compared. The results showed that there were same spectrum curve trends of Carex vegetation (soil moisture content of 51%, 32%, 14% and three regional comparative analysis)reflectance between SOC710VP and ASD FieldSpec 3, including the two reflectance peak of 550nm and 730 nm, two reflectance valley of 690 nm and 970nm, and continuous near infrared reflectance platform. However, The two also have a very clear distinction: (1) The reflection spectra of SOC710VP leaves of Carex Carex leaf spectra in the three soil moisture environment values are greater than ASD FieldSpec 3 collected value; (2) The SOC710VP reflectivity curve does not have the smooth curve of the original spectrum measured by the ASD FieldSpec 3, the amplitude of fluctuation is bigger, and it is more obvious in the near infrared band. It is concluded that SOC710VP spectral data are reliable, with the image features, spectral curve features reliable .It has great potential in the research of hyperspectral remote sensing technology in the development of wetland near earth, remote sensing monitoring of wetland resources.

1. Introduction

In a relatively complex wetland ecological system, the hyperspectral remote sensing plays an important role in the quantitative detection of wetland resources by remote sensing, research on spatial distribution characteristics of wetland types, remote sensing inversion of the wetland vegetation ecosystem and the like by the virtue of its huge advantages, such as thousands of available wave bands, extremely high definition and special sensitivity to spectral features [1]. The wetland vegetation is an important component of the wetland ecosystem. In recent years, domestic scholars have also implemented extensive and deep study to the application of the hyperspectral remote sensing extraction technology with remote sensing into the collection of wetland vegetation information [2]. Lin Chuan and Gong Zhaoning have studied the identification method of the typical wetland vegetation ecological types based on spectral signature variants [3]. Li Fengxiu and others have conducted actual tests and measurements of the reflection rate and chlorophyl density at the Calamagrostis angustifolia shrub layer under different coverage and water depth level to create related vegetation indexes based on the data collection from both visible light - near-infrared band and find out the vegetation indexes and the wave bands with the highest relevance to chlorophyl [4]. Zou Weina analyzes the effects of coverage and canopy water depth to the spectral characteristics of the submerged plant cabomba caroliniana. The research results indicate the basic characteristics of the reflection rate of cabomba caroliniana cluster under different coverage are mostly apparent in the green and near-infrared band[5]. Schmidt and others use GER 3700 wild portable field spectrometer to measure the reflection spectrum of 27 species of wetland vegetation in the coastal saline alkali wetland of southern Holland, apply the statistical tests, continuous removal and distance analysis to finally select out 6 optimal bands for classification[6]. Zomer and others use GER 2600 wild portable spectrometer to measure the spectrum of both the canopy and the leaves of 7 species of wetland vegetation in the Pacheco Creek Salt Marsh in US California. Afterwards, they apply the acquired ground object spectrum curve to create a spectrum bank, providing the scientific bedrock for the classification of wetland vegetation based on hyperspectral remote sensing image. Among such studies most of them are the study and application of the hyperspectral data acquired through aerial imaging and the non-imaging hyperspectral data acquired on the ground without figures. However, the analysis and monitoring techniques about the imaging spectral characteristics of wetland vegetation based on the ground hyperspectral imaging are still quite scarce. The field spectrometer can study a single crop, but cannot do the imaging, and the canopy spectrum collected from the vegetation is mixed large cluster of spectral information. This has restricted the precision of various inversion models to be established. Therefore, this study uses both the imaging spectrum and the non-imaging field spectrometer and selects the water environment which is the most sensitive in the wetland ecosystem, collects the spectral information from the single plant of sedge, which is a wetland vegetation, and compares and analyzes the differences of the hyperspectral band obtained from both equipment and uses the field spectrometer as the standard to verify the reliability of the data acquired from the imaging spectrum; then selects the typical wetland vegetation index to study the spectral differences of the sedge vegetation in the environments with different moisture gradients and provides support to the further study of the wave bands selection for the wetland hyperspectral information acquisition using ground object imaging spectrum.

2. Material and Method

2.1. Test location

The test location of this research selects the Jinjiang river segment in the middle section of Yangtze river located in the Hunan East Dongting Lake in Yueyang city, northeastern Hunan province. The place is one of the wetland natural protection zone designated by the Chinese government and also gets listed in the "Ramsar Convention". The central coordinates of the place are 28°59′52″N and 112°59′52″E. The area belongs to the subtropical humid climate zone with very typical wetland types and wetland vegetation, ample sunshine, abundant precipitation, rich submerged plants, floating plants and emerging plants. The typical wetland vegetation sedge is widely distributed in the eastern Dongting lake wetland protection zone and is a typical plant species of this wetland.

2.2. Data collection instrument

Table.2-1 Comparison of Data Collection Instrument

Comparison of instruments	Imaging spectrum (SOC710VP)	Field spectrometer (ASD FieldSpec 3)
Spectral range:	400-1000 nm	350-2500 nm
Spectral resolution	4.6875 nm	3nm@700nm; 10nm@1400nm,2100n
Wave band characteristics	128 bands	Data interval: 1nm
Sampling features	30 lines/sec Pixel per line: 696; 23.2 sec/cube (696 by 520 cube)	1.4nm@350-1050nm; 2nm@1000-2500nm
Lens/focus	Adjustable (based on the camera used)	Cameras with the view angles of 1°, 5°, 8°

2.3 Test plan and measurement method

2.3.1. Experimental program. Use the soil moisture detector to get the concentrated distributed moisture gradient environment. Take sedge as the example and select three regions with sedge distribution with soil moisture of 51%, 32% and 14% respectively for concentrated measurement. Compare and analyze the test results of single wetland vegetation type scale obtained by the SOC710VP imaging spectrum and field spectrometer ASDFieldSpec 3. The test content is the relevance analysis of the sedge vegetation indexes under different soil moisture level to determine the soil moisture distribution patterns suitable for the ground object classification.

(a)Sedge spectral data acquired by SOC710VP imaging spectrum

(b)the sedge spectral data acquired by ASDFieldSpec 3

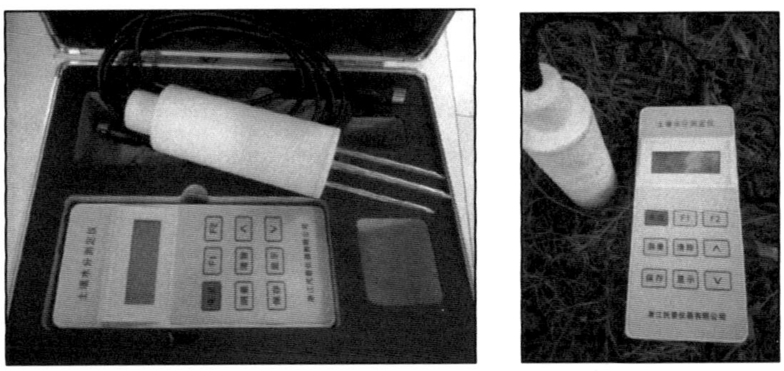

(c)Soil moisture measurement

Figure 2-1 Field Data Collection

2.3.2. Measurement methods. Use the soil moisture meter to measure the sample points of sedge distribution areas under different moisture gradient environments, then collect SOC imaging hyperspectral data, convert the original data into reflection rate, before the conversion implement the dark current removal, wavelength calibration, radiation calibration and other operations at first, afterwards remove the electrical noises which is affected by the environmental factors including the ambient temperature. Therefore, ahead of the start of each test, conduct the Dark measurement and keep the record. Then conduct the wavelength calibration and radiation calibration. The method of reflection rate conversion: To use the actual tested whiteboard reference board, select the target sedge, use the empirical linear algorithm to solve the calibration coefficients, then extract the reflection rate. Sedge reflection rate=(sedge DN value/reference board DN value) × reference board reflection rate. The reflection rate of the reference board is the indoor calibrated value. Finally, the image DN value can be converted into the reflection rate through the processing module of the hyperspectral software.

3. Result and Analysis

3.1 Imaging spectrum data analysis

We can obtain the effective imaging hyperspectral data and non-imaging field spectrometer data in three sedge distribution area with the moisture levels of 51%, 32% and 14% respectively through the aforementioned measurement method with 25 data groups for each sedge area which means 75 samples in total. Administer the post-treatment of the measured data to acquire the pretreated results of the sedge imaging spectrum data and conduct noise reduction to the original imaging spectrum data. The sedge spectral characteristic curves in the visible light and infrared and near infrared bands are all clearly seen.

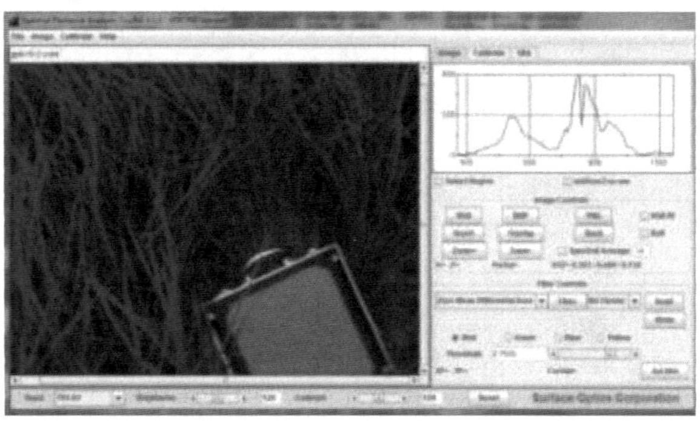

(a) Post-treatment of collected imaging spectrum data

(b) Conversion process of whiteboard reference board reflectivity

(c) The True color RGB combination and False color RGB combination of Carex imaging spectral data

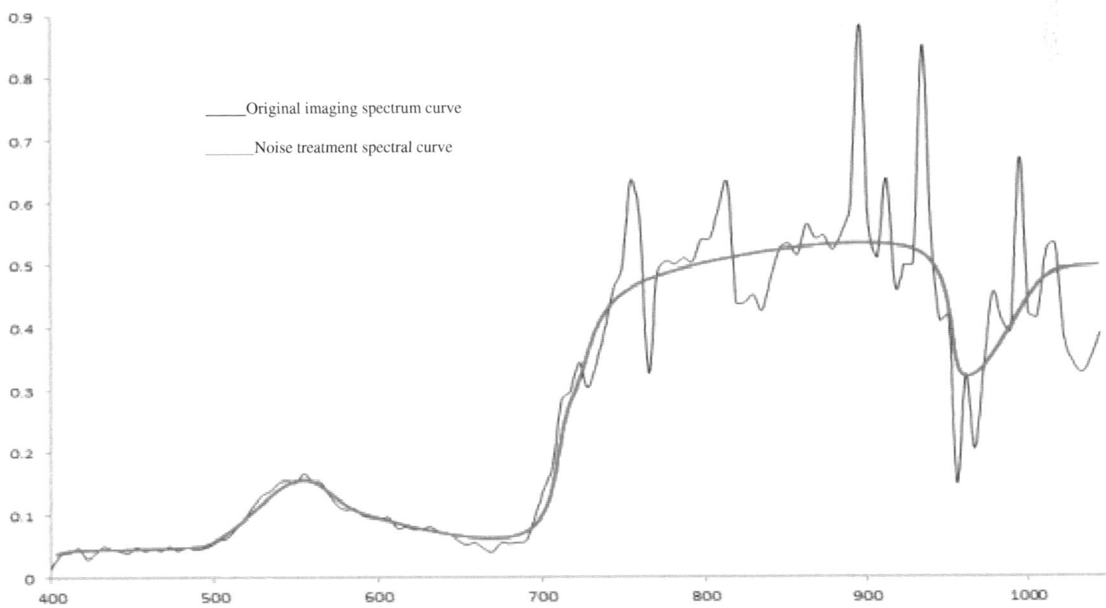

Figure 2-2 Pre-treatment Results of Sedge Imaging Spectrum Data

Based on the data reduction and statistics, we can obtain the spectral signature curve of the three sedge distribution areas with different moisture gradients with the moisture of 51%, 32% and 14% respectively. The non-imaging spectrum data collected by the ASD FieldSpec 3 from the same area are shown in the following diagram:

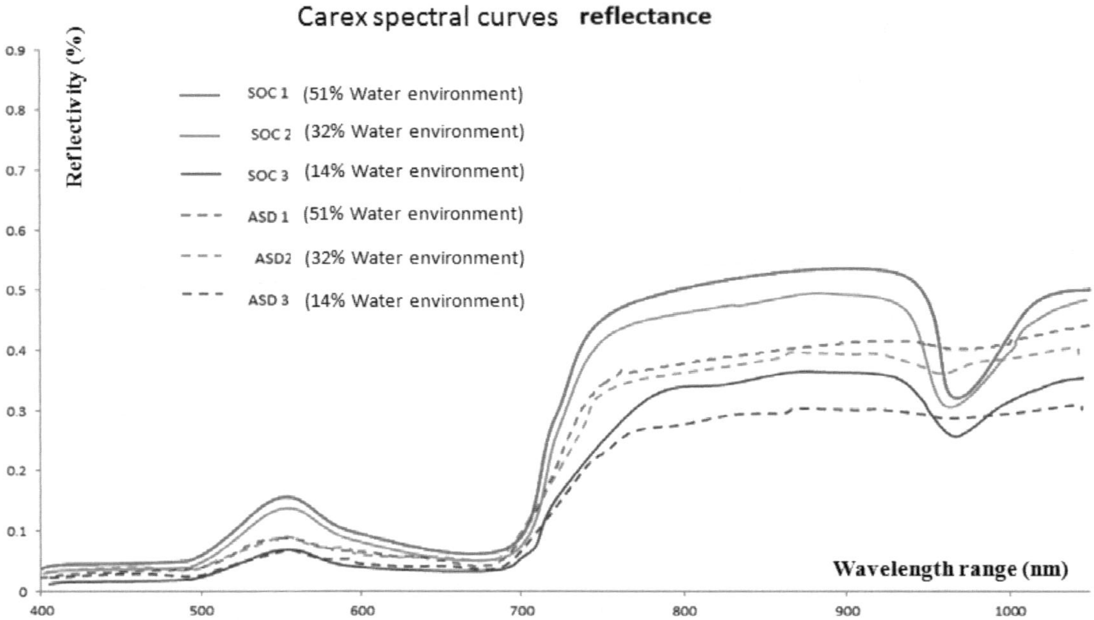

Figure.2-3 Reflection Rate Curve Collected by SOC710VP and ASD FieldSpec 3 under Different Moisture Levels

The spectrum of the sedge in different distribution gradients measured by using the ground imaging spectrum SOC710 and non-imaging ASD FieldSpec 3 respectively shows the strongest reflection in the band between 550nm and 730nm. After 750nm there is a near infrared high reflection platform. The wavelength around 690nm and 970nm witnesses the apparent absorption trough. The sedge spectral characteristic curve measured by two instruments has basically maintained the same development trend, but the two instruments also bear some quite obvious differences;

(1) The reflection rates acquired by two instruments in the same sedge area under the same moisture level are different in value. The spectral data of sedge leaves collected by SOC710VP in all three areas are all bigger than the sedge leaves spectral reflection rate collected by ASD FieldSpec 3.

(2) The original SOC710VP reflection rate curve is not as smooth as the original spectral curve measured by ASD FieldSpec 3. The former has more turbulence, in particular at the near-infrared band. After the noise is diminished, the curve trough of the reflection rate collected by the SOC710VP witnesses more drastic change of magnitude at 970nm. (As shown in Figure 2-3)

The analysis of the reflection rate curve collected by SOC710VP under three different moisture indicates the whole trend completely complies with the sedge vegetation spectral characteristic curve, the difference lies at the characteristic bands (red edge, green edge, near infrared platform) of the sedge spectrum curve under 51% soil moisture all show that the water content directly affects the vegetation spectral reflection characteristics and the interference magnitude is bigger, the change is more stark, which is helpful for the subsequent spectral terminal element clustered analysis.

The results indicate that through the application of spectral data of a single sedge leaf collected by the SOC710VP imaging spectrum under different moisture level and the analysis of the leaf reflection rates under 3 moisture gradients, we can determine that the reflection rate characteristic curve of the SOC710VP spectrometer is similar to the results from the field spectrometer ASD. Both resemble the same pattern which is manifested by the reflection peak at 550nm and 730nm and the absorption trough at 690nm and 970nm as well as the continuous near infrared reflection platform. This result is identical with the precedent conclusions obtained from satellite hyperspectral imaging facilities such as Hyperion, CHRIS, CASI and the like.

3.2 The sedge imaging spectrum vegetation index characteristic analysis under different water content environments

In order to proceed to compare and analyze the moisture's influence to the vegetation spectrum of the single sedge species collected from the imaging spectrum and the non-imaging spectrum, we apply the method of analyzing the vegetation indexes to give an analysis to sedge vegetation information characteristic. This research selects 9 vegetation indexes frequently used by the precedent cases and compare them with the sedge index in the wetland under different moisture condition, and relies on the data collected from both the SOC710VP and the ASD FieldSpec 3 for further analysis of the data from the vegetation imaging spectrum SOC71. Select the following vegetation indexes: Vegetation indexes with visible light: VARIgreen, VARI700, CARI; Visible light/near infrared vegetation index: NDVI, EVI, SR, DVI; Red edge ratio index: VOG3, Carter2.

Table 3-1 The Results of Spectral Vegetation Index of Carex Pure Pixel in Different Water Environment

Vegetation index	Soil moisture content of 51%		Soil moisture content of 32%		Soil moisture content of 14%	
	ASD	SOC	ASD	SOC	ASD	SOC
NDVI	0.718	0.879	0.625	0.789	0.518	0.621
EVI	0.542	0.675	0.508	0.606	0.507	0.555
SR	8.143	9.845	7.104	7.841	7.007	7.889
DVI	0.291	0.478	0.281	0.329	0.235	0.247
CARI	0.041	0.051	0.033	0.039	0.051	0.054
$VARI^{green}$	0.552	0.553	0.425	0.497	0.514	0.573
$VARI^{700}$	1.216	1.611	1.366	1.247	1.839	1.936
VOG3	1.121	1.414	1.787	1.271	1.507	1.455
Carter2	0.187	0.195	0.141	0.165	0.161	0.187

From the Tab 3-1 we can see that the sedge spectrum data measured by SOC710VP and ASD FieldSpec 3 under different water content environment is the representative vegetation index value of that area, among which the data obtained from the imaging spectrum represents the visible light to which the green vegetation is most sensitive/the near infrared light index characteristic NDVI, EVI, SR and DVI value under three moisture environments all manifest very direct relationship. As the decreasing of the soil moisture these indexes also assume a debilitating trend and is slightly higher than the data acquired by the ASD non-imaging spectrum. This is highly identical to the spectral characteristic curve analyzed previously, which proves the data is reliable. From Table 2-1, the results in the visible light band and the red edge ratio area also show very good relevance. In summary with the selection of 9 vegetation indexes from the precedent investigations, especially the selection of 3 visible light vegetation indexes, 4 visible/near infrared vegetation index, 2 red edge ratio indexes, thereby the total quantity of vegetation indexes is 9. The imaging spectrum SOC710VP and non-imaging spectrum obtain the index characteristic of sedge vegetation index under different moisture gradient. The calculation results indicates the changing trends of the selected vegetation indexes are identical which means the imaging spectrum SOC and the field spectrometer ASD possess basically the same and consistent data acquisition.

4. Conclusion

By using the SOC710VP imaging spectrum and the ASD FieldSpec 3 field spectrometer, we collect the spectral reflection data information of the typical wetland vegetation sedge within the wetland, and compare the imaging spectral signature of the sedge pure pixel region and the vegetation index characteristics under different water contents, to understand and command the spectral features of the imaging spectrum on the single sedge area.

IOP Conf. Series: Journal of Physics: Conf. Series **910** (2017) 012045 doi:10.1088/1742-6596/910/1/012045

The study proves that the advantage of imaging spectrum is the ability to select the study scope on the spectral imaging figure, and the data are consistent with the data acquired directly obtained from the biological structure of leaves. From the vegetation indexes obtained from both instruments, the value acquired by the imaging spectrum SOC710VP is more accurate than the data acquired by field spectrometer ASD. From a side view, it means the imaging spectrum is able to determine the selection area which is acquired simultaneously with the biological parameters at the same position, this can enhance the extraction precision of vegetation biological related data. In summary, the data from the imaging spectrum has the advantages of figure-spectrum integrity and spectral characteristic curve reliability. The imaging spectrum has great potential in the hyperspectral remote sensing study of the wetland and expands the technical means of wetland resources remote sensing and monitoring.

5. Acknowledgments

This study was funded by the central public-welfare research institutes basic scientific research special-project "Study on Wetland Vegetation Classification based on Hyperspectral Imaging. (IFRIT201505)".The paper's corresponding author is Hua LIU, (Liuhua@ifrit.ac.cn).

6. References

[1] Qingxi TONG,Bin ZHANG,Lafeng ZHENG.Hyperspectral remote sensing- Principle, Technology and Application.Beijing: Higher Education Press 2006: 136-146.

[2] Yuqing WAN,Kelong TAN,Riping ZHOU.Application of hyperspectral remote sensing. Beijing: Science Press (in Chinese) 2006: 150-152.

[3] Lin C, Gong Z N, Zhao W J, et al.Identifying typical plant ecological types based on spectral characteristic variables: a case study in Wild Duck Lake wetland, Beijing[J]. Acta Ecologica Sinica, 2013,33(4): 1172-1185.

[4] LI Feng-xiu, ZHANG-Bai1, LIU Dian-wei, WANG Zong-ming, SONG Kai-shan. Hyperspectral remote sensing estimation models for chlorophyll a concentration of Calamagrostis angustifolia. [J] 2008,27(07):1077-1083 .

[5] ZOU Wein, YUAN Lin, ZHANG Liquan.Impacts of coverage and canopy water depth on the spectral characteristics for a submerged plant Cabomba caroliniana[J] Acta Ecologica Sinica, 2012, 32(3): 706-714.

[6] Schmidt K S, Skidmore A K. Spectral discrimination of vegetation types in a coastal wetland. Remote Sensing of Environment, 2003, 85(1) :92-108.

[7] Zomer R J, Trabucco A, Ustin S L. Building spectral libraries for wetlands land cover classification and hyperspectral remote sensing. Journal of Environmental Management,2009, 90: 2170-2177

[8] http://www.azup.com.cn/html/2014/PortableHSI_0404/ 58.html SOC710 VP . Infrared imaging spectrum.

[9] tps://www.asdi.com/products-and-services/fieldspec-spectroradiometersASD FieldSpec 3 Spectroradiometer

Application of 3D Laser Scanning Technology in Inspection and Dynamic Reserves Detection of Open-Pit Mine

Zhumin Hu [1,2,3,a], **Shiyu Wei** [1,2,3,b] **and Jun Jiang** [1,2,3,c]

[1] Chongqing Engineering Research Center of Automatic Monitoring for Geological Hazards, 400042 Chongqing, China

[2] Chongqing Institute of Geology and Mineral Resources, 400042 Chongqing, China

[3] Chongqing Huadi Design Institute of Engineering Investigation , 400042 Chongqing, China

Email: [a] hzmsakra@163.com, [b] weishiyu.606@163.com, [c] jiangjun_dyy@163.com

Abstract. The traditional open-pit mine mining rights verification and dynamic reserve detection means rely on the total station and RTK to collect the results of the turning point coordinates of mining surface contours. It resulted in obtaining the results of low precision and large error in the means that is limited by the traditional measurement equipment accuracy and measurement methods. The three-dimensional scanning technology can obtain the three-dimensional coordinate data of the surface of the measured object in a large area at high resolution. This paper expounds the commonly used application of 3D scanning technology in the inspection and dynamic reserve detection of open mine mining rights.

1. Introduction

At present, the mining rights verification and the dynamic reserve detection of the open-pit mine are based on the means that measures the turning point coordinates of mining surface contours through the total station and RTK, so the accuracy of the measurement is too low, and the information of the coordinates is too little. The results are mainly based on the coordinates of the turning points to build model, resulting in low accuracy and being impacted easily by human factors, thus the final results have more interference. Three Dimensional Laser Scanning Technology, also known as "real copy technology", is a high-speed laser scanning measurement method, which has fast access to three-dimensional coordinate data of the measured object surface with large-area and high-resolution [1]. Therefore, mining rights verification and dynamic reserves detection of open-pit mine can use three-dimensional scanning technology [2].

2. 3 D laser scanning mine verification process

According to the area, surface, mining surface and other characteristics, three-dimensional laser scanner station location and the number of stations are planned. The original station scattered point cloud can be gained after the completion of the field work.

After the above pretreatment work, the obtained point cloud data can be used for GIS analysis, such as transboundary analysis, resource reserves calculation and other results. Open-pit mine three-dimensional laser scanning technology process is shown in Fig.1, the main point cloud GIS analysis results are shown in Fig.2.

Figure.1 Three-dimensional laser scanning process of open-pit mine

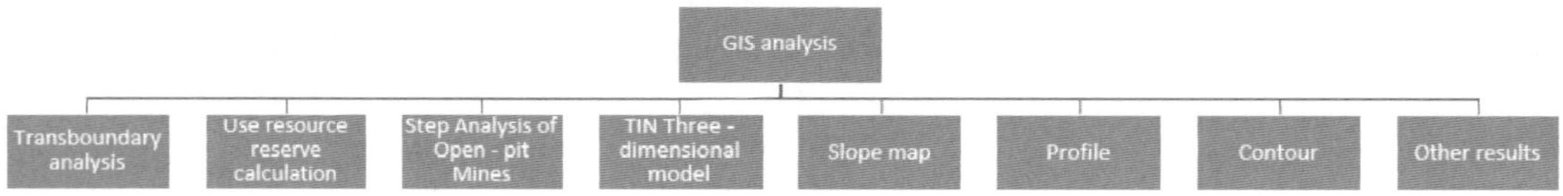

Figure.2 gis analysis results of point cloud

3. Three-dimensional laser scanning mine verification work

The three-dimensional laser scanner pilot work takes a limestone mine in Chongqing as an example. Scanning instrument is the Leica C10 three-dimensional laser scanner. The main processing software is Cyclone7.0, Map3D, and ArcGIS10.0.

3.1 Point cloud preprocessing

Open pit mines generally have more ups and downs and dense stones. Excavators work together, resulting in three-dimensional laser scanner's narrow perspective, and resulting in station-intensive station. Open-pit mine layout of three-dimensional laser scanner station should follow the principles below:

1), the vision of station field should be lager, which can cover a larger area, with the scanning surface angle closer to 90 ° better;

2), the station should be arranged in a more stable area, around no large equipment operations or artificial stripping operations;

3), the station should have common characteristics, such as poles, signs and so on.

Results analysis needs the entire mine as the object, so the various sites point cloud data should be stitched into a complete point cloud. The Scanning adopts the method of no target and station with free way, and it should be noted that the stations must contain a common point between the cloud points, the outside work takes a short time, but inside work would take long time to stitch point cloud of characteristics. Such as limestone mine have a number of security warning signs around the layout, which can be used as a common point between the stations cloud. The characteristic points are shown in Fig.3.

Figure.3 Point cloud stitching feature map

Point cloud can be stitched through the [Create Registration] tool of the Cyclone software. Point cloud stitching can be divided into small parts to stitch, and then the larger parts into a whole point cloud. Point cloud stitching should pay attention to the error column value size of [Constraint List], which shows the two stations point cloud compliance, generally about 0.02m. If the value is large, it should be needed to re-find the characteristics of objects re-stitching.

Constraint ID	ScanWorld	ScanWorld	Type	Status	Weight	Error	Error Vector	Group
Cloud/Mesh 1	Scan-0003.bin (Leveled)	Scan-0002.bin (Leveled)	Cloud: Cloud/Mesh - ...	On	1.0000	0.006 m	aligned [0.015 m]	Ungrouped
Cloud/Mesh 2	Scan-0002.bin (Leveled)	Scan-0004.bin (Leveled)	Cloud: Cloud/Mesh - ...	On	1.0000	0.006 m	aligned [0.014 m]	Ungrouped
Cloud/Mesh 3	Scan-0005.bin (Leveled)	Scan-0002.bin (Leveled)	Cloud: Cloud/Mesh - ...	On	1.0000	0.006 m	aligned [0.014 m]	Ungrouped

Figure.4 Point cloud stitching error map

After stitching, a complete original point cloud is gained, as shown in Fig.5. There are some obstacles in the original point cloud that do not belong to the mine itself, such as staff, vehicles, equipment, and the point cloud trajectory caused by these objects during the scanning process, which need to be removed manually[3].

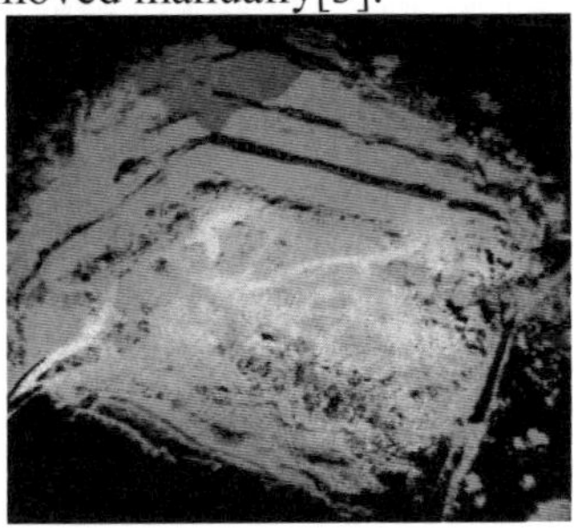

Figure.5 Entire mine point cloud

Figure.6 Feature point cloud

Free local station mode needs to convert the local coordinate system to Xi'an 80 coordinate system. The coordinates of two or more points should be collected by using RTK at the same time, shown in Fig.6.Due to the same vertical direction of two coordinate system, the scale factor is consistent, it only need to give a rotation factor and basis elevation. The entire point cloud coordinate system is converted to Xi'an 80 coordinates through the Cyclone software [Set Coordinate System from Pick Points] tool. The basic available point cloud data for open-pit mine can be accessed by above preprocessing steps.

It is necessary to convert point cloud data which is too large to process directly in ArcGIS into a raster image, the raster image stores coordinates as pixel location and elevations as pixel color. Map3D software is used to convert the gray scale image, because ArcGIS does not provide good support for point cloud data file conversion. Using the [MAPPOINTCLOUDMANAGER] command to load the point cloud file. Using the [MAPSURFACEMANAGER] command to generate, as shown in Fig.8. The blue line is determined by the two mining boundaries, red line for the mine management department to provide the mine turning point coordinates.

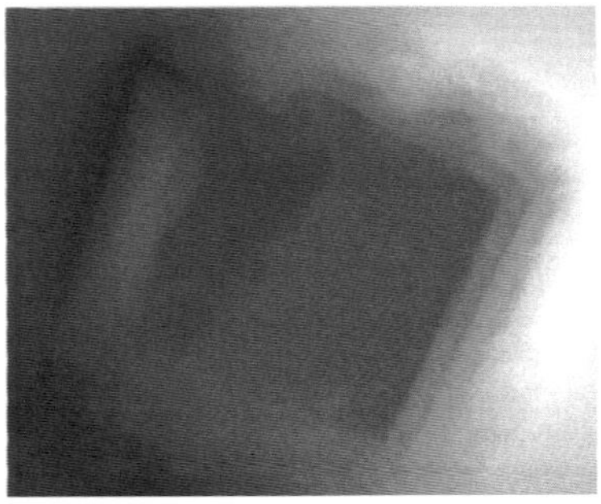

Figure.7 Elevation grayscale

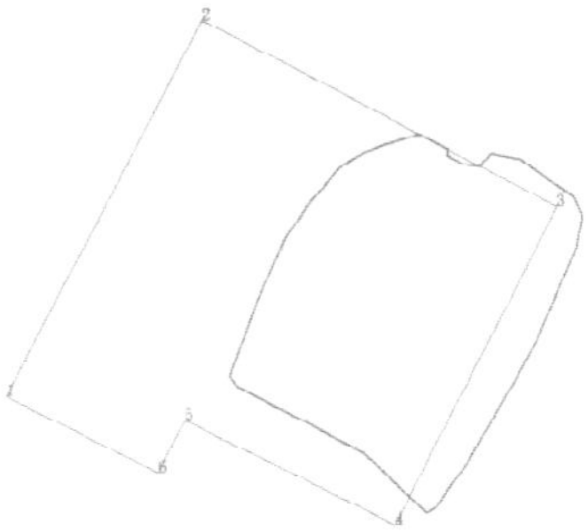

Figure.8 Mining boundary and range

Using the [Extraction] tool in the ArcGIS [Spatial Analyst] tool to extract the elevation gray scale in the excavation boundary extracted in the previous section as a mask to facilitate the subsequent analysis. The first time scan of the TIN digital elevation model is generated by the [Export Raster] tool in the ArcGIS [3D Analyst] tool. Fig.9 is a TIN top view of the excavation boundary after the first time scan.

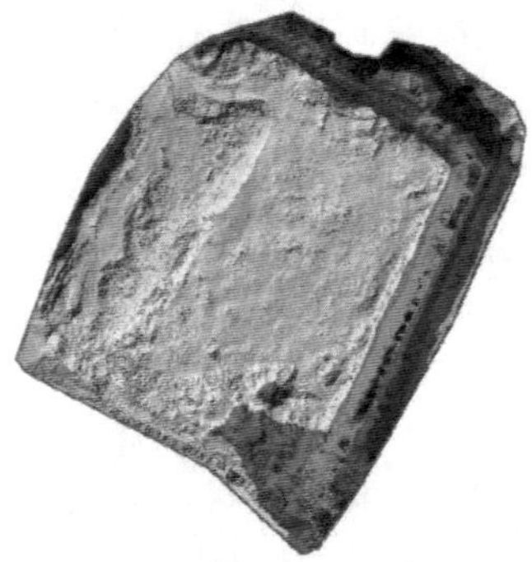

Figure.9 Top view of TIN

Figure.10 TIN model

Using the ArcScene software to open the above generated TIN digital elevation model file, through the [Export TIN] tool to get the scope of the boundary of TIN, according to the scope of the boundary elements of the polyline elevation, stretching the surrounding boundary, the elevation of basic polygon feature should be set 510m.Fig.10 shows the TIN model for the first 3D laser scanner, with the bottom boundary being the lowest mining point (510 m). The total volume of the model is the total remaining amount currently in the current excavation area.

3.2 Transboundary analysis

The current mining boundary is determined according to Fig.8. As shown in Fig. 9, the east side step has been beyond the boundaries of the application area. The maximum value exceeded is 29.8m after the measurement, blue area is dug, and the red area is filled. The color red and blue staggered at the east and north sides of the mine, the elevation has not changed, so the step did not change. The conclusion is that the mine has crossed the border but did not continue.

Fig.12 shows the superposition of the TIN and the lowest mining plane. Obviously, the mining surface has a great distance between the lowest mining plane (510m). The lowest mining elevations of two times were 557m, 547m, vertical space is 37m from lowest mining plane after the measurement. So in the vertical direction is not exceeded.

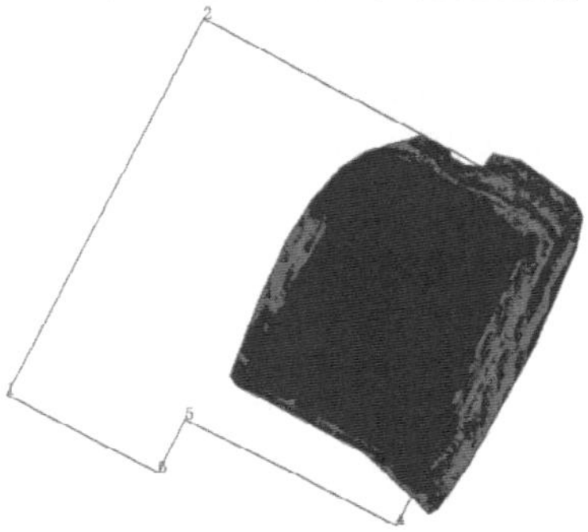

Figure.11 Excavation and mining area

Figure.12 TIN and lowest Mining plane

3.3 Resource reserves calculation [4]

Using the [raster calculation] tool in ArcGIS [3D Analyst], scanning changes in the raster image has been obtained with the two elevation gray scale subtraction, through the raster image, we access to the elevation difference TIN map, As shown in Fig.13. the elevation difference TIN map with the same pond, white for the water, then no elevation changes, white below has been excavated area, the deeper the color, the deeper the dig by configuring the tone. It can get the data cut or filled through [Cut Fill] tool in [Raster Surface] toolset. Fig.14 shows the excavation volume, blue indicates the excavated area, and red indicates the unexcavated area. After the statistics, the total area of the two excavations is 39250m2, the resource reserves in using is 256203m3, the average excavation depth is 6.5m, and the deepest excavation depth is 20.7m.

IOP Conf. Series: Journal of Physics: Conf. Series **910** (2017) 012046 doi:10.1088/1742-6596/910/1/012046

Figure.13 Elevation difference TIN map

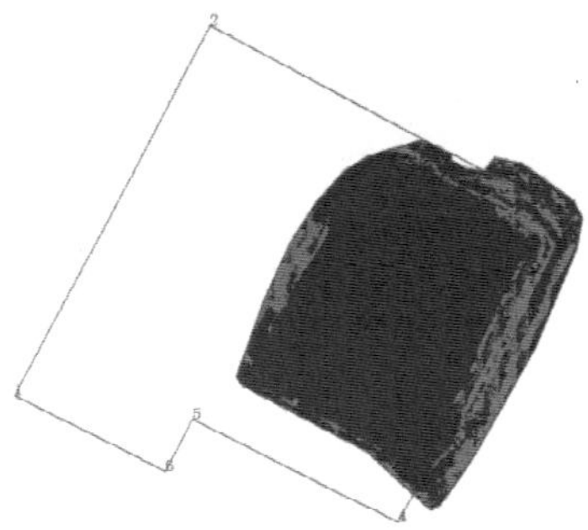

Figure.14 Cut and fill map

3.4 Step analysis

According to the design requirements of the given slope, such as mining slope angle of the limestone mine is 55 °, the slope of the mining area symbol system to "classified" way, as shown in the red area is 55 ° or more area. In this way, it is possible to quickly judge whether the slope angle of the open pit is in compliance with the requirements.

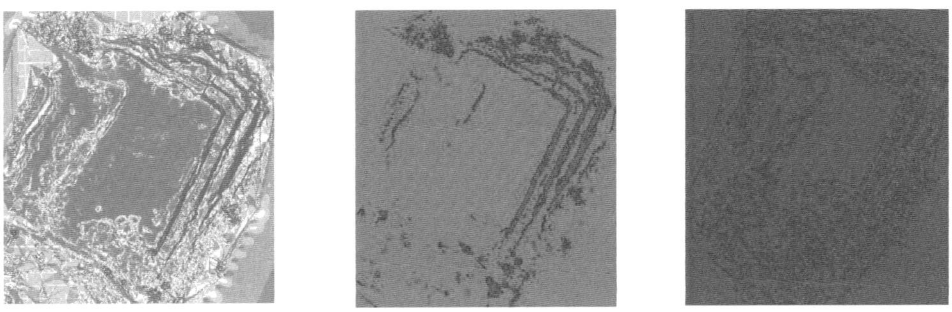

Original slope map 55 °classification 30 °classification

Figure.15 Slope classification map

3.5 Profile

Using the [3D Analyst] tool to obtain a profile of the corresponding position which required, as shown in Fig.16. With elevation or slope as the data source, we can quickly get the required elevation profile and slope profile, elevation distribution of the section line in elevation profile can be saw intuitively.

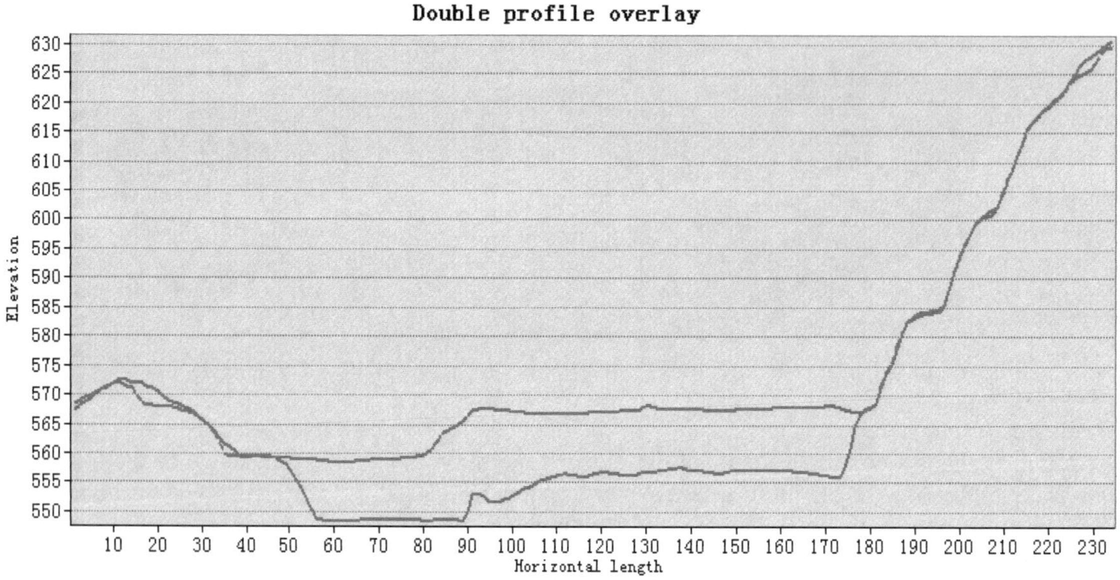

Figure.16 Mine elevation profile

3.6 Other results

Contours can be obtained with opening the front generated elevation gray scale and setting the contour distance through [Contour] tool in [Raster Surface] tool ,then it generate contours automatically, it can be exported to CAD format for the other software to use[5,6] after some other labeling.

4. Conclusion

Through the three-dimensional scanning of the open pit mine, three-dimensional laser scanner method can achieve the open-air mine cross-border determination, step height, width, mining slope analysis, the use of resource reserves calculation, TIN numerical elevation model (three-dimensional model), contours and other results, At the same time we can obtain the following conclusions:

1, High precision. As three-dimensional laser scanner method is to obtain all the information on the surface and the traditional means of measurement is to obtain the contours of the transfer point according to the vertical section method (the mine is divided into multiple sections, cross-section between the ladder, truncated cone or pyramid to estimate) to estimate dynamic reserves, so the accuracy is superior to traditional means of measurement.

2, Good intuition. In the traditional measurement concept, the data obtained by the final output are two-dimensional form of the results, but with the development of digital mines, three-dimensional management model has gradually replaced the two-dimensional, because the two-dimensional results are difficult to accurately describe mine key information.

3, Rich results. Three-dimensional laser scanning technology is the mine surface of the full point of information, so to obtain more rich than the traditional means of the results, such as open-air mine anywhere in the direction and direction of the section, and open-pit mine three-dimensional model.

5. References

[1] Chen Jing; Li Qingquan; Li Bijun. Laser scanning measurement system application research [J]. Surveying and Mapping Engineering .2001 (01).

[2] Su Xiangyi; Wang Wei, etc. Three-dimensional laser scanning technology in the application of mine surveying and mapping [J]. Shandong Industrial Technology. 2016 (22).

[3] Quan Yushu. 3D point cloud data preprocessing research [D]. Northwestern Polytechnic University.

[4] Fan Zhengyue; Xu Maowen. Based on three-dimensional laser scanning of open-pit mine reserves measurement research [J]. Mine survey .2014 (04).

[5] Liu Changjun; Zhao Yu, etc. Research on the key technology of rapid measurement of mine terrain based on 3D laser scanning technology [J]. Survey of Surveying and Mapping.
[6] Mei Wensheng; Zhou Yanfang; Zhou Jun. Based on the ground three-dimensional laser scanning fine terrain mapping [J]. Surveying and mapping .2010 (01).

Independent Component Analysis of Gravity Earth Tide Signal Based on Differential Evolution Algorithm

Lue Gao[1,a] and Haiyan Quan[1,b]

[1]Faculty of Information Engineering and Automation, Kunming University of Science and Technology, Kunming,China

Email: [a]gaolue@foxmail.com, [b]quanhaiyan@163.com

Abstract. To solve the difficulty of extracting orthogonal component of tidal harmonic from gravity earth tide signals, an orthogonal decomposition model for gravity earth tide signal is proposed in this paper .The harmonic component of the gravity earth tide signal is implemented orthogonal decomposition by the model in space, in order to realize the extraction of the harmonic component frequency reflects the independent geophysical information on the space. Thus, a new feature extraction method of gravity earth tide signal is presented by combining independent component analysis and differential evolution algorithm. It can be more effective to get the independent component which is consistent with the gravity solid tidal signal decomposition model. In the experiment, the measured signal is processed and compared with the theoretical value in Kunming area. Experimental result shows that the method can reveal the relationship of tidal harmonic component in gravity earth tide signals and this relationship corresponding to the gravity earth tide signal orthogonal decomposition model. Each spectral information component contained accords with theory value.

1. Introduction

Due to the change in the relative position of the sun and moon, the gravitational force of the sun and moon will cause the earth's gravitational field to change periodically with time[1].According to reflect this change in the vertical direction and numerical of gravity, can be divided into the ground tilt solid tide and the gravity earth tide. However, in this paper, we only study the gravity earth tide signal. The gravity earth tide has not only become an important means of seismic observation, but also is an important part of the study of seismic origin, earthquake precursor and earthquake prediction[2].Rich tidal harmonic components of gravity earth tide signals can be divided into long period wave, diurnal wave, semi diurnal wave three harmonic line[3]. Because these harmonic components are rich in geophysical information, it is significant to analyze the harmonic component of gravity tide signal.

Research of solid earth tide to years is mainly carried out harmonic analysis [4], at present, the Venedikov harmonic analysis method and the ETERNA harmonic analysis method are two standard tidal analysis methods recommended by the international earth tide Center (ICET).In fact, Harmonic analysis is a method based on Fourier transform, which decomposes the observed information into a simple periodic function. However, Fourier transform is a kind of integral transform which is defined by the sine basis function in the infinite period. It can only analyze the stationary signal with time and the global analysis of the signal. But it cannot be analyzed the nonlinear non-stationary signal with time variation and local analysis of the signal. This shows that harmonic analysis cannot fully reveal the modulation relationship between harmonics, and it cannot realize the orthogonal decomposition of the harmonic component of gravity earth tide signal.

In order to effectively extract the geophysical information of the gravity tide signal, the modulation relation between them is revealed. The independent component analysis (ICA) algorithm based on differential evolution (DE) algorithm is introduced in this paper, and an orthogonal decomposition model is proposed for the analysis and treatment of the observed signals. By using this method, in this paper, the corresponding relation between the independent component of tidal harmonic in gravitational earth tide signal and the gravitational tide effect among celestial bodies is revealed. A new method has been found for the study of gravity tides.

2. Gravity Solid Tide Signal Orthogonal Decomposition Model

As shown in Figure 1, an observation point A on earth, which under the influence of tidal force exerted by the earth's rotation, the sun and the moon, triggered a solid tidal F , can be decomposed into gravity earth tide signal F_g . In this work, we only analyze the gravity earth tide signal F_g , which can be decomposed into 2 orthogonal vectors: a signal component F_1 parallel to the equatorial plane, a parallel to the earth's self-axis of the signal component F_2 [7]. And F_1 can be decomposed into two orthogonal signal components F_{11} and F_{12} (in the same plane). Thus, gravity earth tide signal is orthogonal decomposition of three-dimensional orthogonal vector (see Fig 1). According to the model of 3D orthogonal vector decomposition, the F_1 is parallel to the equatorial plane, which means that it is related with the earth's rotation and revolution, it is mainly reflected on the harmonic components of diurnal wave F_{11} and semi diurnal F_{12} . However, F_2 is parallel to the earth's axis of rotation, has nothing to do with the rotation of the earth, its weight in the tidal gravity signals for years wave, half years wave, excluding day wave, semi diurnal wave. In this way, according to the three-dimensional orthogonal decomposition model, the different harmonic components of gravity earth tide signals can be decomposed into three dimensional orthogonal vectors corresponding to each harmonic.

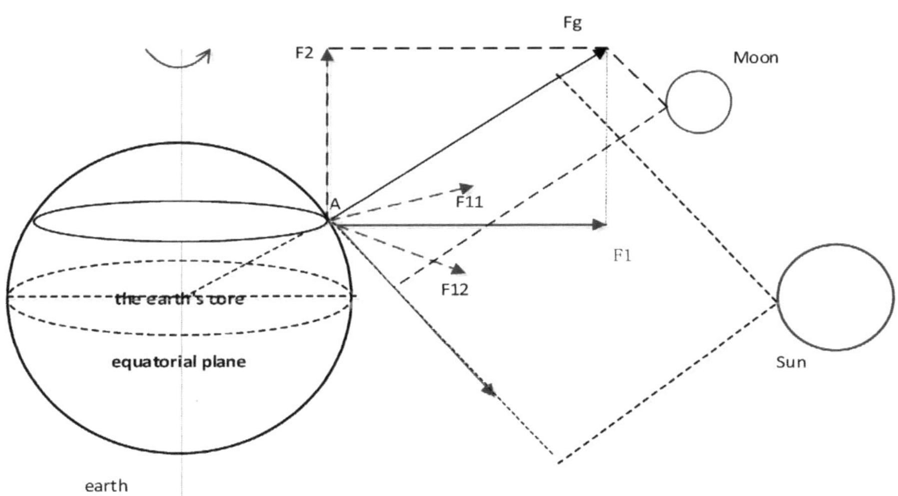

Figure.1 The orthogonal decomposition model of gravity earth tide signal

3. Analysis method

3.1 Independent component analysis (ICA) algorithm

ICA is one method for performing blind signal separation that aims to find de-mixing matrix W_2 of overlapping signal X , so that each component of $Z = W_2Y$ as far as possible independent of each other, in which the independent criterion function is G . ICA mathematical model:

IOP Conf. Series: Journal of Physics: Conf. Series **910** (2017) 012047 doi:10.1088/1742-6596/910/1/012047

Make $\vec{x} = (x_1, x_2,..., x_n)^T$, $\vec{s} = (s_1, s_2,..., s_n)^T$,

$$\vec{x} = A\vec{s} = \sum_{i=1}^{n} a_i s_i, i = 1,2,..., n \quad (1)$$

Where A is an unknown mixing matrix, and $\vec{x}$ represents the latent source signals.

$$A^{-1} = W_1 W_2 \quad (2)$$

Where W_1 is an whitening matrix which is usually computed singular or eigen-value decomposition of the covariance matrix of x , W_2 is de-mixing matrix.

$$\vec{y} = W_1 \vec{x} , \quad (3)$$

Here, $\vec{y}$ is the whitening signals.

De-mixing model:

$$\vec{z} = W_2 \vec{y} = G\vec{s} , \quad (4)$$

Where $\vec{z}$ is independent component, G for global matrix / system matrix. If $\vec{y} = \vec{s}$, the target of separation (or) to recover the source signal is achieved .

In fact, the ICA algorithm is including two aspects:

preprocessing (processing of X to zero mean, spheroidize);

core algorithm (separation matrix processing);

From above, We can get approximate estimates of various components of the source signals . In fact, Its essence is to use the objective function and the corresponding core algorithm (optimization algorithm)that is used to optimize the objective function, and the process of the optimal separation matrix is obtained.

The model of ICA can be depicted by Figure 2.

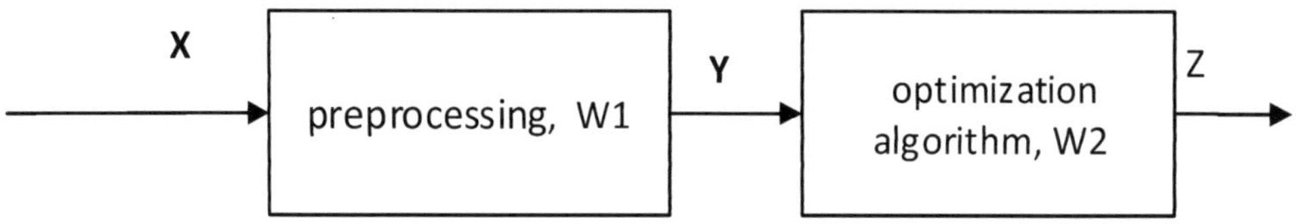

Figure.2 Processing of ICA

In this paper, in order to obtain the optimal separation matrix W_2 , we first use the negative entropy as the objective function of ICA, and then use the differential evolution algorithm to optimize the objective function.

3.2 Differential evolution (DE) algorithm

Differential evolution algorithm is a kind of group evolution algorithm which can remember the individual optimal solution and the information sharing in the population, it can solve the optimization problem by the cooperation and competition among individuals in the population, which means that it is a greedy genetic algorithm based on real number coding with the idea of preserving the best[8]. But compared with the traditional genetic algorithm, it not only retains the global searching strategy based on population, with real encoding, simple mutation differential and a survival strategy based on competition, and reduce the complexity of genetic operations.

One of the most frequently used mutation strategies, named "DE/rand/1/bin" [9],will be employed in this paper. Differential evolution's basic steps can be described as follows:

Step 1: Generating the initial population

$$x_i = (x_{i,1}, x_{i,2},..., x_{i,D}),\ i = 1,2,..., NP\ , \tag{5}$$

where D is the dimension, NP is the population size.

Step 2: Mutation operation

$$V_i(g+1) = x_{r1} + F \cdot (x_{r2}(g) - x_{r3}(g)) \tag{6}$$

where x_{r1}, x_{r2}, x_{r3} are randomly selected vectors, $x_{r2}(g) - x_{r3}(g)$ is differential vector and F is scaling factor.

Step 3: Crossover operation

$$u_{i,j}(g+1) = \begin{cases} v_{i,j}(g+1), & Rand_{i,j}(0,1) \le CR \text{ 或 } j = j_{rand} \\ x_{i,j}(g), & Otherwise \end{cases} \quad j = 1,2,..., D\ ,\ j_{rand} \text{ is } [1,D] \tag{7}$$

Where $Rand_{i,j}$ is a uniform random number within [0,1]. CR denotes the crossover constant with outcome $\in$ [0,1]. j_{rand} is a uniform random integer chosen index $\in$ 1,2,...,n[10].

Step 4: Selection operation

$$x_i(g+1) = x\begin{cases} u_i(g+1), & f(u_i(g+1)) < f(x_i(g)) \\ x_i(g), & Otherwise \end{cases}, \tag{8}$$

where $x_i(g+1)$ is target vector for the next generation. Until the algorithm reaches the maximum number of iteration by doing steps from step 2 to step 4 through.

3.3 ICA and gravity earth tide signals based on DE

In the optimization process, the DE algorithm can be based on the dynamic change of the whole population to adjust the search strategy, with a strong ability of global optimization, so the optimization objective function ICA algorithm to deal with the tidal gravity signal DE algorithm(Figure3).

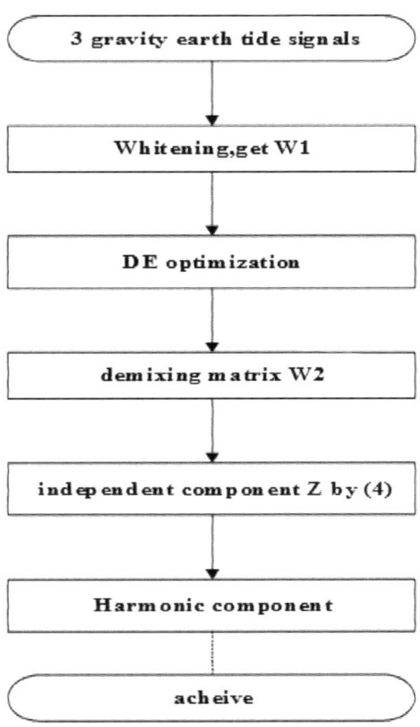

Figure.3 The processing of ICA-DE feature extraction

Step 1: Obtaining the gravity earth tide signal at the observation point. According to the gravity earth tidal model, it is known that the gravity earth tide signal is caused by the tidal force of the earth's rotation, the sun and the moon, so the gravity earth tidal wave is selected as the three signal;

Step 2: Whitening matrix W_1 is obtained for the signal of gravity earth tide;

Step 3: Using DE algorithm to optimize the separation matrix W_2;

(1). Set parameter initial value: population size NP and dimension $nDim$, mutation rate $F0$, the maximum iteration number GM, the crossover rate CR, the parameter value of the initialization algebra G.

(2).Start iteration loop: $i = 1 : Np$, from (5) to derive the variance vector, from (6) to cross the variance vector;

(3).The particle matrix transpose, according to the fitness value function $Fitness = E\{u^4\} - 3(E\{u^2\})^2$ that the kurtosis of the separated signals in non-Gaussian of measure to calculate the particle fitness value of Fitness ;

(4).According to the formula (4) optimization, If $Fitnessnew < Fitnessold(i)$, the termination of circulation, the matrix transpose of the particle, obtain the separation matrix W_2;

Step 4: According to the formula (8), the output signal $\bar{z}$;

Step5: The harmonic component of gravity earth tide signal is extracted, and its spectral analysis is done.

4. Results and discussion

According to the orthogonal decomposition model of gravity solid tidal signals.In order to extract the harmonic component and the geophysical information of the gravity earth tide, in the period of January 2010 to June 2010, three different observation points on the same longitude $(130^0E, 40^0N)$, $(130^0E, 50^0N)$, $(130^0E, 60^0N)$ were selected to carry out ICA-DE analysis on the gravity earth tidal signals.

DE algorithm parameter settings, as shown in Table 1.

Table.1 The parameter of the experiment

Parameter name	mutation rate(F0)	Maximum Iteration number (Gm)	Populat-ion size(NP)	Crossov-er rate (CR)	Initial algebra(G)	Dimensio-n (nDim)
Param-eter values	0.5	10	100	0.9	1	m^2

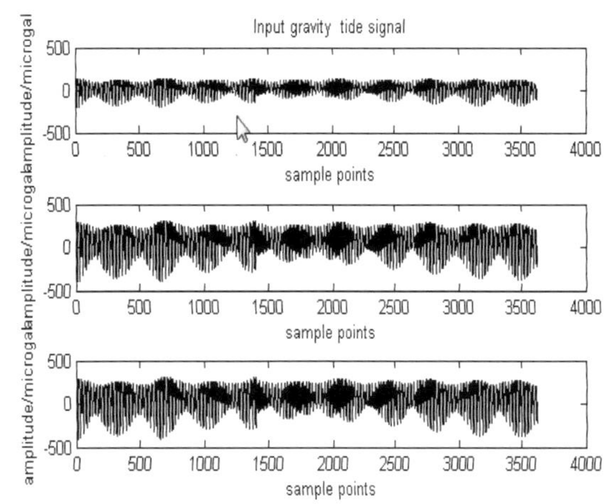

Figure.4. The input gravity earth tide signals

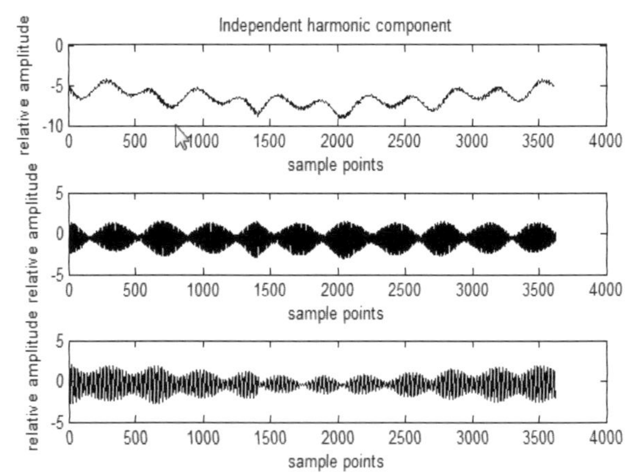

Figure.5. The ICA-DE processing results of the gravity earth tide signal

IOP Conf. Series: Journal of Physics: Conf. Series **910** (2017) 012047 doi:10.1088/1742-6596/910/1/012047

The waveform of gravity earth tide wave at the observation point is shown in Figure 4. The signal waveform after ICA-DE algorithm is processed as shown in Figure 5

To test the experimental results, the spectral analysis of the output signal is carried out (see Figure 6). According to the expansion formula proposed by Du Sen, the angular frequency of gravity earth tide signal is analyzed, and the theoretical value of the main harmonic component of the gravity earth tidal signal is calculated. The harmonic information extracted from the output signal is compared with the theoretical value as shown in Table 2.

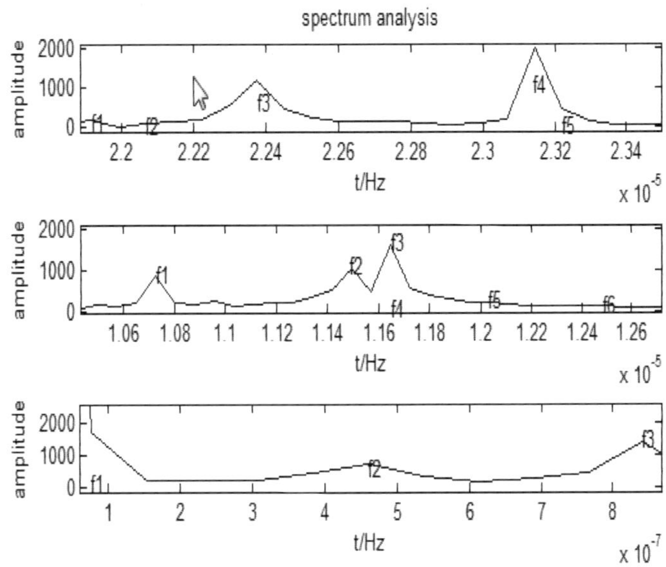

Figure.6. Spectrum analysis of the output signal after processing by ICA-DE algorithm

Table.2 Frequencies of observation points and theoretic values

output signals	Observation point	Observed frequency f(Hz)	Theoretic frequency f(Hz)	Harmonic component	
	f_1	2.192e-05	2.1944e-05		
	f_2	2.207e-05	2.2e-05		
$y_1(t)$	f_3	2.238e-05	2.2364e-05	Semidiurnal waves	
	f_4	2.314e-05	2.3148e-05		
	f_5	2.322e-05	2.3212e-05		
	f_1	1.073e-05	1.0759e-05		
	f_2	1.149e-05	1.1511e-05		
	f_3	1.165e-05	1.1637e-05		
$y_2(t)$	f_4	1.165e-05	1.1606e-05	diurnal waves	
	f_5	1.203e-05	1.2026e-05		
	f_6	1.249e-05	1.2453e-05		
	f_1	7.663e-08	6,33777e-08	half year waves	
$y_3(t)$	f_2	4.598e-07	4.2004e-07	waves month	Long period waves
	f_3	8.429e-07	8.4725e-07	half month waves	

From table 2:

(1): The actual frequency of the 3 channel gravity earth tide signal is in agreement with the theoretical frequency;

(2): Based on the orthogonal decomposition model of gravity earth tide, the ICA-DE algorithm can separate the gravity earth tidal signal in accordance with the harmonic component.

(3).The signal components of the 3 channel gravity earth tide signals are F_{11} , F_{12} and F_{2} respectively, and are not interfered with each other.

(4). Diurnal waves, and semidiurnal waves, harmonic component belongs to the equatorial plane signal components, long period of harmonic component is part of the earth's rotation axis signal components, between them have obvious modulation .

5. Conclusions

Based on the orthogonal decomposition model of gravity earth tide, this paper uses the algorithm of ICA and DE to optimize the gravity earth tide signals, and the results are consistent with the orthogonal decomposition model of gravity earth tidal wave. Then, By means of spectrum analysis, this method can get the spectrum information which is consistent with the generation mechanism of the gravity earth tidal signal. Finally, compared with the actual observed value (after some processing), we can get some information about the earthquake. This proves that the algorithm is effective, and it is more convenient to compare with other methods. In addition, multiplicative relations are still existed in the decomposed components extracted from the gravity earth tidal signals. This provides a good entry point for the next research.

6. Acknowledgment

This work is supported by National Natural Science Foundation of China (No. 41364002), State Science Foundation of Yunnan (No. 2009ZC048M) and Science Grant of Kunming University of Science and Technology (No. KKZ3201103022). The authors are graceful for the support of the grants.

7. References

[1] Guan Ze lin,Ning Jin Sheng.Earth shape and external gravity field[M].Surveying and Mapping Publishing House.1981.

[2] Zhang Li,Fu R S,Zhou Z,et al.Extraction of seismic precursory information from ravity tide at kunming station based on HHT[J]. Acta Seismologica Sinica. 2007

[3] LI Q Y, QUAN H Y. Analysis on the signal of the Gravity E- arth Tide based on ICA and extract the information of tidal harmonics [J].Kunming:Journal of Yun- nan University: Natural Sciences Edition,50(1): 845-850. 2015.

[4] WU Q C, ZHOU Z, LIANG H. Hong,et al.AM-FM model of gravity tide IMF and its non-linear data fitting[J]. Computer Engineering and Applications,45(30):138-142. 2009.

[5] Zhou Z,Shan X M,Zhang L,et al.The gravity tide of Kunming & Xiaguan based on the HHT[J].Chinese Journal Of Geophysics(in Chinese),51(3);836~844.2008.

[6] ZHANG C M.Cha feng jin hua suan fa li lun yu yun yong[M].Beijing: Beijing Institute of Technology press.2014.

[7] Li Y F, Extraction the geopgysical information from the gravity tide based on HHT[D].Kunming: Kunming University of Science and Technology. 2014.

[8] LIU B, WANG L, JIN Y H. Advances in differential evolution [J].Control and Decision,22(7): 722-729.2007.

[9] Chengfu Sun, ZHOU H Y, CHENG L Q. Improved Differential Evolution Algorithms[J].IEEE,142-145. 2012.

[10] LI G Y,LIU M G, The summary of differential evolution algorithm and its improvements[J],V3153-V3156.2010.

[11] ZHU W L, ZHOU J Z, XIAO J et al. An ICA-EMD feature extraction method and its application to vibration signals of hydroelectric generating units[J].Proceedings of the CSEE,33(29): 95-101.

[12] CUI L L, WU C G,WU N. Composite Faul Diagnosis of Rolling Bearings Based on EMD and ICA Algorithm[J].Beijing:JOURN- AL OF BEIJING UNIVERSITY OF TECHNOLOGY ,40(10): 1460-1464. 2014.
[13] IANG L, YANG M H, LI Y F, Hyperspectral remote sensing image classification based on ICA and SVM [J].Spectroscopy and Spectral Analysis30(10): 2725-2728.2010.
[14] SUN H P, XU H Z, CEHN W,et al.Study of Earth`s gravity tide and oceanic loading characteristics in Hong Kong area[J].Chinese Journal Of Geophysics(in Chinese),49(3): 725-734.2006.
[15] Yu X C, Hu D..Theory and appication of blind source separation [D]. Science Press.2011.

Tracking of Thermal Infrared Anomaly before One Strong Earthquake-In the Case of *Ms*6.2 Earthquake in Zadoi, Qinghai on October 17th, 2016

Xuan Zhang[1], Yuansheng Zhang[1,2], Xiufeng Tian[1], Qiaoli Zhang[1], and Jie Tian[1]

[1] Lanzhou Institute of Seismology, CEA, 730000, Lanzhou, China
[2] Lanzhou Base of Institute of Earthquake Forecasting, CEA, 730000, Lanzhou, China
Email: zhangx198565@126.com

Abstract. The detection and tracking process of thermal infrared anomaly before *Ms*6.2 earthquake in Zadio, Qinghai on October 17th, 2016, are reviewed and analyzed; then the different characteristics of thermal infrared brightness temperature data before this earthquake is described in details. According to these characteristics, the tracking process of thermal anomaly is divided into four stages, respectively identification stage, pre-judgment stage, tracking and approaching stage and verification stage. The anomaly forms and turning signals focused in each stage can provide clear indication information for earthquake pre-judgment; finally, the prediction efficiency and technical issues of this method are illustrated and discussed.

1. Introduction

Thermal infrared remote sensing detects the thermal condition on the ground surface by measuring the infrared radiation of earth-atmosphere system, and we can utilize this thermal condition to capture the earthquake precursor information, which can predict the location and time of earthquake occurrence over *Ms*5.5 on a one-month to half-year scale. It is shown by tracking and analysis result of massive thermal infrared data that this method possesses excellent verification rate in short-term earthquake prediction [1-11]. After the *Ms*6.2 earthquake occurred in Zadoi, Qinghai on October 17th, 2016, we traced back the precursor indication and predicting conclusion proposed in pre-earthquake consultation work by applying thermal infrared brightness temperature data, and found that quite complete and typical precursor responses of earthquake thermal anomaly have been formed since September 7th, 2016, when the thermal anomaly related to Zadoi earthquake was proposed in the weekly consultation report. Based on this, on the foundation of thermal infrared anomaly images, the characteristics of thermal anomaly before Zadoi earthquake at different stages are described, and the indication role of anomaly forms in judgment of epicenter in thermal infrared temporal-spatial evolution process is analyzed, and then the prediction efficiency of this method and the technical issues remaining to be improved are discussed.

2. Data source and processing method

The basic data in this paper come from the infrared remote sensing brightness temperature products of China geostationary meteorological satellite FY-2C/E/G (from National Satellite Meteorological Center of China Meteorological Administration). FY-2 satellites are loaded with optional infrared and visible-light scanning radiator, and can provide observation data for at least 24 times every day. In addition, the effective observation range is between 60°N and 60°S, 45°E and 165°E. The earth curvature and unstable factors in satellite orbits will cause the observation data not geometrically calibrated to distort. To avoid complex manual calibration, the geometrically calibrated and scaled data can be directly downloaded from National Satellite Meteorological Center. In everyday work, the one-hour average equivalent blackbody brightness temperature data provided by this center in standard format is mainly adopted. Considering the influence of sunlight radiation, the data from 1am to 5am (Beijing time) in mainland China and its neighboring areas is captured because the morning fog in basins and plains has not yet been formed in this period and the earth-atmosphere can basically reach an equilibrium point. In addition, the brightness temperature data obtained from FY remote sensing data network in selected period is in HDF format, which adopts stratified data structure storage and leads to large data file being generated every day. Faced with such massive data, we carry out channel separation and format transformation on original brightness temperature data. Then, window supplementation method is used for average processing and the processed daily brightness temperature data is supplemented into the original database.

Satellite thermal infrared remote sensing information reflect the integrated influence of ground-surface radiation, atmospheric radiation and remote sensing instrument parameters, which contain the changes of various periodical components, such as sun, cloud, rain, terrain and earthquake preparation. The earthquake thermal infrared anomaly signal is relatively weaker and tends to be submerged into these interference signals and background noises. How to suppress the interference and highlight the useful information is the key of our research. So we utilize the wavelet transformation method, which has been widely applied in digital signal processing in recent years, to separate the signals in different time scales[12-14]. The db8 wavelet basis in Daubechies wavelet system is used in wavelet transformation on brightness temperature data. Considering the different periods of above influential factors, the 7-order low-frequency wavelet component is abandoned to remove the influences from basic earth temperature field, annual temperature field and terrain; the interference from short-term meteorological factors, such as cloud, rain and cold-hot airflow, is removed by abandoning the 2-order high-frequency wavelet component; finally, the 2-order low-frequency component is used to deduct the 7-order low-frequency component to obtain the positive-negative thermal infrared brightness temperature data in the time domain. The earthquake short-impending anomaly information needed by the research is contained in this characteristic period range. Then, Fourier transformation is used to estimate the power spectrum of wavelet data in order to extract the advantageous frequency and amplitude we need in massive thermal infrared brightness temperature data. The specific calculation is as follows: considering that the duration of most earthquake infrared short-impending anomaly ranges between 10 days and 90 days, the window length n is set as 64 days in this paper, and m (m=1, 2, 3, 4, 5, m=1 in this paper) is the sliding window length, then calculates the power spectrum by Fourier transformation. Later, all the power spectra of each pixel is processed by relative amplitude, and the obtained time-frequency spatial data are applied for scanning over the whole time, space and frequency range, thus to analyze the characteristics of thermal anomaly before and after the earthquake[15-16].

3. Tracking of thermal infrared anomaly in Zadoi earthquake

3.1 Introduction to the earthquake event

According to the measurement from China Earthquake Networks, an *Ms*6.2 earthquake occurred in Zadoi County, Yushu State, Qinghai Province at 3:14pm on October 17th, 2016 (Beijing time). The epicenter coordinates are 32.8°N and 94.9°E, and the focal depth is 9km. This earthquake is located inside the Qiangtang Block, and the epicenter is near the Zadoi fault and about 35km from Zadoi County. In addition, this earthquake is strike slip-normal type. Over 1 month before the earthquake, clear and identifiable regional anomaly (red area in Fig.1) has been found in the daily thermal infrared tracking. To ensure the reliability of calculation results and highlight the range and intensity of anomaly response, we screen repeatedly and scan the local thermal infrared anomaly.

3.2 Tracking process of earthquake thermal infrared anomaly

This thermal infrared anomaly is reflected in 1st frequency band with high reliability in the statistical results of typical earthquake cases, and this frequency band has high sensitivity and responsive rate to the thermal anomaly in dry areas. According to the thermal anomaly area covered in Fig.1, we set the local scanning area as 22°~42°N and 85°~110°E, and four stages can be divided by an overview of thermal infrared anomaly tracking process before *Ms*6.2 Zadoi earthquake.

(1) Identification stage: this stage ranges from late August to early September in 2016. It is found by the processing and analysis of thermal infrared data in daily work that there is obvious thermal infrared anomaly within an area ranging among 22°~32°N and 85°~103°E (Fig.1). This anomaly demonstrates as a large-scaled and high-value red response area, mainly involving the region of Sichuan-Qinghai-Tibet border and mid-Yunnan. The anomaly distribution form demonstrates approximate isolated circle shape, and is basically similar to the thermal infrared anomaly forms and intensity in typical earthquake cases. Therefore, it is initially judged as a focus area which is in urgent need of tracking.

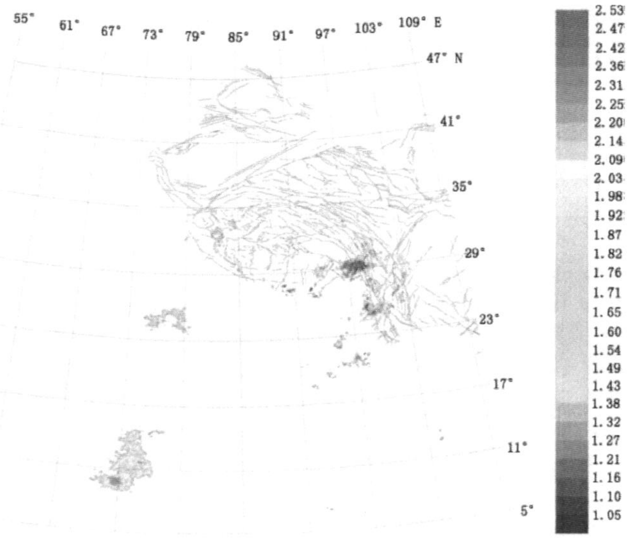

Figure.1. Thermal anomaly region 50 days before Zadoi *Ms*6.2 earthquake

(2) Pre-judgment stage: this stage ranges from early September to mid September in 2016. It is found by the temporal-spatial evolution figure of local thermal infrared brightness temperature power spectrum (Fig.2) that since mid August, there has been large-scale thermal infrared temperature rising area in Sichuan-Qinghai-Tibet border, Sichuan-Yunnan-Tibet border and western Yunnan. In addition, this temperature rising area gradually expands towards different directions. By September 4th, the thermal anomaly area reaches its peak. In the meanwhile, the anomaly forms are evolved into typical Z-shaped distribution. It can be seen from the statistical results of previous earthquake cases that the crossing point position of cross-distributed thermal anomaly forms is where the earthquake will occur frequently. Therefore, it is initially judged that the cross point of Z-shaped distribution, namely the black circle in Fig.3, will be the special area of concern in the next work (above conditions have been exhibited and illustrated in daily consultation conference).

(3) Tracking and approaching stage: this stage is mainly concentrated in mid September to early October. In order to focus the anomaly evolution process, narrow down the possible earthquake occurrence area and determine the earthquake intensity, we also take the local thermal infrared anomaly evolution figure as major tracking object (Fig.4). It is shown that since mid September, this thermal temperature rising area has entered into quick attenuation period. With the attenuation accelerating, the attenuation speed in Sichuan-Qinghai-Tibet border compared to other anomaly areas is the highest. It is shown by massive statistical results of earthquake cases that the epicenter is mainly located in the transition area where the anomaly appears and then quickly disappears. Therefore, we

judge that the possible earthquake occurrence area can be narrowed down to Sichuan-Qinghai-Tibet border (the upper block circle in Fig.3). We carry out time-sequential analysis on the relative power spectrum of the whole anomaly temperature rising area and obtain the change of thermal infrared relative power spectrum in recent 4 years (as shown in Fig.5), which can contributes to judge the time and intensity of earthquakes occurrence in the future. In Fig.5, the anomaly tracking in this area has lasted for above 50 days, and the relative change rate has reached its peak of recent 4 years on August 26th, 2016. Based on the conclusion from previous earthquake cases that the falling stage of anomaly peak is the possible period of time during which the earthquake occurs easily, we can approach the possible occurrence time of the future earthquake. In the meanwhile, combined with the temperature rising area and relative change rate of peaks, we present an empirical potential earthquake magnitude: about *Ms*6.0 (above conditions have been exhibited and illustrated in daily consultation conference).

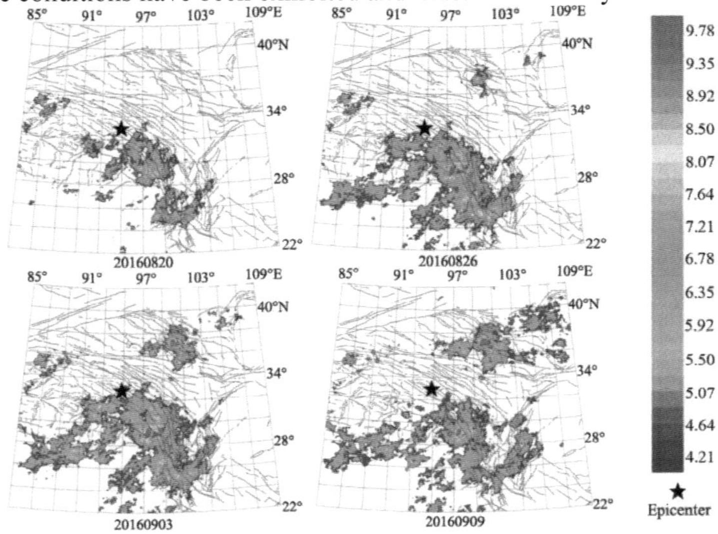

Figure.2. Thermal anomaly evolution before *Ms*6.2 Zadoi earthquake

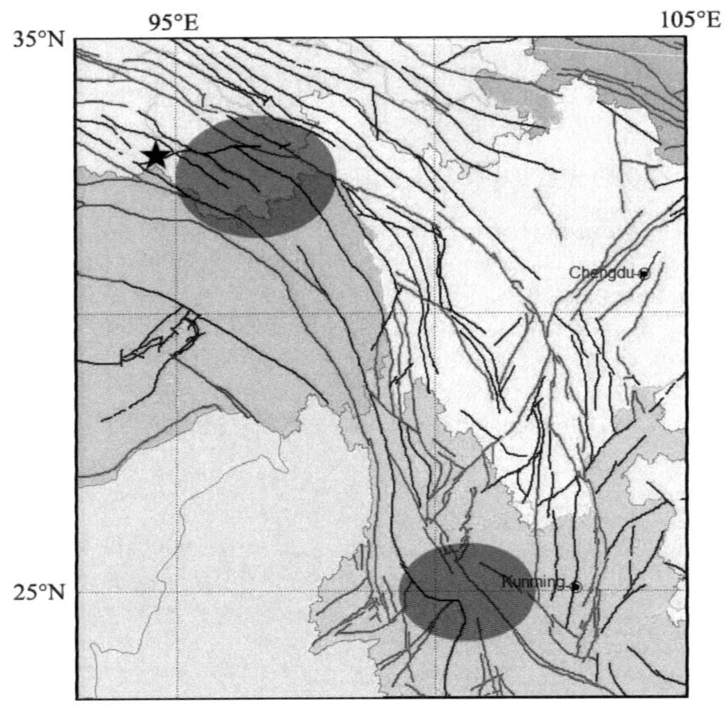

Figure.3. Delineation of thermal anomaly focuses area.

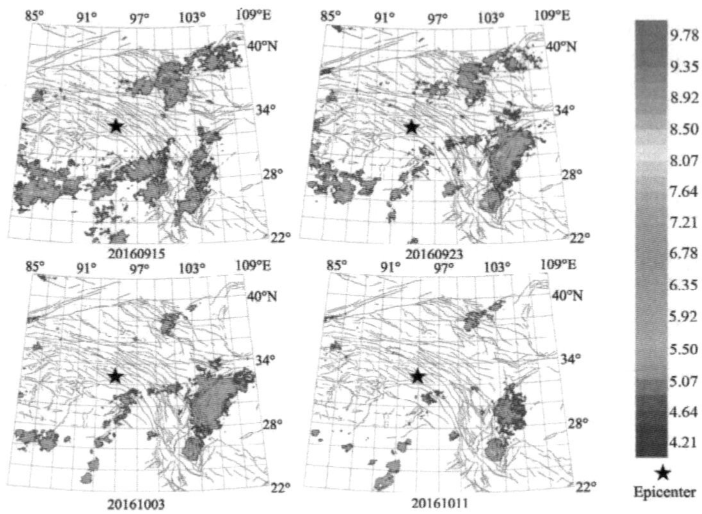

Figure.4. Thermal anomaly evolution before *Ms*6.2 Zadoi earthquake

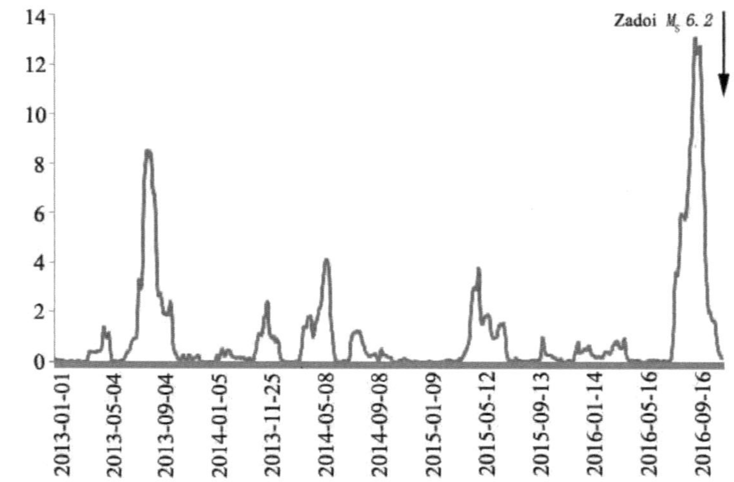

Figure.5. Time-sequential curve of thermal infrared anomaly before *Ms*6.2 Zadoi earthquake

(4) Verification stage: this stage ranges from early October to the date of earthquake occurrence. In this stage, almost all the infrared temperature rising phenomena in the thermal anomaly area have disappeared, and the anomaly has gone through weak-strong-weak evolution process. Based on the development process of anomaly forms, identified characteristics and anomaly turning signals, we speculate that the earthquake occurrence time is gradually approaching. Until 3:14pm on October 17th, 2017, 52 days after the relative power spectrum of brightness temperature reached its peak, an *Ms*6.2 earthquake occurred in Zadoi, Qinghai, and the epicenter was right near the upper black circle in Fig.3 (denoted by ★ in Fig.3), which is basically complying with our focus area. Apart from temporal and spatial correspondence, the intensity of this earthquake is basically close to the magnitude we predicted in approaching stage. Therefore, we believe that the thermal infrared anomaly in this long period can be completely attributed to *Ms*6.2 Zadoi earthquake, and the work of short-impending tracking based on thermal infrared brightness data has produced good reflecting earthquake efficiency in this earthquake case.

4. Conclusion and discussion

In this paper, the earthquake tracking and prediction are carried out by analyzing the thermal infrared remote sensing data. The possible earthquake occurrence regions are found by landmark turning signals in the evolution figure of thermal infrared anomaly. Then, the possible earthquake occurrence period of time and magnitude are determined further by the time sequential curve. It can be seen from this tracking process that the relative power spectrum of thermal infrared data amplified obviously before the Zadoi *Ms*6.2 earthquake, and this anomaly is reflected in the frequency band with high reliability. Above two points are the major judgment criteria and foundation for anomaly tracking process.

Through the thermal infrared anomaly evolution figure before Zadoi earthquake, we find that a large-scale thermal infrared anomaly area was showed up in the southeast of earthquake fault belt about 50 days before the earthquake. Contemporarily, we have provided relevant description of anomaly drawings and delineation of focus area in the weekly consultation work. Later, its anomaly form showed typical expanding and immigrating process. About one and a half month before the earthquake, the thermal anomaly area increased to maximum and then entered quick shrinking and recovering period. In the time domain, this thermal anomaly temperature rising phenomenon occurred about 2 months before the earthquake, and the earthquake occurred after the anomaly reached its peak; in the spatial domain, this thermal anomaly form which had experienced a process of initial appearance, accelerated development and final gradual attenuation, demonstrated a typical Z-shaped distribution, and the epicenter was right near the cross point of this Z-shaped distribution; in the intensity domain, the judgment on the earthquake magnitude was empirical, by comprehensively considering the temperature rising area and relative change rate of peak, the predicted earthquake magnitude is very close to the actual magnitude. Through the above analysis, it can be found that the thermal anomaly tracking effect related to Zadoi *Ms*6.2 earthquake is obvious and significant.

The current research on earthquake prediction and forecasting in China is still in its exploration stage, when we carry out a great deal of work on theoretical model establishment and earthquake case summary. Although massive long-term precursor observation work has been done across a wide range in multiple channels, the most destructive earthquakes occur in areas with relatively weaker earthquake monitoring capability. In other words, the continuous monitoring on various physical parameters in earthquake source region is lacking, and the precursor observation data cannot form an organic system and thus the relevant data cannot be analyzed and compared effectively. The infrared brightness temperature data of geostationary satellite can provide large-scale, quick and stable ground surface information due to their temporal consistency and spatial comparability, so it enjoys obvious advantages in the research of earthquake temporal-spatial characteristics. Currently the judgment of earthquakes based on thermal infrared temporal-spatial parameters is empirical. The prediction reliability on earthquake occurrence time and location is high, and the judgment on these two key elements is the emphasis in the whole tracking process. In Zadoi earthquake, the extraction of time and location elements, especially the precise delineation in location judgment, is obtained by common indicators from typical earthquake cases. On the issue of magnitude judgment, because of the complexity of multiple factors, such as geological environment, atmospheric environment, et al, the earthquake magnitude is not positively proportional to the size of thermal infrared anomaly area. In addition, although there is temporal-spatial correlation between earthquake activity and thermal infrared anomaly, it is still difficult to relate the anomaly to the seismogenic mechanism theoretically. So in the future short-impending tracking work, how to extract information truly related to earthquake precursor from complex thermal anomaly images, distinguish the thermal radiation anomaly characteristics in different regions and geological environments, and propose judgment suggestions and indicators which contain specific indicating information, will be the issue for us to research and explore deeply.

5. Acknowledgments

This work is financially supported by Earthquake Tracking Project of China Earthquake Administration (2017010413), and thanks to National Satellite Meteorological Center of China Meteorological Administration for provision of geostationary satellite data.

6. References

[1] Jin M., Xingjian S.: An attempt to study fault activity using remote sensing technology-a case of the Mani earthquake. Seismology and Geology. 22(3), 210--215 (2003)

[2] Yuansheng Z., Wenrong S., Hui X.: Satellite thermal infrared anomaly before the Xinjiang-Qinghai border *Ms*8.1 earthquake. Northwestern Seismological Journal. 24, 1--4 (2002)

[3] Zuji Q., Changgong L.: Satellite infrared brightness temperature anomaly-Earthquake imminent precursor. Science in China. 28(6), 564--573 (1998)

[4] Xuan Z., Yuansheng Z., Congxin W.: Anomaly of thermal infrared anomaly before the Lushan *Ms*7.0 Earthquake. China Earthquake Engineering Journal. 35(2), 272--277 (2013)

[5] Zhihui D., Yu W., Meihua C.: Satellite infrared anomaly of several strong earthquakes in China mainland. Seismology and Geology. 25(2), 327--337 (2003)

[6] Chunli K., Zhengwei C., Lize C.: Analysis on the satellite infrared anomaly feature before west to Kunlun Mountain Pass *Ms*8.1 earthquake. Northwestern Seismological Journal. 25(1), 12--15 (2003)

[7] Xiudeng X., Xiangmin X., Yu W.: Satellite infrared anomaly before Nantou *Ms*7.6 Earthquake in Taiwan, China. Acta Seismologica Sinica. 22(6), 666--669 (2000)

[8] Yuansheng Z., Xiao G., Congxin W.: The characteristics of seismic thermal radiation of Japan *Ms*9.0 and Myanmar *Ms*7.2 earthquake. Chinese J. Geophys. 54(10), 2575--2580 (2000)

[9] Congxin W., Yuansheng Z., Sahoxing H.: Thermal-Infrared variation of Andaman islands *Ms*7.5 earthquake on August 11, 2009. Journal of Seismological Research. 34(2), 41--45 (2009)

[10] Xuan Z., Yuansheng Z., Jie T.: Anomaly of thermal infrared anomaly in Jinta *Ms*4.7 Earthquake on March 21, 2016. China Earthquake Engineering Journal. 38(S2), 212--217 (2016)

[11] Fan Z., Dan X., Yuyun Z.: Relation between infrared brightness temperature anomalies and earthquake in Zhejiang province and its adjacent areas. China Earthquake Engineering Journal. 38(1), 58--264 (2016)

[12] Shunyun C., Peixun L., LIqiang L.: Wavelet analysis of thermal radiation of land surface and its implementation in the study of current tectonic activities. Chinese J. Geophys. 49(3), 824--830 (2016)

[13] Pan X.: Wavelet-based methods for detecting earthquake information in remote sensing data. Institute of Earthquake Forecasting, CEA (2009)

[14] Xuan Z., Yuansheng Z., Congxin W.: Thermal infrared anomaly before Yiliang of Yunnan *Ms*5.7 earthquake. China Earthquake Engineering Journal. 35(1), 171--176 (2013)

[15] Yuansheng Z., Xiao G., Meijiao Z.: Wenchuan earthquake brightness temperature changes from satellite infrared information. Chinese Science Bulletin. 55(10), 1--8 (2010)

[16] Congxin W.: Research on seismic thermal radiation. Lanzhou Institute of Seismology, CEA (2011)

Service-oriented Software Defined Optical Networks for Cloud Computing

Yuze Liu, Hui Li and Yuefeng Ji

Beijing Advanced Innovation Center for Future Internet Technology, Beijing University of Technology, Beijing, China
395983438@qq.com, lihui@bupt.edu.cn, jyf@bupt.edu.cn

Abstract. With the development of big data and cloud computing technology, the traditional software-defined network is facing new challenges (e.g., ubiquitous accessibility, higher bandwidth, more flexible management and greater security). This paper proposes a new service-oriented software defined optical network architecture, including a resource layer, a service abstract layer, a control layer and an application layer. We then dwell on the corresponding service providing method. Different service ID is used to identify the service a device can offer. Finally, we experimentally evaluate that proposed service providing method can be applied to transmit different services based on the service ID in the service-oriented software defined optical network.

1. Introduction

With the development of new network services (such as high-definition video push service, cloud computing, content delivery networks (CDNs), grids, distributed storage, video on demand), the optical networks are expected to have the characteristics of dynamic flexibility and service adaptability. Meanwhile, Software Defined Optical Network (SDON) and Network Function Virtualization (NFV) has been proposed to maximize the network resource utilization. SDON can be considered as extension of Software Defined Network (SDN) in optical networks. By separation of control plane from date plane, and the standard application programming interface (such as OpenFlow) between these two planes, software defined optical network (SDON) leads to a more flexible control manner for optical networking via software programming [1]. With the help of centralized controller, the SDON controller (e.g., NOX, ODL, RYU) can have a global network view and make strategies in an optimum way. The potential benefit of SDON paradigm is that SDON offers a convenient platform for experimentations of new techniques in optical network.

Network Function Virtualization (NFV) has also attracted a lot of attention in the last years [2] [3]. Traditional telecom services are highly dependent on physical topology graph and vendor specific hardware. NFV uses highly mature virtualization techniques to virtualize the entire classes of network node functions [4] [5]. In this regard, NFV aims to implement network functions that are typically deployed in specialized hardware as software instances running on commodity servers. Besides, with network virtualization, infrastructure providers can offer different virtual optical network (VON) for tenants who have a variety of different demands for physical resources, such as different physical interfaces, bandwidth requirements etc.

The current expansion of cloud computing follows mainly from several recent IT trends, including the "dot-com boom," which started an explosion of interest in outsourcing IT services; the popularity, maturity, and scalability of the present Internet; and the appearance of large data centers developed by companies such as Google, Amazon, and Microsoft [6]. There are three major types of services provided in cloud computing systems: Infrastructure as a service (IaaS), platform as a service (PaaS) and software as a service (SaaS). However, a transport network is an indispensable element of the cloud computing model since it provides connectivity between distributed computing resources. Current transport networks are not efficiently designed to meet the needs of cloud environments for the optical network is still treated as a blind pipe without considering the various cloud computing requirements. Besides that, the separated ownerships for cloud facilities and backbone networks also raise the difficulties of deep cooperation between the two parts.

However, in order to cope with this problem, we can focus on two aspects. First, combining the SDON and NFV technologies to enhance dynamic flexibility. As mentioned above, SDON paradigm offers a unified control plane but it faces the problem of dedicated hardware without a standardized northbound interface. NFV can improve resource utilization by partitioning and abstracting the network infrastructure but it should also be noted that various virtualization network functions may lead to scheduling difficulties [7]. Applied to this context, the combining of SDON and NFV can highly improve the network flexibility in the global network view. With the unified control plane, the virtualization network functions scheduling problem should be processed in the most optimal way. Second, abstract services from the network nodes to extend service adaptability. Different nodes can provide different services. For example, the all-optical switches may have low latency while the flexi-grid optical network devices may have the potential ability to realize the required capacity on demand.

The rest of this paper is organized as follows: Section II presents the proposed service-oriented software defined optical network architecture. Then a service function variable optical node and the service providing method are described in Section III. Section IV is the experimental demonstrate. Finally, Section V concludes the work.

2. Service-oriented Software Defined Optical Networks architecture

Protocol independent transmission method in SDON is proposed in our former work [8]. Based on this method, optical layer functions like Power Equalizer (PE), Dispersion Compensator (DC) and Multicast can be defined as a network services which can be call by upper layers. A service-oriented software defined optical network architecture is proposed based on various optical networks, as shown in Figure. 1. The legacy DWDM and flexi-grid optical network can be regard as transparent network, while the optical consists of O/E/O nodes can be regard as opaque network. In order to realize the service defined optical network, four layers are presented to realize the required function.

The service-oriented software defined optical network architecture include a resource layer, a service abstract layer, a control layer and an application layer. The resource layer consists of all the optical network resources and each optical device logically associated with its own OpenFlow Agent. However, different devices can provide different services. For example, the basic service which the devices in transparent optical network can offer is all-optical switching. But in opaque optical network, with the help of O/E/O convert, the basic service function is the 3R (Reamplification, Reshaping and Retiming). All the information of optical devices will be collected to the service abstract layer through the extended OpenFlow protocol.

Control layer provides different service-oriented virtual optical networks according to the requests from the application layer. The request can be denoted by $R=(s,d,w,q)$, where s is the source node and d is the destination node for the request. w is the bandwidth request by R and q is the service ID. First, according to the service ID, take the $q=2$ for example, the control layer selects nodes from the all-optical switching pool to generate service-oriented virtual optical network. Second, the control layer checks whether the s and d can be mapped into the virtual optical network. Finally, if mapping successful, a path will be calculated for the request based on the w.

The top of the architecture is an application layer. The application layer consists of various third-party applications including graphical user interface (GUI). Cloud computing services requests can be triggered via the Restful API.

The service abstract layer bridges the control layer and resource layer. This layer abstract services from optical nodes via the extended OpenFlow protocol messages. We have defined some services such as all-optical switching, asymmetric spectrum assignment (ASA), low latency, quality of service (QoS), which is shown in Table I. Thus, according to the service ID, the services a node can offer will be collected and a lot of resource pools can be built for the control layer

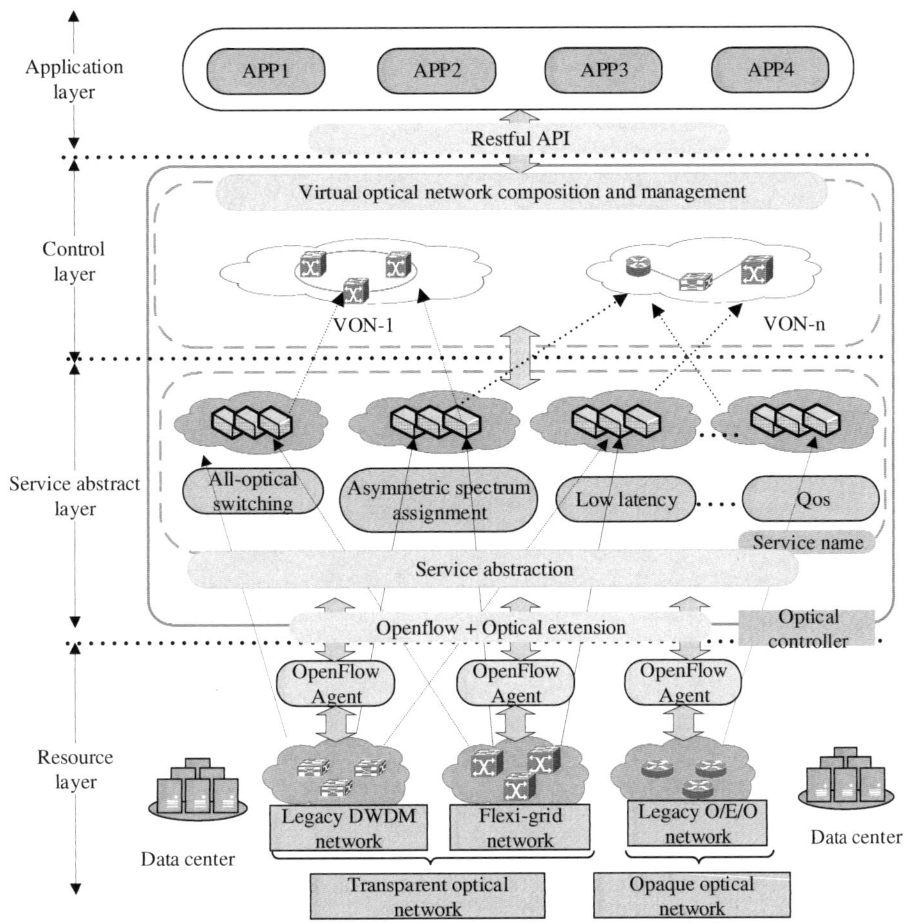

Figure 1 Service-oriented Software Defined Optical Networks architecture

3. Service Variable Optical Node And Service Providing Method

As mentioned above, in service-oriented software defined optical network architecture, it is very important to abstract the services from the hardware. In this section, we designed a service variable optical node (SVON) to provide various services. Its implementation structure is shown in Figure. 2. However, the SVON is composed of three micro-electro-mechanical systems (MEMSs) primarily. Among these MEMSs, various devices are deployed, such as EDFA, DC and optical coupler. As shown in Figure 2, the SVON can provide services function such as PE, DC and Multicast. Only one service can be enabled at a time. A proprietary OVS has been designed for this SVON. In order to realize the service-oriented software defined optical network, a new OFP_SERVICE_MOD message is designed to control devices based on service ID, which is shown in Figure.3. The Service ID is used to identify the service and the Command type will be set to *0x0000ffff* if the service is requested. Spectrum slot ID is designed for the flexi-grid node.

According to the service ID, the SVON can provide different services. For example, when service ID turns to 4, route 4 will be select and route 2 will be choose if all-optical switching is needed. In our early study [9], we have proved that the PE service is an effective solution for adaptive adjustment of transmission performance on optical path.

Table 1 Service Id And Service Name

Service ID	Service
0	None
1	PE without O/E/O
2	All-optical
3	DC
4	Multicast
5	ASA
6	PE with O/E/O
7	Low latency
8	Low BER (QoS)

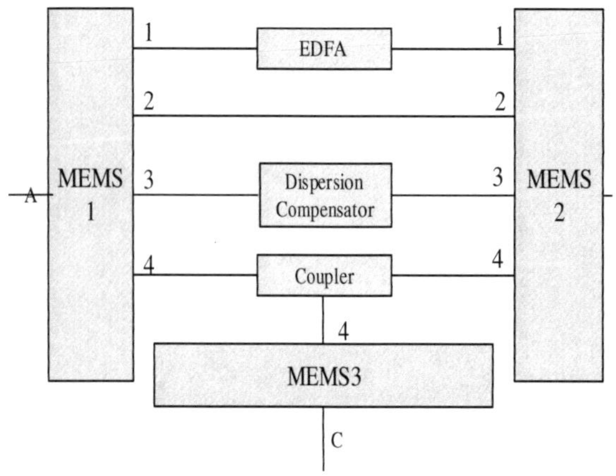

Figure 2 The structure of SVON

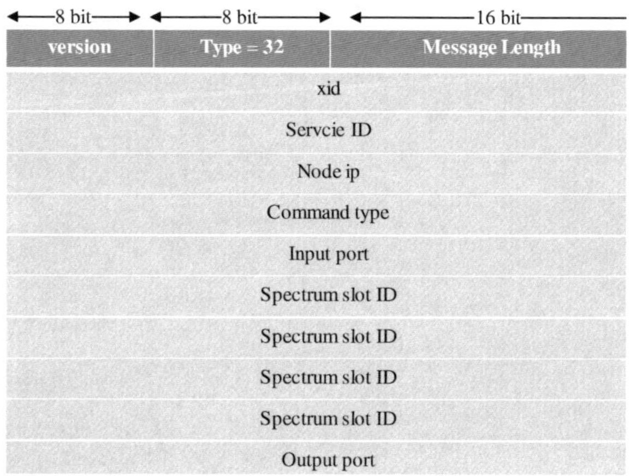

Figure 3 OFP_SERVICE_MOD message

4. Experimental Setup And Results

The experiment testbed is shown in Figure. 4. We construct the optical network based on two OpenFlow-enabled RODAMs (OF-ROADMs), a SVON node, a OpenFlow-enabled OXC and a OpenFlow switch (OF-switch). These nodes offer two paths for traffics. Path1 go through OF-ROADM1~OF-switch~OF-ROADM2 and path2 go through OF-ROADM1~SVON~OF-OXC~OF-ROADM2. All the nodes have the all-optical switching ability except the OF-switch. All the devices are controlled by ODL through OVSs. A High-definition (HD) camera is used to provide the serial digital interface (SDI) high-definition video.

The internet control message protocol (ICMP) is simulated by ping packages between PC1 and PC2 and it is the traffic 1. The carrying wavelength is 1550.67nm. HD camera is used to provide supervisory signal with centre wavelength 1553, 17 nm and it is the traffic 2. However, if a traffic does not apply for any service, the traffic will go through the path1 for path1 is shorter than path2. As traffic2 is a SDI signal without compressing, the traffic cannot pass the OF-switch and the terminal TV cannot receive any signal. Therefore, we set $q=2$ for traffic2. It is means that the all-optical switching service is required and the experiment result is show in Figure. 4, Figure. 5 and Figure. 6.

From the experiment result, we observe that only ICMP signal go through path1 and the SVON offer the all-optical switch service successful, which is shown in Figure. 4. As the SVON is very special and then it does not have the ability to dynamically adjust the bandwidth, only three fields (Service ID, Node ip, Command type) are needed. Figure. 5 shows OFP_SERVICE_MOD message issued by the controller and Figure. 6 shows that both the traffic1 and traffic 2 are fully accepted.

IOP Conf. Series: Journal of Physics: Conf. Series **910** (2017) 012049 doi:10.1088/1742-6596/910/1/012049

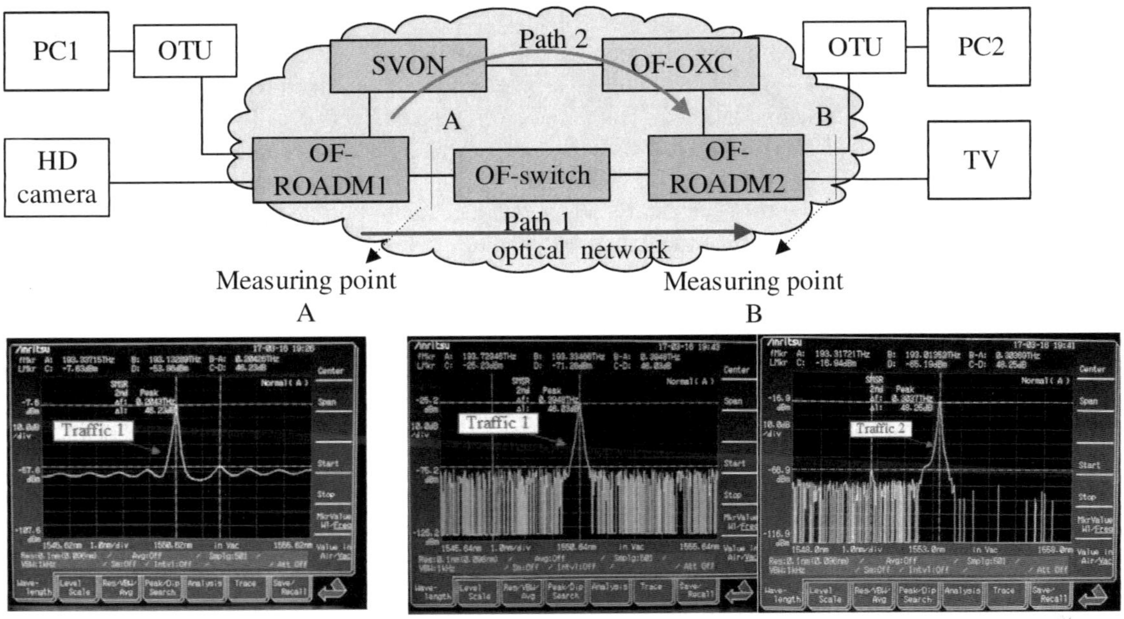

Figure 4 Experiment network architecture

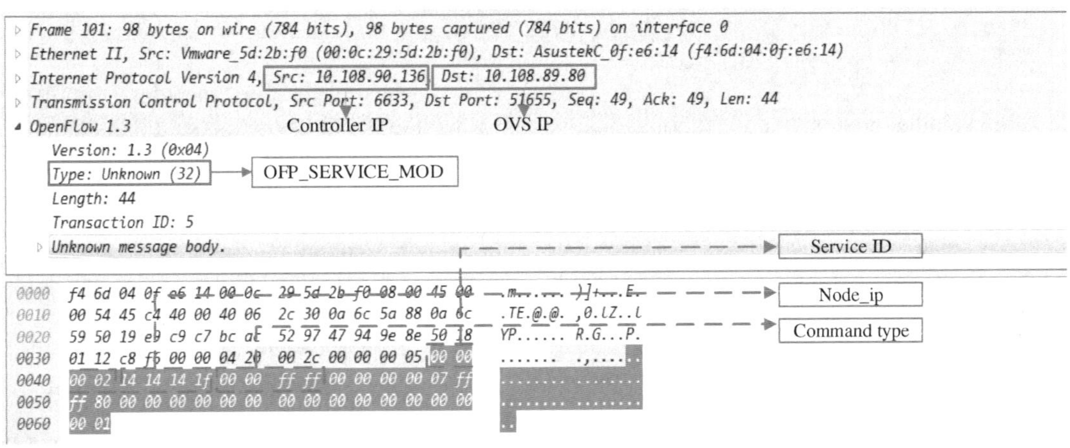

Figure 5 The “OFP_SERVICE_MOD” message issued by the controller and captured by Wireshark

Figure 6 Transmission effect of traffic1 (left) and traffic2 (right)

5. Conclusions

In this paper, we have proposed a service-oriented software defined optical network architecture by utilizing combination of SDN and NFV technologies. The proposed architecture and technologies were experimentally validated and demonstrated. The service-oriented SDON provide a method to enhance the dynamic flexibility and service adaptability. Some optical layer functions such as PE, DC and Multicast are defined as services which can be called by SDN controllers, and as a result, clouding service with different physical interfaces can be adapted directly into optical networks, and less O/E/O transitions are needed which can reduce the network delay and cost shapely.

6. Acknowledgment

This work has been supported by Funds of Beijing Advanced Innovation Center for Future Internet Technology of Beijing University of Technology (BJUT), China.

7. References

[1] W. F. Xia, Y. G. Wen, C. H. Foh, D. Niyato, and H. Xie, “A Survey on Software-Defined Networking,” IEEE Communications Surveys & Tutorials, vol. 17, pp. 27–51, 2015

[2] R. Nejabati, S. Peng, M. Channegowda, B. Guo, and D. Simeonidou, “SDN and NFV Convergence a Technology Enabler for Abstracting and Virtualising Hardware and Control of Optical Networks,” OFC W4J.6, 2015.

[3] K. Shibata, H. Nakayama, T. Hayashi, and S. Ata, “Establishing PDCA cycles for agile network management in SDN/NFV infrastructure,” IFIP, pp 619-625, 2015

[4] R. Vilalta, R. Muñoz, et al, “Transport Network Function Virtualization,” J. Lightwave Technol., vol. 33, no. 8, pp. 1557–1564, 2015.

[5] R. Muñoz, R. Vilalta, et al, “Transport network orchestration for end-toend multi-layer provisioning across heterogeneous SDN/OpenFlow and GMPLS/PCE control domains,” J. Lightwave Technol., vol. 33, no. 8, pp. 1540-1548, 2015.

[6] M. A. Klinkowski, and K. Walkowiak, “On the Advantages of Elastic Optical Networks for Provisioning of Cloud Computing Traffic,” IEEE Netwrok, Vol. 27, no. 6, pp.44-51, 2013.

[7] T. Wood, K. K. Ramakrishnan, J.Hwang, G. Liu, and W. Zhang, “Toward a Software-Based Network: Integrating Software Defined Networking and Network Function Virtualization,” IEEE Netwrok, Vol. 29, no. 3, pp.36-41, 2015.

[8] Y.Z. Liu, H Li, Y.F Hou, Y. J. Qiu and Y. F. Ji, “Protocol Independent Transmission Method in Software Defined Optical Network,” Conference on Optoelectronic Technology and Application, 2016, IPTA15-023

[9] Y. F. Hou, H. Li, W. Zhao, Y. Z. Liu and Y. F. Ji, “QoT-aware adaptive adjustment of optical path transmission performance in softwaredefined optical network,” Electronics letters, vol. 51, no. 15, pp. 1184-1185, 2015.

Design of Xen Hybrid Multiple Police Model

Lei Sun, Renhao Lin and Xianwei Zhu

Zhengzhou Information Science and Technology Institute, ZhengZhou, China
Email: zhjuya_boy@163.com, zhu13939055330@163.com,1056670972@163.com

Abstract. Virtualization Technology has attracted more and more attention. As a popular open-source virtualization tools, XEN is used more and more frequently. Xsm, XEN security model ,has also been widespread concern. The safety status classification has not been established in the XSM,and it uses the virtual machine as a managed object to make Dom0 a unique administrative domain that does not meet the minimum privilege.According to these questions, we design a Hybrid multiple police model named SV_HMPMD that organically integrates multiple single security policy models include DTE,RBAC,BLP. It can fullfile the requirment of confidentiality and integrity for security model and use different particle size to different domain. In order to improve BLP's practicability , the model introduce multi-level security labels. In order to divide the privilege in detail, we combine DTE with RBAC. In order to oversize privilege, we limit the privilege of domain0.

1. Introduction

All In recent years, cloud computing technology enjoys rapid development. With the method of renting calculating resources provided by the cloud service provider, the user can get easy and quick access to software and hardware resources. Virtualization technology[1] is a good way to achieve these benefit that promote the usage of the hardware while decrease the cost of the power.

Considering the issue of security, the direct access to the physical hardware resource is not usually granted to VMs in the virtual machine system. In order to enhance its security, some research scholars have proposed the security model for Xen virtualization platform by referring to the theory and technology of safe operating system.

However, currently so many applications need their multiple parts to communication with each other,which may run on different virtual machines. Consequently, the secure resource sharing manner is needed between related virtual machines, that is to say, the appropriate access control mechanism is needed to enforce the access control between VMs.

In this paper, we introduce a hybrid multiple security police model for Xen named SV_HMPMD(Security Virtualization_Hybrid Multi-Policy Model). There are some advantage as follow: It achieve the minimum privilege to management Dom0 by RBAC and DTE. The single subject of operation of Dom0 is divided into several subjects. In the second place, We improve the BLP model by multi security label. Unlike the original BLP which give each subject and object the current security label, object of SV_HMPMD model are given security label range of (w_min, r_max)and (L_mino, L_maxo) to improve the flexibility of the BLP. The trusted subject is divided to prevent the whole system from the occurrence of a single privileged entity. A number of trusted subjects replace the single trusted subject of Dom0. Privileged domain and privilege state are added to manage the privilege and privilege dynamics.

The rest of this paper is organized as follows. Section2 discuss related work, and the framework design of SV_HMPMD model is discussed in Section 3. In section 4, we discuss the concrete implementation of SV_HMPMD and the resolution of key issues.

2. Related Knowledge Introduction

In the field of secure operating system, domestic and foreign scholars have done a lot of research on access model: Xen's access control model ACM / sKype [2]provides a distributed access control between virtual machines, but it has only a few simple security policies such as STE and the Chinese wall. In order to improve the applicability of the security policy, Xen's security model draws on the mixed model of RBAC and TE in SELinux to manage Dom0 and DomU, and uses the virtual machine as the management unit to carry on the simple role division, it realize the multiple access control strategy but The same management granularity is used for Dom0 and DomU.Ease of Use. But there is no fine-grained division of privilege operations in Dom0 that causes Dom0 privilege to be centralized and violates the principle of system security of minimum privilege. In order to solve this problem, Xenon[3] proposed by the US Navy Laboratory and Qubes[4] proposed by Joanna Rutkowska's divide Dom0 into multiple virtual device domains, Constructing Access Control Model to Realize Dom0 's Fine - grained Management. It can not meet the needs of the classification system. Because it use virtual machine or virtual machine group as a control unit and does not safely classify the system resources. Chuliang wen[] used BLP on a virtual machine system, using virtual machines as access control objects and defining access rules. He use the traditional BLP model, which has * - strict characteristics, did not achieve efficient rights management.

In this paper, we present a hybrid multi-strategy model SV_HMPMD in the face of Xen virtualized environment. Its characteristics are as follows:

(1)We use RBAC and DTE to achieve fine-grained the privileges and access rights of Dom0. The operation of single subject of Dom0 is divided into multiple entities together to complete.

(2) We change the BLP model into a multi-security label BLP. It is different from the original BLP that given the current security label to each subject and object, SV_HMPMD model gives the subject and object with (w_min, r_max), (L_mino, L_maxo) two security label range to improve the flexibility of BLP.

(3) We divide the trusted subject to prevent the occurrence of a single privileged subject to control the entire system. Multiple trusted principals replace Dom0's single trusted subject. Multiple trusted subject replace Dom0's single trusted subject.

3. SV_HMPMD Model

3.1 SV_HMPMD goals

SV_HMPMD security goal is to avoid user of illegal storage of information in Virtualize multi-user sharing information and effective protection system information from by unauthorized users. And the privilege of Dom0 as a trusted subject is divided to achieve fine-grained management.

(1) Due to the multi-user nature of the virtualization platform, the administrator has different permissions from the user's behavior. In order to achieve this goal, particle-size partition is made between the management domain Dom0 and the user domain DomU.

(2) Implement security domain isolation. The model can provide an isolation mechanism that limits the operation of non-trusted domains to ensure that other users' information is not stolen.

(3) It is can restrict the privilege of the trusted subject, because the trusted subject can perform operations that violate the security constraints, and if the trusted privilege can perform too much privilege. it will threaten the confidentiality of the system.

3.2 SV_HMPMD Design

First, DTE provides an easy-to-implement access framework that implements access between subject and object and ensures system integrity. It use the BLP security model to ensure the confidentiality of the virtual domain itself and other domains. Through the trusted subject to specify the Dom0, management domain, violates the BLP security features but the implementation of the system operation allows the action. The privilege interference between the trusted subject and the trusted subject, and the granting of its privilege are achieved by RBAC. RBAC main function is the security label allocation and management in SV_HMPMD. The the access control of the general subject and the privileged subject is realized by the DTE model.

IOP Conf. Series: Journal of Physics: Conf. Series **910** (2017) 012050 doi:10.1088/1742-6596/910/1/012050

4. SV_HMPMD Specific Implementation

From the previous chapter we can see SV_HMPMD through the improved BLP model to achieve the system access control. It achieve the management area Dom0 privilege division and the different users DomU static permissions division though RBAC. Combining it with DTE as a flexible access control model achieve a system integrity and privileged controlled. The principle of the above characteristics will be described later.

4.1 BLP implementation

A multi-tagged BLP model was used to improve the original BLP model as follows:

Using the label range instead of the original label, it do not reduce the confidentiality of the original model while improving the flexibility of the model. SV_HMPMD uses the security tag range that is $(w\_\min, r\_\max)$. w_min is the minimum security label that the subject can write to the object. r_max is the maximum security label that the subject can read on the object. In this range is expressed as ok (s). SV_HMPMD also gives a tag range $(L\_\min_0, L\_\max_0)$ for the object, which is the minimum access security label, the object's maximum access security label. Object security level within this range is ok (o).

BLP safe kilometers are as follows:

a) If a subject can access an object in a read or write manner, the current security level of the subject must control the maximum security level of the object.

$$\forall o \in O, s \in S.(s, o, r \vee a) \in b \Rightarrow f_s(s) \geq L\_\max_0(o)$$

b)*-property

$$(s, o, a) \in b \Rightarrow L\_\max_0 \leq r\_\max_0$$

$$(s, o, r) \in b \Rightarrow L\_\min_0(o) \geq w\_\min_s(s)$$

$$(s, o, w) \in b \Rightarrow ok(o) \subseteq ok(s)$$

c) SV_HMPMD divides the subject into a trusted subject and a general subject. The definition of the general subject is the same as the original BLP model. Part of the trusted subject condition $w\_\min(s) \neq r\_\max(s)$, this part of the subject can only access the object of its security rights. Trusted subject are not restrict to security policy constraints and limit their minimum privileges.

4.2 The Division of the Trusted Subject

In order to limit the authority of Dom0 and prevent Dom0 from controlling the entire system, the Dom0 operations are subdivided into several trusted subsets and assigned to the corresponding "roles" through RBAC. We assign these "roles" to the process of completing the privileged operation and replace the original single trusted user with a trusted user.

4.3 Integrity Implementation

Due to the existence of a credible subject it can violate the security rules. Improper modification of the security information and integrity of the trusted body itself is threatened. The non-passability and inter-domain transformability of DTE's access control relationships provide a better fine-grained access to the model. The idea of its realization is:

The object with the same integrity security requirement is set to the same type. And it uses the Domain Interaction Table (DTT) to specify the access rules of the subject to the object, to prevent the disclosure of the security information to the unauthorized subject, and to limit the privilege of the trusted body with excessive authority in the BLP to ensure the minimum privilege of the trusted body. The isolation of information between domains is achieved through the DIT table of the DTE model, the separation between subject prevent the credibility of subject through the flow of information between the integrity of damage. Integrity access control rules:

a) domain-type access rules

b) domain-domain interaction rules

c) Defines the inter-domain conversion operation, the provisions of the object can occur after the conversion between those domains. The system Domain and channel can be in their specific domain.

IOP Conf. Series: Journal of Physics: Conf. Series **910** (2017) 012050 doi:10.1088/1742-6596/910/1/012050

4.4 Privileged Access Control

The current model for the management of privileged operations is still very small, Most of the management is non-privileged access control, but the actual operation of privileged access seriously affects the security of the system, such as the abuse of privilege. Privileged non-isolated propagation can lead to serious system security issues. So many viruses and malicious code achieve malicious spread by the acquisition of privilege. This paper deals with the problem of dynamic change from privilege scope and privilege. We solve the problem of privilege role checking and privilege calculation. SV_HMPMD first isolates the same role by static role partitioning on the same user, and binds the privileged process to a specific privileged domain. These privilege fields have a specific privilege attribute, that is, each privilege domain specifies the privilege function that the privileged subject of the domain can perform. It limit the scope of the franchise. The privileged state is a sign of the privilege process change. From the above privilege management rules we can see that the privileged process is affected by the role of the process Domain Pr and the privilege field Pd of the process and the shared memory to be executed by the privileged process.

In this paper, (P_i,P_p,P_e) indicates the state of privilege. (f_i,f_e) represents the privileged state of shared memory. It is described in Table 1.

This article uses genetic algorithm that improved by Dragovice[5] to privilege calculations.

$$p_i' = p_i \wedge f_i$$

$$p_p' = (f_i \vee (p_i' \wedge p_p)) \wedge p_r \wedge p_d$$

$$p_e' = p_p' \wedge f_e$$

(p_p', p_i', p_e') is the set of privilege states that the process is executed after. The formula indicates that the privilege state of the inheritable privileged state is the minimum privileged state under the current state of the process privilege state and the shared memory privilege state. P_d is the largest privilege set allowed by the domain where the process is located, The set of privileges that are allowed to execute is determined by the privileges of the current role and the privileged domain and the state of the privileged process. Thus it limit the scope of the privilege set. On the other hand it also achieved the isolation of users of privilege.

Table 1 Privilege process and shared memory

symbol	Privilege set	object	description
P_i	inherit the privilege set	Privileged process	a process can pass privileges to subsequent processes
P_P	Permission privilege set	Privileged process	A maximum privilege set for a privileged operation that a process can perform
P_e	Execute privilege set	Privileged process	A collection of privileges that the current process can perform
f_i	inherit the privilege set	Privileged process	A special collection of privileged operations on shared memory.
f_e	Execute privilege set	Privileged process	When the shared memory is executed, the shared memory maps the privileges of its privileged processes.

4.5 Overall Frame

In SV_HMPMD, the three access control models cooperate with each other in the integrity and confidentiality of functions. This section describes the SV_HMPMD access procedure on a framework. The judgment of privileged as a supplement to BLP confidentiality access runs through the entire access judgment process. Privileged access is divided into supplementary privileged access and managed privileged access. Supplemental privilege access is a supplement to other access policies that are privileged to meet access rules. Managed privileged access is privileged access that violates the access rules of a particular access control model. Such as MAC_OVERRID and DAC_OVERRID, The former can violate MAC access control , The latter is a violation of DAC access control. For the first privileged access, it work with other access rules, so Privileged access takes effect only if the access meets the requirements of the access control model. The second privileged access is common in Dom0, As long as you meet this privilege or other privileged access it will able to take effect. The whole process is shown in Figure 2.

When the subject place an access request, it is first determined whether the domain where the subject is located is allowed to perform the operation by the DTE. If access is allowed, it is judged by MAC and DAC. If it is allowed to access, subject access the object. If it do not allow access, it check whether it meets the appropriate violation of the rule privilege. If it does not allowed to access, it deny to visit.

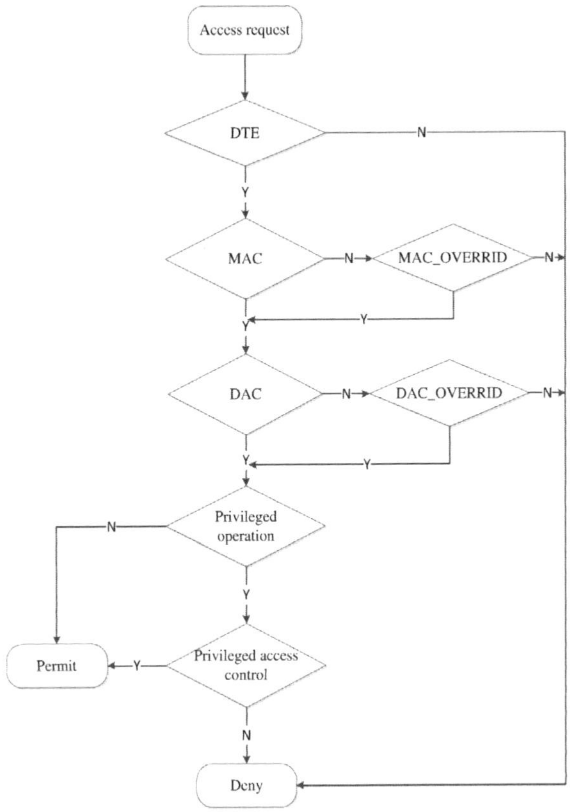

Figure 1 Access control process.

5. Model Rules

5.1 Model Element

User set U; Subject set S={St, SU, P, AR},S_t is a trusted subject, S_u is general subject and S_u=S-S_t; P is a privilege, each privilege depends on the subject. AR is the Subject of access to the operation of its state are {READ, WRITE, APPEND, CREATE, EXECUTE, AUTO};

Object set O={O_t, O_U, M(O), H(O)}; O_t is trusted object, It generate the subject image by executing a trusted object; O_u is general object; M (O) represents the subject's access to the object; H (O) represents the function that the object set maps to the set of set powers. If $O_1 \neq O_2$, then $H(O_1) \cap H(O_2) = \varnothing$.

Domain-type access relationship: $DTT := D \times T \times A$; Domain is D={d_1, d_2, d_3...d_n}; T is the type; A is the way of access $A \subseteq AR$;

Model status output S={subject_P, output}; Subject_P: subject process and privilege with mapping; Output (decision output) value is YES, NO.

The model currently accesses the state. $B = \{b \mid b \in (S, O, AR)\}$;

safety label LEVEL={G, C,≧}; The security level G represents the security level of the subject object. The security category C represents the security scope that the subject and object can perform. ≧ safety label control relationship, High-level security labels can control low-level security labels when security is included.

Label function f={f_s, w_min, r_max, L_min_0, L_max_0};f_s is the maximum security label function of subject. w_min is subject minimum writeable label; r_max is subject maximum readable label; L_min0 is object minimum writeable label; L_max0 object maximum readable label;

5.2 Migration Rules

In order to meet the needs of the actual system development, we give 11 migration rules that the model must meet. These are get_access,execute, release_access, add_user, delete_user, create_object, authorize_role, revoke_role, authorize_dm, authorize_D, authorize_executable_domains. Due to space reasons only three of them to explain and formal proof.

Role1. Domain read and write access get_access(s,d,o,t,p)

If the requested access type is not appropriate, the next state remains unchanged and the decision output is deny; otherwise, check the following conditions:

1.$(d,t,p) \in DTT$

2.$(s,p) \in M(o)$

3.$((p = r) \wedge (L\_\max_o(o) \le v\_max_s(s)))$

$\vee ((p = a) \wedge (L\_\min_o(o) \ge a\_\min_s(s)))$

$\vee ((p = w) \wedge (ran(o) \subseteq ran(s))$

4.$\exists priv \in P, priv \in subject\_privs(s)$, You can get Domain access to p of resource. If the condition 1-3 or the condition 4 is satisfied, the decision output is yes and the next state changes as follows:

$b^* = b \cup \{s,o,p\}$, Other state variables remain unchanged.

Role 2 Domain is executed directly on object O

There are two cases:

a)d=d', If the type of access requested is not appropriate, The next state remains unchanged. Decision output UNDEFINED; Otherwise, check the following conditions:

1.$(d,t,e) \in DTT$

2.$L\_\max_o(o) \le v\_\max_s(s)$

3.$\exists priv \in P, priv \in subject\_privs(s)$

If the condition 1-2 is satisfied or the condition 3 is satisfied, the decision output is YES, The next state changes to follow state:

$b^* = b \cup \{s,o,e\}$

$subject\_images(s^*) = o$

$if \; s \in S_T, then$

$subject\_p(s).P_i = subject\_p(s).P_i \cap$

$a.executable\_p(o).f_i$

$subject\_p(s).P_p = (executable\_p(o).f_i \vee$

$(w.subject\_p(s).P_i \wedge$

$p\_role(subject\_role(s)).P_p \wedge$

$p\_domain(subject\_role(s)))$

$subject\_p(s).P_e = subject\_p(s).P_p \wedge$

$executable\_p(o).f_e$

If the above conditions are not met, the decision output is the NO, the next state remains the same.

b) When the subject and the object are not in the same domain, they need to be domain conversion to complete the operation. If the type of access requested is not appropriate, the next state remains unchanged. Decision output UNDEFINED; otherwise, check the following conditions:

1. $o \in domains\_executables(d^{'})$
2. $(d,t,e) \in DTT$
3. $(d,d^{'}) \in DMD$
4. $role\_domians(d') \subset role\_domains(d)$
5. $L\_max_{o}(o) \leq v\_max_{s}(s)$
6. $\exists priv \in P, priv \in subject\_privs(s)$

If the condition 1-5 is satisfied or the condition 6 is satisfied, the decision output is YES, The next state changes to follow state:

$b^{*} = b \cup \{s, o, e\}$

$subject\_images(s*) = o$

$subject\_domains(s*) = d$

$if\ s \in S_{T}, then$

$(S_{t} = S_{t} \cup \{s\}) \wedge (S_{u} = S_{u} - \{s\}) \wedge$

$subject\_p(s).P_{i} = subject\_p(s).P_{i} \cap$

$execute\_p(o).f_{i}$

$w.subject\_p(s).P_{p} = (executable\_p(o).f_{i} \vee$

$(subject\_p(s).P_{i} \wedge p\_role(subject\_role(s)) \wedge$

$prive\_domain(subject\_role(s))$

$subject\_p(s).P_{e} = subject\_p(s).P_{p} \wedge$

$executable\_p(o).f_{e}$

$if\ (d_{s}, d_{o}) \in DMD$

$o \in domains\_executables(d_{o})$

$role\_domains(d_{o}) \subset role\_domain(d_{s})$

If the above conditions are not met, the decision output is the NO, the next state remains the same.

Role3. Add Domain

A user who has a role r in domain d requests a domain to be added to the system .If the type of access requested is not appropriate, the next state remains unchanged. Decision output UNDEFINED; otherwise, check the following conditions:

1. Add the user's privileges to r
2. d allows users to add privileges and privileges take effect
3. $u \notin U$

If the condition 1-3 is satisfied, the decision output is YES, The next state changes to follow state:

$U^{*} = U \cup \{u\}$

If the above conditions are not met, the decision output is the NO, the next state remains the same.

6. Concolusion

In this paper, according to Xen's isolation, integrity, privilege control, we introduce the hybrid model SV_HMPMD. It guarantees the confidentiality of the system with improved BLP .Through the RBAC and DTE to ensure the integrity of the system and isolation and the privilege management of the system is realized on the this basis. Future work will improve the model and improve the integration of the model to reduce the performance expenditure of the model in actual use.

7. References

[1] Sailer, Reiner, E. Valdez, and T. Jaeger. "sHype : Secure Hypervisor Approach to Trusted Virtualized Systems." Ibm Research Report Rc119.2(2005):331--352.
[2] Freitas, Leo, and J. Mcdermott. "Formal methods for security in the Xenon hypervisor." International Journal on Software Tools for Technology Transfer 13.5(2013):463-489.
[3] I Lu, Qi, Z. X. Cao, and Z. Xue. "Xen-based Qubes OS Technical Architecture and Safety Performance Analysis." Information Security & Technology (2014).
[4] K. Elissa, "Title of paper if known," unpublished.
[5] R. Nicole, "Title of paper with only first word capitalized," J. Name Stand. Abbrev., in press.
[6] Y. Yorozu, M. Hirano, K. Oka, and Y. Tagawa, "Electron spectroscopy studies on magneto optical media and plastic substrate interface," IEEE Transl. J. Magn. Japan, vol. 2, pp. 740–741, August 1987 [Digests 9th Annual Conf. Magnetics Japan, p. 301, 1982].
[7] M. Young, The Technical Writer's Handbook. Mill Valley, CA: University Science, 1989.
[8] Electronic Publication: Digital Object Identifiers (DOIs):
[9] Article in a journal:
[10] D. Kornack and P. Rakic, "Cell Proliferation without Neurogenesis in Adult Primate Neocortex," Science, vol. 294, Dec. 2001, pp. 2127-2130, doi:10.1126/science.1065467.
[11] Article in a conference proceedings: H. Goto, Y. Hasegawa, and M. Tanaka, "Efficient
[12] Scheduling Focusing on the Duality of MPL Representatives," Proc. IEEE Symp. Computational Intelligence in Scheduling (SCIS 07), IEEE Press, Dec. 2007, pp. 57-64, doi:10.1109/SCIS.2007.357670.

IOP Conf. Series: Journal of Physics: Conf. Series **910** (2017) 012051 doi:10.1088/1742-6596/910/1/012051

The Analysis and Processing of Spectra with Splicing Abnormality in LAMOST Based on High Performance Computing Platform

Fanlong Meng, Jingchang Pan* and Jingjing Yu

School of Mechanical, Electrical & Information Engineering, Shandong University at Weihai, Weihai, China

Email: morphy8@163.com, *pjc@sdu.edu.cn, yujingjing2018@163.com

Abstract. Since the completion of LAMOST the observation had produced a large number of data, and officially released data. There are some poor quality spectra in the observation data and even the data released to the outside world. Splicing abnormal spectra is one kind of them. Splicing abnormality is poor continuity spectrum showed in the splicing wavelengths of the red and blue end. This problem can be caused by several factors, such as stability of instrument, observation condition, the response function, and so on. It has important effect on the research and analysis of spectra. It is of great significance to study the automatic identification of the splicing abnormal spectra. The method in this paper we first do the spectrum pretreatment, then the spectral was intercepted into red and blue in order to fit curves. For the two fitting curves, a series of wavelength points are selected to calculate the difference of these points. For the difference set, the statistical characteristics of the difference are analyzed, such as mean and standard deviation. Then we calculate the area of the two curves. Based on these values, we proposed an evaluation function. At the same time, we provided a method for detecting the splicing abnormal spectra based on the high performance computing platform. Through a lot of experiments, the efficiency was greatly improved compared to stand-alone.

1. Introduction

The Large Sky Area Multi-Object Fiber Spectroscopic Telescope [1][2][3] (LAMOST) had been established by Oct, 2008. Since its completion, LAMOST captured a large number of spectral data. In these data, there are still some low quality spectra such as splicing abnormal spectra. With the LAMOST survey going on, thousands of spectra had been captured at each observation night. In the face of such a large amount of data, there is no efficient tool for detecting abnormal splicing spectra and it leads to spectral studies cannot be carried out smoothly. Therefore, it is of great significance to find a method to detect the abnormal spectrum automatically.

In the process of LAMOST survey, the star is divided into red and blue end and captured separately. After observation, the red end data and blue end data are merge together. Because of stability of instrument, observation condition, the response function, some splicing spectra become abnormal. This abnormality in the spectrum is in the splicing region (5600Å -6000Å) spectrum shows flux jumping (Figure.1 (a)), fault (Figure.1 (b)), and trend anomaly (Figure.1 (c)). At present, there is little research on the automatic detection method of LAMOST spectral splicing anomaly. In this paper, an evaluation function to define the spectrum abnormal degree is proposed based on the characteristics of splicing abnormal spectrum. This study not only helps to enhance the detection of abnormal splicing speed spectrum, also contribute to the process of quantification and quality improvement, also it may encourage further research, for example, based on the evaluation we can explore solutions to eliminate anomalies.

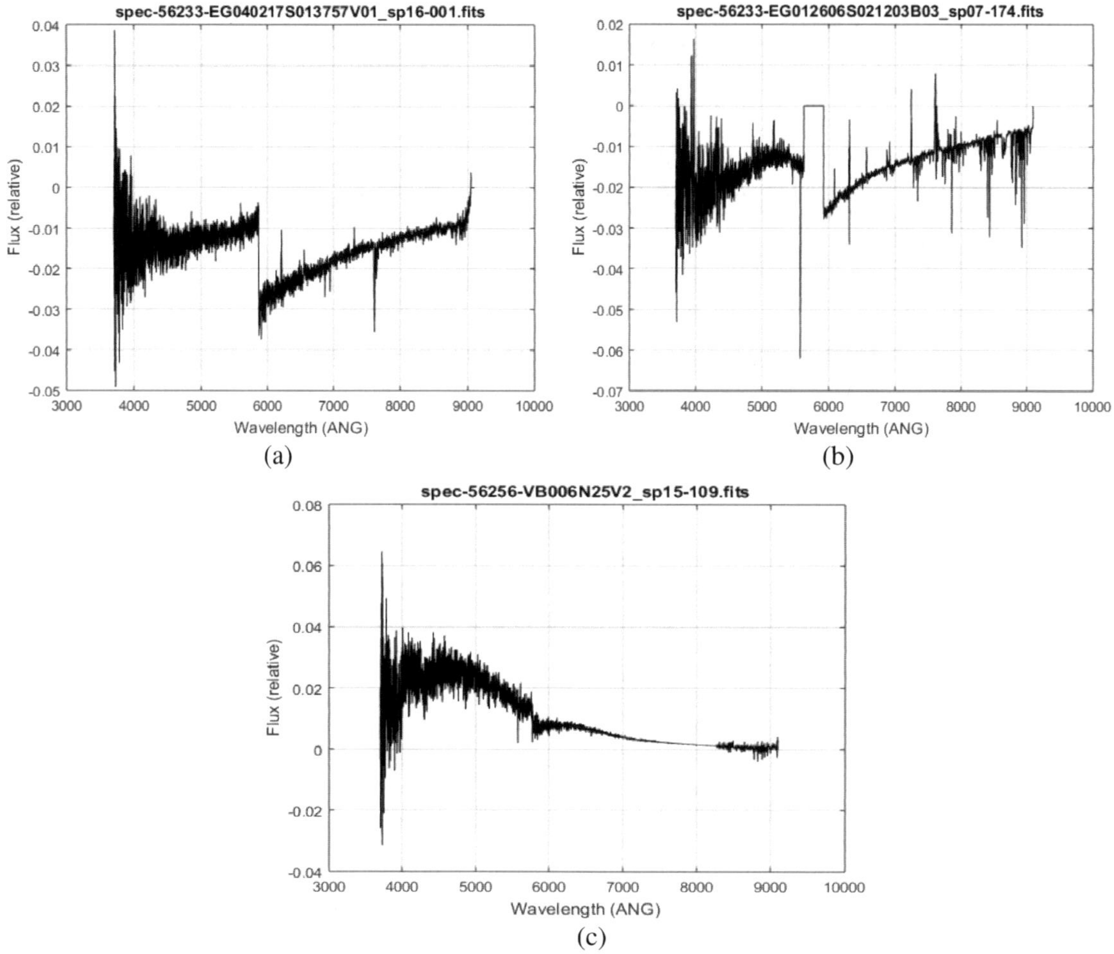

Figure 1 Abnormal splicing spectrum (a) this is the flux jumping. (b) this is the flux fault. (c) this is the trend anomaly

2. Method Detail

The splicing wavelength region is 5600Å-6000Å. After the capture, we mark the overlapped region as the baseline, keep one end motionless, make the other end enlarge or reduce a certain number of times to let the flux coincided in the overlapped wavelength region. Through this, we get the entire curve of full wavelength. Outside the overlapped region, the relative flux is same with origin spectrum and it becomes abnormal under the both effect of red and blue end flux.

In this paper, we propose a piecewise fitting method to analyze the spectrum. Firstly, we intercept the wavelength 5200 Å-6000Å as the blue end, 5600Å-6400Å as the red end. Then, fit the two end separately, so we get two curves. In figure.2, the subgraph upper is the fitting curve of origin spectrum. In this figure, the black line is the origin line, the blue solid line is the fitting curve of blue end and the red solid line is the fitting curve of red end. The two red dotted is the position of 5600Å and 6000Å. The subgraph below is the zoom in detail of the splicing region.

The method of this paper is based on the following principles. The two fitting curves are closely approximate if the spectrum is normal and the flux difference is small. The flux differences distribution is uniform.

In the splicing region, we select a series of points with equal wavelength intervals on the two fitting curves. Then calculate the flux difference of these points. For these differences, we get its mean value and standard deviation and calculate the area surrounded by two lines. If the mean, standard deviation and the area are under the thresholds, we mark it as normal spectrum.

The thresholds of the above three statistical moments (characteristics value) can be got by the analysis of the template spectra. In this paper, the template spectra we use is proposed by Peng Wei in the paper "On the Construction of a New Stellar Classification Template Library for the LAMOST Spectral Analysis Pipeline" [4] for LAMOST. Based on these thresholds, we put forward an evaluation function to define the abnormality degree for the splicing abnormal spectrum.

2.1 Get the Thresholds by Subsection Simulation

In this paper, we define the spectrum abnormal or not by subsection simulation method based on thresholds value, so the thresholds are fundamental. We got the thresholds by analyzing the template spectra because of the template spectra are good splicing spectra. The method to obtain the thresholds is:

The first step, we do the pretreatment, such as flux normalization, sodium line removal. Intercept the flux line of wavelength 5200 Å-6000Å as the blue end (mark the wavelength as bwave, the flux as bflux) , and 5600Å-6400Å as the red end (mark the wavelength as rwave, the flux as rflux).

The second step, we do six order polynomial fitting for the flux line of blue and red end and get red end fitting curve (rp) and the blue end fitting curve (bp).

The third step, for the two fitting curves rp and bp, calculate the flux difference by selecting a series of evenly spaced wavelength points and get flux difference array called df.

The fourth step, for the array df, we calculate its mean value (mark it as MEAN) and standard deviation (mark it as STD).

The fifth step, for the two fitting curves rp and bp, calculate the area surrounded by two lines in the splicing region (mark it as AREA).

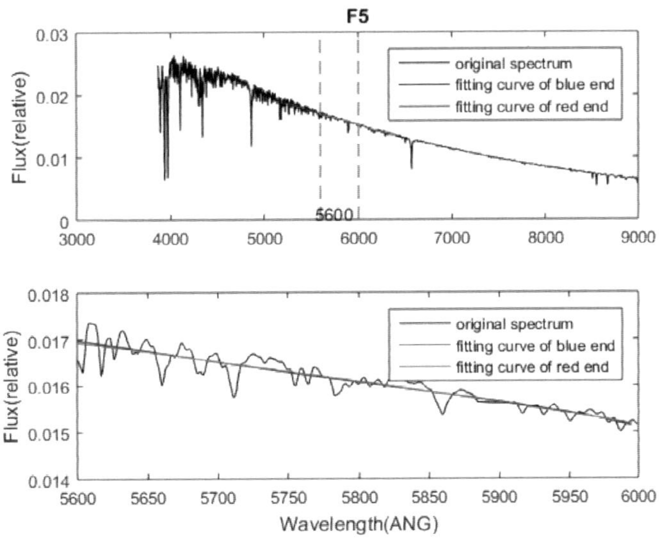

Figure 2 Subsection simulation curve

Table 1 Thresholds of eigenvalues (Note: unit: 10^-4)

TYPE	MEAN	STD	AREA
O	0.86	1.18	325
B	0.08	0.1	30
A	0.15	0.2	80
F	0.17	0.19	78
G	0.25	0.3	100
K	3.7	4.5	1400
M	7.5	9	2900

In the data analysis process, we use 183 template spectra, it contains star type, such as O、B、A、F、G、K、M、double star and carbon star. We only analyze these kinds such as O、B、A、F、G、K、M and ignore other types. We deal with the rest 176 template spectra with the five steps above, so we get 176 groups of characteristic values (it can be expressed as three tuple (MEAN, STD, AREA)). With these statistics, the 176 groups' data can be divided it into 7 groups according to the star type. Each group takes the maximum value of the corresponding statistic as the threshold of this type. Through this, we get the thresholds of the MEAN value, standard deviation and the area. The thresholds in our paper is presented in table.1:

2.2 Remove Spectra of Zero Flux

In the splicing abnormal spectrum, there is a kind of abnormal spectrum of faults, as shown in Figure 1 (b), which the flux is all zero of continuous wavelength points. In this paper, firstly, we judge whether the spectrum exists this fault or not, if yes, we define it as splicing abnormal spectrum.

The specific methods are as follows.

(1) Define a window of "w" points, we set w 10 in this paper.

(2) In the splicing region, we traverse the flux values of all the wavelength. If the flux is zero at one point and all flux are zero in the window of w points from this point, we define this spectrum abnormal.

2.3 Anomaly classification

We use the same subsection simulation method to deal with the test spectrum as the template spectrum. Also, we get the mean value (α′), standard deviation (δ′) and the area surround by the fitting curve (ε′). The method to judge the spectrum is as follows.

The first step, get the star type.

The second step, find the characteristics values according to the star type. The mean value, standard deviation and area threshold are α, δ, ε.

The third step, we define an evaluation function. $p = \sum_{i=1}^{\varepsilon} p_i * \lambda_i . \lambda_1 = \frac{\alpha'}{\alpha}$, $\lambda_2 = \frac{\delta'}{\delta}$, $\lambda_\varepsilon = \frac{\varepsilon'}{\varepsilon}$, p_i is the weight of λ_i. In this paper, p_i (i=1, 2, 3) are 0.4, 0.1 and 0.5 respectively. In the three statistics, the area is the most direct response to the abnormal situation, the weight is set to 0.5, the mean response to the flow of the overall situation, the degree of response to the abnormal situation is also large, set its weight is 0.4, the standard deviation of the difference shows the overall floating situation, the reference is slightly smaller, set the weight of 0.1. Considering these three factors, the weighted average value is used to quantify the degree of splicing abnormal spectra.

The fourth step, judge the degree of splicing abnormal spectra based on the p value calculated by the third step.

Experiments show that the design of the three weights and the evaluation function can be used to judge whether the spectrum is abnormal or not. The value of the evaluation function reflects the degree of abnormal spectrum, and the experimental results show that the evaluation function of the normal spectrum is below 0.9. Through lots of experiments, the abnormal degree is divided into three grades. For the abnormal spectrum, we define the evaluation function value of cut-off point was 5 and 10, namely: $p<=0.9$ is normal spectrum, $0.9<p<=5$ is defined as first level, $5<p<=10$ is defined as second level, $p>10$ is defined as third level. The two level anomaly and the boundary point of the three level anomaly can be adjusted manually according to different needs. The demarcation point between the second and third level can be adjusted manually. Some of the test spectral types are unknown (i.e. Non), and the threshold value is set as the mean of the remaining spectral types

3. The Detecting Method of Splicing Abnormal Spectrum based on High Performance Computing Platform

The high performance computing cluster system of our school uses CPU+GPU+MIC architecture, which is composed of 34 computing nodes and the 1 fat nodes. In addition, there are two graphics processor nodes and two Intel integrated core nodes. The cluster has total of 796 CPU computing core. With the powerful computing capabilities of the platform, we can process fits [5] files rapidly.

In this paper, we use python [6] to invoke mpi [7] method, divide and conquer, a number of fits files are distributed to multiple processes. Each process is executed independently and does not affect each other. The root process gather the result when all processes finish and it will produce standard information stream and standard error stream.

4. Experimental Environment on High Performance Platform

High performance computing platform system uses Red Hat Enterprise Linux Server version of release 6.5. Due to the high performance computing platform does not configure parallel environment based on python, so users need to build their own run environment. The high performance platform has strict permissions, ordinary users can only operate in the current user directory. The software packages can only be manually installed and configured environment variable, in order to build the environment suitable for the requirements of this experiment, in the current user directory, we installed or upgraded some software (package) such as python2.7/jdk1.7/astropy/numpy/scipy/mpi4py/glibc/gmp/mpfr/mpc/gfortran/gcc/jpype.

Python2.7. High performance computing platform comes with a version of Python is 2.6, in order to keep the same with the single experimental environment, we upgrade python from 2.6 to 2.7. Download python2.7 package, unzip and enter the compressed directory, installed by the python setup.py command, due to the current user for ordinary users, it cannot be installed in the default directory, through optional "order prefix" python will be installed below the current user's directory. After installation, when we use python, the system will default to find the python executable file from the system directory and then run it, and did not run the newly installed python, so we need to configure the environment variables. There are two environmental information file ".Bash_profile" and ".Bashrc" under the root directory of current user. The .Bash_profile will be run when a user logs in, then it will call the.Bashrc. We modify the .Bashrc file, configure PYTHONPATH, and add it to the PATH variable. In the search for a python executable file, system will find it in PATH, we put PYTHONPATH before PATH to ensure the system calls the new version of python.

Numpy. Similar to the installation of python, we need to install it in the local directory. We can install the package like Python by downloading the compressed package and do manual installation. We can also use pip to install it. Pip is a python package management tool, it can be used to automatically install the software on PyPI. When using pip, we note that the default software source installation efficiency is low, we can specify the installation source through the "-i" command, such as Tsinghua University, University of Science & Technology China, and Douban. You can specify the installation package version. Pip greatly simplifies the installation process. We will install the package in the current python directory under the site-packages directory, where the installation of third packages put. Then it can be found and imported by the program.

For the software installation, we can configure the environment variables, also we can create a *.pth file in the site-packages directory, when python reads "sys.path", the PTH files will be read to load the configuration directory configured in this file.

There is no need to enumerate all the installation here, such as mpi4py, matplotlib [8], astropy [9] and so on. We will install it in a separate directory, and configure the directory in the PTH file.

For the normal software, we can install it by the three steps, configure (configure), compile (make), install (make install) and then configure environment variable.

5. Analysis of Experimental Results

In this paper, the 4426 splicing abnormal spectra of LAMOST DR1 [10] is used as the test spectra.

About 93% of all splicing abnormal spectra can be found out by this method. The zero flow spectra 1591, first level 1524, second level 608, third level 368. As shown in Figure 3, 4, 5. The scores calculated by the evaluation function are shown in table.2.

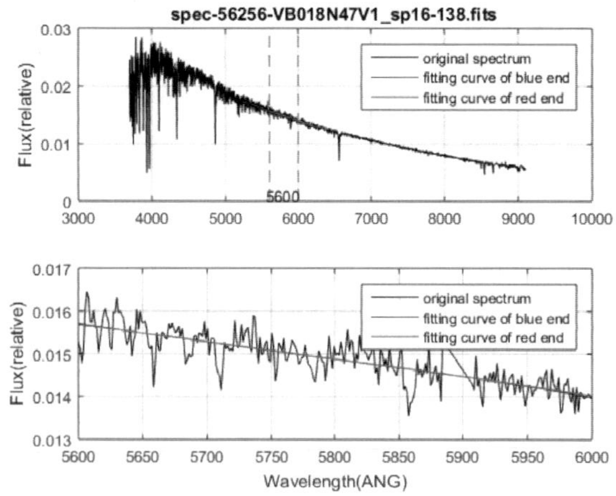

Figure 3 Spectrum judged to be normal splicing

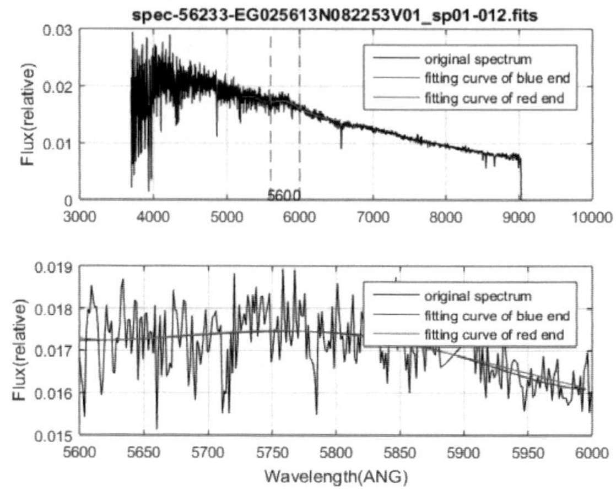

Figure 4 Abnormal splicing spectrum in level I

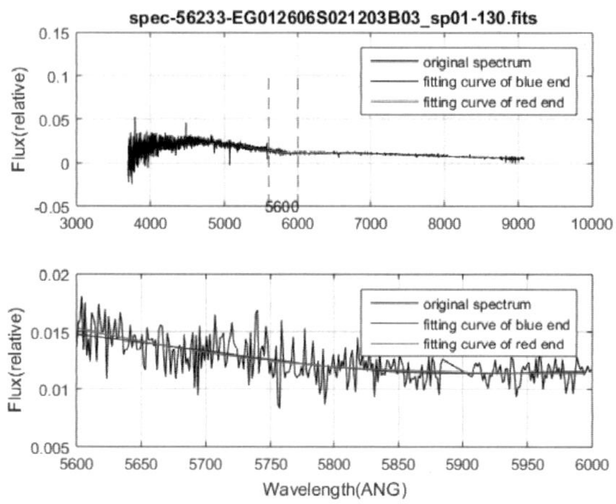

Figure 5 Abnormal splicing spectrum in level II

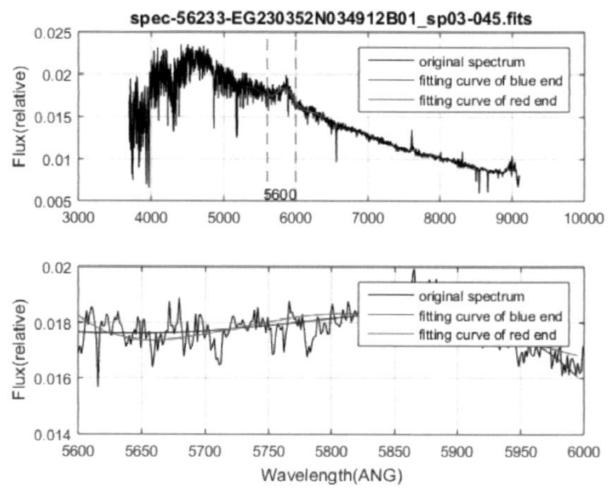

Figure 6 Abnormal splicing spectrum in level III

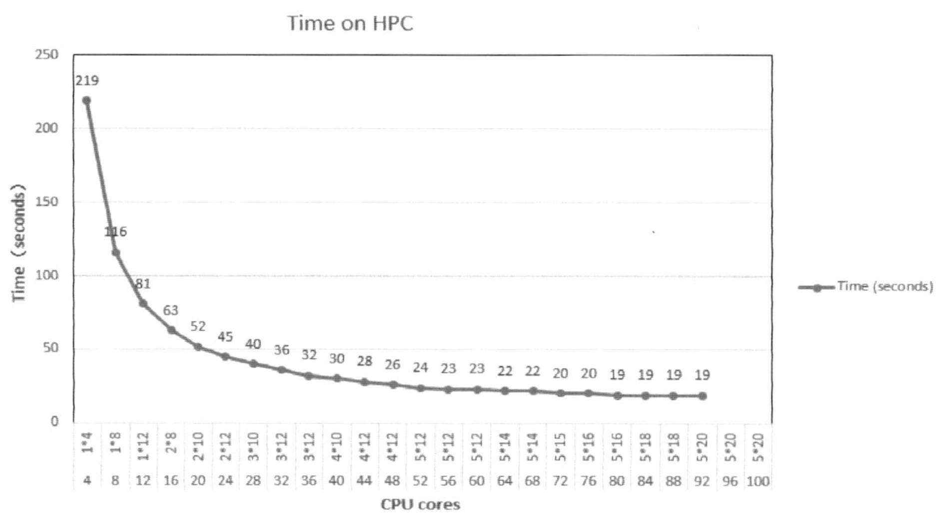

Figure 7 The time change on HPC

Table 2 The score of example spectra

Figure score	Figure 3	Figure 4	Figure 5	Figure 6
score	0.78	2.47	6.24	37.86
result	normal	first	second	third

The experiment analysis is based on the 4426 spectral data. We submit jobs of the same files to the high performance computing platform by specifying different number of CPU core. The run time of each experiment is as follows in Figure.7. The run time on stand-alone is 1471 seconds. It can be seen from the experimental results that the computational efficiency in the high performance environment is obviously higher than that in the stand-alone environment. With the number of CPU core increased, the processing time is shorter, and the efficiency is more and more high. The running time is not linear change. After the core number 80, the running time does not change, the reason is that too many core need a lot of inter process communication and file IO, limit the time shortened.

6. Conclusions

In this paper, a method for automatic detection and classification of splicing abnormal spectra in LAMOST spectra is presented. Firstly the test spectra were pretreated and divided into red and blue ends, then we fitted the flux of red and blue ends. For two fitting curves, we select a series of wavelength interval, calculate the flux difference between these points. After that, we analyze the differences set to get its mean value, standard deviation and the area. In this paper, we propose an evaluation function to deal with the mean, standard deviation and area to determine the degree of abnormal spectrum. A large number of experiments show that the method has a good effect on the detection of splicing abnormal spectra.

7. Acknowledgment

This work was supported by the National Natural Science Foundation of China (U1431102).

8. References

[1] Cui X-Q, Zhao Y-H, Chu Y-Q, et al. The large sky area multi-object fiber spectroscopic telescope (LAMOST)[J]. RAA, 2012, 12(9): 1197.

[2] Gang Zhao, et al. LAMOST spectral survey — An overview. Research in Astronomy and Astrophysics, 12(7):723-734, 2012.

[3] Luo, A., et al. Data release of the LAMOST pilot survey. Research in Astronomy and Astrophysics, 12(9): 1243-1246, 2012

[4] Wei P, Luo A, Li Y, et al. On the Construction of a New Stellar Classification Template Library for the LAMOST Spectral Analysis Pipeline[J]. Journal of Telemedicine & Telecare, 2014,147(147):S129-32.

[5] Donald C. Wells, et al. FITS: A Flexible Imags Transport System. Astronomy and Astrophysics Supplement, 44:363, 1981

[6] Python. https://www.python.org/

[7] OpenMPI. https://www.open-mpi.org/

[8] Ari, Niyazi,Ustazhanov, Makhamadsulton.Matplotlib in python[C].//2014 11th International Conference on Electronics, Computer and Computation: 2014 11th International Conference on Electronics, Computer and Computation (ICECCO 14), Sept. 29 2014-Oct. 1, 2014, Abuja, Nigeria.2014:1-6.

[9] Thomas P. Robitaille,Erik J. Tollerud,Perry Greenfield et al.Astropy: A community Python package for astronomy[J].Astronomy and astrophysics,2013,558(Pt.1):A33-1-A33-1.

[10] LAMOST DR1. http://dr1.lamost.org/

Automatic Identification of LAMOST Spectra with Continuum Problem Based on High Performance Computing

Jingjing Yu, Jingchang Pan* and Fanlong Meng

School of Mechanical, Electrical and Information Engineering, Shandong University, Weihai, China
Email: yujingjing2018@163.com, *pjc@sdu.edu.cn, morphy8@163.com

Abstract. Continuum problem is a phenomenon that the continuum of spectra get off their actual continuum even break off due to interstellar extinction and flux calibration, and it will have negative impact on the subsequent process such as spectral line extraction and so on. Based on this problem and fully considering of the continuum features of the stellar spectra, a method of automatic detection and recognition of the continuum problem in the stellar spectra based on High Performance Computing (HPC) is proposed, which will improve work efficiency greatly compared with the traditional human eyes examination under the condition of maintaining high accuracy. Continuum template matching is used to identify the continuum problem spectra in this paper. The first step goes to fit continuum of the test stellar spectra and the template spectra, then the flux differences of the two continuum we fitted before at every point of wavelength in the continuum spectra will be calculated based on full spectra matching to analyze the features of its distribution. The features we count are average (called β) value and standard deviation (called δ). The percentage of points distributed in range$\beta\pm\alpha*\delta$will be detected to confirm that if there is continuum problem. Experiments was taken to identify the continuum problem spectra on HPC, which demonstrate the validity and efficiency of the method, and this has important reference value to resolve similar massive spectra data processing problems in the future.

1. Introduction

The Large Sky Area Multi-Object Fiber Spectroscopic Telescope [1] (LAMOST) is a National Major Scientific Project built by the Chinese Academy of Science. Since the launch of the formal pilot survey, LAMOST has produced tens of thousands of spectra with each observation, and there are a lot of continuum problem spectra. Continuum problem spectra have negative impact on the subsequent process such as spectral line extraction and so on. The continuum problem spectra were selected by human eyes before, which was time-consuming and labor-intensive and reduced the efficiency and accuracy of spectra analysis. However, it is almost impossible to find the automatic detection of the continuum problem spectra. A method of automatic identification of LAMOST continuum problem spectra based on High Performance Computing (HPC) is proposed after studied the continuum of types of stellar. The flux differences of the two continuum we fitted before at every point of wavelength in the continuum spectra will be calculated based on full spectra matching to analyze the features of its distribution. The features we count are average (called β) value and standard deviation (called δ). The percentage of points distributed in range $\beta\pm\alpha*\delta$ will be detected to confirm that if there is continuum problem.

Fast computation and accurate results are the most outstanding advantages of high performance computing platform compared to traditional computing, and the efficient computing power relies on multi-core computing. Python is used in this paper to realize the parallel computing on HPC by calling

MPI, a plurality of FITS files is divided into subtask that can be executed in parallel across arbitrary numbers of processor cores. And then each processor core is executed in a separate order, the root process is specified to collect the result of each core after the last one is completed, thereby improving the computational efficiency. The experiments on the HPC show that the method we proposed in this paper can achieve high efficiency and accuracy on the identification of continuum problem spectra.

2. The Data

LAMOST has published these spectra online and perform its second data release (DR2) online. In this data release, over 4 million FITS (Flexible Image Transport System) file are published on-line. FITS is a data format for data exchange between different platforms which is used to store images. Each spectrum corresponds to a FITS file. The flux and wavelength information are exist in the data unit of FITS, and some information can be read from FITS directly such as flux, COEFF0 (central wavelength of first pixel), COEFF1 (Log10 dispersion per pixel), NAXIS1 (the number of wavelength array), while the wavelength should be calculated by formula (1), In the formula, "i" is the numbers of sampling points of wavelength, which is in the region of [0, NAXIS1] .

$$wavelength = 10\hat{}(COEFF0 + i*COEFF1) \tag{1}$$

Until July 2014, LAMOST has completed its pilot survey which was launched in October 2011 and ended in June 2012, and the first two years of its regular survey which was initiated on September 2012. LAMOST is a special reflecting Schmidt telescope with 4000 fibers in a field of view of 20 deg2 in the sky. After three-years-survey, it totally obtains 4,136,482 spectra, which consist of stars, galaxies, quasars and other unknown type.

3. Method Introduction

3.1 Stellar Spectra Classification Based on Lick Indices

Lick [2] index is a relatively wide spectral characteristics. Each index in the absorption line with the most prominent, it is named. Lick is an indices that used by astronomers as a line feature. It is simple and feasible to classify stellar spectra by calculating the Euclidean distance of lick indices: on the one hand, 25 dimensional feature matrix can be got by lick indices, which greatly reduces the dimensionality of data and reduces the amount of calculation. On the other hand, lick indices is calculated by the average value of flux so that they have a higher signal-to-noise ratio(S/N) [3] than the original spectra.

3.2 Zero Flux Spectra Detection

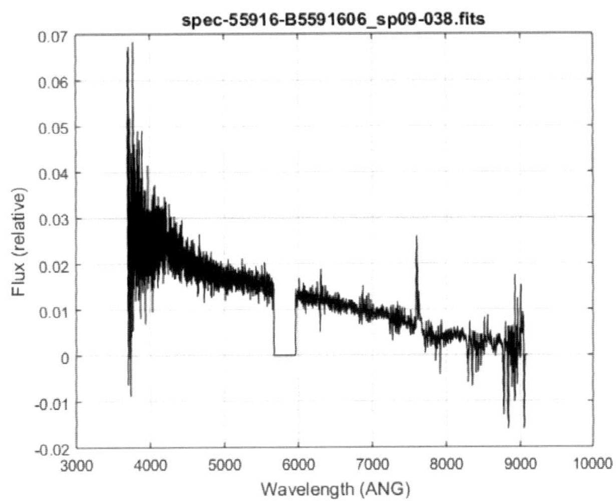

Figure 1 The spectrum which has zero flux at continuity points.

There are many zero flux spectra in continuum problem spectra, as shown in Figure 1 which is the most obvious problem. In this case, not only the continuum fitted inaccurate but the detection efficiency is reduced. To this end, we eliminate the zero flux spectra before all the work. The definite means are as follows:

(1) Set a window size of $window$ =10;
(2) Traverse flux values at all wavelength points of the entire spectrum with the window. If the flow at a certain point is zero, then the spectrum is judged to be zero flux spectrum if all the points in the whole window are 0.

3.3 Fitting the Continuum

There are many methods to fit continuum such as polynomial fitting, wavelet filtering [4], the median filtering [5] and morphological filter and so on. In this paper, we use the method of fitting continuum from Pan Jingchang and Wei Peng, which is mentioned in "A new method for automatic identification of stellar spectra of emission lines [6]". The method is improved on the basis of the polynomial approximation method, when some of the points are selected not all of the points when fitting the continuum, which makes the influence of the wide line and the strong line smaller. To prepare for the continuum fitting, something has been done to deal with the original spectral flux array of stellar spectra in this paper, such as normalizing [7] the flux to region of [0, 1] and flux interpolation. The all steps to fit continuum are as follows:

(1) Take $w(n)$ and $f(n)$ as the original spectral wave and flux of a stellar spectrum respectively;
(2) For test spectra, the flux are interpolated to the wavelength of the template spectra, and the flow value of the test spectra at each wavelength point of the template spectra is obtained, which is $f_i(n)$;
(3) Normalize the flux of $f_i(n)$, and $f_n(n)$ as the normalized spectrum;
(4) Compute the mean flux $f_m(n)$ and standard flux $f_s(n)$ of the normalized flux $f_n(n)$ with the window size of 10;
(5) Take $f_b(n)$ as the points which in the region $[-0.5 * f_s(n) + f_m(n), 0.5 * f_s(n) + f_m(n)]$;
(6) The continuum of the original spectrum is got by six order polynomial fitting to $w(n)$ and $f_b(n)$;

To make the test spectra and template spectra are in the same measurement and preserve the uniqueness of variables, we interpolate the flux of test spectra to the wavelength of the template spectra and normalize the flux of spectra. The comparison of flux between test spectrum and its template spectrum of G5 without flux normalization is shown in Figure 2(a). The measurement of flux is vary greatly in this case. The flux of the two spectrum after flux normalization are shown in Figure 2(b). From these figure, we can know that flux normalization is a crucial step for the continuum template matching.

3.4 Continuum Template Matching

The template spectra we used in this paper is 183 template spectra which is from mentioned in "On the Construction of a New Stellar Classification Template Library for the LAMOST Spectral Analysis Pipeline[8]" by Wei Peng. The continuum of test spectra and template spectra are obtained after steps above. The method based on distance measure to detect continuum problem spectra. The steps of continuum template match method are as follows:

(1) Take $f(1)$, $f(2)$ as the continuum of test spectra and template spectra after steps in part C;
(2) Calculate the distance between $f(1)$ and $f(2)$, we define the distance as an array $Fr = f1_x - f2_x, x = 1,2,\ldots, n$ which represents the flux differences at each wavelength point, n is the dimension of fits files which means the numbers of wavelength points;
(3) Calculate the mean value M and variance S of the array Fr , β is defined as the threshold of flux differences;

IOP Conf. Series: Journal of Physics: Conf. Series **910** (2017) 012052 doi:10.1088/1742-6596/910/1/012052

(4) The test spectrum is defined as continuum problem spectrum if one of the following three conditions is satisfied:

a) Get the number N of the flux differences at a certain wavelength not in the region $[M - \alpha S, M + \alpha S]$, if N is greater than 1 and N accounts for 5% or more of the total wavelength points, then this spectrum has continuum problem;

b) If one of the value in the array Fr is greater than β, then this spectrum has continuum problem;

c) Get the number I of the intersections of the two continuum, if I is greater than or equal to 3, then this spectrum has continuum problem.

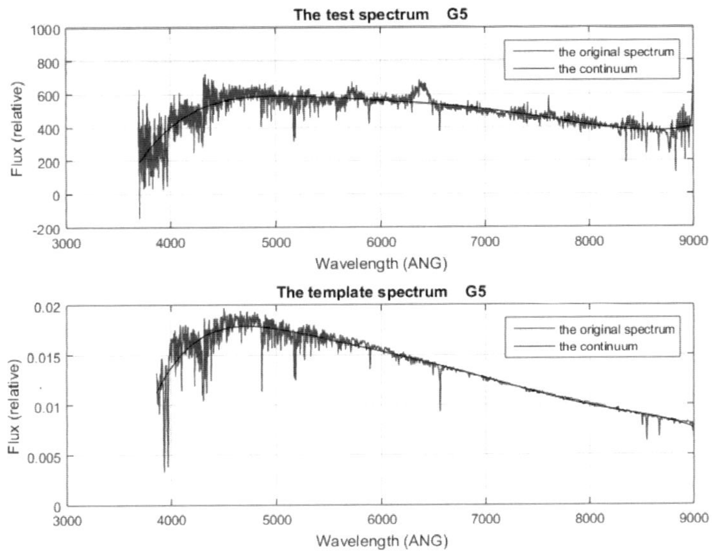

(a)

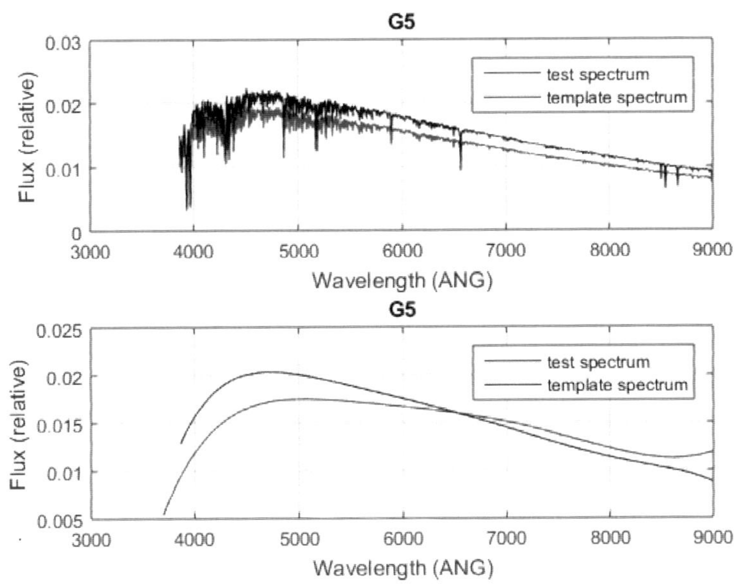

(b)

Figure 2 The comparison of flux.:(a) this is the flux comparison of test spectrum and template spectrum without flux normalization.(b) this is the flux comparison of test spectrum and template spectrum after flux normalization.

In order to determine the continuum problem spectra, the selection of α and β determines the accuracy of the continuum spectra detection. After the experiments, the value of α is 1.5, and the value of β is 0.0025 after normalizing the flux to the same measure. There are some spectra with low signal-to-noise ratio which have the smaller measure than the others that leads to the spectra with continuum problem cannot be detected by the two thresholds previous. In order to ensure that all the continuum problem spectra can be detected, step c came into being.

4. Experiments Based on HPC and results analysis

Python is used in this paper to realize the parallel computing on HPC by calling MPI, a plurality of FITS files is divided into subtask that can be executed in parallel across arbitrary numbers of processor cores. And then each processor core is executed in a separate order, the root process is specified to collect the result of each core after the last one is completed, thereby improving the computational efficiency.

A file with 33,242 FITS files' data is used as the sample file of the test experiments on HPC, 60 processing cores are used to identify the continuum problem spectra. Processing core number increase from 8 to 60 linearly, and the changes of processing time was shown in TABLE.1 to TABLE.4. And the trend of processing time on HPC is shown in Figure.3.

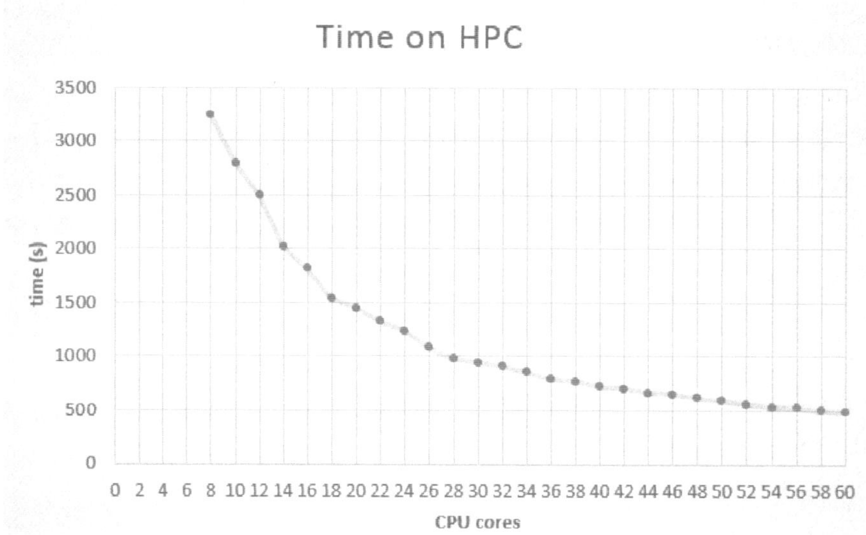

Figure 3 The trend of processing time.

Table 1 The changes of processing time on HPC

cores	8	10	12	14	16	18
time(s)	3238	2787	2490	2015	1816	1538

Table 2 The changes of processing time on HPC

cores	20	22	24	26	28	30
time(s)	1446	1321	1234	1082	973	944

Table 3 The changes of processing time on HPC

cores	32	34	36	38	40	42	44	46
time(s)	908	856	792	768	720	702	660	647

Table 4 The changes of processing time on HPC

cores	48	50	52	54	56	58	60
time(s)	616	590	558	529	521	498	486

Some experiments have been done on the tow platform to identify continuum problem spectra from 33,242 FITS files, the single-machine environment and HPC. The experiments are carried out on the two platforms, and the experimental results show that the continuum problem spectra identified are the same, however, the processing time quite different between the two platforms. In the single-machine environment, it is take 7hours, 51mins for the experiment, and the processing time on HPC is 8mins, 6secs, which greatly improves the processing efficiency.

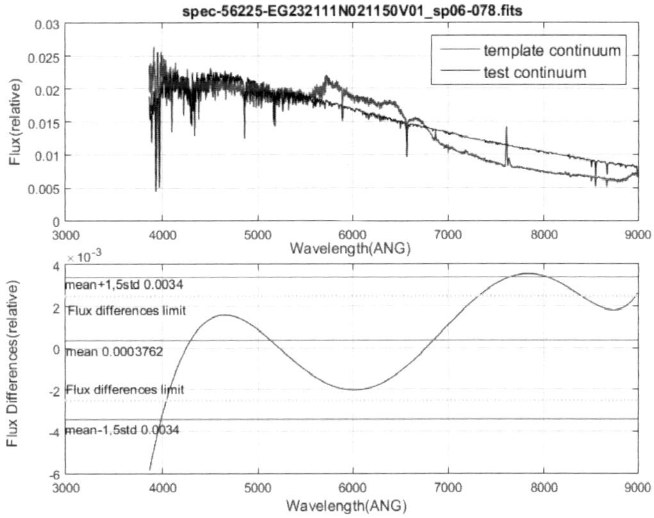

Figure 4 The original spectrum of the continuum problem spectrum, and the distribution of flux differences. The blue curve shows the differences between the two continuum spectra of the original spectra.

A continuum problem spectrum of F5 type which is identified by the method we proposed in this paper is shown as Figure.4. The comparison of the continuum of test spectra and template spectra is shown in the subgraph above of Figure.4. The distribution of the flux differences is shown in the following subgraph of Figure.4. The region $[M - \alpha S, M + \alpha S]$ is identified by three red solid lines, and the thresholds β is identified by two green dashed lines.

The type of the spectrum shown in Figure4 above is F9, which is the late-type star. The feature of the F9 type spectrum is obvious, an absorption bands [9] exist at 4300Å, and the peak value appears at about 4500Å. The red end of the spectrum is smooth, and the metal line begin to strengthen. However, the feature of the continuum problem spectrum above is different from F9's. The spectrum of the red end is obviously deviated from the end of the spectrum, and the blue end is raised, which makes the whole continuous spectrum appear too high in the blue end and too low in the red end. According to the Figure.4 of distribution of flux differences, it is obvious that the distribution of the test spectrum and template spectrum is different. And a numbers of flux differences are found greater than β , which causes the continuum problem.

The experimental results show that the method we proposed in this paper to identify the continuum problem spectra is simple and effective. And the parallel computing on HPC has performed a number of unique advantages in dealing with large amounts of FITS files, which has saved a substantial amount of time.

5. Conclusions

A method of automatic detection and recognition of the continuum problem in the stellar spectra based on HPC is proposed in this paper. The main work is divided into the following three parts:

(1) Fit continuum. The continuum spectra of the test spectrum and the template spectrum with corresponding type are extracted and then we do the process of flux interpolation and normalization in order to analyze data by means of the non-dimensional contrast.

(2) Continuum template matching. In order to make continuum template matching for the test spectrum, the flux differences of the test stellar spectra and template spectra at every point of wavelength in the continuum spectra will be calculated to analyze the features of its distribution. The features we count are average (called β) value and standard deviation (called δ). The percentage of points distributed in range$\beta \pm \alpha * \delta$ will be detected to confirm that if there is continuum problem.

(3) Confirm the subclass of the test stellar spectra. The subclasses of most continuum problem spectra are not defined. The prerequisite of continuum template matching is to determine the subclass of the test spectrum. In this paper, we assign the stellar type of the template spectrum with the minimum distance to the test spectrum by calculating the Euclidean distance between the two lick indices of the template spectrum and the test spectrum.

(4) Automatic identification of LAMOST continuum problem spectra based on high performance computing. We studied parallel computing method based on HPC using python, which by calling MPI to divide many FITS files into a number of processes. The result of each processes will be collected by root when the last processes was completed.

To sum up, the parallel computation method of the automatic recognition of the continuum problem spectrum is realized on the high performance computing platform. It achieves more rapid and efficient recognition effect compared to the single computer.

6. Acknowledgment

This work was supported by the National Natural Science Foundation of China (U1431102).

7. References

[1] BAI Zhongrui. LAMOST 2D pipeline. Intelnational Astronomical Union. 2012.

[2] M Franchini, C Morossi, PD Marcantonio, ML Malagnini, M Chavez. The Lick/SDSS Library. II. [Ca/Fe] and [Mg/Fe] in F, G, and K Stars from SDSS-DR7. Astrophysical Journal, 2011, 730(2):117

[3] Franchini, M., Morossi, C., Di Marcantonio, P., Malagnini, M. L., & Chavez, M. 2011, ApJ, 730, 117

[4] [Starck J.L., R. Siebenmorgen, R. Gredel. The astrophysical journal, 1997, 482: 1011.

[5] J.P.Fitch, E.J.Colye and N.C.Gallagher. Median Filtering by Threshold Decomposition. IEEE Trans. ASSP. Vol.ASSP-32, 1183-1188, 1984

[6] Jingchang Pan, Caiming Zhang, Peng Wei, et al. A New Automated Method to Identify Emission Line Star from Massive Spectra (in Chinese), Spectroscopy and Spectral Analysis, 2012, 32(6): 1689-1693.

[7] JIANG Bin, PAN Jing-chang, GUO Qiang. Journal of Electronic Science and Technology of China, 2007, 36(006): 1292.

[8] P Wei, A Luo, Y Li, L Tu, F Wang. On the Construction of a New Stellar Classification Template Library for the LAMOST Spectral Analysis Pipeline,147(5):101,2014

[9] Jacoby G.H., D.A. Hunter, C.A. Christian. The Astrophysical Journal Supplement Series, 1984, 56: 257-281.

Performance Analysis of Optical Mobile Fronthaul for Cloud Radio Access Networks

Jiawei Zhang, Yuming Xiao, Hui Li and Yuefeng Ji

Beijing Advanced Innovation Centre for Future Internet Technology, Beijing University of Technology (BJUT), Beijing, China
Email: zhangjw1985@gmail.com, ymingxiao_1@sina.cn, lihui@bupt.edu.cn, jyf@bupt.edu.cn

Abstract. Cloud radio access networks (C-RAN) separates baseband units (BBU) of conventional base station to a centralized pool which connects remote radio heads (RRH) through mobile fronthaul. Mobile fronthaul is a new network segment of C-RAN, it is designed to transport digital sampling data between BBU and RRH. Optical transport networks that provide large bandwidth and low latency is a promising fronthaul solution. In this paper, we discuss several optical transport networks which are candidates for mobile fronthaul, analyze their performances including the number of used wavelength, round-trip latency and wavelength utilization.

1. Introduction

The amount of mobile traffic will exceed 500 exabytes by 2020, driven by the proliferation of smart mobile devices and high-definition video application [1]. Small cell and distributed antenna are promising technologies to increase access rate of radio access networks (RANs), but required to be "light" and low power consumption. To satisfy this requirement, RANs architecture is evolving from "distribution" to "cloudification", which means to separate the digital baseband unit (BBU) of conventional cell sites (see Figure. 1(a)) from remote radio heads (RRHs), and pool BBUs to a cloud-centric site (see Figure. 1(b)). With the benefit of the cloud RAN (C-RAN), conventional complicated and power-hungry cell sites can be simplified to RRH only, which reduces the related power consumption and site maintenance.

As the baseband processing moving to the "cloud", a large amount of digitalized in-phase and quadrature-phase (IQ) samples generated by RRHs need to be transported to a BBU pool. The transport network between RRH and BBU pool is so called mobile fronthaul (MFH). Different from the backhaul that connects the BBU pools and mobile evolved packet core (EPC), MFH is a new network segment that appears in the C-RAN. It bridges the RRH and BBU pool through some specific interfaces. Typical MFH interfaces include common public radio interface (CPRI), open base station architecture initiative (OBSAI), and open radio interface (ORI), which cover the areas of digitalized IQ samples transportation, frame control and management issues [2]. With the rapid growth of small cell and distributed antenna, MFH network is attracting much attention. Wavelength division multiplexing (WDM) networks that come naturally with large bandwidth, low latency and green, show great potential as fronthaul solution.

In this paper, we study several possible optical transport solutions for MFH, including passive and active WDM networks. Network performances such as the number of used wavelength, round-trip latency and wavelength utilization ratio are compared and analyzed among different solutions.

2. Mobile Fronthaul Requirements for C-RAN

Firstly, C-RAN architecture brings a huge overhead to fronthaul link. The bit rate has reached the gigabits per second level and presents a multi-rate property according to different cell site technologies and data compression method.

In addition, wireless traffic is time-dependent that require different bandwidth during different time periods. Therefore, MFH should be designed to not only support high-bandwidth capacity, but also support flexible bandwidth allocation. Secondly, since the digitalized IQ data over CPRI link is at the level of physical radio signal, the MFH needs to support strict time delay for immediate signal processing. In addition, advanced C-RAN applications such as coordinated multipoint (CoMP) and virtual BBU migration are time-sensitive, low latency should be guaranteed for these services. The acceptable level of delay for a round-trip is under ~500μs, including the fiber propagation delay and equipment delay [2, 3]. Thirdly, MFH should be able to multiplex CPRI links in either electrical or optical domain, and send them to an available BBU pool. How to coordinate fronthaul resource with radio and BBU resources in an efficient way is a big challenge to mobile operator. In our previous work [4], we proposed a converged optical and wireless network architecture by using software-defined networking (SDN) technology to orchestrate heterogeneous resources in the C-RAN for a resource optimization goal.

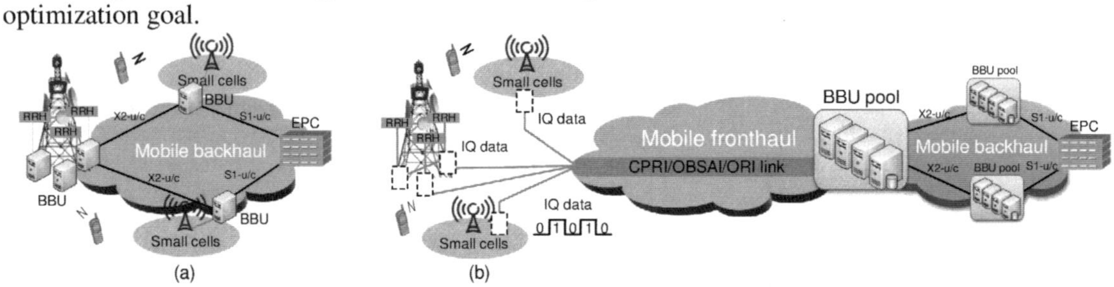

Figure 1 Distributed radio access networks (D-RAN); (b) Cloud based radio access networks (C-RAN)

3. Optical Transport Solutions for Fronthaul Network

WDM networks that come naturally with large bandwidth, low latency and green, show great potential to build MFH networks, and they can be classified in active and passive WDM solutions. As active one, the CPRI signal is encapsulated for example by means of OTN frame or Ethernet frame, and then multiplexed to a WDM wavelength. In this case, optical/electrical/optical (O/E/O) conversions and power supply are required at intermediate nodes. Passive solution is based on the passive multiplexing and de-multiplexing, such as optical splitter (OS) and optical multiplexer (OMUX). No O/E/O conversions and power supply are needed at intermediate nodes in a passive case. For the MFH network topology, point-to-point, tree, and ring are main candidates while taking into account RAN implementation cost. Five possible optical fronthaul solutions are discussed and compared as follows.

Table 1 Optical transport solutions for comparison.

	Fiber required	Latency	Wavelength utilization	Reliability
Dark fiber	***	*	*	*
WDM-PON	**	*	*	**
WDM-Ring	*	*	*	***
TWDM-PON	**	**	**	**
TWDM-Ring	*	***	***	***

Dark fiber is a direct connected fiber-pair between BBU and RRH (see Figure. 2(a)). A RRH has its exclusive fiber link to a BBU pool. Optical transponder at both sides performs E/O or O/E conversion to modulate/de-modulate a CPRI signal onto an optical carrier. Deployment over dark fiber is a straightforward option, and the latency is related to light-propagation delay only. However, it requires more fiber links, and the network reliability and flexibility are rather weak. WDM-PON is a passive

WDM solution (see Figure. 2(b)), CPRI signals are multiplexed in terms of wavelength at optical line terminal (OLT), and duplicated by OS to different optical network units (ONUs).

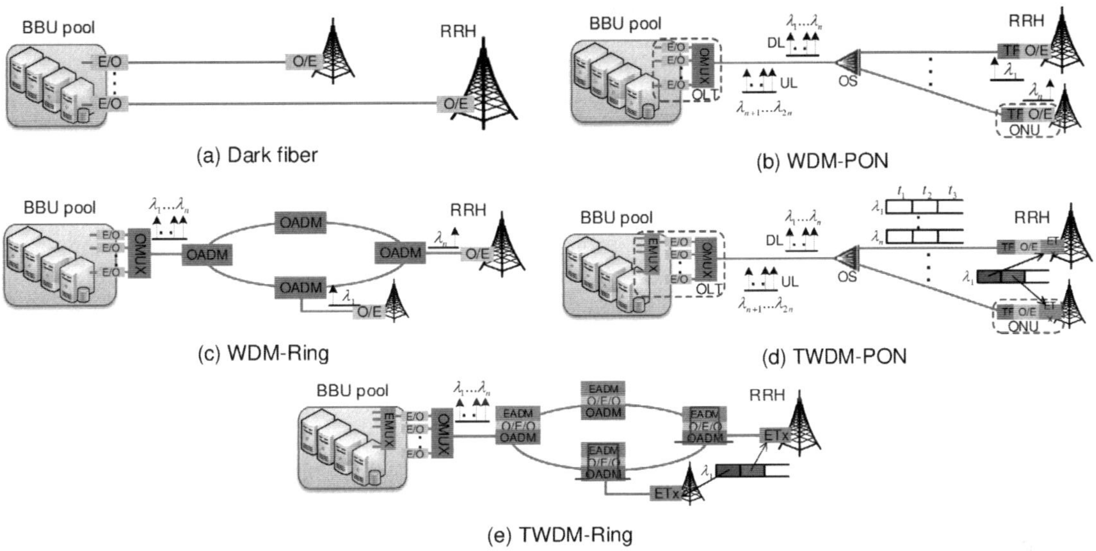

Figure 2 Optical transport solutions for MFH: (a) Dark fiber; (b) WDM-PON; (c) WDM-Ring; (d) TWDM-PON; (e) TWDM-Ring

A tunable filter (TF) will decouple the WDM signals at the ONU site. With the benefit of wavelength multiplexing, WDM-PON saves more fiber links than dark fiber. The latency of WDM-PON depends on propagation. WDM-Ring is similar to WDM-PON (see Figure. 2(c)), but the wavelength selecting function is moved to the intermediate nodes through an optical add-and-drop module (OADM). Compared to WDM-PON, WDM-Ring improves network flexibility and reliability, and further reduces the number of required fibers. However, both WDM-PON and WDM-Ring have low wavelength utilization ratio when mobile traffic rate is far below a wavelength's capacity. To cope with this problem, TWDM technology combines WDM and TDM to not only support high-bandwidth provision but also fine-bandwidth allocation. TWDM-PON solution is shown in Figure. 2(d), CPRI signals are multiplexed in a TDM manner by an electrical multiplexer (EMUX) first, and then multiplexed to a wavelength channel. Therefore, a wavelength contains different RRHs' information, and a TWDM slot which belongs to a given RRH can be extracted at the local ONU site.

It improves wavelength efficiency, but the cost and energy consumption increase because of electrical processing. Besides, dynamic bandwidth reservation for upstream will cause a delay in the TWDM-PON solution. TWDM-Ring is an active WDM solution (see Figure. 2(e)), electrical add-and-drop modules (EADMs) and OADMs are placed at intermediate nodes. Wavelength efficiency is the best among the all, but a large number of O/E/O conversions will increase the cost and energy consumption as well as latency. A qualitative comparison of different MFH solutions are shown in Table I. In the following section, we will introduce some quantitative comparisons from the simulation results.

4. Simulation Results and Analysis

Figure.3 shows the simulation scenario, the vertexes of the regular hexadecagon represent a BBU pool and 15 RRHs. The edge length of the regular hexadecagon is 2.5 kilometers. The blue lines represent the links of ring topology, while the red lines are for PON topology. Table II shows the network parameters in the simulation. The capacity of a wavelength is 10 Gbps. A CPRI basic frame contains 128 bits, and 256 basic frames are to form a CPRI hyper frame. We consider several factors that cause the fronthaul latency. TDM switch latency is to switch TDM signals in a fabric, which is considered in TWDM solutions [3]. Transmission latency is the time duration to transmit a CPRI hyper frame [3]. O-

E-O conversion and FEC coding/decoding occurs at each termination of optical signals [4]. For the upstream of TWDM-PON, dynamic bandwidth allocation (DBA) will cause additional latency [5]. Passive WDM filter and optical splitter introduce no additional delay or a relative small value (~ns), which are ignored in the simulation. We consider two mobile traffic distributions, Nominal and Zipf with the average rate is 2Gbps and ratio is 0.5. Simulation results are shown in Figure.4-6, each data point in the figures is average over 105 experiments.

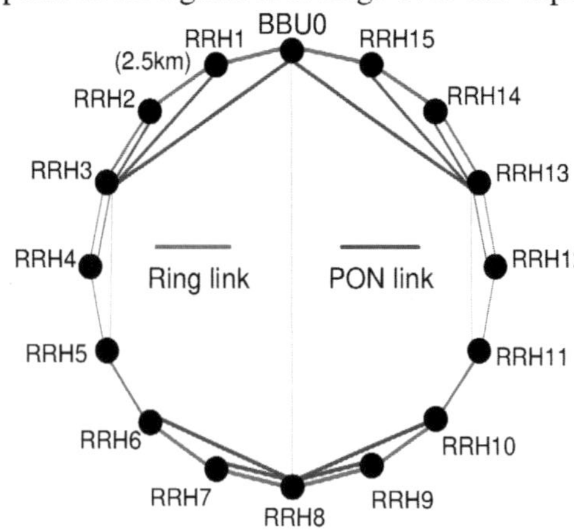

Figure 3 Simulation topology

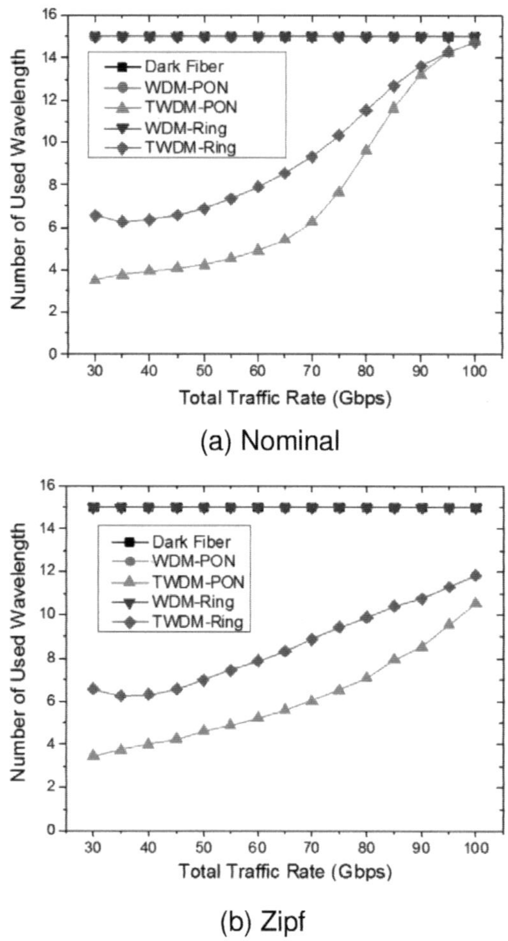

Figure 4 Number of used wavelength vs. total traffic rate.

Figure.4(a) and Figure.4(b) show the number of used wavelength for different MFH solutions. TWDM-PON uses the lowest number of wavelength among the all, because of TDM manner. TWDM-Ring can also multiplex CPRI signals in the time domain, but the wavelengths are added and dropped at intermediate nodes, which causes the wavelength fragment. Wavelength fragment can lead some wavelengths to be not available for some links. TWDM-PON has no wavelength fragment, because all the wavelengths are duplicated and sent to every single ONU.

Figure.5(a) and Figure.5(b) show the round-trip latency. Dark fiber achieves the lowest latency performance, because the fiber is directly connected between BBU and RRH. The latencies of dark fiber as well as WDM-PON and WDM-Ring are caused by light-propagation delay only, and the values are related to the distance between BBU and RRH. TWDM-PON performs higher latency than WDM-PON, because the upstream bandwidth reservation which causes approximate 40μs delay [5]. TWDM-Ring shows the highest latency at low traffic rate because a large number of O/E/O conversions (main latency contributor) occur at intermediate nodes. As traffic rate increases, the curve goes down close to WDM-Ring, because each RRH is assigned to a wavelength to carry CPRI signals.

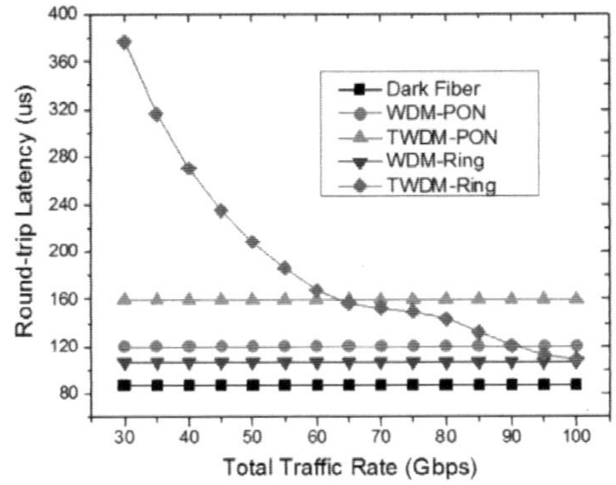

(a) Nominal

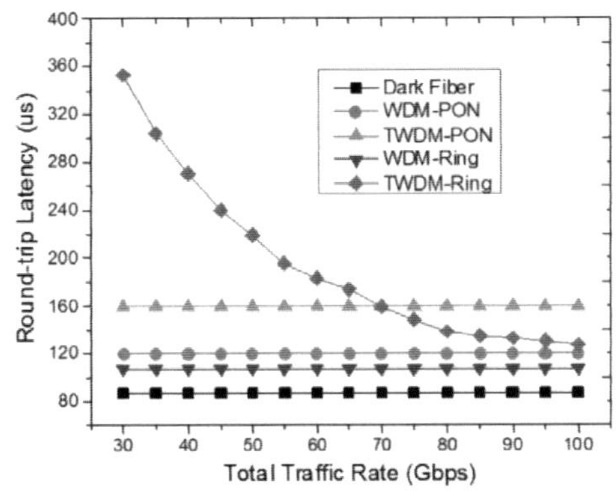

(b) Zipf

Figure 5 Average round-trip latency vs. total traffic rate

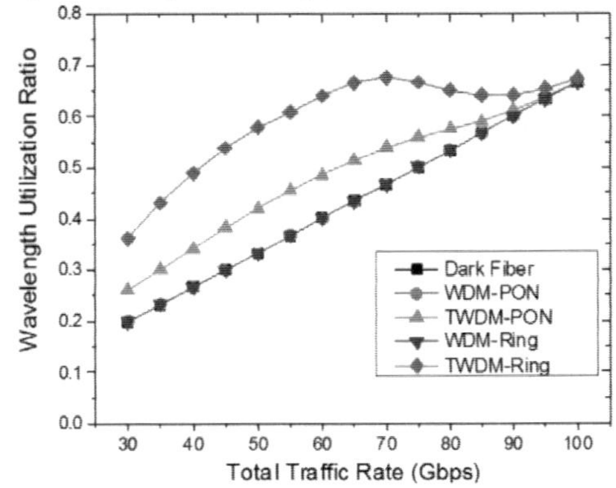

(a) Nominal

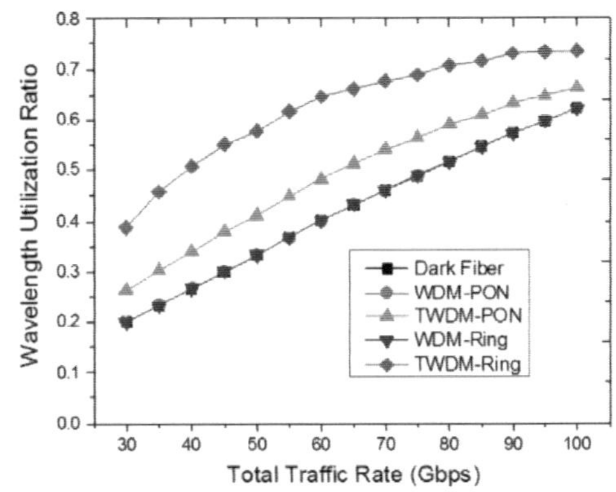

(b) Zipf

Figure 6 Wavelength utilization vs. total traffic rate

Table 2 Network Parameters in Numerical Simulation

Parameters		Values
	Bandwidth of Wavelength	10 Gbps
	CPRI basic frame	128 bits
Latency [2, 3, 5]	TDM switch fabric	~5.2 us
	Transmission time	~4 us
	Propagation time	5 us/km
	Optical-Electrical-Optical conversion	~15 us
	FEC coding/decoding	~5 us
	DBA for PON	~40 us

Figure.6(a) and Figure.6(b) reflect the wavelength utilization efficiency. TWDM-Ring outperforms TWDM-PON. This is because a TWDM slot is duplicated and sent to every single ONU, which leads to the waste of wavelength resource in some links. For example, link (3, 2) of the PON topology contains all the wavelength and time information, but only one TWDM slot is for RRH2.

5. Summary

We discussed various promising optical MFH solutions, and evaluated their performances standing on the network point of view. Network performances, such as the number of used wavelength, round-trip latency and wavelength utilization ratio, were compared and analysis through simulation. From simulation results, we find that TWDM technology can improve the resource efficiency, but introduce higher round-trip latency.

6. Acknowledgment

This work has been supported by fund of Beijing Advanced Innovation Center for Future Internet Technology (BJUT).

7. References

[1] J. G. Andrews, S. Buzzi, W. Choi, S. V. Hanly, A. Lozano, A. C. K. Soong, and J. C. Zhang, "What Will 5G Be?" IEEE J. Select. Areas Commun., vol. 32, no.6, pp. 1065–1082, 2014.

a) Pizzinat, P. Chanclou, F. Saliou, and T. Diallo, "Things You Should Know About Fronthaul," IEEE/OSA J. Lightwave Technol., vol. 33, no. 5, pp. 1077-1083, 2015.

[2] N. P. Anthapadmanabhan, A. Walid, and T. Pfeiffer, "Mobile Fronthaul over Latency-Optimized Time Division Multiplexed Passive Optical Networks," in Proc. of IEEE International Conference on Communications (ICC), Workshop on Backhaul Networks, London, 2015.

[3] J. Zhang, Y. Ji, J. Zhang, R. Gu, Y. Zhao, S. Liu, K. Xu, M. Song, H. Li, and X. Wang, "Baseband Unit Cloud Interconnection Enabled by Flexible Grid Optical Networks with Software Defined Elasticity," IEEE Commun. Mag., vol. 53, no. 9, pp. 90-98, Sep. 2015.

[4] T. Tashiro, S. Kuwano, J. Terada, T. Kawamura, N. Tanaka, S. Shigematsu, and N. Yoshimoto, "A novel DBA scheme for TDM-PON based mobile fronthaul," in Proc. Optical Fiber Communications Conference and Exhibition (OFC), San Francisco, Mar.2014.

IMHRP: Improved Multi-Hop Routing Protocol for Wireless Sensor Networks

Jianhua Huang, Danwei Ruan, Yadong Hong, Ziming Zhao and Hong Zheng

School of Information Science and Engineering, East China University of Science and Technology
Shanghai, China
Email: jhhuang@ecust.edu.cn, rdw1226@163.com, 1534338205@qq.com, 840958663@qq.com, zhenghong@ecust.edu.cn

Abstract. Wireless sensor network (WSN) is a self-organizing system formed by a large number of low-cost sensor nodes through wireless communication. Sensor nodes collect environmental information and transmit it to the base station (BS). Sensor nodes usually have very limited battery energy. The batteries cannot be charged or replaced. Therefore, it is necessary to design an energy efficient routing protocol to maximize the network lifetime. This paper presents an improved multi-hop routing protocol (IMHRP) for homogeneous networks. In the IMHRP protocol, based on the distances to the BS, the CH nodes are divided into internal CH nodes and external CH nodes. The set-up phase of the protocol is based on the LEACH protocol and the minimum distance between CH nodes are limited to a special constant distance, so a more uniform distribution of CH nodes is achieved. In the steady-state phase, the routes of different CH nodes are created on the basis of the distances between the CH nodes. The energy efficiency of communication can be maximized. The simulation results show that the proposed algorithm can more effectively reduce the energy consumption of each round and prolong the network lifetime compared with LEACH protocol and MHT protocol.

1. Introduction

Wireless Sensor Network (WSN) is a distributed sensing network which is composed of many sensor nodes. The sensor nodes cooperate with each other in self-organization [1] and take charge of monitoring the environmental information (such as temperature, vibration, pressure or contaminants) in the WSN [2]. Sensor nodes are typically powered by microcells. Therefore, their computing, storage and communication capacity are limited. It is the first consideration for designing routing algorithm to efficiently make use of node energy to maximize the network lifetime [3]. The routing protocol of the WSN is mainly divided into flat routing protocol and hierarchical routing protocol [4]. The flat routing protocol needs to keep a routing table in each node, which is not suitable for large-scale sensor networks. The hierarchical routing protocol divides the network into clusters. Each cluster has a cluster head (CH) node and the rest are non-cluster head nodes [5]. Each non-cluster head node transmits the data to the CH node. Then the CH node aggregates and transmits the data to the base station (BS) [6]. This approach can meet the scalability of the sensor networks, effectively balance the energy consumption of the sensor nodes and thereby prolong the network lifetime. LEACH [7] is the first proposed energy-efficient clustering routing protocol. The network topology of LEACH is shown in Figure. 1. The LEACH protocol divides the network lifetime into a series of rounds. Each round is divided into two phases: the set-up phase and the steady-state phase. In the set-up phase, each node in the network randomly creates a number between 0 to 1. If the random number of the node is less than the threshold *T(n)*, the node is selected as the CH node. The threshold is calculated by the following equation.

$$T(n) = \begin{cases} \dfrac{p}{1 - p(r \bmod \frac{1}{p})} & \text{if } n \in G \\ 0 & \text{otherwise} \end{cases} \tag{1}$$

Where p is the optimal percentage of cluster heads in the network, r is the current round, and G is the set of nodes in the last $1/p$ round that are not selected as cluster heads.

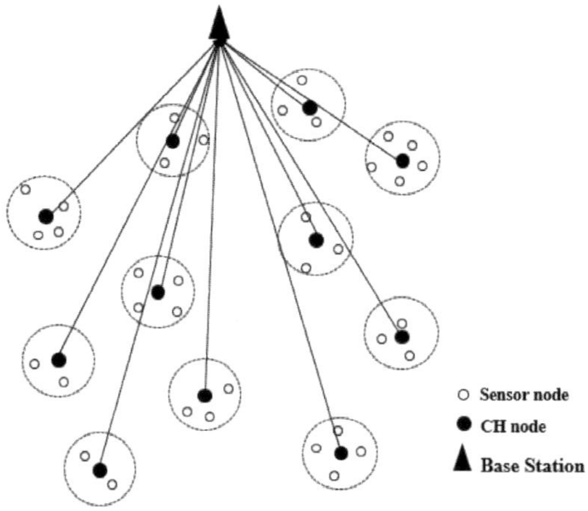

Figure 1 LEACH protocol topology

A large number of clustering routing protocols have been proposed after the LEACH protocol. In recent years, although the hierarchical routing protocol has made great progress, there is still much room for improvement. In this paper, an improved Multi-Hop Routing Protocol (IMHRP) is proposed. The protocol improves the clustering process and routing process. Compared with LEACH and MHT, the lifetime of IMHRP is obviously prolonged, and the energy consumption of nodes is more balanced. The rest of the paper is organized as follows: section 2 presents some of the improvement protocols based on the LEACH protocol as related work. The IMHRP routing model is presented in section 3. The IMHRP protocol is described in detail in section 4. The simulation results are given in section 5 and section 6 is the conclusion.

2. Related Work

In recent years, researchers have proposed a variety of hierarchical routing protocols to balance the energy consumption of WSN to prolong the network lifetime.

MHT [8] is a LEACH based protocol which discusses routing by dividing the network into two layers according to the distances between nodes and BS. However, for large-scale network environment, MHT cannot do a good job. Bhaskar Prince et al. proposed the EEUGCR [9] for large-scale networks. Dividing the network into fixed rectangular clusters in different sizes, EEUGCR selects the nodes with the highest residual energy and the closest distance to the cluster center as the CH nodes. However, the CH nodes can only transfer data to the direct upper layer, instead of based on the energy consumption to choose the next hop. In literature [10], two improved LEACH-based protocols, energy-LEACH and multihop-LEACH, are proposed. energy-LEACH improves LEACH's CH nodes election algorithm. Multihop-LEACH changes single-hop transmission of LEACH to multi-hop transmission. Energy-LEACH chooses the node with the highest residual energy as the CH node, but it cannot guarantee the evenly distributed of the CH nodes. The multihop-LEACH is superior to LEACH only when it is applied in a small-scale network. The EEM-LEACH [11] protocol chooses CH nodes based on the residual energy and the average energy consumption of nodes, which ensures that a CH node transmits the data to the next hop with minimum energy consumption, but it does not guarantee the minimal energy consumption of the total route and network energy consumption is uneven since the uneven distribution of CH nodes. Different from other routing algorithm, DAIC [12] adds gateway CH nodes in the network. In the DAIC, BS compares the distance d_v between the BS to the node and the average distance d_{avg} among between the whole network nodes and the BS. If $d_v<d_{avg}$, the nodes are called sub-layer nodes. Conversely, if $d_v \geq d_{avg}$, the nodes are called main-layer nodes. The DAIC uses the gateway nodes to relay the sub-layer CH nodes and the main-layer CH nodes in order to reduce the energy consumption of the CH nodes, but the d_{avg} in the DAIC is uncertain and the d_{avg} directly affects the energy consumption of nodes. Wendi B. Heinzelman et al. proposed LEACH-C [13] which improves the set-up phase of the LEACH protocol and uses a centralized control algorithm to select the CH nodes. Although LEACH-C can select the optimal number of CH nodes which have the highest residual energy, the uniform distribution of CH nodes cannot be guaranteed. On the basis of EEBCDA, Manoj Pant et al. proposed a multi-hop routing protocol [14]. The protocol makes a more balanced energy consumption of CH nodes which are far away from BS and prolongs the network lifetime, but the load problem of sensor nodes is not taken into account. Xueting Zhang et al. [15] used dynamic weights to classify nodes, but since the distance between nodes with different weights is still likely to be longer, the energy consumed would be more.

3. IMHRP Rounting Model

Most of the existing routing protocols mainly improved electing methods of CH nodes to extend the network lifetime. In addition to the improvement for electing CH nodes during the set-up phase, IMHRP mainly improves the routing algorithm during steady-state phase

3.1 IMHRP Network Model

In this paper, the network model assumes that there are N sensor nodes randomly distributed in the square area of $M*M$. The assumptions of the network environment are as follows:

- All sensor nodes are homogeneous.
- The location of the BS is fixed and positioned outside the square network area.
- After the network is deployed, all sensor nodes are fixed and energy constrained.
- All sensor nodes are able to adjust their own energy consumptions based on the distance to the receiver.

3.2 IMHRP Energy Consumption Model

The energy consumption model of this paper is the same as that in the literature [16] [17] [18], as shown in Figure. 2.

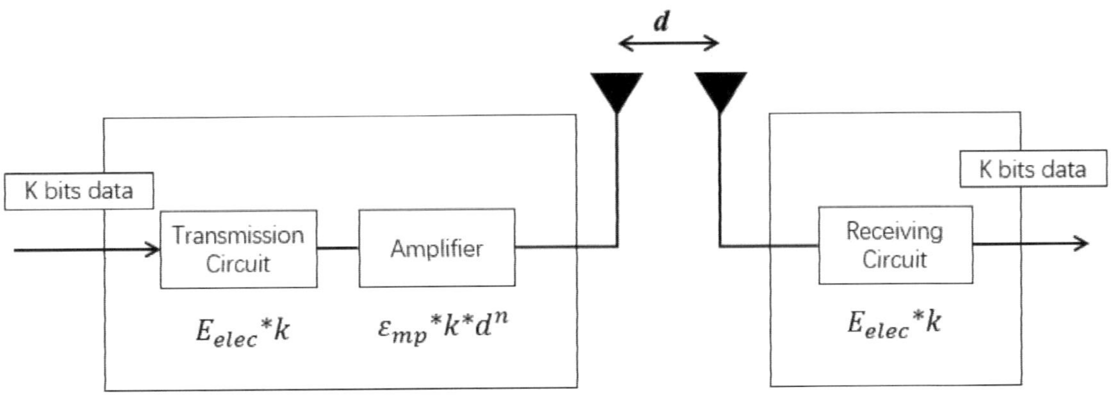

Figure 2 Energy consumption model

The energy consumption of the transmitted data depends on the distance between the sending node and the receiving node. The energy consumed to transmit k-bit data between two nodes with distance d can be calculated as follows:

$$E_{Tx}(k,d)=\begin{cases} k*E_{elec}+k*\varepsilon_{fs}*d^2 & d<d_0 \\ k*E_{elec}+k*\varepsilon_{mp}*d^4 & d\geq d_0 \end{cases} \quad (2)$$

In Eq. (2), E_{elec} is the energy that the sensor node consumes to transmits or receives 1-bit data. The threshold distance d_0 is calculated by the following equation:

$$d_0=\sqrt{\varepsilon_{fs}/\varepsilon_{mp}} \quad (3)$$

Where ε_{fs} and ε_{mp} are two parameters of the amplifier. When $d<d_0$, the energy consumption of the sensor nodes uses the Free-space model, and the amplifier parameter is ε_{fs}. When $d\geq d_0$, the energy consumption of the sensor nodes uses Multi-path fading model, the amplifier parameter is ε_{mp}.

4. IMHRP protocol

At the beginning of the IMHRP algorithm, there is a network initialization phase. The purpose is to initialize the sensor nodes in the network and the relevant parameters of the base station (BS). After initializing the entire network environment, the IMHRP protocol divides the lifetime into a series of rounds and each round is divided into a set-up phase and a steady-state phase.

4.1 Initialization phase

In the initialization phase, the BS broadcasts its own position to the sensor nodes of the whole network. Each sensor node calculates the distance to the BS. The distance between the *i*-th sensor node and the BS is d_{Si}. If $d_{Si}<d_0$, the *i*-th sensor node is an internal node which transmits the data directly to the BS. If $d_{Si}\geq d_0$, the sensor node is an external node which transmits the data to its cluster head. If the external sensor node is the cluster head node, the node transmits the data to the BS according to the routing algorithm mentioned below. The initialized wireless sensor network is shown in Figure. 3.

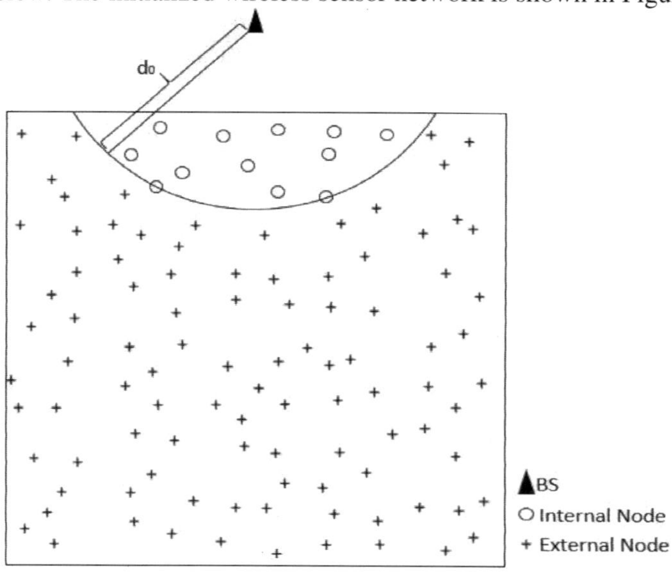

Figure 3 Initialized network layout

4.2 Set-up phase

In the set-up phase, like other clustering protocols, IMHRP divides the network lifetime into rounds. Each node chooses a random number between 0 and 1 in each round. If the random number of a node is less than the threshold *T(n)*, the node would be selected as the CH node. The threshold is calculated by the Eq. (1). In this paper, after selecting a CH node, the nodes with the distance less than d_x to the CH node will not be selected as a CH node. d_x is a constant distance in this paper.

After the CH nodes are selected, all CH nodes broadcast the ADV message to the non-cluster head nodes in the current round. The non-cluster head nodes select a cluster to join according to the signal strength of the received ADV message. Since the greater the signal strength is, the closer the non-cluster head node is to the CH node that issued the message. After selecting a cluster, each non-cluster head node sends an added message to the CH node of the cluster which it belongs to. Finally, the CH node creates a TDMA table for nodes to transmits data and sends it to non-cluster head nodes in the cluster.

4.3 Routing algorithm

According to the distance between the nodes in the network, nodes are divided into internal nodes and external nodes. The multi-hop routing is discussed on the basis of the distance relationships among CH nodes. There are six cases discussed as follows.

In Figure. 4, the tiny black dots represent internal non-cluster head nodes. In Figure.4-Fig.ure 9, the small hollow dots represent CH nodes in the network. Node A is the source CH node. Node B is the internal CH node which is closest to node A and node B_1 is the external CH node which is closest to node A. $d_{BS\text{-}A}$ is the distance between the BS and node A. $d_{BS\text{-}B1}$ is the distance between the BS and node B_1. d_{BA} is the distance between node B and node A. d_{B1A} is the distance between node B_1 and node A. The algorithm selects B_1 as the next hop on the premise that $d_{BS\text{-}A} \geq d_{BS\text{-}B1}$.

1) When $d_{BS\text{-}A} < d_0$, that is, the source node A is an internal node, then the energy consumption of node A to transmit data to the BS is proportional to the square of distance. The multi-hop mechanism cannot save more energy than node A directly connected to the BS. Therefore, node A uses a single-hop communication directly to transmit data to the BS and the internal non-cluster head nodes also directly transmit data to the BS. The internal CH nodes only act as relays for external CH nodes. Routing is shown in Figure. 4.

In Case 2)-Case 6), source node A is an external node, that is $d_{BS\text{-}A} \geq d_0$. According to the energy consumption Eq. (2), the energy consumption for node A to transmit data to the BS is proportional to the biquadrate of the distance. Therefore, it is necessary to consider the multi-hop mechanism in the following cases.

2) When $d_{BS\text{-}A} \geq d_0$, $d_{BA} \geq d_0$, $d_{BS\text{-}A} \geq d_{BA}$, $d_{B1A} \geq d_0$, the energy consumption for transmit data from source node A to node B_1 is proportional to the biquadrate of the distance. Also the energy consumed by node B_1 to transmit data to the BS is proportional to the biquadrate of the distance. The consumed energy of the data transmission between source node A and node B is proportional to the biquadrate of the distance. Since node B is an internal node, the energy that it consumes to transmit data to the BS is in proportion to the square of the distance from node B to the BS. Therefore, source node A transmits the data to the BS via node B, as shown in Figure. 5.

3) When $d_{BS\text{-}A} \geq d_0$, $d_{BA} \geq d_0$, $d_{BS\text{-}A} \geq d_{BA}$, $d_{B1A} < d_0$, the energy consumed by source node A transmit data to node B_1 is proportional to the square of the distance. The energy consumed by node B_1 to transmit data to the BS is proportional to the biquadrate of the distance. The energy consumed by source node A to transmit data to node B is proportional to the biquadrate of the distance between the two nodes. Since the node B is an internal node, the energy consumed by node B to transmit data to the BS is proportional to the square of the distance from the node to the BS. There is no way to directly determine which node to be selected as the next hop. Hence, it is necessary to determine the transmission path of node A by calculation. The routing is shown in Figure. 6.

4) When $d_{BS\text{-}A} \geq d_0$, $d_{BA} \geq d_0$, $d_{BS\text{-}A} < d_{BA}$, $d_{B1A} \geq d_0$, the distance between source node A and node B_1 and the distance between source node A and node B both are greater than the distance between source node A to the BS, so source node A uses a single-hop communication directly to transmit data to the BS. The routing is shown in Figure. 7.

5) When $d_{BS\text{-}A} \geq d_0$, $d_{BA} \geq d_0$, $d_{BS\text{-}A} < d_{BA}$, $d_{B1A} < d_0$, the energy consumed by source node A to transmit data to node B_1 is proportional to the square of the distance. The energy consumed by node B_1 to transmit data to the BS is proportional to the biquadrate of the distance. The distance from source node A to node B is greater than the distance from source node A to the BS. Therefore, source node A chooses node B_1 as a relay for forwarding data to the BS. the routing is shown in Figure. 8.

6) When $d_{BS\text{-}A} \geq d_0$, $d_{BA} < d_0$, the energy consumed by source node A to transmit data to node B and the energy consumed by node B to transmit data to the BS both are proportional to the square of the

distance. At this point, node B_1 is not considered as a relay for source node A, so the data of source node A is transmitted to BS via node B. The routing is shown in Figure 9.

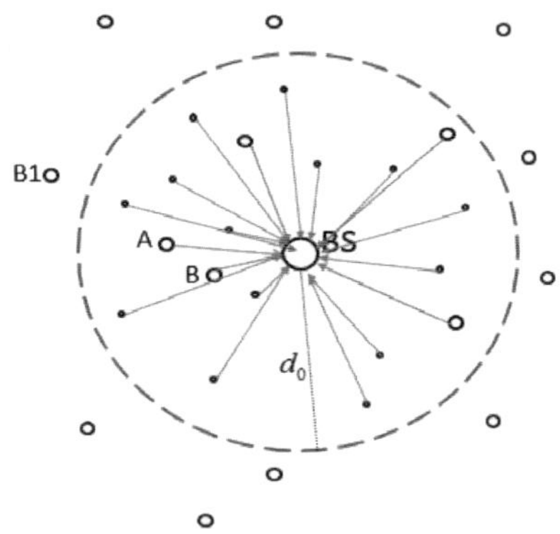

Figure 4 case 1:$d_{BS-A} < d_0$

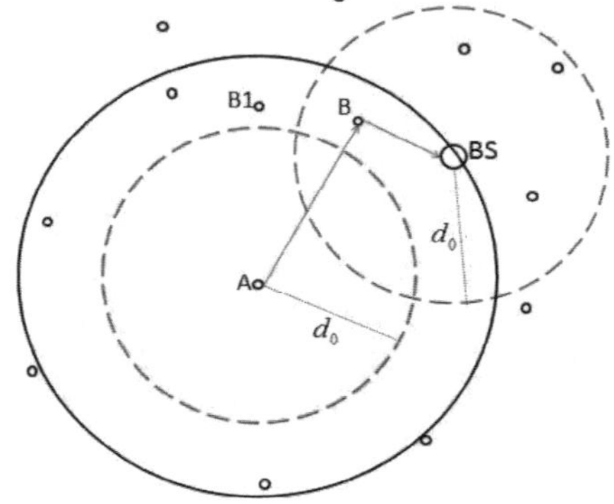

Figure 5 case 2:$d_{BS-A} \geq d_0$,$d_{BA} \geq d_0$,$d_{BS-A} > d_{BA}$,$d_{B1A} \geq d_0$

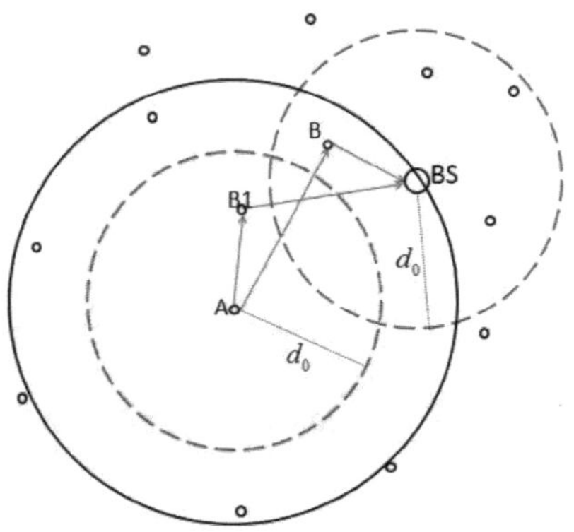

Figure 6 case 3:$d_{BS-A} \geq d_0$,$d_{BA} \geq d_0$,$d_{BS-A} > d_{BA}$,$d_{B1A} < d_0$

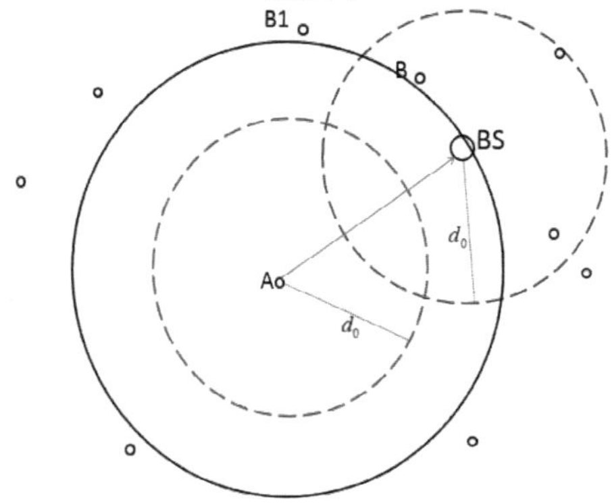

Figure 7 case 4:$d_{BS-A} \geq d_0$,$d_{BA} \geq d_0$,$d_{BS-A} < d_{BA}$,$d_{B1A} \geq d_0$

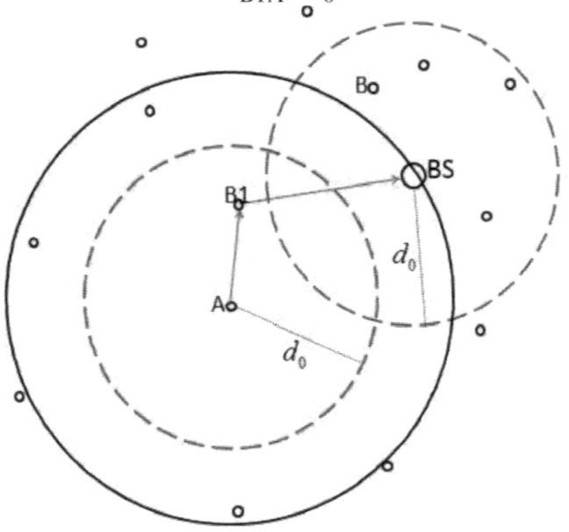

Figure 8 case 5:$d_{BS-A} \geq d_0$,$d_{BA} \geq d_0$,$d_{BS-A} < d_{BA}$,$d_{B1A} < d_0$

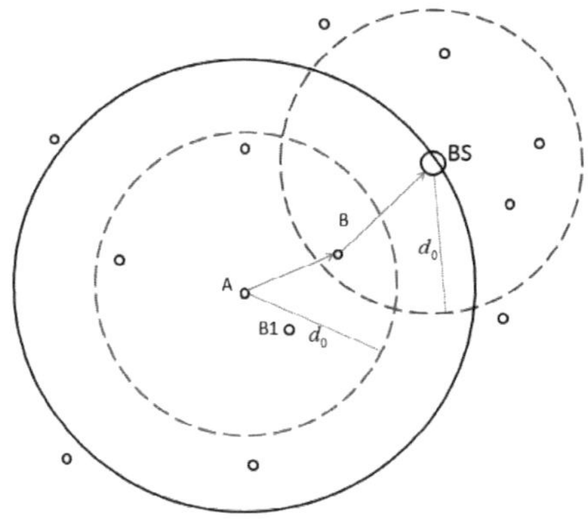

Figure 9 case 6:$d_{BS-A} \geq d_0$,$d_{BA} < d_0$

The routing of CH nodes in the above discussion is based on the distance between CH nodes. The distances from any internal nodes to the BS are less than d_0, so the internal nodes all directly transmit data to the BS. The internal CH node only acts as a relay and does not need to consume energy to aggregate the data collected from non-cluster head nodes. From the CH nodes which are closer to the BS, the source external CH node chooses the next hop node by comparing the energy consumed in

IOP Conf. Series: Journal of Physics: Conf. Series **910** (2017) 012054 doi:10.1088/1742-6596/910/1/012054

transmitting to the nearest external CH node and the energy consumed in transmitting to the nearest internal CH node, thereby achieving a minimum energy consumption of multi-hop transmission routing and extension of the entire network lifetime. The flow chart of the routing algorithm is shown in Figure. 10. According to the routing algorithm we can get the routing process diagram which is shown in Fig.ure 11.

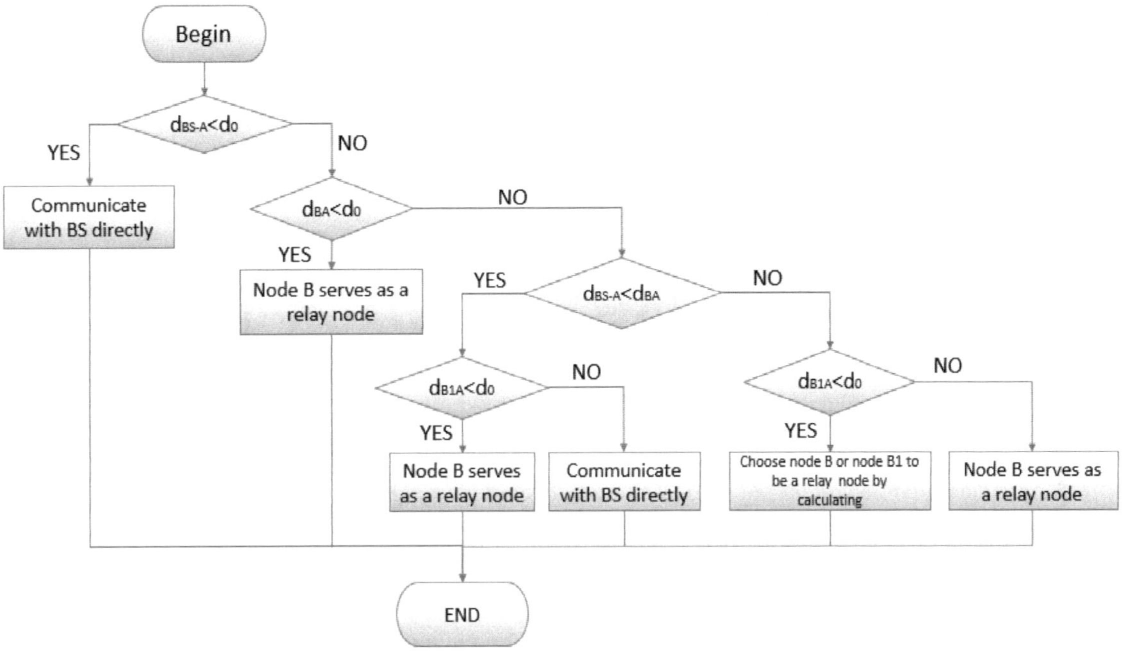

Figure 10 Routing algorithm flow chart

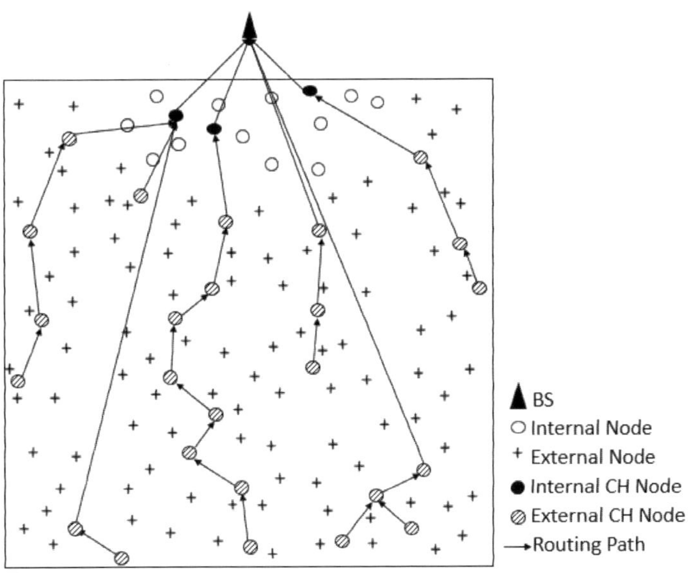

Figure 11 Routing process diagram

4.4 Steady-state Phase

During this phase, if the internal CH node in the network acts as a relay for external CH nodes, the internal CH node aggregates the data of external CH nodes and the data obtained from the environment and transmits the data to the BS directly. If the internal CH node does not act as the relay of external CH nodes, the internal CH node transmits its own environmental data directly to the BS. The internal CH node does not receive the data from the internal non-cluster head nodes. All the internal non-cluster head nodes directly transmit data to the BS. Each external CH node will have an TDMA table for the data sent by the member nodes in the cluster. The external CH nodes use TDMA to allocate the time slots for the cluster members to transmit data. The external CH node fuses the data sent by their member nodes in the same cluster and transfers the data to the BS through the above-mentioned routing protocol. In order to reduce the signal interference between clusters, each cluster of IMHRP has different CDMA spreading codes. All nodes in the same cluster use this code to transmit data to the CH node of the cluster. When the nodes in the cluster receives signals, they would filter out the signals of other clusters.

5. Simulation and Results

We use MATLAB for simulation. In order to simplify the entire simulation process, the paper assumed that the network has an ideal MAC layer. Data link communication is reliable and BS's energy is not restricted. Only the energy of the sensor nodes is considered during the experiment. The parameters of the network area are preprocessed before the algorithm starts. The simulation parameters of sensor nodes are shown in table 1.

Table 1 Sensor node simulation parameters

NO	*Parameter*	*Value*
1	$E_{Tx}/(nJ.bit^{-1})$	50
2	$\varepsilon_{fs}/(pJ.bit^{-1}.m^{-2})$	10
3	$\varepsilon_{mp}/(pJ.bit^{-1}.m^{-4})$	0.0013
4	$E_{DA}/(nJ.bit^{-1})$	5

The simulation parameters of the network area are shown in table 2.

Table 2 Network area simulation parameters

NO	*Parameter*	*Value*
1	Number of Nodes(N)	200
2	Deployment Area(Ω)/m^2	300*300
3	Co-ordinate of Sink(x_0,y_0)/m	(150,380)
4	Initial energy of SN/J	0.5
5	d_x/m	80

dx is obtained after many times of experimentation. The experiment results shown in table 3.

Table 3 The average rounds of IMHRP in the experiments

d_x/m	30	40	50	60	70	80	90
Average rounds	595	591	602	590	605	619	606

The above experiment results indicate that *dx*=80m is more appropriate in the 300m*300m network environment.

The death states of the nodes of LEACH, MHT and IMHRP in the 300m*300m network environment are shown in table 4.

Where FND indicates the round in which the first dead node occurs. LND indicates the round in which the last dead node occurs and the number of rounds of these protocols.

Table 4 The node death status of the network area

	LEACH		MHT		IMHRP	
	FND	*LND*	*FND*	*LND*	*FND*	*LND*
Average rounds	31	383	33	432	75	543

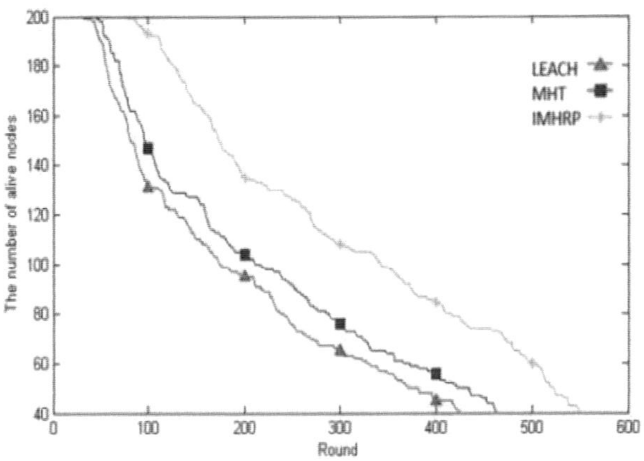

Figure 12 The number of alive nodes

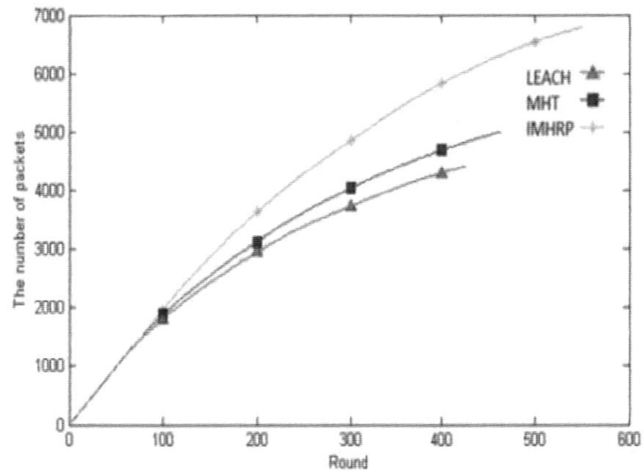

Figure 13 The number of packets received by the BS

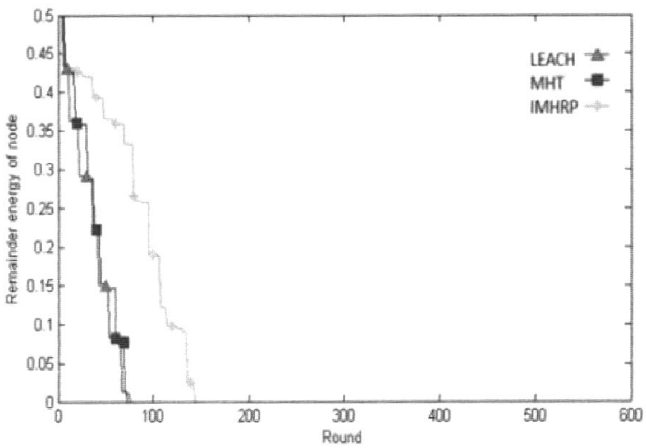

Figure 14 Energy consumption of node at coordinates (150,50)

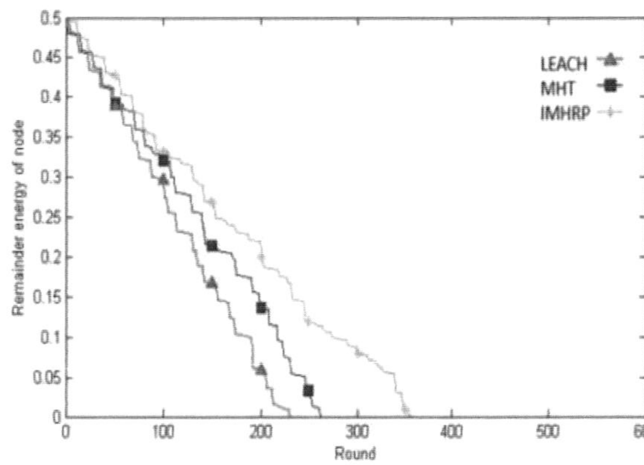

Figure 15 Energy consumption of node at coordinates (150,150)

Simulation experiments were carried out on LEACH, MHT and IMHRP in the corresponding network environment. Figure. 12 shows the changes in the number of alive nodes of the three routing protocols. Figure. 13 shows the number of packets received by the BS in the networks of the three routing protocols. According to table 4, IMHRP protocol can more effectively reduce the energy consumption of node transmission data when considering the distance relationship between CH nodes. In the 300m*300m network environment, the number of rounds of IMHRP is about 41.8% higher than that of LEACH, which is about 26.0% higher than that of MHT, and the round in which first dead node of IMHRP occurs has also delayed significantly, which means that the energy consumption of the whole network of IMHRP is more uniform and the network lifetime is more prolonged.

At the same time, this experiment compares three fixed nodes' energy consumption of LEACH, MHT, IMHRP in the network environment. The results are shown in Figure. 14, 15 and 16.

The three fixed nodes are respectively positioned in (150, 50), (150,150) and (150,300) in the 300m*300m network environment. Comparing IMHRP with LEACH and MHT, it can be concluded that the rate of the energy consumption of each node of IMHRP is slower than the other two protocols which also means that the energy consumption of overall network is more balanced than the other two protocols and the network lifetime is efficiently prolonged.

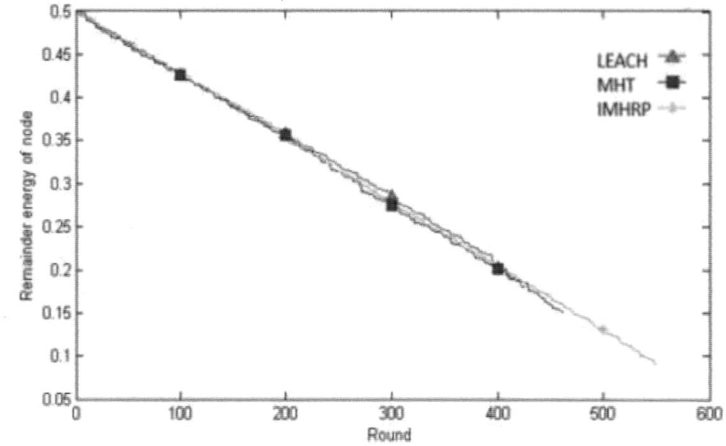

Figure 16 Energy consumption of node at coordinates (150,300)

6. Conclusion

Based on the analysis of various hierarchical routing protocols, an energy efficient hierarchical routing protocol (IMHRP) was proposed. We concentrated on the energy consumption problems like clustering and routing. The IMHRP protocol divides the sensor nodes in the network into internal nodes and external nodes according to the distance from the nodes to the BS and the threshold distance. In the set-up phase, the distance constraints are added so that the distribution of CH nodes would be more uniform. In the steady-state phase, the internal nodes directly transmit data to the BS. The external CH nodes determine their own routes according to the distance relationships with other CH nodes. Each external non-cluster head node transmits data to the CH node in its own cluster. IMHRP achieves multi-hop transmission without fixed hops. In essence, the routing of IMHRP is based on the amount of energy consumed rather than the distance. Simulation experiments showed that compared with MHT and LEACH, the first dead node of IMHRP appears later than the other two protocols. The number of alive nodes in the network reduces at a slower rate and the network lifetime of IMHRP is significantly prolonged. Among the three protocols, the energy consumption of each node of IMHRP is slower than that of the other two protocols. The energy consumption of IMHRP is more balanced than that of the other two protocols.

In future, we will have a more in-depth improvement of the set-up phase to achieve a more even distribution of CH nodes, more balanced energy consumption and more prolonged the network lifetime.

7. Acknowledgment

This work was supported in part by the National Natural Science Foundation of China under Grant 61472139, the key project of Shanghai Science and Technology Commission under Grant 11511504403, and a research grant made to East China University of Science and Technology by Shanghai Education Commission.

8. References

[1] I. F. Akyildiz, W. Su, Y. Sankarasubramaniam, E. Cayirci, "A survey on sensor networks," IEEE Communications Magazine, pp. 102-114, 2002.
[2] Padmavati, T. C. Aseri. "Comparison of routing protocols in wireless sensor network using mobile sink-A survey," Engineering and Computational Sciences (RAECS), 2014: pp. 1-4.
[3] Alotaibi E; Mukherjee B. "A survey on routing algorithms for wireless ad-hoc and mesh Networks," Computers networks,2012,56(2): 940-965.
[4] He LY, Li YM, Wang QD. "Clustering routing for wireless senor network,"Journal of Chongqing University, 2007, 30(1): 50-53

[5] Arumugam G S, Ponnuchamy T. "EE-LEACH: development of energy-efficient LEACH Protocol for data gathering in WSN," Eurasip Journal on Wireless Communications & Networking, 2015, 2015(1):1-9.
[6] Younis, O., Fahmy, S. "HEED: A hybrid, energy-efficient, distributed clustering approach for ad hoc sensor networks," IEEE Trans. Mob. Comput. 2004, 3, 366-379.
[7] Heinzelman W R, Chandrakasan A, Balakrishnan H. "Energy-Efficient Communication Protocol f or Wireless Microsensor Networks," Hawaii International Conference on System Sciences. IEEE, 2000:8020.
[8] EmadAlnawafa, Ion Marghescu. "MHT: Multi-hop technique for the improvement of leach protocol," 2016 15th RoEduNet Conference: Networking in Education and Research,2016, pp. 1-5.
[9] Bhaskar Prince, Prabhat Kumar. "An Energy Efficient Uneven Grid Clustering based Routing Protocol forWireless Sensor Networks," In Proceedings of the 2016 IEEE international conference onwirelesscommunications, signal processing and networking (wispnet), Chennai India, 2016, pp.1580-1584.
[10] F. Xiangning, S. Yulin. "Improvement on LEACH Protocolof Wireless Sensor Network," in International Conference onSensor Technologies and Applications, 2007. SensorComm2007, 2007, pp. 260–264.
[11] A. Antoo, R. Mohammed. "EEM-LEACH: Energy Efficient Multihop LEACH Routing Protocol for Clustered WSNs," IEEE International Conference on Control Instrumentation Communication and Computational Technologies (ICCICCT), pp. 812-818, 2014.
[12] NavinGautam, Jae-Young Pyun. "Distance aware intelligent clustering protocol for wireless sensor networks," Journal of Communications and Networks, 2010, 2010(4):122-129.
[13] W. Heinzelman, A. Chandrakasan, H. Balakrishnan. "An application-specific protocol architecture for wireless microsensor networks," IEEE Trans. Wireless Commun., vol. 1, pp. 660-670, Oct. 2002
[14] Mano j Pant, BiswanathDey, Sukumar Nandi. "A multi-hop routing protocol for wireless sensor network based on grid clustering," Applications and Innovations in Mobile Computing (AIMoC), 2015,pp.137-140.
[15] Xueting Zhang, Yingwu Chen, Jiang Jiang, Bingfeng Ge, Qi Wang. "A routing algorithm based on dynamic weight of the wireless sensor network," 2015 6th IEEE International Conference on Software Engineering and Service Science (ICSESS), 569-572, 2015.
[16] Manjeshwar A, Agrawal D P. "TEEN: a routing protocol for enhanced efficiency in wireless sensor networks," In: Proceedings of 15th Parallel and Distributed Processing Symposium Workshops, San Francisco, USA: IEEE, 2001, 2009-2015.
[17] Ye, Mao, Chengfa Li, Guihai Chen, Jie Wu. "EECS: an energy efficient clustering scheme in wireless sensor networks." In Performance, Computing, and Communications Conference, 2005. IPCCC 2005. 24th IEEE International, pp. 535-540. IEEE, 2005.
[18] Zhijun Gao, Hongyu Wang, Chunyan Xue, Zhonghua Han, "An Energy Efficient Hop-Number-Constrained Multi-hop Routing Algorithm for Heterogeneous Wireless Sensor Networks," In Proceedings of 2012 International Conference on Modelling, Identification and Control, Wuhan, China, June 24-26, 2012.

Research on Credit Evaluation of Mobile Medical APP Interactive Online Consultation Service-Take Haodaifu APP Online Payment Service as an Example

Dandan Xiong and Lei Zhao

School of Management, Shanghai University, Shanghai, China
Email: xiongdandan777@163.com, jolly1949@126.com

Abstract. Mobile medical industry is developing rapidly, but the relevant credit evaluation system is still in the exploratory stage. Haodaifu in the interactive online consulting services more representatives and has a credit evaluation system, but the system is not perfect. There are many problems in the system, such as credits indicators cannot better reflect the results of user's evaluation, the lack of user qualification supervision, evaluation index system and assessment channels are not perfect. On the basis of drawing on the advantages of E-commerce evaluation system and combining with the characteristics of mobile medical itself, two suggestions are put forward: Improve the credit evaluation model (Generate physician static credit index, Change the doctor 's dynamic credit index algorithm, Update the dynamic credit index), Improve the basic mechanism of credit evaluation.

1. Introduction

China's serious shortage of medical resources combined with uneven distribution of resources, leading to "medical difficulties" has become a major problem in today's society. Huge market demand gave birth to a new economic growth point. Now, the Internet develops rapidly, and mobile health came into being. Data show that as of June 2016, the size of Internet users in China reached 7.10 billion, and the Internet penetration rate reached 51.7%, which more than the global average of 3.1 percentage points. China's mobile phone user's scale up to 656 million, of which the use of mobile Internet users accounted for 92.5%, and only through mobile Internet

Users accounted for 24.5%. Internet users further focus on mobile devices[1]. In this context, mobile medical APP mushroomed, there are already more than two thousand related APP come out. Another data show that in March 2016 medical category APP monthly active users (MAU) reached 47.089 million, which is an increase of 547.0%[2]

The development speed of mobile medical APP is faster than imagined, but the quality of development is worrying. Many commercial advertising and bidding model triggered public distrust of mobile medical. The healthy development of the industry is inseparable from the credit management, and now the mobile medical industry credit evaluation system need to be built to standardize corporate behavior, while providing users with valuable basis for judging.

At this stage, the mobile medical APP function for the patient category is divided into the following categories: (1) Non-interactive medical information services. In other words, the interaction between people and intelligent machines. A typical example is the current large number of medical and health program software that with intelligent procedures for patients to make a preliminary diagnosis and to provide patients with treatment recommendations. (2) Interactive online medical counseling services. Specifically, patients and doctors through the Internet for online communication, such as doctors and patients interactive platform(3). Health monitoring guidance(4). Pharmaceutical e-commerce services(5). Medical service process optimization[3]. Among them, interactive online medical advisory services accounted for a large proportion of users. And it's influencing factors, especially the risk factors more than other types.

2. Haodaifu Online Credit Evaluation System Status

Haodaifu online was founded in 2006, which is one of the China's representative medical information and doctors and patients interactive platform. At the beginning of the creation, Haodaifu online focus on providing patients with medical reference information, and it established the first online real-time updated information query system. Today, the Haodaifu online has included 484,186 doctors of 7291 regular hospitals, and of which 142053 experts online Q & A. Data show that in 2016 (as of August 12 data date), Haodaifu online daily average number of active users has been increasing, of which 4, 5, 6 months, the number of active users increased rapidly, and the number of active users after June has remained at 1.5 million[4]. At present, Haodaifu online interactive medical counseling service main function classification as shown in Table 1:

Compared with free online consultation, paid online consultation has a greater advantage in the determination of doctor, counseling and waiting time. So paid online consulting increasingly becomes the main product of online consulting services. The credit evaluation system referred to in this paper is also mainly for paid online consulting services.

Table 1 Haodaifu online interactive advisory service classification table

Free online consultation	Online consultation	Specified doctor's free clinic App arranged for a doctor to answer
Paid online consultation	Online consultation Telephone consultation Video consultation	The patient selected the doctor to answer

Haodaifu online APP has now built an initial credit evaluation system:

The first is information disclosure. The doctor introduction page is mainly composed of the following parts: name, title, hospital and department, patient vote, professional good, practice experience, visiting location, patient evaluation, online inquiry, expert article.

Followed by service assessment. The most intuitive one is the "patient vote", the patient vote to the doctor to assess the situation, and the evaluation page is divided into two parts. The first part is the treatment process and the results of the evaluation, including the name of the disease, the purpose of treatment, treatment, treatment process evaluation, doctor costs, and the current condition. Among them, the treatment process has two main evaluation indicators: treatment results and treatment attitude. The second part is the information collection for medical treatment, including why choose the doctor and the registration route.

In addition, Haodaifu online interactive counseling service has a core evaluation index: "patient recommended heat". This indicator is based on the doctor's medical background, practicing experience, patient voting, doctor reports and the strength of the department where the doctor worked at, similar to the shopping site business credit rating. When the patient to find a doctor, the doctor list by default in accordance with the "patient recommended heat" from high to low, so when the patient choose to seek medical treatment, this indicator has a great impact on patients.

At present, the good doctor online APP evaluation system mainly exist the following deficiencies:

2.1 "Patient Recommended *Heat" Can Not Better Reflect the User's Evaluation Results*

As mentioned earlier, "patient recommended heat" is mainly based on the doctor's background, practice experience, the patient's vote and the strength of the doctor's department, etc. we can see that this indicator is not closely linked with the user's evaluation.

First of all, the impact of "patients recommended heat" factors are mostly doctors static credit indicators, such as medical background, the strength of the department, etc., which shows that the score of patient recommended heat" is more reflected in the doctor's qualifications and background, it does not reflect the user's medical experience evaluation.

Second, in the "patient recommended heat", the only factor associated with user evaluation is "patient vote" that is too general. Patient voting means praise, on the contrary, the patient does not vote that means non-praise. That is, 80% of the satisfaction of the vote and 90% of the satisfaction of the vote on the "recommended heat" has the same effect. For example, in the Haodaifu online APP found such a phenomenon. In the "gynecological department" link, Dr. Hong(attending physician) whose number of patients voting is 681,and whose efficacy and attitude are 100% satisfied and the patient recommended heat score 4.7; Dr. Chen(chief doctor)whose number of patients voting is 361,and whose efficacy and attitude are 99% satisfied, but the patient recommended heat score 5.0. This shows that due to the title or hospital department strength, doctors whose patient satisfaction and the number of votes are high may not have high score in patient recommended heat. This simple hierarchical division makes it difficult to truly and accurately reflect the results of *user's* experience evaluations.

2.2 Evaluation Index System Need to Be Improved

Although the Haodaifu online APP online evaluation contains a number of evaluation indicators, but only have two indicators to evaluate the doctors (medical) services: "treatment result" and "service attitude." This simple evaluation certainly cannot fully reflect the doctor's comprehensive credit situation. On the one hand, these two indicators are too general in the connotation, so patients can not accurately understand the connotation of indicators when they evaluate doctor's treatment. On the other hand, some indicators can also reflect the quality or effect of a doctor's service (treatment) to some extent, but these indicators are not included in the assessment system, such as "rational medication", "treatment" and "waiting time".

2.3 Lack of Supervision of the Evaluator's Qualifications

The Haodaifu online platform does not effectively qualify the evaluator (user or patient) qualification. The current online platform does not require users to authenticate the real name information; the patient simply by phone number can be registered as a user. Although in principle only requires users who have received medical counseling services can evaluate doctor's services (treatment), but the software platform does not set a valid user screening system. In other words, any registered user can evaluate any doctor (whether or not he or she has received the doctor's Treatment or counseling service). This simple authentication system and the lack of regulatory evaluation mechanism lead to such as vicious competition or malicious negative feedback and other phenomena, but also lead to the authenticity of credit evaluation are not guaranteed.

In addition, the influence of the evaluation provided by the users of different identities and qualifications should be treated differently. But in the good doctor online platform system only set the membership system, and the platform did not record the user's credit index. So on the platform, although the evaluator's identity and qualifications are different, but their evaluation of the "patient recommended heat" is the same impact.

2.4 Evaluation Channels Need to Be Improved

The current user evaluation channels mainly exists the following two questions:

First, fewer people involved in the evaluation. From the Haodaifu online doctor introduction page can be seen, comment participation rate is not high. The majority of patients participating in the assessment are less than 10% of the number of online counseling(The Haodaifu online has eight departmental categories, the study selected a branch department in each departmental category randomly and statistics of the top ten doctors evaluation, The results showed that a total of 80 doctors, of which 80% (i.e., 64) of the number of patients assessed less than 10% of the number of online consultation), which cannot give a better reference for other users.

Second, the majority of users involved in the evaluation are patients who received treatment. The evaluation content settled by Haodaifu online have a certain targeted. A lot of content mainly for patients who received treatment face to face, such as the purpose of treatment, treatment, treatment costs, registration path, etc., it can be seen Haodaifu credit evaluation system is still the services evaluation of doctor's traditional offline treatment. But nowadays online medical counseling service,

APP only provides a platform for patients and doctors to communicate, and the doctor's treatment and the APP operator is not linked. The essence of the Haodaifu online evaluation system is still the evaluation of the traditional medical process, and the system does not design the targeted credit evaluation on the basis of mobile medical characteristics.

3. Haodaifu Online APP Credit Evaluation Model Construction

In view of the above situation and shortcomings and reference to the more mature e-commerce credit evaluation system, this study attempts to improve the credit evaluation system of Haodaifu online consulting service, including the generation of "doctors static credit index", change the doctor dynamic credit index Algorithm, and in accordance with the time series to update the credit index. Specific improvement model shown in Figure 1:

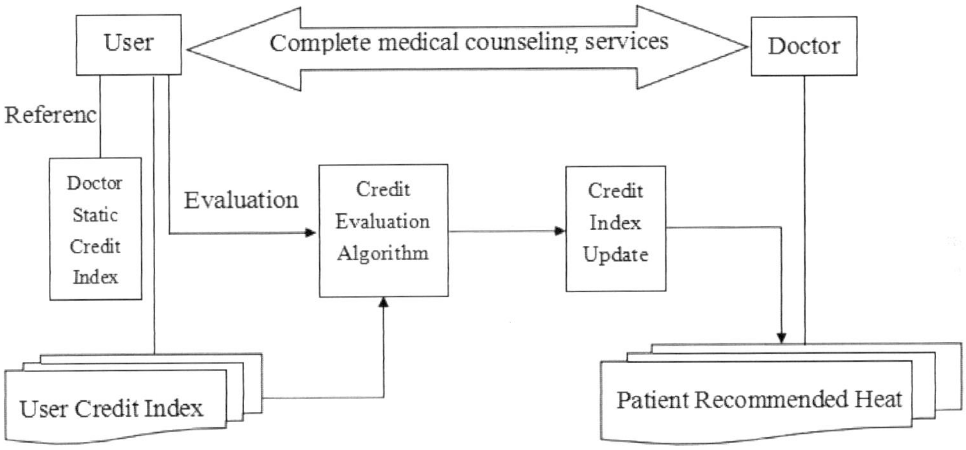

Figure 1 Improved Online Credit Rating Model

3.1 Physician Static Credit Index

The doctor's static credit index is calculated as follows:

$$M=\sum_{i}(Ci*Wi), \quad i=1,2,3\ldots\ldots$$

Note, M: physician static credit index; Ci: score of indicator; WI: weight of corresponding indicator

The above is the specific calculation of doctor static credit index, if you want to ensure the authority of the static index, but also to ensure that the relevant indicators information is true and accurate. Haodaifu online can be in cooperation with the doctor's hospital and the relevant health supervision departments to access to authoritative information, thus the doctor's static credit status can be assessed comprehensively. At the same time, pay attention to the credit index update, you can set the update cycle to update periodically. The renewal process plays a vital role in the improvement of the credit evaluation system, although it is complex and time-consuming and labor-intensive.

3.2 Considering the User's Credit Level

Now the field of e-commerce more use of two-way evaluation model, that buyers and sellers can give each other scoring according to their behavior in the transaction, and they can monitor each other, which is a more scientific evaluation model. The two sides of the transaction personally participate in the evaluation is the basis of the bidirectional evaluation model to ensure its authenticity and validity. Electricity providers can hire more staffs to communicate with customers and participate in the evaluation, but the medical profession is highly professional, and doctor must be involved in the entire transaction process himself. If a two-way evaluation model is used, the physician needs to evaluate each patient, which will spend a lot of time and reduce efficiency, thereby increasing the degree of lack of medical resources. So this model uses a one-way evaluation model, the user credit index by a number of indicators and calculated by the system.

3.3 Change the Doctor Dynamic Credit Index Algorithm

Dynamic credit index algorithm mainly from two aspects to improve: First, the user evaluation using the five-level scoring system, and second, considering the user's credit level.

In practice, you can choose persons who are experienced in this field, and they use the assignment method to quantify the qualitative indicators. And the expert law can be used to determine the corresponding weight of the indicators. This article combines the traditional medical evaluation indicators and mobile medical characteristics and we set a series of indicators as a reference, see Table 2:

Table 2 Physician Static Index Assessment Indicators Detailed Table

Indicator	Index weight	Indicator level division Each grade corresponds to score (Ci)					
Hospital grade	W1	Level 1	Level of first-class	Level 2	Second level of first-class	Level 3	Top 3
		1	2	3	4	5	6
Doctor level	W2	Resident Doctor		Doctor-in-charge	associate senior doctor	senior doctor	
		1		2	3	4	
Time	W3	5 years and below		5-15 years	15-25 years	25 years and above	
		1		2	3	4	
Medical malpractice	W4	The primary responsibility for the first and two level		The secondary responsibility for the first and two level	The primary responsibility for the three and four level	The secondary responsibility for the three and four level	
		-4		-3	-2	-1	

(Note: hospital grade is based on the classification of Haodaifu online platform.)

User credit index formula:

$$T=\sum_{i}(Ai*Wi), \quad i=1, 2, 3......$$

We can calculate the user credit index according to the above indicators and algorithms , and the existing credit index algorithm have been changed, when we calculate the doctor dynamic credit index the user's credit should be considered. That is, different credit levels of user evaluation have different influences on the doctor's credit index, so that the credibility of the doctor's credit index to be ensured.

Doctor dynamic credit index formula:

$$P=\sum_{i}(Ti*Si), \quad i=1, 2, 3......$$

Note, P: physician dynamic credit index; Ti: user credit index; Si: corresponding user to the corresponding doctor's credit score.

Using Five-Level Scoring System

In the calculation of the doctor dynamic credit index using scoring system, great, good, average , poor, very bad corresponding to the score of 5,3,0, -3,5.

The doctor's credit score is calculated as:

$$S=\sum_{i}Mi, \quad i=1, 2, 3......$$

3.4 Credit Index Update

The rapid development of medical technology and the accumulation of clinical experience make the disease cure rate increased rapidly. Therefore, the doctor's treatment and evaluation in different periods have different influence on the user's choice when they consider which doctor is better, generally speaking, the user may refer more to the recent data. Corresponding, different periods of evaluation scores on the doctor's credit index should also be different, so more timeliness. Based on above considerations, the evaluation model of the historical data and the current data need to be calculated differently.

$$P^\wedge=P1*u1+P2*u2+\ldots\ldots+P(n-1)*u(n-1)+Pn$$

Note, P^: a doctor's patients recommended heat in n period, that is, a doctor's total dynamic credit index in n period; Pn: a doctor's dynamic credit index in n period; u (n): the history credit decline coefficient in n period.

Scholars Colin F. Camerer's study[5] have shown that environmental changes, forgetting and other factors have influence on the decline coefficient, and u range of values: $0 <= u <= 1$. If the historical transaction activities and the current transaction activity are relatively high, the decline is less, so the value of u is lager, such as between 0.6-1. Conversely, if the consistency is low, so the decline is large, u can be between 0-0.4.

In modern society people's life is fast, so people can change their concerns in a short time. E-commerce field to Taobao, for example, found that users less concerned about the transaction information before six months, mainly focused on trading activities nearly a month. Products of mobile medical are medical services, the nature of which is special, so users may be more conservative than online shopping. Therefore, this article assumes that when a medical activity occurs in:

A year ago, the credit decline coefficient between 0-0.4value;

More than three months and within a year, the recession coefficient u in the 0.6-1 value;

Within three months, there is no recession, u take 1.

The credit index needs to be updated regularly. According to the time node of u value, the credit index can be updated for three months, and the platform can calculate the accurate patient's recommended heat value.

Note, S: the user to the doctor's credit score; Mi: the user to the specific evaluation index scoring

Note, T: user credit index; Ai: the amount corresponding to the indicator; WI: the weight of the corresponding indicator

User credit index indicators can also be used to quantify the indicators and expert law can also be used to determine the indicator's weight, the specific content in Table 3.

Table 3 User Credit Index Assessment Indicators Details Table

Indicator	Index weight	Index scoring rules		
Number of consultations (paid)	W1	A1: number of paid inquiries		
Accumulated consumption amount	W2	A2: Accumulated Consumption Amount / 50		
Bad comments account for the overall percentage of individual comments (X=The number of bad comments / Individual total number of evaluations) * 100%	W3	$X \geq 60\%$ A3=-4	$X < 60\%$ A3=0	
Registration time	W4	3 years and below A4=2	3-10 years A4=4	10 years and above A4=6

4. Improve the Extensive Support System of Credit Evaluation

The extensive support system of credit evaluation is the basis of credit evaluation model's practicality and scientific. Only the extensive support system to be perfect, users can conduct an effective evaluation, and follow-up credit index accounting and other links can be carried out. The extensive support system to be improved mainly from the following three aspects:

4.1 The Qualifications of Evaluators Should be Supervised Strictly

First, the user real name authentication that is the basis for recording the user's credit index, and it can make the entire evaluation system more standardized. Meanwhile, only users who received a doctor's medical consultation service can evaluate this doctor. The evaluation of who have not personally experienced medical counseling may only be based on "heard", there is a certain degree of evaluation bias, and the evaluation don't have much reference value.

4.2 Subdivide the Evaluator and the Indicator and Establish the Incentive Mechanism

First, subdividing the evaluator. Users will be divided into two categories: understand the doctor information from the APP and then accept the medical diagnose face to face, online consultation users. Two types of users enter their respective evaluation pages and fill in different evaluation information. For these two types of users, the evaluation of different priorities, indicators should also be set differences. Evaluation indicators can be subdivided from two aspects: add new indicators, refine the original indicators. New indicators should be clear, objective and comprehensive, and then the indicator will be graded, each one level of indicators refined into a number of secondary indicators, so that users can understand the meaning of indicators accurately. In addition, you can also add an open topic to encourage users to fill in their own want to fill the indicators, so that enterprises understand customer's needs, and then constantly improve the credit evaluation system according to users' suggestion. Open topics are not included in the credit index accounting. At the same time Haodaifu can establish incentives, give users some gift after they finished the evaluation, such as the gift coupons, good beans or free consultation, so that the enthusiasm of the user evaluation can be improved.

4.3 To Strengthen the Supervision of the Doctor Service Process

Haodaifu online is a intermediate platform to contact the doctor and the patient , if it does not strictly monitor the service process the user's experience will be affected, and the nature of the medical profession is special, so the doctor consultation process is not standardized may cause serious consequences.

Mobile medical industry for the lack of monitoring the main performance in the following two aspects. First, the online consultation cannot guarantee that the doctor involved the consultation himself. Online consultation is different from the traditional diagnose, if the consultation process is monitored non-strictly, phenomenon that doctor assistant or intern instead of the doctor to communicate with users online will happen. Whether the doctor is responsible for consultation, whether to lead the patient to the phenomenon of additional consumption. Whether the doctor is responsible for consultation, whether there is the phenomenon of driving patients to additional consumption. APP operators should carry out the supervision from the online and offline, and establish the feedback complaints mechanism. Strictly monitor the content of online consultation conversations and check frequently. Haodaifu can also set up a monitoring team, and the team members throughout the country, they should often go to the cooperative hospital to patrol. But also be cooperation with the hospital, and the hospital acts as a supervisory role, Haodaifu only need to contact with the hospital to obtain data regularly. At the same time, complaints mechanism can be established, the patient can complain about the behavior of doctors but they should provide evidence.

5. Conclusion

The shortage of medical resources and uneven distribution provide the mobile medical a huge market, mobile medical heat rose again, and the market appears more than two thousand corresponding products. With the exposure of bidding model and Putin, mobile medical industry began to gradually cool, and the user is more calm and rational. Extensive development model has passed, there are so many homogeneous products in the market, if you want to stand out you must enhance the quality of service and user's experience.

Understanding customer needs, and constantly improving the credit evaluation system is the key task of the development of the industry at this stage. Some measures can be considered to carry out, such as supervising doctors and users' qualification strictly, improving the credit evaluation indicators and evaluation channels, optimizing the credit index algorithm, strengthening the service process supervision. With reliable quality, perfect supervision mechanism to enhance customer satisfaction, and to reinforce corporate image, while serving the community, the economic efficiency of enterprises can be strengthened.

The industry can also try to build a third-party credit rating agency in the future, which is independent to the mobile medical platform, to create a more objective credit management environment. Everyone in this industry can try to seek more diverse credit evaluation indicators, to break the status that patient get together to the top three hospitals to accept treatment, to tap more quality doctor resources, to decentralize treatment options, and thus the efficiency of medical services can be improved truly.

6. Acknowledgement

Fund Project: National Social Science Fund Project (16BGL170).

7. References

[1] CNNIC.(The 38th China Internet Development Statistics Report) [R2016.http://www.cnnic.net.cn/hlwfzyj/hlwxzbg/hlwtjbg/201608/P020160803367337470363.pdf.
[2] Quest Mobile, 2016 Spring APP List: Total user volume exceeded 927 million, Wechat, QQ, Taobao, MAU ranking the top three [R].2016.http://www.questmobile.com.cn/blog/blog-39.html.
[3] Fanjin. Telemedicine and Internet Medical [EB/OL]. (2015-03-15) .[2015-12-23].http://health.sohu.com/20150315/n409815923.shtml.
[4] Quest Moblie What if Medical app better than health app?[R].2016.http://www.questmobile.com.cn/blog/blog_56.html.
[5] Colin F. Camerer, Teck-Hua Ho and Juin Kuan Chong. Behavioral Game Theory: Thinking, Learning and Teaching. Advances in Understanding Strategic Behavior pp 120-180.Palgrave Macmillan UK.2004.http://link.springer.com/book/10.1057/9780230523371.

The Analysis of Competitive Strategy and Suggestions for Didi under the Background of the New Car-hailing Regulation

Tingting Wang and Lei Zhao

School of Management, Shanghai University, Shanghai, China
Email: 13962842995@163.com, jolly1949@126.com

Abstract. The emergence of car-hailing service satisfies the need of public travel in Internet era. Didi—the representative of car-hailing service, provides users with cost-effective service and great travel experience and quickly became the leader in the field of mobile travel depending on its price advantage, market segmentation, fast respond and some other competitive strategies. However, the promulgation of the new car-hailing regulation brings many challenges to Didi. After the new regulation, it is hard for Didi to gap away significantly from its competitors in scale and price. Thus the differentiated service is the competitive focus for all platforms. So there is an urgent need for Didi to do something to make difference, such as improving the interface design of the platform and the process of order allocation, establishing exclusive 'station', increasing the interaction between drivers and passengers. By doing so, Didi can reduce the information asymmetry and increase the user engagement and loyalty with high quality service.

1. Introduction

With the development of the Internet, the traditional trip mode has been unable to meet the growing travel demand of the public. The Internet booking taxi is born at the right moment. It establishes a platform on the basis of Internet technology and includes qualified vehicles and drivers. By integrating the supply and demand information, it provides non-parade booking taxi service[1]. The platforms providing car-hailing service, such as Didi tailored car, quickly capture the market depending on its advantage of low-cost and great convenience. However, there are many controversies about the behavior of seizing the market and the potential operational security risks. In this context, the central government promulgated the Interim Measure to the Internet booking taxi service to regulate the platform company, vehicles and drivers, operations and so on. Local governments then made their own management regulations on the basis of realities of situation. The promulgation of the new regulation weakens the original competitive advantage of Didi. As a consequence, Didi and its competitors such as UCAR Inc. and Yidao are almost back to the same starting line. In the new round of competition, Didi must rebuild its core competitiveness in order to maintain its leading position.

2. The Analysis of Competitive Strategy for Didi Tailored Cars Service before the New Regulation

The reason Didi quickly captures the mobile travel market since 2014 when it found is largely due to its cost-effective service .To be specific, its low price advantage, market segmentation, fast respond and some other competitive strategies enable Didi to provides great travel experience to users and develop the consumers' travel habits.

IOP Conf. Series: Journal of Physics: Conf. Series **910** (2017) 012056 doi:10.1088/1742-6596/910/1/012056

2.1 Combine the Low Pricing and High Subsidy to Attract Passengers with Price Advantage

As we can see from the table 1, Didi has an absolute advantage in price compared with its competitors whether in economical kuaiche or comfortable, business and luxurious tailored cars.

In addition to low pricing, Didi also offers passengers with great subsidy. By the calculation of big data, Didi distinguishes different passengers with different sensitivity to price and then gives out red packets and discount coupons specifically. It is wise for Didi to use the price advantage which is a paint point for consumers and attract them with low pricing and high subsidy.

2.2 Devoted to Market Segment and Satisfy the Travel Demands for Different Kinds of Groups

Using market segment, Didi tries to satisfy the travel demands for different kinds of groups: kuciche for general public and tailored cars for high-end businessmen which can divide into the comfort, business and luxury .Passengers is able to make free choices according to their demands, while UCAR only serves high-end businessmen.

In addition to this, Didi continues to mine the potential demands and provides new service. For example, considering it is difficult for the old to use the Internet and mobile devices to call a taxi, Didi provides a new service called Taxi for the old. Passengers confirm the destination for their parents in advance and the old can call a taxi just with a click. The platform automatically recognizes the children's phone number as emergency contact so as to ensure the safety for the old. Another example service is the Enterprise Edition for Didi. It provides intelligent travel solutions for companies. The expense for employee travel is deducted directly from the company account. By doing so, there is no need to issue the invoice and apply for reimbursement. Consequently, it makes the management more efficient.

The comparison in price

Type	Didi			UCAR			Yidao		
	SF	MF	TF	SF	MF	TF	SF	MF	TF
Economy	0	1-2	0.2-0.5	/			6	2.2	0.3
Comfort	10	2.7	0.6-1	15	2.8	0.5	15	2.8	0.5
Business	14	3.8	1.4	20	4.5	0.7	20	4.5	0.6
Luxury	18	3.9	1.6	23	4.6	0.8	25	4.6	0.7

2.3 Provide Great Travel Experience by Fast Respond to Orders

Bringing in lots of private cars to the platform, Didi has the advantage of scale to provide passengers with more choices in the coverage area. In the meantime, the powerful data processing capability of the platform increases the speed of order respond and greatly decreases the waiting time of passengers. The high efficiency meets the public demands for speed.

Since 2014 when Didi provides tailored car service, it has been the leader in the field of mobile travel because of its low price, coverage for all target users and high efficiency.

In December 2015, CNNIC, short for China Internet Network Information Center released the Special Report on the Research of the Development of Tailored Cars Market. The report shows that the current tailored car market has formed a stable pattern where Didi ranks first with the 87.2% rate of utilization while UCAR and Yidao occupy part of the specific market[2]. In March 2016, the global performance management company Nielsen conducted a research on 4000 passengers and 320 drivers in 13 cities in China and released The Report on the Research of the China Internet Mobile Travel Market. As we can see from Fig1, Didi ranks first in user penetration, overall satisfaction, brand preference and brand loyalty[3]. In the same month, I Research released The Report On The Research Of China Mobile Travel Service Market In 2016 which shows that in mobile travel service industry, the rate of active user penetration of the APP of Didi has reached 77.6%, only the proportion of the tailored car users is up to 88.4, and the average order accounting for the 84.1%[4]. In the same year in October, the data of iiMedia Research shows Didi still has the advantage in the proportion of drivers with the ratio 68.4% even after the platform reduced the subsidy for drivers[5]. A month later, Trust Data, a domestic mobile Internet data monitoring organization, released The Report On The Trend Of The Development Of Car-hailing Platforms In China in 2015-2016.The dates show by the end of

September, Didi ranked first with the 36.48 million active users and 25.41 million average times the active users open the APP daily. While the gap between Didi and the following UCAR and Yidao is greatly large[6]. All these indicate Didi has been the absolute leader in the industry.

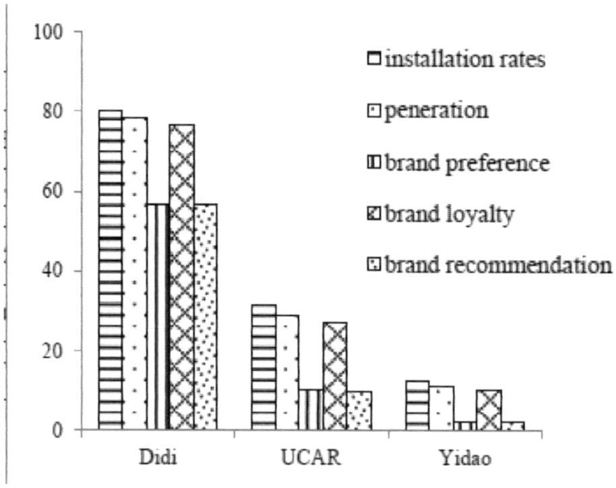

Figure 1 The comparison in consumers' satisfaction with brand

3. The Challenges New Regulation Brings to Didi

On the one hand, the rapid development of the car-hailing service helps to solve the problem in calling taxi and ease traffic congestion. On the other hand, the lack of effective supervision results in the worries about the safety and violation of legal rights .It also becomes a reason for the disorder of the travel market.

3.1 The Strict Limitation on the Vehicles and Drivers Leads to the Sharp Decrease of the Legal Cars in the Platform. There is no Doubt the Travel Experience will be Less Satisfying

The regulation in Beijing and Shanghai are the strictest in all. Both cities require the drivers and cars to have local residence cards and license tags respectively. A group of data from Didi shows that the proportion of cars meeting the demands is only 20%.The data iiMedia Research provide indicate that in Shanghai, there is at most 2% of the cars fit requires considering the limitations on drivers and cars comprehensively. Besides this, some other cities such as Shenzhen, Guangzhou and Nanjing also have a limitation on the year of the cars. That is to say, almost new cars can be used. The requirement of at least 4 SRS in Wuhan makes most cars unable to be used too.

Before the new regulation, the large scale of private cars ensures Didi to have sufficient vehicles in the coverage areas. This helps it to increase the speed of order respond and reduces the passengers' waiting time. According to iiMedia Research, among the improvements of the car-hailing service in 2016, 68.5% of the users hope that it can be easier to call a taxi and 32.1% of users want there are enough cars to satisfy the public needs. As is known to us, the restrict limitation on drivers and cars directly leads to the sharp decrease in cars that can be used for Didi. This is the first challenge the new regulation brings to Didi. So, it is important for Didi to figure out how to use the limited cars to offer equal even value-added service and how to continue to reduce the waiting time and offer more satisfying service.

3.2 There is no price advantage any more. The strategy of low pricing can't support the sustainable development of Didi.

The main reason Didi occupies most of the market just in a few years is the price strategy, which is in the manner of 'burn money'. Because of the large amount of subsidy to drivers and passengers, Didi became the absolute leader. To some extent, cheap has become the pronoun of car-hailing.

But it is regulated in the Interim Measure: Adhere to give priority to develop the urban public and then develop taxis. Develop car-hailing service in accordance with the principle of high quality and differentiated operation. All platform companies must mark the price clearly. They shouldn't obstruct the fair competition and violate the legal rights of passengers as well as the social public interests. Such behaviors like operating with price lower than cost for squeezing other rivals out and monopolizing the market, disturbing the normal market order, operating with high, unfair price to violate the interests of country and other operators are all illegal.

On the one hand, these rules limit the car-hailing service in the more high-ended tailored taxi service and make the private cars Didi bringing in before illegal. On the other hand, the rules indicate that it's impossible for Didi and any other companies to continue to occupy market in the manner of 'burn money' after the regulation. While the data of iiMedia Research show that there are more than 70% users pay attention to the rise of the price after the regulation. This suggests the public demand is not rigid and the price tend to be an important factor they consider when choosing vehicles .So it is not sure whether Didi can maintain its leading role after the new regulation without the advantage of price.

4. Suggestions for Didi after the New Regulation

It is difficult for Didi to gap away significantly from its competitors in price and scale after the new regulation. Thus the differentiated service is the competitive focus for it. Didi is expected to mine the potential needs of passengers and provide them with differentiated service in order to increase the users' engagement and loyalty.

4.1 Improve the Interface Design and the Process of Order Allocation

Now the interface design of Didi is —passengers choose the current position, the App will show them the cars nearby and the system will send the order to drivers. It is up to drivers to decide whether to accept the order or not. However, there are several problems about this design and order allocation.

First, what the App shows is the cars nearby rather than the cars available. This often causes passengers to mistake there are many cars nearby and one of the driver maybe come here quickly and drive him/her to the destination. While this is always not the truth. So it often takes passengers more time to wait after placing an order than expected. So it is necessary for Didi to update the interface design according to specific conditions. For example, differentiate available cars from those nearby but in service. If there are available cars, passengers can decide to call one. If there not, Didi should make sure the system has the ability to provide users with extended informational service, such as information about the amount of the cars which can finish the current service within a period of time ,the average time to wait ,the way to contact drivers and so on. With these informations, passengers can decide whether to continue to place the order or give up.

Second, there is no way for passengers and drivers to make mutual choice now. This can't satisfy some passengers having special require and cause them to cancel the order. Also, the lack of mutual choices always leads drivers to select the orders they want and refuse those they don't prefer. In comparison, the competitor Yidao tries many way to meet the differentiated demands as much as possible. For example, it allows users to choose the types of cars freely and creates information channel for them so that they can collect information about drivers and cars they prefer. So it is feasible for Didi to classify cars on the basis of needs for both drivers and passengers so as to provide mutual choices for them. Specifically, categories cars according to main factors passengers focusing on, such as the type, the gender of drivers, drive experience, invoice, carpooling and so on. Based on these informations, passengers can select freely. For drivers, they can decide whether to accept the order or not according to the requires passengers rise. Of course, it is essential to make a rule that passengers have the preference to choose. That is to say, if a passenger chooses a driver, this driver has the right to refuse, but this will affect the responsively, which is related to their salary.

Third, make use of intellectual technology to optimize the procedures of order acceptance and allocation. According to the current design, many orders can't be allocated reasonably because it doesn't consider the specific situations of drivers. They must click the option 'Don't grab orders temporarily' after they have accepted one. Otherwise, when they are in service, they will often receive voice reminder about orders which not only disturbs passengers but also distracts their attention. If the driver wants to know information about the order or accept the order, they must look away and even manipulate the steering wheel with one hand, this will cause potential risks. So Didi should optimize the procedures to solve problems above. For instance, make use of the technology to help drivers make the intelligent choice. When allocating orders, system should take the drivers' current condition—whether the driver is in service or not, into account, For example, the system won't send orders to drivers who can't finish the current order within 10 kilometers or in 15 minutes .In addition, Didi can improve its supporting facilities. Provide drivers with (Bluetooth) headsets and force them to wear when necessary. Convert cars such as install a wireless Bluetoooth button on the steering wheel to help to make more convenience for drivers. All of these can greatly reduce the security risks during the service as well as improve users' travel experience.

4.2 Establish Exclusive Didi 'station' and Strive to Make Information Equivalency between Users and Drivers

It is probably inevitable for users to have such experience: drivers can't find passengers with the information system provides. No matter the result is the passenger successfully gets to the destination by the longtime of waiting and communication or one party cancels or refuses the order reluctantly, it must be a failure experience for users. In this regard, Didi should improve its positioning technology and establish virtual -physical stations to reduce waiting time and improve travel experience.

Drivers and users have difficulty finding each other is largely because that the existing positioning system isn't precise enough. So it is important for Didi to improve the accuracy of GPS positioning system if it wants to win in the service. With the cooperation with leading companies in the field Didi is able to improve the deficiencies of the current GPS positioning system and provide perfect service combing its powerful big data.

On the other hand, Didi can establish its exclusive stations to change the information asymmetry between drivers and passengers. Outdoor stations have been built in Beijing and Shanghai, but the feedback is not as satisfactory as expected. The public don't know the existence of stations due to insufficient propaganda. In addition, users haven't experienced the convenience yet because the amount of stations is small. Considering these, one feasible solution is to establish virtual-realistic exclusive stations. Make cooperation with entities such as chain convenience stores, fast food shops and shopping malls to post up logos of stations in eye-catching areas. Besides this ,it is necessary to mark these stations accurately in the map and provide these information to passengers using App, so they can see the stations nearby clearly from the map when place an order and decide whether to go to the station or just wait here. With these means, it will takes passengers less time to wait and allows them to get to the destination as soon as possible, especially in busy commercial areas.

4.3 Increase Interactions between Drivers and Users to Improve the Service Quality and Users Engagement

Because of the few interactions between drivers and users, their role definition is simply consumers and service providers. So it may be helpful for Didi to carry out some activities online and off-line to increase interactions and users' engagement on the basis of protecting their legal rights.

In 2015, Didi comprehensively upgraded the anonymous evaluation mechanism for users. If a passenger is not satisfied with the driver or the service, he/she can grade and evaluate the service anonymously after the order. The system will show this information to drivers randomly after a period of time, hiding the information about the user. The scores and evaluations drivers get will be linked to their salary and reward. This evaluation mechanism indeed urges drivers to provide better service and is helpful for users to protect their rights. However, something embarrassing will happen. For example, certain passengers make some unreason able requests or even threat to give low scores and negative evaluation which makes a lot of drivers feel wronged. Therefore, in the next competition, there is need for Didi to upgrade the evaluation mechanism constantly. To be specific, the mechanism should allow drivers to evaluate passengers so that drivers can protect their rights too. It is reasonable to show passengers their scores so as to encourage more civilized behavior. In addition to this, the system can manage the users according to the evaluation and scores they get. For passengers who get lower scores, it is necessary to send them reminders to regulate their behavior. While for those always have improper behaviors, it is considerable to stop their accounts for some time or forbid them to use again. By these means both drivers and passengers can safeguard their legal rights.

Besides the bidirectional evaluation mechanism, Didi can consider to extend interactions beyond the boundary of service. For example, on Thanks-giving Day, Didi made a video about gratitude under the background of its Kuaiche service and put this video on App to propagandize. Users are allowed to send red packets and words via Wechat to show their thanks online. Didi collected thanks offline in the form of drift. Passengers wrote down their thanks voluntarily in the notebook and passed this warm and sweet to the next one. Also, Didi combined the online and off-line activities together. It put those collected thanks words on the body of cars to replace users to express their gratitude. This sincere, equal online and off-line activity has a remarkable effect. On Thanksgiving Day about 10 million people visited the online website and more than 80000 people received red pockets. There are more than 260 messages collected off-line.

This online and off-line interactive activity shows users' great enthusiasm for interaction not just great travel experience. So next, Didi can further extend its interactive pattern both online and off-line, such as theme activities about festivals and a variety of communication channels(including online chatting, send and receive red packet, floating cards and books etc.) to allow more drivers and users to join in. With these, it may be a warm and even surprised travel experience rather than a simple consumption.

These activities will not only further consolidates existing users, make them trust and like Didi ,but also successfully promote to more new users in an effective way so that they may pay attention to this brand and use the service.

5. Conclusion

The competition in car-hailing market is extensive before the new regulation. Platform companies such as Didi captured the market quickly with the advantage of low pricing, high efficiency and coverage for all groups. However, the new regulation, an invisible hand, points out potential problems faced by platform companies and regulates the further development pattern. The time platform companies compete with extensive management has gone. Didi must change its mode from barbaric growth to intensive cultivation in order to face the challenges and take chances the regulation provides. Only when Didi provides high-quality and unique service can it attract more users and maintain the leading role in the future competition.

6. Acknowledgement

* Fund Project: National Social Science Fund Project (16BGL170).

7. References

[1] Ministry of Transport, Ministry of Industry and Information etc. Interim Measure to the Internet booking taxi service[Z].2016-7-27.
[2] China Internet Network Information Center. The Specific Report on The Research of The Development of Tailored Cars Market. [R].Beijing: CNNIC, 2015-12.
[3] Nielsen. The Report on the Research of the China Internet Mobile Travel Market. [R].Shanghai: Nielsen, 2016-3.
[4] Research. The Report on the Research of China Mobile Travel Service Market in 2016[R]. Shanghai:iResearch,2016-3-28.
[5] Media Research. The Surveillance Report On The Mobile Travel Market In 2016. [R].Guangzhou:iiMediaResearch,2016-10-12.
[6] TrustData. The Report On The Trend of The Development of Car-hailing Platforms In China in 2015-2016[R].Beijing:Trustdata,2016-11-4.

Modified PTS-based PAPR Reduction for FBMC-OQAM Systems

Honggui Deng[1], Shuang Ren[2], Yan Liu[2] and Chengying Tang[1]

[1]Guangxi Key Laboratory of Information Materials, Central South University, Changsha, China
[2]School of Physics and Electronics, Central South University, Changsha, China
Email: denghonggui@163.com, 152211025@csu.edu.cn, zndxly@csu.edu.cn, csutcy@163.com

Abstract. The filter bank multicarrier with offset quadrature amplitude modulation (FBMC-OQAM) has been raised great concern in the 5G communication research. However FBMC-OQAM has also the inherent drawback of high peak-to-average power ratio (PAPR) that should be addressed. Due to the overlapping structure of FBMC-OQAM signals, it is proven that directly employing conventional partial transmit sequence (PTS) scheme proposed for OFDM to FBMC-OQAM is ineffective. In this paper, we propose a modified PTS-based scheme by employing phase rotation factors to optimize only the phase of the sparse peak signals, called as sparse PTS (S-PTS) scheme. Theoretical analysis and simulation results show that the proposed S-PTS scheme provides a significant PAPR reduction performance with lower computational complexity.

1. Introduction

5G is a next generation mobile communication beyond 2020, which is characterized by super high mobility, high energy efficiency, high spectrum efficiency, and far ahead of 4G in terms of low latency and high transmission rate. OFDM technique used in 4G is insufficient to meet the requirements of 5G scenarios due to its sensitivity to frequency offset estimation error, time misalignment and high out of band emission. To overcome these shortcomings, a new multicarrier modulation scheme called FBMC (Filter Bank Multicarrier) has increasingly favored by researchers [1~3]. And the combination of FBMC with OQAM modulation can bring low side lobes, achieve maximum bit rate and spectral efficiency, and be equipped with excellent time frequency localization property, which makes FBMC modulation become one of promising candidates for future 5G technique[4~6].

However, an intrinsic drawback of FBMC-OQAM systems is high peak-to-average power ratio (PAPR). Once the high-PAPR FBMC-OQAM signals drive the high-power amplifier (HPA) to operate in the nonlinear region of its characteristics curve, it may cause undesired in-band distortion and out-of-band radiation. Therefore, high PAPR will greatly deteriorate the performance of FBMC-OQAM systems.

Since the adjacent FBMC-OQAM signals are overlapped with each other, researchers have proposed some extended or hybrid methods that are based on conventional methods to tackle the PAPR problem of OFDM systems to deal with the same problem of FBMC-OQAM systems [7~11]. These conventional methods include the selective mapping (SLM) [12], partial transmit sequence (PTS) [13] and tone reservation (TR) [14]. Though these extended or joint scheme combines the advantages of the fundamental PAPR reduction schemes and can achieve better performance, it integrates the weakness of the conventional schemes.

In this paper, in order to ensure that the new scheme causes no distortion and create no out-of-band radiation, we have introduced a novel sparse PTS scheme, termed as S-PTS for simplicity, to reduce the PAPR in FBMC-OQAM systems. For the S-PTS scheme, the first step is to divide the interdependent transmission data into several subblocks, then needs to detect and record the location of random maximum peak of the time domain FBMC-OQAM signals derived from the superposition of the subblocks. And the second step feeds back the recorded location to the subblocks and utilizes phase factors to optimize the phase of each data on the corresponding position of the subblocks. That is to decrease the superposition of the data on that location point but does not modify any data on other positions, so as to reduce the peak power of the FBMC-OQAM signals.

The rest of this paper is organized as follows. Section II presents the FBMC-OQAM system model and the conventional PTS scheme. In section III, we describe the proposed sparse PTS scheme for FBMC-OQAM systems and analyze the computation complexity. In section IV, we study the performance of the S-PTS scheme and mainly compare with C-PTS scheme through simulation. Finally, the conclusion is given in section V.

2. FBMC-OQAM System Model

2.1 FBMC-OQAM Signals

As PHYDYAS mentioned a typical FBMC-OQAM system, which need the OQAM modulation and polyphase network to reduce the interference of adjacent sub-channels and computational complexity [15]. The transmission mode of FBMC-OQAM signals is shown in Figure.1, where the number of the subcarriers is M and the complex input symbols numbers is M, which is defined as:

$$X_n^m = a_n^m + jb_n^m, 0 \le n \le N-1, 0 \le m \le M-1 \tag{1}$$

Where am n and bm n denote the real and imaginary parts of the mth complex symbol Xm n on the nth subcarrier, respectively. For OQAM, the real and imaginary parts of the complex symbols are staggered in time domain by T/2, where T is the symbol period (equivalent to N). Then, the symbols are passed through a bank of transmission filters and modulated with subcarriers modulators where the spacing between every two subcarriers is 1/T. In that way, the mth transmission FBMC-OQAM symbol can be obtained by adding all subcarrier signals, which can be expressed as:

$$S^m(t) = \sum_{n=0}^{N-1} [a_n^m h(t-mT) + jb_n^m h(t-mT-\frac{T}{2})] e^{jn(\frac{2\pi}{T}t+\frac{\pi}{2})} \tag{2}$$

Where h(t) is impulse response of the prototype filter with KT length and K is the overlapping factor[15]. The FBMC-OQAM symbol duration is extended by KT and the transmission symbol rate is still 1/T, which means that K consecutive output symbols overlap with each other in the time domain. The overlapping structure of FBMC-OQAM signals is described in Figure.2. Then the length of FBMC-OQAM signal S(t) that are derived from summation of the M FBMC-OQAM symbols is (M+K-1/2)T, i.e.,

$$S(t) = \sum_{m=0}^{M-1} S^m(t), \quad 0 \le t \le (M+K-\frac{1}{2})T \tag{3}$$

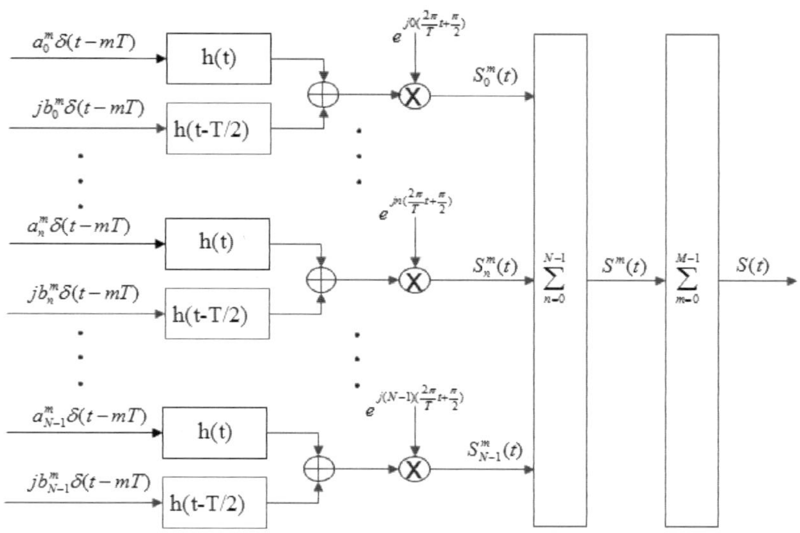

Figure 1 The transmission mode of the FBMC-OQAM signals

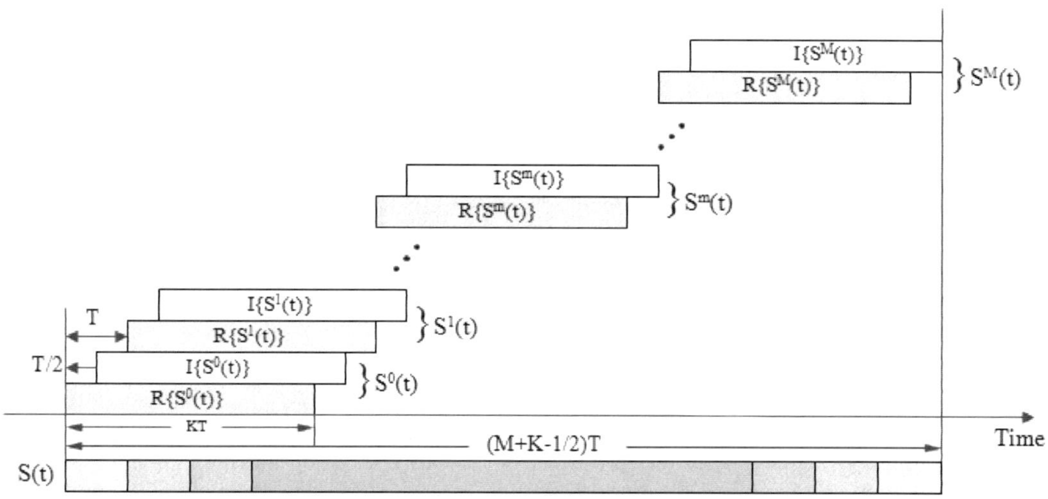

Figure 2 Structure of the FBMC-OQAM signals

2.2 Analysis of PAPR in FBMC-OQAM System

We need to consider multiple overlapped data blocks when reducing the PAPR of FBMC-OQAM systems, rather than each block independently like OFDM system. Thus, the general definition of PAPR for OFDM systems is not applicable to FBMC-OQAM systems. Then, we divide the main value of S(t) into M intervals equally with the time duration T. Thus, the PAPR of each interval is defined by

$$PAPR_i = 10\log_{10}\frac{\max\limits_{iT\le t\le(i+1)T}\left|S(t)\right|^2}{E[\left|S(t)\right|^2]}, i = 0,1,\ldots,M-1 \tag{4}$$

Where $E[|S(t)|^2]$ denotes the expectation, i.e., the average power of FBMC-OQAM signals. Similar to OFDM system, we still utilize complementary cumulative distribution function (CCDF) to measure the PAPR performance of FBMC-OQAM system, which is defined as the probability that PAPR exceeds a given threshold PAPR0.

2.3 Conventional PTS Scheme for FBMC-OQAM

For the conventional PTS (C-PTS) scheme, the subcarriers are divided into V subblocks and each subblock has N/V subcarriers. Then, each disjoint subblock Sm v(t) is multiplied with corresponding phase rotation factor bv, v=1,2,...,V, and the elements in the vector bv are usually selected from the set {-1,1}.

$$S^{m}(t)=\sum_{v=1}^{V} b_{v} S_{v}^{m}(t) \tag{5}$$

The phase vector is chosen for the signals with least PAPR, which is shown as

$$\arg\min_{b_v}\max\left|\sum_{v=1}^{V} b_{v} S_{v}^{m}(t)\right|^{2} \tag{6}$$

In general, an exhaustive searching method is used in Equation (6) to find the optimal solution, and the phase optimization is applied to each data of FBMC-OQAM signals, which is bound to cause a large amount of computation. Moreover, affected by the overlapping structure of consecutive FBMC-OQAM signals, directly applying the C-PTS scheme to reduce the PAPR of FBMC-OQAM systems is not very effective.

In addition, in order to improve the effect of the C-PTS scheme to reduce PAPR, we divide the overlapped filtered signals into several segments and then pick out an optimal phase rotation factors with exhaustive searching method in each segment. And the simulation results will be presented in section IV.

3. PAPR Reduction Based on Sparse PTS Scheme

In this section, we propose a sparse PTS scheme to reduce the PAPR of the FBMC-OQAM signals with better PAPR reduction performance and lower complexity than the conventional PTS scheme.

3.1 Proposed S-PTS scheme

Similar to the conventional PTS scheme, the preliminary procedure for the proposed S-PTS scheme is to transmit the interdependent input signals in packet mode. Thus, the mth original FBMC-OQAM signals can also be expressed as follows:

$$\begin{aligned} S^{m}(t)=\sum_{v=0}^{V-1}\sum_{n=0}^{N/V-1} [a_{n+v(N/V)}^{m} h(t-mT) \\ + jb_{n+v(N/V)}^{m} h(t-mT-\frac{T}{2})]e^{j(n+v(N/V))(\frac{2\pi}{T}t+\frac{\pi}{2})} \end{aligned} \tag{7}$$

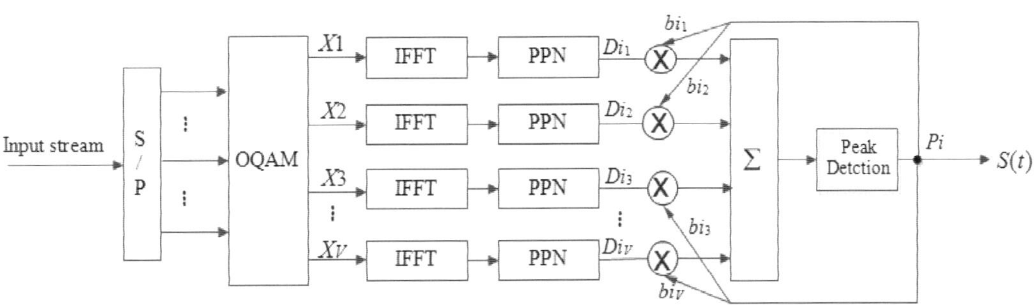

Figure 3 Structure diagram of FBMC-OQAM system combined with S-PTS scheme

Figure.3 illustrate a block diagram. The novel procedure is to detect and record the location of maximum peak of the time domain signals S(t) and employ phase factors to optimize the phase of the data on the recorded position of the subblocks. Then, we set an iteration value I to detect the position of peak signals repeatedly in order to reduce the peak power of FBMC-OQAM signals effectively.

The algorithm of S-PTS scheme involves the following steps:

Detect the location point of maximum peak of FBMC-OQAM signals and record it as Pi, as shown in Figure.3. Assume the total number of iteration as I (i=1,2,...,I).

Look for the data Div (v=1,2,...,V) at the corresponding position Pi in the subblocks and each data is rotated with different phase rotation factors bv∈ {-1,1}, so as to keep these data phase from being consistent at the same time and modify their summation at this position Pi.

Find out a best one (b1, b2,..., bV) from 2V phase rotation factor combinations to achieve the least PAPR for signals. Then multiply the data Div (v=1,2,...,V) by the best combination (b1, b2,..., bV) to obtain the new data (b1Di1, b2Di2,..., bVDiV) with the minimum summation.

Use (b1Di1, b2Di2,..., bVDiV) to update the data at the current position Pi and generate the optimized subblocks.

Repeat the step a to step d until the iteration meets the value of I. Finally, the time domain FBMC-OQAM signals with the least PAPR that are derived from the superposition of the optimized subblocks are selected for transmission.

The difference between the C-PTS and S-PTS schemes lies in that all the data are taken into account to select the optimal phase rotation combination in the C-PTS scheme, while the proposed S-PTS scheme only executes phase optimization operation for few data with high peak, without changing any data on other positions. This means that computational complexity of the proposed S-PTS scheme would be much lower than that of the C-PTS scheme. In addition, since the occurring probability of the maximum peak of FBMC-OQAM signals is relatively low, only a few maximum peaks need to be processed with phase optimization so that its impact on the average power of FBMC-OQAM signals is imperceptible. It is explained theoretically that our proposed S-PTS scheme can make the PAPR of FBMC-OQAM signals decrease significantly.

3.2 Computational Complexity Analysis of S-PTS scheme

In this subsection, we analyze the computational complexity of the proposed S-PTS scheme by comparing with the C-PTS scheme. We mainly analyze the computation of phase optimization operation in both schemes, ignoring the computation of the same and small part. In order to find out the best combination of phase factors from 2V candidates which are derived from exhaustive search, the proposed S-PTS scheme requires IV2V complex multiplications and 2I(V-1)2V complex additions, while the C-PTS scheme requires NMV2V complex multiplications and 2NM(V-1)2V complex additions. Finally, when determining the transmission signals with the least PAPR, the S-PTS and C-PTS schemes require IV and NMV complex multiplications, 2I(V-1) and 2NM(V-1) complex additions, respectively. Then, the numbers of required real multiplications and real additions for the two schemes are listed in Table 1.

Table 1 Comparison of Computational Complexity of S-PTS and C-PTS Schemes

Scheme	Real multiplications	Real additions
S-PTS	$IV(2^V+1)$	$2I(V-1)(2^V+1)$
C-PTS	$NMV(2^V+1)$	$2NM(V-1)(2^V+1)$

From the Table I we can see that the number of iteration I is directly related to the computational complexity in the proposed S-PTS scheme. The probability of maximum peak of FBMC-OQAM signal is relatively low, which means the iteration times I is much less than the value of NM in the C-PTS scheme. Therefore, the computational complexity of the S-PTS scheme will be far less than that of the C-PTS scheme. Similarly, compared to the C-PTS scheme, the recovery of signals phase for the PTF scheme at the receiver also only needs a small amount of calculation.

3.3 Analysis of Side Information in S-PTS Scheme

As we all know, PTS schemes need to send side information (SI) about the phase rotation factors to recover the original phase of signals at the receiving end. For the proposed S-PTS scheme, the side information increases with the increasing of iteration number I, while the side information increases with the increasing of segment number S. Compared with the C-PTS scheme, the superior PAPR reduction performance of the proposed S-PTS scheme is not at the expense of adding additional side information. And the simulation results will be presented in section IV.

4. Simulation Results

In this section, extensive simulations are conducted to assess the PAPR reduction performance of the proposed S-PTS scheme. Table II shows the simulation parameters for FBMC-OQAM system.

Table2 Simulation Parameters

Modulation mode	4OQAM
Number of subcarriers	N=64
Number of sub-blocks	V=4
Phase rotation factor	bi={1,-1}
Number of iteration(S-PTS)	I=6,8,12,16
Number of segment(C-PTS)	S=1,8,16

Figure.4 shows the PAPR reduction of the S-PTS scheme with I=6, 8, 12, 16 and the C-PTS scheme with S=1, 8, 16, respectively. The curve "Original" indicates the FBMC-OQAM signals without any PAPR reduction schemes. It is obvious that increasing I is more effective to decrease the value of PAPR for the S-PTS scheme. If the CCDF is fixed at the value of 10-3, the PAPR values of the proposed S-PTS scheme for I=6, 8, 12, 16 are about 2.5dB, 2.7dB, 3.1dB, 3.4dB smaller than the original FBMC-OQAM signals, respectively. Therefore, the S-PTS scheme could significantly reduce the PAPR of the FBMC-OQAM signals.

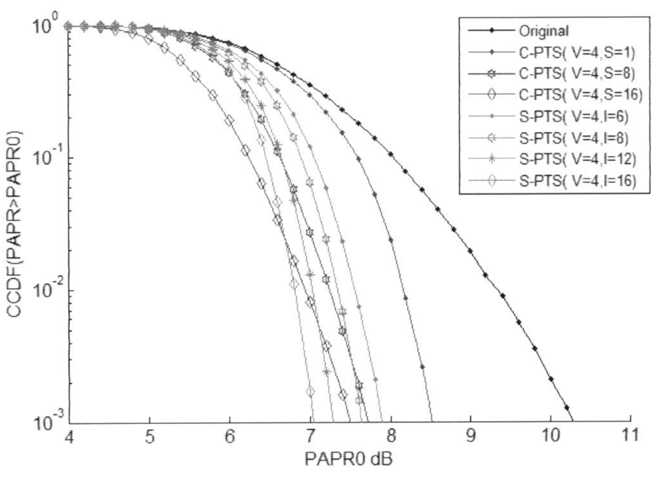

Figure 4 CCDFs of the proposed S-PTS and C-PTS schemes

Furthermore, we compare the PAPR reduction performances of the S-PTS scheme with that of the C-PTS scheme in Figure.4. First of all, we can see that when we directly apply the C-PTS scheme without dividing segments (S=1), it is not very effective to reduce the PAPR of the FBMC-OQAM signals. And the value of PAPR for the C-PTS scheme decrease faster as the increasing of segments S. Nevertheless, at CCDF=10-3, the PAPR of the proposed S-PTS scheme with I=8 is about 7.6dB, smaller than 7.7dB of the C-PTS with S=8. Besides, the PAPR of the proposed S-PTS scheme with

I=16 is about 7.0dB, smaller than 7.5dB of the C-PTS with S=16. That is to say, in the case where the amount of side information is the same, the proposed S-PTS scheme can achieve better PAPR reduction performance than the C-PTS scheme. In addition, the PAPR of the proposed S-PTS scheme with I=12 is 7.3dB, less than 7.5dB of the C-PTS with S=16, which indicates that our proposed S-PTS scheme can achieve superior PAPR reduction performance when it has less side information.

Moreover, we substitute the simulation parameters into the equations in Table I to calculate the numbers of the real multiplications and real additions, as shown in Table III.

Table 3 The Computational Complexity of the Two Schemes

Scheme			Real multiplications	Real additions	PAPR performance
C-PTS	V=4	S=1	69 632	104 448	8.6
		S=8	69 632	104 448	7.7
		S=16	69 632	104 448	7.5
S-PTS	V=4	I=6	408	612	7.8
		I=8	544	816	7.6
		I=12	816	1 224	7.3
		I=16	1 088	1 632	7.0

As can be seen in the Table III, the computational complexity of our proposed S-PTS scheme with various parameters are much smaller than that of the C-PTS scheme.

Figure.5 shows the peak canceling signals of the proposed S-PTS scheme with I=16 and original signals. It can be found that, peak powers of the original signal appear in a few points, majority of points have lower powers, which means that the calculations of our proposed S-PTS scheme will not be too complicated. And the data with the exception of peak signals in black curve coincides with the data in red curve and is covered in the simulation. Obviously, the whole amplitude of the red curve is lower than 2.5, which means our proposed S-PTS can efficiently reduce the PAPR of the FBMC-OQAM signals.

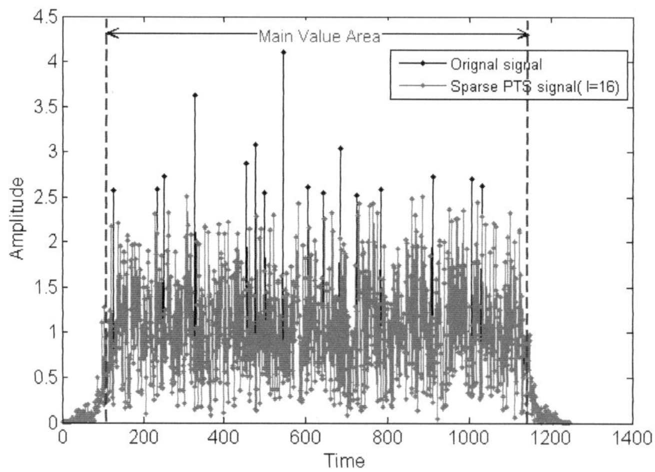

Figure 5 peak canceling signals with PTF scheme and original signals

Therefore, in general, our proposed S-PTS scheme has more excellent PAPR reduction performance and lower computational complexity compared with the C-PTS scheme.

5. Conclusions

In this paper, we propose a novel sparse PTS scheme with low complexity to reduce the PAPR of FBMC-OQAM systems. In order not to cause distortion and waste of frequency resources, the key of this scheme is to use the typical phase rotation factors to optimize the phase of the peak signals derived from peak detection of the filtered signals. Simulation results show that, compared with the C-PTS scheme, our proposed not only achieves better PAPR reduction performance but also has lower computational complexity.

6. Acknowledgment

This work was supported by the Open Science Research Fund of Guangxi Key Laboratory of Information Materials, Guilin University of Electronic Technology.

7. References

[1] Gerhard Wunder, Peter Jung, Martin Kasparick, et al. "5GNOW: non-orthogonal, asynchronous waveforms or future mobile applications," IEEE Communications Magazine, vol.52(2), pp. 97-105, 2014.
[2] B. Farhang-Boroujeny, "OFDM Versus Filter Bank Multicarrier," IEEE Sig. Proc. Mag., vol. 28, May 2011, pp. 92–112.
[3] F. Schaich and T. Wild, "Waveform contenders for 5G OFDM vs FBMC vs UFMC," in Proc. International Symposium on Communications, Control and Signal Processing, Athens, 2014, pp. 457-460.
[4] J. F. Du and S. Signell, "Time frequency localization of pulse shaping filters in OFDM/OQAM systems," in Proc.6th International Conference of Information,Communications & Signal Processing, Singapore, 2007,pp. 1-5.
[5] P. Siohan, C. Siclet, and N. Lacaille, "Analysis and design of OFDM/OQAM systems based on filterbank theory," IEEE Trans. Signal Process., vol. 50, no. 5, pp. 1170–1183, May 2002 .
[6] M. Bellanger, 'Specification and design of prototype filter for filter bank based multi-carrier transmission, IEEE International Conference on Acoustic, Speech and Signal Processing, pp. 2417-2420, May 2001.
[7] D. Qu, S. Lu and T. Jiang, "Multi-Block Joint Optimization for the Peak-to-Average Power Ratio Reduction of FBMC-OQAM Signals",IEEE Trans. on Signal Processing, vol. 61, no. 7, pp. 1605-1613, 2013.
[8] A. Skrzypczak, P. Siohan, and J. P. Javaudin, "Reduction of the peak-to-average power ratio for OFDM-OQAM modulation," in Proc. 63rd IEEE Veh. Technol. Conf., May 2006, pp. 2018–2022.
[9] S. Lu, D. Qu, and Y. He, "Sliding window tone reservation technique for the peak-to-average power ratio reduction of FBMC/OQAM signals," IEEE Wireless Communication Letters, vol. 1, no. 4 , pp. 268-271, April 2012.
[10] V. S. Kumar and S. Anuradha, "Sliding Window tone reservation using smart gradient projection method for PAPR reduction of FBMC signals," in Proc. International Conference on Electrical, Electronics, Signals, Communication and Optimization, Visakhapatnam, 2015,pp. 1-5.
[11] S. Vangala and S. A nuradha, "Hybrid PAPR reduction scheme with selective mapping and tone reservation for FBMC/OQAM," in Proc. 3rd International Signal Processing, Communication and Networking, Chennai,India, 2015, pp. 1-5.
[12] H.-S. Joo, S.-J. Heo, H.-B. Jeon, J.-S. No, and D.-J. Shin, "A new blind SLM scheme with low decoding complexity for OFDM systems," IEEE Trans. Broadcast., vol. 58, no. 4, pp. 669–676, Dec. 2012.
[13] X. Qi, Y. Li, and H. Huang,"A low complexity PTS scheme based on tree for PAPR reduction," IEEE Communication Letters, vol. 16, no. 9, pp. 1486-1488, September 2012.
[14] G. Saeed and A. Ruhallah,"Tone reservation for OFDM systems by maximizing signal-to-distortion ratio,"IEEE Transaction on Wireless Communications, vol. 11, no. 2, pp. 762-770, February 2012.
[15] M Bellanger, D LeRuyet, D Roviras, et al. FBMC physical layer: a primer[J]. PHYDYAS, 2010.

IOP Conf. Series: Journal of Physics: Conf. Series **910** (2017) 012058 doi:10.1088/1742-6596/910/1/012058

PAPR Reduction in OFDM-based Visible Light Communication Systems Using a Combination of Novel Peak-value Feedback Algorithm and Genetic Algorithm

Honggui Deng[1], Yan Liu[2], Shuang Ren[2], Hailang He[3] and Chengying Tang[1]

[1]Guangxi Key Laboratory of Information Materials, Central South University, Changsha, China
[2]School of Physics and Electronics, Central South University, Changsha, China
[3]Department of Information Engineering, Shaoyang University, Shaoyang, China
Email: denghonggui@163.com, zndxly@csu.edu.cn, 601013849@qq.com, 7887302@qq.com, csutcy@163.com

Abstract. We propose an enhanced partial transmit sequence technique based on novel peak-value feedback algorithm and genetic algorithm (GAPFA-PTS) to reduce peak-to-average power ratio (PAPR) of orthogonal frequency division multiplexing (OFDM) signals in visible light communication (VLC) systems(VLC-OFDM). To demonstrate the advantages of our proposed algorithm, we analyze the flow of proposed technique and compare the performances with other techniques through MATLAB simulation. The results show that GAPFA-PTS technique achieves a significant improvement in PAPR reduction while maintaining low bit error rate (BER) and low complexity in VLC-OFDM systems.

1. Introduction

There have been many significant researches on the VLC systems. Compared with conventional radio frequency(RF)communication systems, VLC systems have many advantages such as higher transmission rates, no electromagnetic interference and higher security[1-2]. With respect to all these advantages, VLC has become a promising technology for future wireless communications.

In VLC systems, OFDM technique is employed to meet high-speed transmission requirement and to solve inter symbol interference (ISI) problem[3]. But the disadvantage of high PAPR in OFDM cannot be neglected. If VLC-OFDM signals have high PAPR, systems require all transmitter modules with wide linear operation ranges[4-5]. Otherwise, there will be non-linear distortion of signals when peak value of a signal goes into the nonlinear region of system's components[6]. Light-emitting diode (LED) and power amplifier, which are typically important in transmitter modules, are difficult to design and costly to manufacture to obtain wide linear operation ranges[7]. So the problem of nonlinear distortion caused by high PAPR in VLC-OFDM systems must be solved.

Numerous methods have been investigated to reduce high PAPR of VLC-OFDM systems, including precoding[8], clipping[9-11], companding transform[12], tone reservation [13], selected mapping (SLM)[14], and partial transmit sequences (PTS)[1]. Among them, the partial transmit sequence (PTS) is one of the most potential technique which has no loss in the power and bit error rate (BER), and can significantly improve the PAPR. But it also needs large computational complexity. GA-PTS technique has been proposed to decrease the computational complexity. However, the performance degradation is serious. Reference [1] proposes a modified PTS technique which combines genetic algorithm with hill-climbing algorithm (GH-PTS) to improve PAPR performance. But GH-PTS can only change single phase factor in a phase factor combination in each iteration, so it easily gets stuck at local optimal solution. And the cost for computational amount is too much.

In this paper, we propose a new PTS technique called GAPFA-PTS to solve the above problems which combines genetic algorithm with a novel peak-value feedback algorithm. In this technique, phase factor combination will be continuously changed until it meets the ending condition. Once the optimized signal has lower PAPR than threshold value, the PFA manipulation will be ended no matter whether the current executions equal to the final executions. Compared with GH-PTS, GAPFA-PTS has advantages that it can optimize phase factor combination $[b_1, b_2 ... b_v]$ by many times phase rotation in one PFA manipulation which can combat local optimal solution. The operation will be given in Sec.3 in detail.

The rest of paper is organized as follows. In Sec.2, VLC-OFDM system and PAPR in VLC-OFDM are analyzed. Sec.3 discusses GA-PTS technique and GAPFA-PTS technique in detail. The performance of GAPFA-PTS is analyzed in Sec.4 by comparing GAPFA-PTS with Optimization-PTS, GA-PTS and GH-PTS . In Sec. 5, the conclusion is given.

2. Papr in VLC-OFDM

2.1 System of VLC-OFDM

In VLC systems, signals are transmitted through an LED in form of optical power. In the receiver, a photo detector (PD) is employed to convert optical power signals into electrical signals. A DC bias voltage should be added before transmitting data because baseband-signal in VLC system must be real-valued to meet the requirement of the intensity modulation with direct detection (IM/DD).

VLC-OFDM system block diagram is shown in Figure.1. At the beginning, data are mapped into by using the QPSK mapping. Then, serial data is converted into parallel data. Next, pilot is inserted into the original data. Afterwards, data are processed by Inverse Fast Fourier Transform (IFFT) operation. So far the VLC-OFDM signal without PAPR reduction process has been obtained. It is critical to maintain the stability of the VLC-OFDM system. The following step is to mitigate the high PAPR of signals by applying GAPFA-PTS technique. After digital-to-analog conversion, the signal is emitted by LED emitter and received by a photo detector(PD). At the receiver, the original data is restored through the reverse operation of the transmission scheme.

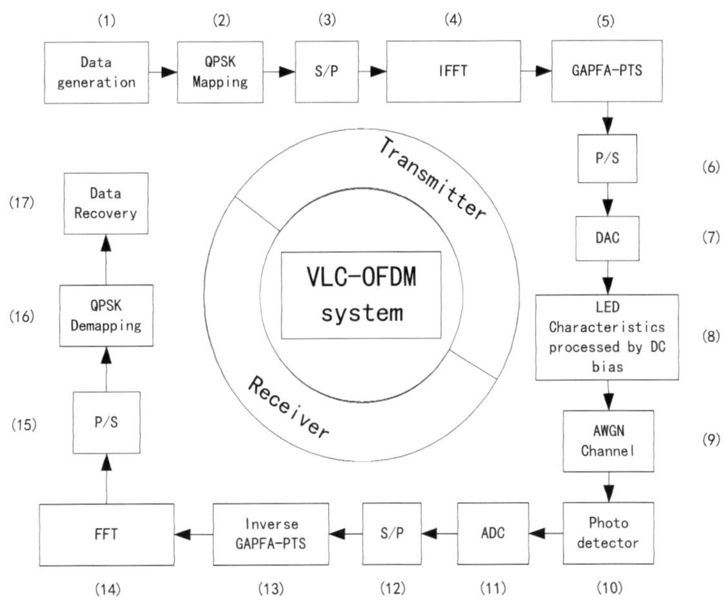

Figure 1. Diagram of VLC-OFDM system.

2.2 PAPR of VLC-OFDM Signal

High PAPR is a common problem for OFDM signals. In VLC-OFDM systems, time domain signals x_n can be expressed as

$$x(n)=\frac{1}{\sqrt{N}}\sum_{k=0}^{N-1}X_k e^{j2\pi kn\frac{1}{N}} \quad n=0,1,...,N-1. \tag{1}$$

Where X_k represents the frequency domain datum, k represents the index number of sub-carrier, N is the number of sub-carriers, and n is the index number of the time domain signal.

PAPR is defined as the ratio of the maximum signal power to average signal power, it can be expressed as

$$PAPR=\frac{P_{\text{peak}}}{P_{\text{average}}}=10\log_{10}\frac{\max\limits_{0\le n\le N-1}\left(\left|x(n)\right|^2\right)}{E\left(\left|x(n)\right|^2\right)}. \tag{2}$$

When the maximum amplitudes of different sub-carriers appear at the same time, the peak power of the signal will appear and the signal has high PAPR.

The pros and cons of a PAPR reduction technique can be evaluated generally by cumulative distribution function (CDF) and complementary cumulative distribution function (CCDF). While, CCDF is more widely used to show PAPR distribution. Given a threshold value z, CCDF represents the probability that the PAPR of an OFDM symbol is greater than z. The CCDF is defined as

$$CCDF=P(PAPR>z)=1-F(z)^N=1-(1-e^{-z})^N. \tag{3}$$

Where N denotes the number of sub-carriers.

By comparing the CCDF in different situations, the advantage of the modified technique can be shown more intuitively.

3. The Proposed GAPFA-PTS Technique

GAPFA-PTS technique is based on genetic algorithm and peak-value feedback algorithm which is a local optimization of GA-PTS technique.

3.1 GA-PTS Technique

GA-PTS is proposed to decrease complexity of PTS technique. The genetic algorithm can be summarized as follows:

1) Select a key parameter G as the largest genetic algebra which determines the system performance and the computational complexity. When the current genetic algebra g is bigger than G, the genetic algorithm meets the ending condition.

2) Randomly choose Q phase factor combinations as parent populations and each population indicates one phase factor combination $[b_1,b_2...b_v]$. The current genetic algebra is set to $g=1$.

3) Introduce the phase factor combination $[b_1,b_2\cdots b_v]$ to rotate disjoint V subblocks and obtain the optimum candidate signal by(4).

$$X=\sum_{m=1}^{V}b_m X_m \tag{4}$$

4) Calculate PAPR of the signal after rotated by each parent population as fitness values. Then obtain offspring populations by genetic manipulation (Figure.2) and calculate its PAPR. Moreover, choose the populations with smaller PAPR in last parent populations and offspring populations as the next genetic algebra's parent populations and set the genetic algebra to $g=g+1$.

5) The step d) is repeated until genetic algebra g meets the condition $g=G+1$ where G is set as the largest genetic algebra in step a). Then choose the best phase factor combination which possesses

the minimum PAPR as the final output $[b_1, b_2 ... b_v]$.

The genetic manipulation makes PTS technique have good global search capability, but its local search capability needs to be improved especially in low complexity.

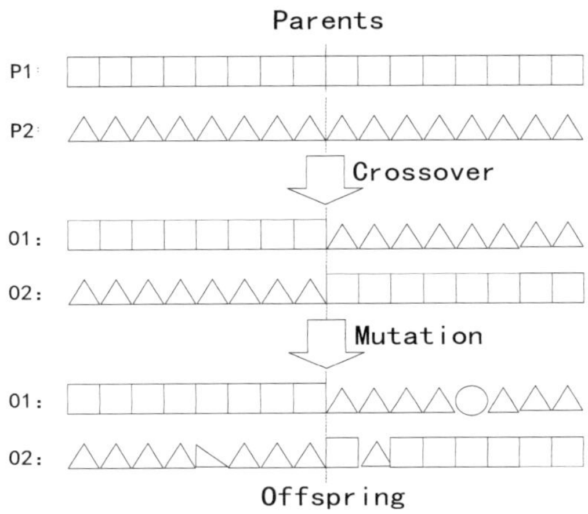

Figure 2. Genetic manipulation.

3.2 GAPFA-PTS Technique

PFA is proposed to gain local search capability of GA-PTS technique. The process of PFA can be summarized as follows: (Setting parameters: set the cumulative number of executions for PFA recorded as c ; set the final number of executions for PFA as C ; set the threshold value as θ .)

1) Through GA-PTS technique, we can find a phase factor combination $[b_1, b_2 ... b_v]$ which makes x have a relatively optimal PAPR. Then we should compare the PAPR with threshold value θ : if PAPR$\leq \theta$, set the cumulative number of executions $c = C + 1$, and regard the current phase factor combination $[b_1, b_2 ... b_v]$ as the final output; if PAPR$> \theta$, continue to find the peak point of the signal and regard the peak point as the sampling point. The cumulative number of executions (c) is attached to the initial value: $c = 1$;

2) Arrange V sub-blocks of the signal in a descending order by their value at the sampling point, and set the sequence number as $v_i (i = 1, 2, 3 ... V)$. At this time, the maximum amplitude of the sub-block corresponds to v_1 ;

3) All the sub-carriers on the v_1 sub-block are multiplied by a phase factor -1, other phase factors are constant. As a result, the combination of the coefficients [$b_1, b_2 ... b_v$] is changed to $[b_1, b_2 ... - b_{v_1} ... b_v]$, then set the changed combination coefficient as $[d_1, d_2 ... d_v]$;

4) Compare the PAPR of the signal before rotated (set as $PAPR_1$) with the PAPR of the signal after rotated(set as $PAPR_2$). If $PAPR_1 > PAPR_2$, it will be regarded as an efficient operation, and cumulative number of executions is set to $c = c + 1$. Otherwise it is an invalid operation, we should multiply the signals of second large block (v_2) by a weighting factor -1, and other phase factors stay unchanged, then set the PAPR as $PAPR_3$; if $PAPR_1 < PAPR_3$, continue to operate the next block of data until we get a smaller PAPR compared with $PAPR_1$; if all the rotated sub-blocks are not satisfied, we will consider the signal satisfied enough to transmit. Then we will set the cumulative number of executions to $c = C + 1$. Obviously parameter $c > C$, it meets the ending condition. So the current phase factor combination $[d_1, d_2 ... d_v]$ is regarded as the best phase factor combination with lowest

PAPR and $[d_1, d_2 ... d_v]$ is as the final output.

5) Repeated Steps b) to d) until the PAPR of the signal after rotated meets the ending condition ($PAPR < \theta$) or the value of c equals to the ending number $C+1$. At this time, the combination of the weighting coefficients $[d_1, d_2 ... d_v]$ can be regarded as the best phase factor combination with lowest PAPR.

The GAPFA-PTS is a further optimization of GA-PTS technique, combining the global search capability of GA-PTS and local search capability of PFA. Overall diagram of GAPFA-PTS is shown in Figure.3 and the superiority of the proposed algorithm are verified in the simulation(see section IV).

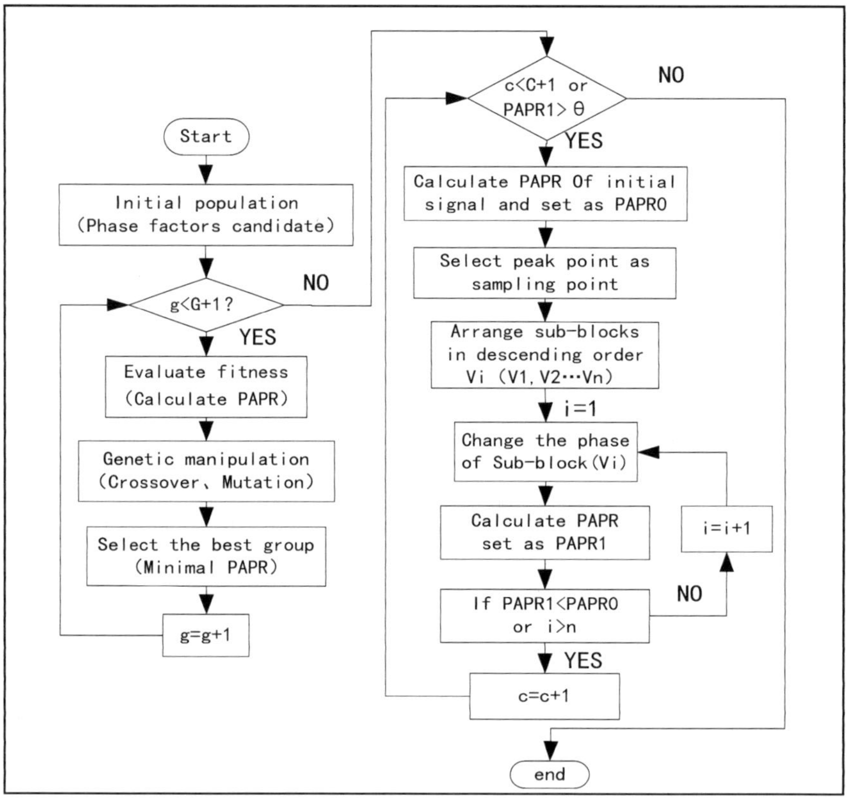

Figure 3. Overall diagram of GAPFA-PTS.

4. Simulation Results and Analysis

4.1 PAPR Performance and Analysis

We simulate 4 techniques where genetic manipulation is completed by applying adjacent divided and multiple-point crossover. Each crossover manipulation needs to select 2 populations within Q populations. So the crossover rate for each population is $2/Q$. Moreover, we will do $Q/2$ times crossover manipulations in each genetic manipulation. So the crossover rate P_{GA} is calculated by (5), in which Q is the number of populations and G is the final genetic algebra.

$$P_{GA} = 1 - (1 - 2/Q)^{Q/2*G} \tag{5}$$

Other parameters of PAPR simulation are summarized in TABLE 1.

Figure.4 shows the CCDF curves, where GAPFA-PTS technique is compared with GA-PTS technique, GH-PTS technique and Optimization-PTS technique.

Table 2 presents a comparison of four techniques which shows the computational amount and performance of PAPR reduction in detail. The threshold value of PFA is set as 7.1dB and the number of rotation factors(set as W) is two as it is selected from -1,1.

Table 1 PAPR performance simulation parameters.

Parameter	number
Modulation mode	QPSK
number of sub-carriers	N =512
Number of sub-blocks	V =16
Number of symbols	1000
Phase rotation factor	b_v =-1,1
Number of populations	Q =10
Number of generations	G =4
Number of PFA	C = 10
Crossover rate	0.988
Mutation rate	0.01

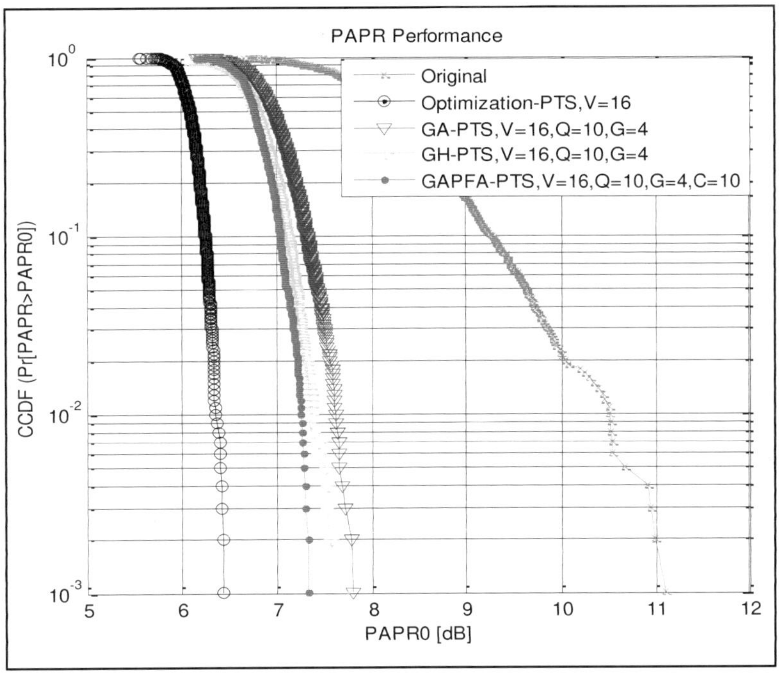

Figure 4. The PAPR CCDF of GAPFA-PTS, GH-PTS, GA-PTS, Optimization-PTS Technique.

Table 2 Comparison of the computational amount and PAPR performance.

Technique	Computational Amount For Each Symbol	PAPR(dB)
Original Signal	0	11.01
Optimization-PTS (W =2, V =16)	W^V =65536	6.45
GA-PTS ($W = 2, V = 16, Q = 10, G = 4$)	$2 \times Q \times G = 80$	7.89
GH-PTS ($W = 2, V = 16, Q = 10, G = 4$)	$2 \times Q \times G + V \times G = 144$	7.64
GAPFA-PTS ($W = 2, V = 16, Q = 10, G = 4$ C =10)	$2 \times Q \times G + 11497/1000 = 92.497$	7.32

As it can be seen from Table 2 and Figure.5, although Optimization-PTS techniques has good

PAPR reduction performance, its computational amount is nearly 700 times as large as both GAPFA-PTS's and GA-PTS's, and the huge computational amount decreases applying feasibility. When $V = 16, Q = 10, G = 4$, the PAPR of signal processed by GA-PTS is 7.89dB, and the computational amount is 80. When $V = 16, Q = 10, G = 4, C = 10$, the PAPR of signal processed by GAPFA-PTS is 7.32dB, and the computational amount is 92.497. As an optimization of GA-PTS technique, GAPFA-PTS technique has better performance of PAPR reduction than GA-PTS technique by 0.57dB with the computational amount only increased by 12.497. When $V = 16, Q = 10, G = 4$, the PAPR value of applying GH-PTS is 7.64dB, and the computational amount of applying GH-PTS is 144. Compared with GH-PTS technique, GAPFA-PTS has better PAPR performance by 0.32dB and decrease 36% computational amount. It suggests that GAPFA-PTS technique is more efficient than GH-PTS technique with better PAPR performance and lower computational amount.

4.2 BER Performance

The BER performance is studied in an AWGN channel using the QPSK Modulation mode. Other parameters of this simulation are listed in TABLE 3.

From Figure.5, we can see that the BER curves of these PTS techniques are very close. Compared with the GA-PTS technique, the signal processed by GAPFA-PTS technique has nearly no extra BER lose. Moreover, the BER performance of GAPFA-PTS technique is better than GH-PTS technique because of its lower computational amount. So we can draw a conclusion that the improvement of PAPR performance costs nearly no extra BER loss for GAPFA-PTS technique.

Table 3 PAPR Performance Simulation Parameters.

Parameter	**number**
Number of data symbols	1000
Number of sub-carriers	N =512
FOV at the receiver	60 [deg]
Surface area of PD	1.0 [cm^2]
Gain of optical filter	1.0
Efficiency of photo electric conversion	0.53 [A/W]

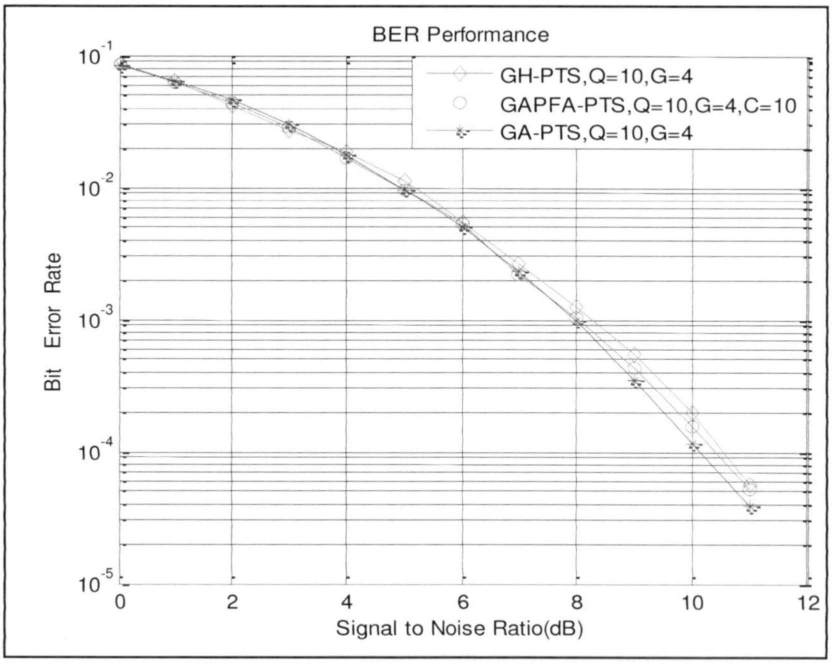

Figure 5 The BER curves of GAPFA-PTS,GH-PTS and GA-PTS.

5. Conclusion

In this paper, we have proposed a new PTS technique called GAPFA-PTS to mitigate high PAPR of OFDM-based VLC systems. This technique can efficiently search for optimum phase factors by applying PFA manipulations which have good local search capability. In these simulations, we have compared the GAPFA-PTS with GA-PTS, GH-PTS and Optimization-PTS for their PAPR performance. Compared with Optimization-PTS, GAPFA-PTS has an advantage in computational complexity. Compared with GA-PTS technique, the proposed GAPFA-PTS has an advantage in PAPR reduction performance. Compared with GH-PTS technique, the proposed GAPFA-PTS has an advantage in both PAPR reduction and computational complexity. While the BER performances of all techniques are close. So, we can draw a conclusion that GAPFA-PTS we proposed has better performance in mitigating PAPR of VLC-OFDM signal with no gain of BER than GA-PTS and GH-PTS technique.

6. Acknowledgment

This work was supported by the Open Science Research Fund of Guangxi Key Laboratory of Information Materials, Guilin University of Electronic Technology. This work was also supported by Research Foundation of Education Bureau of Hunan Province, Project No.17B238.

7. References

[1] Wei Xiao, H. Deng, F. Jiang, K. Zhu, L. Yin, "Peak-to-average power ratio reduction in orthogonal frequency division multiplexing visible light communications system using a combination of genetic algorithm and hill-climbing algorithm," Optical Engineering, vol. 54, pp. 036106-1-9, Oct 2015.

[2] Shan Jiang, Honggui Deng,Wei Xiao, Jie Du, "New nonlinear Yu Zhenhua, R. J. Baxley and G. T. Zhou, "Achievable Data Rate companding technique to decrease peak to average power ratio of the optical orthogonal frequency division multiplexing system based on Hartley transform," Optical Engineering, vol. 54,no. 10, pp. 313-317 Mar 2015.

[3] Analysis of Clipped Flip-OFDM in Optical Wireless Communications," Proc. IEEE Globe com Workshops, vol. 48, no. 11,pp. 1203–1207 2012.

[4] Ji Zhou, Zhenshan Zhang, Tiantian Zhang, Mengqi Guo, Xizi Tang, Zhansheng Wang, Yaojun Qiao "A combined PAPR-reduction technique for asymmetrically clipped optical OFDM system," Optics Communications,vol. 366, pp. 451–456, 2016.

[5] Honggui Deng, Yi Sun, Chen Yang, "A New Method to Reduce Peak to Average Power Ratio in Optical OFDM Communication," Science Journal of Physics, vol. 2013,pp. 1-6, Mar 2013.

[6] Yu Zhenhua, R. J. Baxley and G. T. Zhou,"Brightness Control in Dynamic Range Constrained Visible Light OFDM Systems," IEEE Wireless and Optical Communications Conference (WOCC), 2014, pp. 1–5 .

[7] Y. Pei, S. Zhu, H. Yang, L. Zhao, X. Yi, J. Wang, and J. Li, "LED Modulation Characteristics in a Visible-Light Communication System," Optics and Photonics Journal, vol. 3, pp. 139–142, 2013.

[8] Wu, Tsung-Wei, Chung, Char-Dir, "Correlatively Precoded OFDM With Reduced PAPR,"IEEE Transactions on Vehicular Technology, vol. 65, no. 3, pp.1409-1419, 2016.

[9] Yu Z, Baxley R J, Zhou G T. "Iterative clipping for PAPR reduction in visible light OFDM communication," Proceeding of the IEEE MILCOM, Baltimore, 2014, pp. 1681–1686.

[10] Xu W, Wu M, Zhang H, X You, "ACO-OFDM-specified recoverable upper clipping with efficient detection for optical wireless communications," IEEE Photonics Journal, vol.6,no.5, pp. 1-17, 2014.

[11] Ranjha B, Kavehrad M, "Precoding techniques for PAPR reduction in asymmetrically clipped OFDM based optical wireless system," Proceeding of the SPIE, San Francisco, vol. 8645, no. 6, pp. 540-547, 2013.

[12] Y. Wang, L.-H. Wang, J.-H. Ge, and B. Ai, "An efficient nonlinear companding transform for reducing PAPR of OFDM signals," IEEE Trans. Broadcast., vol. 58, no. 4, pp. 677–684, Dec 2012.

[13] K Park, IC Park, "Low-Complexity Tone Reservation for PAPR Reduction in OFDM Communication Systems," vol. 20, no. 10, pp.1919-1923, 2012.
[14] Amnart Boonkajay, Fumiyuki Adachi, "Selected mapping technique for reducing PAPR of single-carrier signals," Wireless Communications and Mobile Computing, vol. 16, no. 16, pp.2509-2522, 2016.

An Inter-Group Swapping Based Resource Scheduling Scheme for Quasi-Single Frequency Multi-beam Mobile Satellite System

Shengyao Liu[1], Yinuo He[1], Yong Luo[1], Jianjun Wu[1] and Yilei Sun[2]

[1]Institute of Advanced Communications, EECS, Peking University, Beijing, P. R. China

[2]School of Franchise Management, Beijing Normal University (Zhuhai Campus), Zhuhai, P. R. China

Email: amateurliu@pku.edu.cn, just@pku.edu.cn

Abstract. A resource scheduling scheme is proposed in this paper for multi-beam satellite system to eliminate downlink interference as well as to overcome the noise amplification problem caused by Tomlinson-Hiroshima preceding. Firstly, to analyze the noise amplification problem, Tomlinson-Hiroshima preceding (THP) is discussed. Secondly, an inter-group swapping based resource scheduling scheme is proposed to according to the principle of minimizing the maximum noise amplification factor of the same frequency users. Finally, simulations are executed to validate the proposed scheme, which suggest that compared to no-swapping scheme, the proposed scheme can successfully reduce the ratio of low SINR users and improve the minimum signal to interference and noise ratio(SINR).

1. Introduction

With the development of wireless communication technology and diversified wireless communication applications, the Mobile Satellite System (MSS) is becoming more and more popular between research community and industry due to its wide coverage, large capacity, flexible networking and cost independent of distance. In recent years, based on the development of land mobile communication technology, OFDM-based multi-beam satellite mobile communication system has been widely studied and applied.

Usually, the multi-beam satellite system can suppress the inter-beam interference (IBI) [1] by using the traditional frequency-division reuse scheme, which is divided into three-color, four-color and seven-color reuse. But the spectral efficiency is low by using this method. Most frequency bands of the quasi single frequency network are the same frequency reuse among the beams, and a few of the frequency points are reused among the beams, which can effectively improve the spectrum efficiency of the system. In quasi single frequency networked satellite systems, co-channel interference between beams can be effectively controlled or even eliminated by using specific interference cancellation techniques at the transmitter or the receiver, or both [2]. In [4], a method is proposed to eliminate inter-beam interference by using iterations at the receiver, but its drawback is that the design complexity and computational complexity of the receiver are dramatically increased. THP is a simple and effective interference cancellation technique for terrestrial systems. Aiming at the adaptability requirement of THP n uasi single frequency network satellite communication, his paper designs a resource scheduling scheme.

2. System Model

In this paper, a multi-beam satellite mobile communication downlink system model based on quasi single frequency network is proposed. In the downlink of the quasi single frequency network system, most frequency points adopt the same frequency reuse mode among the beams, and for the same frequency reuse users, interference cancellation with precoding is performed at the transmitter side. In the same frequency division reuse mode, the users are grouped in a certain resource scheduling mode, and the users that are pre-encoded at the same time are assigned to the same user group for joint precoding. At the same time, several extra frequency points are reserved for users who do not derive a performance gain from the precoding interference cancellation. These frequency points take a three-color frequency reuse pattern among the beams and not participate in the precoding interference pre-elimination, as is shown in Figure.1.

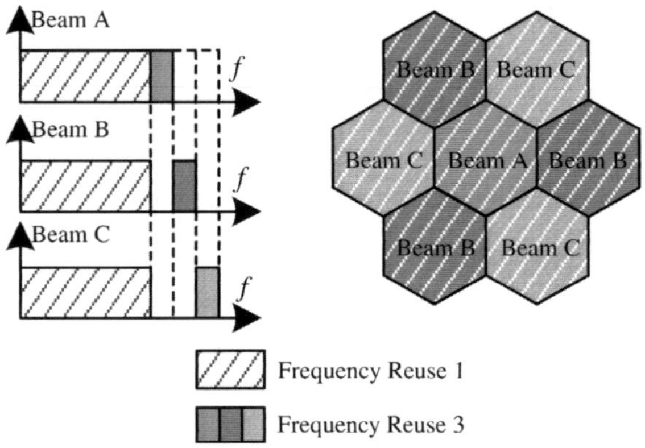

Figure 1 System Model

At this time, system frequency reuse factor FR_{sys} is given by:

$$FR_{sys} = \frac{N + 3 \cdot f_a}{N + f_a} \tag{1}$$

Where N represents the number of users in each beam, and for the sake of simplicity it is assumed that the number of users in each beam is the same, and f_a represents the number of extra frequencies reserved in each beam. When N = 10 and $f_a = 2$, FR_{sys}= 1.33, slightly greater than 1, can achieve quasi single frequency reuse effect.

3. Resource Scheduling Scheme Based on Inter-group Swap

3.1 Interference Pre-cancellation with THP

THP is a kind of precoding technique which is used when the condition of channel state information (CSI) is known at transmitter. The signal is pre-processed by pre-estimation of the channel before the signal is transmitted, so pre-equalization can be achieved to eliminate interference between users, the processing block diagram shown in Figure.2.

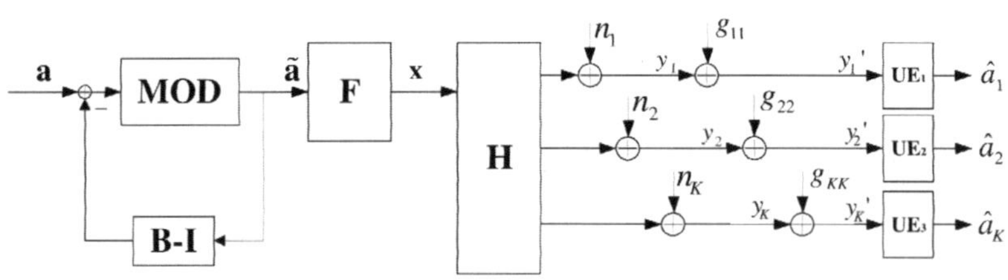

Figure 2 Structure of THP

At receive side, receive signal **y** is represented as

$$\mathbf{y} = \mathbf{HFB}^{-1}\mathbf{v}+\mathbf{n} \tag{2}$$

signal processing matrix is **G**, output is $\mathbf{y}'$, i.e.

$$\begin{aligned}\mathbf{y}' &= \mathbf{Gy} \\ &= \mathbf{GHFB}^{-1} + \mathbf{Gn}\end{aligned} \tag{3}$$

To obey zero-forcing rule, need force

$$\mathbf{GHFB}^{-1} = \mathbf{I} \tag{4}$$

Define

$$\mathbf{HF} = \mathbf{S} \tag{5}$$

F,**S** can be calculated by performing QR factorization on $\mathbf{H}^{\mathbf{H}}$ i.e.

$$\mathbf{H}^{H} = \mathbf{QR} \tag{6}$$

$$\mathbf{F} = \mathbf{Q} \tag{7}$$

$$\mathbf{S} = \mathbf{R}^{H} \tag{8}$$

F is a unitary matrix, **S** is a lower triangular matrix, define

$$\mathbf{G} = \mathrm{diag}[s_{11}^{-1}, s_{22}^{-1}, \ldots, s_{KK}^{-1}] \tag{9}$$

$$\mathbf{B} = \mathbf{GS} \tag{10}$$

Then **B** is a unit lower triangular matrix and is normalized matrix of **S**.

On the other hand, QR factorization can be accomplished by using Schmidt orthogonalization process. Let $\boldsymbol{\alpha}$ represent column vectors of $\mathbf{H}^{\mathrm{T}}$, i.e.

$$\boldsymbol{\alpha}_{\mathrm{i}} = [g_{i1}, g_{i2}, \ldots, g_{iN_T}]^{\mathrm{T}} \tag{11}$$

if $\boldsymbol{\alpha}_{\mathrm{i}}$ is linear independence, after Schmidt orthogonalization:

$$\boldsymbol{\beta}_i = \begin{cases} \boldsymbol{\alpha}_i, i = 1 \\ \boldsymbol{\alpha}_i - \sum_{k=1}^{i-1} \frac{\langle \boldsymbol{\alpha}_i, \boldsymbol{\beta}_k \rangle}{\|\boldsymbol{\beta}_k\|^2} \cdot \boldsymbol{\beta}_k, i \in [2, N_T] \end{cases} \tag{12}$$

$$s_{ii} = \|\boldsymbol{\beta}_i\|, i \in [1, N_T] \tag{13}$$

$$\hat{a}_i = a_i + 1/\|\boldsymbol{\beta}_i\| \cdot n_i, i \in [1, N_T] \tag{14}$$

Therefore, the performance of interference pre-cancellation is determined by the value of $1/\|\boldsymbol{\beta}_i\|(i=[1,N_T])$,only when $1/\|\boldsymbol{\beta}_i\|$ is comparatively small($\|\boldsymbol{\beta}_i\|$ *comparatively large)*,THP can be used for interference pre-cancellation in satellite communication.

Considering the distance between the users, the position of the user in the beam and the coding order of the users, these factors affect the performance of the system through affecting the noise amplification factor $1/\|\boldsymbol{\beta}_i\|$ Users' noise amplification factor has a direct impact on the performance of THP.

Considering all the users' pre-coding output power is the same, when a user's $1/\|\beta_i\|$ is larger, the channel noise will be amplified at the receiving end, resulting in a lower SINR. If there are a lot of users with low signal to noise ratio in the system, the degradation of the communication quality will cause the loss of the capacity and performance of the whole system. On the other hand, when the multi-beam satellite communication system serves a small number of users, it should also avoid the existence of individual users have a very low received signal to interference and noise ratio. From the point of view of practical operation, this paper adopts the principle of minimizing the maximum noise amplification factor of the same frequency users to make user scheduling, so as to determine the more effective THP user grouping method.

3.2 Resource Scheduling Scheme

It is assumed that there are K beams and N users in each beam and N sub-channels are used for the system downlink. Resource scheduling is to be completed from the frequency distribution and power distribution. Select a user from each beam of the K beams, a total of K users in the same group, and for this group of users assign the same sub-channel downlink transmission, the total need to repeat N times, the i-th *(i=[1,N])* sub-channel is assigned to the i-th merged users. Number the beams. The beam at the center is defined as beam 1 and the other beams are numbered from the inner ring to the outer ring in a counter-clockwise spiral order outwardly from the center beam. Let $U_k = \{u_k^1, u_k^{2,\ldots} u_k^P\}$ represent users in k-th beam not allocated before finding users in i-th group (*k=[1,K], P= N-i+1*).

1) Step a: pre-grouping

Firstly, the idea of pre-grouping all users is to find groups of users in each beam which have the same relative coordinates as their beam centers are defined as (0,0)

2) Step b:inter-group swap

For each group of users using the same frequency point, find all users whose SINR<SINR_th*(SINR Threshold,set as 5dB)*, swap their group number with those users closest to the center.

3) Step c:Resource allocation

According to the result of grouping,for the users who stay in the same channel sequence, use the "Best-first" algorithm to sort and allocate the same frequency point for THP interference pre-cancellation.

Table 1 Inter-Group Swap Algorithm

1. Set number of iterations times swap;
2. For all users in kth beam,

$$U_k = \{u_1^{n_1}, u_2^{n_2}, \ldots, u_K^{n_K}\}$$

1) forming a queue in which the distance between the user and its beam center is small to large, $D_k = \{u_{k_1}^{n_{k_1}}, u_{k_2}^{n_{k_2}}, \ldots, u_{k_K}^{n_{k_K}}\}$

2) forming a queue in which the SINR is small to large $S_k = \{u_{k_1'}^{n_{k_1}'}, u_{k_2'}^{n_{k_2}'}, \ldots, u_{k_K'}^{n_{k_K}'}\}$

3) find t users whose SINR<SINR_th $S_{(k,t)} = \{u_{k_1'}^{n_{k_1}'}, u_{k_2'}^{n_{k_2}'}, \ldots, u_{k_t'}^{n_{k_t}'}\}$

4) swap the group number of $S_{(k,t)}$ with the first t users in D_k, er grouping after swapping become:

3. repeat 2 for times swap times.

4. Simulation Results

4.1 Ratio of Low SINR Users

After swapping, change of ratio of low SINR users in system is shown in figure 3.

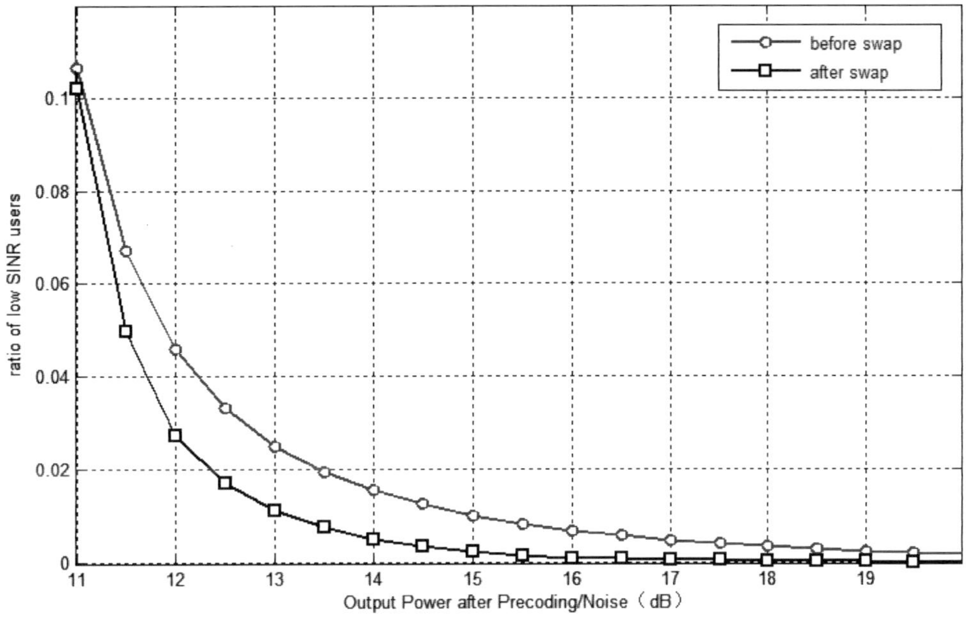

Figure 3 Ratio of low SINR users(before and after swap)

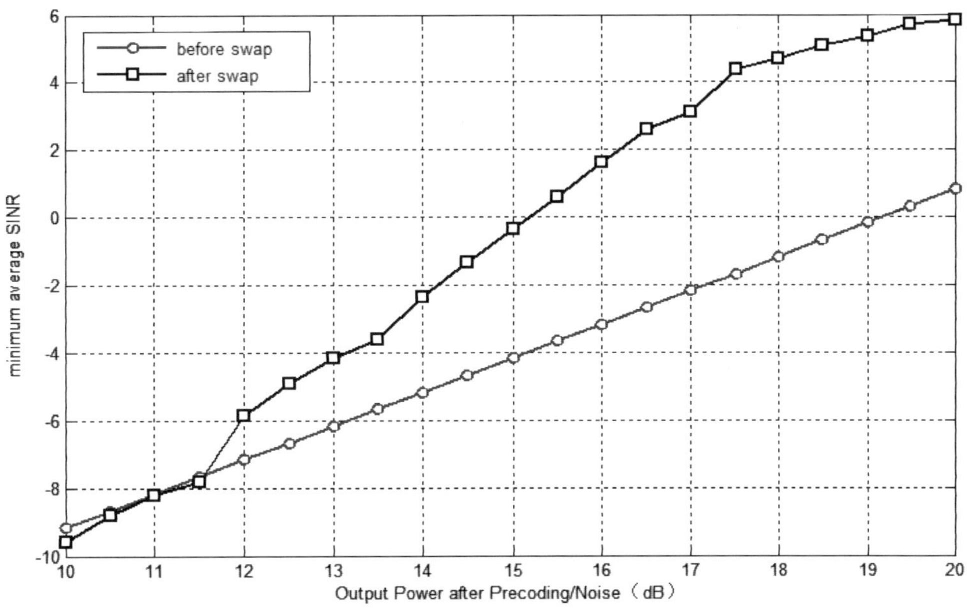

Figure 4 minimum average SINR(before and after swap)

As shown in figure, after swapping, ratio of low SINR sers decrease harply. When precoding output power to channel noise is 12dB, ratio of low SINR users decrease from 4.44% to 2.67% nearly 40% of origin ratio, the larger precoding output power to channel noise increase, the faster the ratio achieves zero.

4.2 Minimum User SINR
In addition to focusing on the ratio of the number of low signal to noise ratio users, in order to ensure the performance of user with poor signal transmission quality, for user with the lowest signal to noise

ratio , whether the performance improves, simulation results as shown above:

The minimum SINR in the system is improved after swapping. When the ratio of precoding output power to hannel noise is 13dB, the minimum SINR in the system increases from -3.5dB to -2.3dB. And when the ratio of the pre-coding output power to the channel noise is increased, the promotion ratio tends to increase.

5. Summary

In this paper, the adaptability of THP in multi-beam downlink is treated as the main research object. Based on the influence factors of user location and coding sequence,

The user packet scheduling model based on inter-group swap is designed. The performance of the proposed method is compared with that without the proposed method, which proves the rationality of the proposed algorithm. Among them, the inter-group swapping scheme significantly improves the fairness of the system without significantly reducing the average signal to interference and noise ratio and the average beam spectrum efficiency of the users in the whole system.

6. Acknowledgment

This paper is supported by the National Natural Science Foundation of China (No.61371073).
Corresponding author: Jianjun Wu; Email: just@pku.edu.cn.

7. References

[1] Kobayashi K, Ohata K, Ueba M, et al. Forward link multiple access scheme for broadband and scalable mobile satellite communication system[C]//Vehicular Technology Conference, 2004. VTC2004-Fall. 2004 EEE 60th. IEEE, 2004, 6: 3997-4001.

[2] D. Gesbert, M. Shafi, D. shan Shiu, P. Smith, and A. Naguib, "From theory to practice: an overview of MIMO space-time coded wireless systems," Selected Areas in Communications, IEEE Journal on, vol. 21, no. 3, pp. 281–302, Apr 2003.

[3] Q. Spencer, C. Peel, A. Swindlehurst, and M. Haardt, "An introduction to the multi-user MIMO downlink," IEEE Commun. Magazine, vol. 42, no. 10, pp. 60–67, Oct. 2004.

[4] B. Beidas, H. El Gamal, and S. Kay, "Iterative interference cancellation for high spectral efficiency satellite communications," IEEE Trans. Commun., vol. 50, no. 1, pp. 31–36, Jan 2002.

Authentication and Authorization of End User in Microservice Architecture

Xiuyu He and Xudong Yang

Department of Computer Science and Technology, Beijing University of Posts of Telecommunications, Beijing, China
Email: xyhe@bupt.edu.com, xdyang@bupt.edu.com

Abstract. As the market and business continues to expand; the traditional single monolithic architecture is facing more and more challenges. The development of cloud computing and container technology promote microservice architecture became more popular. While the low coupling, fine granularity, scalability, flexibility and independence of the microservice architecture bring convenience, the inherent complexity of the distributed system make the security of microservice architecture important and difficult. This paper aims to study the authentication and authorization of the end user under the microservice architecture. By comparing with the traditional measures and researching on existing technology, this paper put forward a set of authentication and authorization strategies suitable for microservice architecture, such as distributed session, SSO solutions, client-side JSON web token and JWT + API Gateway, and summarize the advantages and disadvantages of each method.

1. Introduction

For any mature system, a complete authentication and authorization of end user module is essential. It protects the boundaries of the system. what is the difference between authentication and authorization?

Authentication is the process of verifying who you are. When you log into a PC with a user name and password you are authenticating. Authorization is the process of verifying that you have access to something. For example, some resources only allow the administrator to view but not the ordinary users.

1.1 Auth in Monolithic Application

In a traditional monolithic architecture project, User's requests are handled within a single process in the backend. Put a filter in the system boundary to verify the identity and access, and determine the response or distribution of the request. Figure1 is a typical monolithic architecture. As HTTP is a stateless protocol, it is usually based on the session that the server generates for the client to manage the user's status.

Here is the session control process:

1. The client provides authentication credentials.
2. The server valid the credentials
   - Re-certification if verification failed.
   - Generate session if verification successful.
3. The client requests the resource with session
4. The server gets session of the user by session and response resource if the user has access

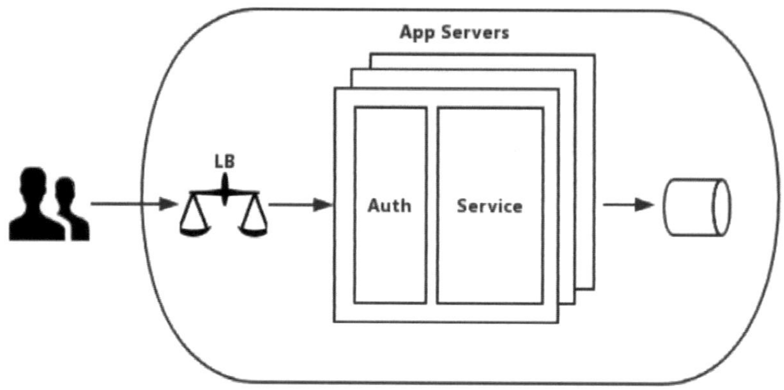

Figure 1 Monolithic auth architecture

The advantages of the session-based system are simple easy to implement and exposed less entrance to the system. But there are many challenges as well. The server needs to save the session in memory, which may cause high memory usage and reduce performance. On the other hand, the authentication feature and other system's features mix together, high coupled results in reducing scalability and flexibility of the system. As the traffic increases, the system needs to deploy multiple nodes to load balance. Share a session in multiple nodes is a problem. Besides, the session-based system uses cookie at most time, so it should deal with some cookie-based attacks from client.

1.2 Distributed Session Management

As the feature of the system becomes more complex and the number of user increases, application deployed in single machine is unable to handle the pressure. From a single node to a cluster, the multiple nodes must share a session when they use session-based auth. There are some solutions about distributed like sticky session, session replication, session centralized management and so on.

- Sticky session ensures that all the subsequent request will be send to the server who handled the first request corresponding to that request.
- Session replication means each server saves session data, and synchronizes through the network. So it will be affected by the network situation.
- Centralized management adds a specific server to manage session. Every service request session from the session server.

In any case, the distributed session is complex to design and difficult to maintain.

1.3 Token-Based Auth

The token-based authentication system allows users to enter their username and password in order to obtain a token which allows them to fetch a specific resource without using their username and password. Once their token has been obtained, the user can offer the token which offers access to a specific resource for a time period to the remote site. The Figure 2 simply shows the process of token-based authentication.

By using token, there is no need to keep a session store; the token is a self-contained entity that conveys all the user information. In addition, the token is stateless so it's easy to deal with server side scalability. The token is adapted to different client like browser and mobile. In the meantime, it can effectively avoid CORS and CSRF attack.

IOP Conf. Series: Journal of Physics: Conf. Series **910** (2017) 012060 doi:10.1088/1742-6596/910/1/012060

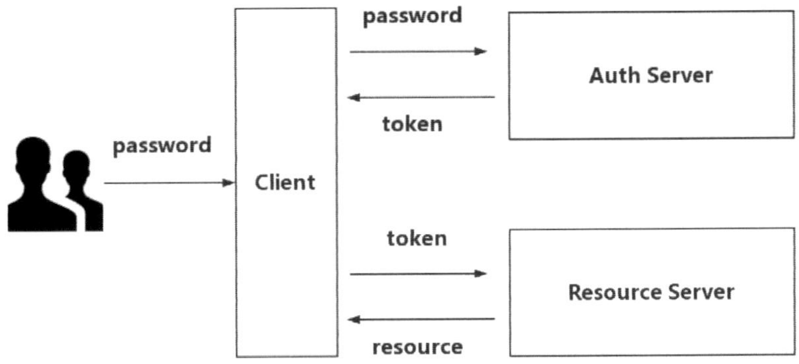

Figure 2 Token-based auth

2. Auth Challenges in Microservice

2.1 What is Microservice Architecture

While there is no standard formal definition of micro service. In the words of Fowler & Lewis [25], a micro service is an approach to developing a single application as a suite of small services, each running in its own process and communicating with lightweight mechanisms, often an HTTP resource API. [1]. Figure 3 is example architecture of microservice.

We like microservice for a bunch of reasons.

- Composing functionality
- Self-contained services
- "Bounded context"
- Independent scaling
- Independent deployment
  - Containers
  - Kubernetes, Mesos
- Localized failures
- Prefer statelessness
  - Not rely on HTTP sessions

2.2 Auth Service in Microservice

There is a simple way to deal with the authentication and authorization in microservice that is to imitate the way of monolithic structure. Each service uses own database or shared database that stores credential data and implements own function to verify the the user independently. This way is easy to understand but has several deficiencies. First, when joining a new service in the system every time, we must re-implement the auth function for the new service. Second, when we use a single shared user database, it will have single point of failure problem. Also, this way is against to the single responsibility principle of micro service.

In order to improve the above architecture, and to adapt to the microservice design principles, we put a separate auth services in the system like Figure 4. The auth service focuses on the authentication and authorization of the user. Other services authenticate the user's identity and authority by interacting with the auth service. As a result, each service is focused on its own business, while improving the scalability and loosely coupled of the system.

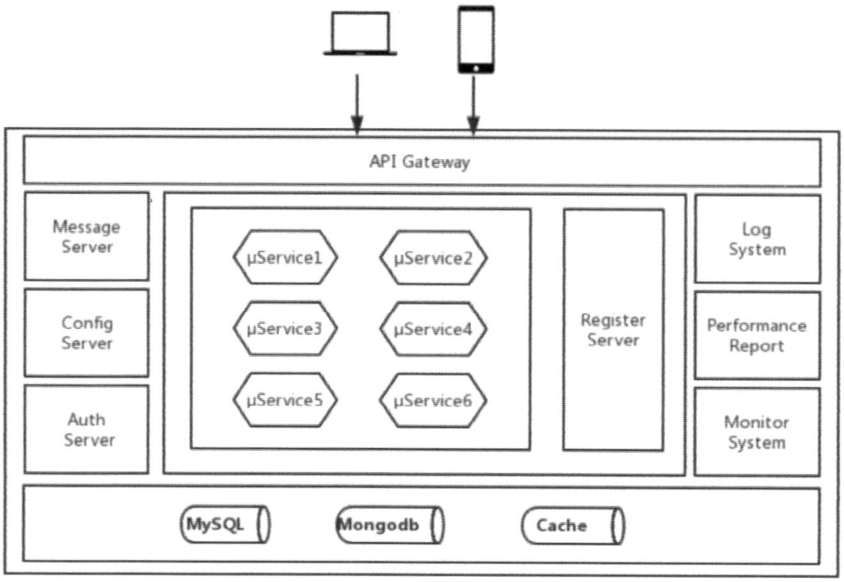

Figure 3 Microservice architecture

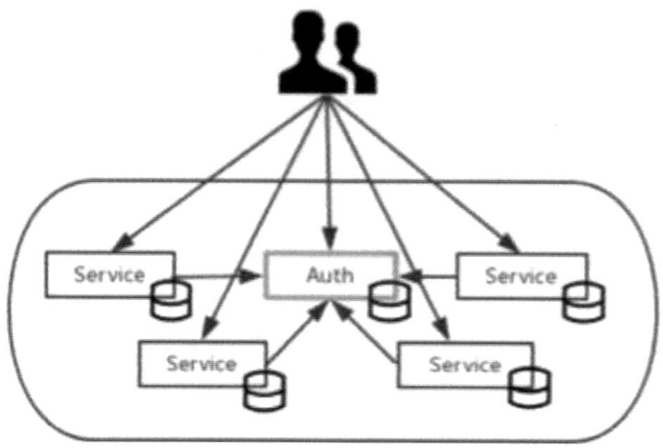

Figure 4 Auth service in microservice system

2.3 What do We Want

- Security

Security means that users are secured identified. The sharing of information within microservice could be a concern depending on how we actually deploy the system.

- Stateless services

We want stateless services with multiple instances in running because we are deploying in the cloud and anything can fail at any time.

- No single point of failure

As mentioned in the previous section, shared databases solution has the problem of SPOF.

- No inherent bottlenecks
- Ability to log out
- Integration with microservices
- Simple to implement

3. Using Distribute Session

In the first part we mentioned the traditional session-based authentication and authorization method and distributed session management. It is well known that microservice is distributed architecture. This section explains how to implement the distribution session solution in micro services. Figure 6 is the architecture.

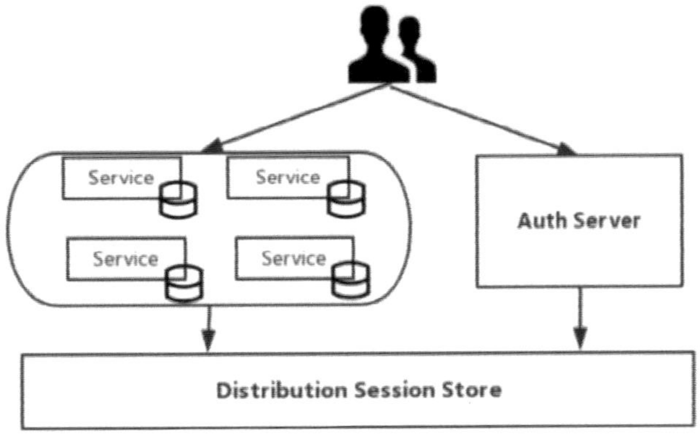

Figure 5 Distribution session solution architecture

User requests access for some services, service redirects to auth server if there are no sessionid in the request. Auth requires the user to provide credentials, generates session for the user and store it in session store, then return sessionid to user. User re-requests access for services with sessionid. Services query session store with sessionid to get user's status and make an appropriate response.

Distribution session solution is feasible but is little complicated to implement as mentioned in first part. The next will introduce several token-based solutions about authentication and authorization in microservice.

4. Using SSO Server

Single sign-on (SSO) is an authentication process that allows a user to access multiple applications with one set of login credentials. In the microservice, each service can be seen as an application. When entering the system, only need to certify the user once.

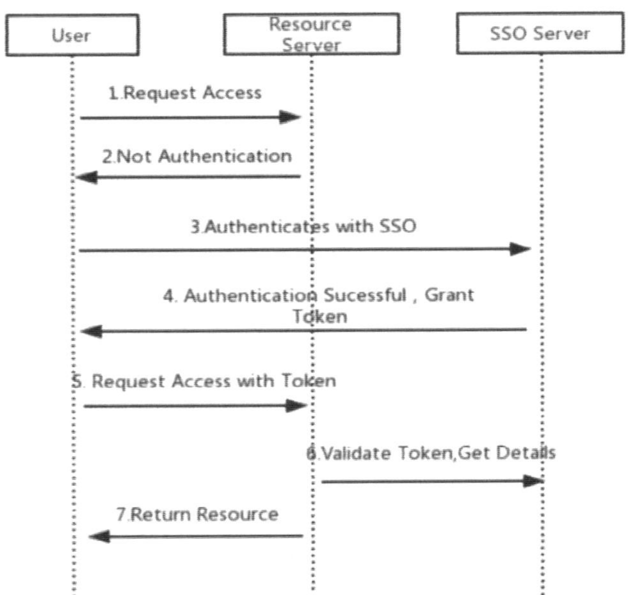

Figure 6 SSO solution timing diagram

Let's take a look at the process of a user want to get resources from microservice. User request access for the resources server. Resource server check if the request carrying a token. If there is a token, valid token from SSO server, and get user details. Judge whether the user has access to the resource and make appropriate response. If the token is not existed nor verified, redirect the request to SSO server. User provides the username and password to get the token and re-visit the resource with token. Figure 5 is the SSO solution timing diagram.

In the SSO solution, “login” state is usually opaque, all requests will finally arrive SSO server to valid the token. This will cause chatty traffic and single point of failure (SPOF) problem in SSO server. when switch the service inside the system, it need to valid the user again with SSO server. We can optimise with local “login” caching to solve the problem.

5. Using JSON Web Token

5.1. JSON Web Token(JWT)

JSON Web Token (JWT)[2] is an information transfer protocol based on an open standard (RFC 7519) that consists of Header, Payload and Signature.

1. Header: The header typically consists of two parts, one of them is token type, namely JWT, and another one is hashing algorithm used, such as HMAC SHA256 or RSA. The format of JSON header is as follow:

```
{
    "alg": "HS256",
    "typ": "JWT"
}
```

Then, this JSON is Base64Url encoded to form the first part of the JWT.

2. Payload: Second part of token which consists of claim. Claim is the statement about an entity (usually users) and supplementary metadata.

```
{
 "exp": 1494428560,
 "name": "John Doe",
 "authorities": [
  "ROLE_ADMIN",
  "VIEW_RESOURCE"]
  ...
}
```

The payload is then Base64Url encoded to form the second part of the JSON Web Token.

3. Signature: This section is to signature the Header and Payload sections according to the algorithm specified in the Header section to prevent modification maliciously.

For example, if you want to use the HMAC SHA256 algorithm, the signature will be created in the following way:

```
HMACSHA256(
 base64UrlEncode(header) + "." +
 base64UrlEncode(payload),
 secret)
```

5.2. Client-Side JWT Solution

When the user sends request to the resources server, the resource server redirect to auth server if there is no token or token is expired. User sends credentials to auth server. Auth server valid the credentials and generate JWT token for the user which contains user’s authority, expired time and so on. User re-requests resource server with JWT token this time. Resource server decodes the token and verifies user permission. Then make an appropriate response. Figure 7 is JWT token solution timing diagram.

Due to the signature, JWT can validation itself rather than request auth server. So this will reduce auth server pressure and chatty traffic. Another benefit is that system is easy to expand for the stateless of the token.

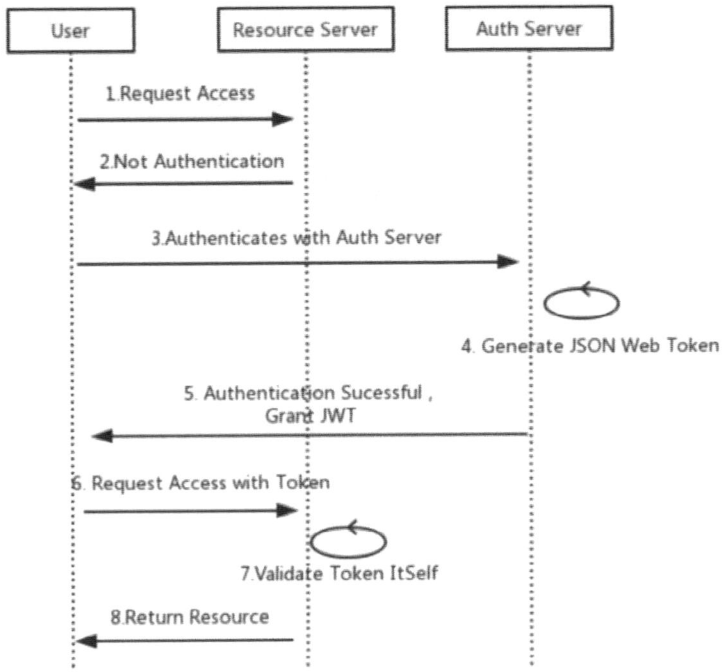

Figure 7 JWT token solution timing diagram

It's a good solution so far. But how to logout. We know that JWT contains an expire time. Before the expire time, the token is always valid. When we want logout, the simple way is to remove the token from client side. But this is not safe, because the server side has not been aware. Another way is periodically check with auth service in the server side. This will cause additional network traffic.

6. Using JWT+API Gateway

In the microservice architecture of Figure 3, we can found there is an API gateway in the boundary of the system. An API gateway provides a single, unified API entry point across one or more internal APIs.

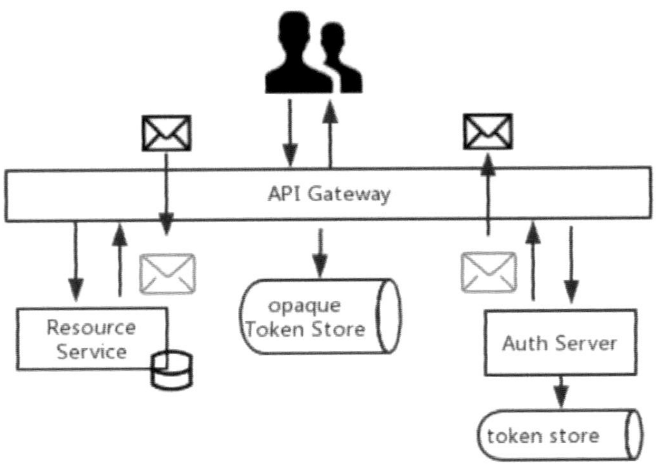

Figure 8 JWT+API gateway solution architecture

The most important thing about logout is to invalidate the token on server side. As every request will go through API gateway, so we can use it to solve the logout problem. The key step is that API gateway translates JWT to an opaque token. Figure 7 is the JWT+ API gate solution architecture and Figure 8 is the timing diagram.

Most steps of JWT+API gateway solution are the same as the solution in V. The only difference is that the gateway has transferred the token. Auth server generate a JSON web token for the user and deliver it to gateway. Gateway transfer the token to an opaque token that only itself can resolve. Then store the relationship of the opaque token and origin token in a token store and send the opaque token to client. When the gateway receives an opaque token from client, transfer it to the origin token, then forward the origin token to the resource service. With that transfer step, we can remove the token relationship in the API gateway token store, So the client need to re-request a token. This is a brilliant solution for the logout issue.

By comparing with the traditional monolithic architecture, this paper explores several solutions of authentication and authorization in microservice architecture. Distributed session solution is similar to the traditional auth way but its complex to implement and maintain. SSO server will cause SPOF and traffic problem. Using API gateway to improve the client-side JWT solution can solve the logout problem. So there is no absolute right solution but only suitable options for your system.

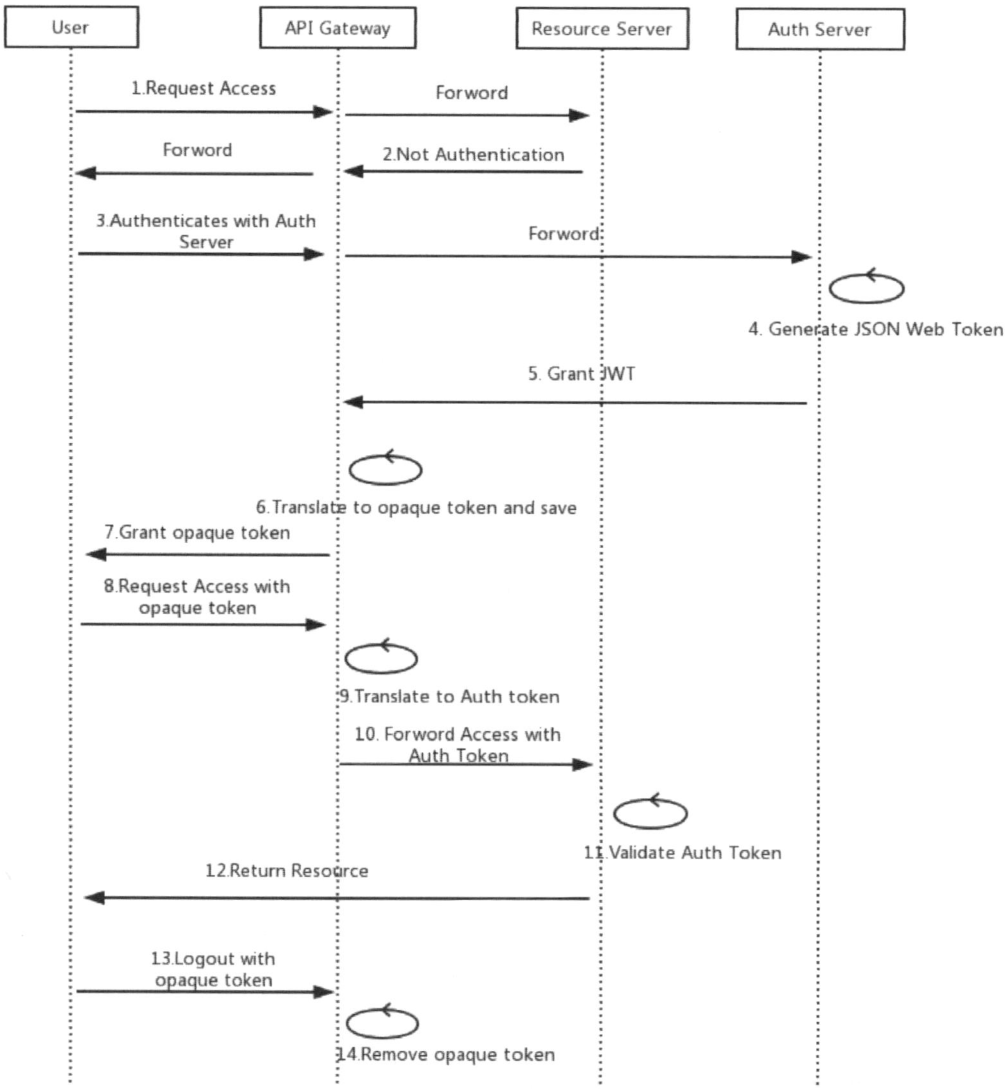

Figure 9 JWT+API gateway solution timing diagrConclusions

7. References

[1] Fowler, M. & Lewis, J. Microservices, 2014 Retrieved From Martin Fowler: http://martinfowler.com/articles/microservices.html.
[2] Bradley J,Sakimura N,Jones M.JSON Web Token(JWT)[Jl,2015.
[3] J. Phys.: Conf. Ser. Flexible session management in a distributed environment, 2010
[4] Peng D,Li C,Huo H.An Extended Username Token-based Approach for REST-style Web Service Security Authentication[C]. Computer SCience and Information Technology,2009.ICCSIT 2009.2nd IEEE International Conference on.IEEE,2009:582—586.
[5] Dmitry Namiot, Manfred Sneps-Sneppe. "On Microservices Architecture".International Journal of Open Information Technologies ISSN: 2307-8 162 vol. 2, no. 9, 20 14. p24-27
[6] Chris Schmidt. "Token Based Authentication - Impelementation Demonstration". Workshop Friend of a Friend, Social Networking and the Semantic Web, 1st-2nd September 2004.
[7] X. Li. "The design of digital campus unified identity authentication system based on web services", Applied Mechanics and Materials Vols. 427-429 pp. 2301-2304, 2013.

AMLSA Algorithm for Hybrid Precoding in Millimeter Wave MIMO Systems

Fulai Liu[1,a], Zhenxing Sun[2,3,b*], Ruiyan Du[1,c] and Xiaoyu Bai[2,d]

[1]Engineering Optimization & Smart Antenna Institute, Northeastern University at Qinhuangdao, Qinhuangdao, China
[2]School of Computer Science and Engineering, Northeastern University, Shenyang, China
[3]Department of Electrical and Information Engineering, Northeast Petroleum University at Qinhuangdao, Qinhuangdao, China
Email: [a]fulailiu@126.com, [b]sunzx-0@163.com, [c]ruiyandu@126.com, [d]bxyu.good@163.com

Abstract. In this paper, an effective algorithm will be proposed for hybrid precoding in mmWave MIMO systems, referred to as alternating minimization algorithm with the least squares amendment (AMLSA algorithm). To be specific, for the fully-connected structure, the presented algorithm is exploited to minimize the classical objective function and obtain the hybrid precoding matrix. It introduces an orthogonal constraint to the digital precoding matrix which is amended subsequently by the least squares after obtaining its alternating minimization iterative result. Simulation results confirm that the achievable spectral efficiency of our proposed algorithm is better to some extent than that of the existing algorithm without the least squares amendment. Furthermore, the number of iterations is reduced slightly via improving the initialization procedure.

1. Introduction

With the development of wireless communication technology, the capacity of wireless networks have to drastically grow to meet the explosive demands for high-data-rate multimedia access. To be specific, the upcoming 5G networks will be expected to achieve 1000X increase in capacity by 2020 [1]. Millimeter-wave (mmWave) communication is a burgeoning technology for future 5G cellular networks [2]. It has the underlying ability to provide gigabit-per-second data rates by developing the large bandwidth available at mmWave frequency range from 30 GHz to 300 GHz [3]. Millimeter-wave MIMO pre-coding can use large-scale antennas at transceivers to provide prominent beamforming gains to combat the path loss and to synthesize highly directional beams, since the small wavelength of mmWave signals.

For a classical MIMO communication system, fully digital processing is made possible by connecting to every antenna dedicated baseband and radio-frequency (RF) hardware. In a mmWave MIMO communication system making use of large arrays, this approach consumes a lot of power due to components such as analog-to-digital converters (ADCs) and RF devices [4]. Hybrid MIMO architectures [5] can overcome this limitation by splitting the MIMO processing between the analog and digital domains. The analog pre-coding stage processes the received signals at the different antenna elements feeding a number of radio-frequency (RF) chains much lower than the array size. A mmWave criterion with a hybrid beamforming architecture has already been designed and tested in [6].

Hybrid precoding is a newly-emerged technique in mmWave MIMO systems [7]. So far the main efforts are on the fully-connected structure [8]. In [9], an optimal hybrid precoder design in a special case was identified, i.e., when the number of RF chains is at least twice that of the data streams. VGA-enabled hybrid precoding is investigated ac-cording to different design criteria[10]. By removing VGAs from the RF domain, low-power analog precoders with phase shifters were also considered, whose phases are heuristically extracted from those of the VGA-enabled solution.

On the other hand, much less attention has been paid on the partially-connected structure. In [11], [12], codebook-based design of hybrid precoders was presented for narrow-band and orthogonal frequency division multiplexing (OFDM) systems, respectively. Although the codebook-based design enjoys a low complexity, there will be certain performance loss, and it is not clear how much performance gain can be further obtained. By utilizing the idea of successive interference cancellation (SIC), an iterative hybrid pre-coding algorithm for the partially-connected structure was proposed in [13]. The algorithm is established based on the assumption that the digital precoding matrix is diagonal, which means that the digital precoder only allocates power to different data streams, and the number of RF chains should be equal to that of the data streams.

The remainder of this paper is organized as follows. We shall introduce the system model and channel model, followed by the problem formulation in Section II. An alternating minimization algorithm with the least squares amendment for the fully-connected structure is depicted in Section III. In Section IV, simulation results will be presented. Finally, we will conclude this paper in Section V.

Notations: the notations below are used throughout this paper. $\mathbf{a}$ and $\mathbf{A}$ stand for a column vector and a matrix, respectively; $\mathbf{A}_{i,j}$ is the entry on the i th row and j th column of $\mathbf{A}$; The conjugate, transpose and conjugate transpose of $\mathbf{A}$ are represented by $\mathbf{A}^*$, $\mathbf{A}^T$ and $\mathbf{A}^H$; $\det(\mathbf{A})$ and $\|\mathbf{A}\|_F$ denote the determinant and Frobenius norm of $\mathbf{A}$; $\mathbf{A}^\dagger$ is the Moore-Penrose pseudo inverse of $\mathbf{A}$; $\mathrm{Tr}(\mathbf{A})$ and $\mathrm{vec}(\mathbf{A})$ indicate the trace and vectorization; Expectation and the real part of a complex variable is noted by $\mathbb{E}[\cdot]$ and $\mathcal{R}[\cdot]$.

2. System Model and Problem Formulation

Firstly, the system model and channel model of the considered mmWave MIMO sys-tem will be described in this section, subsequently the hybrid precoding problem will be formulated to introduce the objective function and constraint condition.

2.1 System Model

Consider a single-user mmWave MIMO system as shown in Figure. 1, where N_s data streams are sent and collected by N_t transmit antennas and N_r receive antennas. The numbers of RF chains at the transmitter and receiver are respectively denoted as N_{RF}^t and N_{RF}^r, which are subject to constraints $N_s \le N_{RF}^t \le N_t$ and $N_s \le N_{RF}^r \le N_r$.

The hybrid precoders consist of an $N_{RF}^t \times N_s$ digital baseband precoder $\mathbf{F}_{BB}$ and an $N_t \times N_{RF}^t$ analog RF precoder $\mathbf{F}_{RF}$. The normalized transmit power constraint is given by $\|\mathbf{F}_{RF}\mathbf{F}_{BB}\|_F^2 = N_s$. Thus, the received signal after decoding processing is given as

$$\mathbf{y} = \sqrt{\rho}\mathbf{W}_{BB}^H\mathbf{W}_{RF}^H\mathbf{H}\mathbf{F}_{RF}\mathbf{F}_{BB}\mathbf{s} + \mathbf{W}_{BB}^H\mathbf{W}_{RF}^H\mathbf{n} \tag{1}$$

where ρ stands for the average received power, $\mathbf{H}$ is the channel matrix, $\mathbf{W}_{BB}$ is the $N_{RF}^r \times N_s$ digital baseband decoder, $\mathbf{W}_{RF}$ is the $N_r \times N_{RF}^r$ analog RF decoder at the receiver, $\mathbf{s}$ is the $N_s \times 1$ symbol vector such that $\mathbb{E}\left[\mathbf{s}\mathbf{s}^H\right] = \frac{1}{N_s}\mathbf{I}_{N_s}$, and $\mathbf{n}$ is the noise vector of independent and identically distributed (i.i.d.) $\mathcal{CN}(0,\sigma_n^2)$ elements. In this paper, we assume that perfect channel state information (CSI) is known at both the transmitter and receiver. The achievable spectral efficiency when transmitted symbols follow a Gaussian distribution can be expressed as

$$R = \log\det\left(\mathbf{I}_{N_s} + \frac{\rho}{\sigma_n^2 N_s}(\mathbf{W}_{RF}\mathbf{W}_{BB})^{\dagger}\mathbf{H}\mathbf{F}_{RF}\mathbf{F}_{BB}\mathbf{F}_{BB}^H \times \mathbf{F}_{RF}^H\mathbf{H}^H(\mathbf{W}_{RF}\mathbf{W}_{BB})\right) \tag{2}$$

Furthermore, the analog precoders are implemented with phase shifters, which can only adjust the phases of the signals. Thus, all the nonzero entries of $\mathbf{F}_{RF}$ and $\mathbf{W}_{RF}$ should satisfy the unit modulus constraints, namely $\left|(\mathbf{F}_{RF})_{i,j}\right| = \left|(\mathbf{W}_{RF})_{i,j}\right| = 1$ for nonzero elements.

According to different signal mapping strategies from RF chains to antennas, the transceiver architecture can be categorized into the fully-connected (as illustrated in Figure. 1) and partially-connected hybrid precoding structures. For the fully-connected structure, which this paper concentrates on researching, the output signal of each RF chain is sent to all the antennas through phase shifters.

Subsequently, we consider the problem of the channel model. Due to high free-space path loss, the mmWave propagation environment is well characterized by a clustered channel model, namely the Saleh-Valenzuela model [14]. This model depicts the mmWave channel matrix as

$$\mathbf{H} = \sqrt{\frac{N_t N_r}{N_{cl} N_{ray}}} \sum_{i=1}^{N_{cl}} \sum_{l=1}^{N_{ray}} \alpha_{il} \mathbf{a}_r(\phi_{il}^r, \theta_{il}^r) \mathbf{a}_t(\phi_{il}^t, \theta_{il}^t)^H \tag{3}$$

where N_{cl} and N_{ray} represent the number of clusters and the number of rays in each cluster, and α_{il} denotes the gain of the l th ray in the i th propagation cluster. Furthermore, $\mathbf{a}_r(\phi_{il}^r, \theta_{il}^r)$ and $\mathbf{a}_t(\phi_{il}^t, \theta_{il}^t)$ represent the receive and transmit array response vectors, where $\phi_{il}^r(\phi_{il}^t)$ and $\theta_{il}^r(\theta_{il}^t)$ stand for azimuth and elevation angles of arrival and departure (AoAs and AoDs), respectively. In this paper, we consider the uniform square planar array (USPA) with $\sqrt{N} \times \sqrt{N}$ antenna elements. Therefore, the array response vector corresponding to the l th ray in the i th cluster can be written as

$$\mathbf{a}(\phi_{il}, \theta_{il}) = \frac{1}{\sqrt{N}}\left[1, \cdots, e^{j\frac{2\pi}{\lambda}d(p\sin\phi_{il}\sin\theta_{il} + q\cos\theta_{il})}, \cdots, e^{j\frac{2\pi}{\lambda}d((\sqrt{N}-1)\sin\phi_{il}\sin\theta_{il} + (\sqrt{N}-1)\cos\theta_{il})}\right]^T \tag{4}$$

where d and λ are the antenna spacing and the signal wavelength, $0 \le p < \sqrt{N}$ and $0 \le q < \sqrt{N}$ are the antenna indices in the 2D plane. While this channel model will be used in simulations, our precoder design is applicable to more general models.

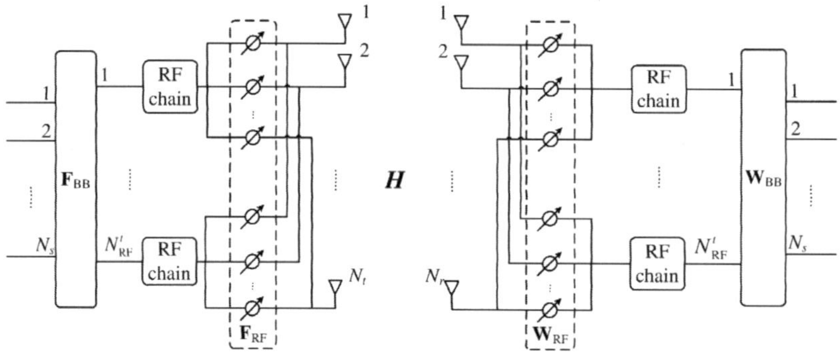

Figure. 1. The transmitter-receiver single user mmWave system architecture

2.2 Problem Formulation

The design of precoders and decoders can be separated into two sub problems, i.e., the precoding and decoding problems. They have similar mathematical formulations except that there is an extra power constraint in the former.

Therefore, we will mainly focus on the precoder design in the remaining part of this paper and the algorithms proposed in this paper can be equally applied for the decoder. The corresponding problem formulation is given by

$$\begin{aligned} &\underset{\mathbf{F}_{RF},\mathbf{F}_{BB}}{\text{minimize}} && \left\| \mathbf{F}_{opt} - \mathbf{F}_{RF}\mathbf{F}_{BB} \right\|_F \\ &\text{subject to} && \begin{cases} \left\| \mathbf{F}_{RF}\mathbf{F}_{BB} \right\|_F^2 = N_s \\ \left| \left(\mathbf{F}_{RF} \right)_{i,j} \right| = 1 \end{cases} \end{aligned} \tag{5}$$

where $\mathbf{F}_{opt}$ stands for the optimal fully digital precoder, while $\mathbf{F}_{RF}$ and $\mathbf{F}_{BB}$ are the analog and digital precoders to be optimized. It has been shown in [17] that minimizing the objective function in (5) approximately leads to the maximization of the spectral efficiency. It is also intuitively true that the optimal hybrid precoders should be sufficiently "close" to the unconstrained optimal digital precoder. In addition, the unconstrained optimal precoder and decoder are comprised of the first N_s columns of $\mathbf{V}$ and $\mathbf{U}$ respectively, which are unitary matrices derived from singular value decomposition (SVD) of the channel, i.e., $\mathbf{H} = \mathbf{U}\mathbf{\Sigma}\mathbf{V}^H$.

In this paper, by decoupling the analog and digital precoders, alternating minimization stands out as an efficient method to obtain an effective solution. With the principle of alternating minimization, we will alternately solve for $\mathbf{F}_{RF}$ and $\mathbf{F}_{BB}$ while fixing the other, which will be the essential idea of this paper.

3. AMLSA Algorithm for the Fully-Connected Structure

For the fully-connected structure, literature [9] has shown that the Frobenius norm in (5) can be made exactly zero under the condition that $N_{RF}^t \geq 2N_s$. This means that the hybrid precoders can achieve the performance of the fully digital precoder in this special case, and the optimal hybrid precoders were obtained in [9]. Therefore, we will focus on the range of $N_s \leq N_{RF}^t \leq 2N_s$ in this paper.

Inspired by [15], this paper will propose an alternating minimization algorithm with the least squares amendment for the fully-connected structure, which is used to minimize the objective function in (5) and obtain the hybrid precoding matrix. The proposed algorithm introduces an orthogonal constraint to the digital precoding matrix $\mathbf{F}_{BB}$, which is amended subsequently by the least squares after obtaining its iterative result. So that the result here is regarded as the final result of $\mathbf{F}_{BB}$. It is shown in simulation results that the achievable spectral efficiency of our proposed algorithm is better to some extent than the existing algorithm without the least squares amendment. Furthermore, the number of iterations is reduced slightly via improving the initialization procedure.

3.1 Hybrid Precoder Design

It is well known that the columns of the unconstrained optimal precoding matrix $\mathbf{F}_{opt}$ are mutually orthogonal in order to cut down the interference between the multiplexed streams. On the basis of this structure of the unconstrained precoding solution, a similar constraint is introduced that the columns of the digital precoding matrix should be mutually orthogonal, i.e.,

$$\mathbf{F}_{BB}^H\mathbf{F}_{BB} = \mathbf{I}_{N_s} \tag{6}$$

where $\mathbf{I}_{N_s}$ is an identity matrix that is of size $N_s \times N_s$. Therefore, the objective function in (5) can be rewritten as

$$\begin{aligned}&\underset{\mathbf{F}_{RF},\mathbf{F}_{BB}}{\text{minimize}} && \left\|\mathbf{F}_{opt}-\mathbf{F}_{RF}\mathbf{F}_{BB}\right\|_F \\ &\text{subject to} && \begin{cases}\left\|\mathbf{F}_{RF}\mathbf{F}_{BB}\right\|_F^2=N_s\\ \left|\left(\mathbf{F}_{RF}\right)_{i,j}\right|=1\\ \mathbf{F}_{BB}^H\mathbf{F}_{BB}=\mathbf{I}_{N_s}\end{cases}\end{aligned} \tag{7}$$

In terms of [15] and [16], the objective function can be further recast as

$$\begin{aligned}&\left\|\mathbf{F}_{opt}-\mathbf{F}_{RF}\mathbf{F}_{BB}\right\|_F^2\\ &=\mathrm{Tr}\left(\mathbf{F}_{opt}^H\mathbf{F}_{opt}\right)-\mathrm{Tr}\left(\mathbf{F}_{opt}^H\mathbf{F}_{RF}\mathbf{F}_{BB}\right)-\mathrm{Tr}\left(\mathbf{F}_{BB}^H\mathbf{F}_{RF}^H\mathbf{F}_{opt}\right)\\ &\quad+\mathrm{Tr}\left(\mathbf{F}_{BB}^H\mathbf{F}_{RF}^H\mathbf{F}_{RF}\mathbf{F}_{BB}\right)\\ &=\left\|\mathbf{F}_{opt}\right\|_F^2-2\mathrm{Tr}\left(\mathbf{F}_{BB}\mathbf{F}_{opt}^H\mathbf{F}_{RF}\right)+\left\|\mathbf{F}_{RF}\mathbf{F}_{BB}\right\|_F^2\end{aligned} \tag{8}$$

Subsequently, the third item of (8) is focused and has the following upper bound

$$\begin{aligned}&\left\|\mathbf{F}_{RF}\mathbf{F}_{BB}\right\|_F^2\\ &=\mathrm{Tr}\left(\mathbf{F}_{BB}^H\mathbf{F}_{RF}^H\mathbf{F}_{RF}\mathbf{F}_{BB}\right)\\ &=\mathrm{Tr}\left\{\begin{pmatrix}\mathbf{I}_{N_s} & \\ & 0\end{pmatrix}\mathbf{K}^H\mathbf{F}_{RF}^H\mathbf{F}_{RF}\mathbf{K}\right\}\\ &\le\mathrm{Tr}\left(\mathbf{K}^H\mathbf{F}_{RF}^H\mathbf{F}_{RF}\mathbf{K}\right)\\ &=\left\|\mathbf{F}_{RF}\right\|_F^2,\end{aligned} \tag{9}$$

Where $\mathbf{F}_{BB}\mathbf{F}_{BB}^H=\mathbf{K}\begin{pmatrix}\mathbf{I}_{N_s} & \\ & 0\end{pmatrix}\mathbf{K}^H$ is the SVD of $\mathbf{F}_{BB}\mathbf{F}_{BB}^H$ and the equality holds when $N_{RF}^t=N_s$. In terms of the mathematical expression above, the expression (8) can be further recast as

$$\begin{aligned}&\left\|\mathbf{F}_{opt}\right\|_F^2-2\mathrm{Tr}\left(\mathbf{F}_{BB}\mathbf{F}_{opt}^H\mathbf{F}_{RF}\right)+\left\|\mathbf{F}_{RF}\right\|_F^2\\ &=\mathrm{Tr}\left(\mathbf{F}_{RF}^H\mathbf{F}_{RF}\right)-2\mathrm{Tr}\left(\mathbf{F}_{BB}\mathbf{F}_{opt}^H\mathbf{F}_{RF}\right)\\ &\qquad+\mathrm{Tr}\left(\mathbf{F}_{BB}\mathbf{F}_{opt}^H\mathbf{F}_{opt}\mathbf{F}_{BB}^H\right)\\ &=\left\|\mathbf{F}_{opt}\mathbf{F}_{BB}^H-\mathbf{F}_{RF}\right\|_F^2\end{aligned} \tag{10}$$

Because of the product form between digital precoding matrix and the analog precoding matrix, the existing algorithm that directly optimizes the objective function (8) will still bring in high complexity. In this context, we plan on adopting the upper bound (10) as the objective function rather than the original one. Thus, by adopting (10) as the objective function, the hybrid precoder design problem is given as

$$\begin{aligned}&\underset{\mathbf{F}_{RF},\mathbf{F}_{BB}}{\text{minimize}} && \left\|\mathbf{F}_{opt}\mathbf{F}_{BB}^H-\mathbf{F}_{RF}\right\|_F^2 \\ &\text{subject to} && \begin{cases}\left\|\mathbf{F}_{RF}\mathbf{F}_{BB}\right\|_F^2=N_s\\ \left|\left(\mathbf{F}_{RF}\right)_{i,j}\right|=1\\ \mathbf{F}_{BB}^H\mathbf{F}_{BB}=\mathbf{I}_{N_s}\end{cases}\end{aligned} \tag{11}$$

Confronted with the expression above, we propose to solve (11) by an alternating minimization procedure. To be specific, since the analog precoding matrix removes the product form with digital precoding matrix, it has a closed-form solution as shown in [15]

$$\arg(\mathbf{F}_{RF}) = \arg\left(\mathbf{F}_{opt}\mathbf{F}_{BB}^{H}\right) \quad (12)$$

where $\arg(\mathbf{F})$ generates a matrix whose entries are the phases of the corresponding entries of matrix $\mathbf{F}$.

Since $\mathbf{F}_{RF}$ is obtained, it can be regarded as fixed, we intend subsequently to solve a digital precoder problem. In terms of [16], OPP tries to minimize $\|\mathbf{A}\mathbf{\Omega} - \mathbf{B}\|_F$, where the optimization variable $\mathbf{\Omega}$ is a square unitary matrix. The mathematical model of OPP is similar to the objective function in (11) if the constraint of $\mathbf{\Omega}$ is relaxed to only columns orthogonality which has the same form of $\mathbf{F}_{BB}$. The solution to this problem, in terms of [16], is given by

$$\mathbf{F}_{BB} = \mathbf{V}_{N_s}\mathbf{U}^{H} \quad (13)$$

where $\mathbf{F}_{opt}^{H}\mathbf{F}_{RF} = \mathbf{U}\mathbf{\Sigma}\mathbf{V}^{H} \approx \mathbf{U}\mathbf{S}\mathbf{V}_{N_s}^{H}$,which is the SVD of $\mathbf{F}_{opt}^{H}\mathbf{F}_{RF}$, $\mathbf{S}$ is a diagonal matrix whose elements are the first Ns nonzero singular values, and $\mathbf{V}_{N_s}$ is the first N_s columns of right singular vector $\mathbf{V}$.

As stated before, The proposed algorithm can alternately optimize the digital precoder and the analog precoder resort to expression (12) and (13), respectively .For better performance, the least squares is used to update $\mathbf{F}_{BB}$ after the last iteration of the algorithm, which is given as

$$\mathbf{F}_{BB} = \left(\mathbf{F}_{RF}^{H}\mathbf{F}_{RF}\right)^{-1}\mathbf{F}_{RF}^{H}\mathbf{F}_{opt} \quad (14)$$

So that the result here is regarded as the final result of $\mathbf{F}_{BB}$ .It is shown in simulation result that the achievable spectral efficiency of our proposed algorithm is better to some extent than the existing algorithm without the least squares amendment. Furthermore, in some existing algorithms, such as [15], the initialization of $\mathbf{F}_{RF}$ has been achieved with random phases, which can increase the number of iterations. The proposed method reduces slightly the number of iterations, which is depicted as follows: first, the initial digital precoding matrix $\mathbf{F}_{BB}$ is generated randomly which is orthogonalised subsequently to guarantee its columns orthogonality. Second, initialize analog precoding matrix $\mathbf{F}_{RF}$ in terms of the expression (12). It is shown in simulation result that the number of iterations is reduced slightly via improving the initialization procedure.

3.2 Algorithm Procedure

As stated before, the algorithm procedure is described as follows:

AMLSA Algorithm

Input: $\mathbf{F}_{opt} \in N_t \times N_s$

1: Generate $\mathbf{F}_{BB}$ randomly and orthogonalize it subsequently to guarantee its columns orthogonality;

2: Initialize $\mathbf{F}_{RF}$ in terms of the expression $\arg(\mathbf{F}_{RF}) = \arg\left(\mathbf{F}_{opt}\mathbf{F}_{BB}^{H}\right)$;

3: Repeat

4: Fix $\mathbf{F}_{RF}$, compute $\mathbf{F}_{opt}^{H}\mathbf{F}_{RF} = \mathbf{U}\mathbf{\Sigma}\mathbf{V}^{H} \approx \mathbf{U}\mathbf{S}\mathbf{V}_{N_s}^{H}$;

5: $\mathbf{F}_{BB} = \mathbf{V}_{N_s}\mathbf{U}^{H}$;

6: Fix $\mathbf{F}_{BB}$, compute $\arg(\mathbf{F}_{RF}) = \arg\left(\mathbf{F}_{opt}\mathbf{F}_{BB}^{H}\right)$;

7: Until a stopping criterion triggers;

8: Update by least squares $\mathbf{F}_{BB} = \left(\mathbf{F}_{RF}^{H}\mathbf{F}_{RF}\right)^{-1}\mathbf{F}_{RF}^{H}\mathbf{F}_{opt}$;

9: For the digital precoder at the transmit end, normalize

$$\hat{\mathbf{F}}_{BB} = \frac{\sqrt{N_s}}{\|\mathbf{F}_{RF}\mathbf{F}_{BB}\|_F}\mathbf{F}_{BB}$$

---

4. Simulation Results

This section will numerically evaluate the performance of our proposed algorithms. Suppose that data streams are transmitted from a transmitter with N_t = 144 to a receiver with N_r = 36 antennas, while both are equipped in the shape of USPA. The channel parameters are set as N_{cl} = 5 clusters, N_{ray} = 10 rays and the average power of each cluster is $\sigma^2_{\alpha,i}$ = 1. The azimuth and elevation angles of AoDs and AoAs follow the Laplacian distribution with uniformly distributed mean angles over [0, 2π) and angular spread of 10 degrees. The antenna elements in the USPA are separated by a half wavelength distance and all simulation results are averaged over 1000 channel realizations. The initial phases of the analog precoder matrix comply with the rule which is depicted in the algorithm procedure. The PE-AltMin algorithm is regarded as the opponent in contrast to our proposed algorithm.

4.1 Spectral Efficiency Evaluation

The spectral efficiency is evaluated when the number of RF chains is complied with the rule $N_s \le N'_{RF} = N^r_{RF} < 2N_s$. Figure. 2 compares the spectral efficiency of different precoding schemes for different N_{RF}. In this case, as shown in Figure 2, for the fully-connected structure, the proposed algorithm achieves higher spectral efficiency in some extent than the existing PE-AltMin algorithm, which benefits from the least squares amendment. Meanwhile, It also has been shown that when compliance with the rule $N'_{RF} = N^r_{RF} = N_s$, it is the worst case since the number of RF chains cannot be smaller than N_s as previously mentioned.

4.2 The Number of Iterations Evaluation

The number of iterations is evaluated when $N_s = 6$ and $N'_{RF} = N^r_{RF} = 9$. Figure. 3 compares the number of iterations of different precoding schemes for different accuracy. In this case, as shown in Figure. 3, for the fully-connected structure, the proposed algorithm experiences less number of iterations slightly than the existing PE-AltMin algorithm, which benefits from reasonable initialization of the analog precoding matrix.

As stated before, the performances including spectral efficiency and the number of iterations, which belong to the proposed algorithm, have been both improved to some extent in contrast to the existing algorithm without the least squares amendment and improved initialization procedure.

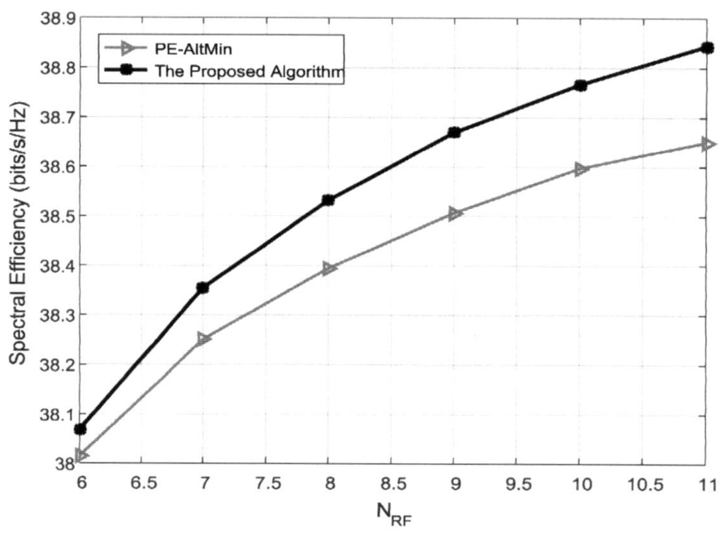

Figure. 2. Spectral efficiency of different precoding schemes for different N_{RF} when $N_s \le N'_{RF} = N^r_{RF} < 2N_s$.

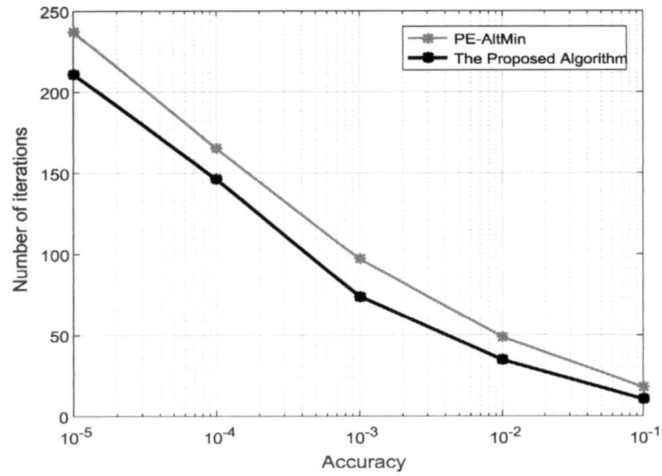

Figure. 3. The number of iterations of different precoding schemes for different accuracy when $N_s = 6$ and $N'_{RF} = N'_{RF} = 9$.

5. Conclusions

In this paper, based on the principle of alternating minimization, an AMLSA algorithm is proposed for hybrid precoding in mmWave MIMO systems. The proposed method can increase the spectral efficiency to some extent, as well as reduce slightly the number of iterations. Simulation results confirm the effectiveness of the proposed algorithm, i.e., the spectral efficiency is better to some extent than the existing algorithm without the least squares amendment, moreover, the number of iterations is reduced slightly via improving the initialization procedure.

6. Acknowledgment

This work was supported by the Program for New Century Excellent Talents in University (NCET-13-0105), and by the Support Program for Hundreds of Outstanding Innovative Talents in Higher Education Institutions of Hebei Province, under Grant No.BR2-259, and by Natural Science Foundation of Hebei Province (No. F2016501139), and by the Specialized Research Fund for the Doctoral Program of Higher Education of China (No.20130042110003), and by the Fundamental Research Funds for the Central Universities under Grant No. N142302001, and by the Fundamental Research Funds for the Central Universities under Grant No.N162304002.

7. References

[1] J. G. Andrews et al., "What will 5G be?", IEEE J. Sel. Areas Commun., vol. 32, no. 6, 2014, pp. 1065–1082.

[2] A. Ghosh et al., "Millimeter-wave enhanced local area systems: A highdata- rate approach for future wireless networks," IEEE J. Sel. Areas Commun., vol. 32, no. 6, 2014, pp. 1152–1163.

[3] M. R. Akdeniz et al., "Millimeter wave channel modeling and cellular capacity evaluation", IEEE J. Sel. Areas Commun., vol. 32, no. 6, 2014, pp. 1164–1179.

[4] A. Alkhateeb, J. Mo, N. González-Prelcic, and R. W. Heath, Jr., "MIMO precoding and combining solutions for millimeter-wave systems", IEEE Commun. Mag., vol. 52, no. 12, 2014, pp. 122–131.

[5] O. El Ayach, S. Rajagopal, S. Abu-Surra, Z. Pi, and R. W. Heath, "Spatially sparse precoding in millimeter wave MIMO systems", IEEE Trans. Wireless Commun., vol. 13, no. 3, 2014, pp. 1499–1513.

[6] W. Roh et al., "Millimeter-wave beamforming as an enabling technology for 5G cellular communications: Theoretical feasibility and prototype results", IEEE Commun. Mag., vol. 52, no. 2, 2014, pp. 106–113.

IOP Conf. Series: Journal of Physics: Conf. Series **910** (2017) 012061 doi:10.1088/1742-6596/910/1/012061

[7] P. Wang, Y. Li, L. Song, and B. Vucetic, "Multi-gigabit millimeter wave wireless communications for 5G: From fixed access to cellular networks", IEEE Commun. Mag., vol. 53, no. 1, 2015, pp. 168–178.

[8] M. Kim and Y. Lee, "MSE-based hybrid RF/baseband processing for millimeter wave communication systems in MIMO interference channels," IEEE Trans. Veh. Technol., vol. 64, no. 6, 2015, pp. 2714–2720.

[9] E. Zhang, C. Huang, "On achieving optimal rate of digital precoder by RF-baseband codesign for MIMO systems," in Proc. 80th IEEE Veh. Technol. Conf. (VTC-Fall), Vancouver, BC, Canada, Sep. 2014, pp. 1–5.

[10] G. Wang and G. Ascheid, "Joint pre/post-processing design for large millimeter wave hybrid spatial processing systems," in Proc. 20th Eur. Wireless Conf., Barcelona, Spain, May 2014, pp. 1–6.

[11] J. Singh and S. Ramakrishna, "On the feasibility of codebook-based beamforming in millimeter wave systems with multiple antenna arrays," IEEE Trans. Wireless Commun., vol. 14, no. 5, 2015, pp. 2670–2683.

[12] C. Kim, T. Kim, and J.-Y. Seol, "Multi-beam transmission diversity with hybrid beamforming for MIMO-OFDM systems," in Proc. IEEE Global Commun. Conf. Workshops (GLOBECOM), Atlanta, GA, USA, Dec. 2013, pp. 61–65.

[13] L. Dai, X. Gao, J. Quan, S. Han, and C.-L. I, "Near-optimal hybrid analog and digital precoding for downlink mmwave massive MIMO systems," in Proc. IEEE Int. Conf. Commun. (ICC), London, U.K., Jun. 2015, pp. 1334–1339.

[14] T. S. Rappaport, R. W. Heath, R. C. Daniels, and J. N. Murdock, "Millimeter Wave Wireless Communications", New York, NY, USA: Pearson Education, 2014.

[15] X. Yu, J. C. Shen, J. Zhang, K. B. Letaief, "Alternating minimization algorithms for hybrid precoding in millimeter wave MIMO systems", IEEE Journal Selected Topics in Signal Processing, vol. 10, no. 3, April 2016, pp. 485-500.

[16] P.H. Schönemann, "A generalized solution of the orthogonal Procrustes problem", Psychometrika, vol. 31, no. 1, 1966, pp. 1–10.

A Simple Deep Learning Method for Neuronal Spike Sorting

Kai Yang[1,2,a], Haifeng Wu[1,2,b] and Yu Zeng[1,2,c]

[1]School of Electrical and Information Technology, Yunnan Minzu University
[2]Key Laboratory of Wireless Sensor Networks of Yunnan Province, Kunming, China
Email: [a]250145249@qq.com, [b]whf5469@gmail.com, [c]yv.zeng@gmail.com

Abstract. Spike sorting is one of key technique to understand brain activity. With the development of modern electrophysiology technology, some recent multi-electrode technologies have been able to record the activity of thousands of neuronal spikes simultaneously. The spike sorting in this case will increase the computational complexity of conventional sorting algorithms. In this paper, we will focus spike sorting on how to reduce the complexity, and introduce a deep learning algorithm, principal component analysis network (PCANet) to spike sorting. The introduced method starts from a conventional model and establish a Toeplitz matrix. Through the column vectors in the matrix, we trains a PCANet, where some eigenvalue vectors of spikes could be extracted. Finally, support vector machine (SVM) is used to sort spikes. In experiments, we choose two groups of simulated data from public databases availably and compare this introduced method with conventional methods. The results indicate that the introduced method indeed has lower complexity with the same sorting errors as the conventional methods.

1. Introduction

Neuronal spikes are basic units of a neural communication system. Neurons in a brain communicate with each other through the spikes. Hence, the study of neuron spikes is a key to understand brain activity. Spike sorting is very important for the study of spikes since different spikes sorted successfully means their neurons identified successfully. For example, neurons about memory process could be identified successfully through spike sorting [1].

Generally, neuronal spikes are collected from electrodes [2]. The aim of spike sorting is to find when the spikes occur and which neuron the spikes belong to. The earliest spike sorting method is matched filtering [3], which can identify a neuron spike from signals collected by a single electrode. The idea of matched filtering is to match the observed signal to a neuron spike template. If the matched errors are not beyond a threshold, the matched spike would be considered as the neuron's spike. Today, the matched filtering is still widely applied in single-cell electrophysiology [3]. When the observed signals are from multi-neuron signals collected by multi-electrodes, however, the method does not work well [4] because the observed signals may be from other neurons. In this case, the matched errors will change and the threshold needs to be manual adjustment for better performance.

A cluster method [2, 5-12] is a popular spike sorting method, which works better in the case of multi-neuron spikes collected by multiple electrodes. The cluster method usually involves three steps [5]. First, segment a raw signal into some chips containing spikes via threshold parameters, such as absolute values [13], square values [6] and Teager energy [14], or some nonlinear operators [15]. Second, extract the eigenvalues of the segmented spikes [2, 6, 7]. Third, cluster the extracted eigenvalues via some clustering method, such as k-means [5], superparamag-netic clustering (SPC) [2] and other hybrid cluster method [8].

With the development of modern electrophysiology technology, however, the number of spikes simultaneously collected by electrodes increases dramatically. Now, the multi-electrode technology has been able to record the activity of thousands of neuronal spikes simultaneously [16, 17]. For such large numbers of spikes sorted, the computational complexity has to be considered. Some conventional methods even spend several days or weeks to complete the data sorted [18-20].

At present, there are several new proposed cluster-based methods to solve the spike sorting [10, 11, 21]. Sparse coding algorithm is an unsupervised learning method, which is used to find a set of "overcomplete" base vectors to represent the sample data more efficiently. In recent years, sparse coding has also been applied to spike sorting [21, 22], and has shown better performance. However, as the number of neurons increases, their complexity would not be overlooked.

This paper introduces a principal component analysis network (PCANet) for spike sorting. PCANet is a deep learning framework proposed in recent years [23]. The learning algorithm could be trained and used to classify different types of data and background. The advantage of PCANet is lower computational complexity and simple parameter modulation. This paper will start from a conventional model and train a PCANet, where some eigenvalue vectors are extracted. Then, support vector machine (SVM) is used to sort spikes. From experiment results, the introduced method has lower complexity than traditional sorting methods.

2. System Model

Before sorting spikes, we need to preprocess the raw voltage trace recorded by electrodes. The preprocessing steps are high-pass filtering, whitening and peaks detecting [10]. After detecting the peaks, we find the time when the neuron fires an action potential. Then, the filtered and whiten voltage trace is segmented into some chips containing the peaks. The segmented spikes are just what we need to sort. If $v(m)$ is denoted as the potential amplitude of the mth sampling point in a segmented spike, it could be expressed as [10]

$$v(m) = \sum_{j=1}^{J} w_j(m) * \delta(m - n_j) + \varepsilon(m) \tag{1}$$

where

$m \in \{1, 2, \cdots, M\}$,

M denotes the length of the spike segment,

$w_j(m)$ denotes the potential amplitude of the jth neural spike template at the mth point,

$\delta(\cdot)$ denotes a unit impulse function,

$*$ represents a convolution,

J denotes the number of neurons,

$n_j \in \{-M+1, -M+2, \cdots, M\}$ denotes the time shifted,

$\varepsilon(m)$ is an additive noise.

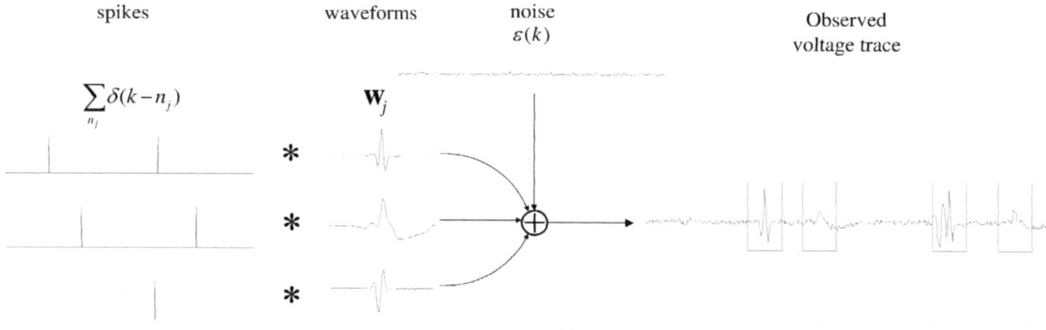

Figure 1 Schematic diagram of spikes signal system model

Eq. (1) means that observed voltage trace can be expressed as the sum of convolutions between several spike template signals and their respective time-shifted impulse functions, shown in Fig. 1. Note that the spike template signal $w_j(m)$ in (1) could be from the center of clusters via methods, e.g. *k*-means. For spikes collected from multiple electrodes, besides, we could constitute them one-by-one into a sequence signal. Here, we only give the case of a single electrode for simplified expression.

3. Algorithm

3.1 Training Data for PCANet

As shown in last section, an observed voltage trace is in fact the superposition of several time-shifted spike templates. We could obtain training data through the time-shifted spike templates and the training data will contain the recorded voltage trace. Then, the parameter of PCANet could be obtained via the training data. Next, we describe how to establish the training data. Firstly, we change (1) into a matrix form as

$$\mathbf{V} = \mathbf{W}\boldsymbol{\Delta} + \boldsymbol{\varepsilon} \tag{2}$$

where

$$\mathbf{V} = [v(1), v(2), \cdots, v(M)]^{\mathrm{T}} \tag{3}$$

and $\boldsymbol{\varepsilon} = [\varepsilon(1), \varepsilon(2), \cdots, \varepsilon(M)]^{\mathrm{T}}$. Since (1) represents convolutions between spike templates and impulse functions, the matrix $\mathbf{W}$ in (2) should be a Toeplitz matrix. Moreover, $\boldsymbol{\Delta}$ is a vector composed of the unit impulse function $\delta(\cdot)$. And, the matrix $\mathbf{W}$ could provide training data that PCANet requires. Here, we will establish the Toeplitz matrix $\mathbf{W}$ as

$$\mathbf{W} = [\overline{\mathbf{W}}_1, \overline{\mathbf{W}}_2, \cdots, \overline{\mathbf{W}}_J] \tag{4}$$

where,

$$\overline{\mathbf{W}}_j = [\mathbf{w}_j(-r_j), \mathbf{w}_j(-r_j+1), \cdots, \mathbf{w}_j(s_j-1)] \tag{5}$$

$$\mathbf{w}_j(m) = \begin{cases} \left[\mathbf{0}_{1\times m}, w_j(1), w_j(2), \cdots, w_j(M-m)\right]^{\mathrm{T}} \\ \quad \text{if } 0 \le m \le s_j - 1 \\ \left[w_j(-m+1), w_j(-m+2), \cdots, w_j(M), \mathbf{0}_{1\times(-m)}\right]^{\mathrm{T}} \\ \quad \text{if } -r_j \le m \le -1 \end{cases} \tag{6}$$

From (4-6), the matrix $\overline{\mathbf{W}}_j$ is composed of several column vectors $\mathbf{w}_j(m)$, $m=-r_j$, $-r_j+1$, $\cdots$, s_j-1, which is from the *j*th neural spike template $w_j(m)$, m=1,2,...,M through left or right time-shifted. Therefore, the Toeplitz matrix $\mathbf{W}$ could be obtained from J time-shifted neural spike templates. Besides, r_j and s_j represent the maximum left and right shifts, respectively. The choice for r_j and s_j is decided by the waveform amplitude of the spike templates and ensure $0 < r_j < M$, $0 < s_j < M$. Finally, the dimension of $\mathbf{W}$ is $M \times N$ if $u_j = r_j + s_j$, $N = \Sigma_{j=1}^{J} u_j$, and the column vectors $\mathbf{w}_j(m)$, $m=-r_j$, $-r_j+1$, $\cdots$, s_j-1 would be the training data for PCANet. Finally, the dimension of $\mathbf{W}$ is $M \times N$ if $u_j = r_j + s_j$, $N = \Sigma_{j=1}^{J} u_j$.

When the training data $\mathbf{W}$ is constructed, it is necessary to transform the *i*-th training sample $\mathbf{W}_i$ into a matrix, and Figure 2 gives an example of its construction.

After the training matrix is constructed, PCANet algorithm is used to obtain the required model parameters. Table I gives the calculation procedure.

When PCANet completes training, the eigenvalues of each segmented spike signals are extracted. Thus, SVM could sort the spikes via the eigenvalues. The details of PCANet applied to spike sorting are given in Table II.

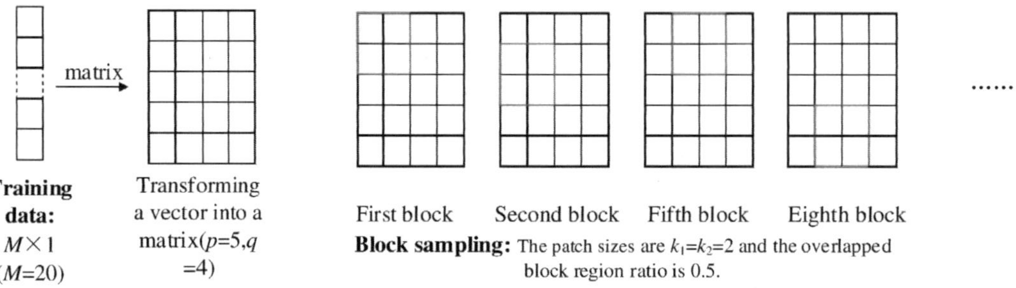

Figure 2 An example of transforming training data into matrices

3.2 Computational Complexity

In this subsection, we will analyze the computational complexity of this sorting method. From (4), the i th column vector $\mathbf{w}_i$ is converted into a $p\times q$ matrix $\mathbf{w}_i$, where $M = p\times q$. In our method, the patch size, i.e. 2D filter size is denoted as $k_1\times k_2$ at all stages. If we let the number of the i th layer filter in PCANet be L_i, the computational complexity of a two-staged PCANet applied to our spike sorting could be shown as [23]

$$O(pqk_1k_2(L_1+L_2)+pq(k_1k_2)^2) \tag{7}$$

4. Experiment

In experiments, we use two groups of simulated data to verify our PCANet method. And, our methods are compared with the traditional k-means, sparse coding [21] and SPC algorithm [2]. All of algorithm codes for this experiments are performed on MATLAB 2012b and the CPU of PC computer is Core i3 3.70GHz.

4.1 Simulated data1

In this sub-section, a simulated data are used to verify the introduced methods and the traditional algorithms. The data denoted by C_Easy1_noise015 are from literature [2], where their waveforms are from real environment and noises are chosen to be able to simulate realistic background activity. Some parameters in this experiment and the algorithms are shown in Table III. More parameters for SPC could be found from [2] and a software, WaveClus version 2.0 downloaded from https://vis.caltech.edu/~rodri/Wave_clus/Wave_clus_home.htm.

The first line of Table IV gives the results for sorting C_Easy1_noise015 data through the four methods, PCANet, SPC, k-means, and sparse coding (denoted by SC), respectively. Except the parameters listed in Table III, we specify some parameters for PCANet as follows. A maximum left shift and right shift r_j, s_j of the Toeplitz matrix in (4-6) are set to r_j=20, s_j=19, j=1,2,3. We specify that, besides, if the difference between spike time in experiment and ground truth is within 4ms, it will be concluded that the both match. When the true spike time is not found to match the experimental one, we regard it as "miss". When the experimental one does not match the true one, on the other hand, we regard it as "false positives". We can see from this group of results that, the misses by SPC is the highest, arriving at 486. The number of missies for PCANet is 246, nearly the same as k-means, 244 and both of them are higher than sparse coding. The reason why the misses by k-means are higher is that, clustering methods will make the eigenvalue point for an overlapped spike be far away from all centers of clusters. For SPC, the highest misses have the following two reasons. One is peak detection because some spikes are not detected. The other is the same as k-means. After all, SPC is also based on clustering methods. In addition, the reason for higher misses of PCANet is overlapping spikes. Finally, it is seen that the false negatives of PCANet is 4, slightly higher than 2 of *k*-means, lower than 18 of SPC and 10 of sparse coding.

Table 1 PCANet Algorithm

Training steps

Input: training data **W**

computation

$$\bar{\mathbf{w}}_i = [\bar{\mathbf{w}}_{i,1}, \bar{\mathbf{w}}_{i,2}, \cdots, \bar{\mathbf{w}}_{i,pq}] \quad \text{(T-1)}$$

$$\tilde{\mathbf{W}} = [\bar{\mathbf{W}}_1, \bar{\mathbf{W}}_2, \cdots, \bar{\mathbf{W}}_N] \in \mathbb{R}^{k_1k_2 \times Npq} \quad \text{(T-2)}$$

$$\hat{\mathbf{W}}_l^1 = mat_{k_1,k_2}(q_l(\tilde{\mathbf{W}}\tilde{\mathbf{W}}^T)) \in \mathbb{R}^{k_1 \times k_2}, l = 1, 2, \cdots, L_1 \quad \text{(T-3)}$$

$$\mathbf{W}_i^l = \mathbf{W}_i * \hat{\mathbf{W}}_l^1 \quad \text{(T-4)}$$

$$\bar{\mathbf{Y}}_i^l = [\bar{\mathbf{y}}_{i,l,1}, \bar{\mathbf{y}}_{i,l,2}, \cdots, \bar{\mathbf{y}}_{i,l,pq}] \in \mathbb{R}^{k_1k_2 \times Npq} \quad \text{(T-5)}$$

$$\mathbf{Y}^l = [\bar{\mathbf{Y}}_1^l, \bar{\mathbf{Y}}_2^l, \cdots, \bar{\mathbf{Y}}_N^l] \in \mathbb{R}^{k_1k_2 \times Npq} \quad \text{(T-6)}$$

$$\hat{\mathbf{W}}_l^2 = mat_{k_1,k_2}(q_l(\mathbf{Y}\mathbf{Y}^T)) \in \mathbb{R}^{k_1k_2}, l = 1, 2, \cdots, L_2 \quad \text{(T-7)}$$

$$O_i^l = \{\mathbf{W}_i^l * \hat{\mathbf{W}}_l^2\}_{l=1}^{L_2} \quad \text{(T-8)}$$

$$\mathcal{T}_i^l = \sum_{l=1}^{L_2} 2^{l-1} H(\mathbf{W}_i^l * \hat{\mathbf{W}}_l^2) \quad \text{(T-9)}$$

$$\mathbf{f}_i = [Bhist(\mathcal{T}_i^1), \cdots, Bhist(\mathcal{T}_i^{L_1})]^T \in \mathbb{R}^{(2^{L_2})L_1B} \quad \text{(T-10)}$$

where

$\bar{\mathbf{w}}_i$ denotes the *i*-th training sample of the training data **W**, which is obtained by mean-removed, block sampling.

$M = p \times q$, k_1 and k_2 are the patch sizes.

L_i denotes the number of filters in layer *i*.

$q_l(\tilde{\mathbf{W}}\tilde{\mathbf{W}}^T)$ denotes the feature vector corresponding to the *l*-th principal component of $\tilde{\mathbf{W}}\tilde{\mathbf{W}}^T$.

$mat_{k_1,k_2}(\mathbf{v})$ is a function that maps **v** to matrix $\hat{\mathbf{W}}_l^1$.

$Bhist(\mathcal{T}_i^l)$ means that after $\mathcal{T}_i^l$ is divided into B blocks, all B histograms are connected into one vector.

Output: filters $\hat{\mathbf{W}}_l^1$ and $\hat{\mathbf{W}}_l^2$, feature vector $\mathbf{f}_i$.

Steps: 1. The matrix $\hat{\mathbf{W}}_l^1$ is obtained from the training data **W** via (T-1) and (T-2).

2. The filter $\hat{\mathbf{W}}_l^1$ of the first layer network is obtained by the formula (T-3).

3. The output $\mathbf{W}_i^l$ of the *l*-th filter of the first layer network is obtained by (T-4).

4. The mean-removed and block sampled matrix $\mathbf{Y}^l$ is obtained by (T-5) and (T-6).

5. The filter $\hat{\mathbf{W}}_l^2$ of the second layer network is obtained by the formula (T-7).

6. The output O_i^l of the second layer network is obtained by the formula (T-8).

7. The final feature vector $\mathbf{f}_i$ is obtained by (T-9) and (T-10).

Table V gives running time of PCANet for the data, C_Easy1_noise015. From the table, PCANet spends about 441.1765 seconds, which is lower than 3905 of sparse coding.

Table 2 PCANet Applied to Spike Sorting

Algorithm steps
1. A raw voltage trace is preprocessed, i.e. whitened, filtered, peaks detected and segmented into a number of signal chips containing the peaks.
2. The eigenvalues of the signal segments are extracted through PCA and clustered into J clusters through *k*-means. Then, the centers of the clusters are used as J spike templates $w_{j(0)}$, $j=1, 2, \cdots, J$.
3. The templates $w_{j(0)}$, $j=1, 2, \cdots, J$ time-shifted constitute a Toeplitz matrix w in (4), and its column vectors $w_{j(0)}$ are used as training data for PCANet.
4. The required model parameters are obtained by Table I, and then SVM is used to classify the signal segments.

Table 3 Some Parameters for the First Group of Data

Parameter	**Value**
Number of neurons	*J*=3
Number of channel	1
Filtered	Cut-off frequency at 250 Hz with a Butterworth high pass filter of order 50.
Whiten	Whitened in time
Peaks detected	A threshold is 6
Length of signal segment	*M*=81
k-means:	
Number of principal component	4
Contribution rate of principal components	90%
Distance measure	Euclidean distance
Clustering repeated	25
Selection of initial centroid	3 centroid randomly selected
Sparse coding:	
Sparse coding method	Non-negative least squares
Prediction method	k-nearest neighbor
A maximum left shift and right shift	$r_j = s_j$ =42, j =1,2,3
I-th largest	*I*=25
Kernel function	Linear
Sparsity threshold	10^{-4}
PCANet algorithm:	
Number of Stages	2
Patch Size	$k_1 = k_2 = 7$
The size of matrix $\bar{w}_i$	*p*=*q*=9
Number of Filters	$L_1 = L_2 = 8$
Size of each block	7,7
overlapped block re-gion ratio	0.92

Next, table VI shows the impact of r_j and s_j on the performance of PCANet. From C_Easy1_noise015, the values of r_j and s_j have a slight impact on the performance of the algithms. In practice, we choose the values having fewest sorting errors, r_j =20, s_j =19, j =1,2,3.

Table 4 Results of Sorting Spike in Two Groups of Data by Four Methods

Experiment data	Number of spikes	PCANet		SPC		*k*-means		SC	
		Misses	FP	Misses	FP	Misses	FP	Misses	FP
C_Easy1_noise015	3477	246	4	486	18	244	2	7	10
Simulation 2	1058	0	11	0	0	0	0	0	0
Average	2268	123	8	243	9	122	1	4	5

Note: SC and FP refer to sparse coding, false positives, respectively. These abbreviations indicate the same meaning in the following tables.

Table 5 Running time (second) of PCANet algorithm in different experimental data

Experiment data	PCANet	SPC	*k*-means	Sparse coding
C_Easy1_noise015	441.1765	16.8100	6.1590	3905
Simulation 2	148.5905	12.6100	8.7420	316.70

Table 6 Impact of Maximum Right Shift and Left Shift on Performance of PCANet

r_1	s_1	r_2	s_2	r_3	s_3	C_Easy1_noise015		Simulation 2	
						Misses	FP	Misses	FP
50	49	47	49	48	47	260	18	3	17
46	46	44	46	44	44	260	18	8	12
44	44	43	44	43	43	260	18	15	18
42	42	42	42	42	42	252	10	5	17
40	40	41	40	41	41	254	12	7	17
38	38	40	38	40	40	257	15	10	17
34	33	32	32	34	34	266	24	1	13
31	30	30	30	31	30	259	17	0	11
28	28	28	27	28	26	261	19	1	12
24	23	24	23	22	21	248	6	4	18
22	21	22	21	21	20	250	8	2	13
20	19	20	19	20	19	246	4	4	15
19	18	19	18	19	18	249	7	1	15
18	17	18	17	19	17	248	6	1	16

4.2 Simulated Data2

In this subsection, we choose another simulated spike data to verify our method and other conventional methods. The simulated data are from literature [24] and availably downloaded from http://www2.le.ac.uk/centres/csn/soft ware. Most of the parameters are the same as Table III, except some of them in Table VII.

The second line in Table IV gives the sorting results of four methods above, where the values of and are chosen as r_1=31, s_1=30, r_2=30, s_2=30, r_3=31, s_3=30. From the table, the misses for all of the four methods are 0 and the false negatives of PCANet are 11, slightly higher than the other three methods.

Table V give the running time for PCANet. From the table, PCANet spends about 148.5905 seconds, which is also lower than sparse conding. Next, Table VI shows the impact of r_j and s_j on the performance of PCANet. From simulation 2, the values of r_j and s_j have a slight impact on the performance of the algithms. In practice, we choose the values having fewest sorting errors, r_1 =31, s_1 =30, r_2 =30, s_2 =30， r_3 =31, s_3 =30.

Table 7 Some Parameters for the Second Group of Data

Parameter	**Value**
k-means:	
Number of principal compon-ent	2.
Sparse coding:	
A maximum left shift and rig-ht shift	$r_1 = r_2 = 18$, $r_3 = 19$ $s_1 = s_2 = s_3 = 17$
I-th largest	*I*=18

5. Conclusion

Our method is based on a conventional model. We firstly adopt *k*-means algorithm to obtain several centers of clusters, through which a Toeplitz matrix is established. Then, the column vectors of the Toeplitz matrix are consi- dered as input training signals of PCANet. In the next step, the output eigenvalue vectors of PCANet would be the input of SVM. The final output of the SVM is sorting result. From experiment results, of course, our method does not show better sorting performance on overlapping spikes. However, the method has less running time of sorting spikes. Therefore, it is probable that a deep learning algorithm, PCANet would have lower computational complexity when sorting multi-electrode spikes.

6. Acknowledgment

This work was supported by the National Natural Science Foundation of China under Grant No. 61262091, The17th batches of Young and Middle-aged Leaders in Academic and Technical Reserved Talents Project of Yunnan Province under Grant No. 2014HB019, Supported by Program for Innovative Research Team (in Science and Technology) in University of Yunnan Province. Corresponding author is Haifeng Wu.

7. References

[1] Hernan Gonzalo Rey, Maryam Ahmadi, and Rodrigo Quian Quiroga,"Single trial analysis of field potentials in perception, learning and memory", Current Opinion in Neurobiology, vol. 31, pp. 148-155, 2015.

[2] Quian Quiroga R, Nadasdy Z, Ben-Shaul Y, "Unsupervised spike detection and sorting with wavelets and superparamagnetic clustering", Neural Computation, vol. 16, no. 8, pp. 1661 - 1687, 2004.

[3] Turin G, " An introduction to matched filters. Information Theory", IRE Transactions on Information Theory, vol. 6, no. 3, pp. 311–329, 1960.

[4] Bar-Gad I, Ritov Y, Vaadia E, Bergman H, " Failure in identification of overlapping spikes from multiple neuron activity causes artificial correlations", Journal of neuroscience methods, vol. 107, no. 1-2, pp. 1–13, 2001.

[5] Lewicki MS, " A review of methods for spike sorting: the detection and classification of neural action potentials", Network Computation in Neural Systems, vol. 9, no. 4, pp. 53–78, 1998.

[6] Rutishauser U, Schuman EM, Mamelak AN,"Online detection and sorting of extracellularly recorded action potentials in human medial temporal lobe recordings, in vivo", Journal of Neuroscience Methods, vol. 154, no. 1-2, pp. 204–224, 2006.

[7] Harris K, Henze D, Csicsvari J, Hirase H, Buzsaki G, " Accuracy of tetrode spike separation as determined by simultaneous intracellular and extracellular measurements", Journal of Neurophysiology, vol. 84, no. 1, pp. 401–414, 2000.

[8] Shoham S, Fellows MR, Normann RA, " Robust, automatic spike sorting using mixtures of multivariate t-distributions", Journal of Neuroscience Methods, vol. 127, no. 2, pp. 111–122, 2003.
[9] Pouzat C, Mazor O, Laurent G, " Using noise signature to optimize spike sorting and to assess neuronal classification quality", Journal of Neuroscience Methods, vol. 122, no. 1, pp. 43–57, 2002.
[10] Pillow J, Shlens J, Chichilsnisky E, Simoncelli EP, " A model-based spike sorting algorithm for removing correlation artifacts in multi-neuron recordings", PLoS ONE, vol. 8, no. 5, article e622123, 2013.
[11] Chen B, Carlson D, Carin L, " On the analysis of multi-channel neural spike data", Advances in Neural Information Processing Systems 24, pp. 936-944, 2011.
[12] Wehr M, Pezaris JS, Sahani M, " Simultaneous paired intracellular and tetrode recordings for evaluating the performance of spike sorting algorithms", Neurocomputing, vol. 26–27, pp. 1061–1068, 1999.
[13] Obeid I, Wolf P, " Evaluation of spike-detection algorithms for a brain–machine interface application", IEEE Transactions on Biomedical Engineering, vol. 51, no. 6, pp. 905–911, 2004.
[14] Choi JH, Jung HK, Kim T, "A new action potential detector using the MTEO and its effects on spike sorting systems at low signal-to-noise ratios", IEEE Transactions on Biomedical Engineering, vol. 53, no. 4, pp. 738–46, 2006.
[15] Rebrik SP, Wright BD, Emondi AA, Miller KD, " Cross-channel correlations in tetrode recordings: implications for spike-sorting", Neurocomputing, vol. 26–27, pp. 1033–1038, 1999.
[16] Dion Khodagholy, Jennifer N Gelinas, Thomas Thesen, et al," NeuroGrid: recording action potentials from the surface of the brain", Nature Neuroscience, vol. 18, no. 2, pp. 310-315, 2014.
[17] Hernan Gonzalo Rey, Carlos Pedreira and Rodrigo Quian Quiroga,"Past, present and future of spike sorting techniques", Brain Research Bulletin, vol. 119, no. Part B, pp. 106-117, 2015.
[18] Rossant C, Kadir S N, Goodman D F M, et al, " Spike sorting for large, dense electrode arrays", Nature Neuroscience, vol. 19, no. 4, pp. 634-641, 2016.
[19] Hilgen G, Sorbaro M, Pirmoradian S, et al, " Unsupervised spike sorting for large-scale, high-density multielectrode arrays", Cell Reports, vol. 18, no. 10, pp. 2521-2532, 2017.
[20] Pachitariu M, Steinmetz N, Kadir S, et al, " Kilosort: realtime spike-sorting for extracellular electrophysiology with hundreds of channels", bioRxiv, 2016: 061481.
[21] Kai Yang, Haifeng Wu, Yu Zeng,"Sparse coding approaches for neuron spike sorting", Advanced Information Management, Communicates, Electronic and Automation Control Conference (IMCEC), pp. 600-604. IEEE Press, Xi An, 2016.
[22] Li Y, Ngom A, " Sparse representation approaches for the classification of high-dimensional biological data", BMC systems biology, vol. 7, no. 4, pp. S6, 2013.
[23] Chan T H, Jia K, Gao S, et al, " Pcanet: A simple deep learning baseline for image classification?", IEEE Transactions on Image Processing, vol. 24, no. 12, pp. 5017-5032, 2015.
[24] Martinez J, Pedreira C, Ison M J, et al, "Realistic simulation of extracellular recordings", Journal of neuroscience methods, vol. 184, no. 2, pp. 285-293, 2009.

IOP Conf. Series: Journal of Physics: Conf. Series **910** (2017) 012063 doi:10.1088/1742-6596/910/1/012063

A Multi-reference Extended Variable Step-size LMS Algorithm for Solving Dynamic Weighting Coefficients of the Satellite Beam-forming Networks

Yao Zhang[a], Bing Wang, Yong Luo[b], Yuxin Cheng and Jianjun Wu

Institute of Advanced Communications, EECS, Peking University, Beijing, P. R. China

Email: [a]zhang_yao@pku.edu.cn, [b]just@pku.edu.cn

Abstract. A Multi-beam satellite communication system can form beam by coherently superimposing the beams generated by multiple feeds in space. In order to form the beam based on dynamic user, a multi-reference extended variable step-size LMS algorithm is introduced in this paper to solve the dynamic weighting coefficients. According to the simulation result, when the mean square error of the synthesized beam is small, this algorithm can effectively improve the beamforming convergence rate.

1. Introduction

Multi-beam mobile communication system is widely used in satellite communication. Its synthesized beam is similar to the wireless cell on the ground, and it can seamlessly cover the region by overlaping beam. The enhanced beamforming method can form the required beam with the coherent superposition of beams from multiple feeds. It can adjust the weighting coefficients of each feed to control the gain and beam shape and finally improve the communication quality and anti-interference performance.

The algorithm in this paper is intended to find optimal weighting coefficients w that can achieve the requirement of beamforming, gain, main-lobe width and side-lobe suppression.

In order to improve the beam service quality, a dynamic user-centered synthesis system is adopted by the multi-beam system to synthesize flexible and variable beams. The corresponding beamforming network not only needs to cover the full region, but also decrease the width of the main-lobe and suppress the side-lobe. For achieving real-time tracking effect and improving the information transfer rate, it is supposed to dynamically adjust the beam according to the user geographical position and channel or generate new beams to provide services. During the beamforming, the algorithm solving dynamic weighting coefficients plays an important role.

SMI, LCMV are the widely used beamforming algorithms. However, due to the complexity of the calculation and the accuracy of beamforming, they can't meet the system demand. Least mean square (LMS) algorithm is a self-adaptive beamforming algorithm based on Minimum Mean Square Error (MMSE) criterion and it is the most widely used beamforming algorithm currently. LMS can get optimal weighting coefficients and higher accuracy with lower complexity by iterative computation. Beams formed with LMS algorithm have better center directivity, stronger main-lobe constraint and greater side-lobe suppression than SMI, LCMV.

The dynamic beamforming network requires higher convergence speed, better precision and more flexibility of algorithm to satisfy dynamic user demands. Intended for this issue, a multi-reference extended variable step-size LMS algorithm is designed to effectively improve the convergence rate when the beamforming accuracy is satisfied.

This paper consists of five parts. First, chapter 1 briefly introduces the background of the research. Then, chapter 2 describes the system model the proposed algorithm is based on. The multi-reference extended variable step-size LMS algorithm is elaborated in Chapter 3. Finally, chapter 4 shows the verification simulation result of the algorithm and the last chapter is a full-text summary.

2. System Model

Consider about the situation dynamicly forming beams based on users shown in Figure 1. When the user's position changes, especially in high-speed, it requires a faster tracking capability of the beam to ensure that user is always in the center of the beam. This makes higher requirements on the convergence, precision and complexity of beamforming networks.

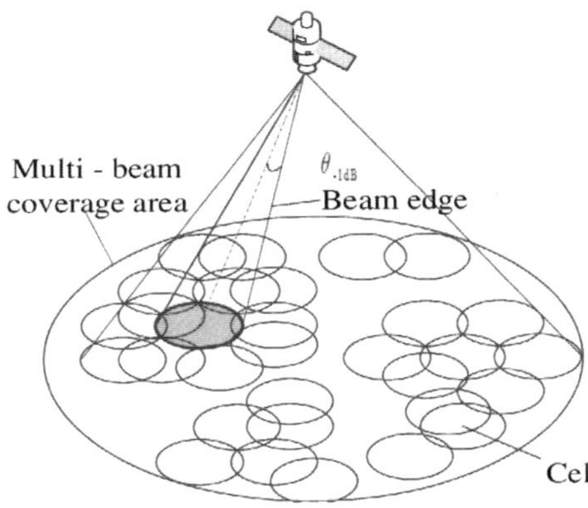

Figure 1. Application Model of Dynamic Multi – beamforming

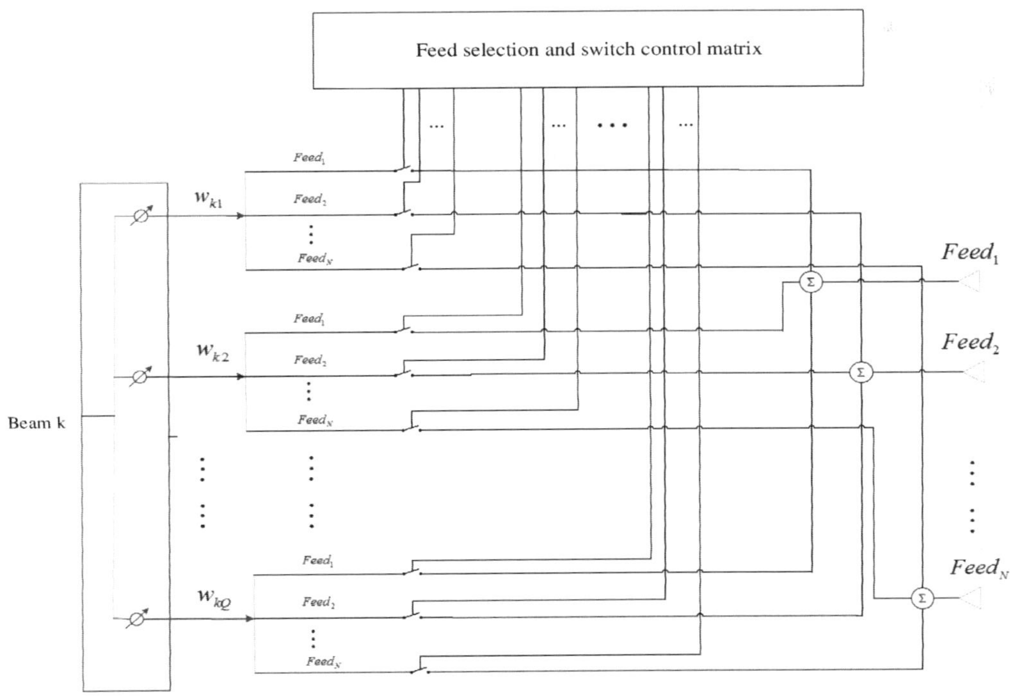

Figure. 2 Dynamic Beamforming Network Framework

The framework of dynamic beamforming network shows in Figure 2. If the beam k serves a user, it will select user's position as the center of the beam at first, then choses corresponding feed combination and its weighting coefficients. Finally form the beam by changing the switch matrix to open or close a feed branch.

The beam needs to follow the user's moving so a new feed combination and its corresponding weighting coefficient need to be calculated. Different with the method finding new feed combination to form fixed beam in traditional beamforming network, the scheme in this paper multiplexes the weighting coefficient network, namely, updates the weighting coefficient network of the original beam k and selects the new feed combination by switch matrix to make the center of the beam move arbitrarily.

3. Multi-reference Extended Variable Step-size LMS Algorithm

3.1 Variable Step-size LMS Algorithm

For the beamforming algorithm, the convergence performance of the traditional LMS algorithm makes great difference on the speed of the dynamic beamforming system to generate adaptive beam [1]. The early analysis about the convergence performance of LMS is studied by Widrow[2][3] and other researchers, they mainly make the theoretical derivation and verification on the convergence performance of the weight vector error. In paper [4] a variable step-size LMS algorithm based on sigmoid function is proposed which can effectively improve the convergence rate when ensuring the convergence.

$$\mu(n) = \beta\left(\frac{1}{1+\exp(-\alpha\,|e(n)|)} - \frac{1}{2}\right) \tag{1}$$

$e(n)$ is the multi-reference error vector, α and β are function parameters,

$$e(n) = d(n) - X^{T}(n)W(n) \tag{2}$$

X(n) and d(n) respectively represent the reference field strength of the current synthesized beam and the target beam.

In the n+1 times iteration, the weighting coefficients are calculated as follow,

$$W(n+1) = W(n) + 2\mu(n)e(n)X(n) \tag{3}$$

In this paper, the algorithm is extended to a multi -reference-point variable step - size dynamic beamforming algorithm which can improve the convergence performance and ensure the accuracy of beamforming at the same time.

3.2 Multi-reference Extended Variable Step-size LMS Algorithm

As a dynamic beamforming system, the requirement of synthetic beam quality is relatively high, especially the main-lobe width and side-lobe suppression. In order to improve the accuracy of beamforming, the variable step-size LMS beamforming algorithm proposed in the previous section is extended to an algorithm with more reference points and less error between the target beam and the actual synthesized beam. At the same time, considering the central position of the synthesized beam needs to follow the user, the beam must can track, so this algorithm is extended to a real-time algorithm the so-called dynamic beamforming algorithm [5].

First, the number of target beam reference points is K which covers the main-lobe and side-lobe. Em (Φ,Ψ,t) represents the secondary radiation field intensity of the synthetic beam m at time t, then we get

$$\hat{d} = [E_m(\Phi_1,\Psi_1,t) \quad E_m(\Phi_2,\Psi_2,t) \quad \ldots \quad E_m(\Phi_K,\Psi_K,t)] \tag{4}$$

Vector $\hat{d}$ represents sampling value of the target beam reference points, Φ,Ψ is the azimuth angle. In order to simplify the ratiocination, we set all feed weighting coefficient to be 1 at time t so that all reference point amplitude and phase can be expressed as

$$X(t) = \begin{pmatrix} \varepsilon_1(\Phi_1,\Psi_1,t) & \cdots & \varepsilon_1(\Phi_K,\Psi_K,t) \\ \vdots & \ddots & \vdots \\ \varepsilon_N(\Phi_1,\Psi_1,t) & \cdots & \varepsilon_N(\Phi_K,\Psi_K,t) \end{pmatrix} \tag{5}$$

After the weighting operation, the synthesized beam at reference point is

$$\hat{y}(t) = \hat{w}(t)X(t) \tag{6}$$

$\hat{w}(t) = [w_1(t), w_2(t), \ldots, w_N(t)]$ is the weighting coefficient vector of the feed array at time t, then the synthetic beam error is

$$\hat{e}(t) = \hat{d}(t) - \hat{y}(t) \tag{7}$$

Mean square error is

$$\hat{\xi}(t) = \left|\hat{e}(t)\right|^2 \tag{8}$$

Because of multiple reference points and each point has relative error with target beam, it is impossible to make all .the errors reach the minimum value at the same time. In this paper, use the synthetic beam optimization criteria which means that the optimization weighting coefficient vector needs to make the mean square error of all reference points minimum at time t.

$$\min\{\xi(t) = \sum_{k=1}^{K} \xi_k(t)\} \tag{9}$$

Among them

$$\xi_k(t) = \left|d_k(t) - \hat{w}(t)X(k,t,:)^T\right|^2 \tag{10}$$

When optimizing the weighting coefficients some reference point error may be relatively large, but the criterion can make sure the overall result so a few reference with large error almost make no difference on beamforming. At the same time, time t is extended to the algorithm which makes the algorithm satisfy the real-time requirement. According to the steepest descent theory, the fastest changing direction for the function is the direction of the gradient, which means the direction of the negative gradient is the fastest decreasing direction of the function. Therefore, in each iteration, the iteration coefficients change along the negative gradient to get the weighting coefficient finally. The weighting coefficients iteration formula can be expressed as

$$w_{k+1}(t) = w_k(t) + \mu(n)[-\nabla_w f(w_k(t))] \tag{11}$$

f(w(t)) is the function of w(t)， ∇w (f(wk (t))) represents the gradient of the function at wk (t). μ(n) is step size which will decrease when iteration increases. After several iterations, the steepest descent method reaches a steady state and the final weighting coefficients can get minimum mean square error. For its gradient

IOP Conf. Series: Journal of Physics: Conf. Series **910** (2017) 012063 doi:10.1088/1742-6596/910/1/012063

$$
\begin{aligned}
\nabla_{\hat{w}}\xi_k(\mathrm{t}) &= \nabla_{\hat{w}}\left|d_k(\mathrm{t}) - \hat{w}\ (\mathrm{t})X\ (k\ ,\mathrm{t},:)^T\right|^2 \\
&= \nabla_{\hat{w}}\{[d_k(\mathrm{t}) - \hat{w}\quad X\ k\quad{}^T\ d_k\quad - w\quad X\ k\quad{}^{T\,*} \\
&= \nabla_{\hat{w}}[\left|d_k(\mathrm{t})\right|^2 - d_k(\mathrm{t})X(k,\mathrm{t},:)^H\hat{w}^* \quad - w\quad X\ k\quad{}^T d_k^* \quad + w\quad X\ k\quad{}^T X(k,\mathrm{t},:)^H\hat{w}^*\ (\mathrm{t}) \\
&= -2[X(k,\mathrm{t},:)^T d_k^*(\mathrm{t}) - X(k,\mathrm{t},:)^T X(k,\mathrm{t}:)^H\hat{w}^*(\mathrm{t})] \\
&= -2X(k,\mathrm{t},:)^T[d_k(\mathrm{t}) - y_k(\mathrm{t})]^* \\
&= -2e_k^* X(k,\mathrm{t},:)^T
\end{aligned}
\tag{12}
$$

$$
\nabla_{\hat{w}}\xi(\mathrm{t}) = \nabla_{\hat{w}}\sum_{k=1}^{K}\xi_k(\mathrm{t}) = \sum_{k=1}^{K}\nabla_{\hat{w}}\xi_k(\mathrm{t}) = \sum_{k=1}^{K} -2e_k^* X(k,\mathrm{t},:)^T = -2\hat{e}^*(n)X(\mathrm{t})^T \tag{13}
$$

Finally get

$$
\hat{w}(n+1,t) = \hat{w}(n,t) + 2\mu(n)\hat{e}^*(\dot{n})X(\mathrm{t})^T \tag{14}
$$

The above equation is the multi-reference extended variable step-size LMS algorithm.

4. Simulation Results

In general, choose α=0.2, β=0.4 for the simulation to compare the convergence and synthesized beam performance between LMS algorithm with variable step-size and fixed step-size LMS.

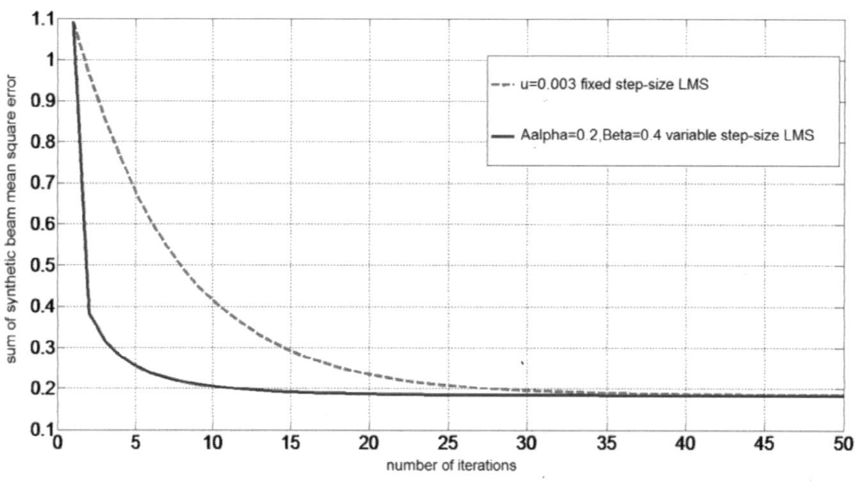

Figure 3. Convergence Performance Comparison of Variable Step-size and Fixed Step-size LMS Algorithm

In Figure 3, the fixed step-size LMS algorithm needs about 40 iterations to achieve mean square error stability, but the multi-reference extended variable step-size LMS algorithm only needs about 20 iterations. The beam convergence rate approximately doubled.

Then compare the beamforming performance between the fixed step-size LMS algorithm with the variable step-size LMS algorithm.

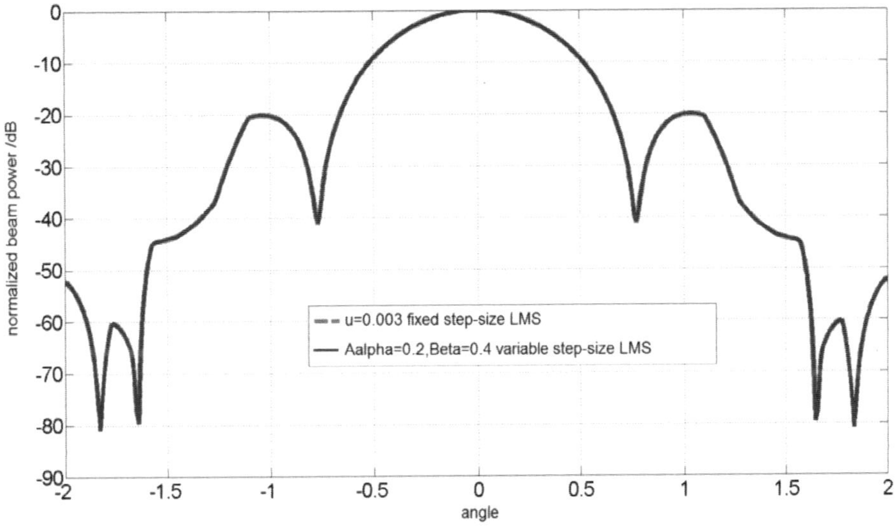

Figure 4. Comparison of Beamforming Performance of Variable Step-size and Fixed Step-size LMS

The simulation results show that the LMS algorithm with the fixed step-size u=0.003 and the variable step-size LMS algorithm with α=0.2, β=0.4 are almost identical in the performance beamforming. Their graph almost coincident and these two algorithm both can achieve the main-lobe width and side-lobe level suppression requirement of dynamic beamforming system. The error vector e(n) and corresponding value μ(n) is larger in the initial iteration but e(n) and μ(n) will decrease over each iteration which make the weighting coefficients more accurate and ensure the performance of the synthesized beam.

5. Conclusion

This paper proposes a multi-reference extended variable step-size LMS algorithm for dynamic beamforming network. The simulation result shows that the algorithm can effectively improve the convergence rate of weighting coefficients and ensure the beamforming performance at the same time. It can effectively track users and meet the requirements of dynamic beamforming.

6. Acknowledgment

This work is supported by the National Natural Science Foundation of China (No. 61371073). Corresponding author: Jianjun Wu; Email: just@pku.edu.cn.

7. References

[1] Zhou Jianying, research and implementation of smart antenna beam forming based on LMS algorithm (in Chinese). Harbin Institute of Technology, navigation, guidance and control, 2008, master's thesis.
[2] Widrow B and Walach E. On the statistical efficiency of the LMS algorithm with nonstationary input. IEEE Trans. Inform.Theory.1 984. IT-30(2): 211-229.
[3] Widrow B, McCool J M, Larimore M G, et a1. Stationary and nonstationary learning characteristics of the LMS adaptive filter. IEEE, 1976, 64(8): 1151-1162.
[4] Gao Ying, Xie Shengli. A variable step size LMS adaptive filtering algorithm and analysis (in Chinese) [C]// Chinese Control Conference. 2000:1094-1097
[5] Li Bin, Research on joint transmission and reception technology of multi beam satellite mobile communication (in Chinese). Peking University, communication and information system, 2013, master's thesis.

A Non-adjacent Feed Selection Method for Beamforming Performance Enhancement for Multi-beam Satellite Antenna

Yongjie Pu[a], Bing Wang, Yong Luo[b], Yuxin Cheng and Jianjun Wu

Institute of Advanced Communications, EECS, Peking University, Beijing, P. R. China

Email: [a]puyongjie@pku.edu.cn, [b]just@pku.edu.cn

Abstract. In recent years, user-centered beamforming systems have attracted increasing interest in the study of multi-beam satellite communication systems. In order to ensure service quality, the beam directivity of targets should have enough flexibility to bear the randomness of user locations. In this paper, a non-adjacent feed selection method is proposed to minimize the mean square error of beamforming and provide enough flexibility of the beam directivity at the same time.

1. Introduction

In recent years, satellite mobile communication has been widely applied in many areas due to its large coverage, long communication distance and flexible communication modes. Owing to the large territory of China, the terrestrial wireless communication system alone cannot guarantee effective Internet access or even telephone communication. In comparison, satellite mobile communication is not restricted by geographical constraints and thus provides good services for remote areas. Meanwhile, satellite mobile communication has played an irreplaceable role in many important fields such as earthquake relief and military defense [1]. Therefore, the development of satellite mobile communication has attracted increasing attention from the government.

The multi-beam satellite communication system was proposed in 1970s to enhance the signal quality in satellite mobile communication [2]. After years of rapid development, this technology is still facing a great challenge.

As beamforming is formed by the weighted superposition of single feed beams, it is inevitable that the Co-channel Interference (CI) exists. The seriousness of the CI increases with smaller central beam intervals, resulting in the decrease of channel quality. Therefore, the performance of beamforming, especially the main lobe bandwidth and the side lobe suppression, should be taken into consideration [3]. In addition, as the demand of users gets stricter, the single total-coverage beam cannot meet the requirements, leading to an increasing beam numbers. The increase of beam numbers results in additional beamforming network complexity, higher hardware resource consumption and more network cost. Meanwhile, more requirements on the performance of the beamforming algorithm are raised [4].

In order to improve the service quality of cell edge users, the concept of single beam for single user is proposed. The satellite communication not only aims to realize a complete coverage of the ground, but also provides higher service quality for the covered district. Researches focus on a user-centered service mechanism and analyze the adaptive phased array antenna technology. Former researchers have done comprehensive analyses on the concrete implementation of the beamforming network, using a most common feed selection method, the adjacent seven feeds [5]. However, the feed selection has not been well discussed, which also has obvious influence on the performance of beamforming. In this paper, a non-adjacent feed selection method is proposed and discussed [6].

The remainder of this paper is organized as follow. The system model is introduced in Section II. In Section III, the drawback of the traditional feed selection method is analyzed and a novel non-adjacent feed selection method is proposed. Simulation results of the performance of the proposed method are given in Section IV. Finally, Section V concludes the paper.

2. System Modle

In the multi-beam satellite communication system, enhanced multi-beamforming technology is often used to improve the gain of the beam and effectively control the shape of the beam. In this technology, the target beam is the weighted sum of several feeds. In the case of the regular feed array, the most common method is the adjacent seven feeds, which is depicted in Figure.1. It can not only ensure the performance of beamforming within a certain range, but also decrease the complexity of feed selection.

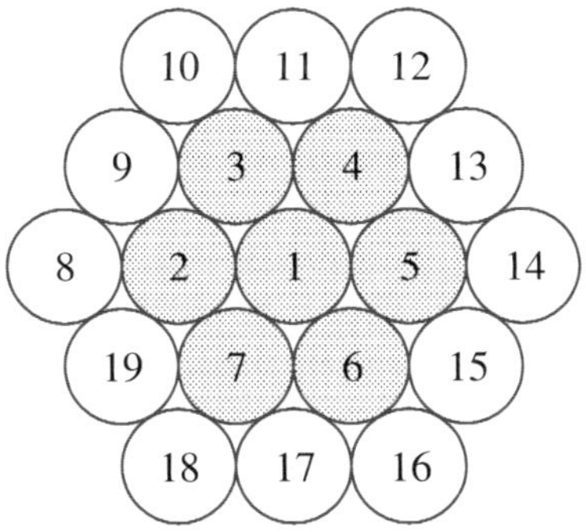

Figure.1. The method of adjacent seven feed sources

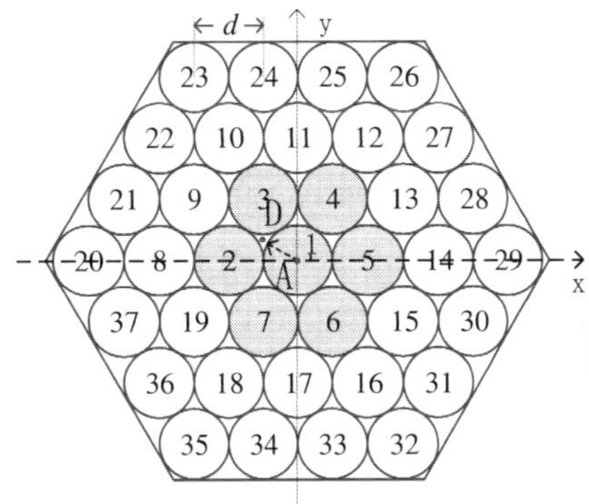

Figure.2. The pattern of the movement of the target beam

However, in the case of user-centered beamforming systems, the beam directivity of targets should have enough flexibility to bear the randomness of user locations. Due to the symmetry of feeds, the levels of the main lobe and side lobes are sometimes hard to control when the traditional adjacent feed selection is applied, leading to serious interface between beams and poor performance of the system.

Consider a case as follow. The pattern of the feed array is given in Figure.2, the same as the single feed direction pattern on the ground. The user starts at the origin point of the coordinates and moves in a direction of 150 degrees along the X-axis. Take a reference point at every 0.1 degree, as the center of the composite beam. Due to the symmetry of the feed array, only the case that the user moves from the origin point to the overlapped marginal location of beam 1, 2 and 3 should be considered. In this paper, four equal spacing points A, B, C and D are picked as the central location of the target beam.

3. The Feed Selection Method

3.1 The Traditional Adjacent Feed Selection Method

In this method, when the center of the beam locates between A and D, the No.1 feed source is the nearest one to the center of the beam according to the basic geometrical relationship, so it is reasonable to choose the No.1 feed source (as the center) and six feed sources (from No.2 to No.7) around it as a combination.

The performance of beamforming in different locations is depicted in Figure.3. Take D as an example. Feed source No.1, No.4, No.5, No.6 and No.7 have greater effect on the right side of the beam, while No.2 and No.3 on the left side. More feed sources mean better performance of beamforming. Thus the right side of the beam at D have lower side lobes while the left side higher. This kind of asymmetry can cause obvious interface between beams.

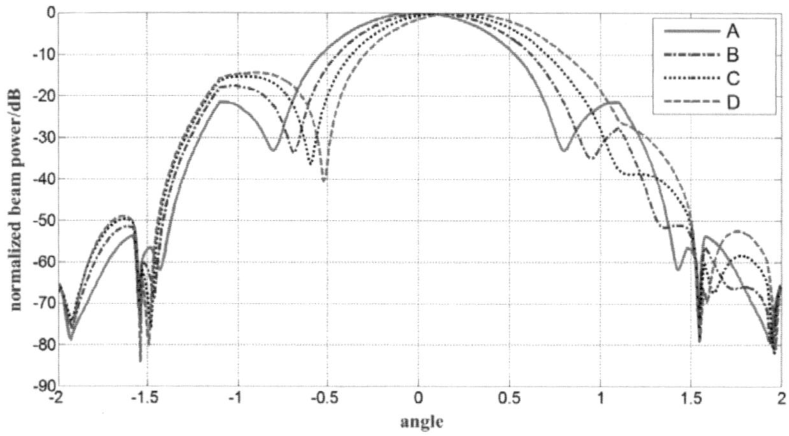

Figure.3.The pattern of the beamforming in different locations

3.2 Novel Non-adjacent Feed Selection Method

Take D as an example again. Assume the set including selected feed sources as ϕ, and then:

- Find the three nearest feed sources from the target beam, usually adjacent feed sources. For D, the No.1, No.2 and No.3 feed sources are selected. Include them into ϕ. They have the biggest effect on beamforming.
- There are nine adjacent feed sources around the three chosen feed sources. Their effect on beamforming is second to the latter. Choose N ($0<N<9$) feed sources from these nine adjacent feed sources and include them into ϕ as well to complete the feed selection.
- Two requirements on the chosen feed source set ϕ are shown as follow. First, the performance of beamforming should meet the requirements, including steering accuracy, -3dB bandwidth of the main lobe and the suppression of side lobe levels. Second, to decrease the complexity of beamforming network and keep amplifiers from nonlinear area, the number of feed sources should be as small as possible.

Therefore, the procedure of feed selection is shown as follow. First, set N=1 before the beamforming algorithm iterates through all possible feed source combinations and calculates the weight. Second, find the combination which minimizes the sum of mean square error. After that, judge whether the performance of beamforming meets the requirements. If so, the expected feed selection is found; otherwise, set N=N+1 and repeat the procedure above until the performance meets the requirements.

4. Simulation Results

Take D as an example. According to the simulation results, the level of side lobes cannot meet the requirements in any feed selection when the number of feed sources is less than seven. So consider the case where seven feed sources are chosen and simulate the performance of the proposed non-adjacent feed selection method.

After the iteration of all those combinations, the best combination is found, that is, No.1, No.2, No.3, No.4, No.6, No.9 and No.19, which is depicted in Figure.4 (a).

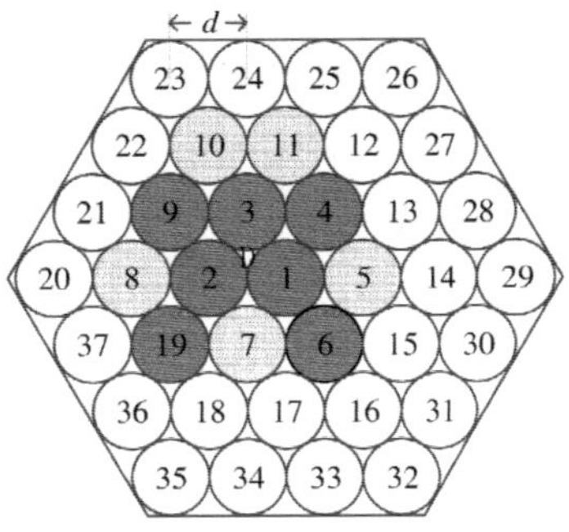

(a)

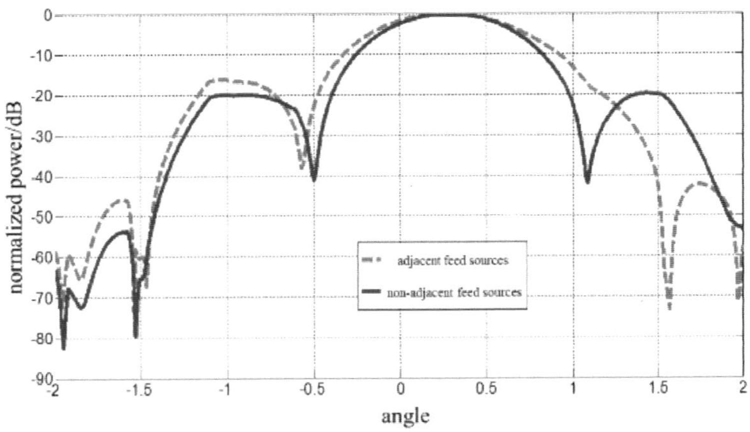

(b)

Figure 4.The simulation result at D. (a) The pattern of the combination of seven non-adjacent feed sources. (b) The pattern of beamforming

The pattern of beamforming in this case is depicted in Figure.4 (b). As in the graph, by applying the proposed method, the performance gets better on the suppression of side lobe levels, which is lower than -20dB. Consequently, the interface between beams can be decreased effectively. Meanwhile, the width of the main lobe is 0.6° which meets the requirements.

The reason for better performance is the symmetry of the combination of feed sources (including No.1, No.2, No.3, No.4, No.6, No.9 and No.19). In this case, the left and right side lobes suffer almost equal influence from these feed sources. Besides No.1, No.2 and No.3 feed sources, No.9 and No.19 have a greater effect on the left side of the beam, which solved the problem in the traditional method, namely the poor suppression of left side lobes.

Then, take C as another example and repeat the procedure. The simulation results are depicted in Figure.5 (a) and Figure.5 (b). As can be seen in Figure.5 (b), compared with the traditional method, the non-adjacent feed selection method can suppress the left side lobes more effectively, leading to a better waveform symmetry. Besides, the requirements of the system are all met, which shows the effectiveness of the new method.

As for the number of feed sources, the seven feeds selection already meets the requirements. Although extra feed sources can further improve the accuracy of beamforming, the change is less meaningful, and it will not be discussed in the paper.

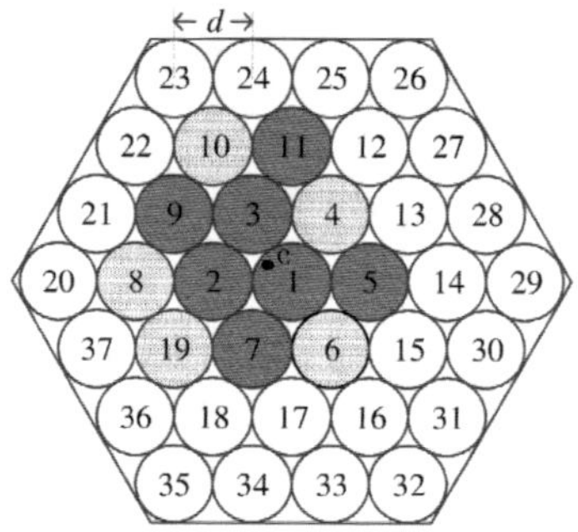

(a)

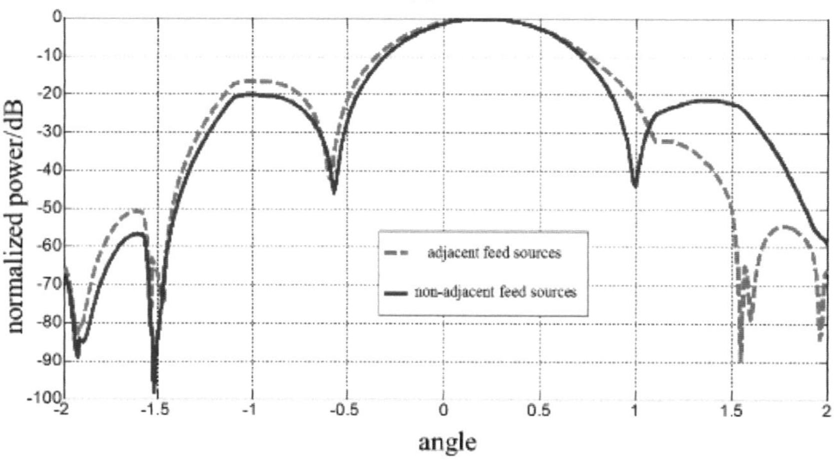

(b)

Figure.5.The simulation result at C. (a) The pattern of the combination of seven non-adjacent feed sources. (b) The pattern of beamforming

5. Conclusion

In this paper, a novel non-adjacent feed selection method is proposed to improve the performance based on traditional adjacent feed selection method. The proposed method decreases the level of side lodes and relieves the interface between beams, resulting in more coverage and beam directivity flexibility while retaining the quality of the beamforming. Results show that the performance of the proposed method meets the requirements of user-centered beamforming systems.

6. Acknowledgment

This paper is supported by the National Natural Science Foundation of China (No.61371073). Corresponding author: Jianjun Wu; Email: just@pku.edu.cn.

7. References

[1] G. W. Beakley. Overview of Commercial Satellite Communications, in IEEE Transactions on Aerospace and Electronic Systems, vol. AES-20, no. 4, pp. 455-464, July 1984.
[2] Hongfen Hu, Xiaohong Bai. Application and Development of Multi-beam Antenna (in Chinese)[J]. Modern Radar,2002,24(4);69-75.
[3] Shaohui Sun, Yingmin Wang, Rakesh Tamrakar, Yongbin Xie. Enhanced Beam-forming Technique[J]. China Communications, 2009,6(2):144-149.
[4] Yang Wang, Danfeng Zhao, Xi Liao. Low Complexity Multiuser Detection Algorithm for Multi-Beam Satellite Systems[J]. Journal of Harbin Institute of Technology,2015,22(5):105-109.

[5] Xinsheng He, Jianfeng Zhu, Baoming Li. Research of Beamforming with Multi-beam Antenna Equiped on GEO Satellite (in Chinese) [J]. Communication Countermeasures,2014,33(1);31-35.
[6] Bin Li. Research on Joint Transmitting and Receiving Technology of Multi - beam Satellite Mobile Communication (in Chinese) [M]. Peking University, Communication and Information System, 2013, Master Thesis.

An Efficient Authenticated Key Transfer Scheme in Client-Server Networks

Runhua Shi and Shun Zhang*

School of Computer Science and Technology, Anhui University, Hefei City, China
Email: shirh@ahu.edu.com, *shzhang27@163.com

Abstract. In this paper, we presented a novel authenticated key transfer scheme in client-server networks, which can achieve two secure goals of remote user authentication and the session key establishment between the remote user and the server. Especially, the proposed scheme can subtly provide two fully different authentications: identity-base authentication and anonymous authentication, while the remote user only holds a private key. Furthermore, our scheme only needs to transmit 1-round messages from the remote user to the server, thus it is very efficient in communication complexity. In addition, the most time-consuming computation in our scheme is elliptic curve scalar point multiplication, so it is also feasible even for mobile devices.

1. Introduction

Secure remote user authentication over insecure channel is an important issue for many distributed applications, especially electronic transactions (e.g., on-line shopping, Internet banking and pay-TV). Furthermore, in order to ensure the session's confidentiality, it needs to establish a secure session key between the remote user and the server over insecure communication channel. Therefore, remote user authentication with key establishment in client-server networks is becoming the focus of widely attentions.

Currently, there appeared lots of novel authenticated key transfer schemes [1-12] in client-server networks. According to the way of authentications among these schemes, there are two well-known but fully different categories: identity-based authentication and anonymous authentication.

With our best knowledge, there is not a scheme that can meet both two different authentication requests, i.e., identity-based authentication and anonymous authentication. However, in some real applications, it needs to provide different authentications according to the different conditions. For example, suppose there is a client-server system, which is composed of a server and multiple remote users. The server manages an important storage space (or database), which is shared (or created) by all legal users. When a remote user wants to upload files (or data) to the storage space (or database) of the server, the server must authenticate the identity of the user to avoid false or illegal files (or data). However, when he downloads files (or data) from the server, the user expects to guarantee his anonymity. In fact, there are lots of similar application scenarios, what's more, e.g., in some private forums, it needs the registered user to publish the information in the real-name way, but it is anonymous to browse the information.

Motivated by these concerns, in this paper, we present a novel authenticated key transfer scheme for client-server networks, which can satisfy both two different authentications under the condition of the user only holding a private key in secret. Obviously, this scheme can not only save system resources, but can also help the server to more efficiently manage the users' private keys.

IOP Conf. Series: Journal of Physics: Conf. Series **910** (2017) 012065 doi:10.1088/1742-6596/910/1/012065

1.1 Related work
In 2009, to achieve two secure goals of remote user authentication and key establishment between the remote user and the server in client-server networks, Yang and Chang [1] proposed an efficient identity-based and authenticated key agreement protocol. However, Yoon and Yoo [2] demonstrated that Yang and Chang's protocol is vulnerable to the impersonation attack and does not provide perfect forward secrecy, and then proposed an improved protocol. Later, He *et al.* [3] again confirmed that Yoon and Yoo's protocol does not provide perfect forward secrecy yet and fails to achieve forward secrecy. In addition, they also pointed out that a special hash function called MapToPoint function which is used to map an identity information into a point on elliptic curve is required in the previous protocols. To improve the efficiency, they presented a new remote user authentication protocol without the MapToPoint function [3]. Subsequently, there appeared more improved protocols of authenticated key establishment for client-server networks [4-7].

The authentications in the protocols mentioned above are all identity-based (i.e., identifiable), which cannot protect user's privacy. For this, Islam and Biswas [8] proposed an anonymous authentication protocol with key establishment, but only against outside attackers. In 2014, Kumari *et al.* [9] proposed a password-based anonymous authentication scheme similarly against the outside adversary. However, Chaudhry *et al.* [10] analyzed that Kumari *et al.*'s scheme is vulnerable to anonymity violation attack, and further proposed an improved anonymous authentication scheme. Later, Liu *et al.* [11] presented certificateless remote anonymous authentication schemes against outside attackers for wireless body area networks. Recently, Shi *et al.* [12] presented a novel anonymous and authenticated key transfer protocol in client-server networks, which perfectly protects the remote user's anonymity against not only outside attackers, but also other legal users and the server. In addition, for secure roaming services in wireless networks, Yang *et al.* [13] presented two efficient anonymous roaming protocols, and defined the notion of Strong User Anonymity that protects user's identity against both eavesdroppers and foreign servers. In the same year, Chang and Tsai [14] presented an anonymous and self-verified mobile authentication scheme with key agreement in large-scale wireless networks. Subsequently, similar anonymous authentication schemes for secure roaming services were proposed [15-18].

1.2 Contributions
In this paper, we present a novel authenticated key transfer scheme for client-server networks. Compared with existing authentication key establishment schemes in client-server networks, our scheme has the following advantages:

(1) In the proposed scheme, by using the same private key, the user can complete two fully different kinds of authentications: identity-based authentication and anonymous authentication.

(2) In order to achieve two secure goals: user authentication and session key establishment between the remote user and the server, the proposed scheme only needs to send 1-round messages. Thus, it is very efficient in communication complexity.

(3) In the proposed scheme, the most time-consuming operation is elliptic curve scalar point multiplication, so it is feasible even for mobile devices.

(4) In the proposed scheme, the server does not need to store the private keys of all authorized users. Thus, it reduces the risk of the leakage of these keys.

(5) The proposed scheme can provide the confirmation of the user's private key.

2. Proposed Scheme
The proposed scheme mainly includes three phases: Initialization, User Registration, and Authentication with Key Transfer, where the last phase includes two different kinds of authentications: identity-based authentication and anonymous authentication. The detailed description is as follows:

2.1 Initialization
The server generates the system parameters as follows: (1) The server chooses an elliptic curve equation $\boldsymbol{E}_p(a,b)$ defined on finite field F_p [19-22], where p is a large prime. (2) The server selects a base point $\boldsymbol{P}$ with the prime order q over $\boldsymbol{E}_p(a,b)$. (3) The server randomly generates its master key

$k_s \in_R \mathbb{Z}_q^*$ and computes the corresponding public key $\boldsymbol{Q_s} = k_s\boldsymbol{P}$. (4) In addition, the server chooses a secure hash function, $H:\{0,1\}^* \times \{0,1\}^* \times \{0,1\}^* \to \mathbb{Z}_q^*$. (5) Finally, the server publishes these system parameters: $\{p, \boldsymbol{E_p}(a,b), q, \boldsymbol{P}, \ \boldsymbol{Q_s}, H(\cdot)\}$.

2.2 User Registration

Step1. Each user u_i $(1 \le i \le n)$ sends his identity, ID_i, to the server. Then the server checks the authenticity and legality of each identity. After confirming the authenticity and legality of the user u_i, the server prepares a smart card including a secret k_i, where $k_i = H(ID_i, ID_s, k_s)$. Here ID_i and ID_s denote the identities of the user u_i and the server, respectively, and k_s is the master key of the server. Then the server privately sends the smart card with the secret k_i to the user u_i.

Step 2. Suppose that in all there are n authorized users who have each obtained their respective secret k_is. The server constructs a polynomial function $f(x)$ by all secrets, k_is, as follows:

$$f(x) = (x - k_1)(x - k_2)\cdots \ (x - k_m) \bmod q \tag{1}$$

The above function can be rewritten as the following expression,

$$f(x) = (x^n + c_{n-1}x^{n-1} + \cdots + c_1x^1 + c_0) \bmod q \tag{2}$$

Furthermore, the server computes and publishes: $\{c_0\boldsymbol{P}, c_1\boldsymbol{P}, c_2\boldsymbol{P}, \ldots, c_{n-1}\boldsymbol{P}\}$ and $\{c_1\boldsymbol{Q_s}, c_2\boldsymbol{Q_s}, \ldots, c_{n-1}\boldsymbol{Q_s}\}$ on his bulletin board. In addition, the server announces two lists of the authorized users and the revoked users. Finally, the server deletes all k_is and c_is (for $i = 1$ *to* n), while he only privately holds the master key, k_s, and the coefficient of the constant term of $f(x)$, c_0, which can be stored in a smart card for security considerations.

Step 3. By the public information of the server, each user u_i verifies the validity of the secret in his smart card as follows.

Compute:

$$Ki = k_i^n P + k_i^{n-1}(c_{n-1}P) + \cdots + k_i^2(c_2P) + k_i^1(c_1P) \tag{3}$$

Verify:

$$Ki = -(c_0P) \tag{4}$$

Otherwise, he broadcasts a complaint about the dishonesty of the server to the whole client-server networks. After a waiting period, if he does not receive any complaint about the server, then the user u_i believes that the server is honest.

2.3 Identity-based Authentication with Key Transfer

Step 1. Suppose that the i-th user u_i requests the server to verify his legality and expects to share a session key between him and the server. The user u_i inserts the smart card into the terminal device, reads out his private key k_i, and further computes $\boldsymbol{U}$, V, w and k by the following equations:

$$\begin{aligned} U &= rP \\ (Q_x, Q_y) &= rk_iQ_s \\ V &= S \oplus Q_x \\ w &= H(S, t, ID_i) \\ k &= H(S, t, 1) \end{aligned} \tag{5}$$

Where $r \in_R \mathbb{Z}_q^*$, $S \in_R \mathbb{Z}_p^*$, Q_x and Q_y denote the x and y coordinates of the point $rk_i\boldsymbol{Q_s}$, t is a timestamp that denotes the current time, and k is a session key for later communications between the user u_i and the server. Furthermore, the user u_i sends the messages $\{ID_i, \boldsymbol{U}, V, w, t\}$ to the server. Finally the user u_i takes out the smart card and deletes r and S from the memory.

Step 2. After receiving the messages $\{ID_i, \boldsymbol{U}, V, w, t\}$ from the user u_i, the server verifies if t is valid. If t is not fresh, the server aborts the process; otherwise, he performs the next step.

Step 3. The server computes:

$$\begin{aligned} k_i &= H(ID_i, ID_s, k_s) \\ (Q'_x, Q'_y) &= k_i k_s U \\ S' &= V \oplus Q'_x \end{aligned} \tag{6}$$

Then he verifies if $w = H(S', t, ID_i)$ holds. If the equation holds, the verification passes successfully. Furthermore, the server computes the session key $k = H(S', t, 1)$. Otherwise, the verification fails.

2.4 Anonymous Authentication with Key Transfer

Step 1. Suppose that the i-th user u_i requests the server to verify his legality and expects to share a session key between him and the server. Then the user u_i inserts the smart card into the terminal device, reads out his private key k_i, and further computes $\boldsymbol{Q}_i$ by the server's public information and his private key k_i:

$$Q_i = k_i^n Q_s + k_i^{n-1}(c_{n-1} Q_s) + \cdots + k_i^2(c_2 Q_s) + k_i^1(c_1 Q_s) \tag{7}$$

Further he computes:

$$\begin{aligned} U &= rP \\ (Q_x, Q_y) &= rQ_i \\ V &= S \oplus Q_x \\ w &= H(S, t, 0) \\ k &= H(S, t, 1) \end{aligned} \tag{8}$$

Where $r \in_R \mathbb{Z}_q^*$, $S \in_R \mathbb{Z}_p^*$, t is a timestamp that denotes the current time, and k is a session key for future communications between the user u_i and the server. Furthermore, the user u_i sends the messages $\{\boldsymbol{U}, V, w, t\}$ to the server. Finally the user u_i takes out the smart card, and deletes r, S and $\boldsymbol{Q}_i$ from the memory.

Step 2. After receiving the messages $\{\boldsymbol{U}, V, w, t\}$ from the user u_i, the server verifies if t is valid. If t is not fresh, the server aborts the process; otherwise, he performs the next step.

Step 3. The server computes and further verifies,

$$\begin{aligned} (Q'_x, Q'_y) &= -c_0 k_s U \\ S' &= V \oplus Q'_x \\ w &= H(S', t, 0) \end{aligned} \tag{9}$$

If the above equation holds, the verification passes successfully. That is, the server confirms that the initiator of this session (i.e., the user u_i) is an authorized user. Finally, the server computes the session key $k = H(S', t, 1)$. Otherwise, the verification fails.

3. Analysis

The security of our scheme mainly relies on the difficulties of solving Elliptic Curve Discrete Logarithm (ECDL) problem (Given two points $\boldsymbol{P}$ and $\boldsymbol{Q}$ over an elliptic curve $\boldsymbol{E}_p(a,b)$, it is computationally infeasible to find an integer k such that $\boldsymbol{Q} = k \cdot \boldsymbol{P}$) and Elliptic Curve Computational Diffie-Hellman (ECCDH) problem (Given three points $\boldsymbol{P}$, $a \cdot \boldsymbol{P}$ and $b \cdot \boldsymbol{P}$ over $\boldsymbol{E}_p(a,b)$, it is computationally infeasible to compute a point $\boldsymbol{W}$ such that $\boldsymbol{W} = ab \cdot \boldsymbol{P}$) [5]. And then, we give Performance comparisons of some related schemes.

3.1 Security Analysis

In this section, we analyze that the proposed scheme can withstand various related security attacks.

3.1.1 Outside Attack. Assume that an outside attacker wants to compute k_i from the public information $\{c_0\boldsymbol{P}, c_1\boldsymbol{P}, c_2\boldsymbol{P}, \ldots, c_{n-1}\boldsymbol{P}\}$ or $\{c_1\boldsymbol{Q}_s, c_2\boldsymbol{Q}_s, \ldots, c_{n-1}\boldsymbol{Q}_s\}$ of the server. On the one hand, based on the difficulty of solving ECDL problem, the attacker can not obtain the corresponding coefficients $c_0, c_1, \ldots, c_{n-1}$, and then directly get k_i by solving the linear equation of $f(x) = 0 mod q$. On the other hand, if the attacker first guesses k and directly computes $f(k)\boldsymbol{P}$ by the public information $\{c_0\boldsymbol{P}, c_1\boldsymbol{P}, c_2\boldsymbol{P}, \ldots, c_{n-1}\boldsymbol{P}\}$, then he can verify if k is a valid secret, that is, he can verify if $f(k)\boldsymbol{P} = \boldsymbol{O}$ holds, where $\boldsymbol{O}$ denotes a point at infinity (i.e., the identity element of the group of points on elliptic curve). But the successful probability of guessing a right secret is not more than $\frac{n}{q}$ (n is far less than q). Thus, it is infeasible to compute k_i only by the public information $\{c_0\boldsymbol{P}, c_1\boldsymbol{P}, c_2\boldsymbol{P}, \ldots, c_{n-1}\boldsymbol{P}\}$. In addition, it is impossible for the attacker to get k_s from the server's public key $\boldsymbol{Q}_s$ based on the difficulty of solving ECDL problem.

Furthermore, suppose that the outside attacker wants to compute the session key by the transmitted messages $\{ID_i, \boldsymbol{U}, V, w, t\}$ (or $\{\boldsymbol{U}, V, w, t\}$) from the remote user to the server. By Eq. (5) (or Eq.(8)), obviously the attacker must first extracts S from $\boldsymbol{U}$ and V, and then can compute $k = H(S, t, 1)$. But, for the attacker, there are three unknowns: r, S and k_i (or $\boldsymbol{Q}_i$) in two equations of respectively computing $\boldsymbol{U}$ and V. Here, we assume that $H(\cdot)$ is a secure hash function. Thus, the attacker cannot get any secret information about S only by the transmitted messages $\{ID_i, \boldsymbol{U}, V, w, t\}$ (or $\{\boldsymbol{U}, V, w, t\}$), and he cannot obtain the session key without S obviously.

Therefore, outside attack is infeasible for the proposed scheme.

3.1.2 Replay attack. In the proposed scheme, the server can verify the freshness of the transmitted messages by the freshness of the timestamp t. Furthermore, the timestamp t is embedded in the hashed message $w = H(S, t, ID_i)$(or $w = H(S, t, 0)$) by the remote user, such that it can guarantee the integrity of the timestamp. Therefore, the proposed scheme can resist replay attack.

3.1.3 Impersonation Attack. Assume that an attacker wants to impersonate a legal user u_i. In order to pass the verification of the server, the attacker has to construct a valid triple $\{\boldsymbol{U}', V', S'\}$, such that the equation of $S' = V' \oplus F_x(k_s k_i \boldsymbol{U}')$ (or $S' = V' \oplus F_x(-c_0 k_s \boldsymbol{U}')$) holds, where $F_x(\boldsymbol{Q})$ denotes the x coordinate of the point $\boldsymbol{Q}$. However, without knowing the secret k_i (or c_0), it is impossible for the attacker to rightly generate the valid triple $\{\boldsymbol{U}', V', S'\}$.

Furthermore, if the attacker intercepts the session messages $\{ID_i, \boldsymbol{U}, V, w, t\}$ (or $\{\boldsymbol{U}, V, w, t\}$) and gets S rightly from w, he can impersonate a legal user to perform an attack as follows: The attacker first computes $F_x(rk_i\boldsymbol{Q}_s) = V \oplus S$ (or $F_x(r\boldsymbol{Q}_i) = V \oplus S$), and then he generates and sends new messages $\{ID_i, \boldsymbol{U}, V_{new}, w_{new}, t_{new}\}$ (or $\{\boldsymbol{U}, V_{new}, w_{new}, t_{new}\}$) to the server, where $V_{new} = S_{new} \oplus (V \oplus S)$, $w_{new} = H(S_{new}, t_{new}, ID_i)$ (or $w_{new} = H(S_{new}, t_{new}, 0)$), S_{new} is random integer, and t_{new} is a new timestamp. Obviously, the fake messages $\{ID_i, \boldsymbol{U}, V_{new}, w_{new}, t_{new}\}$ (or $\{\boldsymbol{U}, V_{new}, w_{new}, t_{new}\}$) can also pass the verification of the server. However, due to the one-way property of hash function, an outsider cannot extract S from $w = H(S, t, ID_i)$ (or $w = H(S, t, 0)$).

In addition, obviously any authorized user cannot impersonate the server to obtain the resulting session key of other authorized user, because it requires the server's master key to compute the session key in Eqs. (6) and (9).

Therefore, the proposed scheme can resist impersonation attack.

3.1.4 Known-key Attack. In our scheme, the session key $k = H(S, t, 1)$ is computed by using a secure hash function. Obviously, the *session* key only depends on the short-term secret S, instead of the long-term secret k_i. Furthermore, each session has different short-term secret S, and S is randomly generated in Z_p^*. Thus, the current session key is independent of the previous session. That is, an outsider cannot compute the current session key even he knows some previous session keys. Therefore, Known-key attack is infeasible for the proposed scheme.

3.1.5 Perfect Forward Secrecy. Perfect forward secrecy means that if a long-term private key is compromised, this does not compromise any earlier session keys. Here, we suppose that the user u_i's secret, k_i, is *compromised* to a attacker, and the attacker further gets the earlier messages $\{ID_i, \boldsymbol{U}, V, w, t\}$ (or $\{\boldsymbol{U}, V, w, t\}$), which was transmitted from the user u_i to the server before his secret is compromised. For this attack, we will analysis in the following two cases.

Case 1 for identity-based authentication with key transfer: Though the attacker gets the user's secret k_i, the attacker does not know the server's secret k_s. Even given $r\boldsymbol{P}$ (i.e., $\boldsymbol{U}$) and $k_ik_s\boldsymbol{P}$ (where $\boldsymbol{Q}_s = k_s\boldsymbol{P}$ is public and k_i is revealed), it is still computationally hard for the attacker to compute $rk_ik_s\boldsymbol{P}$ based on the difficulty of solving ECCDH Problem. Accordingly, it is impossible for him to rightly compute S and then obtain the session key k, since $S = V \oplus F_x(rk_ik_s\boldsymbol{P})$ and $k = H(S, t, 1)$.

Case 2 for anonymous authentication with key transfer: On the one hand, the attacker does not known the server's secrets k_s and c_0, thus he cannot rightly compute $c_0k_s\boldsymbol{U}$. On the other hand, even given $r\boldsymbol{P}$ (i.e., $\boldsymbol{U}$) and $\boldsymbol{Q}_i$ (suppose that $\boldsymbol{Q}_i = y\boldsymbol{P}$, $y \in \mathbb{Z}_q^*$), it is still computationally hard for him to compute $r\boldsymbol{Q}_i$ (i.e., $ry\boldsymbol{P}$) based on the difficulty of solving ECCDH Problem. Without knowing $c_0k_s\boldsymbol{U}$ or $r\boldsymbol{Q}_i$, obviously the attacker cannot get the session key $k = H(S, t, 1)$, where $S = V \oplus F_x(r\boldsymbol{Q}_i)$ or $S = V \oplus F_x(-c_0k_s\boldsymbol{U})$.

Therefore, the proposed protocol can provide perfect forward secrecy.

3.1.6 User Authentication and Anonymity. We analyze two cases of "Identity-based Authentication with Key Transfer" and "Anonymous Authentication with Key Transfer", respectively.

In the case of Identity-based Authentication with Key Transfer, the user's identity is embedded in the hashed message $w = H(S, t, ID_i)$. Thus, any attacker cannot modify the identity ID_i of the sender in the transmitted messages, because the hashed message can provide its integrity. Furthermore, the server can check the legality of the sender by verifying if he knows the secret k_i, which is shared only between the user u_i and the server. Therefore, this protocol can guarantee remote user authentication.

In the case of Anonymous Authentication with Key Transfer, on the one hand, for a legal user u_i, we can easy infer that the following equation holds,

$$\begin{aligned} rQ_i &= r[k_i^n Q_s + k_i^{n-1}(c_{n-1}Q_s)\cdots + k_i^2(c_2Q_s) + k_i^1(c_1Q_s)] \\ &= r[k_i^n + k_i^{n-1}(c_{n-1}) + k_i^{n-2}(c_{n-2})\cdots + k_i^1(c_1)]Q_s \\ &= r[k_i^n + k_i^{n-1}c_{n-1} + k_i^{n-2}c_{n-2}\cdots + k_i^1c_1]Q_s \\ &= r[(f(k_i) - c_0) \bmod q]Q_s \\ &= -rc_0Q_s\,(f(k_i) = 0 \bmod q) \\ &= rc_0k_sP \\ &= -c_0k_s(rP) \\ &= -c_0k_sU \end{aligned} \tag{10}$$

Thus, the server can rightly compute $S = V \oplus F_x(-c_0k_s\boldsymbol{U})$ and then verify the legality of the user. However, for an outside attacker, obviously he cannot forge a fake $\boldsymbol{Q}_i^*$, such that $r\boldsymbol{Q}_i^* = -c_0k_s\boldsymbol{U}$. Thus, the proposed protocol can provide user authentication. On the other hand, for the remote user u_i, obviously there exists the equation of $\boldsymbol{Q}_i = -c_0\boldsymbol{Q}_s$ (Please note that the server does not publish $c_0\boldsymbol{Q}_s$).

That is, all Q_is are equal, which can guarantee the unconditional anonymity of the remote user against outsiders, other authorized users and the server.

3.2 Security and Performance Comparisons

We have analyzed the security of the proposed scheme in the above section. Furthermore, we give security comparisons of our proposed scheme and other related works, as shown in Table 1. In addition, we evaluate the performance of our proposed scheme in terms of the computation and communication costs, and list performance comparisons in Table 2. In Table 1, partial anonymity denotes that it is anonymous only for outsiders, but not for the server; while unconditional anonymity denotes that it is anonymous for anyone, including the server. In Table 2, same as References [5, 7] we assume the timestamp length is 16-bit, the size of p used in the ECC of the scheme is 160-bit, the digest message size of hash function (e.g., SHA-1) is 160-bit, and the identity size is 80-bit. In addition, off-line implies that the computation can be precomputed in advance (e.g., Eq. (7)).

Table 1 Security Comparisons

Security properties	**Yang *et al.*'s scheme [1]**	**Isam *et al.*'s scheme [8]**	**Liu *et al.*'s scheme [11]**	**Shi *et al.*'s scheme [12]**	**Proposed scheme**
Private key confirmation	Not Provided	Not Provided	Provided	Not Provided	Provided
Outside attack	yes	yes	yes	yes	yes
Replay attack	yes	yes	yes	yes	yes
Impersonation attack	no	yes	yes	yes	yes
Known-key attack	yes	yes	yes	yes	yes
Forward secrecy	no	yes	yes	no	yes
User authentication	yes	yes	yes	yes	yes
User anonymity	no	partial	partial	unconditional	unconditional

Table 2. Performance Comparisons

Schemes	**Communication costs**		**Computation costs**		
	Round	**Message size**	**User (on-line)**	**User (off-line)**	**Server**
Yang *et al.*'s scheme [1]	2	1392 bits	4EM+2EA+4H	-	4EM+2EA+4H
Isam *et al.*'s scheme [8]	3	1440 bits	3EM+2EA+4H	-	4EM+2EA+5H
Liu *et al.*'s scheme [11]	2	1296 bits	4EM+2EA+1P+2H+1MAC	-	1EM+1EA+2P+2H+1MAC
Shi *et al.*'s scheme [12]	2	(n+2)160 bits	nM+1H	-	(n+1)mM+1H
Proposed scheme – (Identity-base)	1	736 bits	2EM+2H	-	1EM+2H
Proposed scheme – (Anonymous)	1	656 bits	2EM+2H	nEM+(n-1)EA	1EM+2H

Note: EM, EA, P, H, and M denote the time complexity of Elliptic Curve Scalar Point Multiplication, Elliptic Curve Point Addition, Bilinear Pairing, Hash Function, and Scalar Multiplication, respectively.

According to Table 1 and Table 2, clearly the proposed scheme has the following advantages:

(1) The proposed scheme can withstand all related security attacks.

(2) The proposed scheme can provide the confirmation of the user's private key.

(3) The proposed scheme needs lower communication costs and computation costs, thus it is more efficient in both communication and computation complexity.

(4) The proposed scheme can provide unconditional anonymity for the remote user in client-server networks.

(5) The proposed scheme can subtly provide two fully different kinds of authentications: identity-based authentication and anonymous authentication.

Furthermore, when the server in our scheme wants to revoke a user, he first adds the identity of the revoked user to the revoked user list and then renews the public information, while the secrets of other authorized users remain unchanged. Similarly, when the server authorizes a new user, he first generates a secret key for the new user, adds the identity of the new user to the authorized user list, and then renews the public information, while the secrets of other authorized users remain unchanged. In actual applications, the server may update the public information in batch way, i.e., after a certain period of time, but not whenever new user registering to the server.

In addition, in the proposed scheme, we focus on the remote user authentication, not mutual authentication between the remote user and the server. If achieving mutual authentication, it only requires the server to return $MAC_k(w)$ as the response, and then the remote user checks the integrity of $MAC_k(w)$, where $MAC_k(w)$ denotes Message Authentication Code (MAC) of the message w by using the session key k.

Finally, we consider the extendibility of the proposed scheme. In order to increase the number of the remote users, the proposed scheme can be easily and naturally extended into the multi-level architecture (server – sub-server – user), where the authentication of each level can be implemented in the proposed method.

4. Conclusion

In this paper, we proposed a novel authenticated key transfer scheme to achieve two secure goals: remote user authentication and key establishment between the user and the server. Especially, for the different secure requests in some practical applications (e.g., uploading/downloading files to/from the private forum), the proposed protocol can subtly provide two different kinds of authentications: identity-based authentication and anonymous authentication, while the remote user only holds a single secret, which is written into a smart card for the security considerations. In addition, our scheme only costs 1-round messages transmitted from the remote to the server, few elliptic curve scalar point multiplications and hash functions, thus it is very efficient in the communication and computation complexity.

5. Acknowledgment

This work was supported by National Natural Science Foundation of China (No.61772001), and Talents Youth Fund of Anhui Province Universities (No 2013SQRL006ZD1).

6. References

[1] J.H. Yang, C.C. Chang, "An ID-based remote mutual authentication with key agreement scheme for mobile devices on elliptic curve cryptosystem," Computers & Security, vol. 28, pp. 138-143, May 2009.

[2] E.J. Yoon, K.Y. Yoo, "Robust ID-based remote mutual authentication with key agreement protocol for mobile devices on ECC," Proceeding of 2009 international conference on computational science and engineering, IEEE Computer Society Washington, DC, USA, Aug 2009, Vol. 02, pp 633–640.

[3] D.B. He, J.H. Chen, J. Hu, "An ID-based client authentication with key agreement protocol for mobile client–server environment on ECC with provable security," Information Fusion, vol. 13, pp. 223–230, Jul 2012.

[4] E.J. Yoon, S.B. Choi and K.Y. Yoo, "A Secure and Efficiency Id-based Authenticated Key Agreement Scheme based on Elliptic Curve Cryptosystem For Mobile Devices," International Journal of Innovative Computing, Information and Control, vol. 8, pp. 2637-2653, Apr 2012.

[5] C.H. Chou, K.Y. Tsai, C.F. Lu, "Two ID-based authenticated schemes with key agreement for mobile environments," Journal of Supercomputing, vol. 66, pp. 973–988, Nov 2013.
[6] D. Wang, C.G. Ma, "Cryptanalysis of a remote user authentication scheme for mobile client-server environment based on ECC," Information Fusion, vol. 14, pp. 498-503, Oct 2013.
[7] M.S. Farash, M.A. Attari, "A secure and efficient identity-based authenticated key exchange protocol for mobile client-server networks," Journal of Supercomputing, vol. 69, pp. 395-411, Jul 2014.
[8] S.K. Hafizul Islam, G.P. Biswas, "A more efficient and secure ID-based remote mutual authentication with key agreement scheme for mobile devices on elliptic curve cryptosystem," The Journal of Systems and Software,vol. 84, pp. 1892-1898, Nov 2011.
[9] S. Kumari, M.K. Gupta, M.K. Khan, *et al.*, "An improved timestamp-based password authentication scheme: comments, cryptanalysis, and improvement," Security and Communication Networks, vol. 7, pp. 1921-1932, Nov 2014.
[10] S. A. Chaudhry, M. S. Farash, h. Naqvi, *et al.*, "An enhanced privacy preserving remote user authentication scheme with probable security," Security and Communication Networks, vol.8, pp. 3782-3795, Jun 2015.
[11] J. Liu, Z. Zhang, X. Chen and K.S. Kwak, "Certificateless Remote Anonymous Authentication Schemes for Wireless Body Area Networks," IEEE Transactions on parallel and distributed systems, vol. 25, pp. 332-342, Feb 2014.
[12] R.H. Shi, H. Zhong and L.S. Huang, "A novel anonymous authentication scheme without cryptography," Transactions on Emerging Telecommunications Technologies, vol. 25, pp. 875-880, Sep 2014.
[13] G. Yang, Q. Huang, D.S. Wong, and X. Deng, "Universal Authentication Protocols for Anonymous Wireless Communications," IEEE Transactions on Wireless Communications, vol. 9, pp. 168-174, Jul 2010.
[14] C.C. Chang and H.C. Tsai, "Mobile Authentication with Authenticated Key Agreement for Large-Scale Wireless Networks," IEEE Transactions on Wireless Communications, vol. 9, pp. 3346-3353, Oct 2010.
[15] H.J. Jo, J.H. Paik, and D.H. Lee, "Efficient Privacy-Preserving Authentication in Wireless Mobile Networks," IEEE Transactions on Mobile Computing, vol. 13, pp. 1469-1481, Jul 2014.
[16] Q. Xie, X. Tan, D.S. Wong, *et al.*, "A practical anonymous authentication protocol for wireless roaming," Security and Communication Networks, vol. 7, pp. 1264-1273, Aug 2014.
[17] J.K.. Liu, C.K. Chu, S. S.M. Chow, *et al.*, "Time-Bound Anonymous Authentication for Roaming Networks," IEEE Transactions on Information Forensics and Security, vol. 10, pp. 178-189, Jan 2015.
[18] M.S. Farash, S.A. Chaudhry, M. Heydari, *et al.*, "A lightweight anonymous authentication scheme for consumre roaming in ubiquitous networks with provable security," International Journal of Communication Systems, Jul 2015, DOI:10.1002/dac.3019.
[19] A. C.C. Yao and Y. Zhao, "Privacy-Preserving Authenticated Key-Exchange Over Internet," IEEE Transactions on Information Forensics and Security, vol. 9, pp. 125-140, Jan 2014 .
[20] S.A. Chaudhry, H. Naqvi, T. Shon, *et al.*, "Cryptanalysis and Improvement of an Improved Two Factor Authentication Protocol for Telecare Medical Information Systems,". Journal of Medical Systems, vol. 39, pp. 66, Jun 2015.
[21] S.A. Chaudhry, K. Mahmood, H. Naqvi, *et al.*, "An Improved and Secure Biometric Authentication Scheme for Telecare Medicine Information Systems Based on Elliptic Curve Cryptography," Journal of Medical Systems, vol. 39, pp. 175, Sep 2015.
[22] S.A. Chaudhry, M.S. Farash, H. Naqvi, *et al.*, "A secure and efficient authenticated encryption for electronic payment systems using elliptic curve cryptography," Electronic Commerce Research, vol.16, pp. 113-139, Mar 2016.

IOP Conf. Series: Journal of Physics: Conf. Series **910** (2017) 012066 doi:10.1088/1742-6596/910/1/012066

Distributed Ship Navigation Control System Based on Dual Network

Ying Yao and Wu Lv

Systems Engineering Research Institute, China State Shipbuilding Corporation, Beijing, China
Email: baeng@163.com, lw_all@126.com

Abstract. Navigation system is very important for ship's normal running. There are a lot of devices and sensors in the navigation system to guarantee ship's regular work. In the past, these devices and sensors were usually connected via CAN bus for high performance and reliability. However, as the development of related devices and sensors, the navigation system also needs the ability of high information throughput and remote data sharing. To meet these new requirements, we propose the communication method based on dual network which contains CAN bus and industrial Ethernet. Also, we import multiple distributed control terminals with cooperative strategy based on the idea of synchronizing the status by multicasting UDP message contained operation timestamp to make the system more efficient and reliable.

1 Introduction

In ship navigation system, there are many different devices and sensors responsible for collecting information, computing data, sending, receiving and executing the command. These sensors and devices are usually connected via CAN bus. This scheme is related to the system demand and the development of early technique at that time. As the development of communication technology, it is possible to network the sensors and devices via Ethernet. Under this background, we propose the dual communication network as the data transmission channel in navigation system. CAN bus and Ethernet coexist in the dual communication network. All devices and sensors will be connected via Ethernet. At the same time, some of them will also be connected to CAN bus. In dual communication network, the CAN bus is used as the redundant backups to ensure the data integrity and reliability [1]. The reliability of the system is guaranteed while bringing in the new technology to promote the development of the ship navigation system.

To make full use of the dual communication network, provide comprehensive navigation information for the system devices and make the users know the status of the system in real time, we import the integrated management device to collect the navigation information from CAN bus and Ethernet. After processing the raw information, data will be sent to the certain device to confirm that these devices can obtain all needed information. Also, administrators can use the integrated management devices to send the control command via dual communication network to the devices and sensors. The devices and sensors receive the control command, adjust the working status and report the related information to the integrated management devices.

The control commands can be divided into ordinary commands and important commands. The important commands should be sent by the integrated management device with authority. Deploying more than one integrated management device can make the system more flexible and reliable. To make the control operation correct and effective, especially for the important commands, the navigation system needs some kind of status synchronization strategy. The strategy must guarantee that no more than one integrated management device can send the important commands at any time.

In common situation, any integrated management device can send the ordinary command. For the important command, one integrated management device must lock the control authority. The locked control authority can be canceled by any other integrated management device in the system after verifying operation permission. The integrated management device that cancels the locked authority can also get the authority at the same time. When the control authority is locked, any other integrated management device can emit the ordinary command, but only the authorized one can emit the important command.

To achieve the basic scheme mentioned above, all integrated management devices must exchange some message based on TCP or UDP protocol to synchronize the working status. The synchronization message should be sent at a high frequency to confirm the effectiveness and real-time property. And the message from one terminal should be received by all the other terminals in the communication group. According to the above reasons, UDP protocol is more suitable for the status synchronization of multiple integrated management devices [2].

However, each integrated management device simply multicasts the UDP message with status information may lead to some mistakes in the synchronization process. There may be some problems in the situation of exchanging the operation authority among two integrated management devices. To solve this problem, we add the operation timestamp that represents the execution time of each operation into the synchronization UDP message. The integrated management devices receive two UDP messages with mutually-exclusive operation message can compare the operation timestamp in the messages to determine which one is valid.

In this paper, we introduce the dual communication network in Section II. Then in Section III, we introduce the conflicted problem in the process of synchronization while simply multicasting the UDP message. Also, the cooperative strategy of multiple integrated management devices based on operation timestamp is introduced in this section. Furthermore, the evaluation is shown in Section IV, and a conclusion is followed in Section V.

2 Dual Communication Network

In this section, there is a simple description and comparison of CAN bus and Ethernet to explicit the advantages and disadvantages of each other and describe the characteristics of dual communication network.

CAN bus is a kind of serial communication network which can efficiently support the distributed or real-time control [3]. It is a kind of multiple master bus which is widely used in industrial automation, ship, medical device and industrial device for its features of high performance and reliability in communication. CAN bus has a lot of advantages [4]. First, the coding scheme of CAN bus is data encoding. Multiple kinds of transmission method can be realized based on packet filtering. Second, CAN bus has the functionality of priority and arbitration, multiple control terminals can connect to the CAN bus and realize the functionality of multi-host in local network. Third, the CAN message adopts a short frame structure. The disturbed probability is very low as the transmission time of a message is short. In this way, the error ratio of CAN message is very small. Finally, when the communication node has something terribly wrong, it can disconnect from the CAN bus by itself. Except the advantages mentioned above, CAN bus also has some disadvantages. First, CAN bus cannot connect to the Internet. It is impossible to realize the remote information sharing. Second, it is difficult to connect with the normal computer directly. The related device is much more expensive than the ordinary NIC (Network Interface Card). Third, comparing to the industrial Ethernet, the communication distance and information throughput rate of CAN bus is lower.

The Ethernet based on TCP/IP is a kind of standard open network. The compatibility and interoperability of the information system based on Ethernet is much better [5]. Also, the information sharing ability is better [6]. And it is easy to realize the sharing of control data and resources in the system. Furthermore, the transmission distance of industrial Ethernet is long. The transmission rate is high [7]. Adopting the industrial Ethernet can networking the devices in the system at a low cost. The computers and servers can be connected via the original interface. And the industrial Ethernet has received extensive technical support in recent years. Certainly, the industrial Ethernet also has some disadvantage. It is a kind of non-deterministic network system. There is certain reduction in the reliability of Ethernet [8], especially in the extreme environment.

Based on the above analysis, either CAN bus or industrial Ethernet respectively has its own advantages, disadvantages and applied range. Each one has its own application field of comparative advantage. They will coexist for a long time. However, the development of CAN bus has entered a bottleneck. There are some limitations in the interconnect among each bus. After years of practical application, the developers have gained rich experience about the designation and application of the Ethernet. The cost of developing and application is relatively low. Importing both CAN bus and Ethernet into ship navigation system is consistent with the development trend of technology and the new demand of the system. Utilizing the advantages of real-time and stability of CAN bus can ensure the functionality with high demand of instantaneity and certainty work correctly and effectively. For the application scenario with large data and relatively low demand of instantaneity, the industrial Ethernet is much more suitable. Utilizing the CAN bus and Ethernet simultaneously can ensure the system meet multiple functional and performance requirements.

Connecting the functional devices and sensors via CAN bus and Ethernet simultaneously partly increases the deployment difficulty and complexity. Also, the redundancy of data transmission is raised. However, using two ways to do the system connection can make the data reliability further improved. Data transmitted via different channels can be used for cross validation. In this way, the abnormities of devices or sensors in the system could be discovered timely. Extending the system data transmission by importing the Ethernet can improve the data richness in the system. Each functional device can obtain more information. The functionality of the navigation system will also be enriched. In addition, as the continuous development of storage technology and the improvement of the computer hardware, the navigation system can store the redundant data for a long period of time. Analyzing these data is useful for adjusting the overall system situation in the future and improving the reliability and usability.

In general, adding the Ethernet to connect the devices and sensors is in line with the technology development trend [9][10]. Reserving the CAN bus connection of some devices and sensors is helpful to make the system more reliable. Coexisting of CAN bus and Ethernet can help to build the reliable ship navigation system with the functionality of network connection. The example scenario of dual communication network is shown in Figure. 1.

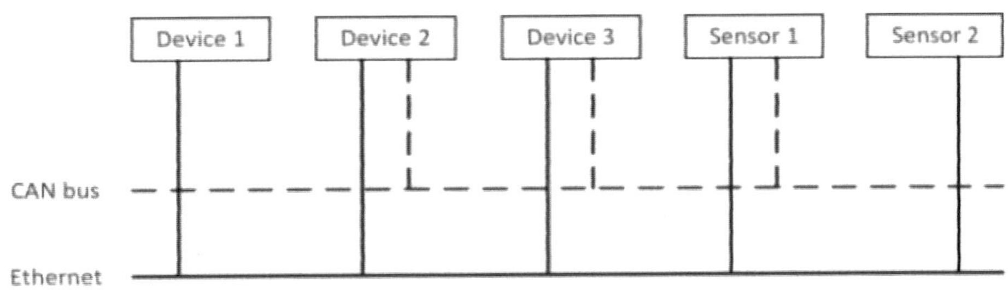

Figure 1. Example scenario of dual communication network

3 The Cooperative Strategy

The dual communication network can make the system more flexible and effective. Multiple integrated management devices can further enhance the reliability and flexibility. To ensure that multiple integrated management devices work correctly, some kind of authority synchronization method is needed. However, some synchronization strategy may lead to the conflicted problem in the system.

3.1 The Conflicted Problem

The simplest way to exchange authority among multiple integrated management devices is to let the one with authority to transfer the operation permission to another one. This way is easy to realize without complicated scheme to synchronize the authority status among the integrated management devices. However, this scheme is not flexible enough as the authority unlock operation must be done at specific integrated management device during some period of time. Allowing the other integrated management devices to unlock the authority by some kind of authentication is much more convenient as it is possible for someone who is far away from the integrated management device with authority to directly finish the operation without leaving current using one.

To achieve the functionality of authority remote unlocking, we add some authentication information in the synchronization UDP message multicast by the integrated management device with authority. Other integrated management devices receive this UDP message and record the authentication information in local storage. The operator must enter the correct information to match the authentication information before changing the authority status.

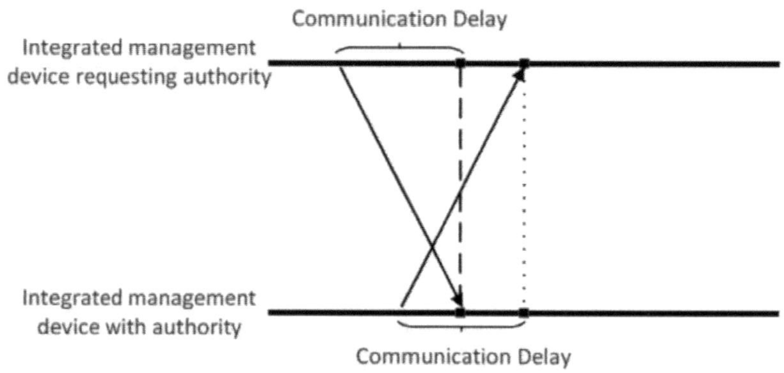

Figure 2 Failed scenario of simplest strategy

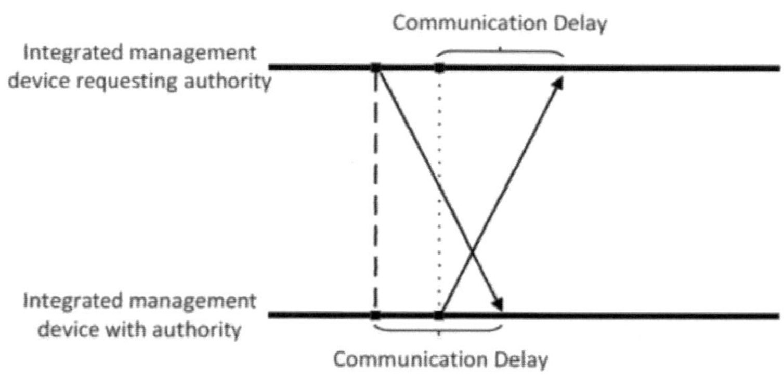

Figure 3. Failed scenario of strategy with ordinary timestamp.

The designation above seems like to make the system work flexibly and correctly. However, there may be some mistakes caused by the tiny time delay in the transmission and processing of the UDP message [11]. After the operator passes the authentication and transfers the authority from original integrated management device to a new one, the later one will multicast the UDP message with authentication and authority information. However, there may be something wrong in the synchronization procession. The example scenario is shown in Figure. 2. The later one might receive the UDP message with authority information from the original one just after emitting its own UDP message with authority information. Then the later one would unlock the authority and believe that the authority is locked by another one. The same situation might happen on the former one again. The status synchronization of two integrated management devices is wrong. Each one would think the authority is locked by the other one. Then they will send the UDP message without authority information. The authority is unlocked. None of the integrated management devices owns the authority. No one can emit the important command. And anyone can lock the authority.

For the problem mentioned above, someone might argue that adding the timestamp information into the UDP message would be helpful to solve the problem. The UDP message with authority whose timestamp is newer will be seen as valid. However, the simple message generated timestamp may not always be useful for the tiny delay between generating and arriving of the UDP message. And the synchronization may not be successful. The example situation is shown in Figure. 3. In the example, the timestamp information in the UDP message with authority information from the integrated management device without authority is smaller than that in the UDP message with authority form one with authority. The authority status would not change.

Someone might think that collecting the delay information and introducing it into the process of synchronization might be helpful. However, the delay consists of transmission delay and propagation delay. The transmission delay is not easy to calculate as the practical situation may always change. The delay is not very specific. Also, the processing of the UDP message is not completely real-time. The synchronization based on delay information will be very complicated.

3.2 The Cooperative Strategy

The insufficiency of ordinary timestamp is that it increases as time goes on. To make the status synchronization among multiple integrated management devices correct and efficient, we propose the cooperative strategy based on the operation timestamp. Different with the ordinary timestamp which is dynamic refreshing, the operation timestamp represents the execution time of each operation recorded in each UDP message. It is stable without refreshing as time goes on. And the operation in the UDP message whose operation timestamp is larger will be seen as valid.

Importing the operation timestamp is very helpful to solve the problem in the process of synchronization mentioned in last subsection. When the integrated management device locks the authority, it records the operation timestamp. Then the UDP message with authority information from this integrated management device will carry the operation timestamp information. When another integrated management device wants to get the authority, it records the operation timestamp after passing the authentication and sends out the UDP message with authority information and operation timestamp. The former integrated management device receives the UDP message from another one. The operation timestamp in newly received UDP message is larger. Then the former one will multicast the UDP message without authority information. Even if the later one received the UDP message with authority information from the former one, the authority synchronization would not be failed as the operation timestamp in the UDP message from the former one is smaller than that recorded locally.

Besides solving the conflicted problem, importing the operation timestamp into the UDP message has some other benefits in some special situations. In the navigation system, if the message from one device or sensor was not heard for a period of time, the device could be seen as disconnected from the communication network. And if one device did not hear any message from any device or sensor, it would see itself as disconnected. When the integrated management device with authority is disconnected, the authority will be unlocked and any other integrated management device can lock the authority. In the special situation, the disconnected integrated management device might be reconnected. The authority might be locked by another one during the time when the original authorized integrated management device was disconnected. The reconnected one would multicast its own UDP message. If it received the UDP message with authority information before multicasting its own, then it would multicast the UDP message without the authority information. If not, it would multicast the UDP message with authority information until receiving the UDP message from another one with authority information. Even if it multicast the UDP message with authority information, there would be nothing wrong in the status synchronization as the operation timestamp from the original one is smaller comparing with the operation timestamp in the later one.

4 Evaluation

We carry out some simulating experiments to demonstrate the effectiveness of the cooperative strategy that solves the problem in the process of synchronization by importing operation timestamp into the UDP message. In the simulated experiments, there are some integrated management devices. One integrated management device is chosen to have the control authority. Then another one is chosen to get

IOP Conf. Series: Journal of Physics: Conf. Series **910** (2017) 012066 doi:10.1088/1742-6596/910/1/012066

the authority from the former one. The simulated experiments mainly calculate the ratio of successful authority transmission in the scenario with different strategy, including strategy without timestamp, strategy with ordinary dynamically refreshing timestamp and strategy with stable operation strategy.

Based on the content in last section, both strategy without timestamp and with ordinary timestamp cannot solve the conflicted problem. The probability of successful authority transmission is mainly related to the sending time of UDP message and the communication delay. The failed scenario of strategy without timestamp and strategy with ordinary timestamp is shown in the grey zone in Figure. 4 and Figure. 5.

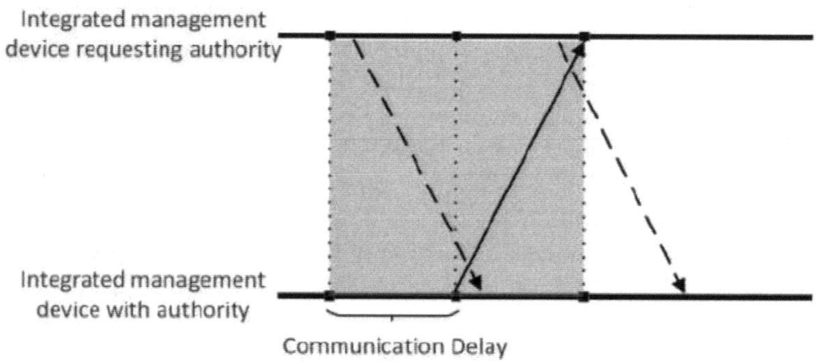

Figure 4. Failed situation in strategy without timestamp.

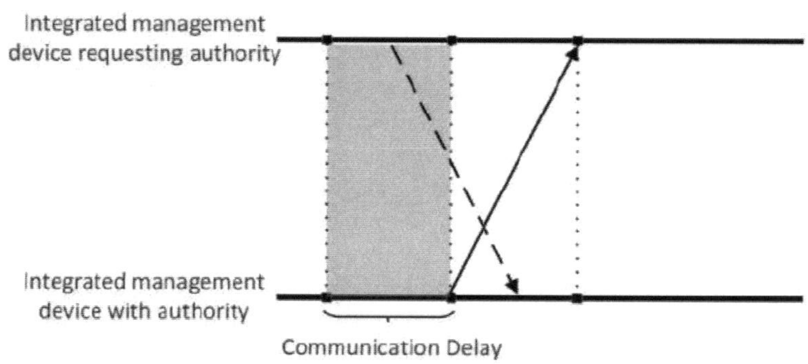

Figure 5. Failed situation in strategy with ordinary timestamp.

We use T_f, T_l and T_d to respectively represent the sending time of UDP message on the integrated management device with authority, the sending time of UDP message on the integrated management device requesting authority and the communication delay between these two devices. It is shown in Figure. 4, in the strategy without timestamp, if these three variables meet the conditions in Eq. (1), the authority synchronization will be failed.

$$Tf - Td < Tt < Tf + Td \quad (1)$$

It is shown in Figure 5, in the strategy with ordinary timestamp, if these three variables meet the conditions in Eq. (2), the authority synchronization will be failed.

$$Tf - Td < Tt < Tf \quad (2)$$

Based on Eq. (1) and Eq. (2), the result of synchronization operation could be judged from the sending time of UDP message on the integrated management devices with authority and requesting authority, and the communication delay between these two devices.

In each group of experiments, the number of integrated management device is different. The synchronization cycle is 1 second. And the sending time of UDP message is set randomly between 0 and 1000 milliseconds. And the delay between two integrated management devices is set randomly in a certain scope. The scope is related to the number of integrated management device. The bigger the number is, the larger the scope is. The experiment with certain number of integrated management device is repeated for many times to calculate the average radio of successful authority transmission. The random seed is different in each experiment.

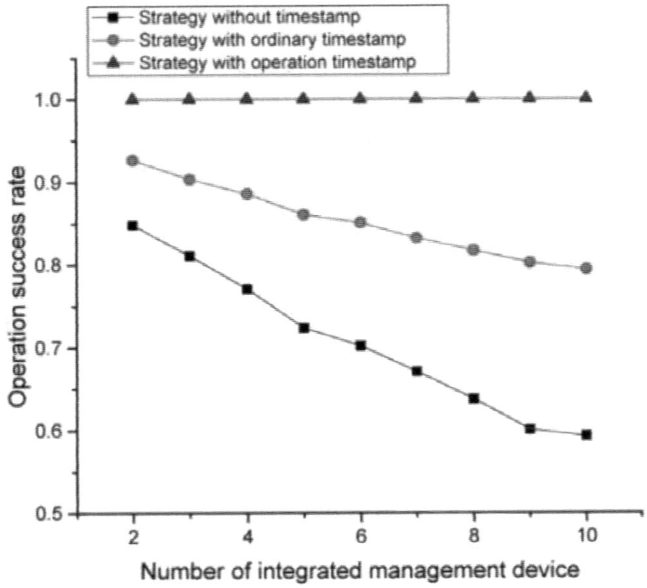

Figure 6. Simulated experiments result.

The experiments result is shown in Figure. 6. It is shown in the picture that the operation successful rate of strategy with operation timestamp is the highest. And the operation successful rate of strategy without timestamp is the lowest. Also, as the random scope of delay between two integrated management devices increases, the operation successful rate gets lower in the experiment with ordinary timestamp and without timestamp.

The navigation system is vital to the normal sailing of a ship. There should not be anything wrong in the authority management of navigation system. From such a perspective, the cooperative strategy with operation timestamp is very suitable for the authority management among multiple integrated management devices in the navigation system.

5 Conclusion

This paper describes the distributed ship navigation control system based on dual communication network which contains the CAN bus and Ethernet. The distributed multiple integrated management devices are introduced to make the system efficient. However, the authority management among these devices faces a conflicted problem which might lead the authority synchronization failed. To solve the conflicted problem and make the authority management among distributed multiple integrated management devices correct and effective, we introduce a kind of cooperative strategy by importing the operation timestamp information into the UDP message. The proposed strategy is easy to realize and effective enough.

6 References

[1] Cao H, Ma J, Zhang G, Zhang J. Marine Main Engine Remote Control System with redundancy CAN bus based on distributed processing technology[C]// International Conference on Intelligent Control and Information Processing. IEEE, 2010:638-640.

[2] Madhuri D, Reddy P C. Performance comparison of TCP, UDP and SCTP in a wired network[C]// International Conference on Communication and Electronics Systems. IEEE, 2017:1-6.
[3] Shi S, Song C, Wu Y. CAN bus performance test technology[C]// IEEE International Conference on Electronic Measurement & Instruments. IEEE, 2015:1395-1399.
[4] Wan X F, Xing Y S, Cai L X. Application and implementation of CAN bus technology in industry real-time data communication[C]// International Conference on Industrial Mechatronics and Automation. IEEE Xplore, 2009:278-281.
[5] Wester C, Adamiak M. Practical applications of Ethernet in substations and industrial facilities[C]// Protective Relay Engineers, 2011, Conference for. IEEE, 2011:67-78.
[6] Inoue M, Owada Y, Hamaguti K, Miura R. Nerve Net: A Regional-Area Network for Resilient Local Information Sharing and Communications[C]// Second International Symposium on Computing and NETWORKING. IEEE, 2015:3-6.
[7] Bencivenni M, Carbone A, Fella A, et al. High rate packet transmission on 10 Gbit/s Ethernet LAN using commodity hardware[C]// Real Time Conference, 2009. Rt '09. Ieee-Npss. IEEE, 2009:167-182.
[8] Gong Z, Liu B, Yang S, Gui X. Analysis of Industrial Ethernet's Reliability and Realtime Performance[C]// International Conference on Reliability, Maintainability and Safety. IEEE, 2009:1133-1136.
[9] Tian M, Sun Z, Zhang L. The simulation research on the ship networked control system based on Ethernet[C]// International Conference on Signal Processing. IEEE, 2015:2348-2353.
[10] Park H C, Lee J S, Jang K W, et al. Development of Network Platform for Integrated Information Exchange on Shipboard[J]. International Journal of Computer Science & Network Security, 2010(1).
[11] Aftab A, Ghani A, Abidi Z U A, Baqar M. Simulation based performance evaluation of TCP variants along with UDP flow analysis of throughput with respect to delay, buffer size and time[C]// International Conference on Open Source Systems & Technologies. IEEE, 2017:36-41.

IOP Conf. Series: Journal of Physics: Conf. Series **910** (2017) 012067 doi:10.1088/1742-6596/910/1/012067

Neighbor Discovery Algorithm in Wireless Local Area Networks Using Multi-beam Directional Antennas

Jin Wang, Wei Peng and Song Liu

College of Computer, National University of Defense Technology, Changsha, Hunan, China

Email: jinwang1206@163.com, wpeng@nudt.edu.cn, liusong13a@nudt.edu.cn

Abstract. Neighbor discovery is an important step for Wireless Local Area Networks (WLAN) and the use of multi-beam directional antennas can greatly improve the network performance. However, most neighbor discovery algorithms in WLAN, based on multi-beam directional antennas, can only work effectively in synchronous system but not in asynchro-nous system. And collisions at AP remain a bottleneck for neighbor discovery. In this paper, we propose two asynchrono-us neighbor discovery algorithms: asynchronous hierarchical scanning (AHS) and asynchronous directional scanning (ADS) algorithm. Both of them are based on three-way handshaking mechanism. AHS and ADS reduce collisions at AP to have a good performance in a hierarchical way and directional way respectively. In the end, the performance of the AHS and ADS are tested on OMNeT++. Moreover, it is analyzed that different application scenarios and the factors how to affect the performance of these algorithms. The simulation results show that AHS is suitable for the densely populated scenes around AP while ADS is suitable for that most of the neighborhood nodes are far from AP.

1. Introduction

Neighbor discovery in WLAN refers to the process where a given node determines that it can communicate directly to the neighbors [1]. Neighbor discovery, which is used to discover its one-hop neighboring nodes in the network initialization, is the first step in the network self-configuration process. The information about one-hop neigh-boring nodes is very important for MAC protocols [2], routing protocols [3] and topology control algorithms.

In traditional wireless network using omnidirectional antennas, a certain broadcast scheme could be used to establish connections among neighboring nodes. However, recent studies have strong interest on the use of multi-beam directional antennas which can increase the throughput of wireless networks [4]. Using multi-beam directional anten-nas in WLAN has many advantages, such as increasing network capacity, extending transmission distances and reducing energy consumption. However, the limited beam width of multi-beam directional antennas poses many challenges to the design of neighbor discovery algorithms due to the following reasons: 1)The limited radial range of beamwidth covers only a fraction of the azimuth, and the limitation requires the neighbor discovery scheme to be repeated in different directions to cover the whole azimuth. 2) Neighboring nodes must know when and where to point their directional beams to discover each other.

According to the features of multi-beam directional antennas, neighbor discovery algorithms can be divided into three categories: omnidirectional transmission and direction-nal reception (OD), directional transmission and directional reception (DD) and directional transmission and omnidirecti-onal reception (DO) [5]. Most neighbor discovery algorithms assume directional transmission and directional reception or directional transmission and omnidirectional reception. Some neighbor discovery algorithms for Ad Hoc Networks are discussed in [5], [6]. [7]-[9] assume successful commun-ication between a pair of nodes can occur only when one node's transmitting beam points in the direction of the other node's receiving beam. And [5], [6], [10] assume directional transmission and omnidirectional reception. However, these algorithms are not applicable for WLAN. Although a few algorithms have been proposed in [11], [12] for WLAN, most of them are discussed in synchronous scenarios. Synchronous algorithms need GPS or other timing devices in networks, which can cause increased costs of networks. Also occasions without timing devices are still common in practice.

In this paper, we propose two asynchronous neighbor discovery algorithms, asynchronous hierarchical scanning algorithm (AHS) and asynchronous directed scanning neighbor (ADS) in the wireless local area networks. The AP needs priori-knowledge about the number of neighboring nodes. No clock synchronization is required between nodes. The nodes in the network are equipped with the same multi-beam directional antennas, and the algorithm adopts directi-onal transmission and omnidirectional reception to complete the neighbor discovery process. AHS algorithm and ADS algorithm reduce the probability of conflict in AP to a certain extent, and the two algorithms are suitable for different application scenarios.

The rest of the paper is arranged as follows. In Section II, we describe current research related to neighbor discovery algorithms using multi-beam directional antennas. Section III presents the system model. Section IV gives a description of two asynchronous algorithms, AHS and ADS, in detail and the synchronous neighbor discovery algorithm proposed in [6] is briefly reviewed. Section V compares the simulation results for all the three algorithms and analyses the influence of parameters on performance and the application scenarios of AHS and ADS algorithm. Finally, we present the conclusions and future work in Section VI.

2. Related Work

A lot of research on neighbor discovery algorithms has been conducted by [5]-[11]. Many algorithms assume directional transmission and directional reception. For example, in [7], Zhang extends the 1-way discovery algorithm to 2-way algorithms, and first proposes three random neighbor discovery algorithms, completely random or probabilistic algorithm with directional transmit and receive (CRA-DD), with directional transmit and omnidirec-tional receive (CRA-DO) and omnidirectional transmit and omnidirectional receive (CRA-OO). But for scan-based algorithms, the paper ignores the collision effect when the beamwidth is large. In [8], the author considers the collision case that more than one node exist in one directional beam and proposed a new Improved Scan-based algorithm (I-SBA) based on the SBA algorithm. The algorithm increases the detection probability when the number of neighbors is small, and reduces the collision probability when the number of neighbors is large. In [13], a new Scan-Based Algorithm with Probabilistic Idle (SBA-PI) is proposed. Each node chooses its mode in every time slots among transmission, reception or idle with probability P_t, P_r and P_i respectively. SBA-PI can address the high collision problem caused by high node density. These algorithms are synchronous, and need a Global Positioning System or other methods to achieve clock synchronous.

There have been some asynchronous neighbor discovery algorithms. In [14], Khatibi proposes an asynchronous pure directional quorum-based neighbor discovery algorithm for the first time. All nodes can find all their potential neighbors in a relatively small time interval. In [15], Feng uses the “Problem of Coloring Balls” model to analyze the one-way handshake-based asynchronous neighbor discovery algori-thm, and proposes a hybrid handshake-based asynchronous algorithm. The feedback scheme is used which allows a better performance. In [9], a multi-carrier asynchronous neighbor discovery algorithm named 3-way MC-NDA is proposed. The algorithm avoids the failure of data transmission due to unnecessary collision and shortens the discovery delay by adding ACK packet. Compared with the two-way neighbor discovery algorithm, it overcomes the disadvantage that the node density has a great influence on the performance of algorithm.

Also many algorithms which assume omnidirectional transmission and directional reception are proposed. The packets sent by the node can be received when the omnidirectional receiving node is within the beam range of the directional sending node. In [10], the author presents a random or probabilistic protocol that called “birthday protocol”, which saves energy during the deployment of the node and makes neighbor discovery more efficient. In [5], a synchronous Direct-Discovery algorithm is proposed. In each time slot, a node selects directional transmission or omnidirectional reception with probability P_t and $1 - P_t$, and the direction angle is randomly selected. When a certain number of nodes are analyzed, the optimal Pt that can maximize the probability of the node discovering its neighbors within a given amount of time. [9] proposes a synchronous neighbor discovery algorithm based on the four-way handshake mechanism, in which the nodes are in the transmit, receive or idle mode with probability of P_t, P_r and P_i. The algorithm can discover not only one-hop neighbors, but also two-hop neighbors. And the ACK packet contains the information of the node itself and its neighbors previously discovered by the node. In [6], a synchronous neighbor node discovery algorithm is proposed in WLAN, which solves the conflict by using a contention-based directional polling mechanism. In this paper, we propose two asynchronous neighbor discovery algorithms in WLAN.

3. System Model

3.1. Multi-beam Directional Antennas Model

In this paper, we assume that the multi-beam directional antenna is composed of a circular antenna array. The cover-age area is divided into M $(1 \leq M)$ non-overlapping sectors, the relationship between the angle for each sector and the number of sectors is shown in (1). The other assumptions about the antenna are described as fallows.

$$\theta = \frac{2\pi}{M} \tag{1}$$

1) The antenna can work in omnidirectional mode and directional mode, the antenna mode can be freely switched and the switching time is ignored.

2) The direction of the beam can be adjusted in the directional mode by changing the antenna array factor, and the time required for the antenna to switch the beam direction can be ignored.

3) The antenna works in half-duplex mode and can only be in omnidirectional mode or directional mode at a certain time.

4) The side lobe gain can be ignored compared to the main lobe.

5) The beam angle of the omnidirectional mode is $\theta = 360°$, the directional mode defines two kinds of beam angle θ and $\theta / 2$. The transmission power P remains constant, and the variation of the beam angle leads to the variation of the range and coverage area of the antenna, as shown in Figure. 1. The relationship between the transmission distance d and the beam angle θ is shown in (2), where d_o is the transmission distance in the omnidirectional mode [6].

$$d = d_o \frac{\sqrt{52525}}{\theta} \tag{2}$$

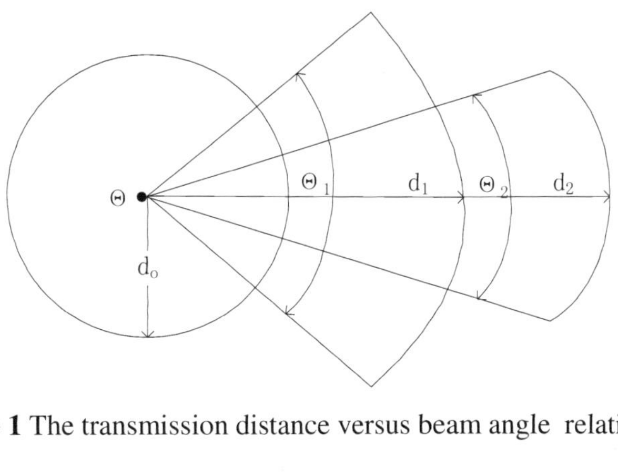

Figure 1 The transmission distance versus beam angle relationship.

3.2. Network Model

We assume that there is an access point (AP) with N nodes around it in the network. Each node is equipped with the same multi-beam directional antenna, all nodes have the same transmission power, and the AP can communicate with all neighboring nodes. The node can adjust the working mode of the antenna, and the node in the omnidirectional receiving state adjusts the antenna to omnidirectional mode for receiving information. The node in the omnidirectional transmission state adjusts the antenna to omnidirectional mode for sending information. There are two directional transmission states of the nodes, namely the directional tran-smission state A and the directional transmission state B respectively, corresponding to the directional mode of beam angle θ and $\theta/2$. Other features of these nodes are shown as follows.

1) Each node has a unique ID.

2) All nodes and the AP work in half-duplex mode. At any time, the AP can be in omnidirectional receiving state, omnidirectional transmission state, or directional transmission state. The nodes can be either in omnidi-rectional receiving state or directional transmission state, but not both.

3) The AP and one node can communicate directly, required AP and the node to be distributed in omnidi-rectional receiving state and directional transmission state or in omnidirectional receiving state and omnidirectional transmission state, and the receiving node is within the beam range of the transmitting node. At the same time, the receiving node can receive a complete packet should be ensured. In other words, the node in receiving state cannot obtain information through parts of a data packet. Figure. 2 shows an example of successful communication between node A and node B.

4) Data packets cannot be recovered when a node receives packets from two or more neighboring nodes at the same time, and that is called collisions. Figure. 3 shows an example of a conflict where it receives packets from node A and node C at the same time.

5) The clock is asynchronous between nodes. All packets have a fixed duration.

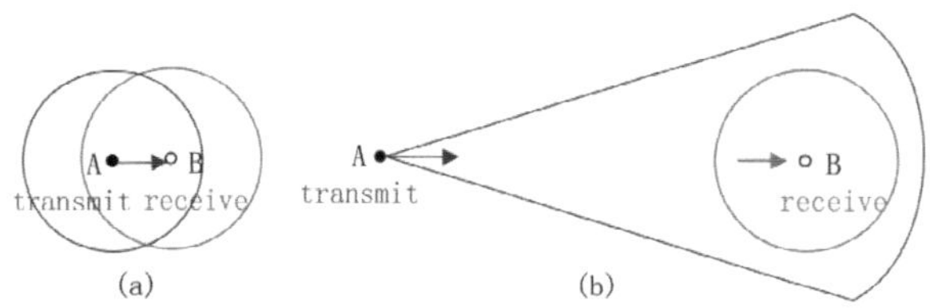

Figure 2 Nodes successfully communicating. (a) Successfully communi- cating between omnidirectional transmission node A and omnidirectional receiving node B. (b) Successfully communicating between directional transmission node A and omnidirectional receiving node B.

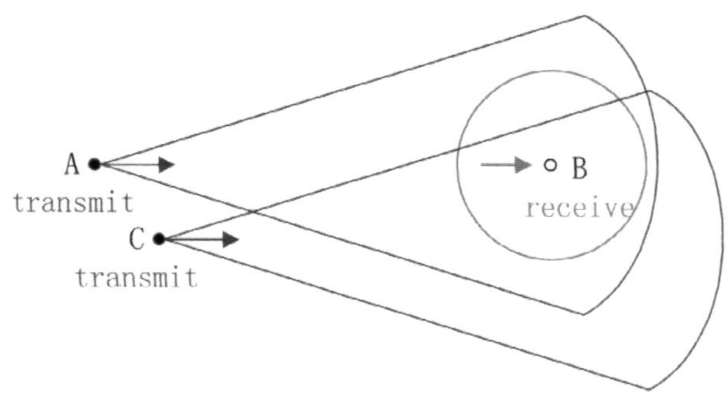

Figure 3 Collisions occur between node A and node C.

4. Neighbor Discovery Algorithm

In this section, we first give a brief description of the synchronous neighbor discovery algorithm (hereinafter referred to as IL algorithm) proposed in [6]. Then, we describe our proposed neighbor discovery algorithms, AHS and ADS. In these two algorithms, the AP makes use of three-way handshaking mechanism to find neighboring nodes and both of them can efficiently reduce collision. AHS uses a hierarchical scanning mechanism to find neighboring nodes in different regions. These regions are divided according to the communication radius of the AP. ADS finds neighbors more quickly using a directional scanning method, in which it divides the communication area into multiple sectors. For ease of description, the symbols are defined as follows.

N: the number of neighboring nodes around the AP, and there are N + 1 nodes totally in the network.

φ: the beam direction angle of multi-beam directional antennas.

t_p: the packet transmission time.

t_d: the delay of packet propagation.

n_A, n_B: the number of beams corresponding to the direc-tional transmission state A and the directional transmission state B respectively.

4.1. Synchronization Neighbor Discovery Algorithm

The IL algorithm assumes time is slotted and all nodes are perfectly synchronized on time slots, each time slot is divided into L mini-slots, as shown in Figure. 4.

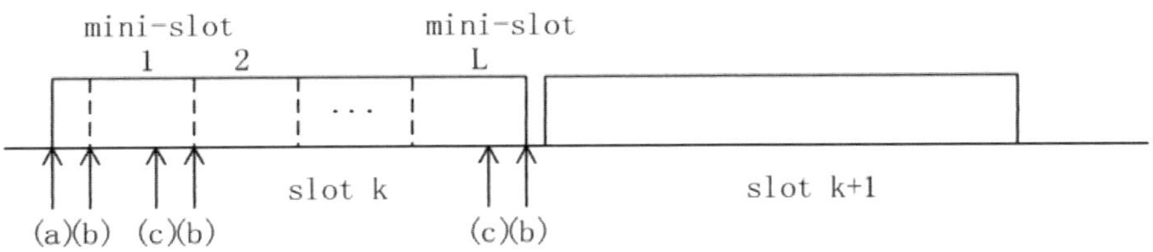

Figure 4 The example of time slots.

Step 1, at the beginning of each slot, that is at time (a), and after all users from the previously formed beam are resolved, the AP selects a new direction to send polling message PM.

Step 2, time (b) is the beginning of mini-slot. And the AP turns to omnidirectional receiving state at the beginning of time (b). If the AP successfully receives an ACK, it happens that the AP only receives an ACK from neighboring nodes (for example, node j).

Step 3, the RACK containing the ID of node j and AP is transmitted at time (c). If the AP receives more than one ACK, there is collision and AP does not transmit anything.

Repeat the step 2 and step 3 until all neighboring nodes in the beam range are resolved. In the L_{th} mini-slot, if the AP receives an ACK, that is there are still unsolved neighboring nodes.

The other nodes are in omnidirectional receiving state at time (a) (for example, node j). If the polling message PM from the AP is received and the RACK containing its own ID has not been previously received, then send an ACK to the AP at time (b), the ACK contains ID of node j. The node j changes to omnidirectional receiving state at time (c). If a RACK containing ID of node j from the AP is received, the node j sends RACK no longer. Otherwise, the ACK is sent with probability p at the beginning of the next mini-slot. Repeat the above process until all the beams have been traversed.

4.2. Asynchronous Hierarchical Scanning Algorithm

The Asynchronous Hierarchical Scanning algorithm (AHS) is based on the three-way handshake mechanism to complete the neighbor discovery process. For the AP, the algorithm is divided into three phases.

Stage 1, the AP is in omnidirectional transmission state, and sends the polling message PM, which is identified by m_1 and contains the ID of the AP. After sending PM, the AP enters omnidirectional receiving state and continues for T_1 time. If an ACK is successfully received from the neighb-oring nodes during the period, then parse the contents of ACK. And after the end of the omnidirectional receiving state, the AP enters the omnidirectional transmission state and transmits the RACK. The ID of the neighboring nodes included in the received ACK and the ID of the AP itself are added to the RACK.

Stage 2, the AP converts to directional transmission state A, randomly selects a beam direction angle φ, and direction-ally sends PM. The PM is identified by m_2 and contains the ID of the AP. After sending PM, the AP enters omnidire-ctional receiving state and continues for T_2 time. If the ACK is successfully received, then analyze it. At the end of omnidirectional reception state, the AP turns into directional transmission state A, and the beam direction angle is still φ. Then directly sends the RACK. After that, the next beam direction angle is selected in the counterclockwise direction based on the beam direction angle φ, and repeats the above process until all the n_A beam is traversed.

Stage 3, the AP is converted to directional transmission state B, randomly selects a beam direction angle φ, the discovery process is similar to stage two. The relationship between the duration time T_i of AP and the selected maximum of random integer k_i by neighboring nodes after receiving PM identified as m_i is as shown in (3).

$$T_i = t_p \times 2 + t_d \times (k_i + 1)(i = 1,2,3) \tag{3}$$

For neighboring nodes around the AP, for example, expressed in node j, node j is in omnidirectional receiving state. When the PM is received, node j does not receive the RACK from the AP or receives the RACK from the AP but the RACK does not contain the ID of node j. And if the PM is identified as m_i, then randomly selects an integer R in the interval $[0,\ k_1]$. After $R \times t_d$ time, node j converts to directional transmission state A, then adds the ID of node j and the AP to the ACK and sends the ACK in the direction of receiving the PM. If the received PM is identified as m_2 or m_3, the algorithm process is similar to the process above. After sending the ACK, node j changes to omnidirectional receiving state. When the RACK from the AP is received, then judges whether or not the RACK contains its own ID. If the ID is included, it indicates that the AP has discovered node j, and node j no longer sends an ACK when it receives the PM from the AP again.

For the AP, the variable $mode_{AP}$ indicates the state of the AP. $mode_{AP}$ has four values, omniTra, omniRec, dirA and dirB, which indicates that the AP is in the omnidirectional transmission state, the omnidirectional receiving state, the directional transmission state A and the directional transmit-ssion state B. After the AP sends PM in the omnidirectional state, the directional transmission state A and the directional transmission state B, the duration time of the omnidirectional receiving state is respectively T_1, $T_2$and T_3. The neighboring nodes around the AP are represented by j. The algorithm process at AP is as follows.

1) After the node starts, it enters to the omnidirectional sending state, $mode_{AP} =$ omniTra;
   a) Send PM identified as m_1;
   b) $mode_{AP} =$ omniRec, and continue for T_1 time;
   c) $mode_{AP} =$ omniTra, if the ACK is successfully received from j in the omnidirectional reception period, then add the ID of j into the RACK and send the RACK;
2) $mode_{AP} =$ dirA, randomly selects a beam direction angle φ;
   a) FOR (i = 1; i < n_A; i + +)
      i) Send PM identified as m_2 at direction angle φ;
      ii) $mode_{AP} =$ omniRec, and continues for T_2 time;
      iii)$mode_{AP} =$ dirA, if the ACK is successfully received from j in the omnidirectional reception period, then add the ID of j into the RACK. If the channel is idle, send the RACK at the direction angle φ, and select the next beam direction angle φ; in counterclockwise order;
      iv) IF i = n_A
      v) $mode_{AP} =$ dirB;
3) $mode_{AP} =$ dirB, randomly select a beam direction angle φ;
   a) FOR (j = 1; j < n_B; j + +)
      i) Send PM identified as m_3 at direction angle φ;
      ii) $mode_{AP} =$ omniRec, and continue for T_3 time;
      iii) $mode_{AP} =$ dirB, if the ACK is successfully received from j in the omnidirectional reception period, then add the ID of j into the RACK. If the channel is idle, send the RACK at the direction angleφ, and select the next beam direction angle φ in counterclockwise order;
      iv) IF j = n_B, the algorithm executes once.

For neighboring nodes around the AP, we take node j as an example. And set a boolean variable isFound for node j. When the ACK from node j is received by the AP and the ID of j is included in the AP's RACK, isFound is set to true, otherwise false. The variable $mode_j$ indicates the state of j. $mode_j$ has three values, omniRec, dirA and dirB. Also we set a variable k, and k has three values as k_1, k_2 and

k_3, respectively the maximum value of random integers is chosen when j receives PM identified as s_1, s_2 and s_3. The algorithm process for node j is described as below.

1) After the node starts, it enters the omnidirectional receiving state, $mode_j$ = omniRec;
2) When the PM sent from the AP is received, if isFound is false, then
   a) IF the identification from PM $\mathrm{s_i} = s_1$
      i) $\mathrm{m} = \mathrm{random}(0,\ k_1)$;
      ii) After waiting for $\mathrm{m} \times t_d$ time, $mode_j =$ dirA, if the channel is idle, send ACK to the AP in the direction of receiving the PM;
   b) ELSE IF the identification from PM $\mathrm{s_i} = s_2$
      i) $\mathrm{m} = \mathrm{random}(0,\ k_2)$;
      ii) After waiting for $\mathrm{m} \times t_d$ time, $mode_j =$ dirA, if the channel is idle, send ACK to the AP in the direction of receiving the PM;
   c) ELSE IF the identification from PM $\mathrm{s_i} = s_3$
      i) $\mathrm{m} = \mathrm{random}(0,\ k_3)$;
      ii) After waiting for $\mathrm{m} \times t_d$ time, $mode_j =$ dirB, if the channel is idle, send ACK to the AP in the direction of receiving the PM;
   d) After the ACK is sent, $mode_j =$ omniRec;
3) When receives the RACK sent by the AP, if isFound is false and the RACK contains the ID of the node j, then set isFound = true.

4.3. Asynchronous Directional Scanning Algorithm

The Asynchronous Directional Scanning Algorithm (ADS) is also based on the idea of three-way handshake of TCP protocol, and utilizes the direction transmission state B of the multi-beam directional antenna to fulfill the neighbor discovery process. At the beginning of ADS, the AP is in the directional transmission state B, randomly selects a beam direction angle φ to send the polling message PM, and then selects the next beam direction angle to send the PM in the counterclockwise order until the PM is transmitted in all directions of the directional transmission state B. Where PM contains the ID of AP and n_i, n_i indicates the number of times that the AP sends PM. For example, if the AP sends PM for the first time, then the value of n_i is 1, and the maximum value of n_i is the number of beams when the AP is in directional transmission state B, namely $n_i = n_B$.

Then the AP turns into omnidirectional receiving state, where the duration of the omnidirectional reception is T, and the relationship between T and the maximum value k of the selected integer after the neighboring nodes receive the PM is shown in (4). If the reply packet ACK from the neigh-boring nodes is received during the period, then the node's ID in the ACK and the direction of the arrival (DOA) of ACK are recorded. At the end of omnidirectional receiving state, the AP is converted to the directional transmission state B, the beam direction angle φ is randomly selected, and sends RACK. The information about the ID of neighboring nodes received at the direction angle φ and the ID of the AP are placed in the RACK. And then selects the next beam direction angle in the counterclockwise order to send a RACK until all beam directions are polled.

$$\mathrm{T} = \mathrm{t_p} \times 2 + \mathrm{t_d} \times (\mathrm{k} + 1) \tag{4}$$

The neighboring nodes around the AP are in the omnidir-ectional receiving state, node j is used as example. When the polling message PM from the AP is received, node j parses the PM and gets the value of n_i. Then node j randomly selects the integer R in the interval [0, k], after $(R + n_B - n_i) \times t_d$ time, sends the ACK in the direction of receiving the PM, where ACK contains the ID of the node j and the ID of the AP. When the RACK from the AP is received, node j judges whether or not the RACK contains the node j's own ID. If the ID is included, it indicates that the node j has been found by the AP.

For the AP, we set the variable n_i to indicate that the number of times that the AP sends the PM. The variable $mode_{AP}$ indicates the state of the AP. $mode_{AP}$ has two values, omniRec and dirB, which indicates that the AP is in the omnidirectional receiving state and the directional transmission state B. The neighboring nodes around the AP are denoted by j. The algorithm process is as follows.

1) After the node starts, it enters the directional transmission state B, $mode_{AP}$ = dirB, randomly selects a beam direction angle φ;
   a) FOR ($n_i = 1; n_i < n_B; n_i + +$)
      i) If the channel is idle, send the PM contains n_i and the ID of AP at the direction angle φ, and select the next beam direction angle φ in the counterclockwise order;
      ii) IF $n_i = n_B$, it enters the omnidirectional recei-ving state;
2) $mode_{AP}$ = omniRec, and continue for T time. If an ACK from node j is received, the ID and the DOA of j are recorded;
3) At the end of the omnidirectional receiving state, $mode_{AP}$ = dirB, randomly select the beam direction angle φ;
   a) FOR ($n_i = 1; n_i < n_B; n_i + +$)
      i) Add the ID of node j received at the angle φ and the ID of the AP to the RACK. If the channel is idle, send a RACK at the direction angle φ and select the next beam direction angleφin counterclockwise order;
   b) IF $n_i = n_B$, the algorithm executes once.

The nodes around the AP is assumed to be node j. We set a boolean variable isFound, and isFound is set to true when node j receives RACK contains the ID of j from the AP, otherwise false. The variable $mode_j$ indicates the state of j. $mode_j$ has two values, omniRec and dirB. The algorithm process for node j is described as below.

1) The node j enters omnidirectional receiving state after it starts, $mode_j$ = omniRec;
2) When the PM sent by AP is received, if isFound is false, then
   a) n = n_i;
      i) m = random(0, k);
      ii) Waiting for $(m + n_B - n) \times t_d$ time, after that, if the channel is idle, $mode_j$ = dirB, then send the ACK to the AP at the direction of receiving PM;
      iii) After the ACK is sent, $mode_j$ = omniRec;
3) When the RACK sent by AP is received, if isFound is false and the RACK contains the ID of the node j, then
   a) isFound = true;

5. Evaluation

5.1. Eexperiment Setup

In this paper, OMNeT++ 5.1.1 is used to simulate the experiment, and we construct an AP-based network. The nodes in the network are equipped with multi-beam directional antennas. The nodes are randomly distributed in a square with an area of 2000m ∗ 2000m. The AP is located in the center of the square. The size of the network varies from 20 nodes to 200 nodes. The transmission rate of the packet namely the transmission delay is certain, and the packet transmission delay is also a fixed value. The parameters used in the simulation experiment are shown in Table 1.

We assume the total number of nodes in the network is 20, 50, 100, 200, that is, the number of neighboring nodes N is 19, 49, 99, 199 and an AP. In these four scenarios, compare the performance of the asynchronous hierarchical scanning algorithm (AHS), the asynchronous directional scanning algorithm (ADS) and the synchronization algorithm (IL). In the AHS algorithm, the corresponding maximum value of random numbers in the optional integer interval is k_1, k_2 and k_3 when the neighboring nodes receive the PM identified as s_1, s_2 and s_3. In the ADS algorithm, the maximum random integer that the neighboring nodes can select when receiving the PM is k. In the IL algorithm, the mini-slots and the probability of the ACK returned by neighboring nodes is represented by L and P. We adjust the values of L and P to make the algorithm as optimal as possible in the case of different number of neighboring nodes.

5.2. Performance Results

When the number of neighboring nodes is certain, the 50 kinds of topological structures of neighboring nodes distributed randomly are generated, and the AHS, ADS and IL algorithm are executed respectively. The time required for all neighboring nodes to be found is compared with the statistical average. When the number of neighboring nodes N is 19, the k_1, k_2 and k_3 of the AHS algorithm is 5, 5, 5, the k of the ADS algorithm is 20, the L and P of the IL algorithm are 6 and 0.5. When number of neighboring nodes N is 49, the k_1, k_2 and k_3 of the AHS algorithm is 5, 5, 5, the k of the ADS algorithm is 50, the L and P of the IL algorithm are 6 and 0.5. When the number of neighboring nodes N is 99, the k_1, k_2 and k_3 of the AHS algorithm is 5, 10, 10, the k of the ADS algorithm is 100, the L and P are 10 and 0.2. When the number of neighboring nodes N is 199, the k_1, k_2 and k_3 of the AHS algorithm is 5, 10, 10, the k is 200, and the L and P are 10 and 0.2. The comparison of algorithm performance is shown in Figure. 5. When the number of neighboring nodes is small, the ADS performs better than AHS and IL, and the AHS requires less time to find neighboring nodes than IL. When the number of neighboring nodes is large, the performance of AHS and ADS is better than that of IL, and it takes less time for ADS than AHS to find all neighboring nodes. When almost 75% of neighbors are found of the total number of neighboring nodes, the AHS and ADS performance is similar. With the increase of the number of neighboring nodes, the difference in performance of AHS, ADS and IL algorithm is more obvious, and two asynchronous neighbor discovery algorithms proposed in this paper are more advantageous.

Table 1 Experimental Parameters

Parameter Name	Parameter Name
area	2000 meters × 2000 meters
number of Neighbor Nodes	20, …, 200
packet transmission delay	60us
packet propagation delay	30us
omnidirectional transmission radius	150m
beam angle of directional mode A	45°
beam angle of directional mode B	22.5°

Neighboring nodes are distributed randomly, and the neighboring nodes are scattered around the AP, that is, the nodes in the area near the AP are few, and most nodes in the area far away from the AP. The experimental results show that ADS is more suitable for such scenes compared to AHS.

We assume that the AP are located at the center of the network area, the neighboring nodes of the AP are distributed around the AP which is similar to the normal distribution pattern. The nodes in the area near the AP are more concentrated, that is, the number of nodes is larger, and the nodes in the area far from the AP are less. In this scenario, we construct 50 kinds of network topologies, and run the AHS and ADS algorithm respectively. According to the statistical average of the simulation results, the line graph of the relationship between the number of nodes found and the required time is drawn, as shown in Figure. 6 (a). Similarly, in this scenario, 50 network topologies with a total number of 100 nodes are constructed and the parameters are adjusted so that the time required for AHS and ADS algorithm to run once is roughly the same. Record the algorithm execution times and the number of neighboring nodes discovered by AP at the end of each algorithm execution. And plot the histogram that is the comparison of run times and the number of neighboring nodes found, as shown in Fig. 6 (b).

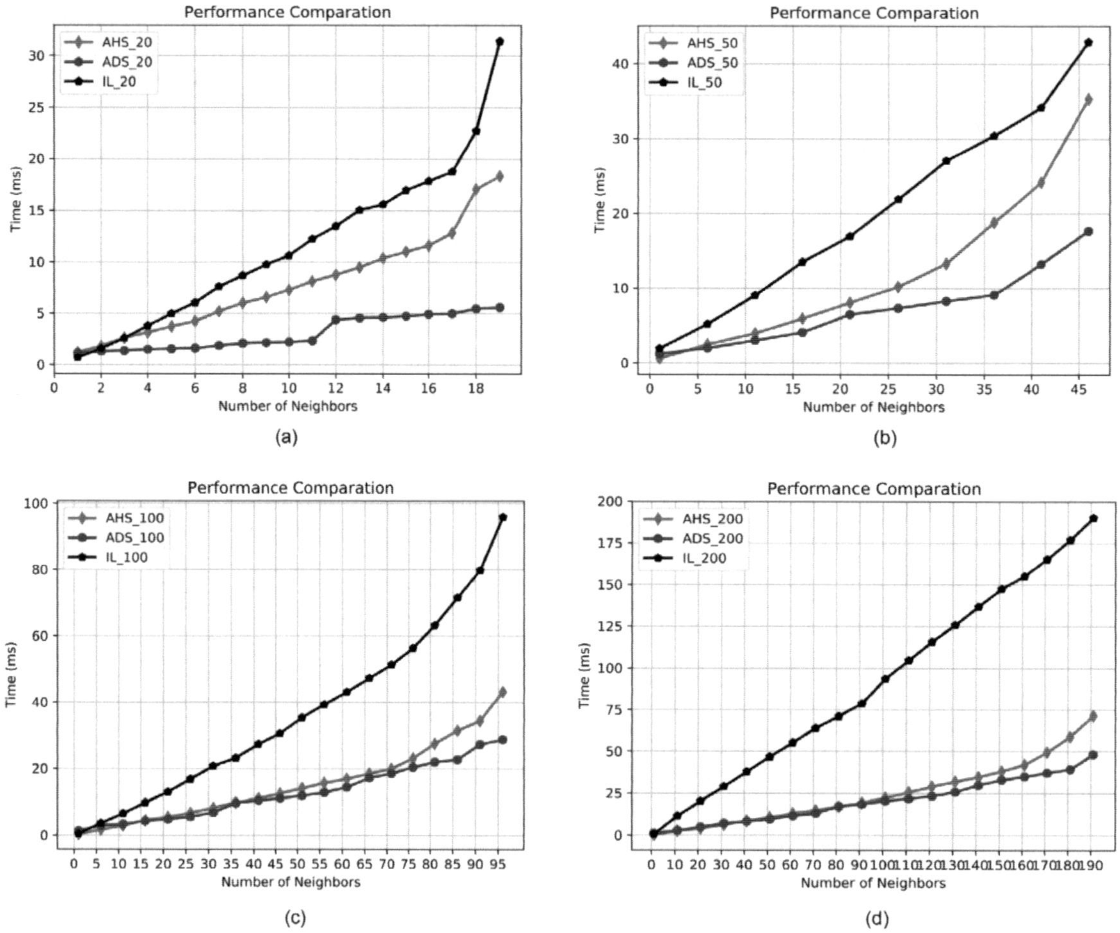

Figure 5 Performance comparison of algorithms. (a) Neighboring nodes in the network is 19. (b) Neighboring nodes in the network is 49. (c) Neighboring nodes in the network is 99. (d) Neighboring nodes in the network is 199.

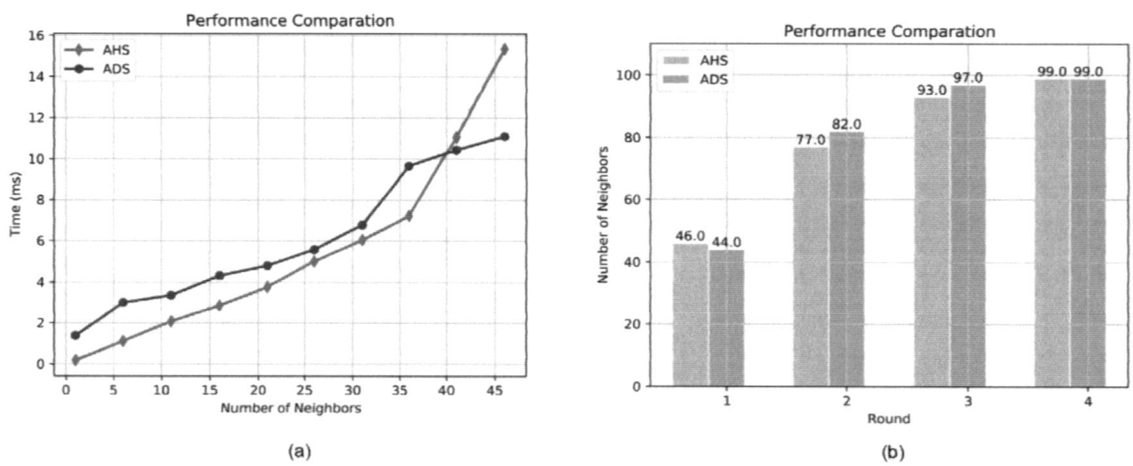

Figure 6 Performance comparison of AHS and ADS. (a) The number of neighboring nodes found and the required time. (b) Run times and the number of neighboring nodes found.

When the number of nodes in the area near the AP is large, the AHS can find more nodes in the same run times than the ADS in the early stage of algorithm execution and the performance is better than that of the ADS. However, at the end of the algorithm execution, AHS takes more time to find the same number of neighboring nodes as ADS. When AHS and ADS spend roughly the same time for each execution, the two algorithms need to execute four times to find all neighbors. However, when the algorithms execute once, the AHS finds more nodes than the ADS. This is because the hierarchical scanning mechanism is used in AHS, and the PM is sent in the area near the AP first, and AP immediately transfers to the receiving state after sending a PM, which can quickly find closer neighbors. While ADS uses the directional scanning mechanism, the AP sends PM at all beam direction and then turns into receiving state, which causes ADS to find neighbors slowly at the beginning. Therefore, the AHS algorithm is more suitable for the scene that most nodes distributed in the area close to the AP and less nodes in the area away from the AP, it can quickly find neighbors. ADS is applicable to scenes that most nodes are in the area away from AP.

5.3. The Effect of Parameter k

We analyze the influence of parameters on AHS algorithm. When the number of neighboring nodes is fixed, this paper analyze the influence of the maximum value of random integers k_1, k_2 and k_3 can be selected by neighbors on the algorithm performance. The number of neighbors is 99, and we construct 20 different network topologies in which nodes are randomly distributed. The k_1, k_2 and k_3 are respectively $k_1 = 5, k_2 = 10, k_3 = 10$ and$k_1 = 10, k_2 = 10, k_3 = 10$ and $k_1 = 5, k_2 = 10, k_3 = 5$and $k_1 = 5, k_2 = 15, k_3 = 10$. The line graph of the number of found neighbors and the time required is shown in Figure. 7 with these four kinds of values, where the performance is best when the value is $k_1 = 5, k_2 = 10, k_3 = 10$. That is because nodes are randomly distributed, and the nodes in the omni-directional transmission range of AP are less, $k_1 = 5$ can avoid the conflict that the neighboring nodes in the range send ACK at the same time. More nodes are located in the range of directional beam of AP, k_2 and k_3 are 5 will increase the probability of conflict generated at AP, thereby the performance degrades, the value of k_2 and k_3 should be increased. While $k_2 = 15$ reduces the conflict probability, it increases the execution time of the algorithm, which makes the algorithm performance worse. Compared to the other two values, $k_2 = 10$ makes the performance better.

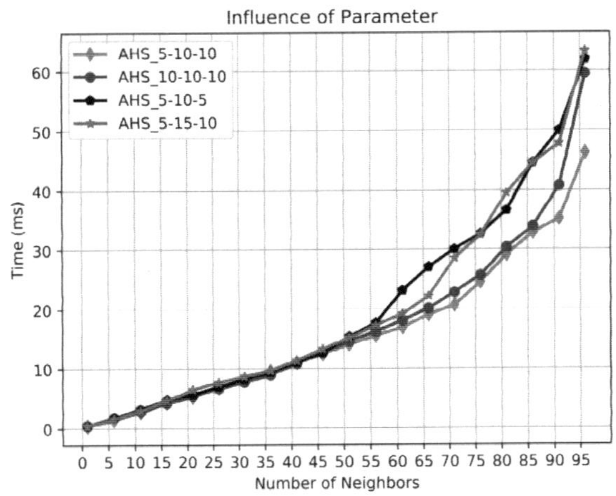

Figure 7 Effect of parameter k on AHS.

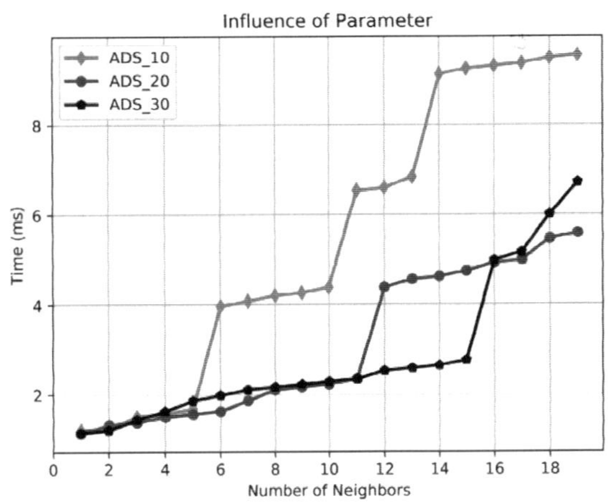

Figure 8 Effect of parameter k on ADS.

Then we analyze the influence of parameters on ADS algorithm. When the number of neighboring nodes is certain, the maximum value k in the random integer interval chosen by the neighbors affects the performance of the algorithm. This paper assumes that the number of neighbors is 19, constructs 20 kinds of network topologies with nodes randomly distributed, k value is 10, 20 or 30. The line graph of the number of found nodes and the time required with different k is shown in Figure 8. Compared with the other two values, the algorithm performance is the best when k is 20. When the k value is 10, the neighboring nodes send the ACK at the same time so that the probability of collision at the AP is large, and the times of algorithms executed to find all neighbors is more, which will take more time. When the k is 30, although the probability of collision at AP is reduced, the times of algorithms that need to be executed to find all neighbors is the same as the number of k values of 20. But it will take longer time to execute the algorithm once, so the algorithm performance is worse.

6. Conclusion and Future Work

This paper proposes two algorithms as AHS and ADS to solve the problem of neighbor discovery with multi-beam directional antennas. AHS makes use of omnidirectional /directional transmission and omnidirectional receiving mode of antennas, and ADS adopts directional transmission and omnidirectional receiving mode. These two algorithms use similar back-off mechanism to reduce the collision probability at AP when more than one neighbor send ACK at the same time, which makes the algorithm more efficient. Also, the performance of AHS and ADS is compared and the applicable scenarios of the two algorithms are analyzed.

In the future works, we can construct the mathematical model of these two algorithms and theoretically analyze the relationship among the number of neighbors, the distribution of neighboring nodes and the parameters of the algorithm to make the algorithm performance better, and analyze the relationship between the number of neighbors and the number of times that the algorithm needs to run to find all neighboring nodes. It is also possible to record the number of nodes found in each beam direction during the neighbor discovery process and use this information to improve the performance of AHS and ADS algorithms, so that a more efficient neighbor discovery algorithm can be designed.

7. Acknowledgment

This work is supported by projects of Natural Science Foundation of China under grant no. 61601483 and 61379147.

8. References

[1] S. A. Borbash, A. Ephremides, and M. J. Mcglynn, "An asynchronous neighbor discovery algorithm for wireless sensor networks," Ad Hoc Networks, vol. 5, no. 7, pp. 998–1016, 2007.

[2] Bao, Lichun, GarciaLunaAceves, and J. J., "A new approach to channel access scheduling for ad hoc networks," International Conference on Mobile Computing and Networking, vol. 63, no. 1, pp. 210–221, 2001.

[3] Z. Zhang, "Pure directional transmission and reception algorithms in wireless ad hoc networks with directional antennas," in IEEE International Conference on Communications, 2005, pp. 3386–3390 Vol. 5.

[4] S. Yi, Y. Pei, and S. Kalyanaraman, "On the capacity improvement of ad hoc wireless networks using directional antennas," in ACM International Symposium on Mobile Ad Hoc NETWORKING & Computing, 2003, pp. 108–116.

[5] S. Vasudevan, J. Kurose, and D. Towsley, "On neighbor discovery in wireless networks with directional antennas," in IEEE Joint Conference of the IEEE Computer and Communications Societies, 2005, pp. 2502–2512 vol. 4.

[6] G. Astudillo and M. Kadoch, "Neighbor discovery and routing schemes for mobile ad-hoc networks with beamwidth adaptive smart antennas," Telecommunication Systems, pp. 1–11, 2017.

[7] Z. Zhang and B. Li, "Neighbor discovery in mobile ad hoc selfconfig-uring networks with directional antennas: algorithms and compari-sons," IEEE Transactions on Wireless Communications, vol. 7, no. 5, pp. 1540–1549, 2008.

[8] H. Cai, B. Liu, L. Gui, and M. Y. Wu, "Neighbor discovery algorithms in wireless networks using directional antennas," in IEEE International Conference on Communications, 2012, pp. 767–772.

[9] S. Zhao, Y. Liu, T. Yang, Z. Feng, Q. Zhang, and C. Gao, "3-way multi-carrier asynchronous neighbor discovery algorithm using directional antennas," wireless communications and networking conference, pp. 1–6, 2016.
[10] M. J. Mcglynn and S. A. Borbash, "Birthday protocols for low en-ergy deployment and flexible neighbor discovery in ad hoc wireless networks," in ACM Interational Symposium on Mobile Ad Hoc NETWORKING and Computing, MOBIHOC 2001, October 4-5, 2001, Long Beach, Ca, Usa, 2001, pp. 137–145.
[11] I. Koutsopoulos and L. Tassiulas, "Fast neighbor positioning and medium access in wireless networks with directional antennas ," Ad Hoc Networks, vol. 11, no. 2, pp. 614–624, 2013.
[12] E. Felemban, R. Murawski, E. Ekici, and S. Park, "Sand: Sectored-antenna neighbor discovery protocol for wireless networks," in IEEE Communications Society Conference on Sensor, Mesh and Ad Hoc Communications and Networks, 2010, pp. 1–9.
[13] B. Liu, B. Rong, R. Q. Hu, and Y. Qian, "Neighbor discovery algorithms in directional antenna based synchronous and asynchronous wireless ad hoc networks," Wireless Communications IEEE, vol. 20, no. 6, pp. 106–112, 2013.
[14] S. Khatibi, M. Dehghan, and M. A. Poormina, "Quorum-based pure directional neighbor discovery in self-organized ad hoc networks," in International Symposium on Telecommunications, 2010, pp. 476–481.
[15] F. Tian, B. Liu, H. Cai, H. Zhou, and L. Gui, "Practical asynchronous neighbor discovery in ad hoc networks with directional antennas," IEEE Transactions on Vehicular Technology, vol. 65, no. 5, pp. 3614–3627, 2016.

An Analysis Model of IEEE 802.11p with Difference Services

Jin Tian[1,a*] and Lu Xu[2,b]

[1]School of Networks and Telecommunications Engineering, Jinling Institute of Technology, Nanjing, China
[2]College of Internet of Things Engineering, HoHai University, Nanjing, China
Email: [a]jim.tian@jit.edu.cn, [b]surrow@hhu.edu.cn

Abstract. Quality of Service (QoS) is a critical issue for the broadcast scheme of IEEE 802.11p systems in Vehicular Ad hoc Networks (VANETs). We present a 3-dimensional (3-D) Markov chain that the extra dimension, which models the deferring period, allows us to accurately solve for the broadcast scheme of 802.11p systems. The 3-D model solutions are validated by key simulations. Our analyses reveal that the lack of retransmission in the 802.11p system results in poor QoS performance during heavy traffic load, particularly for large VANETs.

1. Introduction

Intelligent Transportation System (ITS) is becoming a crucial part in transportation system, because it provides several key advantages compared with the old system. For example, it can detect the real-time vehicle accident problem to the driver and update the information for the related devices to take appropriate measures in time. An important infrastructure part of ITS is VANETs, which focuses on the Vehicular-to-Vehicular (V2V) and Vehicular-to-Roadside (V2R) communications.

The Wireless Access Vehicular Environment (WAVE) [1] standard includes IEEE 802.11p [2] and IEEE 1609 standard families. The IEEE 802.11p includes the physical layer and medium access control (MAC) layer. The EDCA technique prioritizes the transmission channel into four access categories. The biggest category has the highest priority to transmit packets for it has the shortest channel sensing time and the least contention window (CW).

However, EDCA protocol is based on Carrier Sense Multiple Access with Collision Avoidance (CSMA/CA). The vehicles must contend for transmission opportunities. So, the current multi-channel MAC protocol in WAVE standard is not fully applicable for real-time VANETs. Arbitrary interframe spacing (AIFS) has an important influence on QoS performance. Few works have been done to accurately evaluate the performance and reliability of the IEEE 802.11p EDCA broadcast. We propose a 3-D Markov chain queuing model with finite buffer under finite load in order to characterize the IEEE 802.11p QoS performance. The 3rd dimension represents AIFS character of different priority stream. AIFS on IEEE 802.11p has scarcely been researched by Markov chain and the result is not published in the literature.

The remainder of this paper is organized as follows. In section II briefly describes related proposals to analyze the performance of the WAVE. Section III the proposed analytical model is presented. Section IV provides simulation analysis on delay and queuing length. Research results are shown and discussed in Section IV. Finally, conclusions of this paper are drawn in section V.

2. Related Work

Many researches have been taken to improve the performance of WAVE. Many works only have one backoff stage to reflect and only applicable to systems with a single access category (AC) [4], [5].

Gallardo et al. [6], [7] proposed a model of the WAVE broadcast on the CCH but assumed an infinite MAC queue length. Meanwhile, their work did not consider the virtual collision so that all Markov chains of different priority ACs are in 1-D modeling. Yao et al. [8] establish two Markov chain models for ACs with different priorities to analyze the performance and reliability of safety-critical data broadcast on the CCH under both nonsaturated and saturated conditions. Tian et al. [9] proposed a 2-D Markov chain model, which includes backoff counter and remaining frozen time in deferral period.

Malone et al. [10] added post-backoff states to Bianchi's model to analyse the unsaturated traffic networks under a bufferless assumption. Liu et al. [11] proposed a 3-D Markov chain model with finite buffer under unsaturated traffic load. The extra 3th dimension models the queue length. Baozhu Li et al. [3] propose a 2-D Markov chain queueing model with finite buffer under finite load to characterize the IEEE 802.11p broadcast scheme for VANETs. In [3]'s 2-D model, the one dimension represents the backoff counter and the another dimension embodies the queuing length. All including backoff counter, queuing length and AIFS is few taken considerate into model.

3. Model and Mathematical Analysis

In the model, we assume that there are n stations in a network and each station has four ACs and a frame arrival follows the Poisson process. We model the backoff procedure of each AC as the finite state 3-dimensional Markov chain under the assumption that p (the transmission failure probability of the i-th access category AC_i, here only considering collision failure), is independent of the backoff procedure.

Let n_i (i = 0, 1, 2, 3) denote the number of stations in the priority i class. A transmitted frame collides when one more station also transmits during a slot time. The probability p_i, that a station for the priority i class in the backoff stage senses the channel busy, is given as

$$p_i = 1 - \left[\prod_{h=0}^{3} (1-\tau_h)^{n_h} \right] (1-\tau)^{n_i - 1} \left[\prod_{h=i+1}^{3} (1-\tau_h)^{n_h} \right] \tag{1}$$

q and q_T are the probability that a packet arrives while in each backoff state and transmission state respectively. Each state of the Markov chain consists of three parameters, (h, k, l), where h is a queue length, k is a backoff counter, and l is the remaining number of slots in AIFS deferring.

Let $b_{h,k,l}$ be the stationary distribution of the proposed 3-D Markov chains and be the steady probability of state (h, k, l). Then, based on the chain regularities (Fig. 1) we calculate the following steady state probabilities for the i_{th} AC. The priority i class responds to n_i, p_i, et al. To simple denote, the signs denote one priority with no subscript, such as p instead of p_i. From Fig. 1 and from the normalization condition, for all i we have:

Deferring period equation:

$$b_{h,k,l} = q(1-p)b_{h-1,k,l+1} + (1-q)(1-p)b_{h,k,l+1}$$
$$1 \le h \le L, 1 \le l \le d-1 \tag{2a}$$

$$b_{0,k,l} = (1-q)(1-p)b_{0,k,l+1}$$
$$1 \le k \le W-1, 1 \le l \le d-1 \tag{2b}$$

Node equilibrium equation:

$$\frac{1}{W}[q_T b_{h,0,0} + (1-q_T)b_{h+1,0,0}] + qpb_{h-1,k,d} + (1-q)p[b_{h,k,0} + \sum_{l=1}^{d-1} b_{h,k,l}] = [1-p(1-q)]b_{h,k,d} \qquad 2 \le h \le L-1 \tag{3a}$$

$$\frac{1}{W}[qb_{0,0,0} + (1-q_T)b_{1,0,0}] + (1-q)p[b_{0,k,0} + \sum_{l=1}^{d-1} b_{0,k,l}]$$
$$= [1-p(1-q)]b_{h,k,d} \tag{3b}$$

$$\frac{1}{W}[p(1-q)b_{0,0,0}+q_T b_{1,0,0}+(1-q_T)b_{2,0,0}]+qpb_{0,k,d}$$
$$+(1-q)p[b_{1,k,0}+\sum_{l=1}^{d-1}b_{1,k,l}]=[1-p(1-q)]b_{1,k,d} \tag{3c}$$
$$qb_{idle}=(1-q)b_{0,0,0},1\le k\le W-1$$

$$\frac{1}{W}q_T b_{L,0,0}+qpb_{L-1,k,d}+(1-q)p[b_{L,k,0}+\sum_{l=1}^{d-1}b_{L,k,l}]$$
$$=[1-p(1-q)]b_{h,k,d}\quad 1\le k\le W-1 \tag{3d}$$

Backoff period equation:

$$b_{h,k,0}=q(1-p)b_{h-1,k+1,0}+(1-q)(1-p)(b_{h,k,1}+b_{h,k+1,0})$$
$$1\le h\le L,1\le k\le W-2 \tag{4a}$$

$$b_{0,k,0}=(1-q)(1-p)(b_{0,k,1}+b_{0,k+1,0})$$
$$1\le k\le W-2 \tag{4b}$$

$$(1-q)b_{L,k,0}=q(1-p)b_{L-1,k+1,0}+(1-q)(1-p)(b_{L,k,1}+b_{L,k+1,0})$$
$$1\le k\le W-2 \tag{4c}$$

Normalization condition equation:

$$1=b_{idle}+\sum_{k=0}^{W-1}b_{0,k,0}+\sum_{h=1}^{L}b_{h,0,0}+\sum_{h=1}^{L}\sum_{k=1}^{W-1}b_{h,k,0}+\sum_{h=0}^{L}\sum_{k=1}^{W-1}\sum_{l=1}^{d}b_{h,k,l} \tag{5}$$

Above equations allow us to calculate b0,0,0.

For a network with *n stations* we define the transmission probabilities τ for each AC. Since the transmission attempt of AC happens at state (*h, 0, 0*), τ is described as

$$\tau=qb_{0,0,0}+\sum_{h=1}^{L}b_{h,0,0} \tag{6}$$

where L is the maximum queue size of AC.

Packet arrival and transmitting probability in the transmitting state onto 'fan-out' backoff count selectors with Queue length *h* is

$$q_T b_{h,0,0}+(1-q_T)b_{h+1,0,0}$$
$$\text{for } h=1,2,\ldots,L-1 \tag{7}$$

Packet arrival probability in the deferring state is

$$P_{defer}=\sum_{r_1}^{d-f_1}\binom{r_1+f_1}{f_1}(1-q)^{r_1}(1-p)^{r_1+f_1}q^{f_1}$$
$$f_1=0,1,\ldots,d. \tag{8}$$

Packet arrival probability in the back-off state is

$$P_{(1+h+f)(1+h)}=\frac{1}{W}\sum_{k=f}^{W-1}\sum_{r}^{k-f}\binom{r+f}{f}(1-q)^{r}(1-p)^{r+f}q^{f}$$

$$for\begin{cases} h = 0,\dots, L-1, \\ f = 0,\dots, f^{*}, \\ f^{*} = \min(L-1-h, W-1) \end{cases} \tag{9}$$

$$P_{jk} = 0,\ j = h + f + 1, \text{otherwise}, \text{for } j \neq L+1;$$

$$p_{(L+1)k} = \frac{W+1}{2} - \sum_{j=1}^{L} P_{jk}, \text{for } k = 1,\dots,L+1$$

We define $P[Q=h]$ as the probability of the interface queue length, Q, being h at any given time, for $h = 0,\dots, L$. Noting that state transitions are not equally spaced, we use $P_S[Q = h]$, which we define as the probability of the interface queue length, Q, being h during each n-station state transition, for $h = 0, \dots, L$, to calculate $P[Q=h]$. To determine $P_S[Q = h]$, we also have to account for non-backoff contributions to $P_S[Q= h]$.

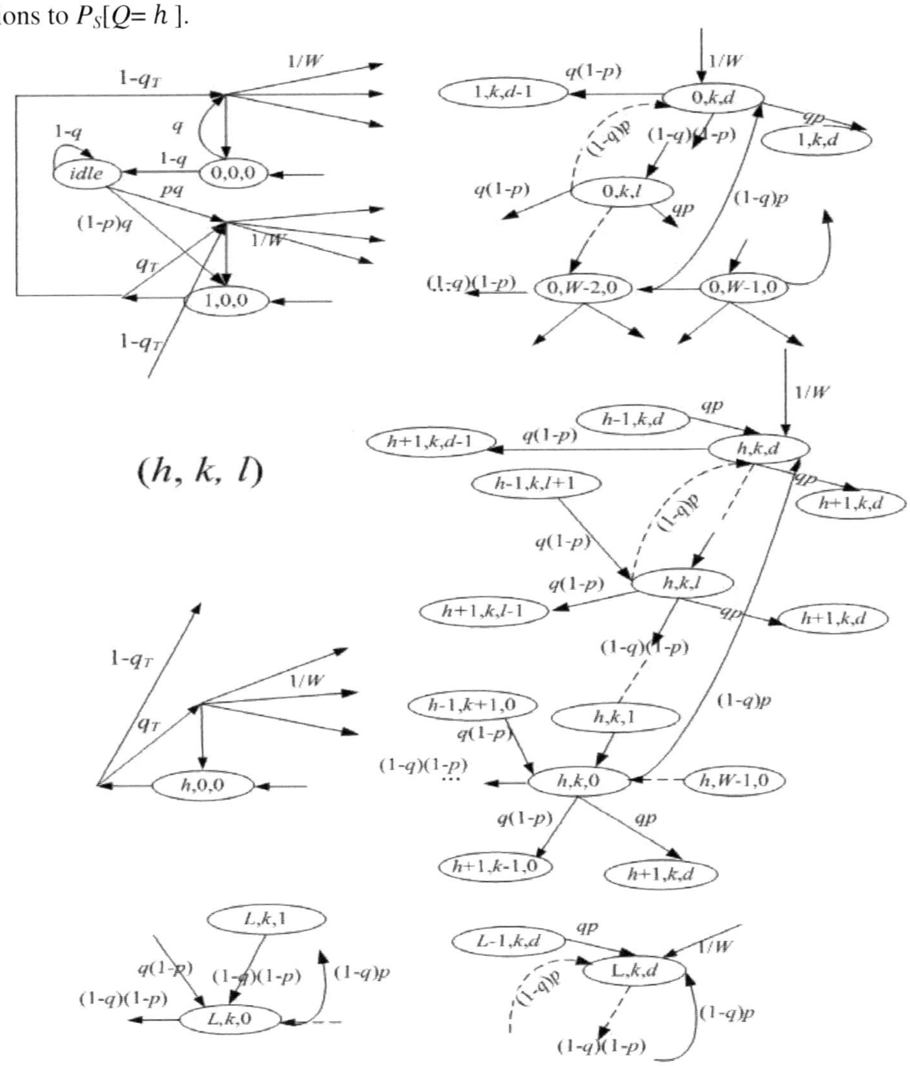

Figure 1 Markov chain state transfer of 802.11p dealing with deferring period.

Then, including non-backoff *contributions* to $P_S[Q= h]$, which comprise being in (*idle*) and transmitting in (1, 0) directly after a packet arrives in (*idle*) and the channel is sensed 'free', we have progressing probability product, which is denotes

IOP Conf. Series: Journal of Physics: Conf. Series **910** (2017) 012068 doi:10.1088/1742-6596/910/1/012068

$$P_s[Q = 0] = qb_{0,0,0} + (1 - q_T)b_{1,0,0} \qquad h = 1,2,3,\ldots,L-1 \tag{10a}$$

$$P_s[Q = h] = P_{defer}\Big\{ P_{(h+1)2}\,p(1-q)b_{0,0,0} + P_{(h+1)1}[qb_{0,0,0} + (1-q_T)b_{1,0,0}] + \sum_{k=2}^{L} P_{jk}[q_T b_{k-1,0,0} + (1-q_T)b_{k,0,0}] \Big\} \tag{10b}$$

$$P_s[Q = L] = P_{defer}\Big\{ P_{(L+1)2}\,p(1-q)b_{0,0,0} + P_{(L+1)1}[qb_{0,0,0} + (1-q_T)b_{1,0,0}] + \sum_{k=2}^{L} P_{jk}[q_T b_{k-1,0,0} + (1-q_T)b_{k,0,0}] + P_{(L+1)(L+1)}q_T b_{L,0,0} \Big\} \tag{10c}$$

$$P_s[Q = h \cap T_x] = qb_{0,0,0},\ h = 0 \tag{11a}$$

$$P_s[Q = h \cap T_x] = b_{h,0,0},\ h = 1, 2, \ldots, L \tag{11b}$$

We seek solutions to the n-station system as a function of the normalised offered load, denoted λ, which is the expected number of packets delivered to all n stations per "time for a successful transmission". Thus:

$$\lambda = n r_q T_s \tag{12}$$

where is the packet arrival rate per second and r_q defines q_T and q as:

$$q_T = r_q T_c \text{ ; and}$$

$$q = r_q E_{s|n-1}$$

To determine time-based system properties, it is necessary to convert the state probabilities into time-based probabilities. The durations are: σ when no station is transmitting; T_s when a successful (single) transmission occurs (from any station); and T_c when there is a collision due to two or more simultaneous transmissions. We will also need the expected time spent per Markov state for states in which the station is not transmitting.

We denote this $E_{s|n-1}$, since, due to station independence, $E_{s|n-1}$ is equivalent to E_s for an $(n-1)$-station system. Thus:

$$E_{s|n-1} = P_{e|n-1}\sigma + P_{s|n-1}T_s + P_{c|n-1}T_c \tag{13}$$

where $P_{e|n-1} = (1-\tau)^{n-1}$; $P_{s|n-1} = (n-1)\tau(1-\tau)^{n-2}$; and $P_{c|n-1} = 1 - P_{e|n-1} - P_{s|n-1}$

Weighting *transmitting* states by T_b and non-transmitting states by $E_{s|n-1}$, and normalizing by E_s, gives $P[Q = h]$ as:

$$P[Q = h] = \frac{\{P_s[Q = h] - P_s[Q = h \cap T_x]\}E_{s|n-1} + P_s[Q = h \cap T_x]T_b}{E_s} \tag{14}$$

Then, finally, $\overline{Q}$ is:

$$\overline{Q} = \sum_{h=0}^{L} hP[Q = h] \tag{15}$$

The MAC delay, Δ, is the average time from a packet first reaching the head-of-the-line till successful transmission. Δ_l denotes the average packet service time from a packet first reaching the head-of-the-line till it leaves the system, either by successful transmission or retransmission failure. According to no retransmission and acknowledge in IEEE 802.11p, we have:

$$\Delta l = \Delta$$

To obtain Δ, we first calculate the average time per state, given a positive queue length, $E_{s|Q\geq 1}$,as:

$$E_{s|Q\geq 1} = \frac{\tau(1-p)T_s + \tau p T_c + P_{e\cap Q\geq 1} E_{s|n-1}}{\tau + P_{e\cap Q\geq 1}} \tag{16}$$

where $P_{e\cap Q\geq 1}$ is the state probability of being in a backoff state with positive queue length, such that:

$$P_{e\cap Q\geq 1} = \sum_{h=1}^{L}\sum_{k=1}^{W-1} b_{h,k,0} + \sum_{h=0}^{L}\sum_{k=1}^{W-1}\sum_{l=1}^{d} b_{h,k,l} \tag{17}$$

where *b0,k,0* is the state probability of being in state (*0*, *k*, *0*).

We then find *the* state probability of a packet leaving the system, by successful transmission, given a positive queue length:

$$P_{l|Q\geq 1} = \tau(1-p)/(\tau + P_{e\cap Q\geq 1}) \tag{18}$$

Dividing (16) by (18) gives:

$$\begin{aligned}\Delta = \Delta_l &= E_{s|Q\geq 1} / P_{l|Q\geq 1} \\ &= T_s + \frac{pT_c}{1-p} + \frac{P_{e\cap Q\geq 1} E_{s|n-1}}{\tau(1-p)}\end{aligned} \tag{19}$$

For the basic channel access, we have service time T_s(successful transmission) and T_c (collision):

$$T_s = \mathrm{H} + L_p + \varepsilon \tag{20}$$

$$T_c = \mathrm{H} + L_p + \varepsilon \tag{21}$$

where *H* is the time occupied by the PHY and MAC headers, L_p is the average time to send a DATA frame, and ε is the propagation delay.

D is the *average* time from a packet first joining the input interface queue till its successful transmission.

$$D = \Delta_l \sum_{h=0}^{L} hP[Q=h] + \Delta \tag{22}$$

The states (*L*, *0*, *0*), … (*L*, *W−1*, *0*) represent that the queue is full. In these states, if a new packet arrives, they are blocked except for (*L*, *0*, *0*); for (*L*, *0*, *0*), the packet being transmitted is removed from the queue so as to allow a new arrival. Thus, the queue blocking probability,

$$P_B = P[Q=L] - b_{L,0,0} T_s / E_s \tag{23}$$

4. Simulation and Performance evaluating

Simulations were conducted in NS2 to validate the accuracy of our 3-D Markov chain deferring model. All nodes are within communication range of each other, and each node independently generates Poisson traffic with specified rate. Each simulation experiment for a particular network configuration was run for 80 seconds after a 10 second initialization. The values of interest were calculated for each subsequent 10-second block. The average of the eight 10-second blocks was taken as one simulation value. The network parameters used in the simulations are given in Table I.

The simulation result is very matched to the proposed theoretical analysis because AIFS is considered. Fig. 2 shows the average queue length varying with normalized offered load for different network sizes. The queue length exhibits an almost bimodal behavior. It is close to zero before saturation, and is almost full to the buffer size when the network is saturated. The switching point from empty to full queue corresponds to the level of offered load where the network reaches saturation. The Fig.2 shows that when the level of offered loads is contrast, the fewer node, the longer the queue.

In theory, the system delay is small if we don't consider AIFS. The simulation result is very matched to the proposed theoretical analysis because AIFS is considered. Fig. 3 shows the average packet delay varying with normalized offered load for different network sizes. Not surprisingly, similar to the queue length in Fig. 2, the packet delay also exhibits an almost bimodal behavior. However, it is interesting that packet delay reaches different saturation levels for different network sizes: large network incur high packet delay. The reason is that there are more nodes to transmit in a large network. In such case, a node in a large network will encounter more freezes in its backoff count down than the node in a small network. Such frequent freezes prolong the packet backoff count and increase the packet media access delay. As a result, the packet delay becomes long for a large network.

Table 1 Simulation Setting

Parameter	*Value*	*Parameter*	*Value*
Slot time, δ	*20 μs*	*Buffer size, L*	*50*
CW, *W*	*32*	*Node numbers, n*	*5, 10, 20*
AIFS, *d*	*9*	*Data payload Lp*	*64 bytes*
Basic rate	*1 Mbit/s*	*Propagation delay, ε*	*1 μs*
Transmission rate	*1 Mbit/s*	*Heads, H*	*(144+48) bits, (30+4) bytes*

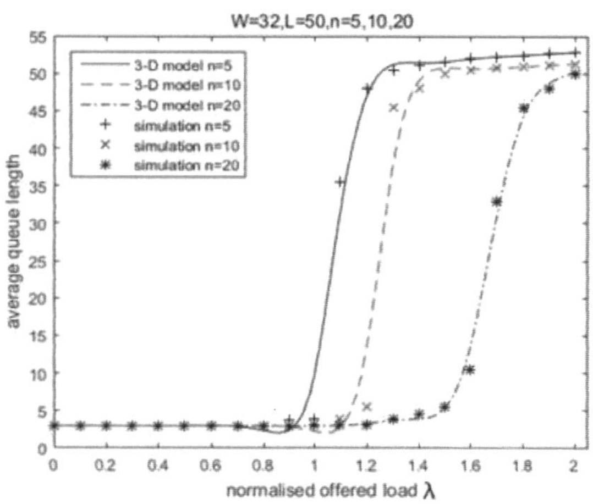

Figure 2 Average queue length vs offered load over different network sizes.

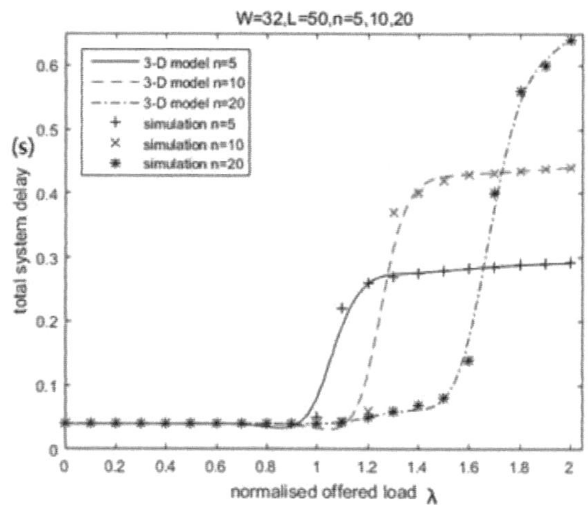

Figure 3 Average packet delay vs offered load over different network sizes.

5. Conclusion

In this paper, we proposed a 3-D Markov chain deferring model for analysing the QoS performance of the IEEE 802.11p broadcast scheme in VANETs. Our 3-D model characterizes the unsaturated traffic condition by including an idle state and post-backoff states, and integrates the queueing process with the EDCA deferring procedure. Important QoS measures were obtained and validated by key simulations. Our analytical results revealed the QoS performance degradation due to the lack of binary exponential backoff and retransmission in VANETs. Such performance deterioration can be avoided by proper traffic control, and our 3-D Markov chain model pointed to a traffic load threshold from which the network QoS starts to degrade. To maintain good QoS performance for VANETs, traffic control guidelines can be regulated by this threshold.

6. Acknowledgement

Nanjing City Sci. & Technol. Program (2012ZD003); scientific research funded by Jinling Institute of Technology (jit-n-201304).

7. References

[1] IEEE Standard for Wireless Access in Vehicular Environments (WAVE)—Multi-channel Operation" IEEE Std 1609.4-2010 (Revision of IEEE Std 1609.4-2006), pp. 1-89.

[2] IEEE Standard 802.11p: Wireless Access in Vehicular Environments (WAVE), draft standard 2008 edition.

[3] Baozhu Li, Bo Hu, Ren Ping Liu, and Shanzhi Chen, "Modeling and QoS analysis of the IEEE 802.11p broadcast Scheme in vehicular Ad Hoc networks", in Proceedings of ICC'2013, pp. 6444-6448, Budapest, Hungary, June 2013.

[4] B. S. Gukhool and S. Cherkaoui, "IEEE 802.11p modeling in NS-2," in Proceedings of 33rd IEEE Conference on Local Computer Networks (LCN'08), pp. 622-626, Montreal, Canada, October 2008.

[5] K. Bilstrup, E. Uhlemann, E. G. Strom, and U. Bilstrup, "Evaluation of the IEEE 802.11p MAC method for Vehicle-to-Vehicle communication," in Proceedings of the 68th Semi-Annual IEEE Vehicular Technology (VTC '08), Calgary, Canada, September 2008.

[6] J. Gallardo, D. Makrakis, and H. Mouftah, "Performance analysis of the EDCA medium access mechanism over the control channel of an IEEE 802.11p WAVE vehicular network," in Proc. IEEE Int. Conf. Commun., Dresden, Germany, 2009, pp. 1–6.

[7] J. Gallardo, D. Makrakis, and H. T. Mouftah, "Mathematical analysis of EDCA's performance on the control channel of an IEEE 802.11p WAVE vehicular network," EURASIP J. Wireless Comm., pp. 1–15, 2010.

[8] Yuan Yao, Lei Rao, and Xue Liu, "Performance and reliability analysis of IEEE 802.11p safety communication in a highway environment," IEEE TRANSACTIONS ON VEHICULAR TECHNOLOGY, VOL. 62, NO. 9, pp. 4198-4212, November 2013

[9] Jin Tian, Ren Ping, Liu, Xu Zhang, "Saturation delay performance analysis by Markov chain for WAVE", International Journal of Advanced Information Science and Technology, 45(45): 93-97, January 2016.

[10] D. Malone, K. Duffy, and D. Leith, "Modeling the 802.11 distributed coordination function in nonsaturated heterogeneous conditions," networking, IEEE/ACM Transactions on, vol. 15, no. 1, pp. 159–172, 2007.

[11] R. P. Liu, G. J. Sutton, and I. B. Collings, "A new queueing model for QoS analysis of IEEE 802.11 DCF with finite buffer and load," IEEE Trans. Wireless Commun., vol. 9, pp. 2664–2675, August 2010.

A Survey of Mainstream Indoor Positioning Systems

Fangzheng Zhou

University of Electronic Science and Technology of China, Chengdu, China
Email: 843550456@qq.com

Abstract. Indoor positioning problems have been one of the most challenging research topics in recent years, which comprise smartphone-based indoor localization, tracking, and navigation. Many positioning systems have been designed to provide such reliable indoor location-based services (LBS). In this paper, we have compared different indoor positioning systems, and discussed its future improvements.

1. Introduction

In recent years, indoor localization has been one of the most popular problems in mobile community. In indoor environments, satellite positioning cannot perform well, since the satellite signal cannot penetrate the building, thus indoor localization technology is needed. Indoor localization technology can be applied in many areas, e.g., rescuing people in fire, tracking suspects, locating special targets in buildings. There are many indoor localization approaches, including infrared-based, ultrasonic-based, Bluetooth-based, RFID-based, UWB-based, and WLAN-based. These technologies use unique signal to replace the satellite signal. Among all these technologies, WLAN-based indoor localization approaches have been the center of attention, due to their low deployment cost, potential for reasonable accuracy and easiness to integrate into the mobile device. In this paper, advantages and disadvantages of different indoor localization approaches will be illustrated, and our focus is on the WLAN-based indoor localization approaches.

2. Motivation

Although so many technologies have been introduced, indoor localization is still facing many challenges, including environmental impact, channel attenuation. And higher accuracy with lower cost is still the center of attention. Since there are such a lot of indoor localization technologies and they all have advantages and disadvantages. It is necessary to distinguish them and to understand their principle, so that we can apply suitable technology to the right situation. In this paper, we will compare different methods in accuracy, cost, portability and anti-jamming capability, and our emphasis is the WLAN-based technology, including fingerprint method, propagation model method and technologies that combine the two methods. Our goal is to give a brief introduction to common indoor localization technologies, that readers can understand the status quo of the development of indoor localization.

The remainder parts of this paper is organized as follows, in Section 3 we will introduce several common indoor localization technologies, including infrared-based, ultrasonic-based, Bluetooth-based, RFID-based, UWB-based. In Section 4 we will focus on the WLAN-based technologies. We will present some future work in Section 5.

3. some common indoor location technology

3.1. Indoor Location Technology Based on Infrared

The infrared location system generally contains two parts, one is the CPE (customer premise equipment) which can transmit the modulated infrared ray (infrared IR mark) outwardly at a fixed frequency, the other is the receivers deployed in the room which can receive the signal from the CPE and locate it. A number of CPE and receivers will be used to form a signal detection network to cover the space to be measured.

The advantage of the system based on infrared is the high accuracy that the error is always below 10 cm. But infrared cannot pass through the obstacle, where the transmission distance is short, and vulnerable to indoor lighting or sunlight interference. And high accuracy needs many receivers, so the cost will be high.

3.2. Indoor Location Technology Based on Ultrasonic

The ultrasonic indoor location systems can be divided into two categories [6].

3.2.1 System based on reflective ranging method. A range finder carried by the user transmits the ultrasonic signal at a fixed frequency, and several (at least 3) receivers are deployed in the building to receive the ultrasonic signal and reflect the signal to the user. The range finder measures the distance to every receiver by the time at which the signal is received until the reflected signal is received, then the user's coordinates can be obtained.

3.2.2 Ultrasound combined with radio frequency technology. The system contains several beacon nodes that *transmit* their own special mark of the RF signal and the ultrasonic signal without any information through the synchronizing channel. The CPE collects the radio frequency signal and the ultrasonic signal from each beacon node, and use the TDOA (time difference of arrival) to calculate the location of the user.

The system based on ultrasonic has a high accuracy and a simple system structure. But ultrasound is susceptible to multipath effects and obstructions, and for reflective ranging method, the cost is high due to the need of a large number of receivers.

3.3. Indoor Location Technology Based on Bluetooth

The key of Bluetooth positioning is the connection between user's Bluetooth device and the Bluetooth anchor nodes deployed in the building. When the connection established, the Bluetooth anchor nodes will collect the RSSI (Received Signal Strength Indicator) and LQ (Link Quality) of user's device, and send the information to the server, then the server will calculate the user's location.

The advantage of Bluetooth location technology is the low cost and low power consumption, Bluetooth module is easy to integrate, and now the mobile communication devices generally have Bluetooth. But the accuracy of Bluetooth system is low with an error more than 1 meter, which is also susceptible to noise signal interference, and the transmission distance is short [7].

3.4. Indoor Location Technology Based on RFID

The system usually consists of a RFID tag carried by user and several RFID readers deployed in the building plane which can receive the radio frequency signal transmitted by the RFID tag. Common RFID positioning system has two kinds.

3.4.1 LANDMARK System [8].. The system has several reference labels, the reference label which has the most similar signal strength with the target position is found by the NNSS (nearest neighbor(s) in signal space) algorithm, and the position of this label is the position of the target to be measured. The measurement accuracy depends on the number of reference labels.

3.4.2 SpotON system [9].. The system uses the relationship between the recognition distance and the signal strength and the triangulation method to find the coordinates of the label to be measured.

System based on RFID has high accuracy and low cost, yet susceptible to environment. The propagation distance is short and the system is not easy to integrate into the mobile device.

3.5. Indoor Location Technology Based on Computer Vision

There are two ways of use computer vision to locate.

By setting the geometric constraints between the feature points in the image and the feature points of the actual objects, a mathematical equation is established to obtain the closed solution of the object position in the camera coordinates, and the position of the object in the world coordinates is obtained by solving the equation [10].

Drawing the road signs on the ceiling of the building, and identifying the landmark information by the ceiling image taken by the camera set on the target to be measured, in order to obtain the location of the target [11].

The advantage of using computer vision positioning is high accuracy, and it is suitable for indoor robot routing. But image positioning is vulnerable to interference (debris in the images), and the need for a camera real-time shooting images makes it not easy to integrate the system into the mobile device.

3.6. Indoor Location Technology Based on UWB

Ultra-wideband(UWB) is a technique for transmitting data by sending and receiving very narrow pulses with nanoseconds.

Here are two common UWB system [5].

3.6.1 Sapphire system. The system consists of a plurality of robots and several (at least 4) receivers and a control center. The robots can transmit a pulse signal. The receiver measures the arrival time of the original signal and sends the result to the control center. Based on four or more data, the control center can calculate the three-dimensional position of the robot.
Unbise system

3.6.2 Unbise system combines RF (radio frequency) and UWB. The system consists of sensors and robots. Multiple sensors deployed in a fixed location using standard Ethernet, which can receive RF and UWB signals. The robot can send the UWB signal and the RF signal containing the identifier, the sensor uses the TDOA and AOA methods to calculate the position of the robot.

UWB signal is not sensitive to the channel weakness, which can go through the obstacles, and which is difficult to be intercepted with high precision, high security and low cost. UWB system almost meets all needs of indoor positioning. There are no obvious shortcomings, the main research direction is to further improve accuracy, reduce costs and improve compatibility.

Table 1 Comparison of Multiple Location Techniques

Technology	*Accuracy*	*Cost*	*Portability*	*Anti-jamming capability*
Infrared	High	High	Integrated into mobile devices	Susceptible to indoor lighting or sunlight
Ultrasonic	High	High	Integrated into mobile devices	Susceptible to multipath effects and obstructions
Bluetooth	Low	Low	Integrated into mobile devices	Susceptible to noise signal interference
RFID	High	Low	Not integrated into mobile devices.	Susceptible to environment
Computer Vision	High	Low	Not integrated into mobile devices	Susceptible to interference(debris in the images)
UWB	High	Low	Not integrated into mobile devices	Anti-interference

All of the approaches in Section 2 need special devices to transmit signal to make it expensive to deploy the system. WLAN-based system uses wireless network signal that has almost covered everywhere in our world, this is an advantage of the WLAN-based system. In next section we will focus on the WLAN-based system.

4. WLAN-based Indoor Positioning System

4.1 Fingerprint - based WLAN Indoor Positioning System

In general, system based on the signal fingerprint positioning is divided into two stages. Take the RADAR system [4] as an example. First, in the offline phase, a client device that can transmit wireless signals is used to record the coordinates of multiple points in the building plane. The combination of the coordinates and the signal strength received by each access point is the fingerprint of a point. Use all fingerprint to plot a radio map. In the online phase, according to the signal strength information received by each access point, the nearest neighbor algorithm map can be used to find the most matching position in the radio, which is the estimated user location.

RADAR as the earliest WLAN signal positioning system is not accurate enough, because in the calculation of signal strength, in order to simplify the model, only consider the influence of user's towards to the signal strength, ignoring some of the wireless channel changes. And then some positioning systems such as HOURS [3], improve accuracy by identifying and positioning wireless channel changes, and use location clustering technology to reduce the amount of calculation.

In practical applications, the accuracy of positioning system based on position fingerprint is depend on the density of position fingerprint in radio map. The higher the fingerprint density, the higher the positioning accuracy. In addition, multiple samples selected in the online phase can also improve accuracy, but the increase in the number of samples will also increase calculation delay, so the number of samples should be dynamically adjusted according to demand.

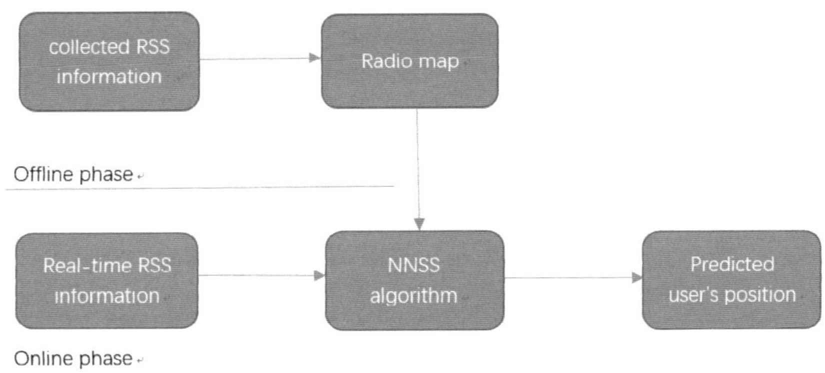

Figure 1 The schematic of the RADAR system

The biggest advantage of fingerprint-based WLAN location system is the small computational complexity and the high accuracy, and now the penetration of WLAN network is high, so no additional wireless signal transmitter is needed. However, the cost of making a high-density radio map is high, and since the attenuation of the signal strength has a strong relationship with the layout of the plane to be measured, it is necessary to redraw the radio map if the layout of the plane changes. In the mall, cinemas and other place where people move around or obstacles move frequently, the measurement results will be a great impact.

4.2 Model-based WLAN Indoor Positioning System

The propagation model method uses the signal strength as the medium of the distance measurement, and converts the signal strength into a distance to determine the position of the point to be measured. The signal strength is inversely proportional to the square of the propagation distance, but in the indoor environment, due to the obstacles, the wireless signal will get multiple refraction, diffraction and

IOP Conf. Series: Journal of Physics: Conf. Series **910** (2017) 012069 doi:10.1088/1742-6596/910/1/012069

reflection, making the propagation model become complex. There are two kinds of propagation model method, which are statistical propagation model method and deterministic propagation model method.

In the statistical propagation model method, all environmental factors are implicitly included in it, so the polynomial regression model can be used as the signal propagation model. Another common model is the path loss model, which is based on the Floor Attenuation Factor propagation model. This model introduces a Wall Attenuation Factor, which mainly considers the affection of wall to the signal attenuation. The statistical propagation model does not need to draw a radio map, only a few samples are needed to calculate the parameters in the propagation model.

The deterministic propagation model is based on the physical nature of the propagation of radio signals and needs to be suitable for all environments. Therefore, the realization of the deterministic propagation model requires a huge database of environmental characteristics and is computationally intensive so it is difficult to implement in practical applications. Therefore, the deterministic propagation model is usually combined with the fingerprint, which is used to reduce the workload of fingerprint sampling.

The biggest advantage of the propagation model method is that it does not need to draw the radio map, nor does it need to arrange special receiving equipment, hence the cost is low, and the flow of personnel within the measured space has little effect on the measurement results, but the measurement accuracy is a bit lower than the fingerprint method and the calculation is also complex. At present the main research direction is still focused on how to build a more accurate wireless signal propagation model.

The following is a simplified propagation model in RADAR.

$$P(d)[dBm] = P(d_o)[dBm] - 10n\log\left(\frac{d}{d_o}\right) - \begin{cases} nW * WAF & nW < C \\ C * WAF & nW \geq C \end{cases}$$

Figure 2 Propagation model in RADAR

In this model n represents the ratio of the path attenuation to the distance increase, P represents the signal strength, d represents the distance between the receiver and the transmitter, C represents the maximum number of obstacles, nW represents the number of obstacles between the receiver and the transmitter. WAF is the wall attenuation factor, which is related to the layout of the building and the wall material. Using the propagation model, the relationship between the signal strength and the distance can be obtained so that the user's position is determined by the actual received signal strength.

4.3 WLAN Indoor Positioning System Combine Propagation Model Method and Fingerprint Method

In the offline phase of the fingerprint method, the propagation model is used to reduce the workload of sampling. Taking the ARIADNE system [2] as an example, the system is divided into a map generation module and a search module. In the map generation module, a propagation model is constructed using ray tracing and simulated annealing algorithm. Combine the propagation model and a set of signal strength information received by the three sniffers arranged in the measured space, a location fingerprint can be found. In the search module, use clustering technology to reduce the amount of calculation and improve accuracy.

Although the propagation model reduces the cost of making radio maps, location fingerprint is still sensitive to environmental changes. And a radio map made by propagation model is accurate, even with five sniffers, the average error is still more than one meter. The accuracy can now be improved by optimizing the propagation model and sniffer distribution, or using historical location information. And a dynamic signal strength map can solve the problem that location fingerprint method sensitive to environmental changes

For example, Modellet [12], an algorithmic approach that optimally approximates the actual radio map by unifying model-based and fingerprint-based approaches. Modellet represents the radio map using a fingerprint-cloud that incorporates both measured real fingerprints and virtual fingerprints, which are computed from models with a local support based on the key concept of the supporting set. A supporting set is a set of real fingerprints from which a derived model can best approximate the local radio map. Using propagation model reduces the amount of measurement data needed to make a high quality radio map and to reduce costs.

5. Conclusion and Future Work

In this paper, we have compared different indoor localization approaches, and we believe WLAN-based technology is the most promising technology. Since WLAN network has a high penetration rate, we can integrate the positioning system into existing WLAN devices, so the cost is low, and the WLAN-based system has potential for reasonable accuracy. The fingerprint method will be the major approach, and the propagation model method can help to make high quality radio map.

The future work of indoor positioning system is to improve accuracy and reduce costs, make the system applicable to a variety of situations, and add more user-friendly features [13-17]. For WLAN-based systems [18, 19], finding a better way to combine the fingerprint method and the propagation model method may significantly increase the accuracy.

6. References

[1] Ying Wang, X. Jia, Hyung Keun Lee: An indoor wireless positioning system based on wireless local area network infrastructure. In University of Technology Dresden Germany & the University of Technology, 2003.
[2] Yiming Ji, Saad Biaz, Santosh Pandey, Prathima Agrawal: ARIADNE: A Dynamic Indoor Signal Map Construction and Localization System. In International Conference on Mobile Systems, 2006.
[3] Moustafa Youssef, Ashok Agrawala: The Horus WLAN Location Determination System. In International Conference on Mobile Systems, 2005.
[4] Paramvir Bahl, Venkata N. Padmanabhan: RADAR: An In-Building RF-based User Location and Tracking System. In Infocom Nineteenth Joint Conference of the IEEE Computer & Communications Societies IEEE, 2000.
[5] LI Xiao-hui, LIU Ya, ZHANG Li-rong: Indoor Positioning System Using Ultra Wideband Radio. In MEASUREMENT & CONTROL TECHNOLOGY, 2007.
[6] HAN Shuang, LUO Haiyong, CHEN Ying, DING Yuzhen: The Design and Implementation of a TDOA-Based Ultrasonic Indoor Localizing System. In CHINESE JOURNAL OF SENSORS AND ACTUATORS, 2010.
[7] Sofía Aparicio, Javier Pérez, Paula Tarrío, Ana M. Bernardos, José R. Casar: An Indoor Location Method Based on a Fusion Map Using Bluetooth and WLAN Technologies. In International Symposium on Distributed Computing and Artificial Intelligence 2008.
[8] Lionel M. Ni, Yunhao Liu, Yiu Cho Lau, Abhishek P.Patil: LANDMARC: Indoor Location Sensing Using Active RFID. Wireless Networks. November 2004.
[9] Jeffrey Hightower, Gaetano Borriello, Roy Want: SpotON: An Indoor 3D Location Sensing Technology Based on RF Signal Strength. In Uw Cse, 2000.
[10] De Xu, Liwei Han, Min Tan, You Fu Li: Ceiling-Based Visual Positioning for an Indoor Mobile Robot with Monocular Vision. In IEEE Transactions on Industrial Electronics, 2009.
[11] LI Yibo, SHEN Xiaoling, LIU Yanmei, WANG Xinxing: Method for indoor airship localization with computer vision. In Computer Engineering and Applications, 2013.
[12] Liqun Li, Guobin Shen, Chunshui Zhao, Thomas Moscibroda, Jyh-Han Lin, Feng Zhao: Experiencing and Handling the Diversity in Data Density and Environmental Locality in an Indoor Positioning Service. In International Conference on Mobile Computing & Networking, 2014.
[13] Ruipeng Gao, Mingmin Zhao, Tao Ye, Fan Ye, Guojie Luo, Yizhou Wang, Kaigui Bian, Tao Wang, Xiaoming Li, "Multi-story Indoor Floor Plan Reconstruction via Mobile Crowdsensing," *IEEE* Transactions on Mobile Computing (IEEE TMC), vol. 15, no. 6, pp. 1427-1442, 2016.
[14] Ruipeng Gao, Yang Tian, Fan Ye, Guojie Luo, Kaigui Bian, Yizhou Wang, Tao Wang, and Xiaoming Li, "Sextant: Towards Ubiquitous Indoor Localization Service by Photo-taking of the

Environment," IEEE Transactions on Mobile Computing (IEEE TMC), vol.15, no.2, pp.460-474, 2016.

[15] Ruipeng Gao, Mingmin Zhao, Tao Ye, Fan Ye, Yizhou Wang, Guojie Luo, "Smartphone-based Real Time Vehicle Tracking in Indoor Parking Structures," IEEE Transactions on Mobile Computing (IEEE TMC), vol. 16, no. 7, pp. 2023-2036, 2017.

[16] Ruipeng Gao, Bing Zhou, Fan Ye, and Yizhou Wang, "Knitter: Fast, Resilient Single-User Indoor Floor Plan Construction," in Proceedings of IEEE INFOCOM, 2017.

[17] Ruipeng Gao, Mingmin Zhao, Tao Ye, Fan Ye, Yizhou Wang, Kaigui Bian, Tao Wang, and Xiaoming Li, "Jigsaw: Indoor Floor Plan Reconstruction via Mobile Crowdsensing," in Proceedings of ACM MobiCom, 2014.

[18] K. Bian, and J. Park. Maximizing Rendezvous Diversity in Rendezvous Protocols for Decentralized Cognitive Radio Networks, IEEE Transactions on Mobile Computing (IEEE TMC), Vol. 12(7):1294--1307, July 2013.

[19] Y Liao, K Bian, L Song, Z Han. Full-duplex MAC protocol design and analysis. IEEE Communications Letters 19 (7): 1185-1188, July 2015.

Efficient Secure and Privacy-Preserving Route Reporting Scheme for VANETs

Yuanfei Zhang, Qianwen Pei, Feifei Dai and Lei Zhang

Shanghai Key Laboratory of Trustworthy Computing, East China Normal University, Shanghai, China

Email: 51151500152@stu.ecnu.edu.cn, 51164500231@stu.ecnu.edu.cn, 51151500011@stu.ecnu.edu.cn, leizhang@sei.ecnu.edu.cn

Abstract. Vehicular ad-hoc network (VANET) is a core component of intelligent traffic management system which could provide various of applications such as accident prediction, route reporting, etc. Due to the problems caused by traffic congestion, route reporting becomes a prospective application which can help a driver to get optimal route to save her travel time. Before enjoying the convenience of route reporting, security and privacy-preserving issues need to be concerned. In this paper, we propose a new secure and privacy-preserving route reporting scheme for VANETs. In our scheme, only an authenticated vehicle can use the route reporting service provided by the traffic management center. Further, a vehicle may receive the response from the traffic management center with low latency and without violating the privacy of the vehicle. Experiment results show that our scheme is much more efficiency than the existing one.

1. Introduction

Nowadays, vehicles bring much convenience for modern life. But with the increasing amount of vehicles, urban transportation system suffers many challenges. Traffic congestion happens in big cities during rush hours everyday, which has negative effects on traffic efficiency of the cities. Air pollution and low traffic efficiency caused by traffic congestion are serious problems for traffic management department. Also the complexity of the traffic environment makes drivers difficult to make decisions and could easily lead to accidents. Therefore, government focuses on technologies that can be used to resolve traffic problems. Vehicular ad-hoc network (VANET) is proposed recently to improve traffic efficiency. It has attracted more and more attention from both industry and academic community. A VANET usually consists of vehicles and infrastructures including roadside units (RSUs). Each entity in VANET is equipped with on-board unit (OBU) which has communication module and computation module. A VANET enables vehicle to vehicle (V2V), vehicle to infrastructure (V2I) and infrastructure to vehicle (I2V) communications based on dedicated short range communication (DSRC) protocol [1].

VANET is a core component of intelligent traffic management system. With the knowledge of traffic information, intelligent traffic management system could be applied to control traffic situation directly or indirectly. For example, according to traffic flow, intelligent traffic management system can adjust traffic light to improve traffic efficiency [2]. As traffic congestion wastes much time of drivers and passengers, route reporting schemes supported by intelligent traffic management system would be a solution. Route reporting schemes require drivers to report their future routes to traffic management center so that possible traffic congestion could be predicted by analysing each vehicle's route information. According to the route information, drivers can change their routes to avoid traffic jam.

Although great convenience is provided by route reporting, it also suffers from several security and privacy issues that must be taken into consideration. In route reporting schemes, vehicles need to broadcast messages which contain their location, future routes, etc. If an attacker may learn a vehicle's location or travel plan, then driver's privacy would be violated. Pseudonyms are generally used to protect vehicles' privacy. But most of the existing route reporting schemes using pseudonyms cannot meet the privacy requirements of drivers, because the reported information can be used to link pseudonyms [3]. In addition, an attacker can forge messages to mislead other vehicles or send fake messages to RSUs. Authentication mechanism should be provided to guarantee the security of the messages in VANETs. Digital signatures can be employed to provide authentication of the messages. To avoid the certificate management problem, identity-based signature is an ideal option.

Besides security and privacy, efficiency is a crucial concern for a route reporting scheme. When vehicles are driving at high speed, the entities in VANETs should react quickly so that vehicles can get response in real-time. A route reporting scheme should have low computation and communication overheads. In existing secure and privacy-preserving route reporting schemes, however, RSUs need to verify a mass of signatures to achieve authentication. If the signatures are verified one by one, it could not achieve expected efficiency requirement. To satisfy the efficiency requirement, an efficient batch verification technique should be employed in route reporting schemes. By using the batch verification technique, a set of signatures could be verified in a short period of time.

Recently, a scheme called privacy-preserving route reporting for infrastructure-based VANETs (PRIB) was proposed [3]. PRIB protects vehicles' privacy by using pseudonyms to hide vehicles' identities and using homomorphic encryption to hide vehicles' travel plans. In PRIB, the traffic management center can only learn the total number of vehicles in each road segment. A vehicle's travel plan will not be leaked to other entities.

1.1. Related Work

The concept of intelligent traffic management system was proposed in [4]. Intelligent traffic management system can be employed to define and apply traffic management decisions. Many intelligent traffic management schemes have been studied recently. Salama et al. proposed an intelligent cross road traffic management system which can control traffic lights by using photoelectric sensors to monitor traffic density [5]. Based on the radio frequency identification (RFID) technology, a scheme that collects and calculates average traffic flow information on each road to get latest congestion messages was proposed in [6]. Recently, Liu et al. proposed an intelligent traffic light control scheme which is secure, privacy-preserving and fog device friendly [2].

Route guidance is a valuable application in intelligent traffic management system. Khosroshahi et al. proposed an optimal route search algorithm for dynamic route guidance [7]. Hajiahmadi et al. proposed a high-level scheme for optimal dynamic route guidance using macroscopic fundamental diagram [8]. Wang et al. proposed a real-time path-planning algorithm based on a hybrid-VANET-enhanced intelligent transportation system, which takes advantages of VANETs and transportation system [9]. Lin et al. introduced a scheme that can be used to predict traffic conditions and determine alternative optimal routes based on shared traffic information [10]. Rabieh et al. proposed a privacy-preserving route reporting scheme, in which attackers cannot link pseudonyms [3].

The security and privacy issues in VANETs have been studied for a long time [11]-[14]. Pseudonyms are commonly applied to achieve vehicles' privacy. Chaurasia et al. proposed a privacy sustaining strategy based on appropriate pseudonym update for VANETs [15]. Chim et al. proposed a VANET-based privacy-preserving navigation scheme based on anonymous credential [16]. Wei et al. proposed an RSU aided beacon-based trust management system which aims to thwart internal malicious attackers by using beacon messages and event messages [17]. Recently, Song et al. realized security requirements in VANETs by using secure multiparty computation and homomorphic encryption [18].

1.2. Our Work

In this paper, we propose an efficient secure and privacy-preserving route reporting scheme. Our scheme has five stages: setup, initialization, route request and reporting, batch verification and route aggregation, traffic guidance. In setup and initialization stages, system parameters and entities' private-public keys are generated. In route request and reporting stage, a vehicle who wants to use the route reporting service has to connect to service provider, i.e., traffic management center, via an RSU. It has to verify the RSU' signature at first and then generate a route request which contains its signature and encrypted route. Batch verification and route aggregation is an important stage which is mainly executed by RSUs. In this stage, RSUs verify signatures from nearby vehicles and aggregate encrypted routes. Compared with a recent proposal, our main improvement is that we reduce the computation overhead of this stage. In traffic guidance stage, the traffic management center receives aggregated routes and generates traffic guidance for the vehicles. Performance evaluation is also given to show the efficiency of our scheme.

The remainder of this paper is organized as follows. Section II is the background which introduces our network model, design goals and related techniques. In section III, we give a high-level review of the PRIB scheme and discussion the efficiency of the scheme. In section IV, the improved scheme is proposed in detail. In section V, the performance evaluation and security analysis are presented. The last section is the conclusion.

2. Background

2.1 Network Model

As shown in Figure. 1, a route reporting scheme consists of vehicles, RSUs, traffic management center (TMC), department of motor vehicles (DMV).

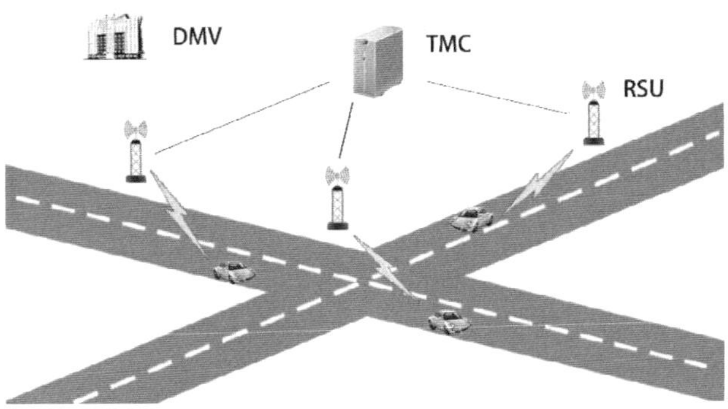

Figure 1 Network model.

- Vehicles: Vehicles run on the road and send route request packets when enter into the communication range of the RSUs. Vehicles also can use computation modules to encrypt and sign messages with issued credentials by DMV.
- RSUs: RSUs are deployed along the roads and also equipped with computation modules. RSUs can communicate with vehicles and TMCs. RSUs are responsible for route aggregation, traffic guidance forwarding, etc.
- TMCs: Each TMC is connected to a group of RSUs which are located in the TMC's monitoring area. TMCs generate traffic guidance for vehicles by using traffic information analysis algorithm.
- DMV: DMV is a trust agency that manages registration for vehicles and RSUs. Typically, DMV is responsible for generating public-private key pairs and certified pseudonyms for vehicles.

2.2 Design Goals

• Vehicle Privacy: The location privacy of the vehicles should not be leaked to other entities. Specifically, if the future route of a vehicle is leaked, an attacker can conduct monitoring which could impact the safety of the driver.

- Authentication: Authentication must be concerned in VANETs. If authentication is not supported, entities cannot detect whether the messages received have been modified or even malicious attackers may forge traffic guidance which would threat the driver's safety.
- Computation efficiency: As we mentioned, computation efficiency is closely related to the latency of the route reporting scheme. Our scheme mainly focuses on the computation overhead to help drivers get route guidance in time.

2.3 Bilinear Maps

Bilinear map has been applied in various VANETs' schemes [19], [20] with its excellent cryptographic properties.

Let $\mathbb{G}_1$, $\mathbb{G}_2$ be two cyclic additive groups of prime order q, and $\mathbb{G}_T$ be a cyclic multiplicative group of the same order. Let P be a generator of $\mathbb{G}_1$, Q be a generator of $\mathbb{G}_2$. A map $\hat{e}: \mathbb{G}_1 \times \mathbb{G}_2 \to \mathbb{G}_T$ is a bilinear map with following properties:

- Bilinearity: $\hat{e}(aP, bQ) = \hat{e}(P, Q)^{ab}$ for all $a, b \in Z_q^*$.
- Non-degeneracy: $\hat{e}(P, Q) \neq 1$.
- Computability: There exists an efficient algorithm to compute $\hat{e}(U, V)$ for any $U \in \mathbb{G}_1, V \in \mathbb{G}_2$.

2.4 Homomorphic Encryption

Homomorphic encryption is a form of encryption that allows perform computation on ciphertext. The decrypted result matches the result that the same operation performed on the plaintext. Here is a form of homomorphic encryption:

$$\mathcal{E}(m_1) \cdot \mathcal{E}(m_2) = \mathcal{E}(m_1 \cdot m_2)$$

Paillier cryptosystem is applied in this paper because of its additive homomorphic properties.

3. The PRIB scheme

In this section, we describe PRIB briefly and give a discussion of PRIB.

3.1. Review of PRIB

In PRIB, roads are divided into segments and each segment has a unique identifier. DMV generates a public-private key pair and a secret for each vehicle. Each TMC generates a public-private key pair and publishes the public key. The RSU broadcasts route request packet to vehicles in its communication range. Vehicle verifies the signature of the RSU then reports its encrypted route if the signature is valid. Vehicle sends back packet containing its encrypted route by using Paillier homomorphic encryption [21]. After the RSU receives a certain number of route packets, it verifies the vehicles' signatures by leveraging batch verification technique [22], [23]. If the signatures are valid, the RSU aggregates the vehicles' routes and send aggregated route ciphertext to the TMC. TMC verifies the signature of aggregated route packet and decrypt the ciphertext. After decryption, the TMC gets the total number of vehicles that will drive in each road segment. Due to the homomorphic encryption, the TMC cannot acquire any vehicle's routes separately. With the number of vehicles in each road segment and traffic information, the TMC sends traffic guidance to vehicles through RSUs. So vehicles can easily avoid traffic congestion by the route reporting.

3.2. Discussion of PRIB

PRIB scheme aims to achieve privacy-preserving in route reporting scheme. To prevent attackers from using reported information to link pseudonyms, PRIB uses homomorphic encryption and route aggregation to protect vehicles' privacy. It's a valid approach for TMCs to analyse traffic condition and generate traffic guidance without leakage of vehicles' route information. However, more aspects should be concerned in traffic management system such as efficiency. Apparently, time overhead in PRIB needs to be low enough so that the drivers can make decisions timely. RSU plays an important role in PRIB, it sends requests, verifies signatures, aggregates routes, forwards routes. However, when RSU verifies the signatures of vehicles' routes, the verification of n vehicles' route reporting packets needs n+1 times of pairing operations in PRIB. So the pairing operations in PRIB increases linearly with the number of route reporting packets. As we know, pairing operations consume much of time in cryptography computation. If many vehicles are sending route reporting packets, drivers would spend much time waiting for traffic guidance. A technique that has lower time overhead should be employed to shorten the latency of traffic guidance.

4. Our Scheme

As shown in Figure. 2, our improved scheme have five stages: setup, initialization, route request and reporting, batch verification and route aggregation, traffic guidance. In setup stage, DMV generates system parameters and publishes them to other entities. In initialization stage, DMV generates keys for vehicles and RSUs. In route request and reporting stage, the RSU sends route request to the vehicles in its communication range. After receiving request, the vehicle generates route reporting packet and sends it to the RSU. In batch verification and route aggregation stage, the RSU verifies route reporting packets using batch verification technique, then sends aggregated routes to the TMC. In traffic guidance stage, TMC decrypts aggregated routes and analyse routes to provide traffic guidance for vehicles.

Table 1 Notations and descriptions

Notations	Descriptions
s	The master private key of DMV
P_0	The master public key of DMV
V_i	The ith vehicle
R_i	The ith RSU
T_i	The ith TMV
PID_i	The pseudo identity of V_i
vs_i	The private key of V_i
rs_i	The private key of R_i
rp_i	The public key of R_i
$\|\|$	Message concatenation operation
$\oplus$	A bitwise XOR operation

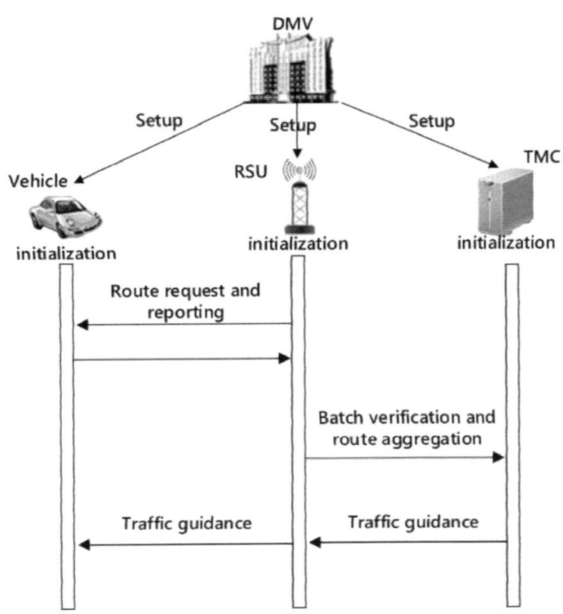

Figure 2 System description.

4.1. Setup

At this stage, DMV generates the public parameters for the whole system. It does the following:

- Choose two cyclic additive groups $\mathbb{G}_1$, $\mathbb{G}_2$ and a cyclic multiplicative group $\mathbb{G}_T$ of prime order q, let $\hat{e}$ be a bilinear map $\hat{e}: \mathbb{G}_1 \times \mathbb{G}_2 \rightarrow \mathbb{G}_T$. P is a generator of $\mathbb{G}_1$.
- Choose cryptographic hash functions $H_0(\cdot): \{0,1\}^* \rightarrow \mathbb{G}_1$, $H_1(\cdot): \{0,1\}^* \rightarrow \mathbb{G}_2$, $H_2(\cdot): \{0,1\}^* \rightarrow Z_q^*$ and $H_3(\cdot): \{0,1\}^* \rightarrow \{0,1\}^l$, where l is the length of the pseudo identities.
- Randomly pick $s \in Z_q^*$ as its master private key, and computes $P_0 = sP$ as its master public key.
- Publish the system parameters as

$$params = (\mathbb{G}_1, \mathbb{G}_2, \mathbb{G}_T, P, P_0, H_0(\cdot) \sim H_3(\cdot))$$

where $params$ are prestored in each entity in the system.

4.2. Initialization

At this stage, DMV generates private keys for vehicles and each RSU generates its private-public key pair. This stage is generally the same as that in PRIB [3].

- DMV generates private key vs_i for vehicle $\mathcal{V}_i$ by computing $vs_i = sH_1(PID_i)$, PID_i is pseudonym of $\mathcal{V}_i$. DMV sends vs_i and PID_i to $\mathcal{V}_i$ through secure channel.
- R_i generates its private key by choosing $rs_i \in Z_q^*$ and generates public key by computing $rp_i = rs_i P$.
- TMC generates x as private key and (m, μ) as public key. TMC publishes the public key.

4.3. Route Request and Reporting

At this stage, the RSU sends route request to vehicles, the vehicle verifies the RSU's signature then sends reporting packet to the RSU.

- R_i generates the signature $\sigma_i = rs_i H_1(M \parallel ID_i \parallel \tau)$, where M is the list of road segments that the RSU queries, ID_i is the RSU's identity, τ is the timestamp.
- The RSU send route request packet $M \parallel ID_i \parallel \tau \parallel \sigma_i$ to the vehicles nearby.
- Each vehicle checks the equation

$$e(P, \sigma_i) = e(rp_i, H_1(M \parallel ID_i \parallel \tau))$$

to verify the signature of the RSU. If the equation holds, go next step, otherwise abort. The following is the proof of the equation.

$$\begin{aligned} e(P, \sigma_i) &= e(P, rs_i H_1(M \parallel ID_i \parallel \tau)) \\ &= e(rs_i P, H_1(M \parallel ID_i \parallel \tau)) \\ &= e(rp_i, H_1(M \parallel ID_i \parallel \tau)) \end{aligned}$$

• For $1 \leq j \leq t$, $L_{i,j} \in \{000, 001\}$ represents whether $\mathcal{V}_i$ will run on the jth road segment. $\mathcal{V}_i$ chooses a random number η_i and encrypts the route by computing $\xi_i = \mu^{(L_{i,1} \parallel L_{i,2} \parallel \cdots \parallel L_{i,t})} \eta_i^m \bmod m^2$.

- $\mathcal{V}_i$ selects a random number α_i and computes $X_i = \alpha_i P'$, $Y_i = \alpha_i P_0' + H_2(\xi_i, X_i) vs_i$, where $P' = \varphi(P)$ and $P_0' = \varphi(P_0)$, φ is a mapping from $\mathbb{G}_1$ to $\mathbb{G}_2$.
- $\mathcal{V}_i$ picks a random number β_i and computes $V_i = \beta_i P$, $C_i = PID_i \oplus H_3(\beta_i rp_i)$.
- $\mathcal{V}_i$ sends route reporting packet $\psi_i = (V_i, C_i, \xi_i, \omega_i)$ to R_i, where $\omega_i = (X_i, Y_i)$.

4.4. Batch Verification and Route Aggregation

At this stage, we modify the verification method of RSUs. When the RSU receives vehicles' reporting routes, the RSU uses batch verification techniques to verify multiple signatures. The underlying signature scheme used here is [24]. Then the RSU aggregates encrypted routes which is same as PRIB's aggregation.

- After R_i receives n route reporting packets, for $1 \leq i \leq n$, it computes $PID_i = C_i \oplus H_3(rs_i V_i)$.
- R_i picks n random numbers $r_1, r_2 \dots r_n \in Z_q^*$, computes $Q_i = r_i H_2(\xi_i, X_i) H_1(PID_i)$, $X_i' = r_i X_i$, $Y_i' = r_i Y_i$.
- R_i verifies the packets by checking the equation

$$e\left(P, \textstyle\sum_{i=1}^{n} Y_i'\right) = e\left(P_0, \textstyle\sum_{i=1}^{n} (X_i' + Q_i)\right)$$

If the equation holds, go next step, otherwise abort. The following is the proof of the equation.

$$
\begin{aligned}
& e(P, \textstyle\sum_{i=1}^{n} Y_i') \\
&= e(P, \textstyle\sum_{i=1}^{n} r_i Y_i) \\
&= e(P, \textstyle\sum_{i=1}^{n} r_i (\alpha_i P_0' + H_2(\xi_i, X_i) v s_i)) \\
&= e(P, \textstyle\sum_{i=1}^{n} r_i (\alpha_i s P' + H_2(\xi_i, X_i) s H_1(PID_i))) \\
&= e(sP, \textstyle\sum_{i=1}^{n} (r_i \alpha_i P' + r_i H_2(\xi_i, X_i) H_1(PID_i))) \\
&= e(P_0, \textstyle\sum_{i=1}^{n} (X_i' + Q_i))
\end{aligned}
$$

- We use PRIB's method to aggregate encrypted routes and generate signature. The aggregated routes is $\Lambda = \prod_{i=1}^{n} \xi_i \mod m^2$. The signature is $\varrho_i = rs_i H_1(\Lambda \parallel ID_i \parallel \tau)$.
- R_i sends aggregated route packet $\Lambda \parallel ID_i \parallel \tau \parallel \varrho$ to TMC.

4.5. Traffic Guidance

At this stage, TMC verifies the signature of aggregated route packet from RSUs and generate traffic guidance for vehicles. This stage is the same as that in PRIB.

- After the TMC T_i receives aggregated route packet, T_i checks
$$e(P, \varrho_i) = e(rp_i, H_1(\Lambda \parallel ID_i \parallel \tau))$$
If the equation holds, go next step, otherwise abort. The following is the proof of the equation.
$$
\begin{aligned}
e(P, \varrho_i) &= e(P, rs_i H_1(\Lambda \parallel ID_i \parallel \tau)) \\
&= e(rs_i P, H_1(\Lambda \parallel ID_i \parallel \tau)) \\
&= e(rp_i, H_1(\Lambda \parallel ID_i \parallel \tau))
\end{aligned}
$$
- T_i computes $(L(\Lambda^x \bmod m^2)/L(\mu^x \bmod m^2)) \bmod m$ to get $\sum_{i=1}^{n} L_{i,1} \parallel \sum_{i=1}^{n} L_{i,2} \parallel \cdots \parallel \sum_{i=1}^{n} L_{i,t}$.
- T_i uses collected traffic information to generate traffic guidance and send it to vehicles through RSUs.

5. Evaluation

5.1. Computation Overhead

The main difference between our scheme and PRIB is signature batch verification of reporting routes. So in this section, we focus the computation overhead of RSUs and show the efficiency of our improved scheme. Our experiments were implemented on a Linux machine with Intel Core i7-4790 at frequency of 3.60 GHz by using MIRACL library.

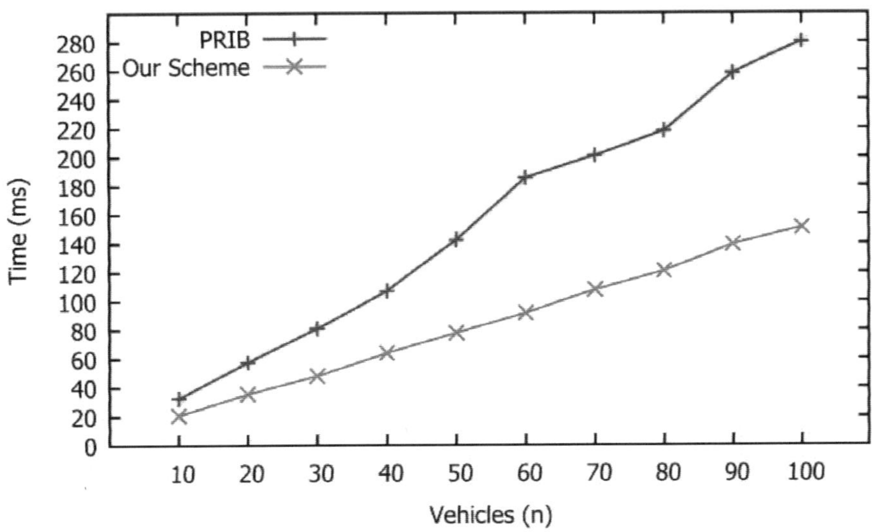

Figure 3 Verification time vs. number of vehicles

As shown in Figure. 3, PRIB's verification time increases faster than ours and always longer than ours. In PRIB, when the RSU verifies the signatures of vehicles' routes, the verification of *n* vehicles' route reporting packets needs *n*+1 times of pairing operations. Pairing operations are time consuming cryptographic computations. However, in our scheme, we reduce the times of pairing operations into only 2. Also, the number of pairing operations in our scheme doesn't increase with the number of signatures. So in Figure.3, the verification time of our scheme is much lower than PRIB's time.

5.2. Security Analysis

In this section, we show that our scheme satisfies security requirements such as privacy-preserving and authentication.

- Vehicle privacy: With homomorphic encryption, each vehicle's future route is transmitted in the form of ciphertext. $\mathcal{V}_i$ encrypts its route under the TMC's public key$m$, so R_i cannot recover $\mathcal{V}_i$'s route. Furthermore, when TMC decrypts aggregated routes which are in form of $\sum_{i=1}^{n} L_{i,1} \parallel \sum_{i=1}^{n} L_{i,2} \parallel \cdots \parallel \sum_{i=1}^{n} L_{i,t}$, it means that only the number of vehicles that will drive on the segment would be achieved. So even TMC cannot get knowledge of each vehicle's future route.
- Authentication: Authentication is achieved by digital signatures. R_i uses batch verification to verify multiple signatures of vehicles. The basic signature scheme in the paper is transformed from Hess's signature scheme [24]. The security of the signature scheme relies on Diffie-Hellman hard problem. The random numbers used in batch verification make the signatures unforgeable.

6. Conclusion

In this paper, we propose an improved route reporting scheme which is more practical than existing one. Our scheme uses the batch verification technique which can verify multiple route reporting packets in a short period of time. Compared to PRIB, our scheme not only achieves required security properties, but the computation overhead has been reduced. Both theoretical analysis and experiments show that our scheme is valid and meets the efficiency for traffic management system in VANETs. We will focus on more techniques to improve traffic management system by enhancing safety and efficiency.

7. Acknowledgment

This work is supported in part by the national key research and development program of China under grant 2017YFB0802004, the NSF of China under Grants 61572198, 61632012, Shanghai high technology field project under grant 16511101400.

8. References

[1] S. Al-Sultan, M. Al-Doori, A. Al-Bayatti and H. Zedan, “A comprehensive survey on vehicular Ad Hoc network, Journal of Network and Computer Applications,” vol 37, pp 380-392, 2014.

[2] J. Liu, J. Li, L. Zhang, F. Dai, Y. Zhang, X. Meng and J. Shen, “Secure intelligent traffic light control using fog computing,” Future Generation Computer Systems, Feb. 2017.

[3] K. Rabieh, M. M. E. A. Mahmoud and M. Younis, “Privacy-Preserving Route Reporting Schemes for Traffic Management Systems,” IEEE Transactions on Vehicular Technology, vol. 66, no. 3, pp. 2703-2713, Mar. 2017.

[4] H. Kirschfink, J. Hernandez, M. Boero and M. Gmbh. “Intelligent traffic management models,” Seventh World Congress on Intelligent Transport Systems, Nov. 2000.

[5] A. Salama, B. Saleh and M. Eassa, “Intelligent cross road traffic management system (ICRTMS),” 2nd International Conference on Computer Technology and Development, pp. 27-31, 2010.

[6] W. Wen, “An intelligent traffic management expert system with RFID technology,” Expert Systems with Applications, vol. 37, no. 4, pp. 3024-3035, 2010.

[7] A. Khosroshahi, P. Keshavarzi, Z. KoozehKanani and J. Sobhi, “Acquiring real time traffic information using VANET and dynamic route guidance,” IEEE 2nd International Conference on Computing, Control and Industrial Engineering, pp. 9-13, 2011.

[8] M. Hajiahmadi, V. Knoop, B. Schutter and H. Hellendoorn, "Optimal dynamic route guidance: A model predictive approach using the macroscopic fundamental diagram," 16th International IEEE Conference on Intelligent Transportation Systems (ITSC 2013), pp. 1022-1028, 2013.
[9] M. Wang, H. Shan, R. Lu, R. Zhang, X. Shen and F. Bai, "Real-Time Path Planning Based on Hybrid-VANET-Enhanced Transportation System," IEEE Transactions on Vehicular Technology, vol. 64, no. 5, pp. 1664-1678, May 2015.
[10] J. Lin, W. Yu, X. Yang, Q. Yang, X. Fu and W. Zhao, "A Real-Time En-Route Route Guidance Decision Scheme for Transportation-Based Cyberphysical Systems," IEEE Transactions on Vehicular Technology, vol. 66, no. 3, pp. 2551-2566, March 2017.
[11] L. Zhang, Q. Wu, B. Qin, J. Domingo-Ferrer and B. Liu, "Practical secure and privacy-preserving scheme for value-added applications in VANETs," Computer Communications, vol. 71, pp 50-60, 2015.
[12] L. Zhang, Q. Wu, J. Domingo-Ferrer, B. Qin and C. Hu, "Distributed Aggregate Privacy-Preserving Authentication in VANETs," IEEE Transactions on Intelligent Transportation Systems, vol. 18, no. 3, pp. 516-526, Mar. 2017.
[13] L. Zhang, "OTIBAAGKA: A New Security Tool for Cryptographic Mix-Zone Establishment in Vehicular Ad Hoc Networks," IEEE Transactions on Information Forensics and Security, DOI: 10.1109/TIFS.2017.2730479.
[14] L. Zhang, Q. Wu, B. Qin, J. Domingo-Ferrer, "APPA: Aggregate Privacy-PreservingAuthentication in Vehicular Ad Hoc Networks," The 14th Information Security Conference (ISC 2011), vol. 7001, pp. 293-308, 2011.
[15] B. Chaurasia, S. Verma, G. Tomar and S. Bhaskar, "Pseudonym Based Mechanism for Sustaining Privacy in VANETs," First International Conference on Computational Intelligence, Communication Systems and Networks, pp. 420-425, 2009.
[16] T. Chim, S. Yiu, L. Hui and V. Li, "VSPN: VANET-Based Secure and Privacy-Preserving Navigation," IEEE Transactions on Computers, vol. 63, no. 2, pp. 510-524, Feb. 2014.
[17] Y. Wei and Y. Chen, "An Efficient Trust Management System for Balancing the Safety and Location Safety and Location Privacy in VANETs," 2012 IEEE 11th International Conference on Trust, pp. 393-400, 2012.
[18] J. Song, C. He, F. Yang and H. Zhang. "A privacy-preserving distance-based incentive scheme in opportunistic VANETs," Security and Communication Networks, vol. 9, no. 15, pp. 2789-2801, Oct. 2016.
[19] B. Liu and L. Zhang, "On the Security of a Privacy-preserving Key Management Scheme for Location Based Services in VANETs," Sixth International Symposium on Foundations & Practice of Security (FPS 2013), vol. 8352, pp. 323-335, 2014.
[20] L. Zhang, Q. Wu, A. Solanas and J. Domingo-Ferrer, "A Scalable Robust Authentication Protocol for Secure Vehicular Communications," IEEE Transactions on Vehicular Technology, vol. 59, no. 4, pp. 1606-1617, May 2010.
[21] P. Paillier, "Public-key cryptosystems based on composite degree residuosity classes," Adv. Cryptology EUROCRYPT, vol. 1592, pp. 223–238, Apr. 1999.
[22] R.Lu, X. Liang, X. Li, X. Lin and X. Shen, "EPPA: An Efficient and Privacy-Preserving Aggregation Scheme for Secure Smart Grid Communications," IEEE Transactions on Parallel and Distributed Systems, vol. 23, no. 9, pp. 1621-1631, Sept. 2012.
[23] L. Zhang, B. Qin, Q. Wu, F. Zhang, "Efficient Many-to-One Authentication with Certificateless Aggregate Signatures," Computer Networks, vol. 54, no. 14, pp. 2482–2491, 2010.
[24] F. Hess, "Efficient identity-based signature schemes based on pairings," in Proc. 9th Annual International Workshop Selected Areas in Cryptography (SAC'02), vol. 2595, pp. 310-324, 2002.

A Systematic Method of Integrating BIM and Sensor Technology for Sustainable Construction Design

Zhen Liu[1, a] and Zhiyu Deng[2, b]

[1] School of Design, South China University of Technology, Guangzhou, P. R. China, 510006
[2] School of Civil Engineering and Transportation, South China University of Technology, Guangzhou, P. R. China 510006
E-mail: [a]liuzjames@scut.edu.cn; [b]dengzy_hnlg@163.com

Abstract. Building Information Modeling (BIM) has received lots of attention of construction field, and sensor technology was applied in construction data collection. This paper developed a method to integrate BIM and sensor technology for sustainable construction design. A brief literature review was conducted to clarify the current development of BIM and sensor technology; then a systematic method for integrating BIM and sensor technology to realize sustainable construction design was put forward; finally a brief discussion and conclusion was given.

1. Introduction

Building Information Modeling (BIM) has received lots of attention of construction field, and has been used in assisting construction management. The application of BIM for construction design mainly focus on construction site layout and construction progress management. There is still grate space for applying BIM in building examine, monitoring, and restructuring, expanding and continued constructions (He et al., 2012). Collecting data from building is an important basis for BIM application in other aspects, and sensor technology is very powerful for acquiring data.

This paper is aimed to develop a method to integrate BIM and sensor technology for sustainable construction design. A brief literature review was conducted to clarify the current development of BIM and sensor technology; then a systematic method for integrating BIM and sensor technology to realize sustainable construction design was put forward; finally a brief discussion and conclusion was given.

2. Literature Review

Various types of sensor can be used in construction field. For structural maintenance and monitoring, Yang et al. (2012) put forward a sensor network to collect the temperature data of the enclosure of buildings [1].

For the selection of sensor types, many researchers developed several of methods to collect different data from buildings. Wu et al. (2001) [2], Gao et al. (2007) [3], and Connolly et al. (2006) [4] put forward that fiber-optic sensors could be used to monitor structure’s stress, strain, crack, vibration and corrosion situations. Wang et al. (2015) developed a method to monitor the structure using Light Detection and Ranging (LiDAR) devices carried by a quadcopter [5]. In other aspects of the combination of sensor and BIM, Fodil et al. (2016) proposed the 3D capture technology combined with BIM, for conduct the life cycle management (LCM) of building [6], the research focuses on photogrammetry and laser scanning technology. Similarly, Costin et al. (2012) proposed that using radio frequency identification (RFID) to carry out the path tracking of objects in the building [7].

For the layout of the sensors, there are many researchers conducted related studies. Ma (2012) proposed that optimizing the sensor arrangement in high-rise building structural health monitoring using complex mathematical method [8]. Papadopoulou (2016) uses a multi-model system identification method to optimize the sensor distribution for monitoring the wind flow around the structure [9].

For the researches about the technical frame of sensor networks, Isikdag (2015) explained in detail about the methods of integrating BIM and information provided by Internet of Things (IoT) and sensors [10]. Jafer et al. (2011) [11] and Huang et al. (2011) [12] proposed that using WSN to monitor indoor environment management; Dorvash et al. (2014) used WSN to monitor the vibration of high rise buildings [13].

At last, for the management and utilization of the data collected by sensors, and the integration of BIM and the data collected by sensors. Deng et al. (2014) raised that using sensor network to collect building big data, and raised several examples of the values of building big data [14].

The integration of sensor technology and BIM to realize sustainable construction design has efficient research basis, and a systematic method can be established according to above methods.

3. A Systematic Method of Integrating BIM and Sensor Technology in Sustainable Construction Design

In above part, a brief review has been conducted to figure out the application status of the integration of BIM and sensor technology in many aspects. This part is aimed to sort out a systematic method of integrating BIM and sensor technology to push forward sustainable construction design, using the methods introduced in above part. The over view of the method is showed in Table 3.1, and the details is described in the follow parts.

Table 1. Overview of the Integrating Method

Integrating method	Sensor Selection	Construction Phase/Requirements	Selection
		Construction plan and design	Laser scanning (Integrating VR)
		Construction safety management	RFID Fiber-optic sensors Others
		Site facilities and materials management	RFID/BLE Others
		Site environment protection	Various site data collection sensors
		Construction progress management	Laser scanning
		Restructure, expansion, and continued construction management	Laser scanning Fiber-optic sensors Others
	Sensor Spatial Layout	No general method, differs according to different goals	
	Data Integration Frame	Aspect	Selection
		Data base selection	NoSQL base
		Data transfer format	JSON
		Graphic presentation	Unity 3D Engine

3.1 Construction Field Management

3.1.1 Construction Plan and Design. In construction plan and design phase, the most practical sensor technology is 3D scanning. Usually laser scanning and photogrammetry will be used to collect the terrain of construction site. This data can integrate in GIS and BIM system to assist managers deploying construction facilities and materials. In BIM integrated with the data collected by 3D scanning facilities, operators could easily recognize and clarify the special relationship among facilities and structural elements. Meanwhile, VR technology can also be integrated into the above system, managers could using VR facilities to directly watch the relationship among completed parts and structural elements needs to be built. Through this system managers can make construction schedules more conveniently, improve the efficiency of facilities and materials' deployment, and reduce the probability of doing wrong works.

According to above parts, a comparison between laser scanning and photogrammetry has been conducted by Fadli et al. and the result shows that laser scanning is better in resolution, precision, range, and data and operation complexity, but photogrammetry is relatively safer and cheap. So it is recommended that photogrammetry could be used to do work which with low accurate requirements, while 3D scanning could be used to acquire mass of accurate points data.

3.1.2 Construction Safety Management. Construction safety management includes 3 aspects: the safety of people, the safety of facilities, and the safety of materials. Among them, the safety of facilities includes the safety of completed structure and the safety of equipment, which will be discussed in restructure, expansion, and continued construction management, and construction facilities & materials management, respectively. This part will focus on the safety management of people.

In construction safety management, positioning and tracing technologies will be essential because it is important to know the real-time positions of workers. RFID is widely used in this aspect. Besides, BLE also belongs to radio frequency technology, and it is also popular in positioning and tracing. By deploying RFID readers in the construction site, implanting RFID tags to workers and combining with the geometry information in BIM, proceeding through a geometry algorithm, workers' position can be located with a relatively high accuracy (about 1.5m). Besides positioning and tracing technology, many other types of sensor can also be used in construction safety management. Some specified examples could still be raised: pressure sensor can be used in underwater construction to detect whether the air pressure is exceed the acceptable range of human; PM 2.5 sensor can be used in a relatively close construction spaces (e.g. underground) with pollution construction machines.

3.1.3 Site Facilities and Materials management. Various sensors could take part in site facilities and materials management. One of them is positioning and tracing technology, in which RFID is most popular. Like positioning and tracing workers in construction site, RFID tags can also planted in facilities and near materials to locate and trace them, with the assist of the structural geometry information from BIM, managers could directly know the position of facilities and materials, and can make deployment schedules, so that a lot of time and labor wasting could be avoid.

In site facilities management, the operation and management of facilities is also an important part and the integration of BIM and sensors can also assist the operation and management of facilities. As an example and case researched by Lee et al, various of sensors were used (e.g. slewing sensors, cable length sensors, boom angle sensors, camera and laser scanning devices) to collect the operation data of a crane, and BIM will provide building and context's 3D information and the detail position of objects need to be lifted, this system is proved to be useful for those cranes have blind spots.

*3.1.4 Site Environment Protection.*In modern construction managements, environment protection is also an important aspect that managers need to consider about. Air pollution and noise pollution are common pollution types in construction site. In order to control the air pollution caused by dust, dustproof cloth is often used to block the diffusion of dust, while PM 2.5 sensors could also be used to collect air quality data from the construction site, with the building information provided by BIM, managers can figure out the diffusion source and reduce the dust emission. Similarly, noise sensors can also detect the noise intensity and make noise reduction measures.

3.1.5 Construction Progress Management. BIM is a practical tool for construction progress management, and it is not strange to use sensors to conduct progress management. With the assist of 3D capture devices (e.g. laser scanning & photogrammetry), the machine can scan the completed structure automatically and compare the scanned model with BIM to assist progress control.

Sensors can also assist construction quality management; it will be introduced in the next part.

3.2 Restructure, Expansion, and Continued Construction Management

Restructure, expansion and continued construction is conducted after the building was finished and be used for a time. Usually designers will check the structural manually to decide whether it is appropriate to be restructured and make construction plan. This method is risky because it is hard to check the inner situation of structures, and some buildings were not used as they were planned. So it is essential for designers to get accurate data of the situation about the structure's health and how did it work.

Using sensors to conduct structure health monitoring is plain and simple. Usually 3D scanning devices were used to examine the surface quality of the structure to check: i) its crack situation, and ii) its deformation, automatically; the stress and strain sensors were planted into or affixed on the concrete to check the inner mechanical situations, also some other mechanical sensors(e.g. velocity sensors, acceleration sensors and seismic sensors) could also be planted into structure; and the temperature sensors could be used to detect the temperature diversifications, which is particularly useful for big volume concrete structures such as dams.

For building mechanical situation detection, optic fiber sensor is recommended for its significant advantages in property (e.g. high sensitivity, small volume, wide range and safety), and extensive usages. A case study of Wuhu Yangtze River Railway Bridge shows that the optical fiber sensors can accurately measure strain and they are suitable for the long-term and automatic monitoring.

For sensor deployment, there is no widely accepted method to deploy various types of sensor, there are only suitable ways to deploy several kinds of sensors. Deploy plan of sensors is still to be determined according to actual situation; a comprehensive deploy method which is suitable for layout various types of sensor is still to be developed.

It is nature that the structural health monitoring system could be used in construction quality management and operation & maintenance phase.

Meanwhile, through the data collected by these sensors, a knowledge base could be found to record the real time operation and usage data of the building, which is useful for scientific research and future design works.

3.3 The Technical Frame of Integrating Sensor Network and BIM

Actually sensor network is a special case of IoT (Internet of Things). And it is a technical problem to realize the communication among sensors, found a database to store those data, integrating sensor data with BIM, and realize the interaction between users and the system. Isikdag et al. has developed a frame, including basic publish-subscribe method for exchanging information among large amounts of IoT nodes; and a message-based cloud update for solving the problems about information updating and saving, etc.

From the research of Khalid et al. (2017) [15], for the selection of database, NoSQL database was recommended for its high scalability, and it was more agility and dynamic for situations that data updates frequently; and in aspect of data transfer format, compared with XML, JSON was recommended because it can be better understood by NoSQL database; Besides, for graphic presentation, Unity 3D game engine was proved to be efficiency in dealing with scenes having lots of vertices.

4. Discussions

This paper developed a method to integrate BIM and sensor technology for sustainable construction design. The method raised above is technically feasible, but without the consideration of economy feasibility, peoples' concept, current management level and enterprises' willing to apply it, which hinted that the cost of applying sensor networks will be high, and current management awareness is lagging in construction field. So the future researches could focus on economic property of sensor networks and promoting information technologies in construction field.

5. Conclusion

This paper developed a method to integrate BIM and sensor technology for sustainable construction design. The current publications shows there are sufficient research about the integration of BIM and sensor technology, the function of sensors can basically meet various of requirements in construction field, the selection of sensor types and the spatial layout is differ according to different goals, and the technical frame for integrating BIM and data collected by sensors is mature.

For policy suggestions, government shall promote BIM applications and applying sensor technologies in construction field; and interdisciplinary research should be encouraged, too.

6. Acknowledgements

The authors are grateful for the support from the South China University of Technology (grant number: j2rs/D6172210), P. R. China, so as its School of Design and the School of the Civil Engineering and Transportation.

7. References

[1] Yang Y., Fang Q., He Z.: Mian xiang jian zhu wei hu jie gou wen du xin xi cai ji de wu xian chuan gan qi wang luo she ji. J. chong qing ke ji xue yuan xue bao (Zi ran ke xue ban) 14(3), 147--150 (2012)
[2] Wu Z., Zhang B., Wang D., Du S.: Guang xian chuan gan qi zai min yong jian zhu jie gou zhong de ying yong. J. Harbin gong ye da xue xue bao 33(4), 469--474 (2001)
[3] Gao, Z., Du, Y., Sun, B., Jin, X.: Strain monitoring of railway bridges using optic fiber sensors. J. Journal of Quality in Maintenance Engineering 13(2), 186--197 (2007)
[4] Connolly, C.: Fibre-optic-based sensors bring new capabilities to structural monitoring. J. Sensor Review 26(3), 236--243 (2006)
[5] Wang, J., Sun, W., Shou, W., Wang, X., Wu, C., Chong, H., Liu, Y., Sun, C.: Integrating BIM and LiDAR for Real-Time Construction Quality Control. J. Journal of Intelligent & Robotic Systems 79(3), 417--432 (2015)
[6] Fadli, F., Barki, H., Shaat, A., Mahdjoubi, L., Boguslawski, P., Zverovich, V.: 3D Capture Techniques for BIM Enabled LCM. In: Bouras, A., Eynard, B., Foufou, S., Thoben, K. (eds.) PLM 2015, IFIP AICT 467, pp. 183--192. Springer, Switzerland (2016)
[7] Costin, A., Pradhananga, N., Teizer, J., Marks, E.: Real-Time Resource Location Tracking in Building Information Models (BIM). In: Luo,Y. (eds.) CDVE 2012, LNCS 7467, pp. 41--48. Springer, Heidelberg (2012)
[8] Ma, S.: Gao ceng jian zhu jie gou jian kang jia ce zhong chuan gan qi you hua bu zhi yan jiu. D. He fei gong ye da xue (2012)
[9] Papadopoulou, M., Raphael, B., Smith, I., Sekhar, C.: Optimal Sensor Placement for Time-Dependent Systems: Application to Wind Studies around Buildings. J. Journal of Computing in Civil Engineering 30(2), 04015024 (2016)
[10] Isikdag, U.: Enhanced Building Information Models. Springer International Publishing (2015)
[11] Jafer, E., Spinar, R., Stack, P., O'Mathuna, C., Pesch, D.: Wireless Sensor Network Deployment for Building Environmental Monitoring and Control. In: Gabrielli, S., Elias, D., Kahol, K. (Eds.) Ambient Media and Systems, 20--27. Springer, Heidelberg (2011)

[12] Huang, Q., Li, X., Shaurette, M., Cox, R.: Novel Sensor Network Architecture for Intelligent Building Environment Monitoring and Management. In: Zhu,Y, Issa, R. (eds.) Computing in Civil Engineering, pp. 347--354. ASCE (2011)
[13] Dorvash, S., Pakzad, S., Naito, C.: In-Service Vibration Monitoring of a Tall Building Structure Using Wireless Sensor Networks. In: Bell, G., Card, M. (eds.) Structures Congress 2014, pp. 2826--2838. ASCE (2014)
[14] Deng, Z., Li, J.: Jian zhu shu ju hua li lun gai shu. J. zhong guo zhu zhai she shi 2014(11), 70--73 (2014)
[15] Khalid, M., Bashir, M., Newport, D.: Development of a Building Information Modelling (BIM)-Based Real-Time Data Integration System Using a Building Management System (BMS). In: Dastbaz, M., Gorse, C., Liu, J., Moncaster, A. (Eds.) Building Information Modelling, Building Performance, Design and Smart Construction. Springer (2017)

Status Quo of Lanzhou Earthquake Early Warning System and Analysis of Its Key Technical Indicators

Xiufeng Tian[1,2], Wenhao Zeng[1], Zhangrong Pan[1], Weidong Zhang[1], Jie Yuan[1], and Xuan Zhang[1]

[1] Institute of Geophysics, China Earthquake Administration, 100081, Beijing, China
[2] Lanzhou Institute of Seismology, China Earthquake Administration, 730000, Lanzhou, China
Email: txf913@163.com

Abstract. The composition, network configuration and system operation condition of Lanzhou earthquake early warning system are introduced in details, and the factors influencing the early warning performance are discussed. Then, in combination of the status quo of system, the key technical indicators are analyzed, and the distribution map of consumed time for earthquake early warning is drawn. The result indicates that the average distance among stations in observation network is large, which casts an obvious influence on the time validity of early warning. Under the set conditions, the radius of blind zone ranges between 30.3km and 39.2km, and the early warning time increases as the epicentral distance enlarges. In addition, the radius of blind zone is influenced by the focal depth. Furthermore, influenced by the station distribution and density, the distribution of consumed time for earthquake early warning is extremely uneven and characteristic of obvious regional difference.

1. Introduction

Based on the fundamental principle that the propagation velocity of P-wave is higher than that of destructive S-wave and surface wave and that the propagation velocity of electromagnetic wave is far higher than that of earthquake wave, earthquake early warning is to utilize the real-time transmission earthquake observation network configured in the early warning target area or potential seismic source zone to determine the basic earthquake parameters and estimate the possible influential field distribution in a very short period after the earthquake occurs as well as then timely send out earthquake alarms and take corresponding emergency handling measures before the destructive earthquake wave reaches the early warning target area, which can effectively alleviate the property loss and personal casualties caused by the earthquake[1-4].

The conception of earthquake early warning was firstly proposed by Dr.Cooper after the earthquake in San Francisco, US in 1868. He assumed to configure earthquake observation stations in Hollister District 100km away from San Francisco, where the earthquake activity is very high. Therefore, once an earthquake occurs, the propagation time difference between electromagnetic wave and earthquake wave can be utilized to sound the alarm in the City Hall of San Francisco in a very short time after the earthquake so that people can take some corresponding emergency rescue measures to reduce casualties [5]. However, this conception was not realized due to limitations of tech level at that time. In recent 20 years, with the quantum leaps in computer technology, data transmission processing technology, earthquake monitoring instruments and earthquake observation method, the conception of Dr.Cooper is gradually evolving into a reality. Currently, in some countries and regions frequent of earthquakes, such as Japan, Mexico, Turkey and Taiwan, multiple earthquake early warning systems aimed at specific facilities, single city or even bigger region, have been established, and remarkable disaster relief effect has been achieved in practical operation; US, Italy and South Korea also have actively carried out relevant real-time tests on earthquake early warning system[6-9].

In mainland China, the relevant research and construction of earthquake early warning started relatively late. In the construction periods of 9th and 10th Five-Year Plan, a large batch of earthquake observation stations were built by China Earthquake Administration and a series of relevant scientific research projects on construction of earthquake early warning system were organized and executed[10-12]. In recent years, relying on its own tech advantages, Fujian Earthquake Agency has built a complete provincial earthquake early warning system; China Earthquake Administration has authorized the execution of China Earthquake Background Field Exploration Project and National Earthquake Social Service Project, and constructed two demonstrative earthquake early warning systems in Beijing and Lanzhou. The construction and operation of these early warning systems have accumulated excellent experience for the development of earthquake early warning in China. Currently, the National Earthquake Intensity Quick Reporting and Early Warning Project has been initiated across a national range and the macroscopic exploration and selection for observation stations have been carried out comprehensively. Before China officially starts the construction of large-scale earthquake early warning systems, it is of great significance to analyze and evaluate the status quo of current early warning network and its capability. Only in this way can we find the corresponding insufficiencies and specify the improvement direction. To address this problem, Lanzhou earthquake early warning system is adopted as research object in this paper; then, in combination of its current network configuration and system operation, the key technical indicators influencing its early warning efficiency is analyzed deeply; next, corresponding improvement directions and suggestions are proposed in the hope of providing a certain reference to the future construction of earthquake early warning system in China.

2. Status quo of Lanzhou earthquake early warning system

As a regional earthquake early warning system specially targeted towards Lanzhou, Lanzhou earthquake early warning system provides earthquake early warning and intensity quick reporting information for Lanzhou in medium or large earthquakes occurring in Lanzhou and its surrounding areas. This system consists of early warning observation network, data processing platform and information distributing platform. Specifically, early warning observation network obtains the original data of earthquake early warning information and mainly accomplishes the real-time detection, transformation and digitization of earthquake signals as well as data packing, communication control and transmission; as the core of the whole system, data processing platform mainly accomplishes real-time monitoring over the operation condition of early warning observation network, identifies and pre-processes the earthquake events, quickly measures the earthquake parameters and outputs earthquake early warning and intensity quick reporting information; as the information terminal of the whole system, information distributing platform faces directly towards the service objects and distributes the early warning information through information distributing terminals and alarming devises. Fig.1 shows the overall structure of above system.

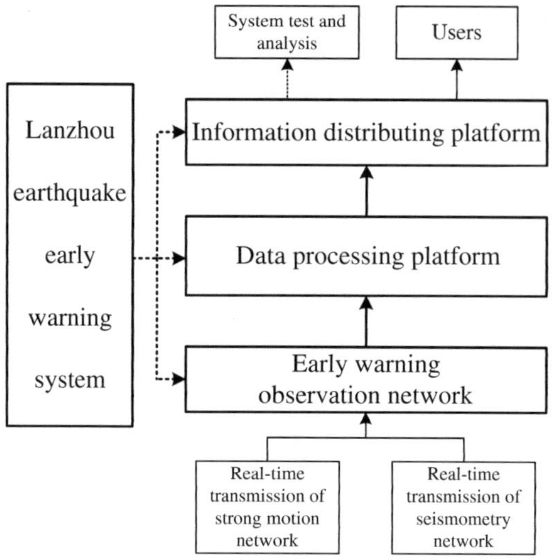

Figure.1. Structure of Lanzhou earthquake early warning system.

2.1 Configuration of early warning observation network

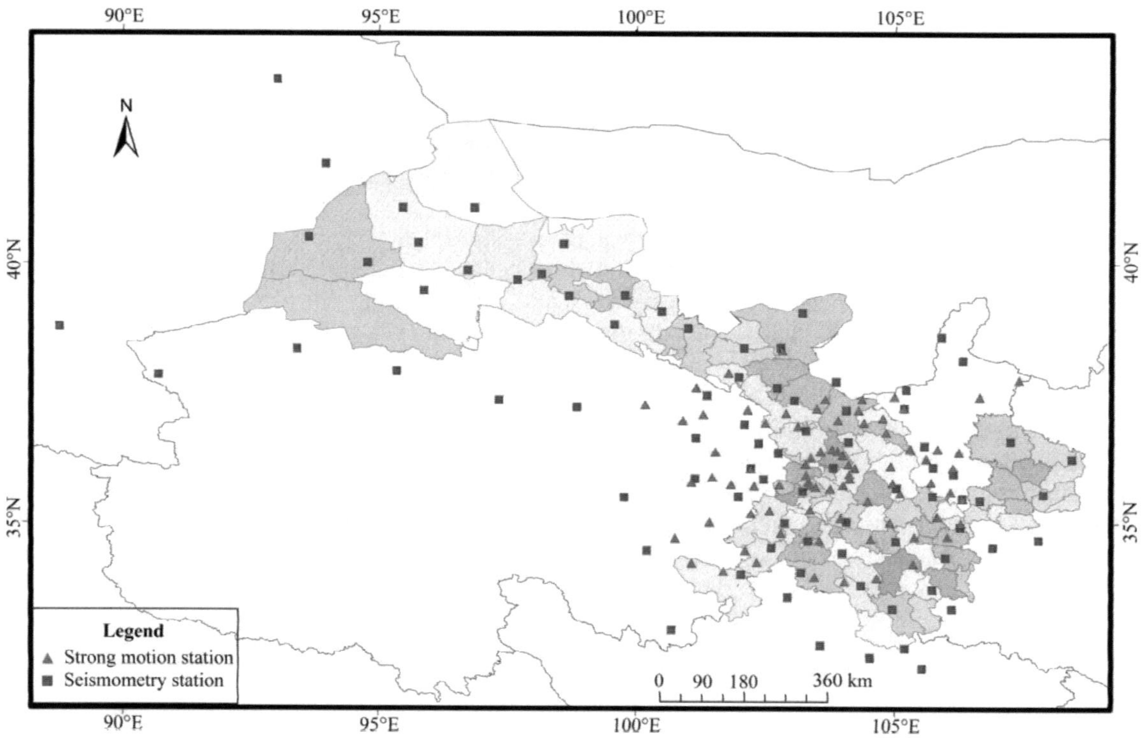

Figure.2. Distribution of early warning observation stations in Lanzhou earthquake early warning system.

The observation network of Lanzhou earthquake early warning system consists of strong motion observation network and seismometry observation network. The monitoring range can cover the entire Gansu and some areas in neighboring provinces. Fig.2 shows the configuration of the whole early warning observation network. Specifically, the strong motion observation network consists of 80 strong motion observation stations, which are centered around Lanzhou and extend outwards in a ring shape. It includes 54 strong motion stations located in Gansu, 15 stations in Qinghai and 11 stations in

Ningxia, and the farthest station is 375km away from Lanzhou. The seismometry observation network consists of 83 seismometry observation stations, which are distributed in a mesh form, including 47 stations in Gansu and 36 data sharing stations in neighboring provinces. In the aspect of instrument category, all the strong motion observation stations adopt the configuration of REFTEK-130REN data collectors and SLJ-100 accelerometer. In addition, the data collectors adopted by seismometry observation stations are mainly EDAS-24 Series, CMG-DM24 and SMART-24, and the sensors are mainly BBVS-60, BBVS-120, CMG-3EMPC and CTS-1. All the above instruments adopt the communication method of real-time transmission by optical fiber.

2.2 Operation status quo of early warning system

Currently, the data processing platform in Lanzhou earthquake early warning system adopts automated processing software, which can identify and process the earthquake events quickly. Within several seconds after the first observation station is triggered, the platform can present the initial earthquake three-element information as well as the early warning time and predicted intensity of early warning target area; within 5-6mins after the earthquake, relevant intensity quick reports, including the instrument intensity figure in the earthquake area and PGA isoline figure, are outputted. In addition, the information distributing platform adopts special database and information reception terminals, which can realize the automatic reception and distribution of early warning information. The information reception terminals currently adopted include digital VHF LED display screen, smart radio, digital VHF loudspeaker and cellphone APP. Since the operation of Lanzhou earthquake early warning system, it has successfully outputted early warning and intensity quick reporting information in several earthquakes in Gansu and its neighboring regions.

3. Key technical indicators of earthquake early warning

3.1 Earthquake early warning modes

According to the difference in working modes, the earthquake early warning can be divided into on-site early warning, off-site early warning and mixed early warning modes. The on-site early warning mode refers to configuring stations in early warning target area and utilizing the arrival time difference between the first-arrival wave (generally P-wave) and the late-arrival destructive wave (S-wave or surface wave) for early warning. Its merits include single target and simple configuration while its shortcomings include little usable information, low accuracy and short early warning time. Compared to on-site early warning mode, off-site early warning mode adopts traditional earthquake monitoring method. In this mode, earthquake observation stations are densely configured in potential seismic source area, which are utilized to record the signals (P-wave or S-wave) to determine the basic earthquake parameters, thus it can estimate the ground motion in early warning target area and send alarms to potential earthquake destruction area. The merits of this mode include long early warning time, rich usable information and accurate earthquake parameter measurement while its shortcomings include the necessity of support from heavily-dense observation network, massive investment and complex system structure[13].

In the early development stage of earthquake early warning, the majority of early warning systems adopt on-site early warning mode due to limitations of observation technology and station density. As the development of observation instruments and data transmission technology as well as station density advance, the difference between on-site and off-site early warning is gradually diluted. The mixed early warning mode combining above two modes is becoming more popular. This mode can integrate the merits of on-site and off-site early warning and make full use of current network resources, but the accompanying problem is that its technical system has become more complex.

3.2 Time validity of earthquake early warning

The time validity of earthquake early warning is generally represented by early warning time and blind zone[13]. The early warning time refers to the time left for the early warning object to take emergency handling measures after it receives the early warning information and before the destructive earthquake wave arrives. The length of early warning time reflects the efficiency of an earthquake early warning system and is directly related to the distance from the early warning object to the earthquake source and the time it takes to distribute early warning information (hereafter called consumed time of early warning). In addition, the early warning blind zone refers to the propagation area of destructive wave (generally S-wave) when the early warning information are distributed. The early warning information cannot be timely delivered to the objects inside the blind zone and thus the early warning efficiency is lost. The range of blind zone is decided by the consumed time of early warning and directly reflects the efficiency of earthquake early warning system. The smaller the blind zone is, the higher the early warning efficiency will be.

Generally, the consumed time of early warning includes the time for P-wave to propagate to stations, data transmission delay, earthquake parameter calculation time, information distribution delay and response delay of reception terminal[14]. Specifically, the time for P-wave to propagate to station is closely related to the focal depth, station density and location of earthquakes; data transmission delay is influenced by instrument and network performance; earthquake parameter calculation time is related to calculation speed and performance of processing software; distribution delay of early warning information and response delay of reception terminal are decided by the software & hardware performance of distributing system and network condition.

4. Analysis of key technical indicators in Lanzhou earthquake early warning system

4.1 Basic parameters of Lanzhou earthquake early warning system

Lanzhou earthquake early warning system is a regional earthquake early warning system adopting off-site early warning mode. It takes Lanzhou as its major early warning object and provides it with earthquake early warning and intensity quick report service. The whole system contains 163 observation stations (as shown in Fig.2) and the average station distance is about 70-80km. In addition, the stations are relatively densely distributed in the east but scarcely distributed in the west. There are fewer observation stations in northwestern Gansu, and the average station distance is correspondingly larger. By comparison, there are relatively more observation stations in mid-eastern Gansu, namely Lanzhou and its surrounding areas, and the station distance is correspondingly shorter. Specifically, 116 observation stations are within 300km away from Lanzhou (including 76 strong motion stations and 40 seismometry stations), and the average station distance is 49.4km. In addition, according to the statistical operation data of Lanzhou earthquake early warning system since its establishment, the transmission delay of the majority of observation stations is within 2s, and the earthquake parameter calculation time is about 5-10s. Considering currently early warning information has not yet been distributed to the public officially, the distribution delay of early warning information and response delay of reception terminal are not considered in this paper.

4.2 Time validity analysis of Lanzhou earthquake early warning system

The theoretical early warning time of Lanzhou earthquake early warning system, T_w , can be expressed as:

$$T_w = T_s - T_{p1} - T_d - T_{pr} = \frac{\sqrt{\Delta^2 + h^2}}{v_s} - \frac{\sqrt{\Delta_1^2 + h^2}}{v_p} - T_d - T_{pr} . \quad (1)$$

where, T_s is the time it takes for S-wave to arrive at the early warning object, T_{p1} is the time it takes for P-wave to arrive at the station closet to the epicenter, T_d is the data transmission delay, T_{pr} is the earthquake parameter calculation time, Δ is the epicentral distance of early warning object, Δ_1 is the epicentral distance of the station closest to the epicenter, h is the focal depth, v_s and v_p are the propagation speeds of S-wave and P-wave.

When $T_w \le 0$, there is no early warning time theoretically. In other words, the early warning objects close to the earthquake source have no time to take emergency handling measures before the destructive earthquake wave arrives due to short epicentral distance, which means they locate in the early warning blind zone.

In this paper, assuming an earthquake occurs at a certain point within 300km away from Lanzhou, considering that the average station distance within this region is 49.4km, we simply assume that the longest epicentral distance of the station closest to the epicenter is 24.7km, namely $0 \le \Delta_1 \le 24.7km$; in addition, we adopt the single-layer crust model in Gansu to calculate the arrival time of earthquake wave[15], where v_s =3.57km/s and v_p =6.10km/s; besides, it is assumed that the focal depth h =15km, T_d =2s and T_{pr} =5s.

When Δ_1 =0km, namely under idea condition, there is one observation station right at the epicenter, assuming T_w =0, we can calculate that Δ =30.3km, that is, the radius of early warning blind zone is 30.3km. When Δ_1 =24.7km, namely that the earthquake occurs right at the middle of two observation stations, at this case we can calculate that the radius of blind zone is 39.2km. So it can be concluded that for this set earthquake, the radius of blind zone in Lanzhou earthquake early warning system ranges between 30.3km and 39.2km.

In addition, we assume that Δ_1 =10km and thus can calculate the change to early warning time T_w with the epicentral distance of early warning object (as shown in Fig.3). It can be seen from Fig.3 that under this setting, the radius of early warning blind zone is 32.2km. When the epicentral distance is 50km, the early warning object has 4.7s early warning time; as the epicentral distance increases, the early warning time is gradually increasing; when the epicentral distance reaches to 100km, the early warning object can obtain 18.4s early warning time.

Likewise, we can draw the changing curve of radius of blind zone with focal depth when Δ_1 =10km (as shown in Fig.4). It can be seen from Fig.4 that when the focal depth ranges between 0 and 30km, the change to radius of blind zone is not obvious and the radius maintains above 30km; when the focal depth exceeds 30km, as the depth increase, the radius of blind zone gradually decreases; when the focal depth increases to 61.4km, the radius of blind zone is 0km, which indicates that at this time the early warning object at the epicenter can also receive the early warning information. Considering that the focal depth of earthquakes in Gansu is mainly between 5km and 20km, therefore, the radius of blind zone in this region is commonly over 30km.

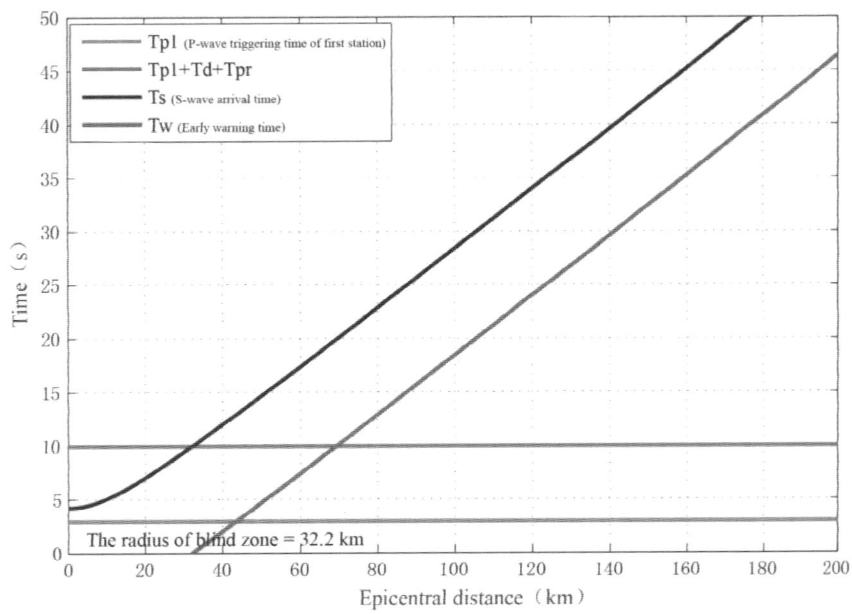

Figure.3. The curve of early warning time.

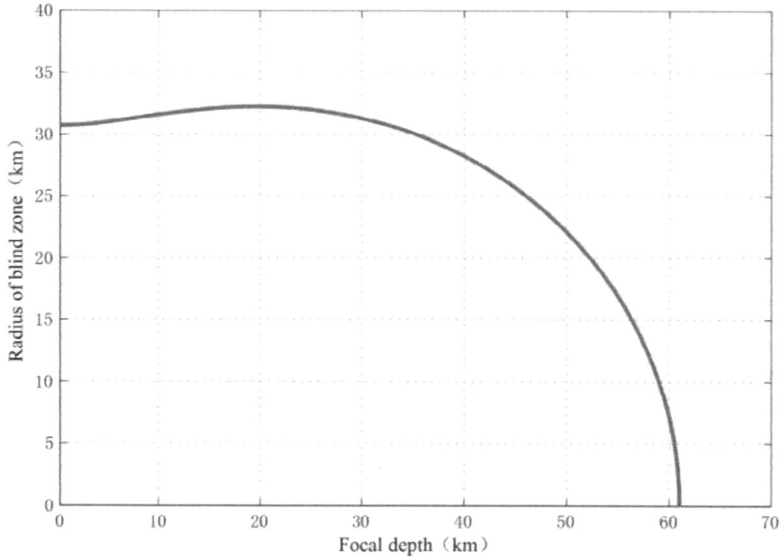

Figure.4. Change of radius of blind zone with focal depth.

4.3 Analysis on consumed time of early warning

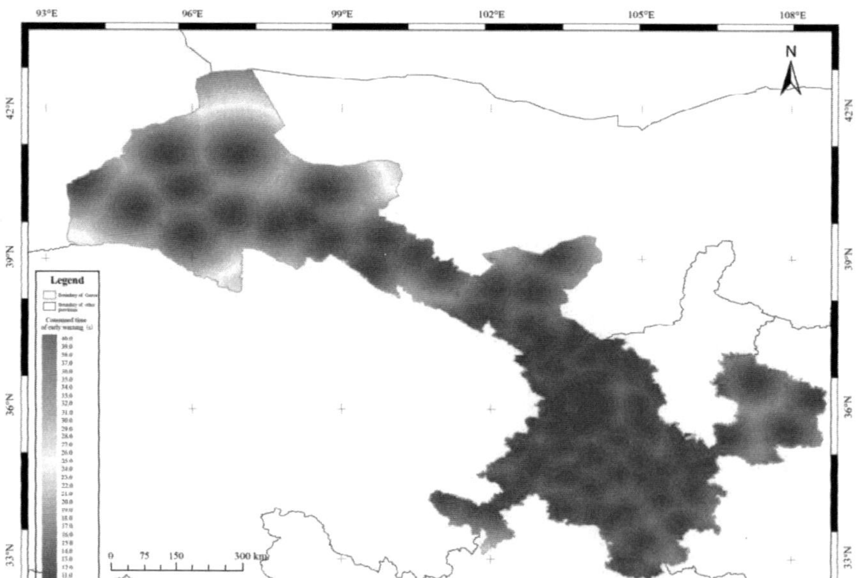

Figure.5. Distribution of consumed time of early warning in Gansu province

The research area ranges from 32°N to 43°N and from 93°E to 109°E, which is just surrounding Gansu province, and then we discretize it into grids based on 0.1°*0.1° unit element. Assuming that one earthquake occurs in each grid point (the focal depth is set as 15km), the system starts calculating, processing and outputting early warning information after the station closest to the epicenter is triggered, and the data transmission delay T_d =2s, the calculation time T_{pr} =5s. Thus, under the current network configuration of Lanzhou earthquake early warning system, the distribution of consumed time of early warning in this area is shown in Fig.5. It can be seen from the figure that the consumed time of early warning is closely related to station distribution and density. The closer the area is to the observation station, the shorter the consumed time will be; the overall consumed time of early warning is correspondingly shorter in areas with higher station density. Based to this influence, the distribution of consumed time of early warning in Gansu is extremely uneven and has obvious regional difference. The shortest consumed time is about 9.0s while the longest is about 40.0s; the overall consumed time is relatively shorter in the mid-eastern region and longer in the northwestern region.

5. Conclusion and discussion

Earthquake early warning system is an enormous and complex project. The deficiency and insufficiency of any link will influence its overall early warning efficiency. The length of early warning time and the size of blind zone are the key to determine if it can play its role in earthquake prevention and disaster reduction. It can be found by the calculation and analysis in the previous sections that:

(1) The average station distance in the observation network of Lanzhou earthquake early warning system is relatively large, which severely influences the time validity of early warning. In mid-eastern Gansu where there are relatively more stations, the average station distance reaches about 50km; in northwestern Gansu, the average station distance is greater. In addition, data transmission and processing delay will cast quite an influence on the early warning time and blind zone radius. In the future, we need to constantly construct more observation stations to improve station density, optimize network configuration and minimize the data transmission and processing delay. Thus, the early warning efficiency of current system can be given to its full play.

(2) Under the conditions set in this paper, the radius of blind zone in Lanzhou earthquake early warning system ranges between 30.3km and 39.2km; the early warning time increases as the epicentral distance of early warning object enlarges; when the epicentral distance is 100km, the early warning time is 18.4s. The research result also indicates that the radius of early warning blind zone is influenced by the focal depth, which changes slightly when the focal depth ranges between 0 and 30km, and basically maintains above 30km; but when the focal depth exceeds 30km, the radius of blind zone is negatively proportional to the focal depth, as the depth increases, the radius will gradually reduce. Considering that the focal depth of the major earthquakes in Gansu is between 5km and 20km, the common radius of blind zone in this region is above 30km. In addition, it is found that due to the influence of station distribution and density, the distribution of consumed time of early warning is extremely uneven in Gansu and has obvious regional difference.

(3) In the calculation and analysis in this paper, the data transmission delay and earthquake parameter calculation time are set as 2s and 5s respectively; due to limitations of network status and waveform quality in working practices, the actual consumed time might be higher than above values. Secondly, the idea of starting earthquake parameter calculation and processing after one station is triggered is adopted in this paper; but in actual application, to ensure the accuracy and reliability of early warning information, the system will only start calculation and processing normally after three stations are triggered. In addition, the distribution delay of early warning information and response delay of reception terminal are not considered in this paper temporarily; however, when the early warning system formally distributes early warning information to public, it is difficult to avoid above delays in information transmission process. All the above factors will directly lead to the increase of consumed time of early warning and finally cause the early warning time to shorten and the blind zone to expand. Therefore, how to perfect the overall performance of system and improve the actual early warning efficiency will be key issues remaining to be solved in our future research.

The work in this paper is fundamental. The analysis of key technical indicators of Lanzhou earthquake early warning system assists us in better understanding the technical level and early warning efficiency of current system as well as exploring its deficiencies and insufficiencies, which can provide necessary reference basis for promotion and application of early warning system in the future.

6. Acknowledgments

This work was financially supported by Fundamental Research Fund of Institute of Earthquake Forecasting, CEA (2016IESLZ03), Earthquake Tech Development Fund of Lanzhou Institute of Seismology, CEA (2015ZB02), High-Speed Railway Joint Fund of NSFC (U1434210) and Lanzhou Earthquake Early Warning System Demonstration Project in National Earthquake Social Service Project.

7. References

[1] Qiang M., Xing J., Shanyou L.: Automatic P-arrival detection for earthquake early warning. Chinese Journal of Geophysics. 56(7), 2313--2321 (2013)
[2] Qiang M.: Stuy and application on earthquake early warning. Institute of engineering mechanics, CEA (2008)
[3] Li L.: Optimization research on seismic array layout and quick evaluation of earthquake damage based on earthquake early warning. Shenyang Jianzhu University (2009)
[4] Hongcai Z.: Study of key technologies in earthquake early warning system. Recent Developments in World Seismology. 4, 44--46 (2014)
[5] Yanwen Z., Xiqiang L., Xuhui H.: Current situation and prospects of the research on the earthquake early warning systems. Recent Developments in World Seismology. 4, 28--34 (2008)
[6] Richard M. A., Gasparini P., Osamu K.: The status of earthquake early warning around the word: An introductory overview. Seismological Research Letters. 80(5), 682--693 (2009)
[7] Espinosa-Aranda J. M., Jiménez A., Ibarrola G.: Mexico city seismic alert system. Seismological Research Letters. 66(6), 42--53 (1995)
[8] Kuyuk H. S., Richard M. A.: Optimal seismic network density for earthquake early warning: a case study from California. Seismological Research Letters. 84(6), 946--954 (2013)
[9] Naichi H., Yihmin W., Li Z.: A new prototype system for earthquake early warning in Taiwan. Soil Dynamics and Earthquake Engineering. 31, 201--208 (2011)
[10] Hongcai Z., Xing J., Jun L.: Review on magnitude estimation methods applied to earthquake early warning. Progress in Geophysics. 27(2), 464--474 (2012)
[11] Qiang M., Shanyou L.: Research on the key technology of earthquake early warning and intensity quick report. China Science and Technology Achievements. 20, 23--25 (2013)
[12] Hongcai Z., Xing J., Jun L.: Progress of research and application of earthquake early warning system. Progress in Geophysics. 28(2), 706--719 (2013)
[13] Hongcai Z.: Study of key technologies in earthquake early warning system. Institute of engineering mechanics, CEA (2013)
[14] Chen Y., Kai G., Suling Z.: Status quo of China earthquake networks and analyses on its early warning capacity. Acta Seismologica Sinica. 37(3), 508--515 (2015)
[15] Cheng Z., Qizhen D.: Inspection of "the travel-time table of seismic wave in Gansu and its nearby areas". Seismological and Geomagnetic Observation and Research. 4, 22--27 (1982)

Institute of Physics
Dirac House, Temple Back
Bristol BS1 6BE UK

ISSN: 1742-6588
ISBN 978-1-5108-5108-5

9 781510 851085

CURRAN ASSOCIATES INC.
proceedings.com